Addison-Wesley
Chemistry

Antony C. Wilbraham

Dennis D. Staley

Candace J. Simpson

Michael S. Matta

Addison-Wesley Publishing Company

Menlo Park, California ▪ Reading, Massachusetts ▪ New York
Don Mills, Ontario ▪ Wokingham, England ▪ Amsterdam ▪ Bonn
Paris ▪ Milan ▪ Madrid ▪ Sydney ▪ Singapore ▪ Tokyo
Seoul ▪ Taipei ▪ Mexico City ▪ San Juan

Reviewers and Consultants

Roy J. Arlotto, Louis E. Dieruff High School, Allentown, PA

E. Patricia Buchanan, Menlo-Atherton High School, Menlo Park, CA

Carolyn Csongradi, Sequoia High School, Redwood City, CA

Marilyn Linner-Luebe, Fulton High School, Fulton, IL

Samuel M. Meyer, Southwest Miami High School, Miami, FL

John Moore, John F. Kennedy High School, Richmond, CA

Margaret Nicholson, Lawrence Hall of Science, Berkeley, CA

Lee Carollo Pforsich, Redwood High School, Larkspur, CA

Roberta Sund, Wichita Falls High School, Wichita Falls, TX

Costella Watson, Tennyson High School, Hayward, CA

Bonnie Williams, Delany High School, Timonium, MD

ISBN 0-201-28239-9

3 4 5 6 7 8 9 10 - VH - 95 94 93

Symbols of Common Elements

Ag	silver	Cu	copper	O	oxygen
Al	aluminum	F	fluorine	P	phosphorus
As	arsenic	Fe	iron	Pb	lead
Au	gold	H	hydrogen	Pt	platinum
Ba	barium	Hg	mercury	S	sulfur
Bi	bismuth	I	iodine	Sb	antimony
Br	bromine	K	potassium	Sn	tin
C	carbon	Mg	magnesium	Sr	strontium
Ca	calcium	Mn	manganese	Ti	titanium
Cl	chlorine	N	nitrogen	U	uranium
Co	cobalt	Na	sodium	W	tungsten
Cr	chromium	Ni	nickel	Zn	zinc

Symbols of Common Polyatomic Ions

$C_2H_3O_2^-$	acetate	$Cr_2O_7^{2-}$	dichromate	NH_4^+	ammonium
ClO^-	hypochlorite	HCO_3^-	hydrogen carbonate (bicarbonate)	NO_3^-	nitrate
ClO_2^-	chlorite			NO_2^-	nitrite
ClO_3^-	chlorate	H_3O^+	hydronium	O_2^{2-}	peroxide
ClO_4^-	perchlorate	HPO_4^{2-}	hydrogen phosphate	OH^-	hydroxide
CN^-	cyanide	HSO_3^-	hydrogen sulfite	PO_4^{3-}	phosphate
CO_3^{2-}	carbonate	HSO_4^-	hydrogen sulfate	SO_3^{2-}	sulfite
CrO_4^{2-}	chromate	MnO_4^-	permanganate	SO_4^{2-}	sulfate

Other Symbols and Abbreviations

α	alpha rays	gmm	gram molecular mass	m	mass
β	beta rays	H	enthalpy	m	molality
γ	gamma rays	H_f	heat of formation	mL	milliliter (*volume*)
Δ	change in	h	hour	mm	millimeter (*length*)
δ^-, δ^+	partial ionic charge	h	Planck's constant	mol	mole (*amount*)
λ	wavelength	Hz	hertz (*frequency*)	mp	melting point
π	pi bond	J	joule (*energy*)	N	normality
σ	sigma bond	K	kelvin (*temperature*)	n^0	neutron
ν	frequency	K_a	acid dissociation constant	n	number of moles
amu	atomic mass unit	K_b	base dissociation constant	n	principal quantum number
(*aq*)	aqueous solution	K_b	molal boiling point elevation constant	P	pressure
atm	atmosphere (*pressure*)			p^+	proton
bp	boiling point	K_{eq}	equilibrium constant	Pa	pascal (*pressure*)
°C	degree Celsius (*temperature*)	K_f	molal freezing point depression constant	R	ideal gas constant
c	speed of light in a vacuum			S	entropy
cm	centimeter (*length*)	K_w	ion product constant for water	s	second
D	density			(*s*)	solid
E	energy	K_{sp}	solubility product constant	SI	International System of Units
e^-	electron	kcal	kilocalorie (*energy*)		
fp	freezing point	kg	kilogram (*mass*)	STP	standard temperature and pressure
G	Gibb's free energy	kPa	kilopascal (*pressure*)		
g	gram (*mass*)	L	liter (*volume*)	T	temperature
(*g*)	gas	(*l*)	liquid	$t_{1/2}$	half-life
gam	gram atomic mass	M	molarity	V	volume
gfm	gram formula mass	m	meter (*length*)	v	velocity, speed

Contents

Science, Technology, Society Features

C	Careers	A	Applications
H	History	I	Instrumentation
R	Research	L	Laboratory Techniques
T	Technology	N	Nature of Science

1 Matter, Change, and Energy

Chapter Preview

1·1 Chemistry
1·A Chemical Technology
1·2 The Scientific Method
1·B Alchemy and the Birth of Chemistry
1·3 Properties of Matter
1·4 The States of Matter
1·5 Physical Changes

1·6 Mixtures
1·7 Elements and Compounds
1·8 Chemical Symbols
1·9 Energy
1·10 Conservation of Energy
1·11 Chemical Reactions
1·12 Conservation of Mass

1·1 Chemistry

Chemists work at an amazing variety of jobs. Some develop new products such as textiles, paints, medicines, or cosmetics. Others may find methods to reduce pollution or to clean up the environment. A specialized chemist may be called on to identify and interpret the evidence found at the scene of a crime. Many chemists teach. Others do analyses of substances or check the quality of manufactured products. What all of these people have in common is a knowledge of chemistry.

Chemistry *is the study of the composition of substances and the changes that substances undergo.* Our world is complex. Chemistry reflects this complexity in the broad areas of interest that it covers. It contributes to other natural sciences including biology, geology, and physics. Chemistry overlaps with agriculture, medicine, and many manufacturing industries as well.

Most of the careers just mentioned apply a knowledge of chemistry to attain specific goals. These are examples of applied chemistry, or chemical technology. Pure chemistry, like other pure sciences, accumulates knowledge for its own sake. Pure science is neither good nor bad. Applied chemistry focuses on the uses of chemical knowledge. Technology can use scientific knowledge in ways that benefit or harm people or the environment. The political and social debates about the uses of scientific knowledge are really debates about technology.

Because chemistry is so diverse, it is usually considered to have five major divisions. With a few exceptions, **organic chemistry** *is the study of*

Chemistry is a natural science.

Figure 1·1
Chemical changes are responsible for many of the spectacular events around us. Here molten lava is being ejected from the Hawaiian volcano Kilauea.

1

Figure 1·2
Glassware is used extensively in a "wet chemistry" laboratory. Most laboratories now use many automated electronic instruments to do chemical analyses.

essentially all substances containing carbon. It was originally the study of substances from living organisms. **Inorganic chemistry** *specializes in substances without carbon*. These are mainly substances from nonliving things. **Analytical chemistry** *is concerned primarily with the composition of substances*. Finding the amount of silver in an ore or minute quantities of a substance in a blood sample requires the practice of analytical chemistry. **Physical chemistry** *specializes in the discovery and description of the theoretical basis of the behavior of chemical substances*. Usually it relies heavily on mathematics. **Biochemistry** *is the study of the composition and changes in composition of living organisms*. Obviously, these five subdivisions of chemistry overlap. For example, one cannot measure a change in an organic or inorganic substance without some proficiency in analytical chemistry.

Figure 1·3
After designing a chemical processing system, a chemical engineer must be able to explain the system's features.

Science, Technology, and Society

1·A Chemical Technology

To meet the needs of our modern world, enormous quantities of chemicals are manufactured every year. These include fertilizers, fuels, pesticides, textiles, plastics, and other materials. Working with ordinary laboratory glassware, chemists develop the processes for producing small amounts of these chemicals. It is the job of the chemical engineer to plan and carry out the same reactions on a large scale.

Chemical engineers are employed by many industries. They work for companies that produce fuels, metals, rubber, cosmetics, drugs, paper, paints, and foods. Chemical engineers must determine whether a reaction can be used for mass production. They plan the layout of a plant, design equipment for it, and supervise its construction and operation. Often they are asked to redesign an existing plant to increase productivity. They may add new safety or pollution control features. Because they are responsible for the efficient operation of the plant, they must always be aware of costs.

Many engineering and technical schools offer degrees in chemical engineering. The course of study emphasizes chemistry, physics, mathematics, economics, writing, and the use of computers. Specialized engineering courses are also required.

1·2 The Scientific Method

Like most fields of human endeavor, science has evolved formal and time-tested methods to solve problems. The scientific method is one important approach. It was through the scientific method that many of the chemical elements were first discovered. *The* **scientific method** *incorporates observations, hypotheses, experiments, theories, and laws. Scientists make* **observations** *when they note and record facts about natural phenomena.* They try to explain their observations by devising hypotheses. **Hypotheses** *are descriptive models for observations.* A hypothesis is useful if it accounts for what scientists observe in many situations.

In order to learn more, *scientists often perform* **experiments** *in which one or more of the conditions are controlled.* An important principle of an experiment is that it can be repeated numerous times. *The observations, which are recorded from an experiment, constitute* **data.** When observations or experimental data do not fit the hypothesis, it must be scrapped or adjusted. The new or refined hypothesis is then subjected to further experimental testing. An important interplay takes place between hypotheses and experiments. The hypothesis guides the design of new experiments. At the same time, experiments guide the rejection or refinement of the hypothesis.

Once a hypothesis meets the test of repeated experimentation, it may be elevated to a theory. *A* **theory** *is a thoroughly tested model that explains why experiments give certain results.* A theory can never be proved. Nevertheless, theories are very useful because they help us to form mental pictures of objects or processes that cannot be seen. Moreover they give us the power to predict the behavior of natural systems under circumstances that are different from those of the original observations.

Another product of scientific research is a law. *A* **scientific law** *is a concise statement that summarizes the results of a broad variety of observations and experiments.* A scientific law is different from a theory in that it only describes a natural phenomenon. It does not attempt to explain it. Scientific laws can often be expressed by simple mathematical relationships. They usually concern natural behaviors that are not immediately obvious. For most people a statement like Boyle's law, "The volume of a gas is inversely proportional to the pressure exerted on it," meets these criteria.

Chemical Connections

Even though scientists may use the same scientific facts, they often disagree about how scientific knowledge should be used by society. For example, one chemist may create a new chemical for use as a pesticide, while another points out the dangers of the same chemical.

When viewing the world, a scientist is interested in both what is happening and why it is happening.

Occasionally a widely held scientific theory must be modified to agree with new experimental evidence. This is usually the cause of much excitement in the scientific community.

Figure 1·4
This outline of the scientific method shows how experimental observations lead to the development of hypotheses and theories. A scientific law summarizes the results of many experiments, but it does not explain why a behavior is observed; that is the role of the hypothesis and the theory. A theory can be modified to account for results of new experiments.

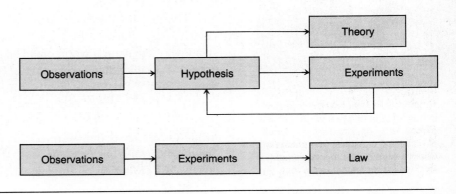

Problem

1. Classify each of the following statements as an example of experiment, hypothesis, theory, or law.
   a. The ashes from a campfire weigh less than the wood that was burned. Therefore mass is destroyed (or lost) when wood is burned.
   b. A body at rest tends to remain at rest.
   c. Water boils at 100°C on your kitchen stove and in the laboratory.
   d. All matter is composed of atoms, which themselves are composed of protons, electrons, and neutrons.

Science, Technology, and Society

1·B Alchemy and the Birth of Chemistry

Long before the science of chemistry existed, people made use of chemical reactions to dye cloth, tan leather, and prepare foods. Eventually people began to search for explanations for the structure and behavior of matter. Greek philosophers 2500 years ago believed that knowledge of the natural world could be achieved through pure reasoning. One group, the "Atomists," proposed that matter was made of tiny indivisible particles called atoms. The famous philosopher Aristotle disagreed. He said that matter was composed of four elements: fire, air, water, and earth. Because of Aristotle's fame and reputation, much of the world agreed with his ideas. As a result, the idea of atoms was not discussed seriously for more than 2000 years.

With the destruction of Greek civilization, European science fell into disrepair and did not reappear until the Middle Ages. This period saw the rise of the alchemists. Their goal was to change common metals into gold. Although the alchemists were unable to succeed in their quest, they did spur the development of science. They developed many experimental procedures and laboratory apparatus. Through trial and error they also developed a wealth of knowledge about the characteristics of substances.

During the thirteenth century, an English Franciscan monk named Roger Bacon introduced a new idea. He held that an understanding of the natural world could be gained through observation and experiment rather than by pure logic. Bacon's ideas were not immediately popular. They were finally put into practice, however, by scientists of the sixteenth and seventeenth centuries. The Englishman Robert Boyle (1627–1691) emphasized the necessity of using experiments to test ideas that were obtained by reason. In his book, *The Sceptical Chymist*, he challenged Aristotle's four "elements." He also advanced a definition of an element very similar to that used today.

Figure 1·5
Alchemists mixed materials together in hundreds of different ways. For this purpose they designed various types of balances, crucibles, and glassware. Chemists later used this equipment for chemical research.

Figure 1·6
This nineteenth century chemical laboratory was named after Lavoisier, the great eighteenth century chemist.

Every substance has a unique set of physical properties.

Antoine Lavoisier (1743–1794) took an important new step in the process of experimentation. This was to make precise measurements of the mass changes in chemical reactions. His experiments transformed chemistry from a science of observation to the science of measurement that it is today. For this reason, Lavoisier is often called the founder of modern chemistry.

1·3 Properties of Matter

The material things around us are all various types of matter. **Matter** *is anything that takes up space and has mass*. Aluminum, water, air, glass, and you are different kinds of matter. Ideas, light, and heat are not matter. They do not take up space or have mass. *The amount of matter that an object contains is its* **mass.** A golf ball has more mass than a table tennis ball. It contains more matter.

Some materials are composed of numerous types of matter; others consist of a single kind. *A* **substance** *is a particular kind of matter that has a uniform and definite composition*. Table sugar is a substance. It is 100% sucrose. Substances are often called pure substances. Lemonade is not a substance. It contains different amounts of sugar and lemon juice in water.

All samples of a substance have identical properties. A **physical property** *is a quality or condition of a substance that can be observed or measured without changing the substance's composition*. It can be specified without reference to any other substance. Some physical properties of matter include color, solubility, mass, odor, hardness, density, electrical conductivity, magnetism, melting point, and boiling point. Physical properties help chemists identify substances (Table 1·1). A colorless odorless liquid in which both salt and sugar dissolve and that boils at 100°C is probably water. Another colorless liquid that has a distinctive odor, that evaporates quickly when placed on your skin, and in which salt will not dissolve is most certainly not water.

Table 1·1 Physical Properties of Some Common Substances				
Substance	State (at normal conditions)	Color	Melting point (°C)	Boiling point (°C)
Oxygen	Gas	Colorless	-218	-183
Bromine	Liquid	Red-brown	-7	59
Water	Liquid	Colorless	0	100
Sulfur	Solid	Yellow	113	445
Sodium chloride (table salt)	Solid	White	801	1413
Iron	Solid	Silver-white (freshly cut)	1535	2750

1·4 The States of Matter

Matter can exist in three different physical states: solid, liquid, and gas. The physical state of a substance is a physical property of that substance. Certain characteristics summarized in Table 1·2 distinguish each state of matter.

A **solid** *is matter that has a definite shape and volume.* The shape of a solid does not depend on the shape of the container (Figure 1·7). Solids usually have a high density, and they expand only slightly when heated. They are almost incompressible. Coal, sugar, bone, ice, and iron are examples of solids.

A **liquid** *is a form of matter that flows, has a fixed volume, and takes the shape of its container.* Liquids are generally less dense than solids. They expand slightly when heated and are almost incompressible. Examples of liquids are water, milk, and blood.

A **gas** *is matter that takes both the shape and the volume of its container.* Gases expand without limit to fill any space and are easily compressed. The term *gas* is limited to those substances that exist in the gaseous state at room temperature. For example, air is a mixture of gases including oxygen and nitrogen. *The word* **vapor** *describes a substance that, although in the gaseous state, is generally a liquid or solid at room temperature.* Steam, the gaseous form of water, is a vapor. Moist air contains water vapor. The words *gas* and *vapor* should not be used interchangeably; there is a difference.

Safety

Gases expand dramatically when heated. Never heat a tightly closed vessel, as the build-up in gas pressure may cause an explosion.

A fourth state of matter, plasma, exists only at very high temperatures. Most of the matter in the universe is plasma because stars are in a plasma state.

Problem

2. What is the physical state of each of the following at room temperature? **a.** silver **b.** gasoline **c.** helium **d.** paraffin wax **e.** rubbing alcohol

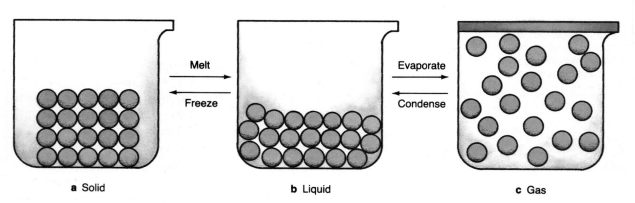

a Solid **b** Liquid **c** Gas

Figure 1·7
The three states of matter differ because of the arrangement of their particles.
a A solid has its own shape and volume. **b** A liquid conforms to the shape but not the volume of its container. **c** A gas will occupy all the volume of its container.

Table 1·2	Important Properties of the States of Matter		
Property	Solid	Liquid	Gas or vapor
Mass	Definite	Definite	Definite
Shape	Rigid	Indefinite	Indefinite
Volume	Definite	Definite	Indefinite
Temperature increase	Very small expansion	Moderate expansion	Large expansion
Compressibility	Almost incompressible	Almost incompressible	Compressible

1·5 Physical Changes

Physical changes in the state of matter can result from changes in temperature.

Matter can be changed in many ways. *A* **physical change** *will alter a substance without changing its composition.* Cutting, grinding, or bending a material will cause a physical change. The substance remains the same. A change in temperature may also bring about a physical change. The melting of ice, the freezing of water, the conversion of water to steam, and the condensation of steam to water are all examples of physical changes. We know that these physical transformations do not change the identity of the water. The physical properties of water are the same for water that has been frozen and melted as they are for water that has been converted to steam and then condensed.

Words like boil, freeze, melt, condense, break, split, crack, grind, cut, crush, and bend usually signify a physical change. Table salt (sodium chloride) is a white solid at room temperature. It can also exist in each of the three physical states. However, the temperatures at which the changes of state occur in sodium chloride are much higher than for the corresponding changes in water. Sodium chloride melts at 801°C and boils at 1413°C.

Figure 1·8
Physical changes can take place as a result of temperature changes. The melting point of the metal gallium is 30°C. Body temperature is about 37°C.

1·6 Mixtures

Most matter consists of mixtures of substances.

Mixtures *consist of a physical blend of two or more substances.* They differ from pure substances because they have a variable composition. Most materials found in nature are mixtures. We all recognize beef stew as a mixture of meat, vegetables, and gravy. On the other hand, the identification of air or brine as mixtures is much less obvious. (Air is a mixture of gases. Brine is a mixture of salt and water.) The component parts of these mixtures cannot be distinguished even under a microscope.

Mixtures can be heterogeneous or homogeneous. *A* **heterogeneous mixture** *is not uniform in composition.* If we were to sample one portion of the mixture, its composition would be different from another sample. Soil is a heterogeneous mixture. It contains bits of decayed material along with sand, silt, and/or clay. By contrast, *a* **homogeneous mixture**

Greek: *heteros* = different
genos = kind
homo = the same

Figure 1·9
All these items are mixtures. The toothpaste is a homogeneous mixture. It has a uniform composition. The sand in water, fruit salad, granite, and soil are heterogeneous mixtures. They consist of a number of phases that are not evenly distributed.

Chemical Connections

Dental fillings are an example of a special kind of mixture made by dissolving silver and tin in mercury. This kind of mixture of mercury and one or more metals is called an amalgam.

Figure 1·10
This is a microscopic view of the cross section of a leaf. Like most foods and all living things, a leaf is a complex heterogeneous mixture of substances.

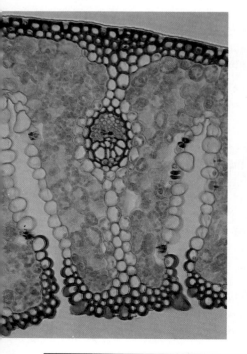

has a completely uniform composition. Its components are evenly distributed throughout the sample. Salt water from the ocean is a homogeneous mixture. It is the same throughout a sample.

One important characteristic of both heterogeneous and homogeneous mixtures is that their compositions may vary. The composition of air in a forest differs from that near an industrial city, particularly in the amounts and kinds of pollutants it contains. Blood is a mixture of water, various chemicals, and cells. Blood composition varies somewhat from one individual to another. Moreover, each person's blood composition varies with health, nutrition, and activity.

Homogeneous mixtures are so important in chemistry that chemists give them the special name of solutions. *A **solution** is a homogeneous mixture.* As shown in Table 1·3, solutions may be gases, liquids, or solids. If we were to sample any portion of a sugar solution, we would find that it has the same composition as any other portion. *Any part of a system with uniform composition and properties is called a **phase.*** Thus a homogeneous mixture consists of a single phase, and a heterogeneous mixture consists of two or more phases. Vinegar and oil dressing is a heterogeneous mixture with two phases.

Some mixtures can be separated into their various components by simple physical methods. We could use a spoon to separate beef stew into meat, vegetables, and gravy. As a slightly more sophisticated example, consider the gray-colored heterogeneous mixture of powdered yellow

Table 1·3	Some Common Types of Solutions
System	Examples
Gas–gas	Carbon dioxide and oxygen in nitrogen (air)
Liquid–gas	Water vapor in air (moist air)
Gas–liquid	Carbon dioxide in water (soda water)
Liquid–liquid	Acetic acid in water (vinegar)
Solid–liquid	Sodium chloride in water (brine)
Solid–solid	Copper in silver (sterling silver, an alloy)

Sulfur

Iron

Mixture of iron and sulfur

Figure 1·11
Physical methods can be used to separate a mixture. The iron filings are attracted to a magnet, but the powdered sulfur is not.

 Issues in Chemistry

Many communities face water shortages. The oceans of the world contain 98% of the earth's water. Can ocean water be used to satisfy the ever increasing demand for fresh water? Why or why not?

sulfur and black iron filings. The individual particles of sulfur and iron can be readily distinguished from one another under a microscope. The mixture is easy to separate. The iron filings can be removed from the mixture with a magnet, leaving the sulfur behind (Figure 1·11). Both the sulfur and the iron are unchanged in composition.

Tap water is a homogeneous mixture of water plus other substances that are dissolved in it. One method used to purify water is distillation. *During* **distillation** *a liquid is boiled to produce a vapor that is then condensed again to a liquid.* When water is distilled it is heated to form steam, which is then condensed to give water (Figure 1·12). The water is collected in a receiver. The solid substances originally dissolved in the water remain in the distillation flask. Distilled water is pure except for the dissolved gases it contains. Water from which even the dissolved gases are removed is a pure substance. Water has a unique set of properties: It freezes at 0°C and boils at 100°C.

Example 1

How would you separate the following mixtures? **a.** iron filings from aluminum filings **b.** sawdust from sand **c.** sand from salt

Solution

a. Use a magnet to attract the iron filings.
b. Add water to the mixture. The sawdust floats, and sand sinks.
c. Add water to dissolve the salt. Pour the mixture onto a piece of cloth. The sand remains on the cloth, and the salt solution goes through. Use evaporation to remove the water and leave solid salt.

Problem

3. Classify each of the following as homogeneous or heterogeneous mixtures. **a.** blood **b.** chocolate-chip ice cream **c.** brass (copper and zinc) **d.** homogenized milk **e.** a cup of coffee

Figure 1·12
A solution of impure water is being distilled. The water is boiled to steam that is passed through a water-cooled condenser. The steam condenses to distilled water that is collected in a receiver.

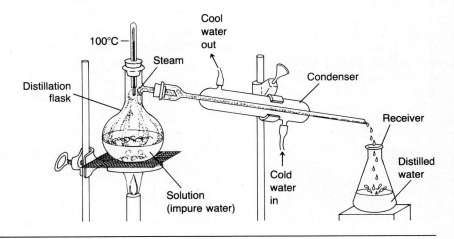

1·7 Elements and Compounds

Every substance is either an element or a compound.

No two substances have identical physical and chemical properties.

Figure 1·13
Some elements and compounds are common household items. The elements shown here are aluminum, gold (in the ring), and mercury (in the thermometer). The compounds are sodium chloride (table salt) and sodium bicarbonate (baking soda).

Safety

Never taste any chemical in the laboratory, even if you are sure that it is safe.

By separating mixtures into their component parts we obtain pure substances. As you remember, a substance has a uniform and definite composition. Substances are divided into two groups, elements and compounds. Because you will work with elements and compounds in the laboratory, you should know how to distinguish between them. **Elements** *are the simplest forms of matter that can exist under normal laboratory conditions.* They cannot be separated into simpler substances by chemical reactions. Elements are the building blocks for all other substances. Two or more elements can combine with one another to form compounds. **Compounds** *are substances that can be separated into simpler substances only by chemical reactions.* The constituent elements in a compound are always present in the same proportions. Every element and every compound has its own unique set of properties.

The difference between elements and compounds can be demonstrated by heating refined cane sugar, a compound. Upon heating, the sugar undergoes a chemical reaction. It decomposes to carbon and water vapor. This experiment shows that cane sugar is a compound, not an element. It is separated by a chemical reaction into two substances, carbon and water. Now, what about the carbon and water? The water that comes from the breakdown of the sugar can be broken down into hydrogen and oxygen by another chemical reaction. Thus water too is a compound. Carbon, hydrogen, and oxygen cannot be broken down into simpler substances. They are elements.

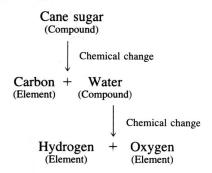

In general, the chemical and physical properties of compounds are quite different from those of their component elements. For example, sugar is a white sweet solid, but carbon is a black tasteless solid. Water is a colorless liquid, but oxygen and hydrogen are colorless gases. Table salt (or sodium chloride, to use its chemical name) is a harmless compound in most instances. It is composed of the elements sodium and chlorine. Sodium is a soft metal that reacts explosively with water. Chlorine is a pale yellow-green poisonous gas (Figure 1·14).

Much of what we have said about elements, compounds, and mixtures is summarized in Figure 1·15. Deciding whether a sample of matter is a pure substance or a homogeneous mixture can be difficult. After all, a homogeneous mixture looks like a pure substance. As a help, ask

Figure 1·14
The compound sodium chloride is common table salt. It is composed of the elements sodium (a solid) and chlorine (a gas). As a safety precaution, sodium is stored under oil to prevent it from reacting explosively with moisture in the air.

Chlorine gas (Cl_2)

Sodium metal (Na)

Salt Sodium chloride (NaCl)

yourself: "Is there more than one kind of this material?" For example, how would you classify gasoline? Based on its physical appearance you might conclude that gasoline is a compound. It must be a mixture, however, since there are so many different grades of gasoline (at so many different prices). You should also conclude that gasoline is a homogeneous mixture because each grade is uniform throughout.

Example 2

When a blue solid is heated in the absence of air a colorless gas and a white solid are formed. Which of these substances are elements and which are compounds? Explain.

Solution

The blue solid was separated into two substances by heating. Therefore it must be a compound. The two resulting substances may be elements or compounds.

Problem

4. A clear liquid in an open container is allowed to evaporate. After three days a solid residue is left. Was the original liquid an element, compound, or mixture? How do you know?

Figure 1·15
Any sample of matter can be classified as an element, compound, or mixture.

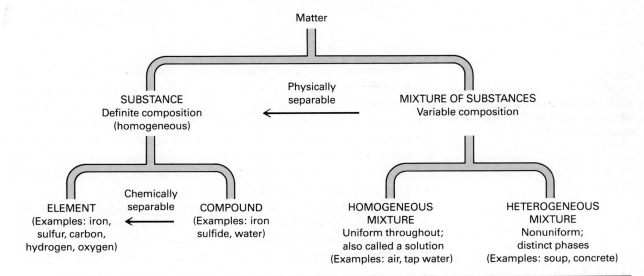

1·8 Chemical Symbols

Elements are represented by chemical symbols.

Symbols of common elements are given in Table A·3 on page 672.

Only about 90 of the elements are known to occur naturally on earth. The other known elements have been created in the laboratory.

Carbon, hydrogen, oxygen, sodium, and chlorine are only a few of the approximately 100 known elements. All matter in the universe is composed of these elements. *Each element is represented by a* **chemical symbol.** The symbols for most elements consist of the first one or two letters of the name of the element (Table A·3). The first letter is always capitalized. If a second letter is needed, it must be lowercase. Some element symbols are derived from older Latin names. In those cases, the symbol is not always consistent with the common name. Table 1·4 lists these exceptions.

Each element has properties in some way different from those of every other element. For example, helium (He) is a chemically inert gas that is lighter than air. Iron (Fe) is a silver-gray solid that conducts electricity and rusts in moist air.

Chemical symbols are used to write chemical formulas of compounds. The familiar compound water is composed of the elements hydrogen and oxygen. The formula for water is H_2O. The formula for sucrose (table sugar) is $C_{12}H_{22}O_{11}$. Sucrose is composed of the elements carbon, hydrogen, and oxygen. Baking soda, sodium bicarbonate, has the formula $NaHCO_3$. Baking soda is made of four elements: sodium, hydrogen, carbon, and oxygen. The numbers in chemical formulas represent the proportions of the various elements in the compound. Every compound is always made-up of the same elements in the same proportions. Thus the formula for each of these compounds is always the same.

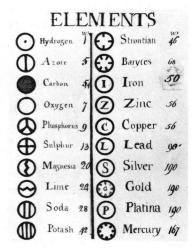

Figure 1·16
Several chemists devised symbols to represent the elements. Dalton's symbols are shown in this reproduction of one of his lecture diagrams.

ChemDirections

Four elements make up most of the human body. Other elements are essential in trace amounts. To learn more about the elements required for life, turn to **ChemDirections** *Mineral Nutrients*, page 652.

Example 3

Write the chemical symbol for each of these elements.   **a.** mercury   **b.** gold   **c.** iodine   **d.** magnesium   **e.** calcium   **f.** barium

Table 1·4	Symbols and Latin Names for Some Elements	
Element	Symbol	Latin name
Sodium	Na	*Natrium*
Potassium	K	*Kalium*
Antimony	Sb	*Stibium*
Copper	Cu	*Cuprum*
Gold	Au	*Aurum*
Silver	Ag	*Argentum*
Iron	Fe	*Ferum*
Lead	Pb	*Plumbum*
Mercury	Hg	*Hydrargyrum*
Tin	Sn	*Stannum*
Tungsten	W	From Wolfram (not Latin)

Solution

a. Hg **b.** Au **c.** I **d.** Mg **e.** Ca **f.** Ba

Problems

5. What chemical elements are represented by the following symbols?
a. Sn **b.** Cu **c.** N **d.** Cd **e.** P **f.** Cl

6. Classify each of the following as an element, compound, or mixture.
a. argon **b.** ethyl alcohol (C_2H_5OH) **c.** grape juice
d. cheese **e.** zinc

1·9 Energy

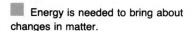
Energy is needed to bring about changes in matter.

Energy *is the capacity for doing work.* Nearly all chemical and physical changes in nature involve the absorption or emission of energy. Energy may exist in any one of several forms. It includes chemical, nuclear, electrical, radiant, mechanical, and thermal energies.

Light is *radiant energy.* The use of light by plants is an example of how one form of energy may be converted to another. Green plants use the radiant energy of sunlight to do the work of making carbon dioxide and water into complex chemicals. This process is called photosynthesis. The chemicals store *chemical energy* that can be used by plants for life processes. Green plants, therefore, survive by acquiring energy in one form and using it in another. Stored chemical energy is a form of **potential energy,** *the energy of position.* Chemical energy depends on the composition of a substance. When we buy gasoline to power our cars, we are buying chemical potential energy. When the gasoline is burned in the car's engine, this potential energy is converted into **kinetic energy,** *the energy of motion.*

Figure 1·17
Describe the energy in a tomato by using terms introduced in Section 1·9. What is the source of its energy?

As a second example of the interconvertibility of forms of energy, consider what can happen when natural gas is burned in air. This chemical process releases the chemical energy stored in the chemical structure of the gas. This chemical energy could be lost as **heat,** *which is energy that is transferred from one body to another because of a temperature difference.* The heat could also be used to raise the *thermal energy* of water, changing it to steam. The steam can drive a turbine (*mechanical energy*), and the turbine can generate *electrical energy* (Figure 1·18). The electrical energy can then be used to power an electric drill (mechanical energy).

Problem

7. One type of bicycle light is powered by a generator located on the rear wheel of the bicycle. Trace the interconversion of energy from the sun to the light produced by this bicycle light.

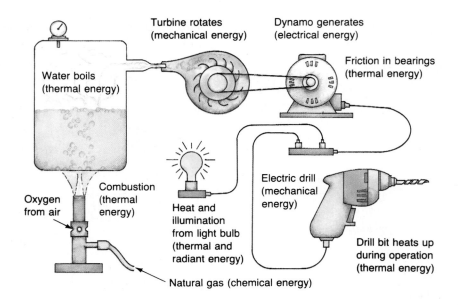

Turbine rotates (mechanical energy)

Dynamo generates (electrical energy)

Friction in bearings (thermal energy)

Water boils (thermal energy)

Electric drill (mechanical energy)

Oxygen from air

Combustion (thermal energy)

Heat and illumination from light bulb (thermal and radiant energy)

Drill bit heats up during operation (thermal energy)

Natural gas (chemical energy)

1·10 Conservation of Energy

The forms of energy can be interconverted, but energy is neither created nor destroyed in ordinary physical and chemical changes.

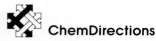

ChemDirections

Solar cells convert the energy of sunlight to electrical energy. Only about 25% of the sunlight that strikes a solar cell is changed into electricity. To find out more about solar cells, turn to **ChemDirections** *Solar Cells*, page 655.

It is not hard to see that various forms of energy may be interconverted. An important question is: What happens to the amount of energy as it is changed from one form to another? When steam is used to drive a turbine, the kinetic energy of the moving turbine will be less than the energy used to heat the water to steam. Does this mean that some of the energy is destroyed? No, because some of the energy escapes as heat. The connections between moving parts of the turbine become hot from friction. The air surrounding the turbine becomes warm. When this thermal energy is taken into account, then the energy input equals the energy output.

The **law of conservation of energy** *states that in any chemical or physical process, energy is neither created nor destroyed.* The energy may be converted from one form to another. Often it is converted to a form that is not useful. In any chemical or physical process, however, energy is conserved. All of the energy involved can be accounted for as work, stored energy, or heat.

1·11 Chemical Reactions

Chemical changes in matter are called chemical reactions.

Energy changes occur whenever a chemical reaction takes place. *In a* **chemical reaction** *one or more substances are changed into new substances.* For example, when refined cane sugar is heated it undergoes a process called caramelization. The sugar melts to a colorless liquid that soon develops a brown color and a caramel odor as the temperature is raised. At this point the sugar is breaking down, and a chemical change is taking place. The important point is that when the hot liquid is allowed to cool, the sugar has lost its identity. It does not turn back into sugar, because the sugar has been changed to other substances. The change in

Mixture of iron and sulfur

Iron sulfide

Figure 1·19
A chemical change occurs when a mixture of powdered sulfur and iron filings is heated. The reactants are converted into a product that has chemical and physical properties different from those of the reactants.

composition of the sugar is the result of a chemical reaction. *In a chemical reaction, the starting substances are called* **reactants,** *and the new substances are the* **products.** Chemists use an arrow as a shorthand form of the phrase "are changed into."

$$\text{Reactants} \longrightarrow \text{products}$$

To help distinguish between physical and chemical changes, recall how sulfur and iron filings may be separated unchanged from a mixture. This separation is an example of a physical change. If the mixture of these two substances is heated, however, a chemical change takes place. The sulfur and iron are changed into a nonmagnetic substance, iron sulfide (Figure 1·19). Using the shorthand expression, this change can be written as follows.

$$\underset{\text{Reactants}}{\text{Iron + sulfur}} \xrightarrow{\text{heat}} \underset{\text{Product}}{\text{iron sulfide}}$$

Several occurrences are common indicators of a chemical change. Energy is usually absorbed or evolved in chemical reactions. Burning coal evolves heat; cooking food absorbs heat. Energy is also absorbed or evolved, however, in physical changes of state. A color change, as in leaves turning in the fall, or an odor change, as in meat rotting, often accompanies a chemical change. The production of a gas or a solid can be the result of a chemical change. *A solid that separates from a solution is called a* **precipitate.** Gas bubbles rising from the bottom of a stagnant lake are formed by the chemical decay of plant matter on the lake floor. Gas or vapor formation can also be the result of a change of state. Boiling water is an example. Finally, it is sometimes most helpful to consider whether the change can be easily reversed. Physical changes, especially those involving a change of state, are usually reversible. Ice can be melted and then the water refrozen. In contrast, most chemical changes are not easily reversed. Once iron has reacted with oxygen to form rust, as it often does on a car, you cannot easily reverse the process.

Figure 1·20
The formation of a gas or of a solid precipitate is a common indication that a chemical change may be taking place. On the left, iron reacts with dilute sulfuric acid to release hydrogen gas. On the right, red silver chromate forms when a solution of sodium chromate is added to a solution of silver nitrate.

Just as every substance has physical properties, it also has chemical properties. *The ability of a substance to undergo chemical reactions and to form new substances constitutes its* **chemical properties.** Chemical properties describe chemical changes. For example, when iron is exposed to water and oxygen it corrodes and produces a new substance called iron oxide (rust). Rusting is a chemical property of iron. Chemical properties are observed only when a substance is undergoing a change in composition. Words like rot, rust, decompose, ferment, grow, decay, and sprout usually signify a chemical change.

Problem

8. Classify the following changes as physical or chemical.

a. Food spoils.	**e.** Oil is pumped out of a well.
b. Water boils.	**f.** Bread is baked.
c. A nail rusts.	**g.** Sugar dissolves in water.
d. A firefly emits light.	**h.** A snowflake melts.

1·12   Conservation of Mass

The total mass of substances involved in a physical or chemical change is conserved.

Combustion, or burning, is an example of one of the most familiar chemical changes. When we burn a piece of coal, atmospheric oxygen combines with the carbon in the coal. The products are carbon dioxide gas, water vapor, and a large residue of ash. With careful measurements we find that the mass of the reactants (the coal and oxygen consumed) equals the mass of the products (the carbon dioxide, water vapor, and ash). During the chemical reaction, the quantity of matter is unchanged. Similarly, when 10 grams of ice melts, 10 grams of water is obtained. Again, in this physical process, mass is conserved. Similar observations have been made for millions of chemical and physical changes. Thus *the* **law of conservation of mass** *states that in any physical or chemical reaction, mass is neither created nor destroyed; it is conserved.* In every case, the mass of the products is equal to the mass of the reactants.

The law of conservation of mass does not hold in nuclear reactions. In these reactions matter and energy are interconverted.

Figure 1·21
This experiment demonstrates the law of conservation of mass. On the left a candle is being burned in a sealed flask of oxygen. As the candle burns it combines with the oxygen within the flask. At the same time, it produces water and carbon dioxide. Throughout this process the mass of the flask does not change.

Key Terms

analytical chemistry	$1 \cdot 1$	law of conservation	
biochemistry	$1 \cdot 1$	of mass	$1 \cdot 12$
chemical property	$1 \cdot 11$	liquid	$1 \cdot 4$
chemical reaction	$1 \cdot 11$	mass	$1 \cdot 3$
chemical symbol	$1 \cdot 8$	matter	$1 \cdot 3$
chemistry	$1 \cdot 1$	mixture	$1 \cdot 6$
compound	$1 \cdot 7$	observation	$1 \cdot 2$
data	$1 \cdot 2$	organic chemistry	$1 \cdot 1$
distillation	$1 \cdot 6$	phase	$1 \cdot 6$
element	$1 \cdot 7$	physical change	$1 \cdot 5$
energy	$1 \cdot 9$	physical chemistry	$1 \cdot 1$
experiment	$1 \cdot 2$	physical property	$1 \cdot 3$
gas	$1 \cdot 4$	potential energy	$1 \cdot 9$
heat	$1 \cdot 9$	precipitate	$1 \cdot 11$
heterogeneous		product	$1 \cdot 11$
mixture	$1 \cdot 6$	reactant	$1 \cdot 11$
homogeneous		scientific law	$1 \cdot 2$
mixture	$1 \cdot 6$	scientific method	$1 \cdot 2$
hypothesis	$1 \cdot 2$	solid	$1 \cdot 4$
inorganic chemistry	$1 \cdot 1$	solution	$1 \cdot 6$
kinetic energy	$1 \cdot 9$	substance	$1 \cdot 3$
law of conservation		theory	$1 \cdot 2$
of energy	$1 \cdot 10$	vapor	$1 \cdot 4$

Chapter Summary

Chemistry is a natural science that deals with the composition of matter and the changes it undergoes. Matter is anything that has mass and occupies space. Matter exists in three states: solid, liquid, and gas. Chemists use the scientific method to learn how matter can be changed and how energy is involved in these changes.

A physical combination of two or more substances is a mixture. A mixture has a variable composition and may be identified as heterogeneous or homogeneous. A mixture can be separated into its components by physical methods. Homogeneous mixtures (solutions) have uniform properties throughout. Solutions may be gases, liquids, or solids. Like all other mixtures, solutions can have variable composition.

A pure substance is either an element or a compound. A pure substance is identified by its physical properties. Elements are the building blocks for all compounds. Elements are always present in the same ratio in a given compound. The properties of a compound are usually quite different from those of the elements of which it is composed. Chemical methods are required to separate compounds into their constituent elements.

A change in the properties of a substance without a change in composition is a physical change. If there is a change in the composition of a substance, however, a chemical change is indicated. In a chemical change (chemical reaction) reactants are converted to products. In any physical or chemical change, both mass and energy are conserved.

Practice Questions and Problems

9. Define chemistry. $\qquad$ $1 \cdot 1$

10. List the five major divisions of chemistry. $1 \cdot 1$

11. Distinguish between a theory, hypothesis, and law. $1 \cdot 2$

12. What is the purpose of an experiment as part of the scientific method? $1 \cdot 2$

13. Which of the following are examples of matter?
a. concrete **b.** acetone vapor **c.** heat $1 \cdot 3$
d. sound **e.** air

14. List four physical properties of an iron nail. $1 \cdot 3$

15. In which state of matter do the following exist at room temperature and atmospheric pressure? $1 \cdot 4$
a. diamond $\qquad$ **d.** mercury
b. oxygen $\qquad$ **e.** clay
c. cooking oil $\qquad$ **f.** neon

16. Match each state of matter with the terms on the left. More than one state can match each term. *1·4*
  a. incompressible   **1.** gas
  b. indefinite shape   **2.** liquid
  c. definite volume   **3.** solid
  d. flows

17. What physical properties would you use to separate these mixtures? *1·5*
  a. iron filings and salt   **b.** salt and water

18. Name two physical properties that could be used to distinguish between these substances. *1·5*
  a. water and rubbing alcohol
  b. gold and aluminum
  c. helium gas and oxygen gas

19. What is the difference between a heterogeneous and a homogeneous mixture?

20. How can the various components of a mixture be separated? *1·6*

21. Identify each of the following samples of matter as homogeneous or heterogeneous. *1·6*
  a. milk   **d.** river water
  b. glass   **e.** cough syrup
  c. table sugar   **f.** nitrogen

22. What are two ways to distinguish between an element and a compound? *1·7*

23. Classify each of the samples of matter in Problem 21 as an element, compound, or mixture. *1·7*

24. Write the chemical symbols for each of the following elements. *1·8*
  a. copper
  b. oxygen
  c. phosphorus
  d. silver
  e. sodium
  f. helium

25. Name the elements found in each compound.
  a. ammonium chloride, NH_4Cl *1·8*
  b. potassium permanganate, $KMnO_4$
  c. isopropyl alcohol, C_3H_7OH
  d. calcium iodide, CaI_2

26. Distinguish between potential energy and kinetic energy. *1·9*

27. List five forms of energy. *1·9*

28. State the law of conservation of energy. *1·10*

29. Classify each of the following as a physical or chemical change. *1·11*
  a. bending a piece of wire   **c.** cooking a steak
  b. burning of coal   **d.** cutting grass

Mastery Questions and Problems

30. Devise a way to separate sand from a mixture of charcoal, sand, sugar, and water.

31. How do you know that each of these is a chemical change?
  a. Food spoils.
  b. A foaming antacid tablet fizzes in water.
  c. A ring of scum forms around your bathtub.
  d. Iron rusts.
  e. A firecracker explodes.

32. When a small amount of a red powder is heated, it darkens and then changes into a shiny silvery liquid. Is the red powder an element or a compound? Explain. Can you classify the shiny liquid with certainty? Explain.

33. Identify each of the following as a mixture or a compound. For the mixtures, classify as homogeneous or heterogeneous.
  a. soda   **e.** an egg
  b. candle wax   **f.** ice
  c. fog   **g.** blood
  d. ink   **h.** gasoline

Critical Thinking Questions

34. Choose the term that best completes the second relationship.
  a. seed:plant   data: _____
     (1) theory   (3) hypothesis
     (2) law   (4) scientific method
  b. cell:organism   atom: _____
     (1) matter   (3) compound
     (2) molecule   (4) element
  c. person:female   mixture: _____
     (1) substance   (3) sea water
     (2) compound   (4) solution

35. Compare the relationships between individual particles in the three states of matter.

36. A snow skier begins at the top of a tall mountain. How does her energy change from the top to the bottom of the mountain? How does her energy escape?

Challenging Questions and Problems

37. Five elements make up 97.9% of the mass of the human body. These elements are oxygen (64.8%), carbon (18.1%), hydrogen (10.0%), nitrogen (3.1%), and calcium (1.9%). Compare this data to that in the pie graph below, which shows the five most abundant elements by mass in the earth's crust, oceans, and atmosphere.

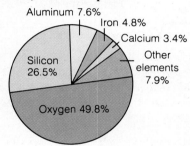

a. Which elements are abundant in both the earth's surface and the human body?
b. Which elements are abundant in the earth's surface but not the human body?
c. Why is the mass percent of oxygen the greatest in each case?
d. Are the compounds making up the human body different from those found in rocks, water, and air? Explain your answer based on the evidence in the pie graph and the data above.

38. Energy is conserved. It cannot be created or destroyed. Trace the origin of the energy in these bodies.
a. a falling raindrop
b. a gallon of gasoline
c. a piece of steak

39. We now know that heat is a form of energy. In the late 1700s, however, it was widely thought by scientists that heat was a weightless invisible fluid called caloric. In defense of these early scientists, can you think of some simple observations or experiments that would seem to support the *incorrect* theory that heat was a fluid?

Research Projects

1. What chemical processes were used by people in primitive cultures or in early societies?

2. Trace the history of alchemy. What did the alchemists contribute to the growth of chemistry? In your opinion, were the alchemists scientists? Why or why not?

3. What are some assumptions of the scientific method? What limitations result from these assumptions?

4. How is scientific research carried out on phenomena which cannot be manipulated or controlled? A few examples are tidal waves, apparent weightlessness in space, volcanic eruptions, solar fusion, and "the greenhouse effect."

5. Devise a controlled experiment to show that the rate of evaporation of water depends on the amount of exposed water surface area.

6. Devise a "system" for separating one of these mixtures into its components: garden soil, ocean water, "natural" (cloudy) apple juice, a carbonated soft drink.

7. Discuss the efficiency of the interconversion of energy in various "alternative" energy sources, such as solar, wind, geothermal, and tidal.

Readings and References

Corrick, James A. *Recent Revolutions in Chemistry.* New York: Watts, 1986.

Gardner, Robert. *Energy Projects for Young Scientists.* New York: Watts, 1987.

Kauffman, George B., and H. Harry Szmant (eds.) *The Central Science: Essays on the Uses of Chemistry.* Fort Worth: Texas Christian University Press, 1984.

Moravcsik, Michael J. *How to Grow Science.* New York: Universe, 1980.

Rosin, Jacob. "The Dark Science." *ChemMatters* (October 1985), p. 10.

Rossotti, Hazel. *Introducing Chemistry.* New York: Penguin, 1975.

2 Scientific Measurement

Chapter Preview

2·1 The Importance of Measurement

2·A Lavoisier: The Founder of Modern Chemistry

2·2 Accuracy and Precision

2·3 Significant Figures in Measurements

2·4 Significant Figures in Calculations

2·5 The Metric System

2·6 Units of Length

2·7 Units of Volume

2·8 Units of Mass

2·9 Measuring Density

2·10 Specific Gravity

2·11 Measuring Temperature

2·B Choosing a Thermometer

2·12 Measuring Heat

2·13 Specific Heat Capacity

Figure 2·1
Chemists use a wide variety of measuring devices. Balances are used to measure mass and thermometers are used to measure temperature. Graduated glassware is used to measure volume.

Everyone makes and uses measurements. You decide how to dress in the morning based on the measured outside temperature. You measure out the ingredients for your favorite cookie recipe. If you were building a cabinet for your stereo system, you would carefully measure each piece of wood.

Measurements are also fundamental to the experimental sciences. The reference standards used by scientists are those of the metric system. This is a system based on multiples of 10. The understanding of scientific concepts is often based on measurements. It is important to be able to make measurements and to decide whether a measurement is good or bad.

2·1 The Importance of Measurement

■ Measuring is a fundamental skill.

One of your most important skills for chemistry is making measurements. **Quantitative measurements** *give results in a definite form, usually as numbers.* **Qualitative measurements** *give results in a descriptive nonnumeric form.* Some measurements can be made either qualitatively or quantitatively. Consider, for example, a sick person with a fever. If we touch a person's forehead we might decide, "Yes! This person does feel feverish." This is a qualitative and subjective evaluation. If we need to be objective, we can use a thermometer. It will give us a numerical, or

21

Figure 2-2
By lifting the watermelon, the child finds it is heavy. This is a qualitative observation. The weight of the watermelon can also be determined with a spring scale. This is a quantitative observation.

quantitative, measurement. We get a specific number, the temperature, that can be recorded for future reference.

If several people touch the person's forehead, they might or might not agree with the original observer. Personal biases, as well as people's own temperatures could influence their decisions. By using a thermometer we can eliminate personal bias. Then everyone will report the same number. Remember, we have not questioned whether the thermometer gives us the correct temperature. Any instrument whose calibration is in doubt must be checked or discarded. A measurement can be no more reliable than the measuring instrument. Before looking at some common types of measurement, we will discuss the accuracy and precision of measurements.

History, Nature of Science

Figure 2-3
When Lavoisier heated mercury in the container at left, some red oxide of mercury formed. At the same time, the volume of air in the bell jar at the right decreased.

Science, Technology, and Society

2-A Lavoisier: The Founder of Modern Chemistry

Antoine-Laurent Lavoisier (1743–1794) was a brilliant French scientist and dedicated public servant. He is most famous for his work on combustion, which brought about the overthrow of the phlogiston theory. This erroneous theory had dominated chemistry for nearly 300 years.

Scientists before Lavoisier believed that all burnable substances contained phlogiston ("fire stuff"). When a substance burned, it was thought to have lost phlogiston. The ash, of course, was the substance minus its phlogiston. Lavoisier performed many experiments in which he made extremely careful measurements of the mass of a substance before and after combustion. He proved that many ashes weigh more than their original substance. This fact could not be explained by the phlogiston theory.

In 1772 Lavoisier began his experiments on combustion. In one of these experiments, he sealed mercury in a glass vessel and carefully determined the mass of the vessel with its contents. He then heated the vessel in a furnace. After 4 days of heating, flecks of red

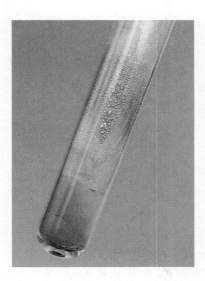

Figure 2-4
The red compound shown here is mercury(II) oxide. When heated, it decomposes into mercury and oxygen. You can see tiny drops of liquid mercury that have condensed on the inner surface of the test tube.

Scientific measurements should be both accurate and precise.

powder appeared on the surface of the mercury. After 12 days of continuous heating, most of the mercury had been converted to the red powder. Lavoisier again weighed the sealed vessel and found the mass unchanged. When the seal was broken, however, he heard the sound of air rushing into the vessel. He reasoned that during the change of mercury to its red powder, the mercury metal must have used up some component of the air sealed in the vessel. This created a partial vacuum. When the vessel was opened air rushed in to fill it. After this air had entered, Lavoisier weighed the resealed vessel again and found it heavier. He deduced that the gain in mass was the mass of the component of air used up.

By 1777 Lavoisier had completed a series of experiments with the red powder of mercury. When this powder was heated to higher temperatures, it gave off a gas that proved to be oxygen. This gas had been discovered in 1773 by Carl Scheele and again in 1774 by Joseph Priestley. (They were working independently.) In many tests, Lavoisier found that nothing could be burned in the absence of oxygen. He concluded that *combustion is the combination of a burnable substance with oxygen*. Younger chemists accepted his conclusions, and the mysterious phlogiston theory soon passed into history.

During the eighteenth century, very few women were involved in scientific work. Nevertheless, Marie Ann Lavoisier actively assisted her husband in scientific research. She recorded measurements and took notes in the laboratory. She learned both Latin and English, which her husband did not know. This allowed her to write translations and condensations of scientific articles. Marie Ann Lavoisier also learned to draw. She illustrated several books which Antoine-Laurent authored.

Like many other scientists of this time, Antoine-Laurent was an aristocrat. He was also a member of the despised royal taxation commission, a position he took to finance his scientific work. Although he had dedicated much of his time and money to improving the lives of the common people, Lavoisier was hated during the French Revolution. He refused to flee France even though he knew his life was in danger. In 1794 he was arrested, tried, and beheaded. A Frenchman has written, "Until it is realized that the gravest crime of the French Revolution was not the execution of the King, but of Lavoisier, there is no right measure of values, for Lavoisier was one of the greatest three or four men France had produced."

2-2 Accuracy and Precision

The words *accuracy* and *precision* mean the same thing to many people. To scientists, however, their meanings are different. **Accuracy** *is how close a measurement comes to the actual dimension or true value of whatever is measured*. **Precision** *is concerned with the reproducibility of the measurement*. Darts stuck in a dart board can be used to illustrate

Figure 2·5
The distribution of the darts illustrates the distinction between accuracy and precision. **a** The ideal combination is good accuracy and good precision. **b** Poor accuracy and good precision is probably the worst situation because it can be misleading. **c** Poor accuracy and poor precision point to the need for improvements.

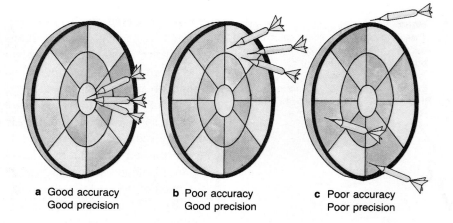

a Good accuracy
Good precision

b Poor accuracy
Good precision

c Poor accuracy
Poor precision

accuracy and precision in making measurements. Let the bull's-eye of the dart board represent the true value of what we are measuring. Then, the closeness of a dart to the bull's-eye is a measure of our accuracy. It represents how close we come to the true value. Let's consider three possible outcomes of tossing three darts at the board (Figure 2·5).

1. All the darts stick in the bull's-eye. Each dart in the bull's eye represents an accurate measurement of the true value. Thus, our measurements are very accurate. All the measurements are also closely grouped together. Therefore our measurements are very reproducible and very precise.

2. All the darts are closely grouped, but they are not in the bull's-eye. Our precision is excellent because our tosses are very reproducible. The tosses are inaccurate, though, because they do not reflect the true value.

3. The darts are spread randomly around the target. Our shots are neither accurate nor precise. We have missed the target, and we have failed to toss the darts reproducibly.

Note that the precision of a measurement depends on more than one measurement. By contrast, an individual measurement might be accurate or inaccurate. The accuracy of real measurements often depends on the quality of the measuring device. Precision often depends on the skill of the person making the measurement. We must usually assume that the measuring devices we use in the laboratory are as accurate as their manufacturers specify. Our job is to use these devices with enough care to obtain good precision.

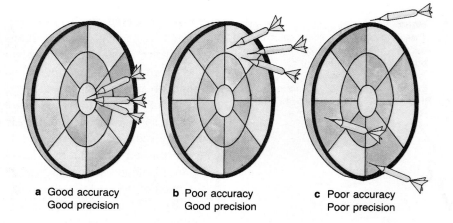

Issues in Chemistry
Most dentists agree that certain levels of fluoride in drinking water retard tooth decay. Why is it important to measure precisely the level of fluoride in public water supplies?

Before reading this section you should review your knowledge of exponential notation by doing Section B·1 of the Math Review in Appendix B.

2·3 Significant Figures in Measurements

Scientists report measurements in significant figures. *The **significant figures** in a measurement include all the digits that can be known precisely plus a last digit that must be estimated.* When you are taking a temperature using a thermometer calibrated in 1° intervals, it is easy to report the temperature to the nearest degree. With this thermometer you can also estimate the temperature to the nearest tenth of a degree. Suppose you estimate a temperature that lies between 35° and 36° to be 35.8°. This number has three significant figures. The first two digits (3 and 5) are known with

Significant figures are digits in a measurement that have meaning.

Figure 2·6

Three different meter sticks could be used to measure the length of a board. What measurement do you get in each case? Is there a difference in the number of significant figures in the three measurements? Why or why not?

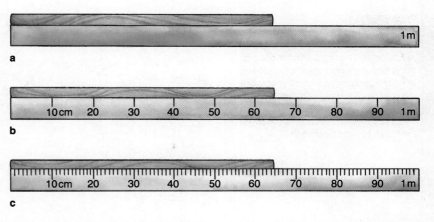

certainty. The last digit (8) has been estimated. It involves some uncertainty. You are not sure that the true temperature is $35° + 0.8°$.

By estimating the last digit in a measurement, scientists are able to get additional information. The digits retained in a measurement are all significant figures, but the last digit is uncertain. Suppose you have a thermometer that is calibrated in 0.1° intervals. You can easily report the temperature to the nearest 0.1° and estimate it to the nearest 0.01°. You might report the temperature as 35.82°. This measurement has four significant figures, and the last digit (2) is uncertain.

The rules for determining which digits in a measurement are significant are as follows.

1. Every nonzero digit in a recorded measurement is significant. The measurements 24.7 m, 0.743 m, and 714 m all express a measure of length to three significant figures. The abbreviation, m, stands for the unit of length, meter.

2. Zeros appearing between nonzero digits are significant. The measurements 7003 m, 40.79 m, and 1.503 m all have four significant figures.

In scientific writing, commas are not used to separate groups of digits. This is because a comma is used as a decimal point in many countries.

3. Zeros appearing in front of all nonzero digits are not significant. They are acting as place holders. The measurements 0.0071 m, 0.42 m, and 0.000 099 m all have two significant figures.

4. Zeros at the end of a number and to the right of a decimal point are significant. The measurements 43.00 m, 1.010 m, and 9.000 m all have four significant figures.

5. Zeros at the end of a measurement and to the left of the decimal point can be confusing. They are not significant if they just serve as place markers to show the magnitude of the number. The zeros in the measurements 300 m, 7000 m, and 27 210 m are probably not significant, but some of them may be. We cannot tell any difference. *If these zeros were measured then they are significant.* To avoid ambiguity, the measurements should then be written in standard exponential form: 3.00×10^2 m, 7.000×10^3 m, and 2.7210×10^4 m. In these examples the number of significant figures is three, four, and five, respectively.

Figure 2·7
Quality control chemists determine the range of characteristics that indicate whether the quality of a product is acceptable.

Significant figures in measurements are used to correctly round off calculated answers.

Example 1

How many significant figures are in each of these measurements?

a. 123 m
b. 0.123 m
c. 40 506 m
d. 9.8000×10^4 m

e. 4.5600 m
f. 0.078 m
g. 0.070 80 m
h. 98 000 m

Solution

All nonzero digits are significant. Use rules 2–5 to decide about zeros.

a. 3 (rule 1)
b. 3 (rule 3)
c. 5 (rule 2)
d. 5 (rules 4 and 5)

e. 5 (rule 4)
f. 2 (rule 3)
g. 4 (rules 2, 3, 4)
h. 2 (rule 5)

Problem

1. Determine the number of significant figures in each of the following measurements.

a. 5.730×10^{-2} m
b. 8765 m
c. 0.000 73 m
d. 40.007 m

e. 3000 m
f. 0.010 m
g. 50 700 m
h. 0.070 020 m

2·4   Significant Figures in Calculations

When calculations are done with scientific measurements, we sometimes end up with an answer with more digits than can be justified as significant. Such numbers must be rounded off to make them consistent with the data they represent. *An answer cannot be more precise than the least precise measurement.* To round off a number, we must first decide how many significant figures it should have. Our decision will depend on the given measurements and on the arithmetic operation used. Once we know the number of significant figures our answer should have, we count that many digits starting on the left. *If the digit immediately following the last significant digit is less than 5, all the digits after the last significant place are dropped. If the digit is 5 or greater, the value of the digit in the last significant place is increased by 1.* Rounding off 56.212 m to four significant figures gives us 56.21 m, and 56.216 m becomes 56.22 m.

Example 2

Round off each of these measurements to the number of significant figures shown in parentheses. Write the answer in standard exponential form.

a. 314.721 m (4)
b. 0.001 775 m (2)

c. 64.32×10^{-1} m (1)
d. 8792 m (2)

Solution

The arrow points to the digit immediately following the last significant digit. (Significant figures are shown in parentheses.)

a. 314.721 m; 2 is less than 5; 314.7 m (4) = 3.147×10^2 m
↑

b. 0.001 775 m; 7 is greater than 5; 0.0018 m (2) = 1.8×10^{-3} m
↑

c. 64.32×10^{-1} m; 4 is less than 5; 60×10^{-1} m (1) = 6 m
↑

d. 8792 m; 9 is greater than 5; 8800 m (2) = 8.8×10^3 m
↑

Problem

2. Round off each measurement to three significant figures and then to one significant figure. Write your answers in standard form.

a. 87.073 m   **c.** 0.01552 m   **e.** 1.7777×10^{-3} m
b. 4.3621×10^8 m   **d.** 9009 m   **f.** 629.55 m

Addition and Subtraction. The answer of an addition or subtraction can have no more digits to the right of the decimal point than are contained in the measurement with the least number of digits to the right of the decimal point.

Example 3

Do the following operations and give the answer to the correct number of significant figures.

a. 12.52 m + 349.0 m + 8.24 m   **b.** 74.626 m − 28.34 m

Solution

a. Align the decimal points and add the numbers.

$$\begin{array}{r} 12.52 \text{ m} \\ 349.0 \text{ m} \\ \underline{8.24 \text{ m}} \\ 369.76 \text{ m} \end{array}$$

The answer must be rounded off to one digit after the decimal point. The answer is 369.8 m or 3.698×10^2 m.

b. Align the decimal points and subtract the numbers.

$$\begin{array}{r} 74.626 \text{ m} \\ \underline{-28.34 \text{ m}} \\ 46.286 \text{ m} \end{array}$$

The answer must be rounded off to two digits after the decimal point. The answer is 46.29 m or 4.629×10^1 m.

Problem

3. Do the following operations and give your answer to the correct number of significant figures.
   a. 61.2 m + 9.35 m + 8.6 m   **c.** 1.36 m + 10.17 m
   b. 9.44 m − 2.111 m   **d.** 34.61 m − 17.3 m

Multiplication and Division.   In calculations involving multiplication and division, the answer must contain no more significant figures than the measurement with the least number of significant figures. The position of the decimal point has nothing to do with the number of significant figures.

Example 4

Do the following operations and give the answer to the correct number of significant figures.
   a. 7.55 m × 0.34 m   **c.** 2.4526 m ÷ 8.4
   b. 2.10 m × 0.70 m   **d.** 0.365 m ÷ 0.0200

Solution

The calculated answer is given, then rounded off.
   a. 2.567 m^2 = 2.6 m^2 (0.34 m has two significant figures)
   b. 1.47 m^2 = 1.5 m^2 (0.70 m has two significant figures)
   c. 0.291 976 m = 0.29 m (8.4 has two significant figures)
   d. 18.25 m = 18.3 m (both numbers have three significant figures)

Problem

4. Do the following problems and give your answer to the correct number of significant figures.
   a. 8.3 m × 1.22 m
   b. $(1.8 \times 10^{-3}$ m$) \times (2.9 \times 10^{-2}$ m$)$
   c. 8432 m ÷ 12.5
   d. 5.3×10^{-2} m ÷ 0.255

2·5   The Metric System

A measurement depends on a standard for reference. *The standards of measurement used by scientists are those of the* **metric system.** Other systems of measurement are still in use, but they are gradually being phased out. They are less useful because there is no systematic digital relationship among the units. The metric system is important because of its simplicity and convenience. All the units are based on 10 or multiples of 10. As a result, conversions between units are easy to do. The metric

Figure 2·8
This calculator was used to multiply the measurements 3.24 cm and 1.78 cm. When correctly rounded off, the product is 5.77 cm².

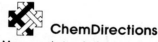

ChemDirections

Many pocket calculators rely on energy from the sun rather than from dry-cell batteries. To discover how these work read **ChemDirections** *Solar Cells*, page 655.

The metric system, or the SI, is the preferred system of scientific measurement.

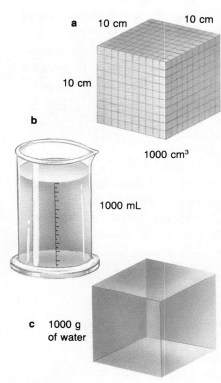

Figure 2·9
The metric units of mass, length, and volume are related by definition. **a** The volume of a 10 cm cube is 1000 cm³. **b** The cube can hold 1000 ml of water. **c** This volume of water has a mass of 1000 grams.

Table 2·1 SI Base Units of Measurement		
Quantity measured	Unit	SI symbol
Length	meter	m
Mass	kilogram	kg
Time	second	s
Electric current	ampere	A
Thermodynamic temperature	kelvin	K
Amount of substance	mole	mol
Luminous intensity	candela	cd

system was originally established in France in 1790. *The* **International System of Units** (abbreviated **SI,** after the French name Le Système International d'Unités) *is a revised version of the metric system.* It was adopted by international agreement in 1960. The SI has seven base units (Table 2·1). From these, other SI units of measurement such as volume, density, and pressure are derived.

It is possible to report all measured quantities in SI units. Sometimes, however, non-SI units are preferred for convenience or practical reasons. Table 2·2 lists some SI and non-SI units of measurement that are used in this text.

When a measurement is made, the numerical value must be assigned the correct units. Without the correct units it is not possible to communicate the resulting measurement. Imagine the confusion that would ensue if you were instructed to "heat the solution for 20." Your immediate response would be "Twenty what? Seconds, minutes, hours, or days?" The units of length, volume, and mass will be discussed in the three sections that follow.

Table 2·2 Some Units of Measurement Used in This Text		
Quantity	SI base unit or SI derived unit	Non-SI unit
Length	meter (m)	
Volume	cubic meter (m³)	liter (L)
Mass	kilogram (kg)	
Density	grams per cubic centimeter (g/cm³)	
	grams per milliliter (g/mL)	
Temperature	kelvin (K)	degree Celsius (°C)
Time	second (s)	
Pressure	pascal (pa)	atmosphere (atm)
		millimeter of mercury (mm Hg)
Energy	joule (J)	calorie (cal)

2·6 Units of Length

The meter is the basic metric unit of length.

A meter is defined as the distance that light travels in a vacuum in 1/299 792 458 of a second.

Figure 2·10
A meter stick is divided into 100 divisions. Each division is a centimeter. Each centimeter is divided into 10 millimeters. How many millimeters are in a meter?

Size is an important property of matter. *In the metric system the basic unit of length, or linear measure, is the* **meter** (*m*). All measurements of length can be expressed in meters. (The length of a page in this book is about one-fifth of a meter.) For very large and very small lengths, however, it is more convenient to use the appropriate metric prefix with the base unit meter. Table 2·3 lists the prefixes in common use in the metric system. As an example, this dash - is about 0.001 m long. Compare the length of this dash with the section of meter stick reproduced in Figure 2·10. The dash measures about 1 millimeter (mm), or 1/1000 of a meter. Both 1 mm and 0.001 m measure the same length.

For large distances, it is most appropriate to express a measurement in kilometers (km). The prefix *kilo-* means 1000 times larger; 1 km is equal to 1000 m. The marathon distance race of about 42 000 m is more conveniently expressed as 42 km (42 × 1000 m). Table 2·4 summarizes the relationships among units of length in the metric system.

Table 2·3	Prefixes in Common Use in the Metric System	
Prefix	Symbol	Meaning
kilo	k	1000 times larger than the unit it precedes (1000 or 10^3)
deci	d	10 times smaller than the unit it precedes (1/10 or 10^{-1})
centi	c	100 times smaller than the unit it precedes (1/100 or 10^{-2})
milli	m	1000 times smaller than the unit it precedes (1/1000 or 10^{-3})
micro	μ	1 million times smaller than the unit it precedes (1/1 000 000 or 10^{-6}) (μ is the lowercase Greek letter "mu")
nano	n	1000 million times smaller than the unit it precedes (1/1 000 000 000 or 10^{-9})

Table 2·4	Metric Units for Length		
Unit	Symbol	Relationship	Examples (approximate values)
Kilometer	km	$1\ km = 10^3\ m$	Length of about five city blocks ≈ 1 km
Meter (base unit)	m		Height of door knob from the floor ≈ 1 m
Decimeter	dm	$10^1\ dm = 1\ m$	Diameter of large orange ≈ 1 dm
Centimeter	cm	$10^2\ cm = 1\ m$	Width of shirt button ≈ 1 cm
Millimeter	mm	$10^3\ mm = 1\ m$	Thickness of dime ≈ 1 mm
Micrometer	μm	$10^6\ \mu m = 1\ m$	Diameter of bacterial cell $\approx 1\ \mu$m
Nanometer	nm	$10^9\ nm = 1\ m$	Thickness of RNA molecule ≈ 1 nm

Problem

5. Measure the following dimensions using metric units with the appropriate metric prefix. **a.** the height of this letter I **b.** the width of Table 2·4 **c.** the height of this page **d.** the distance between these brackets [   ]

2·7 Units of Volume

The space occupied by matter is called **volume.** It is a derived unit. Volume is calculated from linear measures of length, width, and height. The SI unit of volume is the cubic meter (m^3). This is the amount of space occupied by a cube that is 1 m on each side. A more convenient metric unit of volume is the liter (L) (Figure 2·11). A **liter** *is the volume of a cube*

Figure 2·11
The volume of a cube 10 cm on each edge is 1000 cubic centimeters (cm^3). This is a liter. One liter contains 1000 milliliters (mL). Thus 1 mL and 1 cm^3 have the same volume. The face of the cube is shown actual size.

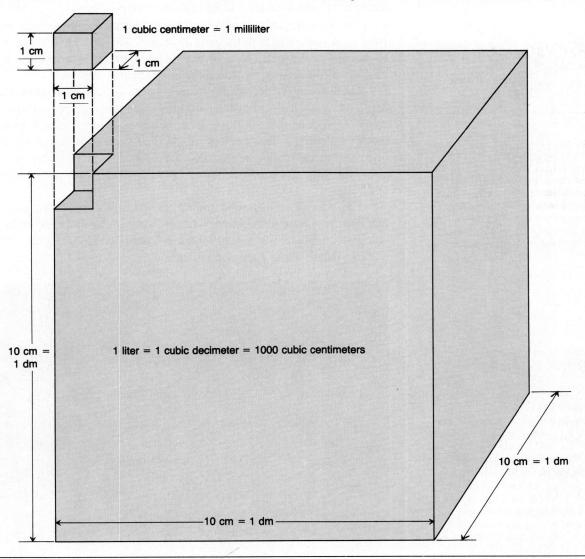

1 cubic centimeter = 1 milliliter
1 cm
1 cm
1 cm
10 cm = 1 dm
1 liter = 1 cubic decimeter = 1000 cubic centimeters
10 cm = 1 dm
10 cm = 1 dm

Table 2·5	Metric Units for Volume		
Unit	Symbol	Relationship	Examples
Liter (base unit)	L		Quart of milk $\approx$ 1 L
Milliliter	mL	10^3 mL = 1 L	20 drops of water $\approx$ 1 mL
Cubic centimeter	cm³	1 cm³ = 1 mL	Cube of sugar $\approx$ 1 cm³
Microliter	μL	10^6 μL = 1 L	Crystal of table salt $\approx$ 1 μL

The liter is the most commonly used metric unit of volume.

Figure 2·12
One cubic centimeter of water is shown actual size. This much water has a mass of 1 g at 4°C. Twenty drops of water equals approximately 1 mL.

that is 10 centimeters (cm) on each edge (10 cm × 10 cm × 10 cm = 1000 cm³ = 1 L). A smaller unit of volume is the milliliter (mL), which is 1/1000 part of a liter. Because a liter is defined as 1000 cm³, the milliliter and cubic centimeter must have the same volume. These two units are used interchangeably. Table 2·5 summarizes the most commonly used relationships among units of volume.

Note that the prefixes *milli-* and *micro-* are used with both the unit of volume, a liter, and with the unit of length, a meter. This is one of the advantages of the metric system. Each prefix retains its same meaning regardless of the unit of measurement to which it is attached.

There are many devices for measuring the volume of a liquid (Figure 2·13). A graduated cylinder is useful for dispensing approximate volumes. A pipet or buret must be used, however, when accuracy is important. A volumetric flask contains a specified volume of liquid when it is filled to the calibration mark. Volumetric flasks are available in many sizes. A syringe is used to measure small volumes of liquids for injection.

The volume of any solid, liquid, or gas will change with temperature. For this reason, accurate volume-measuring devices are calibrated at a given temperature. Usually they are calibrated at 20 degrees Celsius (20°C). This is about room temperature.

Figure 2·13
These five types of glassware are used for measuring liquid volumes. The four beakers at the left and the two Erlenmeyer flasks and tall graduated cylinder at the right are used to measure approximate volumes. Small volumes are measured accurately with a pipet (on the left). Larger volumes are measured accurately with a volumetric flask (in the center).

2·8 Units of Mass

Mass is a physical property that measures the amount of matter.

Figure 2·14
An astronaut's weight on the moon is 1/6th as much as it is on earth. How does his mass on the moon compare with his mass on the earth?

Mass is the quantity of matter an object contains. Mass is not weight. **Weight** *is a force*. It is a measure of the pull on a given mass by the earth's gravity. Although the weight of an object can change with its location, its mass remains constant regardless of its location. A block of wood on earth has a weight that is six times its weight on the moon. This is because the force of gravity on earth is about six times that on the moon. On a passage from the earth to the moon, the weight of the wood decreases. Whether on earth, on the moon, or in space, however, the mass of the wood does not change.

The mass of an object is measured by comparing it to a standard mass of 1 kilogram (kg), the basic SI unit of mass. *A* **kilogram** *is the mass of 1 L of water at 4°C.* (Water freezes at 0°C.) Imagine a cube of water at 4°C measuring 10 cm on each edge. This cube would have a volume of 1 L and a mass of 1000 grams (g), or a kilogram. *A* **gram** *is defined as the mass of 1 cm³ of water at 4°C* (Figure 2·12).

The relationships among units of mass are shown in Table 2·6. The mass of an object is measured with a balance. An object of unknown mass is placed on one side, and standard masses are added to the other side until the beam is in a position of balance (Figure 2·15). When the

Figure 2·15
A platform balance compares an unknown mass to a known mass. **a** The pans are empty, and the balance is balanced. **b** An object of unknown mass is placed on the left-hand pan. The balance is disturbed. **c** Standard masses are placed on the right-hand pan to restore the balance. The mass of the unknown object is the same as the sum of the standard masses added to the right-hand pan. In this case the mass is 150 g.

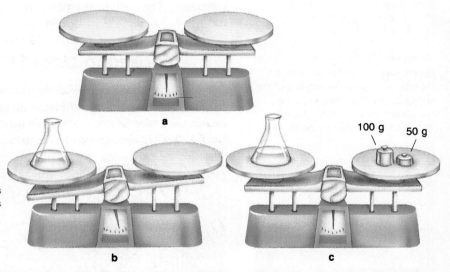

Table 2·6	Metric Units for Mass		
Unit	Symbol	Relationship	Examples
Kilogram (base unit)	kg	$1 \text{ kg} = 10^3 \text{ g}$	Small textbook $\approx 1 \text{ kg}$
Gram	g		Dollar bill $\approx 1 \text{ g}$
Milligram	mg	$10^3 \text{ mg} = 1 \text{ g}$	Ten grains of salt $\approx 1 \text{ mg}$
Microgram	μg	$10^6 \mu\text{g} = 1 \text{ g}$	Particle of baking powder $\approx 1 \mu\text{g}$

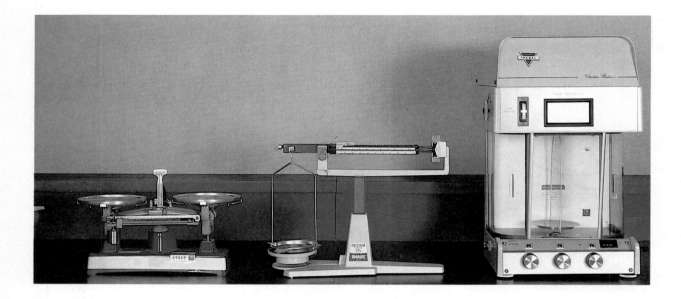

Figure 2·16
It is possible to determine mass to the nearest 0.1 g on a platform balance (left) and to the nearest 0.01 g on a triple beam balance (center). Even greater precision is possible with an analytical balance (right).

Before reading this section you should review your knowledge of algebraic equations by doing Section B·2 of the Math Review in Appendix B.

█ The relationship between mass and volume is called density.

Chemical Connections

Automobile manufacturers use low density materials whenever possible. Lighter cars use less fuel than heavier cars. Many car parts are now made of aluminum or plastic rather than steel.

beam is balanced, the unknown mass is equal to the sum of the standard masses. Several balances are commonly used in laboratories (Figure 2·16). They range from very sensitive instruments with a maximum capacity of only a few milligrams to devices for measuring quantities in kilograms. The analytical balance is used to measure masses of less than 100 g.

2·9 Measuring Density

The statement "Lead is heavier than wood" has no precise meaning. Most people take it to mean that if pieces of lead and wood of the same volume are weighed, the lead has a greater mass than the wood. It would take a much larger volume of wood to equal the mass of the lead. What all this brings us to is the idea that there is an important relationship between an object's mass and its volume. This relationship is called density.

Density *is the ratio of the mass of an object to its volume.*

$$\text{Density} = \frac{\text{mass}}{\text{volume}}$$

When mass is measured in grams, and volume in cubic centimeters, density has units of grams per cubic centimeter. As the density of a substance increases, the volume of a given mass of that substance decreases (Table 2·7). The densities of materials are extremely important. If bones were as dense as lead, for example, you would collapse under your own weight. Table 2·8 gives the densities of some common substances.

The density of a substance usually decreases as its temperature increases. This is because its volume usually increases as the temperature is increased. Meanwhile, its mass remains the same. As we will see in Chapter 15, water is an important exception. Ice floats because it is less dense than water.

Table 2-7   Relationship Between Volume and Density for Identical Masses of Common Substances

Substance	Cube of substance (face shown actual size)	Mass (g)	Volume (cm³)	Density (g/cm³)
Wood		10	20	0.50
Water		10	10	1.0
Aluminum		10	3.7	2.7
Lead		10	0.88	11

Figure 2-17
Solid paraffin (left) sinks in liquid paraffin, but solid water (ice) floats in water. What does this tell you about the density of these substances?

Table 2·8 Densities of Some Common Substances

Substance	Density at 25°C (g/cm³)	Substance	Density at 25°C (g/cm³)
Gold	19.3	Water	0.997
Mercury	13.5	Water	1.00 (4°C)
Lead	11.4	Ice	0.917 (0°C)
Iron	7.87	Gasoline	0.66–0.69
Aluminum	2.70	Wood	0.11–1.33
Bone	1.7–2.0	Air (dry)	0.00119 (1 atm)

Example 5

A copper penny has a mass of 3.1 g and a volume of 0.35 cm³. What is the density of copper?

Solution

We can calculate the density of a substance if we know the mass and volume of a sample of the substance.

$$\text{Density} = \frac{\text{mass}}{\text{volume}} = \frac{3.1 \text{ g}}{0.35 \text{ cm}^3} = 8.8571$$
$$= 8.9 \text{ g/cm}^3 \text{ (two significant figures)}$$

Problems

6. A student finds a shiny piece of metal that she thinks is aluminum. In the lab she determines that the metal has a volume of 245 cm³ and a mass of 612 g. Is the metal aluminum?

7. A plastic ball with a volume of 19.7 cm³ has a mass of 15.8 g. Would this ball sink or float in a container of gasoline?

8. Would the density of a person be the same on the surface of the earth and on the surface of the moon? Explain.

2·10 Specific Gravity

▮ Specific gravity is the ratio of two densities.

Specific gravity *is a comparison of the density of a substance to the density of a reference substance, usually at the same temperature.* Water at 4°C, which has a density of 1 g/cm³, is a convenient reference and is commonly used for this measurement.

$$\text{Specific gravity} = \frac{\text{density of substance (g/cm}^3)}{\text{density of water (g/cm}^3)}$$

The units in the equation cancel. Thus specific gravity has no units.

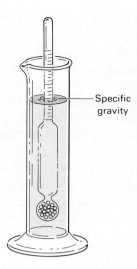

Figure 2·18
A hydrometer is a sealed tube with a weight in the bottom. It is used to measure the specific gravity of a liquid. A hydrometer can be used to check the antifreeze solution in an automobile radiator or the amount of acid in a storage battery. For these purposes, it is enclosed in a glass tube fitted with a rubber bulb. The liquid to be tested is drawn up into the tube and the hydrometer reading is taken.

The specific gravity of a liquid can be measured with a **hydrometer** (Figure 2·18). The depth to which the hydrometer sinks depends on the specific gravity of the liquid. The calibration mark on the hydrometer stem at the surface of the liquid indicates the specific gravity.

The use of measurements of specific gravity is fairly common. A physician uses the measured specific gravity of a patient's urine to help diagnose certain diseases, such as diabetes. A service station attendant checks the condition of the antifreeze of your car by measuring the specific gravity of the solution in the radiator.

Problem

9. Use the values in Table 2·8 to calculate the specific gravity of the following substances at 25°C.   **a.** aluminum   **b.** mercury   **c.** ice

2·11   Measuring Temperature

Temperature *is the degree of hotness or coldness of an object.* Our bodies are sensitive to temperature differences but not to specific temperatures. When we hold a glass of hot water, we feel hot because heat transfers from the hot glass to our hand. If we hold a piece of ice, however, heat transfers from our hand to the ice. Temperature determines the direction of heat transfer. **Heat transfer** *occurs when two objects at different temperatures contact each other.* Heat transfers from the object at the high temperature to the object at the low temperature.

Almost all substances expand with an increase in temperature. They also contract as the temperature decreases (a very important exception is water). These properties are the basis for the common mercury-in-glass thermometer. Until recently this was the only type of thermometer in

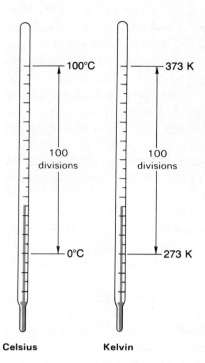

■ Temperatures are measured on either the Celsius or Kelvin scales.

Figure 2-19
These thermometers show a comparison of the Celsius and Kelvin temperature scales.

general use. The mercury in the thermometer expands and contracts more than the volume of the bulb that holds it.

Several temperature scales have been devised. Two readily determined temperatures, the freezing and boiling points of water, are used as reference temperature values. The Celsius scale of the metric system is named after the Swedish astronomer Anders Celsius (1701–1744). *The* **Celsius scale** *takes the freezing point of water as 0°C and boiling point as 100°C.* The space between these two fixed points is divided into 100 equal intervals, or degrees.

Another temperature scale used in the physical sciences is the Kelvin scale, or absolute scale. It is named after Lord Kelvin (1824–1907), a Scottish physicist and mathematician. *On the* **Kelvin scale,** *the freezing point of water is 273 K, and the boiling point is 373 K.* Notice that with the Kelvin scale the degree sign is not used. A change of 1° on the Kelvin scale is the same as that on the Celsius scale. *The zero point on the Kelvin scale, 0 K or* **absolute zero,** *is −273°C.* The relationship between a temperature on the Celsius scale and one on the Kelvin scale is given by these equations.

$$K = {}^\circ C + 273 \text{ or } {}^\circ C = K - 273$$

Example 6

Normal human body temperature is 37°C. What is your temperature in K?

Solution

$K = {}^\circ C + 273$

$\quad = 37{}^\circ C + 273$

$\quad = 310 \text{ K}$

Problems

10. Surgical instruments may be sterilized by heating at 170°C for 1.5 hr. Convert 170°C to K.

11. The boiling point of the element argon is 87 K. What is the boiling point of argon in °C?

Figure 2-20
Liquid-in-glass thermometers commonly contain mercury or colored alcohol. The bulb contains a large amount of liquid which expands and contracts with temperature changes.

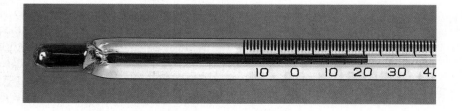

2·B Choosing a Thermometer

A teaspoon of sugar dissolves easily in a cup of hot tea, but lies undissolved in the bottom of a glass of iced tea. A tightly closed container explodes when it is heated. Food stays fresh for days or weeks in a cold refrigerator. By contrast, it spoils rapidly at room temperature. The solubility of solids, the pressure of a gas, and the rate of chemical reactions are just a few of the properties of matter that change with a change in temperature. In fact, all thermometers work because of the change of some property of matter with a change in temperature.

The thermometers used in a school laboratory are usually sealed glass tubes filled with either mercury or alcohol. Both the glass and the liquid inside of it expand when the thermometer is heated, but the liquid expands more. The scale on the thermometer translates the height of the liquid column into a temperature reading. Most mercury thermometers measure temperatures between $-20°C$ and $150°C$, with a precision of $0.2°C$. One type, the Beckmann thermometer, has a precision of $0.001°C$. The chief disadvantage of mercury thermometers is that they are easily broken. This could expose people in the laboratory to toxic mercury metal.

The thermometer in your home thermostat is probably a bimetallic thermometer. Narrow ribbons of two different metals are laid on top of each other to make a layered metal strip. Different metals expand different amounts when heated. Thus the metal strip will curl as the temperature changes. The amount of curvature is translated into a temperature reading. The range of a bimetallic thermometer depends on what metals are used in the strip. Thermometers of this type are available to cover temperatures from $-180°C$ to $425°C$.

Safety

Remember that mercury is toxic and slowly vaporizes. If you break a mercury thermometer, report it to your teacher immediately so that the mercury can be cleaned up thoroughly.

Figure 2·21
Bimetallic strips are often used in home thermostats. When warmed, the strip bends. Why?

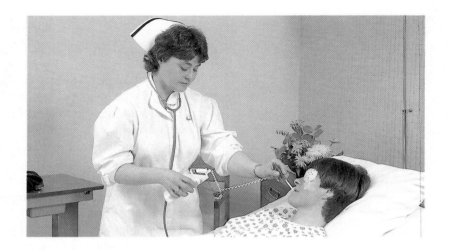

Figure 2·22
The electronic thermometer used in hospitals senses temperature with a thermistor. A thermistor is a metal oxide chip whose resistance to current flow changes with temperature.

Figure 2·23
The liquid crystals in the plastic strip change color at different temperatures. Different strips must be used to measure room temperature and body temperature.

Electronic thermometers are commonly used in hospitals and laboratories. They make use of another temperature measuring device, the thermistor. The thermistor is a bead, rod, or disc made of a mixture of metal oxides. These compounds are semiconductors. Their ability to conduct electricity increases as the temperature increases. The thermistor is connected to an electric current. When the temperature of the thermistor changes, the amount of current that can pass through it changes. This change is translated into a temperature reading. Thermistors are very fast and accurate. Some can record a change as small as $0.0005°C$.

Perhaps you have seen plastic strips or blocks that display the temperature in numbers that change color as the temperature changes. These thermometers contain a liquid crystal. Such a substance flows like a liquid, but has an orderly arrangement of its molecules, like a solid crystal. The flexible strips used to take a person's temperature contain a liquid crystal that changes color dramatically at different temperatures. As the strip heats up, the color of the liquid crystal changes. At each temperature, a different reading is visible on the scale. These devices are used mainly to measure body temperature or room temperature. This is because they are only useful over a small range of temperatures. The liquid crystal remains in its special "in-between" state only when it is very close to its melting point. It will turn into a solid at lower temperatures, and into a true liquid at higher ones. A liquid crystal thermometer is not as accurate as the common mercury thermometer. The former is very easy to use, however, and it is virtually unbreakable.

Very high temperatures, like those encountered in steel making, can be measured very accurately by a thermocouple. This device contains two wires of different metals, joined at two points. One point, called the cold or reference junction, is kept at a known temperature. It may, for example, be cooled in ice water to keep it at $0°C$. The other point, called the sensing junction, is located where the temperature measurement is to be taken. If the temperature at the

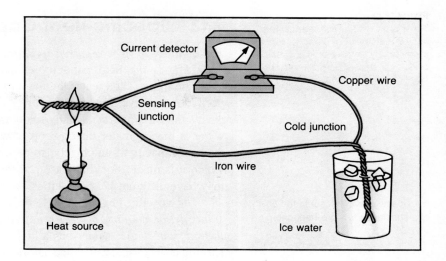

Figure 2·24
Temperatures as high as 1400°C are measured accurately by a thermocouple.

Current detector

Copper wire

Sensing junction

Cold junction

Iron wire

Heat source

Ice water

two junctions is different, an electric current is generated in the wires. The amount of current generated depends on the size of the temperature difference. The current is measured and translated into a temperature reading. The cold junction and the current detector can be located at a considerable distance from the sensing junction with no loss of accuracy. This is very important because this device is used to measure temperatures as high as 1400°C.

2·12 Measuring Heat

Chemical and physical changes involve a change in heat energy.

Heat can be measured in units such as the joule (J) or the calorie (cal). The **joule** *is the SI unit of energy.* It is named after the English physicist James Prescott Joule (1818–1889). *One* **calorie** *is the quantity of heat that raises the temperature of 1 g of pure water 1°C.* The kilocalorie (kcal) is used to measure large amounts of heat; 1 kcal = 1000 cal. Conversions between calories and joules can be carried out by using the following relationships.

A joule is a rather small unit of energy. A television set might use a million joules of energy in one hour.

$$1 \text{ J} = 0.239 \text{ cal} \quad and \quad 1 \text{ cal} = 4.18 \text{ J}$$

The term *Calorie* used in nutrition is actually the kilocalorie. Notice that the dietary Calorie is written with a capital C. The statement "10 g of sugar has 41 Calories" means that 10 g of sugar will release 41 kcal of heat if it is burned completely to carbon dioxide and water by fire. The same amount of heat is also released when 10 g of sugar is "burned" to carbon dioxide and water by your body.

Problem

12. Using the definition of a calorie, how much heat energy does 32.0 g of water absorb when it is heated from 25°C to 80°C?

2·13 Specific Heat Capacity

The quantity of heat required to change an object's temperature by exactly 1°C is the **heat capacity** *of that object.* Heat capacity depends partly on mass. The greater the mass of the object, the greater its heat capacity. A massive steel girder, for example, has a much larger heat capacity than a small steel nail. Likewise, the heat capacity of a cup of water is much greater than a drop of water. Besides varying with mass, the heat capacity of an object also depends on the kind of substance from which it is made. For this reason, different substances of the same mass may have different heat capacities.

The **specific heat capacity,** *or simply the* **specific heat,** *of a substance is the quantity of heat required to raise the temperature of 1 g of the substance 1°C.* Water has a high specific heat, and metals have low specific heats (Table 2·9). One calorie of heat raises the temperature of 1 g of water by 1°C. The same amount of heat raises the temperature of 1 g of iron by 9°C. Therefore iron has a specific heat only one-ninth that of water. Specific heat is important because it can be used to calculate how many calories are required to heat a known mass of substance from one temperature to another. The following equations relate specific heat to the changes in heat and temperature.

$$\text{Specific heat} \left(\frac{J}{g \times {}^{\circ}C} \right) = \frac{\text{heat (J)}}{\text{mass (g)} \times \text{change in temperature (}^{\circ}C)}$$

Using the unit calories, the equation is as follows.

$$\text{Specific heat} \left(\frac{cal}{g \times {}^{\circ}C} \right) = \frac{\text{heat (cal)}}{\text{mass (g)} \times \text{change in temperature (}^{\circ}C)}$$

■ Specific heat measures the ability of a substance to store heat energy.

Safety

Because of its high heat capacity, hot water can cause very serious burns. Use care when heating water in the laboratory.

Figure 2·25
Which food will warm you up most when you are cold: warm soup or hot toast? Why?

Table 2·9 Specific Heat Capacities of Some Common Substances

Substance	Specific heat capacity	
	[cal/(g × °C)]	[J/(g × °C)]
Water	1.00	4.18
Grain alcohol	0.58	2.4
Ice	0.50	2.1
Wood	0.42	1.8
Steam	0.40	1.7
Chloroform	0.23	0.96
Aluminum	0.21	0.90
Glass	0.12	0.50
Iron	0.11	0.46
Silver	0.057	0.24
Mercury	0.033	0.14

Figure 2·26
Water must give off a lot of heat in order to freeze. The high specific heat of water makes the climate warmer in the winter and cooler in the summer near large bodies of water.

Example 7

The temperature of a piece of copper with a mass of 95.4 grams changes from 25.0°C to 48.0°C when the metal absorbs 849 J of energy. What is the specific heat of copper?

Solution

Use the equation for specific heat and substitute in the given values. The change in temperature is $48.0°C - 25.0°C = 23.0°C$.

$$\text{Specific heat} = \frac{J}{g \times °C} = \frac{\text{heat (J)}}{\text{mass (g)} \times \text{change in temperature (°C)}}$$

$$= \frac{849 \text{ J}}{95.4 \text{ g} \times 23.0°C}$$

$$= 0.387 \frac{J}{g \times °C}$$

Problem

13. When 435 J of heat energy is added to 3.4 g of olive oil at 21°C, the temperature increases to 85°C. What is the specific heat of olive oil?

2 Scientific Measurement

Key terms

absolute zero	2·11	kilogram (kg)	2·8
accuracy	2·2	liter (L)	2·7
calorie	2·12	meter (m)	2·6
Celsius temperature		metric system	2·5
scale	2·11	precision	2·2
density	2·9	qualitative	
gram (g)	2·8	measurement	2·1
heat capacity	2·13	quantitative	
heat transfer	2·11	measurement	2·1
hydrometer	2·10	significant figures	2·3
International System of		specific gravity	2·10
Units (SI)	2·5	specific heat	2·13
joule (j)	2·12	temperature	2·11
Kelvin temperature		volume	2·7
scale	2·11	weight	2·8

Chapter Summary

Measurements can be qualitative or quantitative. The accuracy of a measurement describes how close a measurement comes to the true value. By contrast, the precision of a measurement depends on its reproducibility.

The metric system, or International System of Units (SI), is a system of measurement used by scientists. The SI has seven base units from which all other SI units of measurement are derived.

In the metric system the basic unit of length is the meter. The space occupied by matter is called volume. A cubic decimeter (liter) has a volume of 1000 cm^3. One cubic centimeter has the same volume as 1 mL.

The quantity of matter an object contains is its mass. The mass of an object is constant and independent of gravity. A balance is used to measure mass. Weight is not the same as mass. Weight is the measure of the pull of gravity on an object of given mass.

The ratio of the mass of an object to its volume is its density. The unit of density is grams per cubic centimeter. Specific gravity is the ratio of the density of a substance to the density of water. Specific gravity has no units. The specific gravity of a liquid is commonly measured with a hydrometer.

Temperature difference determines the direction of heat flow between two bodies. Heat flows from a hot body to a cold body. Two temperature scales are used: Celsius and Kelvin.

Heat is a form of energy. Heat is measured in joules or calories. One calorie is the quantity of heat required to raise the temperature of 1 g of pure water 1°C.

Heat capacity is the quantity of heat required to raise the temperature of a substance 1°C. For any given substance, the greater the mass of an object, the larger its heat capacity. The heat capacity of 1 g of a substance is the specific heat capacity, or specific heat, of that substance.

Practice Questions and Problems

14. Identify the following as quantitative or qualitative measurements. *2·1*
   a. A flame is hot.
   b. A candle weighs 90 g.
   c. Wax is soft.
   d. A candle's height decreases 4.2 cm per hr.

15. Distinguish between the accuracy and precision of a measurement. *2·2*

16. Comment on the accuracy and precision of these basketball free-throw shooters. *2·2*
   a. 99 of 100 shots are made.
   b. 99 of 100 shots hit the front of the rim and bounce off.
   c. 33 of 100 shots are made; the rest miss.

17. Three students made multiple weighings of a copper cylinder, each using a different balance. The correct mass of the cylinder had been previously determined to be 47.432 g. Describe the accuracy and precision of each student's measurements. *2·2*

	Mass of cylinder		
	Jim	*Leigh Anne*	*Bob*
Weighing 1	47.13	47.45	47.95
Weighing 2	47.94	47.39	47.91
Weighing 3	46.83	47.42	47.89
Weighing 4	47.47	47.41	47.93

18. How many significant figures are in each of these measurements? *2·3*
  a. 143 g   **d.** 1.072 km
  b. 0.074 cm   **e.** 10 800 cal
  c. 8.750×10^{-2} ng   **f.** 5.00 dm^3

19. Round off each of these measurements to three significant figures. *2·4*
  a. 98.473 L   **d.** 12.17°C
  b. 0.000 763 21 cg   **e.** $0.007\,498\,3 \times 10^4$ mm
  c. 57.048 m   **f.** 1764.9 mL

20. Write each of the rounded-off measurements in Problem 19 in standard exponential form. *2·4*

21. Round off each of the answers correctly. *2·4*
  a. 8.7 g + 15.43 g + 19 g = 43.13 g
  b. 4.32 cm $\times$ 1.7 cm = 7.344 cm^2
  c. 853.2 L − 627.443 L = 225.757 L
  d. 38.742 kg ÷ 0.421 = 92.023 75 kg
  e. 5.40 m $\times$ 3.21 m $\times$ 1.871 m = 32.431 914 m^3
  f. 5.47 m^3 + 11 m^3 + 87.300 m^3 = 103.770 m^3

22. Express each of the rounded-off answers in Problem 21 in standard exponential form. *2·4*

23. Water with a mass of 35.4 g is added to an empty flask with a mass of 87.432 g. The mass of the flask with the water is 146.72 g after a rubber stopper is added. Express the mass of the stopper to the correct number of significant figures. *2·4*

24. Express the answer to each problem in standard exponential form. *2·4*
  a. $5.3 \times 10^{+4} + 1.3 \times 10^{+4} =$
  b. $\dfrac{7.2 \times 10^{-4}}{1.8 \times 10^3} =$
  c. $10^{+4} \times 10^{-3} \times 10^{+6} =$
  d. $9.12 \times 10^{-1} - 4.7 \times 10^{-2} =$
  e. $\dfrac{4.8 \times 10^{-5}}{(2.0 \times 10^{-2}) \times (6.0 \times 10^3)} =$
  f. $(5.4 \times 10^{+4}) \times (3.5 \times 10^{+9}) =$

25. List at least two advantages to using the metric system (or SI). *2·5*

26. List the SI base unit of measurement for these quantities. *2·5*
  a. time   **c.** temperature
  b. length   **d.** mass

27. What is the symbol and meaning of these common metric prefixes? *2·6*
  a. milli   **c.** deci
  b. nano   **d.** centi

28. Match the approximate volume with each item.
  a. an orange   **(1)** 30 m^3 *2·7*
  b. a basketball   **(2)** 200 cm^3
  c. a van   **(3)** 20 L
  d. an aspirin tablet   **(4)** 200 mm^3

29. List these units in order from largest to smallest. *2·7*
  a. 1 dm^3 **c.** 1 mL   **e.** 1 cL
  b. 1 μL   **d.** 1 L   **f.** 1 dL

30. Astronauts in space are apparently weightless. Explain why it is incorrect to say that they are massless. *2·8*

31. Does the density of an object depend upon its size? Explain. *2·9*

32. A shiny, gold-colored bar of metal weighing 57.3 g has a volume of 4.7 cm^3. Is the metal bar pure gold? *2·9*

33. A weather balloon is inflated to a volume of 2.2×10^3 L with 37.4 g of helium. What is the density of helium in grams per liter? *2·9*

34. What is the unit of measure of specific gravity? Explain. *2·10*

35. Which of these substances has the highest specific gravity? Which has the lowest specific gravity? *2·10*
  a. mercury   **d.** iron
  b. wood   **e.** lead
  c. ice   **f.** aluminum

36. The element silver melts at 960.8°C and boils at 2212°C. Express these temperatures in Kelvin. *2·11*

37. Liquid nitrogen boils at 77.2 K. What is this temperature in Celsius? *2·11*

38. What are the two common units of heat energy? *2·12*

39. How much heat energy is absorbed when 88.0 g of water is heated from 5°C to 37°C? *2·13*

40. A piece of gold weighing 35.0 g absorbs 185 J of heat energy when its temperature increases by 41°C. What is the specific heat of gold? *2·13*

41. The temperature of two substances with the same mass increases by 20°C when heated. Which absorbs the most energy, the substance with the higher or lower specific heat? Explain. *2·13*

Mastery Questions and Problems

42. List two possible reasons for precise, but inaccurate, measurements.

43. Rank these numbers from smallest to largest.
a. 5.3×10^4 **d.** 0.0057
b. 57×10^3 **e.** 5.1×10^{-3}
c. 4.9×10^{-2} **f.** 0.0072×10^2

44. Criticize this statement: "When two measurements are added together, the answer can have no more significant figures than the measurement with the least number of significant figures."

45. Which is larger?
a. a centigram or a milligram
b. a liter or a centiliter
c. a calorie or a kilocalorie
d. a millisecond or a centisecond
e. a microliter or milliliter
f. a mm^3 or a dm^3

46. Criticize this statement: "When a number is rounded off, the last significant figure is dropped if it is less than 5."

47. The human body is approximately 60% water. Explain how this high percentage of water helps the body handle rapid changes in outside temperatures.

Critical Thinking Questions

48. Is it possible for experimental data to have good accuracy but poor precision?

49. Choose the term that best completes the second relationship.
a. heat:calories   money: _____
    (1) hot   (3) coins
    (2) dollars   (4) spending
b. temperature:thermometer
    volume: _____
    (1) meter stick   (3) balance
    (2) graduate cylinder (4) scale
c. foot:inch   meter: _____
    (1) millimeter   (3) kilometer
    (2) liter   (4) kilogram

50. You are hired to count the number of ducks on three northern lakes during the summer. In the first lake you estimate 500 000 ducks, in the second 250 000 ducks, and in the third 100 000 ducks. You write down that you have counted 850 000 ducks. As you drive away, you see 15 ducks fly in from the south and land on the third lake. Do you change the number of ducks you report? Justify your answer.

51. You have been introduced to the concepts of heat and temperature in this chapter. Are the ideas of hot and cold more closely associated with heat or with temperature? Do you think it is possible for something to be hot and cold at the same time?

Review Questions and Problems

52. What are the correct symbols for each element?
a. sodium   **d.** copper
b. aluminum   **e.** sulfur
c. chlorine   **f.** strontium

53. Classify each of the following as a chemical or physical change.
a. grass growing
b. sugar dissolving in water
c. crushing a rock
d. cooking potatoes
e. bleaching clothes
f. boiling water

54. How would you separate a mixture of ground glass and salt?

55. How can you distinguish between a homogeneous mixture and a compound?

56. Describe at least five energy transformations that are occurring in this scene.

57. What is a general goal of the scientific method?

Challenging Questions and Problems

58. The mass of a cube of iron is 355 g. What is the mass of a cube of lead that has the same dimensions?

59. A swimming pool measures 10 m by 25 m by 3.5 m. Assuming the pool is full of water, how many kcal of heat energy are needed to raise the temperature of the water from 18°C to 33°C?

60. Plot this data that shows how the mass of sulfur increases with an increase in the volume. Determine the density of sulfur from the slope of the line.

Mass of sulfur (g)	Volume of sulfur (cm³)
23.5	11.4
60.8	29.2
115	55.5
168	81.1

Research Projects

1. How did Lavoisier's work benefit people during his lifetime?

2. Trace the development of the liquid-in-glass thermometer.

3. Discuss the history of this country's conversion from English customary units to metric units.

4. Write an account of your daily routine. Use metric units to describe the objects and processes in your life.

5. Compile a pictorial dictionary of objects to demonstrate metric units. (The objects should be easily recognized as having a standard size.)

6. What does the standard deviation of a set of measurements represent? How is it used in research?

7. What are the toxic effects of mercury? Would the mercury in a broken thermometer be enough to contaminate the air in a classroom above acceptable levels?

8. Place an egg in a large glass and cover it with 400 mL of water. Determine how many teaspoons of table salt (sodium chloride) must be dissolved in the water to get the egg to float. Repeat the experiment with sugar, sodium carbonate (washing soda), magnesium sulfate (Epsom salts), and sodium sulfate (Glauber's salts). Compare the results.

9. Collect and determine the density of at least five different samples of rocks.

Readings and References

Frank, Sylvia. "Archimedes." *ChemMatters* (October 1987), p. 17.

Grey, Vivian. *The Chemist Who Lost His Head: The Story of Antoine Laurent Lavoisier*. New York: Coward, McCann & Geoghegan, 1982.

Morrison, Philip and Phylis, and The Office of Charles and Ray Eames. *Powers of Ten: A Book about the Relative Size of Things in the Universe and the Effect of Adding Another Zero*. New York: Scientific American (dist. by Freeman), 1982.

3 Problem Solving in Chemistry

Chapter Preview

3·1 Word Problems
3·2 Techniques of Problem Solving
3·A Problem Solving at Work: Chemical Technicians
3·3 Conversion Factors
3·4 Dimensional Analysis

3·5 Converting Between Units
3·6 Multistep Problems
3·B How Do Your Variables Vary?
3·7 Converting Complex Units

Chemistry is an experimental science aimed at problem solving. As you study chemistry and learn to think like a chemist, you will become a better problem solver. In this course you will encounter new concepts. You will be asked to develop skills that may be unfamiliar to you. As you gain an understanding of these new concepts and learn to apply these new skills, you will improve your ability to solve problems.

3·1 Word Problems

Real world problems are word problems. We deal not with numbers, but with measurements and the relationships between measurements.

Do you like word problems? You know the kind: "Sally is twice as old as Sara, who is three years younger than Suzy. What will Sally's age be one year from now if Suzy is one-third as old as twelve-year-old Sonja will be nine years from now?" (Answer: Nine years old.) This problem is probably more difficult than any problem you will encounter in this class. Even with much simpler word problems, however, you might feel overwhelmed. We hope this is not your situation, but if it is, consider the following analogy.

Persons who like playing tennis generally fall into one of two categories. Either they are already good tennis players (at least in their own minds), or they feel they are getting better through practice. Persons who try tennis and then give up are generally not good players. Neither do they want to put in the time needed to become a good player.

In much the same way, if you like word problems you are probably already a fairly proficient problem solver. Most students who dislike word problems have not yet become good problem solvers. If you dislike word problems you should not follow the example of the tennis player

Figure 3·1
The day-to-day work of chemists involves solving many different types of problems.

Figure 3·2
Developing a skill takes practice.

who quits. You can become a better problem solver. Like the tennis player who gets better with practice, *you can become a better problem solver by doing problems.*

3·2 Techniques of Problem Solving

After you have read a lengthy, complex word problem, the first thought that is likely to cross your mind is "Where do I begin?" Often there are several ways to solve a problem. No one method will work in every situation. Nevertheless, most word problems can be solved by using the following guidelines.

Approaching problem solving in an organized consistent fashion will ensure your success at problem solving.

1. **Identify the unknown.** Be certain that you know what the problem is asking. Read the problem carefully. If the problem is long you might need to read it a number of times. If the problem will have a numerical answer, be aware of the unit of the answer.

 Issues in Chemistry

Not all problems have numerical solutions. Controversial issues such as passive smoking and the "rights" of smokers and nonsmokers need to be resolved. Which of the steps in problem solving suggested in this section might apply to the issue of passive smoking?

2. **Identify what is known or given.** This usually includes a measurement and one or more relationships between measurements. Facts might also be given in a problem that are not needed to solve the problem. Learn to recognize extra information as such, and do not be misled.

3. **Plan a solution.** This is obviously the "heart" of problem solving: getting from the known to the unknown. Sketching a picture of the problem often helps you see a relationship between the given and the unknown. It might also suggest a way to break down a complex problem into two or more simpler problems. At this point you might need to use resources such as tables or figures to find other facts or relationships. For example, you might need to look up a constant or an equation that relates a known measurement to an unknown measurement. The solutions to the simpler problems can then be combined to form the solution to the more complex problem.

4. **Do the calculations.** This is usually a straightforward step if you have done a good job on Step 3. It may involve solving an equation for the unknown, substituting in known quantities, and doing the arithmetic.

Figure 3·3
The problem solving steps in this section are also helpful for solving everyday problems. When fixing a car you need to know how the car should work and what is currently wrong. Then you need to plan how to fix it, do the work, and check that it is fixed before you drive away.

In some problems you might be asked to express a measurement in a different form. In such a case, you must use relationships correctly to move from the given quantity to the unknown.

5. **Finish up.** The answer to a problem should always be expressed to the correct number of significant figures. Where appropriate the answer should be written in standard exponential form. Most importantly, **you must check your work.** Have you found what was asked for? Reread the problem. Did you copy down the given facts correctly? Check your math and check the units. Does your answer make sense? Is it reasonable? You can often estimate an approximate answer as a quick check.

Example 1

What is the volume, in cubic centimeters, of a sample of cough syrup that has a mass of 50.0 g? The density of cough syrup is 0.950 g/cm^3.

Solution

Step 1. The unknown is the volume, in cubic centimeters.

Step 2. The knowns are the mass and density.
a) mass = 50.0 g
b) density = 0.950 g/cm^3

Step 3. An equation that relates mass, density, and volume is this.

$$\text{Density} = \frac{\text{mass}}{\text{volume}}$$

This equation can be solved for the unknown, the volume.

Step 4. Solve for volume.

$$\frac{\text{volume}}{\text{density}} \times \text{density} = \frac{\text{mass}}{\text{volume}} \times \frac{\text{volume}}{\text{density}}$$

$$\text{Volume} = \frac{\text{mass}}{\text{density}}$$

Substitute values for the knowns.

$$\text{Volume} = \frac{50.0 \text{ g}}{0.950 \text{ g/cm}^3} = 52.632 \text{ cm}^3$$

Step 5. Volume = 52.6 cm^3 (three significant figures)

Problems

1. A small piece of gold has a volume of 0.87 cm^3. **a.** What is the mass if the density of gold is 19.3 g/cm^3? **b.** What is the value of this piece of gold if the market value of gold is $12 per gram?

2. How many joules are required to raise the temperature of 10.0 g of water from 10.0°C to 25.0°C?

3. What is the final temperature of a 140-g piece of glass at 21°C that absorbs 8.5×10^2 cal of heat when exposed to sunlight?

4. A volume of 15.0 mL of mercury is added to a beaker that has a mass of 56.7 g. What is the mass of the beaker with the added mercury?

5. The density of silicon is 2.33 g/cm³. What is the volume of a piece of silicon that weighs 62.9 g?

Science, Technology, and Society

3·A Problem Solving at Work: Chemical Technicians

Behind every successful chemist there is probably a well-trained chemical technician. While chemists invent new chemical procedures to produce a desired product, chemical technicians have the job of carrying out the chemists' ideas. Their job is generally practical rather than theoretical. They must have a good understanding of the procedures they are expected to follow. By contrast, the chemist must understand the theory behind the procedure.

Chemical technicians work in every type of chemical laboratory and factory. They may be part of a team that is engaged in research, design, development, production, analysis, or testing. Some technicians work with standard laboratory glassware. Others work with very sophisticated instruments. Technicians are expected to maintain accurate records of their work. They compute and analyze results and make reports. Some technicians may supervise other employees.

Preparation for a chemical technician position includes considerable on-the-job training. The technician must become familiar with the particular procedures, apparatus, and instruments used. For some

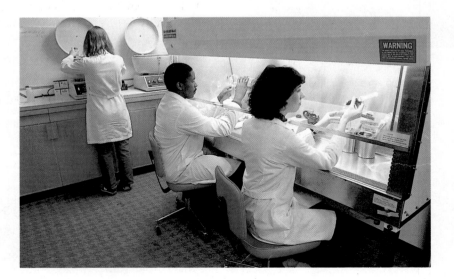

Figure 3·4
A successful year of chemistry in high school can qualify you for a summer job as a chemical technician.

jobs, a strong background in high school science and mathematics may be sufficient education. Many other jobs require some training beyond high school. Community colleges, technical schools, and some universities offer two-year programs for chemical technicians. In general, a person who is willing to continue his or her education will find more opportunities for promotion and challenging work.

3·3   Conversion Factors

The same quantity can usually be measured or expressed in many different ways. One dollar = 4 quarters = 10 dimes = 20 nickels = 100 pennies. These are all expressions, or measurements, of the same amount of money. One meter = 10 decimeters = 100 centimeters = 1000 millimeters. These are different ways to express the same length. Whenever two measurements are equal, or equivalent, a ratio of these two measurements will equal unity, or one.

$$\frac{100 \text{ cm}}{1 \text{ m}} = 1 \quad and \quad \frac{1 \text{ m}}{100 \text{ cm}} = 1$$

▇ Conversion factors show the relationship between two measurements.

Each of these ratios of equivalent measurements is called a **conversion factor.** In a conversion factor, the measurement in the numerator (on the top) is equivalent to the measurement in the denominator (on the bottom). The conversion factors above are read "one-hundred centimeters per one meter" and "one meter per one-hundred centimeters." As we shall see in the next section, conversion factors are very useful in solving problems in which a given measurement must be expressed in some other unit of measure. When a measurement is multiplied by a conversion factor, the value of the measurement remains the same. Although the *numerical value* of the measurement is changed, the change in the unit compensates for this. Consider, for example, 1 g and 10 dg. Even though the numbers in these two measurements differ, both measurements represent the same mass. Conversion factors are defined quantities. Therefore they generally have an unlimited number of significant figures.

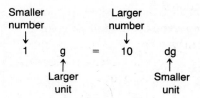

Figure 3·5
The two parts of a conversion factor are equal. The smaller number is part of the quantity with the larger unit. The larger number is part of the quantity with the smaller unit.

Example 2

Write the two possible conversion factors for each pair of units.
a. kilograms and grams   **b.** liters and milliliters

Solution

a. Since 1000 g = 1 kg, the conversion factors are: $\frac{1000 \text{ g}}{1 \text{ kg}}$ and $\frac{1 \text{ kg}}{1000 \text{ g}}$.

b. Since 1 L = 10^3 mL, the conversion factors are: $\frac{1 \text{ L}}{10^3 \text{ mL}}$ and $\frac{10^3 \text{ mL}}{1 \text{ L}}$.

Problems

6. What are the two conversion factors for each pair of units?
   a. meters and micrometers   **b.** grams and centigrams   **c.** grams of water and milliliters of water   **d.** joules and kilojoules

7. Write six possible conversion factors involving these units of measure: 1 g = 100 cg = 10^6 μg.

3·4 Dimensional Analysis

No one method is best for solving every type of problem. Several good approaches are available, however, and one of the best is dimensional analysis (also called the factor-label, or unit factor method). As the name implies, *in* **dimensional analysis** *we use the units (dimensions) that are a part of measurements to help solve (analyze) the problem*. The best way to explain this problem-solving technique is to use it to solve an everyday-type problem and then apply it to chemistry.

The use of units of measurements is a key factor in solving problems by dimensional analysis.

Your school club has sold 600 tickets to a chili supper fund-raising event. You, unfortunately, volunteered to make the chili. You have a very large pot and a chili recipe serving ten persons that calls for two teaspoons of chili powder. How much chili powder do you need for this pot of chili?

Let's work this problem using dimensional analysis and the problem-solving techniques we discussed in Section 3·2.

Step 1. The unknown is the number of teaspoons of chili powder.

Step 2. The knowns are as follows.
   a) 600 servings of chili must be made.
   b) 10 servings require two teaspoons (2 t.) of chili powder.

Step 3. To solve the problem we start with the known measurement, 600 servings of chili. We can use the relationship between servings of chili and teaspoons of chili powder from the recipe to write two conversion factors.

$$\frac{10 \text{ servings}}{2 \text{ t. chili powder}} \quad and \quad \frac{2 \text{ t. chili powder}}{10 \text{ servings}}$$

We always use the form of the conversion factor that has the unit of the known in the denominator. This allows us to cancel the known unit. It also gives us an answer that has the units of the unknown. Our plan then is to change the unit "servings" into the unit "teaspoons of chili powder": servings → teaspoons of chili powder.

Step 4. The solution can now be calculated.

$$\frac{600 \text{ servings}}{1} \times \frac{2 \text{ t. chili powder}}{10 \text{ servings}} = \frac{600 \times 2}{10} = 120 \text{ t. chili powder}$$

Notice that the unit of the known has cancelled and that we are left with the correct unit for the answer.

Figure 3·6
The conversion of a family recipe to a fund-raising-supper recipe can easily be done using dimensional analysis.

Step 5. What have we done in solving this problem? We have taken a known measurement, 600 servings, and multiplied it by a conversion factor (which equals unity). This gives us another measurement: 120 teaspoons of chili powder. According to this recipe, 120 teaspoons of chili powder is needed to make 600 servings of chili.

Now you have a problem of a different sort. Do you really take the time to measure out 120 teaspoons of chili powder or do you just estimate and dump in the whole can? The first option would be tedious; the second could be dangerous! Why not measure out the chili powder by the cup? This should be much quicker and still give a good-tasting product. The question then becomes, how many cups are 120 teaspoons of chili powder?

Step 1. The unknown is the number of cups of chili powder.

Step 2. The known is 120 t. chili powder.

Step 3. We start with the known, 120 t. chili powder. To solve this problem we need to know how many teaspoonsful of chili powder are in a cup of chili powder. A cookbook gives us this information.

3 teaspoons (t.) = 1 tablespoon (T.)

16 tablespoons = 1 cup

We can use the first relationship to write a conversion factor that allows us to express 120 t. as tablespoons. The conversion factor must be written with the unit teaspoons in the denominator so the known unit will cancel. We can then use the second conversion factor to change the unit tablespoons into the unit cups. This conversion factor must be written with the unit tablespoons in the denominator.

The overall plan then is to change the unit teaspoons into the unit tablespoons; then change the unit tablespoons into the unit cups: teaspoons → tablespoons → cups.

Figure 3·7
Metric measuring cups hold 250 mL. A metric tablespoon is 15 mL and a metric teaspoon is 5 mL.

Step 4. The solution can now be calculated.

$$\frac{120\ \cancel{t.}}{1} \times \frac{1\ \cancel{T.}}{3\ \cancel{t.}} \times \frac{1\ cup}{16\ \cancel{T.}} = \frac{120 \times 1 \times 1}{1 \times 3 \times 16} = 2.5\ cups$$

Notice that the units in the numerator of the solution are the same as the units in the plan: t. → T. → cups. *In every problem it is important to check to make sure that the units cancel and that the numerator and denominator of each conversion factor are equal to one another.* You should probably do this *before* you do the actual arithmetic to get an answer.

Step 5. We have shown that 2.5 cups of chili powder is the same amount as 120 teaspoons of chili powder. We have found the answer by multiplying the known measurement by two conversion factors, each of which is equal to unity. In doing this we have calculated a new measurement that is equivalent to the known measurement. The solution to this problem could have been shortened by combining the two relationships given in Step 3 into one: 48 teaspoons = 1 cup. We in effect did this in our solution in Step 4. Look at the product of the two conversion factors in the solution in Step 4.

At the beginning of this section we said that there is usually more than one way to solve a problem. As you read through the examples, you may have been thinking about a different way to approach some of the problems. For instance, the example in this section could also be worked using the ratio-and-proportion method. Some problems are most easily worked with simple algebra. You may already be able to use these other methods of problem solving. If so, you should learn how to use dimensional analysis as well. It will broaden your problem-solving skills. Then you will be able to choose the best way to solve each problem.

Of course you will work to get the correct answer to a problem. At the same time though you should try to explain the reasoning behind the steps to your solution. A negative aspect of solving problems by dimensional analysis is that, for some students, it becomes an almost mechanical procedure. Use the units to help you set up the solution to the problem. More importantly though, *think* the solution through. Try to understand why you are using particular conversion factors in progressing from the *known* to the *unknown*. Consider every problem to be unique. Becoming proficient at a particular method of problem solving will be helpful. It will never, however, replace your ability to read and interpret a problem. With perseverance you can learn to apply all your problem-solving skills toward the solution of problems.

Figure 3·8
How many apples can you buy for $4.40? Work this problem using the ratio-and-proportion method and by dimensional analysis. Notice the similarities in the solutions.

Example 3

The directions of an experiment ask each student to weigh out 1.84 g of copper (Cu) wire. The only copper wire available is a spool weighing 50.0 g. How many students can do the experiment before the copper runs out?

Solution

Step 1. The unknown is the number of students.

Step 2. The knowns are as follows.
a) the mass of copper, 50.0 g
b) each student uses 1.84 g of copper

Step 3. The conversion factor is 1 student/1.84 g Cu.

Step 4. The solution can now be calculated.

$$\frac{50.0 \text{ g Cu}}{1} \times \frac{1 \text{ student}}{1.84 \text{ g Cu}} = \frac{50.0}{1.84} = 27.174 \text{ students}$$

Step 5. The answer is 27 students. The rules for rounding off would allow us to give the answer as 27.2 students, but this does not make any sense.

Problems

8. A 1° increase on the Celsius scale is equivalent to a 1.8° increase on the Fahrenheit scale. If the temperature decreased 85.4°F, what is the corresponding temperature drop on the Celsius scale?

9. One of the first dental amalgams, used for tooth fillings, consisted of 26.0 g of silver, 10.8 g of tin, 2.4 g of copper, and 0.8 g of zinc. How much silver is in a 15.0-g sample of this amalgam?

3·5 Converting Between Units

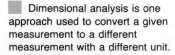

Dimensional analysis is one approach used to convert a given measurement to a different measurement with a different unit.

When doing chemistry it is often necessary to express a measurement in a different unit. The directions for a laboratory experiment may ask you to measure out 145 cg of magnesium metal. The balance in the laboratory is calibrated in grams. How many grams of magnesium do you want? This

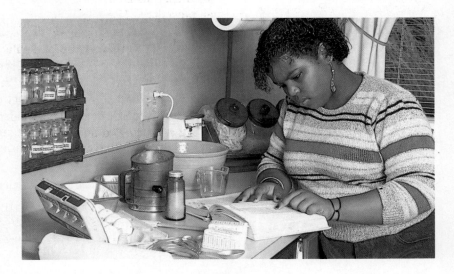

Figure 3·9
Understanding how to use conversion factors can make cooking easier. Some recipes call for ingredients by weight. You can look up the volume for a given weight in a table. Then you can measure the ingredient by volume.

is an example of a typical *conversion problem*. The need here is to express a given measurement in a different unit. As we have seen, conversion problems are easily solved using dimensional analysis.

In working the examples we will use the problem-solving steps discussed in Section 3·2.

Figure 3·10
Architects use conversion factors to accurately scale a building.

Example 4

Express 145 cg in grams.

Solution

Step 1. The unknown is the number of grams.

Step 2. The known is 145 cg.

Step 3. The desired conversion is: centigrams → grams. The expression relating the units is 100 cg = 1 g. The conversion factor is 1 g/100 cg. (We want to cancel the centigram unit and be left with the unit grams.)

Step 4. The solution is as follows.

$$\frac{145 \text{ cg}}{1} \times \frac{1 \text{ g}}{100 \text{ cg}} = \frac{145}{100} = 1.45 \text{ g}$$

Step 5. We see that the known unit (cg) cancels, and the answer has the correct unit, grams.

Example 5

What is the volume of a silver coin that has a mass of 14.0 g? The density of silver (Ag) is 10.5 g/cm^3.

Solution

Step 1. The unknown is the volume of the coin.

Step 2. The knowns are as follows.
a) the mass of the coin = 14.0 g
b) the density of silver = 10.5 g/cm^3

Step 3. This problem can be solved algebraically, but it can also be solved as a conversion problem. We need to convert the given mass of the coin into an equivalent volume. A density measurement gives a relationship between mass and volume: 1 cm^3 Ag = 10.5 g Ag.

Step 4. The solution can now be calculated.

$$\frac{14.0 \text{ g Ag}}{1} \times \frac{1 \text{ cm}^3 \text{ Ag}}{10.5 \text{ g Ag}} = \frac{14.0}{10.5} = 1.33333 \text{ cm}^3$$

Step 5. The answer is 1.33 cm^3.

Problems

10. Using tables from Chapter 2, make the following conversions.
- **a.** 0.73 km to meters
- **b.** 1.75 g to milligrams
- **c.** 4.7 nm to meters
- **d.** 0.072 m to centimeters
- **e.** 53 cm^3 to cubic decimeters
- **f.** 8.34 J to kilojoules
- **g.** 6.7 ms to seconds
- **h.** 45.3 kcal to calories

11. Use the given densities to make the following conversions.
- **a.** 23.6 g of boron to cm^3 of boron (density of boron = 2.34 g/cm^3)
- **b.** 1.4 L of argon to g of argon (density of argon = 1.78 g/L)
- **c.** 8.96 g of mercury to cm^3 of mercury (density of mercury = 13.5 g/cm^3)

3·6 Multistep Problems

When converting between units it is often necessary to use more than one conversion factor. For example, you probably don't know how many seconds are in a day. This is something that you can easily calculate using dimensional analysis.

Figure 3·11
When playing a complex piece of music for the first time, a pianist may play only the upper or lower portion of the music. In a similar way, you should break down a complex problem into simpler parts.

Combining the solutions to two or more simple one-step conversions will give the solution to a multistep conversion problem.

Example 6

How many seconds are in one day?

Solution

Step 1. The unknown is the number of seconds.

Step 2. The known is one day.

Step 3. The desired conversion is: day → seconds. This conversion can be carried out by the following sequence of conversions: day → hours → minutes → seconds. The expressions relating the units are: 1 day = 24 hr, 1 hr = 60 min, and 1 min = 60 s.

Step 4. The solution can now be calculated.

$$\frac{1 \text{ day}}{1} \times \frac{24 \text{ hr}}{1 \text{ day}} \times \frac{60 \text{ min}}{1 \text{ hr}} \times \frac{60 \text{ s}}{1 \text{ min}} = 1 \times 24 \times 60 \times 60 \text{ s}$$

$$1 \text{ day} = 86\ 400 \text{ s}$$

Each conversion factor is written so that the unit in the denominator cancels the unit in the numerator of the previous factor.

Step 5. The answer in standard exponential form is 8.64×10^4 s. *Remember to check that the numerator and denominator of each conversion factor are equivalent. When the units cancel you should be left with the unit of the unknown.*

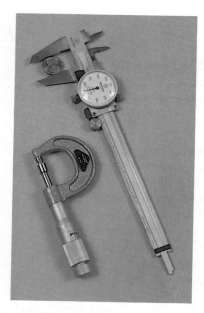

Figure 3·12
Micrometers are tools that are used to measure small distances very accurately. What units might be used for such measurements?

Example 7

How many micrometers is 0.073 cm?

Solution

Step 1. The unknown is the number of micrometers.

Step 2. The known is 0.073 cm.

Step 3. The desired conversion is: centimeters → micrometers. We do not have an expression that relates centimeters to micrometers. However, we know that 100 cm = 1 m and 1 m = 10^6 μm. The problem can be solved in two steps. Change centimeters to meters, then change meters to micrometers: centimeters → meters → micrometers.

Step 4. The solution is as follows.

$$\frac{0.073 \text{ cm}}{1} \times \frac{1 \text{ m}}{100 \text{ cm}} \times \frac{10^6 \text{ } \mu\text{m}}{1 \text{ m}} = \frac{7.3 \times 10^{-2} \times 10^6}{10^2} = 7.3 \times 10^2 \text{ } \mu\text{m}$$

Step 5. Check to make sure that units cancel, that conversion factors are correct, and that the correct unit remains on the answer.

Problems

12. Make the following conversions.
    - **a.** 0.57 μg to centigrams
    - **b.** 27.6 μL to milliliters
    - **c.** 583 nm to millimeters
    - **d.** 0.0023 kcal to decicalories
    - **e.** 678 cJ to kilojoules
    - **f.** 5.75×10^5 cg to nanograms

13. To raise the temperature of a mass of water from 25°C to 50°C requires 7.5 kcal. Calculate the mass of the water. **a.** grams **b.** kilograms.

Science, Technology, and Society

3·B How Do Your Variables Vary?

Imagine you are on a cross-country automobile trip. You decide to do an experiment to see whether your driving speed has any effect on how much gas your car uses. You fill up the gas tank. Holding the car's speed at 80 kilometers per hour, you drive for 3 hours. It takes 20 liters to refill the tank. You now drive at 100 kilometers per hour for a similar time period. It now takes 30 liters to refill the gas tank. You can now calculate the number of kilometers driven per liter of gas consumed at each speed. The results are 12 kilometers per liter at 80 kph and 10 kilometers per liter at 100 kph.

In this experiment, there are two variables. The speed of the car is changing because you are manipulating it, that is, you are deliberately making it change. The speed is called the independent (or manipulated) variable. Your fuel efficiency (kilometers per liter) also

Figure 3·13
On a graph, the independent variable is shown on the horizontal axis. The dependent variable is shown on the vertical axis. Controlled variables are not generally shown. As this graph shows, the lighter-mass subcompact cars are more fuel efficient at all road speeds than the more massive luxury-size car. The best fuel efficiencies are obtained at intermediate speeds.

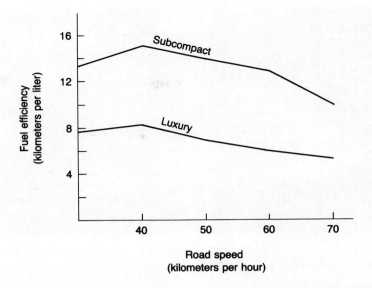

changes, because it depends on the speed. The fuel efficiency is called the dependent (or responding) variable. Any other possible variables should not have been allowed to change during the experiment. For example, the terrain the car is passing through will affect the fuel consumption. You could not tell how much of the mileage change was due to different terrains, and how much was due to the change in speed. You should guard against this problem by driving the same route for each of the two speeds. In this way the terrain would become one of the controlled variables. All other variables should also be controlled. Some of these are the type of gasoline used, the presence of winds, and the use of an air conditioner.

The conclusion that the fuel efficiency is greater at the higher speed is not supported by other experiments. This should lead you to think that some other variable must account for the unexpected results. The experiment must be repeated, with better control over the variables.

In a similar way, chemistry experiments involve manipulated (independent) variables, responding (dependent) variables, and controlled variables. Some examples of variables that must be controlled in sensitive chemical experiments are temperature, pressure, and acidity. The concentration of reactants, degree of mixing, and purity of chemicals must also be controlled.

3·7 Converting Complex Units

Many common measurements are expressed as a ratio of two units. We measure how fast we drive our car in kilometers per hour. We measure the densities of solids and liquids in grams per cm^3. If we use dimensional analysis, converting these types of measurements is just as easy as converting measurements with a single unit.

Figure 3·14
Different units of density would be used in these two situations. In an industrial setting the density is often stated in kg/m³. In the laboratory, it is often stated in g/cm³.

Example 8

The density of manganese is 7.21 g/cm³. What is the density of manganese expressed in units kg/m³?

Solution

Step 1. The unknown is the density in kg/m³.

Step 2. The known is a density of 7.21 g/cm³.

Step 3. The desired conversion is $\dfrac{g}{cm^3} \longrightarrow \dfrac{kg}{m^3}$

We have two conversions to carry out. In the numerator, the mass unit must be changed from grams to kilograms. The relationship is 10^3 g = 1 kg. In the denominator, the volume unit must be changed from cm³ to m³. Since 10^2 cm = 1 m, then $(10^2 \text{ cm})^3 = (1 \text{ m})^3$, or $10^6 \text{ cm}^3 = 1 \text{ m}^3$.

Step 4. The calculations can now be done.

$$\frac{7.21 \, g}{1 \, cm^3} \times \underbrace{\frac{1 \text{ kg}}{10^3 \, g}}_{\substack{\text{(mass} \\ \text{conversion)}}} \times \underbrace{\frac{10^6 \, cm^3}{1 \text{ m}^3}}_{\substack{\text{(volume} \\ \text{conversion)}}} = 7.21 \times 10^3 \text{ kg/m}^3$$

Step 5. Check that the units cancel, the conversion factors are correct, and the answer has the correct ratio of units.

Conversions of measurements with ratios of units are conveniently accomplished by converting each unit of the ratio in sequence.

Problem

14. Make the following conversions.
   a. 4.65 kg/L to grams per cubic centimeter
   b. 0.74 kcal/min to calories per second   milliLiter
   c. 1.42 g/cm³ to milligrams per square millimeter

Key Terms

conversion factor   3·3
dimensional analysis   3·4

Chapter Summary

Problem solving is a skill learned through practice. To solve a word problem you should identify both the known and the unknown. A solution is then planned and calculations are done. An answer to any problem should always be properly rounded off and checked.

Any two measurements that are equal to one another can be written as a ratio. This ratio of equivalent measurements is equal to unity and is called a conversion factor. Conversion factors are used in the problem-solving technique of dimensional analysis. Problems in which you are asked to express a measurement in some other unit (conversion problems) are easily solved using dimensional analysis. In this technique, the units are used to help write the solution to the problem.

No problem-solving method can replace the need for you to *read carefully* and to *think* as you work problems.

Practice Questions and Problems

15. Which of these statements correctly complete this sentence? Good problem solvers . . .   3·2
 a. read a problem only once.
 b. check their work.
 c. use every fact given in the problem.
 d. do as much of the work as possible in their head.

 e. break complex problems down into one or more simpler problems.
 f. look for relationships between pieces of information.

16. What conversion factor would you use to convert between these pairs of units?   3·3
 a. seconds and minutes
 b. grams of water and cm^3 of water
 c. nanograms and grams
 d. cubic decimeters and milliliters

17. A 5.00-kg sample of bituminous coal is composed of 3.20 kg of carbon, 0.500 kg of ash, 0.150 kg of moisture, and 1.15 kg of volatile (gas-forming) material. How many kilograms of ash are in 425 kg of this coal?   3·4

18. Make the following conversions. Express your answers in standard exponential form.   3·5
 a. 74 cm to meters
 b. 8.32×10^{-3} kg to grams
 c. 555 L to cubic centimeters
 d. 0.00527 cal to kilocalories
 e. 8.6 g to micrograms
 f. 9.62 kJ to joules
 g. 9.52 m to decimeters

19. Which of the following linear measures is the longest?   3·5
 a. 4×10^3 cm
 b. 4×10^5 mm
 c. 0.04 km
 d. 4×10^8 nm

20. The density of dry air at 25°C is 1.19×10^{-3} g/cm^3. What is the volume of 20.0 g of air?   3·5

21. An atom of gold weighs 3.271×10^{-22} g. How many atoms of gold are in 1.00 g of gold?   3·5

22. Make the following conversions. Express your answers in standard exponential form.   3·6
 a. 5.0×10^4 mm to kilometers
 b. 3.3×10^5 ng to milligrams
 c. 18.8 kcal to joules
 d. 6.29×10^{-6} dL to microliters

23. Complete this table so that the measurements in each horizontal line have the same value. *3·6*

cg	kg	g	μg
___	___	43.5	___
1.2×10^2	___		___
___	4.4×10^{-2}	___	___

24. How large a container, in liters, would you need to hold 7.2 kg of gasoline? (Take the density of gasoline to be 0.68 g/cm^3.) *3·6*

25. In a primitive barter society the following rates of exchange exist: 1 fot = 5 vum, 2 sop = 3 tuz, 4 bef = 3 tuz, and 9 fot = 2 bef. A man has 4 sop and wants to convert all his possessions into vum. How many vum can he trade for? *3·6*

26. The food that the average American consumes in a day provides 2.0×10^3 kcal of energy. How many calories/second is this? *3·7*

27. A cheetah has been clocked at 112 km/hr over a 100-m distance. What is this speed in m/s? *3·7*

Mastery Questions and Problems

28. A tank measuring 14.3 cm $\times$ 73.0 mm $\times$ 0.72 m is filled with olive oil that weighs 8.2×10^3 g. What is the density of olive oil in kg/L?

29. The specific heat of diamond (carbon) is 5.0×10^{-4} kJ/(g $\times$ °C). Express the specific heat in units cal/(g $\times$ °C).

30. What is the mass of a silver ring if its temperature changes from 20°C to 320°C when it absorbs 4.00×10^2 cal of heat energy?

31. A watch loses 0.15 s every minute. How many minutes will the watch lose in 1 day?

32. An apple can provide 110 kcal of energy. What mass of water could this same amount of energy raise from the freezing point to the boiling point?

33. A quantity of heat equal to 4.44 cal is added to 5.40 g of mercury at 25.0°C. What is the final temperature of the mercury?

34. A flask that can hold 158 g of water at 4°C can hold only 127 g of ethyl alcohol. What is the density of ethyl alcohol?

35. How many kilojoules are required to raise the temperature of 170.0 g of grain alcohol from 10.0°C to 45.0°C?

Critical Thinking Questions

36. Choose the term that best completes the second relationship.
   a. journey:route   problem: ___
      (1) unknown   (3) known
      (2) plan   (4) calculate
   b. meter:100 cm   gram: ___
      (1) kg   (3) 1000 mg
      (2) 100 cm   (4) 100 kg

37. You have solved many word problems up to this point. Review the techniques for solving word problems. Which step is most difficult for you? What kind of problems do you find most difficult?

38. Why are units so important in working word problems?

39. In the lab you add an equal amount of sugar to three beakers of hot water. You cool the three beakers to three different temperatures and examine the amount of sugar that precipitates.
   a. What are the independent and dependent variables?
   b. Why was an equal amount of sugar added to each beaker?

Review Questions and Problems

40. The melting point of gold is 1064°C. Express this temperature in Kelvin.

41. Classify each of the following as an element, compound, or mixture.
   a. milk shake   **c.** dry ice (CO_2)
   b. lipstick   **d.** iron powder

42. Describe how the law of conservation of mass and the law of conservation of energy apply to a burning campfire.

43. List three physical properties of each of the following objects in the photo below.
  a. the glass   **c.** the ice cube
  b. the soda   **d.** the bubbles

44. Identify the larger quantity in each of these pairs of measurements.
  a. nanogram and microgram
  b. calorie and joule
  c. cubic decimeter and milliliter
  d. centimeter and millimeter
  e. decigram and kilogram

45. How many significant figures are in each of these measurements?
  a. 5.432 g   **d.** 0.00304 mm
  b. 9.00×10^6 cal   **e.** 145.6 K
  c. 0.00045 dm^3 **f.** 34.563 s

46. Round each of the measurements in Problem 45 to two significant figures.

47. Name three physical and three chemical changes that you have seen today.

Challenging Questions and Problems

48. When 121 g of sulfuric acid was added to 400 mL of water, the resulting solution's volume was 437 mL. What is the specific gravity of the resulting solution?

49. How tall is a rectangular block of balsa wood measuring 3.2 cm wide and 2.5 cm deep that has a mass of 85.0 g? The specific gravity of balsa wood is 0.20.

50. Seawater contains 8.0×10^{-1} cg of the element strontium per kilogram of seawater. Assuming that all the strontium could be recovered, how many grams of strontium could be recovered from a cubic meter of seawater? Assume the density of seawater is 1.0 g/mL.

51. If 50.0 g of iron at 75°C is placed in 250 g water at 25°C in an insulated container and both are allowed to come to the same temperature, what will the temperature be? The specific heat of iron is 0.106 cal/(g × °C).

52. There are 6.02×10^{23} atoms of mercury in 201 g of mercury. The density of mercury is 13.5 g/mL. What is the volume, in μm^3, occupied by one mercury atom?

53. Different volumes of the same liquid were added to a flask on a balance. After each addition of liquid the mass of the flask with the liquid was measured. Graph the data using mass as the dependent variable. Use the graph to answer these questions.
  a. What is the mass of the flask?
  b. What is the density of the liquid?

volume (mL)	14	27	41	55	82
mass (g)	103.0	120.4	139.1	157.9	194.1

Research Projects

1. Design an experiment to investigate how one of the physical properties of a substance varies under different conditions. Identify the variables that are manipulated, controlled, and responding.

2. Write a computer program to do a routine calculation such as to calculate density given the mass and volume.

3. Find out what kind of mathematical problems are common in one of the five major divisions of chemistry listed in Chapter 1.

Readings and References

Loebel, Arnold B. *Chemical Problem-Solving by Dimensional Analysis: A Self-instructional Program.* Boston: Houghton Mifflin, 1978.

Zaugg, Harold E. "Polywater," *ChemMatters* (December 1987), pp. 10–13.

4 Atomic Structure

Chapter Preview

4·1   Atoms
4·A   John Dalton and the Role of Theories
4·2   Electrons, Protons, and Neutrons
4·B   The Discovery of the Nucleus
4·3   The Structure of the Atom

4·C   New Discoveries of Subatomic Particles
4·4   Atomic Number
4·5   Mass Number
4·6   Isotopes
4·7   Atomic Mass
4·D   The Mass Spectrometer

By the late 1800s scientists were convinced that atoms are the fundamental units of which all matter is composed. They soon discovered, however, that the atom is not indivisible. It is composed of smaller particles: electrons, protons, and neutrons. In this chapter we will begin with Dalton's atomic theory and see how our present knowledge of atomic structure was developed.

4·1 Atoms

A typical atom has a diameter of 10^{-8} cm and a mass of 10^{-23} g.

Greek: *atomos* = indivisible

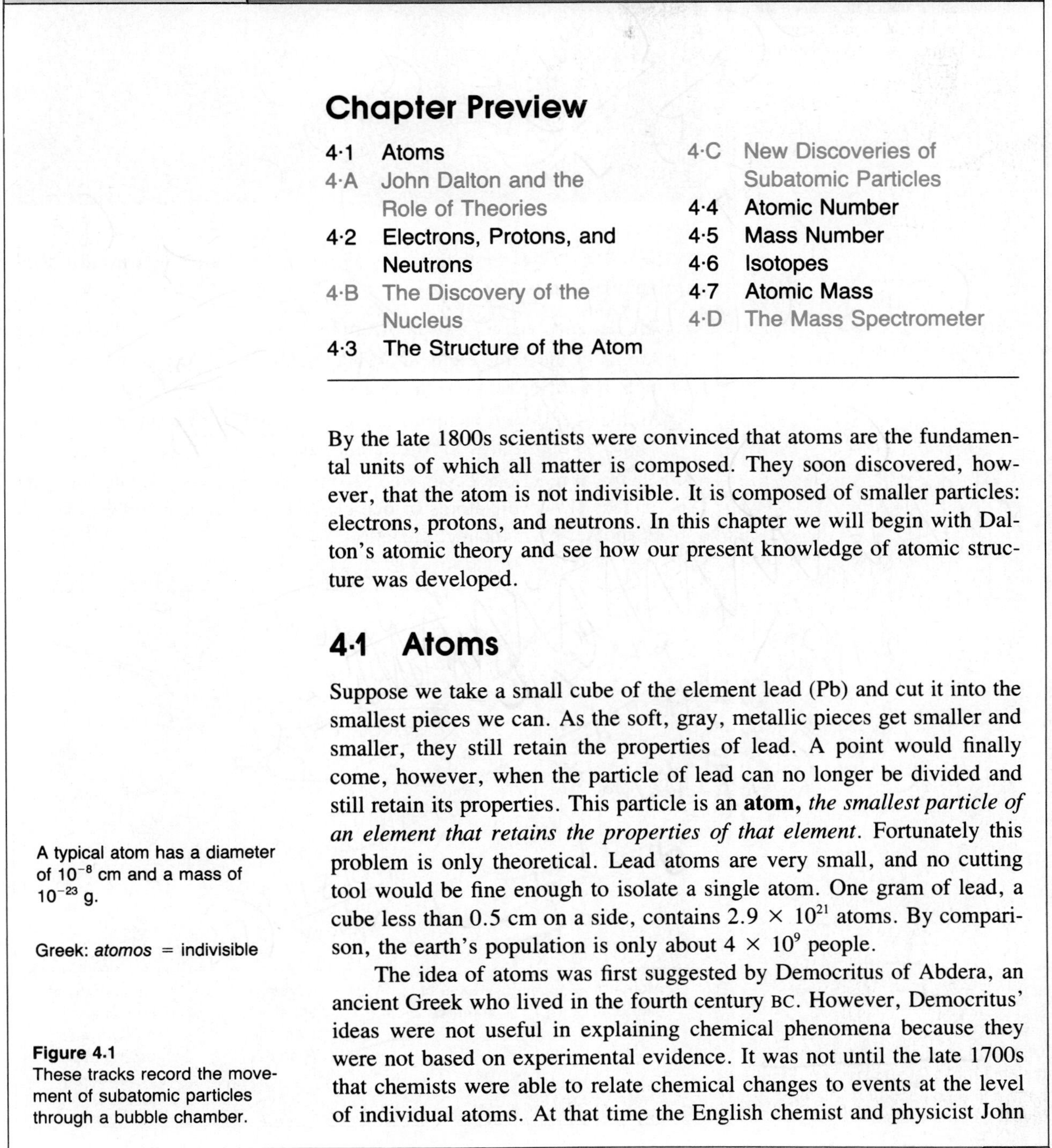

Figure 4.1
These tracks record the movement of subatomic particles through a bubble chamber.

Suppose we take a small cube of the element lead (Pb) and cut it into the smallest pieces we can. As the soft, gray, metallic pieces get smaller and smaller, they still retain the properties of lead. A point would finally come, however, when the particle of lead can no longer be divided and still retain its properties. This particle is an **atom,** *the smallest particle of an element that retains the properties of that element.* Fortunately this problem is only theoretical. Lead atoms are very small, and no cutting tool would be fine enough to isolate a single atom. One gram of lead, a cube less than 0.5 cm on a side, contains 2.9×10^{21} atoms. By comparison, the earth's population is only about 4×10^9 people.

The idea of atoms was first suggested by Democritus of Abdera, an ancient Greek who lived in the fourth century BC. However, Democritus' ideas were not useful in explaining chemical phenomena because they were not based on experimental evidence. It was not until the late 1700s that chemists were able to relate chemical changes to events at the level of individual atoms. At that time the English chemist and physicist John

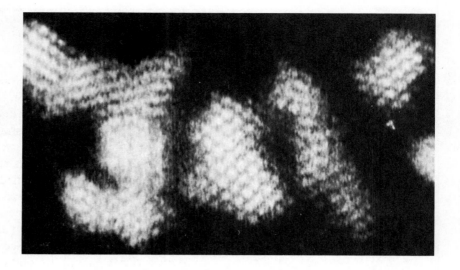

Figure 4·2
Clusters of uranium atoms appear in this photograph taken with a scanning transmission electron microscope. The individual uranium atoms are the white dots on the dark carbon surface. The magnification is ×500 000.

Dalton (1766–1844) first stated his atomic theory. **Dalton's atomic theory** included the following ideas.

An element is composed of many atoms that behave alike.

1. All elements are composed of tiny indivisible particles called atoms.
2. Atoms of the same element are identical. The atoms of any one element are different from those of any other element.
3. Atoms of different elements can combine with one another in simple whole number ratios to form compounds.
4. Chemical reactions occur when atoms are separated, joined, or rearranged. However, atoms of one element are not changed into atoms of another by a chemical reaction.

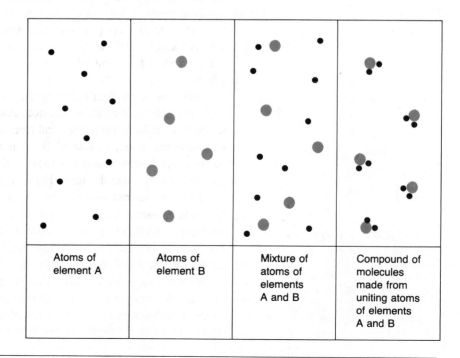

| Atoms of element A | Atoms of element B | Mixture of atoms of elements A and B | Compound of molecules made from uniting atoms of elements A and B |

Figure 4·3
According to Dalton's atomic theory, an element is a large collection of atoms and a compound is a large collection of molecules.

Figure 4-4
John Dalton is known for his interpretation of the experimental results of other chemists. His atomic theory helped chemists understand why many chemicals behave in certain ways.

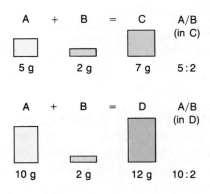

Figure 4-5
Dalton's law of multiple proportions relates to when two elements can combine to form more than one compound. The masses of the one element that combine with a fixed mass of the other element are in small whole-number ratios. In this example, the ratio of the mass of A in C to the mass of A in D (for a fixed mass of B) is 5:10 or 1:2.

4·A John Dalton and the Role of Theories

Two thousand years of error and oversight ended with Dalton's proposal of an atomic theory to explain chemical behavior. Dalton was only twelve years old when he took his first job as a school teacher in his own village in England. Throughout his life he earned his living as a teacher, first in high schools, later in a college, and finally as a private tutor. His real love and his real genius, however, were for science. During his lifetime he studied many different topics, including the aurora borealis, the tradewinds, and color blindness.

Dalton was particularly curious about how substances combined. He discovered that when two substances, A and B, combined to make a new substance, C, they seemed to do so in a fixed proportion by mass. Furthermore, if A and B could also combine to make a second new substance, D, the ratio of the masses of A and B in D would be a simple multiple of the ratio of their masses in C. For example, if substance C consisted of 5 grams of A for every 2 grams of B, substance D might consist of 10 grams of A for every 2 grams of B. Dalton demonstrated this "law of multiple proportions" and proposed an explanation for it in 1803.

Dalton concluded that all matter was made of atoms, which were indivisible, indestructible particles. They could be separated from each other or combined to form new substances. The atoms of different substances had different masses and were packed closely together even in a gas. This explained the differing densities of gases. Dalton assumed that if two elements combined to make only one compound, that compound consisted of one atom of each element. If two compounds were possible, one would consist of one atom of one element and two of the other. Because of this simple ratio of atoms, the masses of the combining elements were in simple ratios to each other.

We now know that some of Dalton's assumptions were not correct. The particles of a gas are not close together. There is no guarantee that the simplest compound formed from A and B will be made of one atom of A and one of B. Water, for example, is made of 2 atoms of hydrogen to one of oxygen. Finally, in the last century, science has shown that the atom is far from indivisible.

Dalton was not known as a great experimenter. He based most of his conclusions on the work of other chemists. He was, however, a brilliant theoretician. His atomic theory was a historic step in the understanding of chemical behavior. It is also an excellent example of the way in which theories bring about progress in scientific work. In spite of its flaws, Dalton's theory inspired a generation of chemists to conduct experiments. These clarified and refined our picture of the fundamental particles of matter.

4·2 Electrons, Protons, and Neutrons

Most of Dalton's atomic theory is still accepted. One revision concerns his idea that atoms are indivisible. This belief was held until about a century ago. Today we know that matter can be broken down into particles more fundamental than atoms. With the help of increasingly powerful atom smashers, physicists have found dozens of subatomic particles. In fact, so many particles have been found that no single theory of atomic structure can account for all of them. In chemistry we concern ourselves with only three of these particles: electrons, protons, and neutrons.

Electrons *are negatively charged subatomic particles.* They were discovered by scientists whose main interests were in electricity rather than chemistry. These scientists studied the flow of electric current through gases at low pressure. They contained the gases using a closed glass tube with metal disks called electrodes at each end (Figure 4·6). When connected to a source of high-voltage, the tube glows. One electrode, the anode, becomes positively charged. The other electrode, the cathode, becomes negatively charged. *The glowing beam, which travels from the cathode to the anode, is called a* **cathode ray.**

Atoms are divisible into electrons, protons, and neutrons.

Safety

Under some conditions, cathode tubes emit X-rays. Care should be used in operating this equipment.

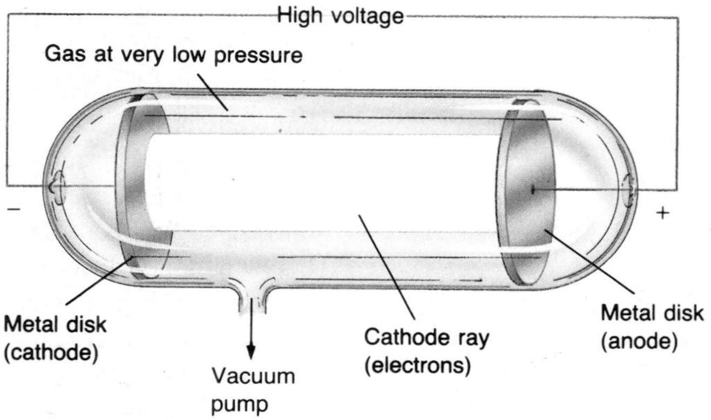

Figure 4·6
In a cathode ray tube, electrons travel as a ray from the cathode ($-$) to the anode ($+$). A television tube is a specialized type of cathode ray tube.

Chemical Connections

The picture tube in your television set is a cathode ray tube. The signal received by the TV antenna activates an electron gun in the tube. The gun fires a beam of electrons at the TV screen, forming the picture you see.

The English physicist Sir Joseph J. Thomson (1856–1940) experimented with cathode rays. In 1897 he found that cathode rays could be deflected either by magnets or by electrically charged plates (Figure 4·7). Thomson showed that a cathode ray is a collection of very small negatively charged particles, all alike, moving at high speed. He named these particles electrons. Moreover, he observed that cathode rays are always composed of electrons. Thomson reasoned that electrons must be a part of the atoms of all the elements. Thomson determined the ratio of the electric charge to mass of the electron. Later, R.A. Millikan (1868–1953) determined the charge of an electron and was therefore able to calculate its mass; it was about 1/2000 the mass of a hydrogen atom.

Shortly after electrons were discovered, scientists began to think about the particles left over when a hydrogen atom loses an electron. Since atoms are electrically neutral, researchers reasoned that the leftover

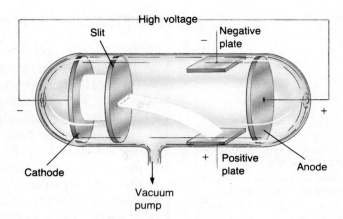

Figure 4·7
Cathode rays are attracted by a positively charged plate. This deflection shows the negatively charged character of the particles.

Greek: *protos* = first

particle should have a positive charge. Experimental evidence for such particles, protons, was soon found. *Researchers named this positively charged subatomic particle a* **proton.** It carries a single unit of positive charge and is 1840 times heavier than an electron.

In 1932 the English physicist Sir James Chadwick (1891–1974) confirmed the existence of yet another subatomic particle: the neutron. **Neutrons** *are subatomic particles with no charge,* but their mass nearly equals that of the proton. Thus the fundamental building blocks of all atoms are the electron, the proton, and the neutron. Table 4·1 summarizes the properties of these subatomic particles.

Figure 4·8
In 1886, Goldstein observed rays traveling in the opposite direction of the cathode ray. This occurred when he used a cathode with holes in it. He called these rays "canal rays." Later, the canal rays were found to be made up of positively charged particles. If the cathode ray tube contained hydrogen gas, the canal rays were made up of protons.

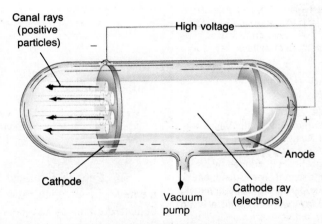

Table 4·1	Properties of Subatomic Particles			
Particle	Symbol	Relative electrical charge	Approximate relative mass (amu)*	Actual mass (g)
Electron	e^-	1−	1/1840	9.11×10^{-28}
Proton	p^+	1+	1	1.67×10^{-24}
Neutron	n^0	0	1	1.67×10^{-24}

*1 amu = 1.66×10^{-24} g.

Safety

Proper shielding should always be used with radioactive emissions such as alpha particles.

An atom has a dense nucleus but is mostly empty space.

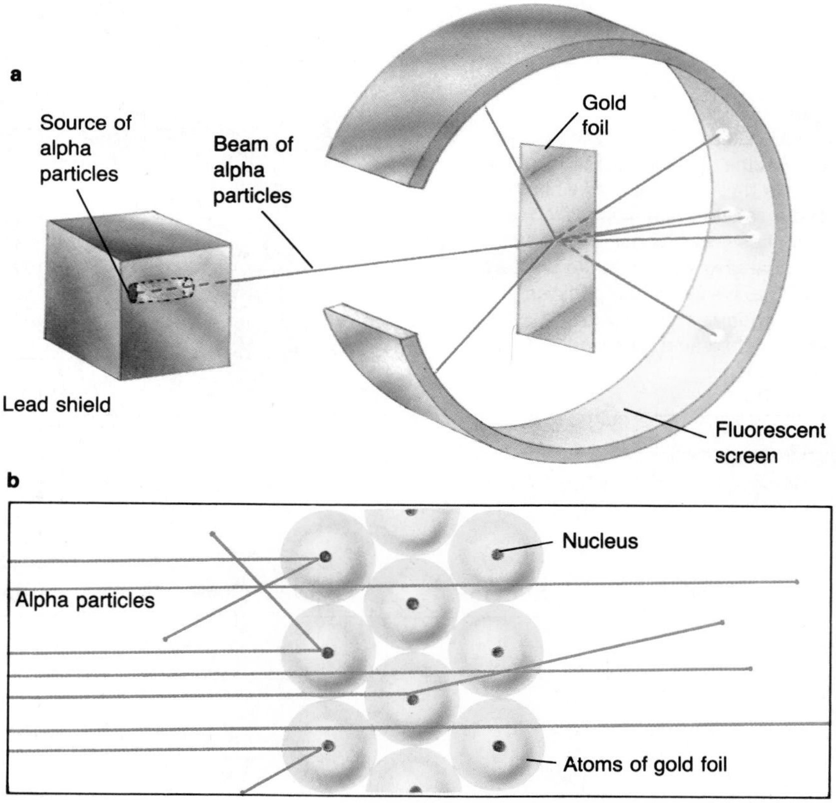

=== Science, Technology, and Society ===

4·B The Discovery of the Nucleus

In 1911 Baron Ernest Rutherford (1871–1937) decided to test the prevailing theory of atomic structure. In the process of doing so, he and his coworkers at the University of Manchester, England, were able to develop a more refined picture of the atom. The current theory was that the protons and electrons were evenly distributed throughout the volume of an atom. To test this theory, they directed a beam of particles at a very thin sheet of gold. The particles they chose for this were alpha particles, which are positively charged helium atoms that each lack two electrons. The expectation was that the alpha particles would pass through the thin sheet of metal unhindered.

To everyone's surprise, a very small fraction of the alpha particles bounced back (Figure 4·9). From these results, Rutherford proposed that the mass of the atom and the positive charge are concentrated in a small region. He called this region the nucleus. He thought of the rest of the atom as more or less empty space. The electrons were in that area, but they were so small that they did not interfere with the movement of the alpha particles. Rutherford later recollected: "It was quite the most incredible event that has ever happened to me in my life. It was almost as if you fired a 15-inch shell into a piece of tissue paper and it came back and hit you."

Figure 4·9
a To learn more about the nature of the atom, Rutherford and Marsden aimed a beam of alpha particles at a piece of gold foil surrounded by a fluorescent screen. They found that most of the particles passed through the foil. A few particles were deflected.
b Rutherford concluded that most of the alpha particles pass through the gold foil because the mass and positive charge are concentrated in a small region of the atom. He called this region the nucleus. Particles that pass near or approach the nucleus directly are deflected.

a

Source of alpha particles

Beam of alpha particles

Gold foil

Lead shield

Fluorescent screen

b

Alpha particles

Nucleus

Atoms of gold foil

4·3 The Structure of the Atom

Even before neutrons were discovered, scientists were wondering how electrons and protons were positioned within an atom. This was difficult to determine since atoms are such small particles. Using a rather ingenious experiment, Rutherford discovered the nucleus, a basic feature of atomic structure. *The **nucleus** of an atom is composed of protons and neutrons.* The nucleus has a positive charge, and it occupies a very small part of the volume of an atom. By contrast, the negatively charged electrons occupy most of the volume of the atom. They exist outside the nucleus. Since protons and neutrons have a much greater mass than electrons, the nucleus of an atom has a very high density.

Most atoms are composed of electrons, protons, and neutrons. How then are atoms of one element different from those of another element?

In an atom, negatively charged electrons move about the positively charged nucleus.

Most atoms of hydrogen have only a proton and an electron; they have no neutrons.

Figure 4·10
If an atom were the size of this stadium, then its nucleus would be about the size of a marble.

Table 4·2	Atoms of the First Ten Elements					
			Composition of the nucleus			Number
Name	Symbol	Atomic number	Protons	Neutrons	Mass number	of electrons
Hydrogen	H	1	1	0	1	1
Helium	He	2	2	2	4	2
Lithium	Li	3	3	4	7	3
Beryllium	Be	4	4	5	9	4
Boron	B	5	5	6	11	5
Carbon	C	6	6	6	12	6
Nitrogen	N	7	7	7	14	7
Oxygen	O	8	8	8	16	8
Fluorine	F	9	9	10	19	9
Neon	Ne	10	10	10	20	10

The answer is simple. The atoms of one element differ from the atoms of another element because their nuclei contain different numbers of protons. For example, atoms of carbon (C) have 6 protons, whereas fluorine (F) atoms have 9 protons, lead (Pb) atoms have 82 protons, and uranium (U) atoms have 92 protons.

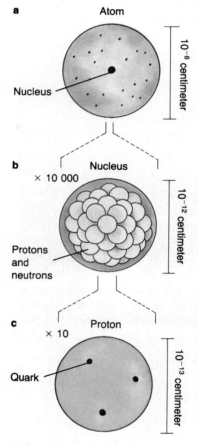

Figure 4·11
a The atom consists of a nucleus surrounded by electrons.
b The nucleus is composed of protons and neutrons.
c Each proton and neutron is composed of three quarks. Notice the large size difference between the atom and the nucleus compared to the size difference between the nucleus and the proton.

4·C New Discoveries of Subatomic Particles

The discovery that all atoms were made of just three types of particles showed a great simplicity in the structure of matter. Unfortunately, this simplicity was soon proven false. In the decades following the discovery of the neutron, scientists discovered more subatomic particles. These particles were not stable; they did not exist in ordinary matter. Instead they turned up when atomic nuclei were struck by high-energy particles or radiation. A target nucleus would shatter, giving rise to one or more new particles. Such collisions happened very infrequently in nature.

In order to have more collisions to study, scientists built cyclotrons, synchrotrons, and linear accelerators. These devices differ in the manner by which the particle "bullets" are accelerated. As a result of these new tools, however, hundreds of new subatomic particles have been discovered. This huge family of particles can now be explained in terms of more truly elementary particles.

Physicists now refer to two families of particles. The first, the leptons, includes the electron, the mu-meson (or muon), the tau-meson, and three types of neutrino. Each neutrino is matched to each of the three particles mentioned. Neutrinos have no charge and no detectable mass. These six particles are grouped as a family because they are affected only by the electroweak force. This is a force that affects charged particles and is also involved in certain kinds of radioactive decay. Leptons seem to be truly elementary. No evidence yet exists that leptons are made of smaller particles.

The second family of particles is called the hadrons. These particles are affected by the strong force that holds the nucleus of an atom together. Hundreds of hadrons are known, including the familiar proton and neutron. These particles are not elementary. Physicists have proposed that hadrons consist of smaller particles called quarks. The six types, or "flavors," of quarks are called up and down, top and bottom (or truth and beauty), and strange and charm. Each type of quark can have one of three "colors." These are not colors in the normal sense but they are more like a kind of charge. Up, top, and charm quarks have a charge of $+2/3$. The remaining three quarks have a charge of $-1/3$. The proton is made of two up quarks and one down quark; thus its charge is $+1$. The neutron is made of two down quarks and one up; its charge is 0.

Each lepton and quark also has a corresponding "antiparticle." For example, the antielectron is called the positron. It has the same mass as the electron but a charge of $+1$. The antidown quark has the same mass as a down quark but the opposite charge and color. When a particle collides with its antiparticle, both are annihilated. A burst of energy is released, and new particles are formed.

The study of the forces within atoms is equally important and fascinating. Physicists have identified the particles that carry the atomic forces. The photon, which has no charge and no mass, carries the electromagnetic force. Other particles, called bosons, carry the strong and weak forces mentioned earlier. The massless particles that carry the strong force, holding quarks together, are called gluons. Those that carry the weak force are known as the W+, W-, and Z particles. Unlike photons and gluons, these particles are very heavy. Their mass is as much as 100 times the mass of the proton.

Again the number of "elementary particles" has grown. They now include 6 types of quarks, 6 leptons, 24 antiparticles, 8 gluons, the W and Z bosons, and the photon. The physicists' hope for simplicity is not satisfied by these numerous subatomic particles. Hence, the search for truly fundamental particles continues.

4·4 Atomic Number

In Table 4·2 an atomic number is listed for each element. *The* **atomic number** *of an element is the number of protons in the nucleus of the atom of that element.* Hydrogen (atomic number 1) contains one proton in its nucleus. Oxygen (atomic number 8) contains eight protons in its nucleus. Atoms are electrically neutral. Thus the number of protons in the nucleus of an atom must equal the number of electrons around its nucleus. An atom of hydrogen has one electron around its nucleus, and an oxygen atom has eight electrons.

The atomic number of each element is given in the periodic table (Figure 12·4). The atomic number is the whole number that increases as you read across each row of the periodic table from left to right.

> The number of protons in an atomic nucleus identifies the element.

Problems

1. Element carbon is atomic number 6. How many protons and electrons are in a carbon atom?

2. The atomic number of an element is 11. What is the element?

4·5 Mass Number

A glance at Table 4·1 shows that the actual mass of protons and neutrons is very small: 1.67×10^{-24} g. Even compared with this small mass, the mass of an electron is negligible: 9.11×10^{-28} g. These values are useful information, but they are inconvenient for us to work with. In most instances chemists use relative comparisons of the masses of atoms. The unit of comparison is the **atomic mass unit** (amu), *which is defined as one-twelfth the mass of a carbon atom that contains six protons and six neutrons*. For all practical purposes we can consider the mass of a single proton or a single neutron as 1 amu.

Most of the mass of an atom is concentrated in its nucleus. *Thus the total number of protons and neutrons in the nucleus is the* **mass number** *of an atom*. Table 4·2 gives the composition of atoms of the first 10 elements. Hydrogen (atomic number 1) has a mass number of 1. Helium (atomic number 2) has a mass number of 4. From these data we can conclude that an atom of helium is four times as heavy as an atom of hydrogen. Likewise, we can see that an atom of neon (mass number 20) is five times heavier than an atom of helium.

■ The mass number is the approximate mass of an atom.

The composition of the nucleus of an atom is calculated from the atomic number and the mass number.

Example 1

How many protons, electrons, and neutrons are in the following atoms?

		Atomic number	Mass number
a.	Beryllium (Be)	4	9
b.	Neon (Ne)	10	20
c.	Sodium (Na)	11	23

Solution

a. For an atom, the number of electrons equals the number of protons and is given by the atomic number. Be has 4 protons and 4 electrons. We get the number of neutrons by subtracting the atomic number from the mass number. Be has $9 - 4 = 5$ neutrons.

b. Ne has 10 protons, 10 electrons, and $20 - 10 = 10$ neutrons.

c. Na has 11 protons, 11 electrons, and $23 - 11 = 12$ neutrons.

Problem

3. Complete this table.

Atomic number	Mass number	Number of protons	Number of neutrons	Number of electrons	Symbol of element
7	——	——	7	——	——
——	——	9	10	——	——
——	39	——	——	19	——
——	59	27	——	——	——

4·6   Isotopes

Most of Dalton's atomic theory (Section 4·1) is still accepted today. We now know, however, that all atoms of the same element are not identical. The nuclei of the atoms of a given element must all contain the same

■ Most elements in nature are a mixture of two or more isotopes.

Greek: *isotopes* = equal place

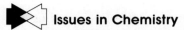 **Issues in Chemistry**

One of the isotopes of uranium is used to produce nuclear energy. What issues must be considered in deciding whether or not to increase our use of nuclear power plants? Which of these issues do you think carries the most weight? Why?

number of protons, but the number of neutrons may vary. *Atoms that have the same number of protons but different numbers of neutrons are called* **isotopes.** Since isotopes of an element have different numbers of neutrons, they also have different mass numbers and different atomic masses. Despite these differences, isotopes are chemically alike. This is because they have identical numbers of protons and electrons. These subatomic particles are responsible for the characteristic behavior of each element.

To symbolize the composition of an isotope we use the chemical symbol with two additional numbers written to the left of it. The mass number is written as a superscript (a number slightly above). The atomic number is written as a subscript (a number slightly below). Atoms of hydrogen with a mass number of 1 are designated hydrogen-1. Atoms of helium with a mass number of 4 are designated helium-4. They are represented as shown.

Mass number $\underset{\text{Atomic number}}{\overset{\text{Mass number}}{}}$ $^{1}_{1}\text{H}$ $^{4}_{2}\text{He}$ Mass number / Atomic number

Hydrogen-1   Helium-4

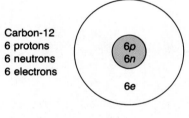

Carbon-12
6 protons
6 neutrons
6 electrons

Carbon-13
6 protons
7 neutrons
6 electrons

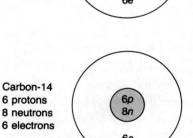

Carbon-14
6 protons
8 neutrons
6 electrons

Figure 4·12
Carbon-12, carbon-13, and carbon-14 are three isotopes of carbon. Do they differ in the number of protons, the number of neutrons, or the number of electrons?

Example 2

Two of the isotopes of carbon are carbon-12 and carbon-14. Give the chemical symbol for each.

Solution

Carbon is atomic number 6. All atoms of carbon have six protons.

Carbon-12, $^{12}_{6}\text{C}$ Carbon-14, $^{14}_{6}\text{C}$

Example 3

How many neutrons are in the following atoms?
a. $^{18}_{8}\text{O}$ **b.** $^{32}_{16}\text{S}$ **c.** $^{108}_{47}\text{Ag}$ **d.** $^{80}_{35}\text{Br}$ **e.** $^{207}_{82}\text{Pb}$

Solution

Recall that the superscript is the mass number and the subscript is the atomic number. The mass number minus the atomic number equals the number of neutrons.
a. 10 **b.** 16 **c.** 61 **d.** 45 **e.** 125

Problems

4. Two isotopes of oxygen are oxygen-16 and oxygen-18. Write the chemical symbol for each.

5. Using the periodic table, determine the number of neutrons in these atoms. **a.** ^{12}C **b.** ^{15}N **c.** ^{226}Ra

Table 4·3 Natural Isotopes of Some Familiar Elements

Name	Symbol	Mass (amu)	Natural percent abundance	"Average" atomic mass
Hydrogen	^1_1H	1.0078	99.985	
	^2_1H	2.0141	0.015	1.0079
	^3_1H	3.0160	negligible	
Helium	^3_2He	3.0160	0.0001	
	^4_2He	4.0026	99.9999	4.0026
Carbon	$^{12}_6C$	12.000	98.89	
	$^{13}_6C$	13.003	1.11	12.011
Nitrogen	$^{14}_7N$	14.003	99.63	
	$^{15}_7N$	15.000	0.37	14.007
Oxygen	$^{16}_8O$	15.995	99.759	
	$^{17}_8O$	16.995	0.037	15.999
	$^{18}_8O$	17.999	0.204	
Sulfur	$^{32}_{16}S$	31.972	95.00	
	$^{33}_{16}S$	32.971	0.76	
	$^{34}_{16}S$	33.967	4.22	32.064
	$^{36}_{16}S$	35.967	0.014	
Zinc	$^{64}_{30}Zn$	63.929	48.89	
	$^{66}_{30}Zn$	65.926	27.81	
	$^{67}_{30}Zn$	66.927	4.11	65.37
	$^{68}_{30}Zn$	67.925	18.57	
	$^{70}_{30}Zn$	69.925	0.62	

Before reading this section you should review your knowledge of percents by doing Section B·3 of the Math Review in Appendix B.

The atomic mass of an element is not always a whole number.

Your grade in a class is often calculated as a weighted average. Major exams, quizzes, homework assignments, and laboratory work may all have a different "weight" (or degree of importance) when your grade is averaged.

4·7 Atomic Mass

The mass of an atom is concentrated in the nucleus. You know that the mass of a proton is 1 amu, as is the mass of a neutron. Thus it is reasonable to expect that the mass of an atom expressed in atomic mass units should be a whole number. This view, however, ignores the existence of isotopes. *The **atomic mass** of an element is the weighted average of the masses of the isotopes of that element.* A weighted average reflects both the mass and the relative abundance of the isotopes as they occur in nature.

The percent abundances of the natural isotopes of some familiar elements are listed in Table 4·3 along with their atomic masses. How is the atomic mass of an element calculated? *Most elements occur as two or more isotopes in nature.* For example, chlorine has two isotopes, both of which have 17 protons in their atomic nuclei. One isotope has 18 neutrons and an atomic mass of 35 amu. This isotope is chlorine-35. The

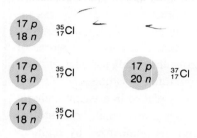

Ratio of chlorine atoms in natural abundance: three $^{35}_{17}$Cl to one $^{37}_{17}$Cl

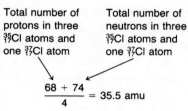

| Total number of protons in three $^{35}_{17}$Cl atoms and one $^{37}_{17}$Cl atom | Total number of neutrons in three $^{35}_{17}$Cl atoms and one $^{37}_{17}$Cl atom |

$$\frac{68 + 74}{4} = 35.5 \text{ amu}$$

Average mass of one atom

Figure 4·13
The average atomic mass of an element is calculated as a weighted average of the masses of its naturally occurring isotopes.

ChemDirections

The radiation given off by the unstable isotope cobalt-60 may be used in a method of food preservation. The process is described in **ChemDirections** *Irradiated Foods*, page 657.

other isotope is chlorine-37. It has 20 neutrons. In nature the isotopes of chlorine occur in a nearly 3 : 1 ratio. What is their average atomic mass? Figure 4·13 shows how the calculation is made. The average atomic mass that is calculated by this method is 35.5 amu. In the periodic table the atomic mass is 35.453 amu. The difference between these numbers is that the ratio of the isotopes is not exactly 3 : 1. In addition, the masses of a proton and a neutron are not exactly 1 amu. In calculating average atomic masses, we must remember to take into account the relative abundance of each isotope.

Example 4

Element X has two natural isotopes. The isotope with a mass number of 10 has a relative abundance of 20%. The isotope with a mass number of 11 has a relative abundance of 80%. Estimate the average atomic mass for the element from these figures. What is the true identity and atomic number of element X?

Solution

The mass that each isotope contributes toward the average atomic mass is equal to the product of its mass multiplied by its relative abundance. Remember to express percent as so many parts per 100 parts: $20\% = 20/100 = 0.20$.

$$
\begin{array}{lll}
^{10}\text{X} & 10 \text{ amu} \times 0.20 = & 2.0 \text{ amu} \\
^{11}\text{X} & 11 \text{ amu} \times 0.80 = & \underline{8.8 \text{ amu}} \\
& \text{Total} & 10.8 \text{ amu}
\end{array}
$$

The average atomic mass of element X is 10.8 amu. Element X is boron, atomic number 5.

Problem

6. The element copper is found to contain the naturally occurring isotopes $^{63}_{29}$Cu and $^{65}_{29}$Cu. The relative abundances are 69.1% and 30.9%, respectively. Calculate the average atomic mass of copper.

Each of the three known isotopes of hydrogen has one proton in the nucleus. The most common hydrogen isotope has no neutrons. It has an atomic mass of 1 amu and is called hydrogen-1 ($^1_1$H), or hydrogen. The second isotope has one neutron and an atomic mass of 2 amu. It is called either hydrogen-2 ($^2_1$H), or deuterium. The third isotope has two neutrons and an atomic mass of 3 amu. This isotope is hydrogen-3 ($^3_1$H), or tritium. Tritium nuclei are unstable and slowly disintegrate. Almost all hydrogen in nature is hydrogen-1. The other two isotopes are present in only trace amounts. Thus the average atomic mass for hydrogen is 1.0079 amu. You will see in Chapter 24 that many isotopes with unstable nuclei such as iodine-131 and cobalt-60 are useful in the diagnosis and treatment of disease.

4·D The Mass Spectrometer

One of the most useful and versatile instruments found in the modern chemistry laboratory is the mass spectrometer. This device was invented early in the twentieth century by F. W. Aston, an English physicist. Aston needed an instrument that could separate atoms of slightly different masses in order to prove that many samples of naturally occurring elements were really mixtures of isotopes.

In a mass spectrometer the sample is placed in a vacuum chamber. The vaporizing sample leaks through a very small hole into the instrument. There the vaporized sample is bombarded by high-energy electrons coming from a heated filament. When a particle in the sample is hit by one of these electrons, it may lose an electron of its own and become positively charged. The charged particles are passed through a slit to give a narrow beam of particles. Then they are accelerated to a high speed by an electric field. Because of their charge, the particles can be deflected by electric and magnetic fields. The amount of deflection depends on the size of the fields and the speed, charge, and mass of the particles.

An analogy may help to explain how this deflection takes place. Imagine an electric fan set up to blow *across* a table top. Now imagine that you are rolling table tennis balls *along* the table top, across the path of the fan. The balls will roll in a straight line until they feel the effect of the breeze from the fan. Then they will be deflected from their path. If you roll the balls faster, the deflection will not be so much. If you increase the speed of the fan, the deflection will be more. If you were to use golf balls instead of table tennis balls, the deflection would be less.

By measuring the deflection of the particles in the mass spectrometer, their mass can be determined. If a sample contains particles of several different masses, each different mass corresponds to a different deflection.

Chemical Connections

Forensic chemists use the mass spectrometer (MS) to identify small amounts of unknown materials. For example, some drugs have highly specific MS patterns. A chemist matches the MS pattern of the unknown material against the patterns of known materials to find the identity of the unknown.

ChemDirections

The mass spectrometer is often used in drug identification. For more information read **ChemDirections** *Drug Testing*, page 653.

Figure 4·14
In a mass spectrometer, the ionized sample is deflected by variable electric and magnetic fields. High mass ions are deflected less than lower mass ions. The detector measures how many particles are present when the beam is deflected by a given amount.

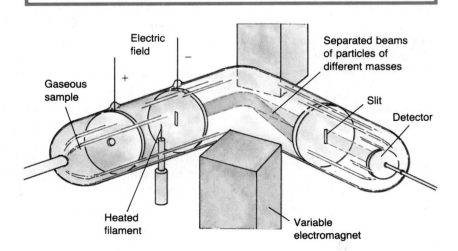

4 Atomic Structure
Chapter Review

Key Terms

atom	4.1	electron	4.2
atomic mass	4.7	isotope	4.6
atomic mass unit	4.5	mass number	4.5
atomic number	4.4	neutron	4.2
cathode ray	4.2	nucleus	4.3
Dalton's atomic		proton	4.2
theory	4.1		

Chapter Summary

The basic building blocks of matter are elements. More than 100 elements exist, and each is composed of atoms. The atoms of a given element are different from the atoms of all other elements.

Atoms are exceedingly small. Dalton theorized that atoms were indivisible, but the discovery of the electron changed this theory. Besides negatively charged electrons, atoms contain positively charged protons and electrically neutral neutrons. The proton has a mass nearly 2000 times the mass of an electron. A proton and a neutron are nearly identical in mass.

The nucleus of the atom is composed of protons and neutrons. The nucleus contains most of the mass of the atom in a very small volume. The electrons surround the nucleus and occupy most of the volume of the atom.

The number of protons in the nucleus of the atom is the atomic number of that element. Atoms are electrically neutral. Thus an atom has the same number of protons and electrons. The sum of the protons and neutrons is the mass number. The atoms of a given element all contain the same number of protons, but the number of neutrons may vary. Atoms with the same number of protons but different numbers of neutrons are isotopes.

The atomic mass of an element is expressed in atomic mass units (amu). An atom of any element has an atomic mass that is approximately a whole number. This is because protons and neutrons each have a mass of about 1 amu. The atomic mass in the periodic table is a weighted average of all the naturally occurring isotopes of that element. For this reason, the atomic mass of most elements is generally not a whole number.

Practice Questions and Problems

7. What is an atom? $\quad 4.1$

8. State the main ideas of Dalton's atomic theory. $\quad 4.1$

9. What are the charges and relative masses of the three subatomic particles that are of most interest to chemists? $\quad 4.2$

10. Describe the composition of the nucleus of the atom. $\quad 4.3$

11. What does the atomic number of each atom represent? $\quad 4.4$

12. How many protons are in the nuclei of the following atoms? $\quad 4.4$
a. sulfur   **c.** phosphorus   **e.** calcium
b. hydrogen   **d.** cadmium

13. What is an atomic mass unit? $\quad 4.5$

14. Complete this table. $\quad 4.5$

Elements	Number of protons	Mass number	Number of electrons	Atomic number	Number of neutrons
Se	——	——	——	——	45
——	——	16	8	——	——
——	14	28	——	——	——
——	——	1	——	——	——

15. Name three ways that isotopes of an element differ. *4·6*

16. Complete this table. *4·6*

Isotope	Symbol	Atomic number	Mass number	Number of protons	Number of neutrons
hydrogen-2	$^{2}_{1}\text{H}$	——	——	——	——
——	$^{83}_{38}\text{Sr}$	——	——	——	——
——	——	92	——	——	146
——	C	——	——	——	8
——	——	——	201	80	——

17. List the number of protons, neutrons, and electrons in each of the following atoms. *4·6*
  a. $^{36}_{18}\text{Ar}$ **c.** $^{3}_{1}\text{H}$ **e.** $^{79}_{35}\text{Br}$
  b. $^{61}_{29}\text{Cu}$ **d.** $^{17}_{8}\text{O}$

18. What is the atomic mass of an element? *4·7*

19. Uranium has three isotopes with the following percent abundances: $^{234}_{92}\text{U}$ (0.0058%), $^{235}_{92}\text{U}$ (0.71%), $^{238}_{92}\text{U}$ (99.23%). What do you expect the atomic mass of uranium to be in whole numbers? Why? *4·7*

20. The table below shows some data collected by Rutherford and his colleagues during their gold foil experiment. *4·8*
  a. What percent of the alpha particle deflections were 5° or less?
  b. What percent of the deflections were 15° or less?
  c. What percent of the deflections were 60° or greater?

Angle of deflection (degrees)	Number of deflections
5	8 289 000
10	502 570
15	120 570
30	7 800
45	1 435
60	477
75	211
>105	198

Mastery Questions and Problems

21. Explain why the atomic masses for most elements are not whole numbers.

22. Compare the relative size and relative density of an atom to its nucleus.

23. How are the three isotopes of hydrogen alike? How are they different?

24. What parts of Dalton's atomic theory no longer agree with our current picture of the atom?

25. The four isotopes of lead are shown below, each with its percent by mass abundance and the composition of its nucleus. Using this data and the approximate atomic mass of each isotope, calculate the atomic mass of lead.

204 amu 82 p 122 n	206 amu 82 p 124 n	207 amu 82 p 125 n	208 amu 82 p 126 n
1.37%	26.26%	20.82%	51.55%

Critical Thinking Questions

26. Choose the term that best completes the second relationship.
  a. female:male   proton: _____
      (1) atom   (3) electron
      (2) neutron   (4) quark
  b. cow:horse   neutron: _____
      (1) proton   (3) atom
      (2) nucleus   (4) quark
  c. proton:quark   house: _____
      (1) school   (3) planet
      (2) nucleus   (4) brick

27. How could you modify Rutherford's experimental procedure to determine the relative sizes of different nuclei?

28. Rutherford's atomic theory proposed a massive nucleus surrounded by very small electrons. This implies that atoms are composed mainly of empty space. If all matter is mainly empty space, then why is it impossible to walk through walls or pass your hand through your desk?

Review Questions and Problems

29. How many significant figures are in each of these measurements?
 a. 56.902 cg
 b. 0.0045 km
 c. 5.200×10^{-3} mL
 d. 10.04 kJ

30. Round each of the measurements in Problem 29 to two significant figures.

31. What is the mass of 2.47 cm^3 of platinum? The density of platinum is 22.5 g/cm^3.

32. A piece of coal was burned and emitted 245 kJ of heat energy. How many calories is this?

33. Oxygen and hydrogen react explosively to form water. In one reaction 4 g of hydrogen combines with oxygen to form 36 g of water. How much oxygen was used?

34. Classify each of the following as an element, compound, or mixture.
 a. silver   **d.** water
 b. salad oil   **e.** cardboard
 c. apple   **f.** perfume

35. An aquarium measures 45.0 cm × 1.10 m × 60.0 cm. How many cm^3 of water will this aquarium hold?

Challenging Questions and Problems

36. Lithium has two naturally occurring isotopes. Lithium-6 has an atomic mass of 6.015 amu; lithium-7 has an atomic mass of 7.016 amu. The atomic mass of lithium is 6.941 amu. What is the percentage of lithium-7 in nature?

37. When the masses of the particles that make up an atom are added together, the sum is always larger than the actual mass of the atom. The "missing" mass, called the mass defect, represents the matter converted into energy when the nucleus was formed from its component protons and neutrons. Calculate the mass defect of a chlorine-35 atom by using the data in Table 4·1. The actual mass of a chlorine-35 atom is 5.81×10^{-23} g.

Research Projects

1. How was the proton discovered?

2. How was the neutron discovered?

3. Measure the effect of distance on the attraction between oppositely charged particles.

4. How are radioactive isotopes isolated? How are they created?

5. What is a cloud chamber? How was it invented, and what discoveries did it make possible?

6. Report on the experiments that are being done to try to find a "free quark" or to record the decay of a proton. Why are these experiments important?

7. Visit a laboratory that has a mass spectrometer. Report on the instrument and its applications.

Readings and References

Berger, Melvin. *Atoms, Molecules, and Quarks.* New York: Putnam, 1986.

Boslough, John. "Worlds Within the Atom." *National Geographic,* (May 1985), pp. 634–663.

Davenport, Derek A. "John Dalton's First Paper and Last Experiment." *ChemMatters* (April 1984), pp. 14–15.

Janos, Leo. "Timekeepers of the Solar System." *Science 80,* (May/June 1980), pp. 44–55.

Taubes, Gary. "Detecting Next to Nothing." *Science 85,* (May 1985), pp. 58–66.

Taubes, Gary. "The Ultimate Theory of Everything." *Discover,* (April 1985), pp. 52–59.

Trefil, James S. *From Atoms to Quarks: An Introduction to the Strange World of Particle Physics.* New York: Scribner, 1980.

ロム酸カリウム
K_2CrO_4

重クロム酸カリ
$K_2Cr_2O_7$

5 Chemical Names and Formulas

Chapter Preview

ChemDirections
Only about 25 of the elements known to exist are necessary for life. To learn what some of these elements are and how they function in the body, turn to **ChemDirections** *Mineral Nutrients*, page 652.

The periodic table groups the elements according to similarities in their properties.

Figure 5·1
The language of chemistry, which you will learn in this chapter, is an international language.

Only about 100 types of atoms, the elements, exist. Even so, a tremendous variety of combinations of atoms can be formed. With such a diversity of substances, we need an orderly way in which to communicate chemistry. We could identify a particular substance by listing its physical and chemical properties. However, this would be very time-consuming at best. Instead, each substance is identified by its own unique name and chemical formula. This identification is the language of chemistry and the topic of this chapter.

5·1 The Periodic Table

The elements are arranged in rows and columns on the **periodic table,** *according to similarities in their properties.* On the next page you can see that the elements are listed in order of increasing atomic number (Figure 5·2). Hydrogen, the smallest lightest element, is in the top left corner. Helium, He, is atomic number 2 at the top right. Lithium, Li, is atomic number 3 at the left side of row two.

A column of elements in the periodic table is known as a **group.** Each group is designated by a number–letter combination. For example, the first vertical column on the left is Group 1A. It is composed of the elements H, Li, Na, K, Rb, and Cs. The next vertical column to the right,

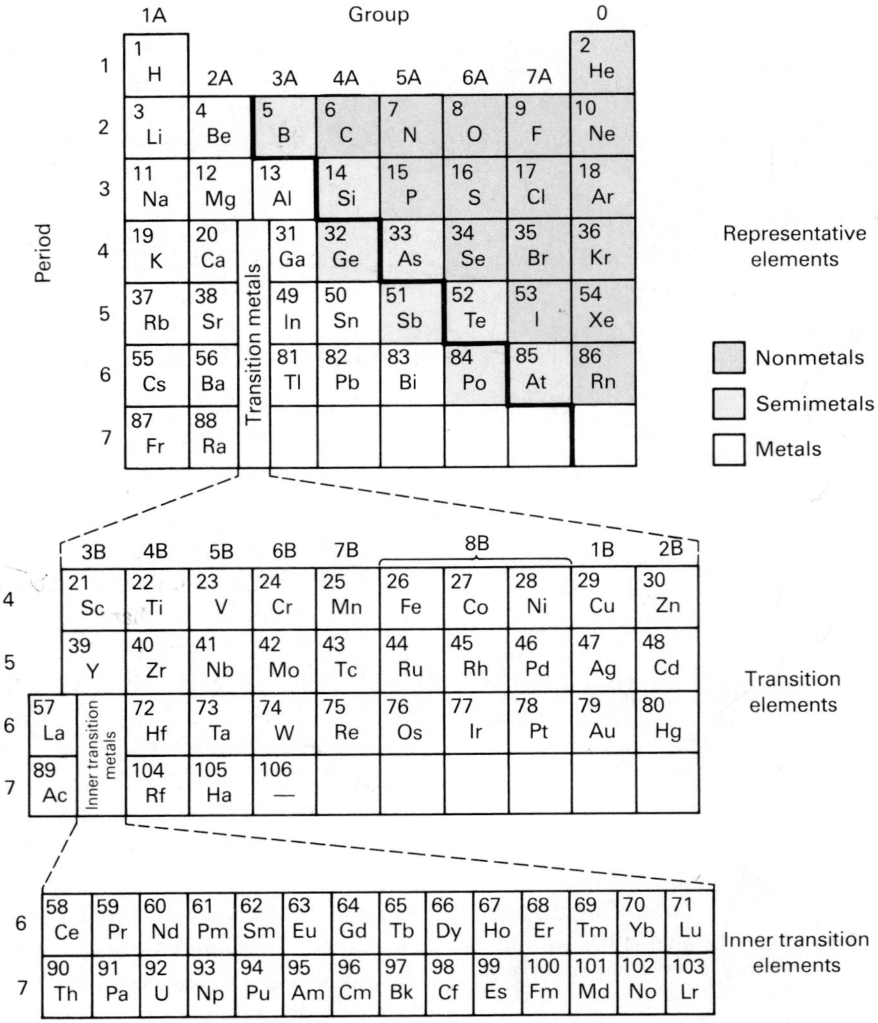

Figure 5·2
The short form of the periodic table shows three major divisions. In what area are the nonmetals grouped?

Figure 5·3
Mercury, a transition metal, is the only metallic element that is a liquid at room temperature. It is used in thermometers and barometers.

Figure 5·4
Sulfur is a nonmetallic element that occurs as a crystalline solid or in the amorphous state. Sulfur is used primarily in the manufacture of sulfuric acid.

starting with Be, is Group 2A, and so forth. *The Group A elements in the periodic table are known as the* **representative elements.** They are called this because they illustrate the entire range of chemical properties. The representative elements include both metals and nonmetals. They do not include the transition metals or the inner transition metals.

The metallic elements are grouped on the left side of the periodic table. **Metals** *are elements that have a high luster when clean and a high electrical conductivity.* They are ductile (can be drawn into wires) and malleable (can be beaten into thin sheets). Most of the elements are metals. They include the **transition metals,** *which are the B group elements,* and the inner transition metals, which are called the rare earths.

The nonmetallic elements are grouped on the right side of the periodic table. **Nonmetals** *are elements that are nonlustrous and are poor conductors of electricity.* Some of these elements are gases, others are brittle solids, and one, bromine, is a liquid at room temperature. Notice that hydrogen is a special case. It is a nonmetal in Group 1A.

Figure 5·5
The semimetal boron is usually found as boric acid in volcanic spring water. It is a component of borax, a cleanser and water softener.

Atoms lose or gain one or more electrons when forming ions.

Ca is a calcium atom. Ca^{2+} is a calcium cation.

The **semimetals,** *or* **metalloids,** *are elements with the properties of both metals and nonmetals.* Silicon is a basic component of transistors and chips because it is a semiconductor. It conducts electricity in a special way.

Problems

1. Identify the following elements as metals, semimetals, or nonmetals.
   a. potassium **b.** boron **c.** molybdenum **d.** iodine **e.** uranium

2. Which of the above elements are part of the representative elements?

5·2   Atoms and Ions

We know that elements are composed of atoms of the same kind. An atom is electrically neutral because it has an equal number of protons and electrons. Sodium is atomic number 11. A sodium atom has 11 positively charged protons, 12 neutrons, and 11 negatively charged electrons. Because neutrons have no charge, the net charge on a sodium atom is zero $[+11 + (-11) = 0]$. In forming a chemical compound, an atom of an element can gain or lose one or more electrons. When the number of electrons is no longer equal to the number of protons, then the atom becomes an ion. **Ions** *are atoms or groups of atoms that have a positive or negative charge.* An ion is formed when an atom or group of atoms loses or gains electrons.

Atoms of the metallic elements tend to form positive ions by losing one or more electrons. *A* **cation** *is any atom or group of atoms with a positive charge.* Compared with an electrically neutral atom, a cation has fewer electrons. For example, a *sodium cation* is formed by the loss of one electron from a sodium atom. Since a sodium ion has 11 protons but only 10 electrons, it has a charge of $1+$ (Figure 5·6). An ionic charge is written with a number followed by a sign. If the number is 1, it can be omitted; Na^{1+} and Na^+ are equivalent. Similarly, a magnesium atom forms a magnesium ion by the loss of two electrons. A magnesium ion therefore has a charge of $2+$ because it has 12 protons but only 10 electrons.

There are many important chemical differences between metals and their ions. Sodium metal, for example, reacts explosively with water. By

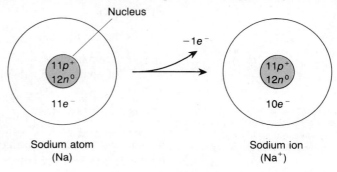

Figure 5·6
A sodium atom loses an electron to become a positively charged sodium ion.

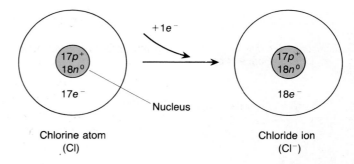

Figure 5·7
A chlorine atom gains an electron to become a negatively charged chloride ion.

Chlorine atom
(Cl)

Chloride ion
(Cl⁻)

Nucleus

Safety

All the alkali metals (lithium, sodium, potassium, rubidium, and cesium) are stored under oil to prevent them from contacting air or water. Their reactions can be very violent. They should be used only with extreme caution and under direct supervision of the teacher.

The names of some common anions are fluoride, chloride, bromide, iodide, oxide, sulfide, nitride, and phosphide.

contrast sodium ions are a component of table salt, a compound that we know is stable in water.

When forming ions, atoms of the nonmetallic elements tend to gain one or more electrons. In this way they form **anions,** *which are atoms or groups of atoms with a negative charge.* In comparison to an electrically neutral atom, an anion has more electrons. For example, when a chlorine atom gains one electron, it forms a *chloride ion.* A chloride ion has 17 protons and 18 electrons. It is therefore an anion with an ionic charge of $1-$ (Figure 5·7). Another common anion is the oxide ion, O^{2-}. The ionic charge of the oxide ion is $2-$ because an oxygen atom gains two electrons in forming its common ion.

Example 1

Write the symbol and name of the ion formed.   **a.** A strontium atom loses two electrons.   **b.** An iodine atom gains one electron.   **c.** A nitrogen atom gains three electrons.   **d.** A hydrogen atom loses one electron.

Solution

An atom that loses electrons forms a positively charged ion (cation). The numerical value of the charge equals the number of electrons lost.

An atom that gains electrons forms a negatively charged ion (anion). The numerical value of the charge is the number of electrons gained. The names of anions of nonmetallic elements end in *-ide*.
a. Sr^{2+}, strontium ion   **b.** I^{1-} or I^-, iodide ion   **c.** N^{3-}, nitride ion
d. H^{1+} or H^+, hydrogen ion

Problem

3. Complete this table.

	Symbol of element	Change in electrons	Formula of ion	Name of ion
a.	Ca	2 electrons lost	_____	_____
b.	F	_____	F^-	_____
c.	_____	_____	Al^{3+}	_____
d.	Se	2 electrons gained	_____	_____

5·3 Compounds

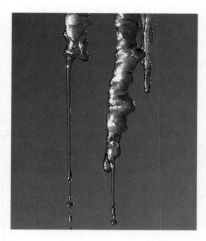

Figure 5·8
Molecular compounds can be solids, liquids, or gases. Water commonly exists in all three phases.

Atoms in a compound are held together by forces called chemical bonds (Chapters 13 and 14).

■ Compounds, which may be molecular or ionic, are formed from elements.

Compounds are pure substances that differ from elements because they contain more than one kind of atom. Compounds are formed when atoms of two or more different elements combine chemically. Compounds obey the **law of definite proportions.** *In any chemical compound the elements are always combined in the same proportion by mass*. Magnesium sulfide is a compound. It is produced by the combination of two kinds of atoms, magnesium atoms and sulfur atoms. If we take 100.00 g of magnesium sulfide and break it down into its elements, we always obtain 43.13 g of magnesium and 56.87 g of sulfur. How the magnesium sulfide was formed, or the size of the sample, will not change this mass ratio. For example, 10.000 g of magnesium sulfide will be composed of 4.313 g of magnesium and 5.687 g of sulfur. The law of definite proportions is explained by Dalton's atomic theory. If atoms combine in simple whole number ratios as Dalton postulated, then their proportions by mass must always be the same.

In many compounds, the atoms are bound together in molecules. *A* **molecule** *is a neutral group of atoms that act as a unit*. All the molecules of any given compound are identical. The molecules of one compound are different, however, from those of any other compound.

Compounds that are composed of molecules are **molecular compounds.** They tend to have relatively low melting and boiling points. Many molecular compounds exist as gases or liquids at room temperature. Most are composed of two or more nonmetallic elements. Water and carbon dioxide are examples of molecular compounds (Figure 5·9).

Not all compounds are composed of molecules. **Ionic compounds** *are composed of positive and negative ions*. The ions are arranged in an orderly three-dimensional pattern. Each positive ion is between two or more negative ions. At the same time, each negative ion is between two or more positive ions. Ionic compounds are electrically neutral although they are composed of ions. Most ionic compounds are crystalline solids at room temperature. They are usually formed from a metallic and a nonmetallic element. A familiar example of an ionic compound is table salt, or sodium chloride.

Chemical Connections

Rubies and sapphires are made up of aluminum oxide, an ionic compound. In a red ruby, 1–2% of the aluminum ions are replaced by chromium ions. In a blue sapphire, 1–2% of the aluminum ions are replaced by iron and titanium ions.

5·4 Chemical Formulas

Chemists have identified more than four million chemical compounds. Some are molecular compounds, others are ionic compounds. No two of these compounds have identical properties. Fortunately, the composition of each of these chemical substances can be represented by a chemical formula. *A* **chemical formula** *shows the kinds and numbers of atoms in the smallest representative unit of the substance*.

The chemical formula of a molecular compound is called a molecular formula. *A* **molecular formula** *shows the number and kinds of atoms present in a molecule of a compound*. The number of atoms of each kind is indicated by a subscript written after the symbol. Water is a molecular

■ The formula of a molecular compound represents a molecule. The formula of an ionic compound represents a formula unit.

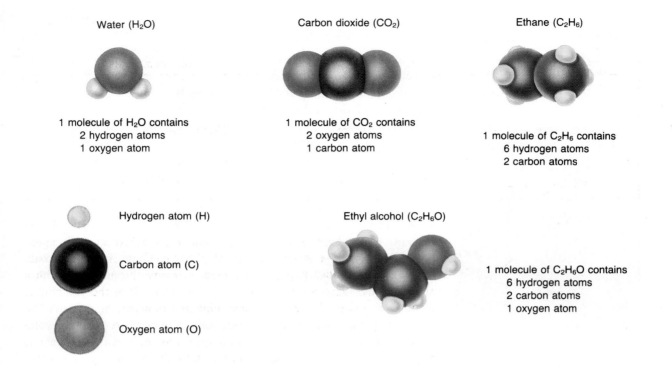

Water (H_2O)

1 molecule of H_2O contains
2 hydrogen atoms
1 oxygen atom

Carbon dioxide (CO_2)

1 molecule of CO_2 contains
2 oxygen atoms
1 carbon atom

Ethane (C_2H_6)

1 molecule of C_2H_6 contains
6 hydrogen atoms
2 carbon atoms

Hydrogen atom (H)

Carbon atom (C)

Oxygen atom (O)

Ethyl alcohol (C_2H_6O)

1 molecule of C_2H_6O contains
6 hydrogen atoms
2 carbon atoms
1 oxygen atom

Figure 5·9
The formula of a molecular compound indicates the numbers and kinds of atoms in a molecule of the compound. The arrangement of the atoms within a molecule is called the molecular structure.

Safety

Ethyl alcohol is extremely flammable. Never use a flame when others are working nearby with flammable liquids.

compound. A water molecule is a tightly bound unit of two hydrogen atoms and one oxygen atom. Its molecular formula is H_2O. The molecular formula for the molecular compound carbon dioxide is CO_2. This formula represents a molecule containing two oxygen atoms and one carbon atom. Ethane, a component of natural gas, is also a molecular compound. Its molecular formula is C_2H_6. This tells us that a molecule of ethane is composed of two carbon atoms and six hydrogen atoms. Another molecular compound is ethyl alcohol, C_2H_6O. What do you know about its composition?

Although a molecular formula gives us the composition of a molecule, it tells us nothing about the molecular structure. We need a diagram to show the arrangement of the atoms within the molecule. Figure 5·9 shows chemical formulas and molecular structures of some molecular compounds.

Chemical formulas can also be written for ionic compounds. In this case, though, the formula does not represent a molecule. Sodium chloride, table salt, is an ionic compound. It is composed of equal numbers of sodium ions, Na^+, and chloride ions, Cl^-. They are arranged in an orderly pattern. In order to represent this compound, should the formula be Na_2Cl_2, Na_3Cl_3, or NaCl? Chemists use the simplest of these formulas, a formula unit, to represent an ionic compound. *A **formula unit** is the lowest whole-number ratio of ions in an ionic compound.* For sodium chloride, the lowest whole-number ratio of the ions is 1:1 (1 Na^+ to 1 Cl^-). Thus the formula unit is NaCl. (Once we have determined the correct ratio of ions, we ignore the charges in writing the formula unit.) Another ionic compound is magnesium chloride. It contains magnesium ions,

Table 5·1   Characteristics of Ionic and Molecular Compounds

Characteristic	Ionic compound	Molecular compound
Representative unit	Formula unit (balance of oppositely charged ions)	Molecule
Type of elements	Metallic combined with nonmetallic	Nonmetallic
Physical state	Solid	Solid, liquid, or gas
Melting point	High (usually above 300°C)	Low (usually below 300°C)

Mg^{2+}, and chloride ions, Cl^-. In magnesium chloride the ratio of magnesium ions to chloride ions is 1:2 (one Mg^{2+} to two Cl^-). Thus the formula unit is $MgCl_2$. Notice that there are twice as many chloride ions (with a 1− charge) as magnesium ions (with a 2+ charge). Thus the grouping is electrically neutral. There is no such thing as a molecule of sodium chloride or magnesium chloride. Instead, sodium chloride consists of a collection of positively and negatively charged ions. As in ionic compounds in general, the ions are arranged in repeating three-dimensional patterns (Figure 5·10).

Much of your success at naming compounds and writing formulas of compounds will hinge on your ability to recognize compounds as either ionic or molecular. Table 5·1 summarizes some of the differences between ionic and molecular substances.

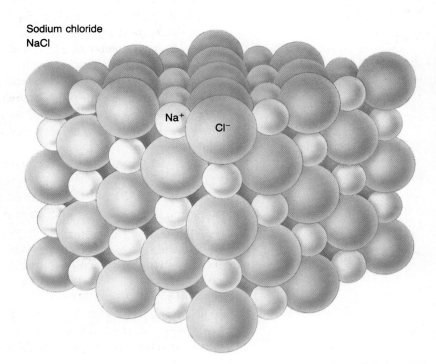

Sodium chloride
NaCl

Na^+ Cl^-

Figure 5·10
Sodium ions and chloride ions form a three-dimensional array in sodium chloride, NaCl. How many ions are part of a single formula unit for this compound?

Figure 5·11
Hydrogen peroxide and water both contain atoms of only hydrogen and oxygen. Nevertheless, they have different chemical and physical properties. Hydrogen peroxide **(a)** bleaches most dyes while water **(b)** does not.

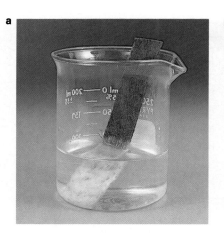
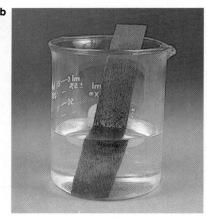

5·5 The Law of Multiple Proportions

■ The law of multiple proportions is explained by atomic theory.

Atoms of two elements can often combine in more than one way. For example, the elements hydrogen and oxygen combine to form the familiar compound water, H_2O. These same two elements can also form the compound hydrogen peroxide, H_2O_2. The compound water and the compound hydrogen peroxide each obey the law of definite proportions. In every sample of water the mass ratio of oxygen to hydrogen is 8 to 1. That is, 8.0 g of oxygen are present for each 1.0 g of hydrogen. In hydrogen peroxide the mass ratio of oxygen to hydrogen is always 16 to 1. For a given mass of hydrogen (1.0 g), the ratio of the masses of oxygen in these two compounds is a simple whole-number ratio.

$$\frac{16 \text{ g O (in } H_2O_2)}{8 \text{ g O (in } H_2O)} = \frac{2}{1}$$

Safety

Concentrated solutions of hydrogen peroxide enhance the combustion of other materials, and can cause explosions. They also cause burns. Handle with extreme care.

In the early 1800s Dalton and his contemporaries studied similar pairs of compounds. In each case, the two compounds were composed of the same two elements. Using the results from these experiments, Dalton put forth the **law of multiple proportions.** *Whenever two elements form more than one compound, the different masses of one element that combine with the same mass of the other element are in the ratio of small whole numbers.*

Example 2

Carbon reacts with oxygen to form two compounds. Compound A contains 2.41 g carbon for each 3.22 g oxygen. Compound B contains 6.71 g carbon for each 17.9 g oxygen. What is the lowest whole-number mass ratio of carbon that combines with a given mass of oxygen?

Solution

For each compound find the grams of carbon that combine per 1.00 g of oxygen.

$$\text{Compound A} \quad \frac{2.41 \text{ g C}}{3.22 \text{ g O}} = \frac{0.748 \text{ g C}}{1.00 \text{ g O}}$$

$$\text{Compound B} \quad \frac{6.71 \text{ g C}}{17.9 \text{ g O}} = \frac{0.375 \text{ g C}}{1.00 \text{ g O}}$$

The mass ratio of carbon per gram of oxygen in the compounds is 2:1.

$$\frac{0.748 \text{ g C (in compound A)}}{0.375 \text{ g C (in compound B)}} = \frac{2}{1}$$

Problem

4. Lead forms two compounds with oxygen. One compound contains 2.98 g of lead combined with 0.461 g of oxygen. The other compound contains 9.89 g of lead combined with 0.763 g of oxygen. What is the lowest whole-number mass ratio of lead that combines with a given mass of oxygen?

5·6 Ionic Charges of the Elements

The periodic table can be used to determine the charge on the ions of many elements.

In order to write chemical formulas, we need to know the types of ions that atoms tend to form. In other words, we need to know the ionic charges of the elements. For the representative elements (those in Group A), the ionic charges can easily be determined using the periodic table.

The metals in Groups 1A, 2A, and 3A lose electrons when they form ions. The ionic charge is positive and is numerically equal to the group number. Lithium, sodium, and potassium in Group 1A form cations. They have a 1+ charge (Li^+, Na^+, and K^+), as do all the other Group 1A ions. Magnesium and calcium are Group 2A metals. They form cations with a 2+ charge (Mg^{2+} and Ca^{2+}), as do all the other Group 2A metals. Aluminum is the only common Group 3A metal. As expected, it forms a 3+ cation (Al^{3+}).

The numerical charge of an ion of a Group A nonmetal is determined by subtracting the group number from 8. The sign of the charge is minus because nonmetals gain electrons and form anions. For example,

Table 5·2	Ionic Charges of Representative Elements						
1A	2A	3A	4A	5A	6A	7A	0
						H^-	
Li^+	Be^{2+}			N^{3-}	O^{2-}	F^-	
Na^+	Mg^{2+}	Al^{3+}		P^{3-}	S^{2-}	Cl^-	
K^+	Ca^{2+}				Se^{2-}	Br^-	
Rb^+	Sr^{2+}					I^-	
Cs^+	Ba^{2+}						

Issues in Chemistry

Radon (atomic number 86) is a Group 0 element. It is a colorless, tasteless, odorless, radioactive gas found in the atmosphere. What are the health hazards associated with radon exposure? What are the steps you can take to minimize your exposure?

You need to be able to recognize ions named with the older system. You will find these names on containers of old chemicals and in the older scientific literature.

Safety

Many transition metals are toxic. Make a habit of washing your hands thoroughly after handling any chemicals.

the elements in Group 7A form ions with a $1-$ charge: fluoride, F^-, chloride, Cl^-, and so forth. When hydrogen reacts with Group 1A metals, it also forms an anion with a $1-$ charge, the hydride ion, H^-. Anions of nonmetals in Group 6A have a $2-$ charge $(8 - 6 = 2)$, for example, oxide, O^{2-}. The three nonmetals in Group 5A can form ions with a $3-$ charge $(8 - 5 = 3)$, for example, nitride, N^{3-}.

The elements in the two remaining representative groups, 4A and 0, do not commonly form ions. The elements in Group 0 seldom form compounds. The two nonmetals in Group 4A, carbon and silicon, are ordinarily found in molecular compounds. The ionic charges of representative elements that can be obtained from the periodic table are summarized in Table 5·2.

Unlike the Group A representative metals, many of the transition metals have more than one common ionic charge. This is also a characteristic of metals in Groups 4A and 5A. For example, the transition metal iron has two common ions, Fe^{2+} and Fe^{3+}. Two methods of naming such ions are used. The preferred method is called the Stock system. A roman numeral in parentheses is used after the name of the element to indicate the numerical value of the charge. For example, the cation Fe^{2+} is the iron(II) ion. This is read as the "iron two" ion.

An older, less preferred method of naming these ions uses a root word and suffixes. The classical name of the element is used as the root word. An -ous ending is used for the name of the ion with the lower of the two ionic charges. An -ic ending is used with the higher of the two ionic charges. In the example of iron, Fe^{2+} is the ferrous ion, and Fe^{3+} is the ferric ion. Table 5·3 lists some of the elements that have two common ionic charges. Note that you can usually identify what may be unfamiliar classical names by looking for the symbol of the element in the name. *Fe*rrous (Fe) is iron; *Cu*prous (Cu) is copper; and *S*tan*nous (Sn) is tin. A major disadvantage of using classical names is that the name does not indicate the charge of the ion. The name only tells you that the ion is either the smaller (-ous) or the larger (-ic) charge of the pair.

A few transition metals have only one ionic charge. The names of these cations will not have a roman numeral. You should know about three of these "exceptions." Silver ions always have a $1+$ charge (Ag^+). Cadmium and zinc ions always have a $2+$ charge $(Cd^{2+}$ and $Zn^{2+})$.

Example 3

What is the charge of the ion of the following elements? (For transition metals with more than one common ionic charge, the number of electrons lost is indicated.) **a.** sulfur **b.** lead, four electrons lost **c.** zinc **d.** argon **e.** bromine **f.** copper, one electron lost

Solution

a. $2-$ (in Group 6A) **b.** $4+$ **c.** $2+$ **d.** 0 (in Group 0)
e. $1-$ (in Group 7A) **f.** $1+$

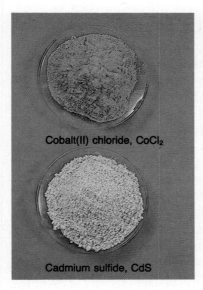

Cobalt(II) chloride, CoCl₂

Cadmium sulfide, CdS

Figure 5·12
Many transition metals form brightly colored compounds. Cobalt(II) chloride and cadmium sulfide are shown here. Some artists' pigments are named after the metals involved, for example: cobalt blue and cadmium yellow.

Table 5·3   Formulas and Names of Common Metal Ions with More than One Ionic Charge

Formula	Stock name	Classical name
Cu^{1+}	Copper(I) ion	Cuprous ion
Cu^{2+}	Copper(II) ion	Cupric ion
Fe^{2+}	Iron(II) ion	Ferrous ion
Fe^{3+}	Iron(III) ion	Ferric ion
$*Hg_2^{2+}$	Mercury(I) ion	Mercurous ion
Hg^{2+}	Mercury(II) ion	Mercuric ion
Pb^{2+}	Lead(II) ion	Plumbous ion
Pb^{4+}	Lead(IV) ion	Plumbic ion
Sn^{2+}	Tin(II) ion	Stannous ion
Sn^{4+}	Tin(IV) ion	Stannic ion
Cr^{2+}	Chromium(II) ion	Chromous ion
Cr^{3+}	Chromium(III) ion	Chromic ion
Mn^{2+}	Manganese(II) ion	Manganous ion
Mn^{3+}	Manganese(III) ion	Manganic ion
Co^{2+}	Cobalt(II) ion	Cobaltous ion
Co^{3+}	Cobalt(III) ion	Cobaltic ion

*A diatomic elemental ion.

Example 4

Name the ions in Example 3. Classify each as a cation or an anion.

Solution

The names of nonmetallic anions end in *-ide*. Metallic ions take the name of the metal. If the metal has more than one common ionic charge, a roman numeral is used.   **a.** sulfide ion, anion
b. lead(IV) or plumbic ion, cation   **c.** zinc ion, cation   **d.** no ion formed   **e.** bromide ion, anion   **f.** copper(I) or cuprous ion, cation

Problems

5. Write the symbol for each ion. Be sure to include the charge.
   a. iodide ion   **d.** barium ion
   b. mercury(II) ion   **e.** stannic ion
   c. phosphide ion   **f.** silver ion

6. Name the following ions. Use Table 5·3 if necessary.
   a. Cu^{2+} **d.** Pb^{2+}
   b. O^{2-} **e.** F^-
   c. Li^+ **f.** H^+

5·7 Polyatomic Ions

As the name implies, **polyatomic ions** *are tightly bound groups of atoms that behave as a unit and carry a charge.* An example of a polyatomic ion is the sulfate ion, SO_4^{2-}. A sulfate polyatomic ion is composed of one sulfur atom and four oxygen atoms. These five atoms together form a unit. This ion has a $2-$ charge.

You should memorize the names and formulas of the common polyatomic ions listed in Table 5·4. There are others, of course, and your teacher may wish to expand this list. Observe that the names of most polyatomic ions end in *-ite* or *-ate*. You should note three important exceptions to this. The ammonium ion (NH_4^+) is the one common polyatomic ion that is postively charged. The two anions that end in *-ide* are the cyanide (CN^-) and the hydroxide (OH^-) ions. *Compounds that give off OH^- ions when dissolved in water are known as* **bases.**

It is important to be able to recognize polyatomic ions. The hydrogen carbonate (HCO_3^-), hydrogen phosphate (HPO_4^{2-}), and the dihydrogen phosphate ($H_2PO_4^-$) ions are essential components of living systems. The cyanide ion (CN^-) is extremely poisonous to living systems because it blocks the cell's means of producing energy. Most laundry bleaches contain the hypochlorite ion (ClO^-).

Notice the relationship among the polyatomic ions for which there is an *-ite/-ate* pair. The *-ite* ending will always indicate one less oxygen atom than the *-ate*. The charge on each ion in the pair is the same. Notice that the ending does not tell you how many oxygen atoms are in the ion.

$$\text{-}ite \quad SO_3^{2-} \quad NO_2^- \quad ClO_2^-$$

$$\text{-}ate \quad SO_4^{2-} \quad NO_3^- \quad ClO_3^-$$

A polyatomic ion whose formula begins with H (hydrogen) can be viewed as a hydrogen ion (H^+) combined with another polyatomic ion. For example, HCO_3^- is a combination of H^+ and CO_3^{2-}. The charge on the new ion is just the algebraic sum of the ionic charges.

$$H^+ + \underset{\text{carbonate}}{CO_3^{2-}} \longrightarrow \underset{\text{hydrogen carbonate}}{HCO_3^-}$$

$$H^+ + \underset{\text{phosphate}}{PO_4^{3-}} \longrightarrow \underset{\text{hydrogen phosphate}}{HPO_4^{2-}}$$

$$H^+ + \underset{\text{hydrogen phosphate}}{HPO_4^{2-}} \longrightarrow \underset{\text{dihydrogen phosphate}}{H_2PO_4^-}$$

Safety

Figure 5·13
These models show the arrangement of atoms in four common polyatomic ions.

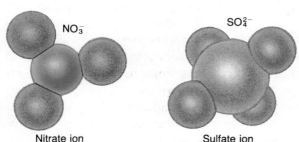

Nitrate ion (NO_3^-)

Sulfate ion (SO_4^{2-})

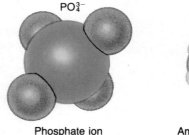

Phosphate ion (PO_4^{3-})

Ammonium ion (NH_4^+)

Table 5·4 Common Polyatomic Ions

1− charge		2− charge		3− charge	
Formula	Name	Formula	Name	Formula	Name
$H_2PO_4^-$	Dihydrogen phosphate	HPO_4^{2-}	Hydrogen phosphate	PO_4^{3-}	Phosphate
$C_2H_3O_2^-$	Acetate	$C_2O_4^{2-}$	Oxalate	PO_3^{3-}	Phosphite
HSO_3^-	Hydrogen sulfite (bisulfite)	SO_3^{2-}	Sulfite		
HSO_4^-	Hydrogen sulfate (bisulfate)	SO_4^{2-}	Sulfate	**1+ charge**	
				Formula	Name
HCO_3^-	Hydrogen carbonate (bicarbonate)	CO_3^{2-}	Carbonate	NH_4^+	Ammonium
NO_2^-	Nitrite				
NO_3^-	Nitrate	CrO_4^{2-}	Chromate		
CN^-	Cyanide				
OH^-	Hydroxide	$Cr_2O_7^{2-}$	Dichromate		
MnO_4^-	Permanganate	SiO_3^{2-}	Silicate		
ClO^-	Hypochlorite				
ClO_2^-	Chlorite				
ClO_3^-	Chlorate				
ClO_4^-	Perchlorate				

The large number of known compounds requires that we use a systematic method of naming.

Figure 5·14
Dentists often administer dinitrogen monoxide (N_2O) to help patients relax while the dentist works.

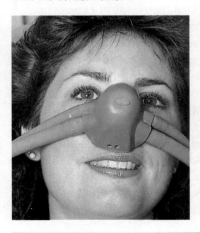

5·8 Common and Systematic Names

As the science of chemistry developed, it became apparent that some systematic method of naming chemical compounds was needed. In the early days of chemistry, the discoverer of a new compound often named the compound. It was not uncommon for the name to describe some physical or chemical property of the substance or the source of the compound. A common name for potassium carbonate, K_2CO_3, is potash. The name evolved because the compound was separated by boiling wood *ash*es in iron *pots*. Laughing gas is the common name for a gaseous compound, dinitrogen monoxide (N_2O). It causes people to laugh when they inhale it. These and other common names are very descriptive. Unfortunately, they do not tell us anything about the chemical composition of the compound. What are the components of baking soda, plaster of paris, quicksilver, or lye? Table 5·5 gives the formulas and systematic names for some substances with common names that you might recognize.

A few common names are used even by scientists. Water (for H_2O) is preferred to the systematic name dihydrogen monoxide. It is important, however, that we use a systematic method of naming compounds. The rest of this chapter is devoted to learning about this system for inorganic compounds. These are compounds that are not carbon-based. We will look first at binary ionic compounds.

Table 5·5 Common and Systematic Names Plus Formulas for Some Familiar Substances

Common name	Formula	Systematic name
Water	H_2O	Dihydrogen monoxide
Lime	CaO	Calcium oxide
Slaked lime	$Ca(OH)_2$	Calcium hydroxide
Lye	$NaOH$	Sodium hydroxide
Potash	K_2CO_3	Potassium carbonate
Table salt	$NaCl$	Sodium chloride
Laughing gas	N_2O	Dinitrogen monoxide
Baking soda (bicarbonate soda)	$NaHCO_3$	Sodium hydrogen carbonate
Muriatic acid	$HCl(aq)$	Hydrochloric acid
Hypo	$Na_2S_2O_3$	Sodium thiosulfate

Chemical Connections

Most drugs have three names: a scientific name, a generic name, and a brand name. The compound with the scientific name of hydroxyacetanilide, for example, is also known by the generic name of acetaminophen and the brand names of Tylenol®, Datril®, Liquipirin®, Tempra®, and Trilium®.

Figure 5·15
In the eighteenth century, pharmacists both prescribed and prepared remedies.

Science, Technology, and Society

5·A Pharmacy from Scheele to the Present

Over 5000 years ago, Chinese physicians treated respiratory diseases with a tea brewed from a common herb *Ephedra sinica*. In the seventh and eighth centuries AD Arabian alchemists extracted and purified the essential ingredients in such herbs. Today millions of allergy patients find relief by taking ephedrine. The Chinese herb contained this chemical which is now used in prescription and over-the-counter drugs.

One pharmacist who dramatically changed his profession was the Swedish scientist Carl Wilhelm Scheele. In 1756 when he was barely 15 years old, Scheele was apprenticed to a pharmacist. At that time pharmacists both prepared remedies and prescribed them for patients. Scheele and others were constantly making new remedies. He used new combinations of materials and reacted them to form novel substances. As a result, Scheele conducted thousands of experiments and made important chemical discoveries.

In Scheele's era, air, fire, water, and earth were considered the elements of which all matter was composed. Even as a teenager, Scheele opposed this view. Later his discoveries directly or indirectly led to the isolation of many true elements. These included chlorine, manganese, molybdenum, and oxygen. The names that Scheele used for several common chemicals reveal some of the mystery faced by pioneer chemists. Oxygen and nitrogen were called fire air and foul air. Carbon dioxide was known as aerial acid, and calcium oxalate was called rhubarb earth.

Perhaps the last item represents Scheele's greatest contribution to pharmacy and chemistry. By combining sugar and nitric acid he

Figure 5·16
Today's pharmacists must have a broad and thorough knowledge of how various medications affect the body. This requires an understanding of the chemistry of drugs.

Metals combine with nonmetals to form binary ionic compounds.

Monatomic ions are formed from individual atoms.

Although composed of positively and negatively charged ions, ionic compounds themselves are electrically neutral.

produced a substance that was identical to an acid obtained from the rhubarb plant. Until this work, scientists were sure that substances produced by plants and animals could not be produced in the laboratory. Scheele's experiment disproved this. He opened the door to the field of pharmaceutical chemistry, which is the study of the analysis, synthesis, and description of drugs. Today most drugs, including ephedrine, are partially or wholly synthesized.

Many specialists now contribute to the work of controlling disease. Pharmaceutical chemists work toward the synthesis of effective new drugs. Pharmacologists focus on the effects of drugs on various parts of the body. This information is used by physicians and pharmacists in deciding which drugs to prescribe. A major responsibility of hospital and drugstore pharmacists is to dispense drugs. Physicians and patients also rely on pharmacists to have a thorough understanding of the effects of medications. Thus a degree program in pharmacy includes five or six years of course work. Chemistry, physiology, microbiology, and pharmacy courses are emphasized.

5·9 Writing Formulas for Binary Ionic Compounds

Binary compounds *are composed of two elements*. The components of a binary ionic compound are a monatomic cation and a monatomic anion. Monatomic cations such as Na^+, Fe^{3+}, and Ca^{2+} are formed from metallic elements. The name of a monatomic cation is just the name of the element followed by the word ion. Monatomic anions such as O^{2-} and F^- are formed from nonmetallic elements. The name of a monatomic anion ends in *-ide*.

Ionic compounds are electrically neutral. In writing a formula for an ionic compound, we must exactly balance the positive charge of the cation by the negative charge of the anion. Stated another way, the net ionic charge of the formula must be zero. The binary ionic compound potassium chloride is composed of potassium ions, K^+, and chloride ions, Cl^-. In this compound, potassium and chloride ions must combine in a 1 : 1 ratio. For each K^+ there is one Cl^-. The formula for potassium chloride then is KCl. The net ionic charge of the formula is zero. In the formula of an ionic compound, the cation is always written first.

Calcium bromide is composed of calcium ions, Ca^{2+}, and bromide ions, Br^-. In this compound the ions must combine in a 1 : 2 ratio. Each calcium ion with its 2+ charge must combine with (or be balanced by) two bromide ions, each with a 1− charge. The formula for calcium bromide is $CaBr_2$.

What is the formula of iron(III) oxide? The roman numeral tells us the charge of the metal ion. Thus this compound consists of Fe^{3+} ions combined with O^{2-} ions. Writing a balanced formula for this compound may not seem as obvious as in the previous two examples. One way to reason through the problem is on the next page.

Figure 5·17
A common name for iron(III) oxide is rust. It is a binary ionic compound.

Action	Result	Observation
1. Start with the ions.	Fe^{3+} O^{2-}	More minus charge is needed.
2. Add an additional O^{2-}.	Fe^{3+} O^{2-} O^{2-}	More positive charge is needed.
3. Add an additional Fe^{3+}.	Fe^{3+} O^{2-} Fe^{3+} O^{2-}	More negative charge is needed.
4. Add another O^{2-}.	Fe^{3+} O^{2-} Fe^{3+} O^{2-} O^{2-}	The charges now balance.
5. Sum the ions.	$2Fe^{3+}$ $3O^{2-}$	
6. Write the balanced formula.	Fe_2O_3	Check that the net ionic charges balance. $2(3+) + 3(2-) = 0$

This is a rather lengthy process. A much quicker way to the same answer is to use what can be called the crisscross method. In this method, the numerical charge of each ion is crossed over and used as a subscript for the other ion. The signs of the numbers are dropped.

$$Fe^{3+} \quad O^{2-}$$
$$Fe_2O_3$$
$$2(3+) + 3(2-) = 0$$

This method can also be used with the previous examples.

$$K^{1+} \quad Cl^{1-} \qquad Ca^{2+} \quad Br^{1-}$$
$$KCl \qquad\qquad CaBr_2$$

Do not forget that formulas for ionic compounds are the lowest whole-number ratio of ions. The formula for calcium sulfide is CaS, not Ca_2S_2.

$$Ca^{2+} \quad S^{2-}$$
$$Ca_2S_2 \text{ reduces to CaS}$$

Of course, if the magnitudes of the charges of the cation and anion are the same, there is no reason to use the crisscross method. The ions will combine in a $1:1$ ratio because the charges will balance.

Example 5

Write formulas for these binary ionic compounds.
a. copper(II) sulfide   **b.** potassium nitride

Solution

Write down the formula (symbol and charge) of each ion. Balance the formula with appropriate subscripts.

a. Cu^{2+} ⤬ S^{2-} **b.** K^{1+} ⤬ N^{3-}

$$CuS \qquad\qquad K_3N$$
$$(2+) + (2-) = 0 \qquad 3(1+) + (3-) = 0$$

Problem

7. Write formulas for compounds composed of these pairs of ions.
   a. Li^+, S^{2-} **b.** Sn^{4+}, O^{2-} **c.** H^+, Cl^- **d.** Mg^{2+}, N^{3-}

5·10   Naming Binary Ionic Compounds

The names of binary ionic compounds have this form: (cation name) (anion name).

Binary ionic compounds are named by writing the name of the cation followed by the name of the anion (-*ide* ending). When a cation has more than one common ionic charge, it is important to use a roman numeral in the name. Giving the compound CuO the name "copper oxide" is not correct. This is because copper commonly forms two ions: Cu^{1+} and Cu^{2+}. The correct cation name of a transition metal can be determined by effectively working the formula backwards. The oxide ion always has a $2-$ charge. What must the charge of the copper ion be if the copper ion and oxide ion combine in a $1:1$ ratio? Obviously it must be copper(II) oxide. Cu_2O is copper(I) oxide.

You need to know which transition metals have two common ionic charges. It should be apparent from this discussion, however, that there is not much advantage to memorizing the charges of the transition metals in Table 5·3. You may know, for example, that tin forms ions with $2+$

Figure 5·18
Some transition metals can exist in more than one ionic state. When combining with another element, these metals can form more than one type of compound. Two pairs of examples for iron compounds are shown here.

Iron(II) sulfate, FeSO₄   Iron(III) sulfate, Fe₂(SO₄)₃

Iron(II) chloride, FeCl₂   Iron(III) chloride, FeCl₃

and 4+ charges. It will still be necessary, however, to determine which of these charges tin has in a particular compound. The name of SnO_2 is tin(IV) oxide. The tin(IV) ion, Sn^{4+}, will balance two oxide ions, each with a 2− charge.

Roman numerals are used on an "as needed" basis only. It is incorrect to use roman numerals with the name of a representative (Group A) metal. This is because these metals form ions with only one common ionic charge. For example, we would never write sodium(I) chloride or magnesium(II) oxide.

Example 6

Name these binary ionic compounds.   **a.** CoI_2 **b.** Cs_2O

Solution

Name the cation followed by the name of the anion.

a. cobalt(II) iodide, or cobaltous iodide

b. cesium oxide

8. **a.** aluminum fluoride
   b. iron(II) oxide or ferrous oxide
   c. copper(I) sulfide or cuprous sulfide
   d. calcium selenide

Problem

8. Write names for these binary ionic compounds.
   a. AlF_3 **b.** FeO   **c.** Cu_2S **d.** CaSe

5·11   Ternary Ionic Compounds

■ Ternary ionic compounds are easily named if we recognize the common polyatomic ions.

Ternary compounds *contain atoms of three different elements*. Ternary ionic compounds usually contain one or more polyatomic ions (Table 5·4). The procedure for writing the formula of a ternary ionic compound is the same as that for binary ionic compounds. This same procedure is also used for ionic compounds that have more than three different elements. First write down the formulas (symbol and charge) of the ions. Then balance the charges. An *-ate* or *-ite* ending on the name of a compound indicates that the compound contains a polyatomic anion. The two common exceptions are the hydroxide and cyanide polyatomic ions.

What is the formula of calcium nitrate? This compound is composed of calcium ions, Ca^{2+}, and nitrate ions, NO_3^-. To balance the formula, two nitrate ions, each a minus one, are needed to balance each calcium ion. Parentheses are used around the nitrate ion in the formula because two nitrate ions, NO_3^{1-}, are needed.

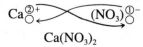

$$Ca(NO_3)_2$$

Figure 5·19
Calcium carbonate ($CaCO_3$) is a ternary ionic compound that oysters produce to form both their shells and pearls.

Whenever more than a single polyatomic ion is needed to balance a formula, parentheses must be used. This is the only time they are used.

Example 7

Write formulas for these compounds.   **a.** potassium sulfate
b. magnesium hydroxide   **c.** ammonium sulfide

Solution

a. K^{1+} $(SO_4)^{2-}$ **b.** Mg^{2+} $(OH)^{1-}$

K_2SO_4 $Mg(OH)_2$

The formula for magnesium hydroxide must have the parentheses.
$MgOH_2$ would be incorrect because there would not be two hydroxide
ions (instead only one oxygen atom and two hydrogen atoms).

c. $(NH_4)^{1+}$ S^{2-}

$(NH_4)_2S$

Problem

9. Write the formulas for these compounds.   **a.** barium sulfate
   b. aluminum hydrogen carbonate   **c.** sodium hypochlorite
   d. lead(IV) chromate

When naming a ternary ionic compound, be sure to note the polyatomic ion in the formula. Like a binary ionic compound, this compound
is named by naming the ions, cation first. The compound $NaC_2H_3O_2$ is
composed of sodium and acetate ions. Its name is sodium acetate. The
compound $K_2Cr_2O_7$ is potassium dichromate (two K^+ combined with one
$Cr_2O_7^{2-}$).

Example 8

Name these compounds.
a. $LiCN$ **c.** $(NH_4)_2C_2O_4$
b. $Sr(H_2PO_4)_2$ **d.** $Fe(ClO_3)_3$

Solution

a. lithium cyanide   **c.** ammonium oxalate
b. strontium dihydrogen phosphate   **d.** iron(III) chlorate

Problems

10. Write names for these compounds.   **a.** $Cr(NO_3)_2$ **b.** $Mg_3(PO_4)_2$
    c. Cu_2HPO_4 **d.** Li_2CrO_4

11. Write the name or formula, as appropriate.   **a.** aluminum hydrox-
    ide   **b.** K_2SiO_3 **c.** LiF **d.** potassium chlorite   **e.** $Sn_3(PO_4)_2$
    f. zinc hydrogen sulfate

5·12   Binary Molecular Compounds

When two nonmetals combine they form a binary molecular compound.

Binary molecular compounds are composed of two nonmetallic elements. Two characteristics of binary molecular compounds affect naming and writing formulas for these compounds. First, because these compounds are composed of molecules, the ionic charges of the representative elements are *not* used in writing formulas for the compounds. Second, when two nonmetallic elements combine, they often do so in more than one way. For example, the elements carbon and oxygen combine to form two different gaseous compounds and an ion, CO, CO_2, and CO_3^{2-}. They each have different physical and chemical properties.

At first glance it might seem satisfactory to give the name "carbon oxide" to a binary compound formed by the combination of carbon and oxygen atoms. This could have some severe consequences, however. Sitting in a room with moderate amounts of "carbon oxide" (CO_2) in the air would not present any problems. We exhale CO_2 as a product of our metabolism. Thus it is normally present in the air we breathe. If, however, the "carbon oxide" were the same amount of CO, we could die of asphyxiation. CO is a poisonous gas that interferes with our blood's ability to transport oxygen to body cells. Obviously, we need to distinguish between these two compounds when naming them.

The names of binary molecular compounds have this form:
(prefix + element name)
(prefix + element root + *-ide*).

Prefixes are used to show how many atoms of each element are present in each molecule (and formula) of a binary molecular compound. These prefixes are listed in Table 5·6. The two compounds of carbon and oxygen then are named carbon monoxide (CO) and carbon dioxide (CO_2). Note that the second element in the name is written with an *-ide* ending. *Thus all binary compounds, both ionic and molecular, end in -ide.* Note also that the vowel at the end of the prefix *mono-* is dropped when the name of the element begins with a vowel: monoxide, not monooxide. This prefix is normally omitted if there is just a single atom of the *first* element in the name.

Remember, prefixes are used in naming compounds composed of two nonmetals.

a

b

Figure 5·20
The binary molecular compounds of carbon and oxygen have very different properties. **a** Carbon monoxide (CO) is a poisonous gas found in automobile exhaust. **b** Carbon dioxide (CO_2) is used to put bubbles into soft drinks.

Table 5·6   Prefixes Used in Naming Binary Molecular Compounds

Prefix	Number
mono-	1
di-	2
tri-	3
tetra-	4
penta-	5
hexa-	6
hepta-	7
octa-	8
nona-	9
deca-	10

Example 9

Name the compounds.   **a.** N_2O **b.** PCl_3 **c.** $AlCl_3$ **d.** SF_6

Solution

When both elements are nonmetals, the compound is molecular, and prefixes are used to indicate the number of each atom.   **a.** dinitrogen monoxide   **b.** phosphorus trichloride   **c.** aluminum chloride (not trichloride! This is an ionic compound formed between a metal and a nonmetal)   **d.** sulfur hexafluoride

Formulas of binary molecular compounds should be among the easiest to write. The prefixes tell you the subscript of each element in the formula. You only need to write down the correct symbols for the two elements with the appropriate subscripts. For example, the formula for tetraiodine nonoxide is I_4O_9; for sulfur trioxide, SO_3; for phosphorus pentafluoride, PF_5.

Problems

12. Name these compounds.   **a.** CBr_4 **b.** Cl_2O_7 **c.** N_2O_5 **d.** BCl_3 **e.** $CrCl_3$

13. Write formulas for these compounds.   **a.** carbon disulfide   **b.** dinitrogen tetrahydride   **c.** carbon tetrachloride   **d.** diphosphorus trioxide

5·13   Acids

Acids are identified by formulas in which the element hydrogen appears first.

Acids are a group of compounds that are given special treatment in naming. You will see that acids are defined in several different ways when you look at the chemistry of acids in more detail in Chapter 18. For now it is sufficient to know that **acids** *are compounds that give off hydrogen ions when dissolved in water.*

The formulas of acids are of a general form HX, where X is a monatomic or polyatomic anion. We have previously named the compound HCl, hydrogen chloride. When the compound HCl is dissolved in water, it is named as an acid. Other compounds, such as HNO_3, exist only in water solution. They are always named as acids.

Consider the acid HX as dissolving in water. The acid can be named using three rules that focus on the ending of the anion of the acid (Table 5·7).

1. When the anion (X) ends in *-ide,* the acid name begins with the prefix *hydro-.* The stem of the anion has the suffix *-ic* and it is followed by the word *acid.* Thus HCl (X = chloride), dissolved in water, is named *hydro*chlor*ic acid.* H_2S (X = sulfide) is *hydro*sulfur*ic acid.*

Table 5·7 Naming Acids

Anion ending	Example	Acid name	Example
-ide	Cl^- chloride	hydro-(stem)-ic acid	hydrochloric acid
-ite	SO_3^{2-} sulfite	(stem)-ous acid	sulfurous acid
-ate	NO_3^- nitrate	(stem)-ic acid	nitric acid

2. When the anion ends in -ite, the acid name is the stem of the anion with the suffix -ous, followed by the word acid. H_2SO_3 (X = sulfite) is sulfurous acid.

3. If the anion ends in -ate, the acid name is the stem of the anion with the suffix -ic, followed by the word acid. Thus HNO_3 (X = nitrate) is nitric acid.

Not all compounds which contain hydrogen are acids. For example methane, CH_4, is not an acid.

Example 10

Name these compounds as acids.   **a.** HClO   **b.** HCN   **c.** H_3PO_4

Solution

a. hypochlorous acid (rule 2)   **b.** hydrocyanic acid (rule 1)
c. phosphoric acid (rule 3)

Formulas of acids are most easily written by using the preceding three rules in a reverse fashion. For example, what is the formula of chloric acid? As rule 3 shows, chloric acid (-ic ending) must be a combination of hydrogen ion (H^+) and chlorate ion (ClO_3^-). The formula of chloric acid is $HClO_3$. Hydrobromic acid (hydro- prefix and -ic suffix), according to rule 1, must be a combination of hydrogen ion and bromide ion (Br^-). The formula of hydrobromic acid is HBr. Hydrogen ion and phosphite ion (PO_3^{3-}) must be the components of phosphorous acid (rule 2). The formula of phosphorous acid is H_3PO_3.

Problems

14. Name these compounds as acids.   **a.** HF   **b.** $HC_2H_3O_2$
 c. H_2SO_4 **d.** HNO_2

15. Write formulas for the following acids.   **a.** chromic acid
 b. hydroiodic acid   **c.** chlorous acid   **d.** perchloric acid

5·14 Summary of Naming and Formula Writing

A systematic approach will lead to success in naming compounds and writing formulas.

In this chapter we have discussed two skills: writing chemical formulas and naming chemical compounds. If this is your first exposure to these concepts, you may feel a little overwhelmed at this point. When do you use prefixes or roman numerals in a name and when should you not use them? Should a compound's name end in -ate, -ide, or -ite? For example, what is the name of each of these compounds: $FeCl_3$, PCl_3, and $AlCl_3$? Each compound is composed of one element combined with the nonmetal chlorine in a 1:3 ratio. On first glance, then, it may look as if these compounds should be named the same way. But the names differ.

$FeCl_3$ is iron(III) chloride. The name must include a roman numeral because iron is a transition metal.

PCl_3 is phosphorus trichloride. The compound is named with prefixes because phosphorus is a nonmetal.

$AlCl_3$ is aluminum chloride. Neither a roman numeral nor prefixes are needed in the name since aluminum is a Group A metal.

Figure 5·21 is a flowchart designed to help you name compounds correctly. By following the flowchart, you are led to the directions for

Figure 5·21
This flowchart will help you to name chemical compounds. P and Q in the general formula, P_xQ_y, can be atoms, monatomic ions, or polyatomic ions.

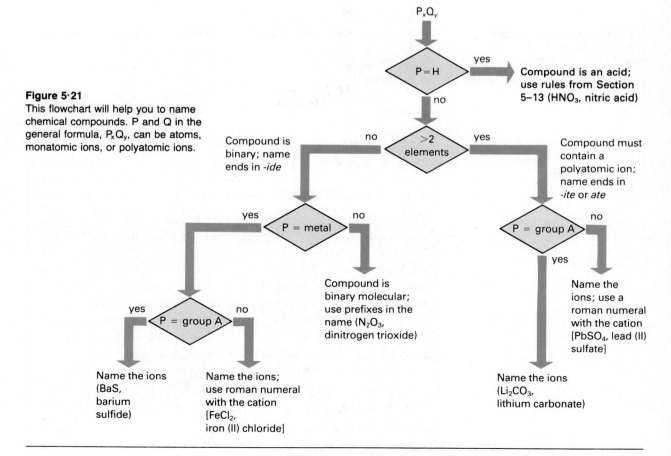

naming a particular compound. Use the flowchart while working exercises to increase your skill at naming compounds. With practice you will be able to mentally ask yourself the questions that will lead you to the correct name without having the flowchart in front of you.

In writing a chemical formula from a chemical name, it is helpful to remember the following.

1. In an ionic compound, the net ionic charge is zero.
2. An -ide ending indicates a binary compound.
3. An -ite or -ate ending means the formula has a polyatomic ion.
4. Prefixes in the name indicate a molecular compound. They show the number of each atom in the formula.
5. A roman numeral shows the ionic charge of the cation.

Problems

16. Name these compounds, using Figure 5·21 as an aid if necessary.

a. $CaCO_3$ f. Mg_3P_2
b. $KMnO_4$ g. $(NH_4)_2SO_4$
c. $PbCrO_4$ h. ICl
d. $CaHPO_4$ i. HNO_2
e. $SnCr_2O_7$ j. $SrBr_2$

17. Write formulas for these compounds.

a. tin(II) hydroxide   f. aluminum hydrogen carbonate
b. barium fluoride   g. sodium phosphate
c. tetraiodine nonoxide   h. potassium perchlorate
d. iron(III) oxalate   i. disulfur trioxide
e. hydrosulfuric acid   j. magnesium nitrate

Science, Technology, and Society

5·B   Chemical Data Banks

The job of naming chemical compounds and recognizing formulas may seem like an overwhelming task. What you have seen so far, however, is much less than the tip of the iceberg. There are 217 different compounds that have the simple formula C_6H_6. Each of these has a unique name. In the field of organic chemistry, hundreds of thousands of new compounds are made, and named, every year. Recognizing the names of all these compounds is a tremendous challenge. Being able to look up the chemical, biochemical, and physical properties of all these substances is important as well.

Chemists today are using computers to manage this evergrowing body of data. This allows them to answer questions that would otherwise take weeks of library and laboratory work. Some systems store information on the names, properties, and reactions of compounds.

Figure 5·22
Chemical data banks are used in poison control centers and by pollution control agencies. They are also widely used in basic research.

Others can draw all the possible structures of a given formula, such as all 217 versions of C_6H_6. One program will "look at" a molecule or part of a molecule drawn by a chemist and then search for it among all the compounds in its memory. A similar program can search for all reactions involving the molecule that has been drawn.

One of the largest chemical data banks is the Chemical Information System. This system, known as CIS, was developed in 1971. It contains data on hundreds of thousands of chemicals. It also is connected to other computer systems that can give bibliographies relevant to each substance. The system is used by government agencies, universities, and industries throughout Canada, the United States, and in many other countries. These customers pay a fee and can gain access to the system by telephone.

A user can type in the name or formula of a chemical and get information on its structure and properties. If the properties of an unknown substance are provided, the computer will list the compounds it might be. If an unknown chemical is released into the environment, an emergency team equipped with a portable computer can determine what the chemical is. Information may also be available on how to clean up the site, how to control any resulting fires, and how to treat any people or animals exposed to the chemical.

CIS also includes most of the specialized programs described earlier and is used in research. Hospitals and poison control centers use the system to identify poisons and choose proper medical treatment. Pollution control agencies can identify and manage pollutants in water supplies. It is an important tool for today's chemists.

Key Terms

acid	$5 \cdot 13$	metal	$5 \cdot 1$
anion	$5 \cdot 2$	metalloid	$5 \cdot 1$
base	$5 \cdot 7$	molecule	$5 \cdot 3$
binary compound	$5 \cdot 9$	molecular	
cation	$5 \cdot 2$	compound	$5 \cdot 3$
chemical formula	$5 \cdot 4$	molecular formula	$5 \cdot 4$
formula unit	$5 \cdot 4$	nonmetal	$5 \cdot 1$
group	$5 \cdot 1$	periodic table	$5 \cdot 1$
ion	$5 \cdot 2$	polyatomic ion	$5 \cdot 7$
ionic compound	$5 \cdot 3$	representative	
law of definite		element	$5 \cdot 1$
proportions	$5 \cdot 3$	semimetal	$5 \cdot 1$
law of multiple		ternary compound	$5 \cdot 11$
proportions	$5 \cdot 5$	transition metal	$5 \cdot 1$

Chapter Summary

The elements in the periodic table are arranged in vertical columns called groups. Groups 1A through 7A and Group 0 make up the representative elements. Metals are on the left and lower sides of the periodic table. Nonmetals are on the right and upper sides.

Every substance is either an element or a compound. An element consists of one kind of atom. A compound consists of more than one kind of atom. A compound is either molecular or ionic in nature.

Molecular compounds are composed of two or more nonmetals. The representative particle of a molecular compound is a molecule. A molecular formula shows the number and kinds of atoms present in a molecule of a compound.

Ionic compounds are composed of oppositely charged ions (cations and anions) combined in electrically neutral groupings. The chemical formula of an ionic compound is a formula unit. A formula unit gives the lowest whole-number ratio of ions in the compound.

The charges of the ions of the representative elements can be determined by the position of these elements in the periodic table. Most transition metals have more than one common ionic charge. A polyatomic ion is a group of atoms that behave as a unit and has a charge.

As the number of known chemical compounds increased, a systematic naming method was devised. Binary (two-element) ionic compounds are named by writing the name of the cation followed by the name of the anion. Binary compounds end in -ide. For example, KBr is potassium bromide.

When a cation has more than one ionic charge, a roman numeral is used in the name. The compound FeO is named iron(II) oxide. Ternary ionic compounds contain at least one polyatomic ion. The names of these compounds generally end in -ite or -ate. The compound KNO_3 is potassium nitrate.

Binary molecular compounds are composed of two nonmetallic elements. The name of a binary molecular compound ends in -ide. Prefixes are used to show how many atoms of each element are present in a molecule of the compound. For example, N_2O_4 is named dinitrogen tetroxide.

Compounds with the formula HX, where X represents an anion, are named as acids when they are dissolved in water.

Practice Questions and Problems

18. Classify each element as a metal, nonmetal, or metalloid. **a.** aluminum **b.** silver **c.** silicon **d.** helium **e.** zinc $5 \cdot 1$

19. In which groups are the elements in Problem 18 found? $5 \cdot 1$

20. State the number of electrons either lost or gained in forming each ion. **a.** S^{2-} **b.** K^+ **c.** Cl^- **d.** Ba^{2+} **e.** Li^+ **f.** H^- $5 \cdot 2$

21. Name each of the ions in Problem 20. Identify them as anions and cations. *5·2*

22. Would you expect the following compounds to be ionic or molecular? *5·3*
a. CO **b.** KBr **c.** C_3H_8 **d.** SO_3

23. List three characteristics each for ionic compounds and molecular compounds. *5·3*

24. State whether each formula in Problem 22 represents a molecule or a formula unit. *5·4*

25. How is the law of multiple proportions explained by Dalton's atomic theory? *5·5*

26. Using only the periodic table, name and write the formulas of the ions of these representative elements. *5·6*
a. lithium **d.** fluorine
b. oxygen **e.** potassium
c. barium **f.** neon

27. Without consulting Table 5·4, name the following ions. *5·7*
a. H^+ **e.** $H_2PO_4^-$
b. CN^- **f.** Sn^{4+}
c. Cr^{3+} **g.** SO_3^{2-}
d. $Cr_2O_7^{2-}$ **h.** Se^{2-}

28. Write the formula and charge of each of the following ions. *5·7*
a. magnesium ion **e.** hydroxide ion
b. lead(IV) ion **f.** iron(II) ion
c. chromate ion **g.** ammonium ion
d. nitrite ion **h.** copper(I) ion

29. Write formulas for compounds formed from these pairs of ions. *5·9*
a. Sr^{2+}, Se^{2-} **c.** Ca^{2+}, N^{3-}
b. K^+, O^{2-} **d.** Co^{2+}, I^-

30. Write formulas for these compounds. *5·9*
a. silver sulfide **c.** sodium nitride
b. stannic chloride **d.** strontium iodide

31. Name these binary ionic compounds. *5·10*
a. ZnO **c.** Cu_2O
b. NaI **d.** $CaBr_2$

32. When are parentheses used in writing a chemical formula? *5·11*

33. Write formulas for these compounds. *5·11*
a. mercury(II) bromide
b. ammonium dichromate
c. lithium hydrogen sulfate
d. chromium(III) nitrite

34. Complete this table by writing correct formulas for the compounds formed by combining positive and negative ions. Then name each compound. *5·11*

	SO_4^{2-}	Cl^-	PO_4^{3-}	S^{2-}
Ca^{2+}	_____	_____	_____	_____
NH_4^+	_____	_____	_____	_____
Al^{3+}	_____	_____	_____	_____
Pb^{4+}	_____	_____	_____	_____

35. What are the components of a binary molecular compound? *5·12*

36. What prefix is used to indicate each of the following number of atoms in a formula of a molecular compound? *5·12*
a. 3 **c.** 2 **e.** 5
b. 1 **d.** 7 **f.** 9

37. Name these binary molecular compounds.
a. OF_2 **b.** Cl_2O_8 **c.** SO_3 **d.** P_4O_{10} *5·12*

38. Assume that each of these is dissolved in water, and name each as an acid. *5·13*
a. $H_2C_2O_4$ **c.** $HClO_2$
b. HF **d.** H_2CO_3

39. Write formulas for these compounds. *5·13*
a. nitrous acid **c.** phosphoric acid
b. hydroselenic acid **d.** acetic acid

Mastery Questions and Problems

40. Identify each set of elements listed below by their numbered location on the periodic table that follows. Some of the sets are made up of a combination of several numbered areas.
a. representative elements
b. inner transition elements
c. nonmetals
d. transition metals
e. metals

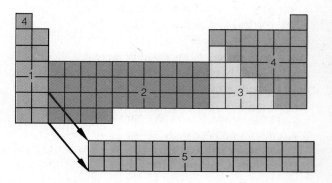

41. Two compounds can be formed when phosphorus reacts with oxygen. One compound is prepared by reacting 8.29 g of oxygen with 6.43 g of phosphorus. The other can be prepared by reacting 2.64 g of oxygen and 3.41 g of phosphorus. Does this pair of compounds confirm the law of multiple proportions?

42. Thousands of different kinds of inorganic chemicals are produced in the United States. The amounts produced of the top ten inorganic chemicals in 1987 are shown in the data table below.
a. What percent of the total production of the top ten did lime (calcium oxide) account for?
b. Three diatomic gases are on the list. What are their names and what was their total production in 1987?
c. What percent of the total production of the top ten did the three acids account for?
d. Write formulas for the top ten inorganic chemicals.

Chemical	Amount produced (billions of kg)
Sulfuric acid	35.2
Nitrogen	21.5
Ammonia	14.7
Oxygen	14.7
Lime	13.8
Sodium hydroxide	10.5
Chlorine	10.0
Phosphoric acid	9.5
Sodium carbonate	8.1
Nitric acid	6.5

43. Name these compounds.
a. AlF_3
b. SnO_2
c. $Fe(C_2H_3O_2)_3$
d. $KHSO_4$
e. CaH_2
f. $HClO_3$
g. Hg_2Br_2
h. H_2CrO_4

44. Write formulas for these compounds.
a. phosphorus pentabromide
b. carbon tetrachloride
c. potassium permanganate
d. calcium hydrogen carbonate
e. dichlorine heptoxide
f. trisilicon tetranitride
g. sodium dihydrogen phosphate

45. Classify each of the ions in Problem 27 as a monatomic or a polyatomic ion.

46. Classify each of the ions in Problem 28 as an anion or a cation.

47. Name each of the following substances. Use Figure 5·21 if necessary.
a. CaO
b. $Ba_3(PO_4)_2$
c. I_2
d. $BaSO_4$
e. $NaC_2H_3O_2$
f. $HClO_4$
g. Cl_2O
h. HgF_2

48. Write formulas for these compounds.
a. calcium carbonate
b. sodium bromide
c. ferric sulfate
d. magnesium sulfide
e. nitrogen
f. barium hydroxide
g. hydrobromic acid
h. sulfite ion

49. Name each compound.
a. KOH
b. HF
c. PI_3
d. $Be(NO_3)_2$
e. K_2CO_3
f. N_2H_4
g. ZnO
h. $Mg(MnO_4)_2$

50. Write formulas for these compounds.
a. silver chloride
b. aluminum carbide
c. lithium hydride
d. acetic acid
e. sodium silicate
f. calcium oxide
g. hydrocyanic acid
h. tin(IV) cyanide

Critical Thinking Questions

51. Choose the term that best completes the second relationship.
a. black:white   cation:_____
(1) anion   (3) sodium ion
(2) metal   (4) iron ion
b. plant:oak tree   anion:_____
(1) sodium ion   (3) cation
(2) magnesium ion   (4) chloride ion
c. acid:hydrogen ions   cloud:_____
(1) base   (3) storm
(2) rain   (4) cumulus

52. Why is it important for chemists to have a systematic system of writing chemical names and formulas?

53. Summarize the rules that chemists use for naming ionic compounds. What is the purpose for each rule?

Review Questions and Problems

54. Make the following conversions.
  a. 5.46 cg of Ag to milligrams of silver
  b. 189 cal to kilojoules
  c. $-25°C$ to K

55. A student finds that 2.62 g of a substance occupies a volume of 1.05 mL. What is the density of the substance?

56. How many protons and electrons are in each of these ions?
  a. bromide ion
  b. aluminum ion
  c. calcium ion
  d. oxide ion

Challenging Questions and Problems

57. The *Handbook of Chemistry and Physics* is a reference work that contains a wealth of information about elements and compounds. Two sections of this book that students often use are the "Physical Constants of Inorganic Compounds" and the "Physical Constants of Organic Compounds."

To familiarize yourself with this work, make a table with these headings: Name, Formula, Crystalline form or color, Density, Melting point (°C), Boiling point (°C), and Solubility in water. Enter these substances in the body of the table: ammonium chloride, barium, barium sulfate, bromine, calcium carbonate, chlorine, copper(II) sulfate pentahydrate, iodine, iron(II) sulfate pentahydrate, mercury, potassium carbonate, and sulfur. Use the *Handbook* to complete the table.

58. Use the table you prepared for Problem 57 to answer the following questions.
  a. You have two unlabeled bottles, each containing a white powder. You are told that one of the substances is calcium carbonate and the other is potassium carbonate. Describe a simple physical test you would carry out to distinguish between these two compounds.
  b. How would you distinguish between samples of copper(II) sulfate pentahydrate and iron(II) sulfate pentahydrate?
  c. A bottle contains a mixture of ammonium chloride and barium sulfate. What method could you use to separate these two compounds?

  d. List the *elements* in the preceding table in order of increasing density. Identify the elements as metals or nonmetals.
  e. List the *compounds* in the table in order of decreasing density.
  f. Calculate the mass of 132 cm³ of mercury.
  g. Calculate the volume of 56 g of sulfur.
  h. How would you distinguish among the halogens listed in this table?

Research Projects

1. Describe the function of various ions in our blood. Some of the most important ones are Na^+, K^+, Ca^{2+}, Mg^+, HCO_3^-, and Cl^-.

2. How do pharmaceutical chemists develop drugs?

3. What is a computer data bank and how is one constructed or maintained? Use chemical examples in explaining how a data bank works.

4. Construct a chemical data bank for one group of elements. Include information on how to identify them, toxicity, common reactions, etc.

5. Scheele's early death is sometimes attributed to his exposure to toxic chemicals. What other famous scientists died as a result of the unknown hazards they faced in their work?

6. Collect samples of metallic elements and devise a way to compare their electrical conductivity.

7. Research the art of growing large crystals. Grow crystals of one or more of the following ionic compounds: potassium sulfate, magnesium sulfate, sodium chloride, and sodium sulfate.

Readings and References

Meadows, Robin. "Fossil Molecules." *ChemMatters* (April 1988), pp. 4–7.

Parker, Sybil P. *McGraw-Hill Encyclopedia of Chemistry*. New York: McGraw-Hill, 1983.

Weast, Robert C., ed. *CRC Handbook of Chemistry and Physics: A Ready Reference Book of Chemical and Physical Data*, 67th ed. Boca Raton, FL: Chemical Rubber Co., 1986.

6 Chemical Quantities

Chapter Preview

Many scientists spend their time answering the question "How much?" How many grams of iron are in a kilogram of iron(II) oxide? How many grams of the elements hydrogen and oxygen must be combined to make 20 g of water? Both these typical chemistry questions emphasize that chemistry is a quantitative science. Chemists, and chemistry students, analyze the composition of samples of matter. They must also do chemical calculations relating quantities of reactants and products to chemical equations. To do any of these problems, you must be able to measure the amount of matter you have.

6·1 Measuring Matter

Counting is one way to measure how much of something we have. For example, we count how many students are in a class or how many pins we knock down when bowling. Another way to measure the amount of substance is by weighing. We buy potatoes by the pound or kilogram and silver and gold by the ounce or gram. We also measure matter by volume. We buy gasoline by the gallon or liter, and we take cough medicine by the teaspoon or milliliter. Some items are sold by all three of these ways: a count, a mass, and a volume. For example, strawberries may be sold for 10¢ each. A basket of strawberries may be sold by the quart or by the pound. Chemists also measure the amount of matter in these three

Figure 6·1
Do you recognize these elements? They are (clockwise from the bottom) sulfur, carbon, cobalt, aluminum, bromine, mercury, and copper (center). Although they have different masses, all the quantities shown here contain an equal number of atoms.

115

■ A mole is an amount of substance.

Figure 6·2
Matter is counted in convenient quantities: 12 eggs equal a dozen; 144 pencils equal one gross; and 500 sheets of paper equal one ream.

Latin: *moles* = heap or pile

■ A mole represents Avogadro's number of representative particles of any substance.

Mole is a word which is used to stand for an exact quantity, just as dozen stands for 12, ream stands for 500, score stands for 20, and gross stands for 144.

ways. In fact, each of these units of measurement is related by chemists to a single quantity called the mole. The mole, which is one of the seven base SI units, measures the "amount of substance."

6·2 The Mole

We know that matter is made up of different kinds of particles. *The term* **representative particle** *refers to whether a substance commonly exists as atoms, ions, or molecules.* The representative particle of most elements is the atom. Iron is composed of iron atoms. Helium is composed of helium atoms. Seven elements, however, normally exist as diatomic molecules (H_2, N_2, O_2, F_2, Cl_2, Br_2, and I_2). The representative particle of these elements and of molecular compounds is the molecule. The molecular compounds water and sulfur dioxide are composed of H_2O and SO_2 molecules respectively. The formula unit is the representative particle of ionic compounds. Calcium chloride, an ionic compound, has the formula unit $CaCl_2$. Calcium and chloride ions are present in a one to two ratio in this formula unit.

Because substances are composed of particles, one way to measure the amount of a substance is to count the number of representative particles of that substance. Atoms, molecules, and ions are very small. Thus the number of particles in even a relatively small sample of any substance is very large. Counting the particles is not practical or even possible. We can "count" particles, however, if we introduce a term that represents a certain number of particles. Just as a dozen eggs represents 12 eggs, *a* **mole** *of a substance represents 6.02×10^{23} representative particles of that substance.* This experimentally determined number, 6.02×10^{23}, is *called* **Avogadro's number.** It is named in honor of Amedeo Avogadro di Quarenga (1776–1856) who was a nineteenth-century Italian scientist and lawyer. His work (Section 6·B) made the calculation of this number possible. We usually talk about moles of elements or compounds. There is nothing wrong, however, with calling 6.02×10^{23} pencils a mole (1 mol) of pencils. The relationship between representative particles and moles of substances is summarized in Table 6·1.

Example 1

How many moles of magnesium is 3.01×10^{22} atoms of magnesium?

Solution

Step 1. The unknown is the number of moles.

Step 2. The known is the number of atoms: 3.01×10^{22} atoms.

Step 3. The desired conversion is: atoms $\longrightarrow$ moles. The expression relating the units is 1 mol Mg = 6.02×10^{23} atoms Mg. The conversion factor is 1 mol Mg/6.02×10^{23} atoms Mg. (We want to cancel the unit atoms and be left with the unit moles.)

Step 4. The solution is as follows.

$$3.01 \times 10^{22} \text{ atoms Mg} \times \frac{1 \text{ mol Mg}}{6.02 \times 10^{23} \text{ atoms Mg}}$$

$$= 0.500 \times 10^{-1} \text{ mol Mg}$$

Step 5. In standard form the answer is 5.00×10^{-2} mol Mg. The answer should have three significant figures. Since the given number of atoms was less than Avogadro's number, we would expect to have less than 1 mol of atoms.

Problems

1. How many moles are 1.20×10^{25} atoms of phosphorus?

2. How many atoms are in 0.750 mol of zinc?

A mole refers to Avogadro's number of representative particles of any substance. A mole of a diatomic element such as nitrogen contains 6.02×10^{23} *molecules* of nitrogen. A mole of any molecular compound is composed of Avogadro's number of *molecules* regardless of the size of the molecule. A mole of dinitrogen tetroxide, N_2O_4, and a mole of sucrose, $C_{12}H_{22}O_{11}$, both contain 6.02×10^{23} molecules of these compounds. In a similar fashion, a mole of any ionic compound is composed of Avogadro's number of *formula units*. A mole of NaI and a mole of $(NH_4)_3PO_4$ each contains 6.02×10^{23} formula units.

> The number of representative particles in a mole does not depend on the kind of particles being counted.

Example 2

How many molecules are there in 4.00 mol of glucose, $C_6H_{12}O_6$?

Solution

Step 1. The unknown is the number of molecules of $C_6H_{12}O_6$.

Step 2. The known is the number of moles: 4.00 mol of $C_6H_{12}O_6$.

Step 3. The desired conversion is: moles $\longrightarrow$ molecules. Thus we need a conversion factor that gives molecules/mole. Glucose is a molecular compound. We know that 6.02×10^{23} molecules of $C_6H_{12}O_6$ is 1 mol of $C_6H_{12}O_6$.

Step 4. In doing the calculation, make sure that the units cancel.

$$4.00 \text{ mol } C_6H_{12}O_6 \times \frac{6.02 \times 10^{23} \text{ molecules } C_6H_{12}O_6}{1 \text{ mol } C_6H_{12}O_6}$$

$$= 24.08 \times 10^{23} \text{ molecules}$$

Step 5. The answer must be rounded to three significant figures and put in standard form: 2.41×10^{24} molecules glucose.

Table 6-1   Representative Particles and Moles

Substance	Representative particle	Chemical formula	Representative particles in 1.00 mol
Atomic nitrogen	Atom	N	6.02×10^{23}
Nitrogen gas	Molecule	N_2	6.02×10^{23}
Water	Molecule	H_2O	6.02×10^{23}
Calcium ion	Ion	Ca^{2+}	6.02×10^{23}
Calcium fluoride	Formula unit	CaF_2	6.02×10^{23}
Sucrose	Molecule	$C_{12}H_{22}O_{11}$	6.02×10^{23}

Problems

3. How many molecules are in 0.400 mol N_2O_5?

4. How many moles are contained in 1.20×10^{24} molecules CO_2?

How many atoms are in a mole of a compound? We cannot answer this question unless we know how many atoms are in a representative particle of the compound. Consider this analogy. A dozen packages of pencils may contain four, eight, or twelve *dozen* pencils depending on the size of the package. If each package contains eight pencils, a dozen packages will contain 96 (8×12) pencils. Similarly, a mole of carbon dioxide, CO_2, contains Avogadro's number of CO_2 molecules. Each CO_2 molecule is composed of three atoms. Hence, a mole of carbon dioxide contains three times Avogadro's number of atoms. *To find the number of atoms in a mole of a compound, we must determine the number of atoms in a representative formula of that compound.*

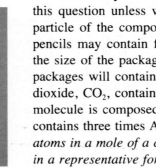

Figure 6-3
A dozen packages of marbles contain more than a dozen marbles. Similarly, a mole of molecules contains more than a mole of atoms. If each molecule consists of 6 atoms, how many atoms will be in a mole of molecules?

Example 3

How many fluoride ions are in 1.46 mol of aluminum fluoride?

Solution

Step 1. The unknown is the number of F^- ions.

Step 2. The known is 1.46 mol of aluminum fluoride.

Step 3. The desired conversion is moles $\longrightarrow$ formula units $\longrightarrow$ ions. We know that 1 mol = 6.02×10^{23} formula units. The first conversion factor will be formula units/mole. To convert formula units to ions we need to know how many F^- ions are part of each formula unit. Thus we need to write the formula for aluminum fluoride: AlF_3. We find that this compound has three F^- ions per formula unit. The second conversion factor will be: three F^- ions/one formula unit AlF_3.

Step 4. The calculations can be done in a single step.

$$1.46 \text{ mol AlF}_3 \times \frac{6.02 \times 10^{23} \text{ formula units AlF}_3}{1 \text{ mol AlF}_3}$$

$$\times \frac{3 \text{ F}^- \text{ ions}}{1 \text{ formula unit AlF}_3} = 26.3676 \times 10^{23} \text{ ions}$$

Step 5. The answer should be rounded to three significant figures and expressed in standard form: 2.64×10^{24} F$^-$ ions.

Problems

5. How many ammonium ions are in 0.036 mol ammonium phosphate, $(NH_4)_3PO_4$?

6. How many carbon atoms are in a mixture of 3.00 mol acetylene (C_2H_2) and 0.700 mol carbon monoxide?

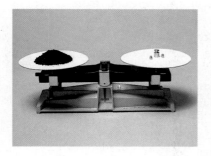

Figure 6·4
Twelve grams of charcoal is a mole of carbon. It contains 6.02×10^{23} atoms.

It would take over a billion (10^9) years to spend Avogadro's number of dollars if you spent a billion dollars each minute.

=== Science, Technology, and Society ===

6·A   The Size of a Mole

Just how large is Avogadro's number? Twelve grams of charcoal contains a mole of carbon, or 6.02×10^{23} atoms of carbon. Of course, these atoms are very, very small, but what if each atom were as large as the period at the end of this paragraph? How big would a mole of carbon atoms be? Would it fill up the room you are in, the building you are in, the Superdome, Lake Erie? Let's assume that the effective volume of a spherical period (carbon atom) is 0.333 mm^3. Now we can calculate the volume of Avogadro's number of carbon atoms in cubic meters.

$$6.02 \times 10^{23} \text{ atoms} \times \frac{0.333 \text{ mm}^3}{1 \text{ atom}} \times \frac{(1 \text{ m})^3}{(10^3 \text{ mm})^3} = 2.00 \times 10^{14} \text{ m}^3$$

Lake Erie has a volume of 4.55×10^{11} m^3. Thus our pile of carbon atoms is much larger than Lake Erie. In fact, we would need the volume of 440 Lake Eries to hold a mole of our period-sized carbon atoms!

6·3   The Gram Formula Mass

You learned in Section 4·7 that the atomic mass of an element (the mass of a single atom) is given in atomic mass units (amu). The atomic masses can be found in the periodic table. Remember that atomic masses of atoms are relative values. A carbon atom with an atomic mass of 12.0 amu is 12 times as heavy as a hydrogen atom with a mass of 1.0 amu. One hundred carbon atoms are 12 times as heavy as 100 hydrogen atoms (Figure 6·5). In fact, any group of carbon atoms is 12 times heavier than

■ The gram atomic mass of an element is the mass of one mole of those atoms. It is numerically equal to the atomic mass in grams.

the same number of hydrogen atoms. Therefore we can be confident that 12.0 g of carbon atoms and 1.0 g of hydrogen atoms contain the same number of atoms.

Chemists always work with large numbers of atoms. One hundred atoms would be a very small amount of a substance. Several grams of atoms would be a much easier amount to measure. *Chemists have defined* **the gram atomic mass** *as the number of grams of an element that is numerically equal to the atomic mass in amu.* For carbon the gram atomic mass (gam) is 12.0 g. For hydrogen the gram atomic mass is 1.0 g. By checking the periodic table you can see that for oxygen, the gram atomic mass is 16.0 g.

You will remember that 12.0 g of carbon atoms and 1.0 g of hydrogen atoms contain the same number of atoms. In a similar way, *the gram atomic masses of any two elements contain the same number of atoms.* If we were to compare 12.0 g of carbon atoms with 16.0 g of oxygen atoms, we would find they contain the same number of atoms. How many atoms are contained in the gram atomic mass of an element? This is a quantity with which you are already familiar. The gram atomic mass of a monatomic element contains one mole of atoms (6.02×10^{23} atoms).

The size of a mole is defined as the quantity of substance that contains as many representative particles as the number of atoms in 12.0 g of carbon-12. As you remember, 12.0 g is the gram atomic mass of carbon-12. Thus the size of a mole is the number of atoms in the gram atomic mass of a monatomic element. Stated another way, *the gram atomic mass is the mass of one mole of atoms of a monatomic element.* Since 24.3 g is the gram atomic mass for magnesium, we know that 24.3 g is also 1 mol of magnesium. This is a quantity that contains 6.02×10^{23} atoms of magnesium.

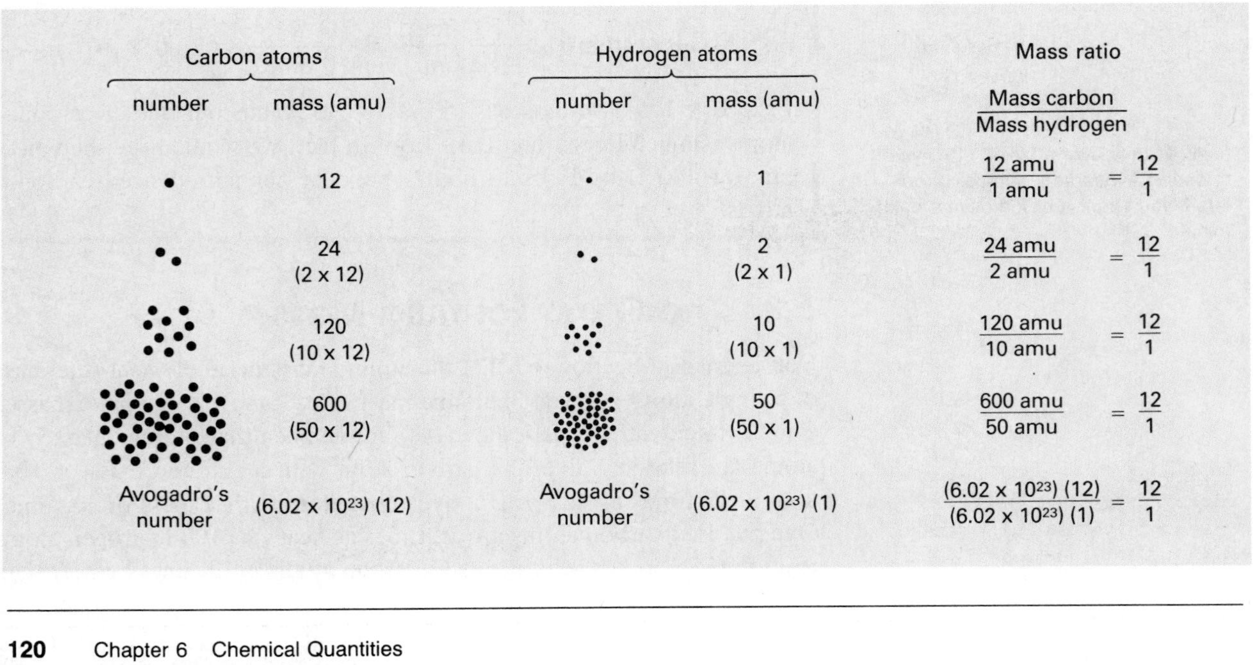

Figure 6·5
The mass ratio of equal numbers of carbon atoms to hydrogen atoms is always 12 to 1.

Carbon atoms		Hydrogen atoms		Mass ratio
number	mass (amu)	number	mass (amu)	$\dfrac{\text{Mass carbon}}{\text{Mass hydrogen}}$
•	12	•	1	$\dfrac{12 \text{ amu}}{1 \text{ amu}} = \dfrac{12}{1}$
••	24 (2 x 12)	••	2 (2 x 1)	$\dfrac{24 \text{ amu}}{2 \text{ amu}} = \dfrac{12}{1}$
(10 dots)	120 (10 x 12)	(10 dots)	10 (10 x 1)	$\dfrac{120 \text{ amu}}{10 \text{ amu}} = \dfrac{12}{1}$
(50 dots)	600 (50 x 12)	(50 dots)	50 (50 x 1)	$\dfrac{600 \text{ amu}}{50 \text{ amu}} = \dfrac{12}{1}$
Avogadro's number	$(6.02 \times 10^{23})(12)$	Avogadro's number	$(6.02 \times 10^{23})(1)$	$\dfrac{(6.02 \times 10^{23})(12)}{(6.02 \times 10^{23})(1)} = \dfrac{12}{1}$

Figure 6·6
One gram atomic mass (gam) for each of three elements is shown. Each of these quantities contains 6.02×10^{23} atoms. Note that the color of sulfur can vary from creme to bright yellow.

1 mole of mercury atoms
200.6 g Hg = 1 gam Hg

1 mole of iron atoms
55.8 g Fe = 1 gam Fe

1 mole of sulfur atoms
32.1 g S = 1 gam S

Problem

7. What is the mass of 1 mol of each of these monatomic elements?
a. sodium **b.** arsenic **c.** uranium

What is the mass of a mole of a compound? The formula of a compound tells us the number of atoms of each element in a representative particle of that compound. For example, the formula of the molecular compound sulfur trioxide is SO_3. This shows us that a molecule of SO_3 is composed of one atom of sulfur and three atoms of oxygen. We can calculate the molecular mass of a molecule of SO_3 by adding together the atomic masses of the atoms making up a molecule of SO_3. The atomic mass of sulfur (S) is 32.1 amu. The mass of three atoms of oxygen is three times the atomic mass of oxygen (O): 3×16.0 amu = 48.0 amu. Thus, the molecular mass of SO_3 is 32.1 amu + 48.0 amu = 80.1 amu.

Issues in Chemistry

"Better Living Through Chemistry" was a popular advertising slogan during the 1960s. Which of the items in your kitchen, laundry room, medicine cabinet, or garage are hazardous if used or disposed of improperly? You may want to check containers for safety information.

1 mole of paradichlorobenzene molecules (moth crystals)
147.0 g $C_6H_4Cl_2$ = 1 gmm $C_6H_4Cl_2$

1 mole of sucrose molecules (table sugar)
342.0 g $C_{12}H_{22}O_{11}$ = 1 gmm $C_{12}H_{22}O_{11}$

1 mole of water molecules
18.0 g H_2O = 1 gmm H_2O

Figure 6·7
One gram molecular mass (gmm) for each of three molecular compounds is shown. Each of these quantities contains 6.02×10^{23} molecules.

You must know the chemical formula of a compound to calculate the number of grams in a mole of the substance.

Now if we substitute the unit, grams, for atomic mass units, we obtain the gram molecular mass of SO_3. *The **gram molecular mass** (gmm) of any molecular compound is the mass of one mole of that compound.* It is equal to the molecular mass expressed in grams. Thus 1 mol of SO_3 has a mass of 80.1 g. Gram molecular masses may be calculated directly from gram atomic masses. For each element in the compound we find the number of grams of that element per mole of the compound. Then we sum the masses of the elements in the compound.

Example 4

What is the gram molecular mass of hydrogen peroxide, H_2O_2?

Solution

The unknown is the number of grams in the gram molecular mass. The known is the molecular formula: H_2O_2. This tells us the number of moles of each element in a mole of compound: 2 mol of hydrogen atoms and 2 mol of oxygen atoms.

We can convert moles of atoms to grams by using conversion factors (g/mol) based on the gram atomic mass of each element: 1 mol H = 1.0 g H and 1 mol O = 16.0 g O. We will sum the mass of each element to get the gram molecular mass.

$$2 \text{ mol H} \times \frac{1.0 \text{ g H}}{1 \text{ mol H}} = 2.0 \text{ g H}$$

$$2 \text{ mol O} \times \frac{16.0 \text{ g O}}{1 \text{ mol O}} = 32.0 \text{ g O}$$

$$\text{Gram molecular mass of } H_2O_2 = \overline{34.0 \text{ g}}$$

Problem

8. Find the gram molecular mass of each compound. **a.** CCl_4 **b.** PCl_3 **c.** C_8H_{18} **d.** N_2O_5

It is inappropriate to calculate the gram molecular mass of potassium iodide, KI, because it is an ionic compound. The representative particle of an ionic compound is a formula unit, not a molecule. *The mass of 1 mol of an ionic compound is the **gram formula mass** (gfm).* It is equal to the formula mass expressed in grams. A gram formula mass is calculated the same way as a gram molecular mass. Simply sum the atomic masses of the atoms that are in the formula of the compound. The atoms are actually present as ions. For example, the gram formula mass of potassium iodide, KI, is the gram atomic mass of potassium plus the gram atomic mass of iodine.

$$39.1 \text{ g K} + 126.9 \text{ g I} = 166.0 \text{ g KI}$$

There are 166.0 g of KI in 1 gfm or 1 mol of KI.

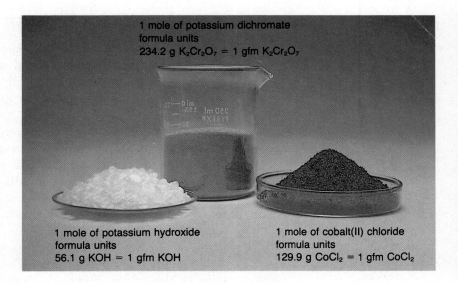

1 mole of potassium dichromate formula units
234.2 g K₂Cr₂O₇ = 1 gfm K₂Cr₂O₇

1 mole of potassium hydroxide formula units
56.1 g KOH = 1 gfm KOH

1 mole of cobalt(II) chloride formula units
129.9 g CoCl₂ = 1 gfm CoCl₂

Figure 6·8
One gram formula mass (gfm) for each of three ionic compounds is shown. Each of these quantities contains 6.02×10^{23} formula units.

Example 5

What is the gram formula mass of ammonium carbonate $(NH_4)_2CO_3$?

Solution

The unknown is the number of grams in a gram formula mass. The known is the formula unit. We know that a mole of this ionic compound is composed of 2 mol of nitrogen atoms, 8 mol of hydrogen atoms, 1 mol of carbon atoms, and 3 mol of oxygen atoms.

We will convert moles of atoms to grams by using conversion factors based on the gram atomic masses. Then we will sum the masses of each element to get the gram formula mass.

$$2 \text{ mol N} \times \frac{14.0 \text{ g N}}{1 \text{ mol N}} = 28.0 \text{ g N}$$

$$8 \text{ mol H} \times \frac{1.0 \text{ g H}}{1 \text{ mol H}} = 8.0 \text{ g H}$$

$$1 \text{ mol C} \times \frac{12.0 \text{ g C}}{1 \text{ mol C}} = 12.0 \text{ g C}$$

$$3 \text{ mol O} \times \frac{16.0 \text{ g O}}{1 \text{ mol O}} = 48.0 \text{ g O}$$

$$\text{Gram formula mass of } (NH_4)_2CO_3 = 96.0 \text{ g}$$

Problem

9. Find the gram formula mass of each of these compounds.
a. strontium chloride, $SrCl_2$
b. sodium carbonate, Na_2CO_3
c. aluminum sulfate, $Al_2(SO_4)_3$
d. calcium cyanide, $Ca(CN)_2$

6·4 The Molar Mass of a Substance

In the last section we introduced three terms: gram atomic mass (gam), gram molecular mass (gmm), and gram formula mass (gfm). Each term is used to represent a mole of a particular kind of substance. The gram *atomic* mass of an element contains a mole of *atoms*. The gram *molecular* mass of a molecular compound contains a mole of *molecules*. The gram *formula* mass of an ionic compound contains a mole of *formula* units. There is nothing wrong, however, with using the term *gram formula mass* to refer to a mole of an element, molecular compound, or ionic compound. In a broad sense, the gram formula mass is the mass in grams of the formula of the substance. *Some people use the term **molar mass** in place of gram formula mass to refer to the mass of a mole of any element or compound.*

Can you see the one advantage to using one of the three specific terms instead of the general term? What is the gram *formula* mass of oxygen? Is it any different from the gram *molecular* mass of oxygen? In the latter question, we have a reminder that oxygen is diatomic and is composed of molecules. The gram molecular mass of oxygen is 32.0 g (2×16.0 g). The gram formula mass of oxygen is 16.0 g *or* 32.0 g. This depends on whether we are talking about a mole of oxygen atoms (16.0 g/mol) or a mole of oxygen molecules (32.0 g/mol). We will use the three specific terms when necessary. Where there is no danger of ambiguity, we will use the term *gram formula mass* in its broadest sense.

■ The gram formula mass, or molar mass, is a mole of any substance.

When we refer to a mole of an element or compound we really mean a mole of atoms of the element or a mole of molecules of the compound.

6·5 Mole–Mass Conversions

The gram formula mass of an element or compound is used to convert grams of a substance into moles. It can also be used to convert moles of a substance into grams.

■ Conversions between moles and mass of a substance are done using the gram formula mass.

Example 6

How many grams are in 7.20 mol of dinitrogen trioxide?

Solution

From a known number of moles of a compound we want to calculate the unknown number of grams of the compound: mol → grams. First we need to write the formula: N_2O_3. If we calculate the gram formula mass we will know the number of grams in 1 mol of the compound.

$$2 \text{ mol N} \times \frac{14.0 \text{ g N}}{1 \text{ mol N}} = 28.0 \text{ g N}$$

$$3 \text{ mol O} \times \frac{16.0 \text{ g O}}{1 \text{ mol O}} = 48.0 \text{ g O}$$

$$1 \text{ mol } N_2O_3 = 1 \text{ gfm } N_2O_3 = 76.0 \text{ g}$$

Now we can use the gram formula mass to write a conversion factor. We will convert moles of N_2O_3 to grams of N_2O_3.

$$7.20 \; \text{mol } N_2O_3 \times \frac{76.0 \text{ g } N_2O_3}{1.00 \text{ mol } N_2O_3} = 547.2$$

$$= 5.47 \times 10^2 \text{ g } N_2O_3$$

Problem

10. Find the mass in grams of each quantity.
    a. 10.0 mol Cr
    b. 3.32 mol K
    c. 2.20×10^{-3} mol Sn
    d. 0.720 mol Be
    e. 2.40 mol N_2
    f. 0.160 mol H_2O_2
    g. 5.08 mol $Ca(NO_3)_2$
    h. 15.0 mol H_2SO_4
    i. 4.52×10^{-3} mol $C_{20}H_{42}$
    j. 0.0112 mol K_2CO_3

Example 7

You must know the chemical formula of a compound to calculate the number of moles in a given mass.

Find the number of moles in 922 g of iron(III) oxide, Fe_2O_3.

Solution

From a known number of grams of a compound we want to calculate the unknown number of moles of the same compound: grams $\longrightarrow$ moles.

First we calculate the gram formula mass (to get a conversion factor).

$$2 \; \text{mol Fe} \times \frac{55.8 \text{ g Fe}}{1 \text{ mol Fe}} = 111.6 \text{ g Fe}$$

$$3 \; \text{mol O} \times \frac{16.0 \text{ g O}}{1 \text{ mol O}} = \underline{48.0 \text{ g O}}$$

$$\text{gfm } Fe_2O_3 = 159.6 \text{ g}$$

The number of grams can now be converted to moles of Fe_2O_3 by using the gfm as a conversion factor.

$$922 \; \text{g } Fe_2O_3 \times \frac{1.00 \text{ mol } Fe_2O_3}{159.6 \text{ g } Fe_2O_3} = 5.776$$

$$= 5.78 \text{ mol } Fe_2O_3$$

Problem

11. Find the number of moles in each quantity.
    a. 72.0 g Ar
    b. 3.70×10^{-1} g B
    c. 187 g Al
    d. 333 g SnF_2
    e. 7.21×10^{-2} g He
    f. 27.4 g TiO_2
    g. 5.00 g hydrogen molecules
    h. 0.000 264 g Li_2HPO_4
    i. 11.0 g CH_4
    j. 847 g $(NH_4)_2CO_3$

6·6 The Volume of a Mole of Gas

You may have noticed in Figure 6·7 that the volumes of moles of different substances can be different. For example, a mole of table sugar is much larger than a mole of water. The volume of a mole of gas is more predictable than the volume of a mole of a liquid or solid.

A mole of any gas at STP occupies a volume of 22.4 L.

The volume of a gas is usually measured at a **standard temperature and pressure** (STP). *Standard temperature is 0°C. Standard pressure is 1 atmosphere (atm).* At STP, 1 mol of any gas occupies a volume of 22.4 L. *This quantity, 22.4 L, is known as the* **molar volume** *of a gas.* A mole of any substance contains Avogadro's number of particles. Thus 22.4 L of any gas at STP contains 6.02×10^{23} representative particles of that gas.

The molar volume (22.4 L) is about the size of a neat stack of 13 of these chemistry books.

Would 22.4 L of one gas also have the same mass as 22.4 L of another gas at STP? Probably not. A mole of a gas (22.4 L at STP) has a mass equal to the gram formula mass. Only gases with equal gram formula masses would have equal masses for equal volumes at STP.

Example 8

Determine the volume, in liters, of 0.600 mol of SO_2 gas at STP.

Solution

The known is the number of moles and the unknown is the number of liters of SO_2. We know that 1.00 mol SO_2 = 22.4 L of SO_2. We can express this equality as a conversion factor and use it to convert moles to liters (mol $\longrightarrow$ L).

$$0.600 \text{ mol } SO_2 \times \frac{22.4 \text{ L } SO_2}{1.00 \text{ mol } SO_2} = 13.4 \text{ L } SO_2$$

Example 9

Determine the number of moles in 33.6 L of He gas at STP.

Solution

Here we want to convert liters to moles of He gas: L $\longrightarrow$ mol. Again, we use a conversion factor based on the molar volume.

$$33.6 \text{ L He} \times \frac{1.00 \text{ mol He}}{22.4 \text{ L He}} = 1.50 \text{ mol He}$$

Problems

12. What is the volume at STP of these gases?
    a. 5.40 mol O_2 b. 3.20×10^{-2} mol CO_2?   c. 0.960 mol CH_4?

13. Assuming STP, how many moles are in these volumes?
    a. 89.6 L SO_2 b. 1.00×10^3 L C_2H_6 c. 5.42×10^{-1} mL Ne

Figure 6·9
Balloons this size at 0°C would each contain a mole of gas. At STP, a mole of any gas occupies 22.4 L and contains Avogadro's number of representative particles. Note however that moles of different gases have different masses.

The gram formula mass of a gas can be determined experimentally by measuring the density of the gas.

6·7 Gas Density and the Gram Molecular Mass

The density of a gas is usually measured in the units g/L. The experimentally determined density of a gas at STP is used to calculate the gram formula mass of that gas. The gas can be an element or a compound.

Example 10

The density of a gaseous compound of carbon and oxygen is 1.964 g/L at STP. Determine the gram formula mass of the compound. Is the compound carbon dioxide or carbon monoxide?

Solution

At STP the gram formula mass of any gas occupies a volume of 22.4 L. The density is used as a conversion factor to convert liters to mass: L ⟶ g.

$$22.4 \text{ L} \times \frac{1.964 \text{ g}}{1 \text{ L}} = 44.0 \text{ g}$$

A volume of 22.4 L of this gas has a mass of 44.0 g. This mass must be compared with the gram formula masses of carbon monoxide and carbon dioxide.

gfm CO: 12 g/mol (C) + 16 g/mol (O) = 28 g/mol

gfm CO_2:   12 g/mol (C) + 2(16 g/mol)(O) = 44 g/mol

The compound is carbon dioxide.

Problem

14. The densities of gases A, B, and C are 1.25 g/L, 2.86 g/L, and 0.714 g/L, respectively. Calculate the gram formula mass of each of these substances. Identify each substance as ammonia (NH_3), sulfur dioxide, chlorine, nitrogen, or methane (CH_4).

6·8 Converting Between Units with Moles

The mole is the most useful way for chemists to express the amount of matter.

We have now defined a mole in terms of particles, mass, and volume (the last for gases at STP). These relationships are summarized in Figure 6·10. This diagram shows the importance of the mole. To change from one unit to another, the mole is used as an intermediate step. The form of the conversion factor will depend on whether you are going *from* moles or *to* moles.

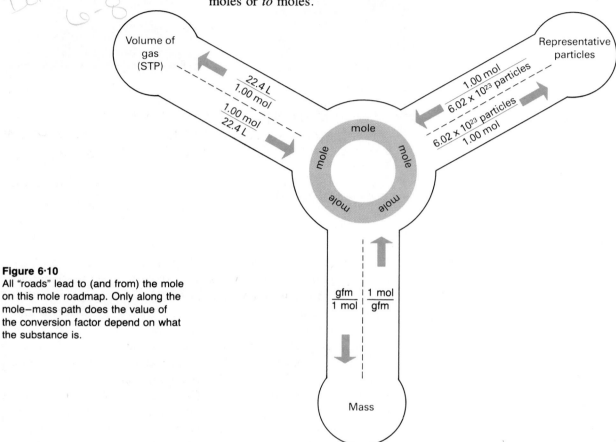

Figure 6·10
All "roads" lead to (and from) the mole on this mole roadmap. Only along the mole–mass path does the value of the conversion factor depend on what the substance is.

Example 11

How many carbon atoms are in a 50.0-carat diamond that is pure carbon? Fifty carats is the same as 10.0 g.

Solution

We want to convert from a known mass to an unknown number of atoms. Figure 6·10 shows that we will need to use moles in this process: grams $\longrightarrow$ moles $\longrightarrow$ atoms. Starting with a given mass, we can use the gram formula mass to find the number of moles.

$$10.0 \text{ g C} \times \frac{1.00 \text{ mol C}}{12.0 \text{ g C}} = 0.833 \text{ mol C}$$

Then we can use Avogadro's number to find the number of particles.

$$0.833 \ \cancel{\text{mol C}} \times \frac{6.02 \times 10^{23} \text{ atoms C}}{1.00 \ \cancel{\text{mol C}}} = 5.02 \times 10^{23} \text{ atoms C}$$

We can also set up the problem in "one solution."

$$50.0 \ \cancel{\text{carats}} \ C \times \frac{10.0 \ \cancel{\text{gram C}}}{50.0 \ \cancel{\text{carats C}}} \times \frac{1.00 \ \cancel{\text{mol C}}}{12.0 \ \cancel{\text{grams C}}}$$

$$\times \frac{6.02 \times 10^{23} \text{ atoms C}}{1 \ \cancel{\text{mol C}}} = 5.02 \times 10^{23} \text{ atoms C}$$

Problems

15. What is the mass in grams of an atom of nickel (Ni)?

16. How many molecules are in a 6.00-L balloon (at STP) filled with carbon dioxide? Would your answer change if the gas was carbon monoxide?

Science, Technology, and Society

6·B Who Was Avogadro?

The first determination of Avogadro's number was made by a German physicist in 1865, nine years after Avogadro's death. Avogadro himself had never made any attempt to calculate it. This important number was given his name, however, to honor him for contributions that were not recognized until after his death.

Count Amedeo Avogadro was a lawyer who turned to the study of mathematics and physics. He soon became a professor of these subjects. His most famous work was his explanation of experiments done by the French chemist Joseph Gay-Lussac. Gay-Lussac had discovered that when two gases combine to give a new substance, they do so in a simple volume ratio. For example, one liter of oxygen combines with two liters of hydrogen to form two liters of gaseous water. One liter of nitrogen combines with three liters of hydrogen to form two liters of ammonia gas. Gay-Lussac called his discovery the law of combining volumes.

Gay-Lussac's results did not seem to be compatible with Dalton's recently published atomic theory. Dalton thought that when any two gases combined to make a single new substance, one atom of each was used. He also thought that atoms of a gas were packed closely together. If these two ideas were correct, it seemed that one liter of oxygen should react with one liter of hydrogen, not two. Even if one atom of oxygen reacted with two of hydrogen, it would seem that together they should form three liters of water, not two.

Avogadro was able to reconcile Gay-Lussac's data with Dalton's atomic theory. In the process he introduced an essential new concept

Figure 6·11
One hundred years after Avogadro published his hypothesis, the Italian Academy of Science had a coin minted to commemorate the occasion. The back of the coin reads: "Amedeo Avogadro. This commemorates the centennial of his molecular theory. September 24, 1911."

to chemistry, the concept of molecules. He disagreed with Dalton on two important points. He said that the particles of a gas were not atoms at all, but molecules. These were clusters of atoms that could be divided or created during chemical reactions. Second, these molecules were not packed closely together in the gas. They were separated by enough distance that they were not much affected by each other. Thus each gas, no matter what its composition, contained the same number of molecules per liter, if the temperature and pressure were the same. This theory, first published in 1811, is still known as Avogadro's hypothesis. It can be stated as follows: *Equal volumes of different gases, at the same temperature and pressure, contain an equal number of molecules.*

Gay-Lussac's data could now be easily explained. Take the case where one liter of oxygen reacts with two liters of hydrogen to give two liters of gaseous water. It follows from Avogadro's hypothesis that *one molecule* of oxygen reacts with *two molecules* of hydrogen to give *two molecules* of water. If one molecule of oxygen can give two of water, the molecule of oxygen must be dividing into halves. Since atoms are indivisible, the molecule must consist of two atoms of oxygen. Based on other experiments, Avogadro was able to demonstrate that hydrogen molecules were also diatomic. Therefore, since two molecules of hydrogen made two molecules of water, each molecule of water must contain one molecule (two atoms) of hydrogen. Thus the formula of water was H_2O, not HO as Dalton had assumed. Using similar reasoning Avogadro established correct formulas for many compounds. These included nitrous oxide, ammonia, carbon monoxide, carbon dioxide, sulfur dioxide, ethyl alcohol, ether, and others.

Avogadro's ideas were a breakthrough in understanding chemical reactions. Unfortunately, however, his work was generally ignored for several reasons. Avogadro did not correspond with other chemists, and his hypothesis was not supported by much experimental work. In addition, the most influential chemist of the day, Berzelius, did not accept Avogadro's conclusions. As a result, progress in chemistry was greatly handicapped for half a century.

Figure 6·12
Avogadro's hypothesis explains why one liter of oxygen and two liters of hydrogen combine to make two liters of gaseous water. Since equal volumes of different gases at STP contain the same number of molecules, then the oxygen molecules must split into two parts. This suggests that oxygen and hydrogen molecules are diatomic and that the molecular formula for water is H_2O.

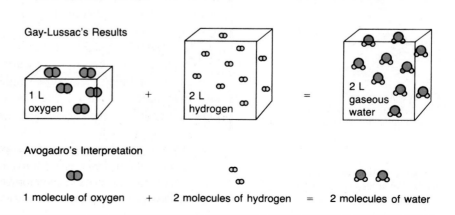

Before reading this section you should review your knowledge of percents by doing section B·3 of the Math Review in Appendix B.

Safety

Potassium chromate is a common laboratory reagent. It has been shown to cause mutations in animals. Always handle all chemicals with caution.

■ The percent by mass of each element in a compound is the percent composition of the compound.

Chemical Connections

Percent composition is of concern when extracting metals from their ores. The ore called hematite, for example, has long been the most important source of iron, partly because it contains a high percentage of the metal.

6·9 Calculating Percent Composition

When a new compound is made in the laboratory, we need to determine its formula. One of the first steps in doing this is to determine the **percent composition,** *the percent by mass of each element in a compound.* The percent composition includes as many percents as there are elements in the compound. For example, the percent composition of K_2CrO_4 is 40.3% K, 26.8% Cr, and 32.9% O. These percents must add up to 100% (40.3% + 26.8% + 32.9% = 100%).

The percent by mass of an element in a compound is the number of grams of the element divided by the grams of the compound, multiplied by 100%.

$$\% \text{ mass} = \frac{\text{grams of element}}{\text{grams of compound}} \times 100\%$$

To calculate the percent by mass we need to know the mass of each element that combined to form the compound.

Example 12

An 8.20-g piece of magnesium combines completely with 5.40 g of oxygen to form a compound. What is the percent composition of this compound?

Solution

Since we know the mass of each element, we can add them to find the mass of the compound.

$$8.20 \text{ g} + 5.40 \text{ g} = 13.60 \text{ g}$$

Now the percent mass of each element can be calculated. The mass of each element is divided by the mass of the compound.

$$\% \text{ Mg} = \frac{\text{mass of Mg}}{\text{mass of compound}} \times 100\% = \frac{8.20 \text{ g}}{13.60 \text{ g}} \times 100\%$$
$$= 60.3\%$$

$$\% \text{ O} = \frac{\text{mass of O}}{\text{mass of compound}} \times 100\% = \frac{5.40 \text{ g}}{13.60 \text{ g}} \times 100\%$$
$$= 39.7\%$$

As a check, the percents of the elements must add up to 100%: 60.3% + 39.7% = 100%.

Problem

17. Calculate the percent composition of the compounds that are formed from these reactions.
  a. 29.0 g of Ag combines completely with 4.30 g of S.
  b. 9.03 g of Mg combines completely with 3.48 g of N.
  c. 222.6 g of Na combines completely with 77.4 g of O.

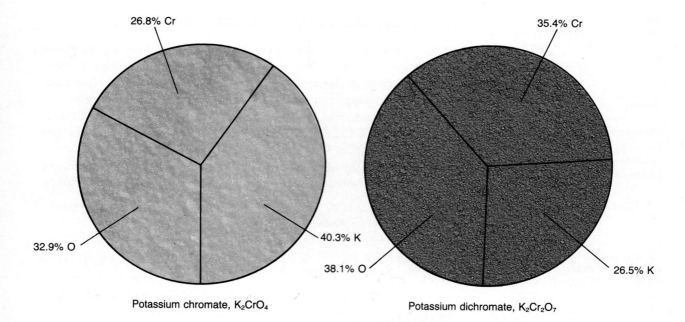

Potassium chromate, K_2CrO_4

26.8% Cr

32.9% O

40.3% K

Potassium dichromate, $K_2Cr_2O_7$

35.4% Cr

38.1% O

26.5% K

Figure 6·13
Potassium chromate, K_2CrO_4, is composed of 40.3% potassium, 26.8% chromium, and 32.9% oxygen. How does this differ from the percent composition of potassium dichromate, $K_2Cr_2O_7$?

Occasionally we may want to calculate the percent composition of a known compound. To do this, we use the chemical formula to calculate the gram formula mass. This gives us the mass of one mole of the compound. Then for each element, we calculate the percent by mass in one mole of the compound. To calculate the percent by mass we divide the mass of the element in one mole by the gram formula mass and multiply this by 100%.

$$\% \text{ mass} = \frac{\text{grams of element in 1 mol of compound}}{\text{gfm of compound}} \times 100\%$$

Example 13

Calculate the percent composition of ethane, C_2H_6.

Solution

Given the formula, we can calculate the percent composition by using the gram formula mass: formula $\longrightarrow$ gfm $\longrightarrow$ % composition. First calculate the gram formula mass of ethane.

$$2 \text{ mol C} \times \frac{12.0 \text{ g C}}{1 \text{ mol C}} = 24.0 \text{ g C}$$

$$6 \text{ mol H} \times \frac{1.0 \text{ g H}}{1 \text{ mol H}} = \underline{6.0 \text{ g H}}$$

$$\text{gram formula mass of } C_2H_6 = 30.0 \text{ g}$$

Now the percent mass for each element can be calculated. The mass of each element is divided by the mass of the compound.

$$\% \text{ C} = \frac{\text{grams of C}}{\text{gfm of C}_2\text{H}_6} \times 100\% = \frac{24.0\ g}{30.0\ g} \times 100\% = 80.0\% \text{ C}$$

$$\% \text{ H} = \frac{\text{grams of H}}{\text{gfm of C}_2\text{H}_6} \times 100\% = \frac{6.0\ g}{30.0\ g} \times 100\% = 20.0\% \text{ H}$$

As a check, we can see that $80.0\% + 20.0\% = 100\%$.

Problem

18. Calculate the percent composition of each of these compounds.
    a. propane, C_3H_8 b. sodium bisulfate, $NaHSO_4$ c. calcium acetate, $Ca(C_2H_3O_2)_2$ d. hydrogen cyanide, HCN

Percent composition can be used to calculate the number of grams of an element in a specific amount of a compound. The mass of the compound is multiplied by a conversion factor that is based on the percent composition.

Example 14

Calculate the mass of carbon in 82 g of ethane, C_2H_6.

Solution

Given the number of grams of ethane, C_2H_6, we want to calculate the number of grams of carbon, C: grams $C_2H_6 \longrightarrow$ grams C. To do this, we use a conversion factor based on the percent mass of carbon in ethane. From Example 13 we know the percent composition is 80% C and 20% H.

$$82\ g\ C_2H_6 \times \frac{80\ g\ C}{100\ g\ C_2H_6} = 66\ g\ C$$

Problem

19. Using the results of Problem 18, calculate the amount of hydrogen in the following amounts of these compounds.
    a. 350 g C_3H_8 b. 20.2 g $NaHSO_4$ c. 124 g $Ca(C_2H_3O_2)_2$
    d. 378 g HCN

Figure 6·14
Lawn and garden fertilizers are labeled with a series of three numbers. These numbers represent the mass percent of the elements nitrogen, phosphorus, and potassium, respectively. One hundred kilograms of a 15-10-5 fertilizer would contain 15 kg of nitrogen (as N), 10 kg of phosphorus (as P_2O_5), and 5 kg of potassium (as K_2O).

Empirical formulas can be determined by chemical analysis.

6·10   Calculating Empirical Formulas

Once a new compound has been made in the laboratory, we can usually determine its percent composition experimentally. From the percent composition data, we can calculate the empirical formula of the compound. *The **empirical formula** gives the lowest whole-number ratio of the elements in a compound.* If we use the mole interpretation, an empirical formula is the lowest whole-number ratio of moles of atoms in a compound. An empirical formula may or may not be the same as a molecular formula. For carbon dioxide, the empirical and molecular formulas are the same, CO_2. Dinitrogen tetrahydride (molecular formula, N_2H_4) has an empirical formula of NH_2 because this is the simplest ratio of nitrogen to hydrogen.

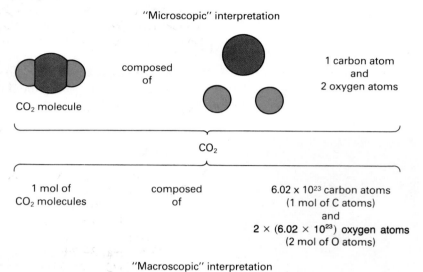

"Microscopic" interpretation

CO_2 molecule — composed of — 1 carbon atom and 2 oxygen atoms

CO_2

1 mol of CO_2 molecules — composed of — 6.02×10^{23} carbon atoms (1 mol of C atoms) and $2 \times (6.02 \times 10^{23})$ oxygen atoms (2 mol of O atoms)

"Macroscopic" interpretation

Figure 6·15
A formula can be interpreted on a microscopic level in terms of atoms or on a macroscopic level in terms of moles of atoms.

Example 15

What is the empirical formula of a compound that is 25.9% nitrogen and 74.1% oxygen?

Solution

The unknown is the empirical formula. Thus we want to calculate the lowest whole-number ratio of moles of nitrogen atoms to oxygen atoms. The known is the percent composition. This tells us the ratio of masses of nitrogen to oxygen atoms in the compound. In 100.0 g of the compound there are 25.9 g N and 74.1 g O.

We can change the ratio of *masses* to a ratio of *moles* by using conversion factors based on the gram atomic mass of each element. (The gram atomic mass tells us the number of grams per mole.)

$$25.9 \text{ g N} \times \frac{1 \text{ mol N}}{14.0 \text{ g N}} = 1.85 \text{ mol N}$$

$$74.1 \text{ g O} \times \frac{1 \text{ mol O}}{16.0 \text{ g O}} = 4.63 \text{ mol O}$$

We now know the mole ratio of nitrogen to oxygen: $N_{1.85}O_{4.63}$. This is not the correct empirical formula because it is not the lowest whole-number ratio. We can reduce the ratio by dividing through by the smallest number of moles. This will give us a "1" for the smallest number of moles. It may not give us whole numbers for the others.

$$\frac{1.85 \text{ mol N}}{1.85} = 1 \text{ mol N}$$

$$\frac{4.63 \text{ mol O}}{1.85} = 2.50 \text{ mol O}$$

Is the final answer $N_1O_{2.5}$? No, because this is not the lowest *whole*-number ratio. We must now multiply each part of the ratio by a number (in this case 2) to convert the fraction to a whole number.

$$1 \text{ mol N} \times 2 = 2 \text{ mol N}$$

$$2.5 \text{ mol O} \times 2 = 5 \text{ mol O}$$

The empirical formula is N_2O_5.

Problems

20. Calculate the empirical formula of each compound with the following percent composition.
   a. 79.8% C, 20.2% H
   b. 67.6% Hg, 10.8% S, 21.6% O
   c. 94.1% O, 5.9% H
   d. 17.6% Na, 39.7% Cr, 42.7% O
   e. 27.59% C, 1.15% H, 16.09% N, 55.17% O

21. Calculate the empirical formula for each compound in Problem 17 without using the percent composition.

6·11   Calculating Molecular Formulas

The molecular formula of a compound is either the same as its experimentally determined empirical formula, or it is some whole-number multiple of it. Consider the two series of compounds in Table 6·2. Acetylene and benzene have the same empirical formula, CH. Glucose, acetic acid, and formaldehyde all have the same empirical formula, CH_2O. This formula is also the molecular formula of formaldehyde. Each of the compounds in these series has a different gram formula mass.

Table 6·2   Comparison of Empirical and Molecular Formulas

Formula (name)	Classification of formula	Gram formula mass
CH	Empirical	13
C_2H_2 (acetylene)	Molecular	26 (2×13)
C_6H_6 (benzene)	Molecular	78 (6×13)
CH_2O (formaldehyde)	Empirical and molecular	30
$C_2H_4O_2$ (acetic acid)	Molecular	60 (2×30)
$C_6H_{12}O_6$ (glucose)	Molecular	180 (6×30)

An empirical formula and a molecular mass determine the molecular formula of a compound.

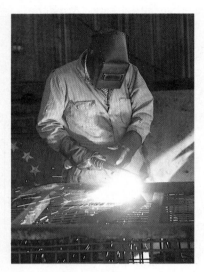

Figure 6·16
Acetylene, C_2H_2, is a gas used in welder's torches. Benzene, C_6H_6, is a flammable liquid. These two compounds have the same empirical formula.

 ChemDirections

Hazardous chemicals like acetylene and benzene are transported on the nation's highways. You will learn to recognize these materials in transit by studying **ChemDirections** *Trucking Chemicals*, page 659.

Other methods to determine the gram formula mass must be used if the substance is not a gas. One such method involves measuring the freezing-point depression (Section 16·10).

We can determine the molecular formula of a compound if we know its empirical formula and its gram formula mass. From the empirical formula we calculate the empirical formula mass. This is simply the gram formula mass of the empirical formula. The known gram formula mass is then divided by the empirical formula mass. This gives the number of empirical formula units in a molecule of the compound.

Example 16

Calculate the molecular formulas of the following compounds.

	Gram formula mass	Empirical formula
a.	60 g	CH_4N
b.	78 g	NaO
c.	181.5 g	C_2HCl

Solution

First calculate the empirical formula mass (efm). Then divide the empirical formula mass (efm) into the gram formula mass (gfm). It takes this many of the empirical formula units to make up the molecular formula of the compound.

	Empirical formula	efm	$\dfrac{\text{gfm}}{\text{efm}}$	Molecular formula
a.	CH_4N	30	$\dfrac{60}{30} = 2$	$C_2H_8N_2$
b.	NaO	39	$\dfrac{78}{39} = 2$	Na_2O_2
c.	C_2HCl	60.5	$\dfrac{181.5}{60.5} = 3$	$C_6H_3Cl_3$

You may wonder how the gram formula mass could be determined without knowing the molecular formula. If the compound is a gas, then we can measure the mass of 22.4 L of the compound at STP. This is the molar volume. The mass of one mole of any compound is the gram formula mass.

Problems

22. The compound methyl butanoate smells like apples. Its percent composition is 58.8% C, 9.8% H, and 31.4% O. If its gram molecular mass is 102 g/mol, what is its molecular formula?

23. You find that 7.36 g of a compound has decomposed to give 6.93 g of oxygen. The rest of the compound is hydrogen. If the molecular mass of the compound is 34.0 g/mol, what is its molecular formula?

Figure 6·17
A salesperson for a chemical company must have a good knowledge of the language of chemistry.

6·C Knowing Your Product: Chemical Sales

A college degree in chemistry can be the basis of a rewarding career on the road or in an office as well as in a laboratory. The field of chemical marketing and sales includes many positions from field sales representative to marketing executive.

A chemical company generally sells chemicals that are used in the manufacture of other products. In addition to needing raw materials, manufacturing companies often need special chemicals for specific purposes. A cosmetics manufacturer may need a preservative that is safe for use in hand lotions. A paint manufacturer may need an additive to improve the spreading quality of paint. An agricultural chemicals company may need an anticaking agent for fertilizers. The job of a sales representative is to find customers that have a need for his or her product and to provide information about it. If the salesperson has accurately assessed the need of the customer, the product will sell itself.

To succeed at this task, a salesperson must know the properties, limitations, and applications of his or her product. He or she must be able to communicate this information to potential customers. Often sales representatives will be asked to train customers in the use of the product. A college degree in chemistry, courses in business, and skill in oral and written communication are important qualifications for someone considering a career in chemical sales.

A sales engineer's job is to help customers solve technical problems that involve the use of the company's products. He or she may also provide technical data to a salesperson and participate in determining prices. Sales representatives and sales engineers who are willing to continue their education often advance to management positions.

6 Chemical Quantities
Chapter Review

Key Terms

Avogadro's number	6·2
empirical formula	6·10
gram atomic mass (gam)	6·3
gram formula mass (gfm)	6·3
gram molecular mass (gmm)	6·3
molar mass	6·4
molar volume	6·6
mole (mol)	6·2
percent composition	6·9
representative particle	6·2
standard temperature and pressure (STP)	6·6

Chapter Summary

The SI unit that measures the amount of substance is the mole. A mole of any substance is composed of Avogadro's number (6.02×10^{23}) of representative particles. The representative particle of most elements is the atom. A molecule is the representative particle of diatomic elements and molecular compounds. The representatve particle of ionic compounds is a formula unit.

The formula mass of an atom, ion, molecule, or formula unit expressed in grams is its gram formula mass. One gram formula mass of any substance is a mole. One mole of any substance contains the same number of representative particles as one mole of any other substance. Thus one gfm of any substance contains 6.02×10^{23} representative particles of that substance.

If we want to determine the number of moles of a gas, it is easier to measure its volume than its mass. One mole of any gas at STP (1 atm pressure and 0°C) occupies a volume of 22.4 L.

The percent composition is the percent by mass of each element in a compound. An empirical formula is the simplest whole-number ratio of atoms of the elements in the compound. An empirical for-

mula can be calculated from a compound's percent composition. A molecular formula will be the same as, or some simple multiple of, an empirical formula. For example, dinitrogen tetroxide has the empirical formula NO_2 and the molecular formula N_2O_4.

Practice Questions and Problems

24. List three common ways that matter is measured, giving examples of each. *6·1*

25. Name the basic SI unit that measures the "amount of substance." *6·1*

26. Name the representative particle (atom, molecule, or formula unit) of each of the following substances. *6·2*
  a. hydrogen
  b. iron(II) hydroxide
  c. copper
  d. carbon monoxide

27. How many hydrogen atoms are in a representative particle of each of these substances? *6·2*
  a. $Ca(OH)_2$ **c.** $(NH_4)_2HPO_4$
  b. C_3H_8O **d.** HC_2H_3O

28. What is the total number of *ions* in a formula unit of each of these substances? *6·2*
  a. $BaCl_2$ **c.** Na_2CO_3
  b. $(NH_4)_2SO_4$ **d.** $FeCl_3$

29. Find the number of representative particles in each of these substances. *6·2*
  a. 5.00 mol Fe
  b. 0.200 mol NaI
  c. 34.0 mol SO_3
  d. 5.25×10^{-4} mol K_2S

30. How many moles is each of the following? *6·2*
  a. 6.02×10^{22} molecules Br_2
  b. 4.81×10^{24} atoms Li
  c. 1.50×10^{23} molecules NH_3
  d. 1 billion (10^9) molecules O_2

31. Calculate the mass of 1.00 mol of each of these substances. *6·3*
  a. H_2
  b. $Fe(OH)_3$
  c. Co
  d. SiO_2

32. Calculate the gram formula mass of each of these substances. *6·3*
  a. $CaCO_3$
  b. $(NH_4)_2SO_4$
  c. C_3H_6O
  d. I_2
  e. H_3PO_4
  f. N_2O_5

33. Find the mass of each of these substances. *6·5*
  a. 2.40 mol $NaOH$
  b. 3.21×10^{-3} mol Ni
  c. 0.780 mol $Ca(CN)_2$
  d. 7.00 mol H_2O_2

34. How many moles is each of the following? *6·5*
  a. 0.800 g Ca
  b. 79.3 g Cl_2
  c. 5.96 g KOH
  d. 937 g $Ca(C_2H_3O_2)_2$

35. Calculate the volume of each of these gases at STP. *6·6*
  a. 9.6 mol He **b.** 4.8 mol N_2 **c.** 3.2 mol CO_2

36. What is the density of each of these gases at STP?
  a. NO **b.** Ar **c.** C_2H_6 *6·7*

37. Find each of the following quantities. *6·8*
  a. the number of molecules in 60.0 g of NO_2
  b. the volume, in liters, of 3.24×10^{22} molecules (STP)
  c. the mass of 18.0 L of CH_4 (STP)
  d. the volume, in liters, of 835 g of SO_3 (STP)
  e. the mass of a molecule of aspirin, $C_9H_8O_4$
  f. the number of atoms in 5.78 mol of NH_4NO_3

38. Calculate the percent composition of each of these compounds. *6·9*
  a. Na_3PO_4
  b. H_2S
  c. $(NH_4)_2C_2O_4$
  d. $Mg(OH)_2$

39. Using your answers from Problem 38, calculate the number of grams of these elements. *6·9*
  a. phosphorus in 804 g of Na_3PO_4
  b. sulfur in 3.54 g of H_2S
  c. nitrogen in 25.0 g of $(NH_4)_2C_2O_4$
  d. magnesium in 97.4 g of $Mg(OH)_2$

40. Which of the following compounds has the highest iron content? *6·9*
  a. Fe_2O_3
  b. $Fe(OH)_3$
  c. $Fe(C_2H_3O_2)_2$
  d. FeO

41. Determine the empirical formula of each of these compounds. *6·10*
  a. 42.9% C, and 57.1% O
  b. 71.72% Cl, 16.16% O, and 12.12% C
  c. 32.00% C, 42.66% O, 18.67% N, and 6.67% H

42. Classify each of these formulas as an empirical or a molecular formula. *6·11*
  a. K_2SO_4
  b. $C_6H_{12}O_6$
  c. $C_{17}H_{19}NO_3$
  d. $(NH_4)_2CO_3$
  e. S_2Cl_2
  f. $C_6H_{10}O_4$

43. What is the molecular formula for each of these compounds? Each compound's empirical formula and gram formula mass is given. *6·11*
  a. CH_2O, 120 g/mol
  b. $HgCl$, 472.2 g/mol
  c. $C_3H_5O_2$, 146 g/mol

44. Determine the molecular formula for each of these compounds. *6·11*
  a. 40.0% C, 53.4% O, and 6.6% H; gfm = 120 g
  b. 94.1% O and 5.9% H; gfm = 34 g
  c. 54.5% C, 13.6% H, and 31.8% N; gfm = 88 g

Mastery Questions and Problems

45. Which of the following contains the largest number of atoms?
  a. 82.0 g Ar
  b. 0.420 mol C_3H_8
  c. 5.00 g H_2

46. What is the total mass of a mixture of 3.5×10^{22} formula units of Na_2SO_4, 0.500 mol of H_2O, and 7.23 g of $AgCl$?

47. Find the empirical formula for each compound from its percent composition.
  a. 72.4% Fe and 27.6% O
  b. 65.2% Sc and 34.8% O
  c. 52.8% Sn, 12.4% Fe, 16.0% C, and 18.8% N

48. What is the total number of moles of atoms in a mixture of 20.0 g of H_2O_2, 1.20×10^{24} molecules of SiO_2, and 3.70 mol of I_2?

49. A typical virus particle is 5×10^{-6} cm in diameter. If Avogadro's number of these virus particles were laid in a row, how many kilometers long would the line be?

50. Explain what is wrong with each of the following statements.
a. One mole of any substance contains the same number of atoms.
b. The gram atomic mass of a compound is the atomic mass expressed in grams.
c. One gram molecular mass of CO_2 contains Avogadro's number of atoms.

51. How many formula units of magnesium nitride, Mg_3N_2, can be made from 60 magnesium ions and 60 nitride ions?

52. Calculate the empirical formula for each of these compounds.
a. The compound consists of 0.25 mol of Fe per 0.50 mol of Cl.
b. The compound has 4 atoms of carbon for each 12 atoms of hydrogen.

53. An imaginary "atomic balance" is shown below. Fifteen atoms of boron on the left side of the balance are being balanced by six atoms of the unknown element E on the right side.
a. What is the atomic mass of element E?
b. What is the identity of element E?

Critical Thinking Questions

54. Choose the term that best completes the second relationship.
a. dozen:pencils   mole: _____
   (1) atoms   (3) size
   (2) 6.02×10^{23} (4) grams
b. gam: element
   gmm: _____
   (1) formula unit   (3) ionic compound
   (2) molecule   (4) molecular compound
c. mole:Avogadro's number
   molar volume: _____
   (1) mole (2) water (3) STP (4) 22.4 L

55. One mole of any gas at STP is equal to 22.4 L of that gas. It is also true that different elements have different atomic volumes, or diameters. How can you reconcile these two statements?

56. Examine the pictures in this chapter that show one mole of a solid substance. Why does the volume of one mole of a solid substance vary between substances? What does this imply about the distance between atoms in a solid?

57. Why does one mole of carbon have a smaller mass than one mole of sulfur? How is the atomic structure of these elements different?

Review Questions and Problems

58. How many moles is 30.0 cm^3 of copper? The density of copper is 8.92 g/cm^3.

59. How does an atom differ from a molecule?

60. Do each of these conversions.
a. 56.3 cal to kilojoules
b. 5.9×10^5 cm/s to km/hr
c. 0.34 mg to decigrams

61. How many protons, electrons, and neutrons are in each of these isotopes?
a. hydrogen-2   **c.** bromine-79
b. mercury-201   **d.** antimony-113

62. Write formulas for these compounds.
a. sodium carbonate   **c.** iron(III) hydroxide
b. carbon tetrachloride   **d.** ammonium iodide

63. Show that this series of the oxides of nitrogen obeys the law of multiple proportions: N_2O, NO, N_2O_3, and N_2O_5.

64. Name these compounds.
a. KNO_3 **d.** AgF
b. CuO   **e.** $Al_2(SO_4)_3$
c. Mg_3N_2 **f.** H_3PO_4

Challenging Questions and Problems

65. How many molecules are in a sample of water that requires 8.40 kcal of heat energy to raise the temperature by 34°C?

66. Nitroglycerine contains 60% as many carbon atoms as hydrogen atoms; three times as many oxygen atoms as nitrogen atoms; and the same number of carbon and nitrogen atoms. The number of moles in a gram of nitroglycerine is 0.00441. What is the molecular formula?

67. The specific gravity of lead is 11.3. How large a cube, in cm^3, would contain 5.00×10^{24} atoms of lead?

68. Dry air is about 20.95 percent oxygen by volume. Assuming STP, how many oxygen molecules are in a 40.0-g sample of air? The density of air is 1.19 g/L.

69. The table below gives the gram formula mass (gfm) and density of seven gases at STP.
   a. Plot this data, with density on the x-axis.
   b. What is the slope of the straight-line plot?
   c. What is the gfm of a gas at STP that has a density of 1.10 g/L?
   d. A mole of a gas at STP weighs 56.0 g. Use the graph to determine the density of this gas.

Substance	Gram formula mass (gfm)	Density (g/L)
Oxygen	32.0	1.43
Carbon dioxide	44.0	1.96
Ethane	30.0	1.34
Hydrogen	2.00	0.0893
Sulfur dioxide	64.1	2.86
Ammonia	17.0	0.759
Fluorine	38.0	1.70

70. The element gold has properties that have made it much sought after through the ages. A cubic meter of ocean water contains 6×10^{-6} g gold. If the total mass of the water in the oceans of the world is 4×10^{20} kg, how many kilograms of gold are distributed throughout the oceans? (Assume that the density of seawater is 1 g/cm^3.) How many liters of seawater would have to be processed to recover a kilogram of gold (which had a value of about $11 000 at 1989 prices)? Do you think this recovery operation is feasible?

71. Have you ever wondered how Avogadro's number was determined? Actually, Avogadro's number has been independently determined by about 20 different methods. In one approach Avogadro's number is calculated from the volume of a film of a fatty acid floating on water. In another method it is determined by comparing the electric charge of an electron to the electric charge of a mole of electrons. In a third method the spacing between ions in an ionic substance can be determined by using a technique called X-ray diffraction. In the X-ray diffraction of sodium chloride it has been determined that the distance between adjacent Na^+ and Cl^- ions is 2.819×10^{-8} cm (Figure 5·10). The density of solid sodium chloride is 2.165 g/cm^3. Thus by calculating the gram formula mass to four significant figures, you can determine Avogadro's number. What value do you obtain?

Research Projects

1. Develop an illustration or a model to make clear to fellow students the magnitude of Avogadro's number.

2. Look up the fertilizer requirements for several kinds of plants. Record the price and percent composition of various brands of fertilizers. Calculate which ones would be the most economical for specific plant types.

3. Make data tables that list various properties of several compounds with identical empirical formulas. Can you find any patterns that hold true for several series of empirical formulas? How would you explain these patterns?

4. Describe one method of determining Avogadro's number using enough detail so that one of your classmates could reproduce the determination.

5. Devise a method to determine the density of carbon dioxide gas. Carbon dioxide can be conveniently produced by dropping an Alka-Seltzer tablet in water.

Readings and References

Dean, J. A. *Lange's Handbook of Chemistry,* 12th ed. New York: McGraw-Hill, 1978.

O'Connor, Rod, and Charles Mickey. *Solving Problems in Chemistry: With Emphasis on Stoichiometry and Equilibrium.* New York: Harper & Row, 1974.

Taylor, L. B., Jr. *Chemistry Careers.* New York: Watts, 1978.

Young, Jay A. "The Exploding Tire." *ChemMatters* (April 1988), pp. 12–14.

7 Chemical Reactions

Chapter Preview

7·1 Writing Chemical Equations
7·2 Balancing Chemical Equations
7·A Passing the Torch for Chemistry: St. Elmo Brady
7·3 Combination Reactions
7·4 Decomposition Reactions
7·5 Single-Replacement Reactions
7·6 Double-Replacement Reactions
7·B Qualitative and Quantitative Analysis
7·7 Combustion Reactions
7·C What's in a Flame?
7·8 Summary of the Types of Reactions
7·D The Chemistry of Firefighting

In Chapter 5 you learned the vocabulary of chemistry by learning how to write chemical formulas and name chemical compounds. We will now use this vocabulary to write chemical sentences called chemical equations. Chemical equations describe chemical reactions. To conform to the law of conservation of mass, equations must be balanced. While practicing your skills in balancing equations you will learn about five general types of reactions. Knowing these reaction types will allow you to predict what will happen when substances undergo a chemical change.

7·1 Writing Chemical Equations

Chemical reactions are continually taking place in us and around us. When our bodies digest food, a whole series of complex chemical reactions occur. A useful chemical reaction also takes place in a car battery to produce energy to start the car. At the same time, an unwanted chemical reaction may be taking place between the iron in the car fender and the oxygen in the air. Rust is being produced. All chemical reactions, whether simple or complex, involve changes in substances. One or more starting substances, the reactants, are changed into one or more new substances, the products.

$$\text{Reactants} \longrightarrow \text{products}$$

Figure 7·1
Copper reacts with nitric acid to produce nitric oxide (a colorless gas) and copper(II) nitrate (in solution). The nitric oxide reacts with the oxygen in air to produce nitrogen dioxide, the brown gas shown here.

143

Figure 7·2
The ingredients for baking are reactants. They undergo physical and chemical changes and combine to form the product, in this case, blueberry muffins.

In a chemical reaction the ways in which atoms are joined together are changed. Bonds are broken and new bonds are formed as reactants are converted into products. The atoms are not created or destroyed. They are just rearranged.

Chemical reactions can be described in different ways. For the example of rusting we could say: "Iron reacts with oxygen to produce iron(III) oxide (rust)." Alternatively, we could identify the reactants and product in this reaction by writing a word equation.

$$\text{Iron + oxygen} \longrightarrow \text{iron(III) oxide}$$

In a word equation, the reactants are written on the left, and the products are written on the right. They are connected by an arrow ($\rightarrow$) that is read as "yields" or "reacts to produce." Other forms of the yields sign and their uses are shown in Table 7·1.

Table 7·1	Symbols Used in Equations
Symbol	**Explanation**
$+$	Used to separate two reactants or two products
$\longrightarrow$	"Yields," separates reactants from products
$=$	An alternative to $\longrightarrow$
$\rightleftarrows$	Used in place of a $\longrightarrow$ for reversible reactions (Chapter 17)
(s)	Designates a reactant or product in the solid state; placed after the formula
$\downarrow$	Alternative to (s); used only for a solid product (precipitate)
(l)	Designates a reactant or product in the liquid state; placed after the formula
(aq)	Designates an aqueous solution; the substance is dissolved in water
(g)	Designates a reactant or product in the gaseous state; placed after the formula
$\uparrow$	Alternative to (g); used only for a gaseous product
$\xrightarrow{\Delta \quad \text{heat}}$	Indicates that heat is supplied to the reaction
$\xrightarrow{\text{Pt}}$	A formula written above or below the yield sign indicates its use as a catalyst (in this example, platinum)

■ A chemical equation is a concise representation of a chemical reaction.

Chemists generally do not use equations in skeleton form. Writing a skeleton equation is only the first step in obtaining a correctly balanced chemical equation.

Many industrial processes require the use of a catalyst. For example, catalysts are used to convert certain otherwise unusable portions of petroleum into efficient fuels for motor vehicles.

A chemical equation shows the rearrangement of atoms that takes place as a result of chemical change.

Word equations are cumbersome. To communicate more effectively, chemists use chemical formulas for writing equations. *In a **chemical equation,** the formulas of the reactants (on the left) are connected by an arrow with the formulas of the products (on the right).* For instance, the chemical equation for rusting is shown here.

$$Fe + O_2 \longrightarrow Fe_2O_3$$

Such equations, which show just the formulas of the reactants and products, are skeleton equations. *A **skeleton equation** is a chemical equation that does not indicate the relative amounts of the reactants and products.*

The physical state of a substance in a reaction can be indicated in the equation. Symbols are used after each formula: (s) for a solid, (l) for a liquid, (g) for a gas, and (aq) for a solution in water.

$$Fe(s) + O_2(g) \longrightarrow Fe_2O_3(s)$$

In many chemical reactions, a catalyst is employed. *A **catalyst** is a substance that speeds up a reaction without being used up.* Because a catalyst is neither a reactant nor a product, it is written above the arrow in a chemical equation. For example, manganese dioxide (MnO_2) catalyzes the decomposition of an aqueous solution of hydrogen peroxide (H_2O_2).

$$H_2O_2(aq) \xrightarrow{MnO_2} H_2O(l) + O_2(g)$$

To write a skeleton equation, you must write the correct formulas of the reactants and products. Put the reactants to the left of the yield sign and the products to the right. Many of the symbols that are commonly used in writing chemical equations are listed in Table 7·1.

a

b
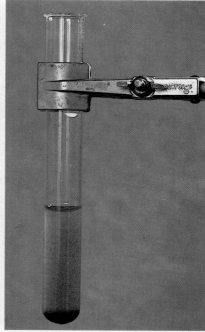

Figure 7·3
Hydrogen peroxide slowly decomposes to form water and oxygen. **a** Manganese dioxide (MnO_2), a black powder, speeds up this reaction, causing bubbles to form. **b** Since it is not used up in the reaction, MnO_2 is a catalyst.

Example 1

Write a skeleton equation for this chemical reaction: When calcium carbonate is heated, calcium oxide and carbon dioxide are produced.

Solution

$$CaCO_3(s) \xrightarrow{\Delta} CaO(s) + CO_2 \uparrow$$

Problem

1. Write a skeleton equation for each of these chemical reactions.
   a. Aluminum metal reacts with oxygen in the air to form aluminum oxide.
   b. When solid mercury(II) sulfide is heated with oxygen, liquid mercury metal and gaseous sulfur dioxide are produced.
   c. Oxygen gas can be made by heating potassium chlorate in the presence of the catalyst manganese dioxide. Potassium chloride is left as a solid residue.

Example 2

Write a sentence that completely describes the chemical reaction shown in this skeleton equation.

$$NaHCO_3(s) + HCl(aq) \longrightarrow NaCl(aq) + H_2O(l) + CO_2 \uparrow$$

Solution

Solid sodium hydrogen carbonate reacts with hydrochloric acid to produce aqueous sodium chloride, water, and carbon dioxide gas.

Problem

2. Write sentences that completely describe each of the chemical reactions shown in these skeleton equations.
   a. $BaCl_2(aq) + H_2SO_4(aq) \longrightarrow BaSO_4 \downarrow + HCl(aq)$
   b. $NH_3(g) + O_2(g) \xrightarrow{Pt} NO \uparrow + H_2O(g)$
   c. $N_2O_3(g) + H_2O(l) \longrightarrow HNO_2(aq)$

7·2 Balancing Chemical Equations

All the equations we have written thus far have been correct in a qualitative sense. They have shown what has happened in terms of reactants changing to products. They have not been correct in a quantitative sense, however. This is because the equations have not shown the amounts of reactants and products. To represent chemical reactions correctly, equations must be balanced so that they are quantitatively correct. *In every*

■ A balanced chemical equation obeys the law of conservation of matter.

balanced equation, *each side of the equation has the same number of atoms of each element.* This is necessary to be consistent with the law of conservation of mass. Remember, in a chemical reaction, atoms are not created or destroyed; they are simply rearranged.

Sometimes when we write the formulas for the reactants and products in an equation, the equation is already balanced. This is true of the equation for the burning of carbon in the presence of oxygen to produce carbon dioxide.

$$C(s) + O_2(g) \longrightarrow CO_2(g)$$

The equation is balanced. One carbon atom is on each side of the equation, and two oxygen atoms are on each side. Carbon can also react with oxygen to produce carbon monoxide.

$$C(s) + O_2(g) \longrightarrow CO(g)$$

This is a correct skeleton equation but it is not balanced. It does not obey the law of conservation of mass.

Another equation we have seen is the reaction of hydrogen with oxygen to produce water.

$$H_2(g) + O_2(g) \xrightarrow{\text{Pt}} H_2O(l)$$

This equation is not balanced even though the formulas for all the reactants and products are correct (Figure 7·4). There are two oxygen atoms on the reactant (left) side of the equation and only one oxygen atom on the product (right) side. The equation does not obey the law of conservation of mass. What can we do? Many chemical equations can be balanced by trial and error. A few guidelines, however, will speed the process. Let's look at these rules and then use them to balance the equation for the formation of water.

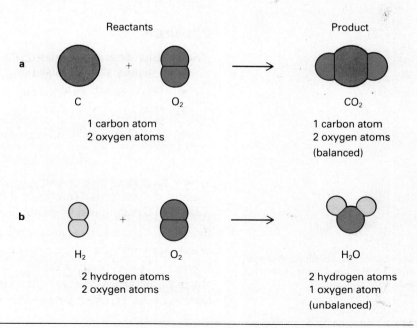

Figure 7·4
a A balanced chemical equation has the same number and type of atoms on each side. **b** An unbalanced equation does not have the same number and type of atoms on each side. Which kind of atom is not balanced in **b**?

Rules for Balancing Equations

1. Determine the correct formulas for all the reactants and products in the reaction.

2. Write the formulas for the reactants on the left and the formulas for the products on the right with an arrow in between. If two or more reactants or products are involved, separate their formulas with plus signs.

3. Count the number of atoms of each element in the reactants and products. A polyatomic ion appearing unchanged on both sides of the equation is counted as a single unit.

4. Balance the elements one at a time by using coefficients. *A coefficient is a small whole number that appears in front of a formula in an equation.* When no coefficient is written, it is assumed to be 1. It is best to begin with an element other than hydrogen or oxygen. These two elements often occur more than twice in an equation. *You must not attempt to balance an equation by changing the subscripts in the chemical formula of a substance.*

5. Check each atom or polyatomic ion to be sure that the equation is balanced.

6. Finally, make sure that all the coefficients are in the lowest possible ratio.

Now let's use these rules to balance the equation for the formation of water from hydrogen and oxygen.

Example 3

When hydrogen and oxygen react, the product is water. Write a balanced equation for this reaction.

Solution

Since the chemical formulas of the reactants and products are known, we can write a skeleton equation.

$$H_2(g) + O_2(g) \longrightarrow H_2O(l)$$

Hydrogen is balanced but oxygen is not. If we put a coefficient of 2 in front of H_2O, the oxygen becomes balanced.

$$H_2(g) + O_2(g) \longrightarrow 2H_2O(l)$$

Now there are twice as many hydrogen atoms in the product as there are in the reactants. To correct this, put a coefficient of 2 in front of H_2. The equation is now balanced.

$$2H_2(g) + O_2(g) \longrightarrow 2H_2O(l)$$

Check the coefficients. They must be in their lowest possible ratio: $2(H_2)$, $1(O_2)$, and $2(H_2O)$. Figure 7·5 illustrates this balanced chemical equation.

Coefficients in a balanced chemical equation that are 1's are usually not written.

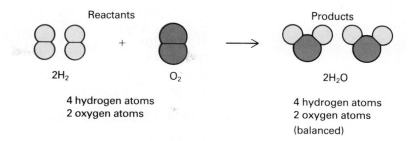

Figure 7·5
A balanced chemical equation obeys the law of conservation of mass. The same number of each kind of atom is on each side of the equation.

Reactants

$2H_2$ O_2

4 hydrogen atoms
2 oxygen atoms

Products

$2H_2O$

4 hydrogen atoms
2 oxygen atoms
(balanced)

When the same polyatomic ion is present as a reactant and a product, it is easier to balance the polyatomic ion than the individual atoms.

Figure 7·6
Copper reacts with silver nitrate to form copper nitrate and silver. Notice the silver plating on the copper strip and the characteristic blue-green color of the copper(II) ion in solution.

Never change the formula of a substance when balancing an equation.

Example 4

Balance the following equation.

$$AgNO_3(aq) + Cu(s) \longrightarrow Cu(NO_3)_2(aq) + Ag(s)$$

Solution

Since this reaction involves the polyatomic nitrate ion, we can save time if we consider the nitrate ion as a unit. Put a coefficient of 2 in front of $AgNO_3$ to balance the nitrate ion.

$$2AgNO_3(aq) + Cu(s) \longrightarrow Cu(NO_3)_2(aq) + Ag(s)$$

By inspection we see that the silver is not balanced. Put a coefficient of 2 in front of Ag.

$$2AgNO_3(aq) + Cu(s) \longrightarrow Cu(NO_3)_2(aq) + 2Ag(s)$$

Example 5

Balance the following equation.

$$Al(s) + O_2(g) \longrightarrow Al_2O_3(s)$$

Solution

First balance the aluminum by adding a coefficient of 2 in front of Al.

$$2Al(s) + O_2(g) \longrightarrow Al_2O_3(s)$$

We now have a situation that occurs quite frequently in balancing equations. We might call it the even-odd problem. Any whole-number coefficient placed in front of the O_2 will give an even number of oxygen atoms on the left. This is because we are always multiplying the coefficient by the subscript 2. The problem is that the odd number of oxygen atoms on the right has to be made even to be able to balance. The simplest way to do this is to multiply the formula with the odd number of oxygen atoms by 2.

$$2Al(s) + O_2(g) \longrightarrow 2Al_2O_3(s)$$

Now there are six oxygens on the right. Balance the oxygens on the left with a 3 and rebalance the aluminum on the left with a 4.

$$4Al(s) + 3O_2(g) \longrightarrow 2Al_2O_3(s)$$

Suppose we had written the equation for the formation of aluminum oxide as follows.

$$8Al(s) + 6O_2(g) \longrightarrow 4Al_2O_3(s)$$

This equation is balanced but incorrect because the coefficients are not in their lowest possible ratio. Each of them can be divided by 2.

Problems

Be sure to count atoms correctly. $3Fe_2(SO_4)_3$ indicates 6 Fe atoms, 9 S atoms, and 36 O atoms.

3. Balance the following equations.
   a. $Al + N_2 \longrightarrow AlN$
   b. $NaCl + H_2SO_4 \longrightarrow Na_2SO_4 + HCl$
   c. $Al + CuSO_4 \longrightarrow Al_2(SO_4)_3 + Cu$
   d. $P + O_2 \longrightarrow P_4O_{10}$
   e. $Fe(OH)_3 \longrightarrow Fe_2O_3 + H_2O$

4. Rewrite these word equations as balanced chemical equations.
   a. sodium + water $\longrightarrow$ sodium hydroxide + hydrogen
   b. hydrogen + sulfur $\longrightarrow$ hydrogen sulfide
   c. iron(III) chloride + calcium hydroxide $\longrightarrow$
                         iron(III) hydroxide + calcium chloride
   d. carbon + oxygen $\longrightarrow$ carbon monoxide
   e. potassium nitrate $\longrightarrow$ potassium nitrite + oxygen

Figure 7·7
St. Elmo Brady's research centered on the chemistry of plants and their products.

=== Science, Technology, and Society ===

7·A Passing the Torch for Chemistry: St. Elmo Brady

In 1916 St. Elmo Brady became the first black man to be awarded a doctorate in chemistry. In that year he left the University of Illinois to begin a career dedicated to inspiring young people in the study of science. Dr. Brady studied plant chemistry under Dr. George Washington Carver at Tuskegee Institute. This remained one of his research interests, along with other branches of organic chemistry. In the course of his career, he developed departments of chemistry at Tuskegee Institute, Howard University, and Fisk University. After his retirement from Fisk in 1952, he went to Tougaloo College in Mississippi to start yet another department of chemistry. Dr. Brady was known for his dedication to research and his love of teaching.

Much of the research in chemistry is done by university professors, assisted by their students. Many brilliant men and women combine scholarship in chemistry with dedication to teaching. University teachers do laboratory research, write scholarly papers, teach several courses, and advise undergraduate and graduate students. They may also have administrative duties. University teaching generally requires a doctorate in chemistry.

ChemDirections

■ The reaction of two or more substances to form a single product is a combination reaction.

Combination reactions usually liberate energy as they take place.

7·3 Combination Reactions

We can only know with certainty what the products of a chemical reaction are by carrying out that reaction in the laboratory. The reactants must be allowed to react and the products of this chemical reaction must be identified. Carrying out each reaction in the laboratory is the ideal, but it is both time consuming and costly. It is possible, however, to predict the products of some chemical reactions. To achieve this, you must be able to recognize various types of reactions. Five general types of reactions are: combination, decomposition, single-replacement, double-replacement, and combustion.

As the name implies, *in a* **combination reaction** *two or more substances react to form a single substance* (Figure 7·8). The reactants of most common combination reactions are either two elements or two compounds. One of the reactants is often water. The product of a combination (or synthesis) reaction must be a compound. Two nonmetals can often combine in more than one way. Thus for combination reactions involving nonmetals you will usually need to be told what the product is. Combination reactions usually liberate energy as they take place.

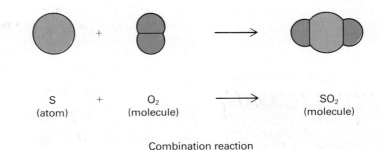

| S (atom) | + | O_2 (molecule) | → | SO_2 (molecule) |

Combination reaction

Figure 7·8
The elements sulfur and oxygen *combine* to form the compound sulfur dioxide, SO_2.

Figure 7·9
When it is ignited, sulfur will combine with oxygen to form sulfur dioxide, a colorless gas.

A general equation for a combination reaction is:

$$R + S \longrightarrow RS$$

This reaction takes place when sulfur-containing fuels are burned. Sulfur trioxide in the atmosphere causes acid rain.

 Issues in Chemistry

Times Beach, Missouri, was a residential community with no industrial plants but dioxin contamination made it a ghost town. How is the issue of hazardous waste disposal handled in your community? What rights (and responsibilities) do you have in decisions involving hazardous waste disposal?

Example 6

Complete the following combination reactions.
a. $Al(s) + O_2(g) \longrightarrow$
b. $S(s) + O_2(g) \longrightarrow$
c. $Cu(s) + S(s) \longrightarrow$
d. $SO_3(g) + H_2O(l) \longrightarrow$
e. $CaO(s) + H_2O(l) \longrightarrow$

Solution

a. Write the correct formula for the compound aluminum oxide by using ionic charges. Then balance the equation.

$$Al(s) + O_2(g) \longrightarrow Al_2O_3(s) \text{ (unbalanced)}$$

$$4Al(s) + 3O_2(g) \longrightarrow 2Al_2O_3(s)$$

b. One reaction between sulfur and oxygen gives SO_2.

$$S(s) + O_2(g) \longrightarrow SO_2(g) \text{ (balanced)}$$

Another possible reaction is the formation of SO_3.

$$S(s) + O_2(g) \longrightarrow SO_3(g) \text{ (unbalanced)}$$

$$2S(s) + 3O_2(g) \longrightarrow 2SO_3(g)$$

c. Two reactions are possible because copper has more than one common ionic charge (Section 5·6).

$$Cu(s) + S(s) \longrightarrow CuS(s) \text{ [copper(II)]}$$

$$2Cu(s) + S(s) \longrightarrow Cu_2S(s) \text{ [copper(I)]}$$

d. Reactions between water and nonmetal oxides usually give an acid. Often we can get the formula of the acid by "adding" the two molecules together.

$$SO_3(g) + H_2O(l) \longrightarrow H_2SO_4(aq)$$

e. A metal oxide reacts with water to give a metal hydroxide called a base. Use ionic charges to write the formula of the compound.

$$CaO(s) + H_2O(l) \longrightarrow Ca(OH)_2(aq)$$

Problem

5. Write balanced chemical equations for the following combination reactions. Refer to the solutions in Example 6 to help you determine the products.
   a. $Ca + S \longrightarrow$
   b. $Fe + O_2 \longrightarrow$ iron(II) oxide
   c. $P + O_2 \longrightarrow$ tetraphosphorus decoxide
   d. $N_2O_5 + H_2O \longrightarrow$
   e. $Na_2O + H_2O \longrightarrow$
   f. $Mg + O_2 \longrightarrow$

7·4 Decomposition Reactions

A single substance is broken down into two or more simpler substances in a decomposition reaction.

Products of decomposition reactions are elements such as O_2 and H_2 or simple compounds such as H_2O and CO_2.

In a **decomposition reaction** *a single compound is broken down into two or more simpler products* (Figure 7·10). These products can be any combination of elements and compounds. It is usually very difficult to predict the products of decomposition reactions. When a simple binary compound breaks down, however, you know the products will be the constituent elements. Most decomposition reactions require energy in the form of heat, light, or electricity.

Figure 7·10
The compound calcium carbonate, $CaCO_3$, *decomposes* into simpler compounds: calcium oxide, CaO, and carbon dioxide, CO_2. This reaction is used to make lime (calcium oxide) from limestone (primarily calcium carbonate).

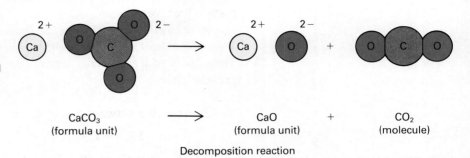

CaCO₃ (formula unit) ⟶ CaO (formula unit) + CO₂ (molecule)

Decomposition reaction

A general equation for a decomposition reaction is:

$$RS \longrightarrow R + S$$

Example 7

Write balanced equations for each of these decomposition reactions.

a. $H_2O(l) \xrightarrow{\text{electricity}}$

b. mercury(II) oxide $\xrightarrow{\Delta}$

Solution

a. Water, a binary compound, breaks down into its elements. Reminder: hydrogen and oxygen are both diatomic.

$$H_2O(l) \xrightarrow{\text{electricity}} H_2(g) + O_2(g) \quad \text{(unbalanced)}$$

$$2H_2O(l) \xrightarrow{\text{electricity}} 2H_2(g) + O_2(g)$$

b. $HgO(s) \xrightarrow{\Delta} Hg(l) + O_2(g) \quad \text{(unbalanced)}$

$$2HgO(s) \xrightarrow{\Delta} 2Hg(l) + O_2(g)$$

Problem

6. Write balanced chemical equations for each of these decomposition reactions.

a. nickel(II) carbonate $\xrightarrow{\Delta}$ nickel(II) oxide + _____

b. $Ag_2O \xrightarrow{\Delta}$

c. ammonium nitrate $\xrightarrow{\Delta}$ dinitrogen monoxide + water

7·5 Single-Replacement Reactions

One element displaces another element from a compound in a single-replacement reaction.

In a **single-replacement reaction** *atoms of an element replace the atoms of a second element in a compound* (Figure 7·11). These reactions are also called displacement reactions. Whether one metal will replace another metal from a compound can be determined by the relative reactivities of the two metals. *The* **activity series of metals** (Table 7·2) *lists metals in order of decreasing reactivity*. A reactive metal will replace any metal found below it in the activity series. Thus sodium would replace zinc or silver from a compound. By contrast, sodium would not replace lithium or calcium.

A nonmetal can also replace another nonmetal from a compound. This replacement is usually limited to the halogens (F_2, Cl_2, Br_2, and I_2). The activity of the halogens decreases as you go down Group 7A on the periodic table.

$$\text{Decreasing activity} \quad \left| \begin{array}{l} F_2 \\ Cl_2 \\ Br_2 \\ I_2 \end{array} \right. \downarrow$$

General equations for single-replacement reactions are:

$$T + RS \longrightarrow TS + R$$

or

$$U + RS \longrightarrow RU + S$$

Example 8

Write balanced chemical equations for each of these single-replacement reactions.

a. $Zn(s) + H_2SO_4(aq) \longrightarrow$
b. $K(s) + H_2O(l) \longrightarrow$
c. $Sn(s) + NaNO_3(aq) \longrightarrow$
d. $Cl_2(g) + NaBr(aq) \longrightarrow$

Solution

a. According to the activity series of metals, zinc will displace hydrogen from an acid and take its place. Hydrogen as an element is diatomic.

$$Zn(s) + H_2SO_4(aq) \longrightarrow ZnSO_4(aq) + H_2 \uparrow$$

Figure 7·11
Iron is above copper in the activity series. The iron *replaces* the copper ions, forming a new compound, iron(II) sulfate, plus copper metal.

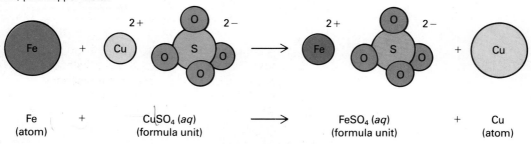

Single-replacement reaction

Figure 7·12
The alkali metals such as potassium (shown here) displace hydrogen from water in a *single-replacement* reaction. The heat of the reaction then ignites the hydrogen. Because these metals react so violently with water, they must be stored in mineral oil or kerosene.

Table 7·2	Activity Series of Metals	
	Name	Symbol
	Lithium	Li
	Potassium	K
	Barium	Ba
	Calcium	Ca
	Sodium	Na
	Magnesium	Mg
	Aluminum	Al
	Zinc	Zn
	Iron	Fe
	Nickel	Ni
	Tin	Sn
	Lead	Pb
	(Hydrogen)	(H)*
	Copper	Cu
	Mercury	Hg
	Silver	Ag
	Gold	Au

Decreasing activity (arrow pointing down, left side of table)

*Metals from Li to Na will replace H from acids and water; from Mg to Pb they will replace H from acids only.

b. Potassium will displace hydrogen from water. It is helpful to write water as HOH and visualize it as being made of H^+ and OH^-.

$$K(s) + H^+OH^-(l) \longrightarrow KOH(aq) + H_2 \uparrow \quad \text{(unbalanced)}$$

$$2K(s) + 2HOH(l) \longrightarrow 2KOH(aq) + H_2 \uparrow$$

c. No reaction occurs because tin is less reactive than sodium.
d. Chlorine is more reactive than bromine and will displace bromine from a compound. Bromine is diatomic and a liquid.

$$Cl_2(g) + NaBr(aq) \longrightarrow NaCl(aq) + Br_2(l) \quad \text{(unbalanced)}$$

$$Cl_2(g) + 2NaBr(aq) \longrightarrow 2NaCl(aq) + Br_2(l)$$

Problem

7. Use the activity series of metals to write balanced chemical equations for each of these single-replacement reactions.
   a. $Al + H_2SO_4 \longrightarrow$
   b. $Cl_2 + KI \longrightarrow$
   c. $Cu + FeSO_4 \longrightarrow$
   d. $Li + H_2O \longrightarrow$

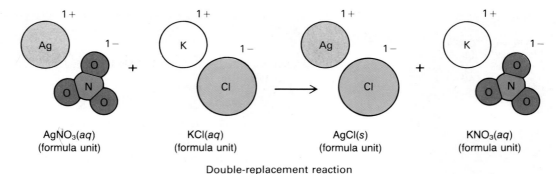

| $AgNO_3(aq)$ (formula unit) | $KCl(aq)$ (formula unit) | $AgCl(s)$ (formula unit) | $KNO_3(aq)$ (formula unit) |

Double-replacement reaction

Figure 7·13
The silver and potassium ions exchange places in this *double-replacement* reaction. Two new compounds are formed: silver chloride and potassium nitrate. The reaction occurs because silver chloride forms a precipitate.

Two ionic compounds exchange cations (or anions) in a double-replacement reaction.

A general equation for a double-replacement reaction is:

$$RS + TU \longrightarrow RU + TS$$

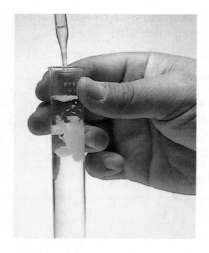

Figure 7·14
Solutions of barium chloride and potassium carbonate react in a double-replacement reaction to form barium carbonate, the white precipitate. Potassium chloride, the other product of the reaction, remains in solution.

7·6 Double-Replacement Reactions

Double-replacement reactions *involve an exchange of positive ions between two compounds* (Figure 7·13). These reactions generally take place between two ionic compounds in aqueous solution. For a double-replacement reaction to occur, one of the following statements is usually true concerning at least one of the products of the reaction. (a) It is only slightly soluble and precipitates from solution. (b) It is a gas that bubbles out of the mixture. (c) It is a molecular compound such as water.

Example 9

Write balanced chemical equations for each of these double-replacement reactions.
a. $NaOH(aq) + H_2SO_4(aq) \longrightarrow$
b. $BaCl_2(aq) + K_2CO_3(aq) \longrightarrow$
c. $FeS(s) + HCl(aq) \longrightarrow$

Solution

a. This reaction takes place because of the formation of water as one of the products.

$$NaOH(aq) + H_2SO_4(aq) \longrightarrow Na_2SO_4(aq) + H_2O(l)$$
(unbalanced)

$$2NaOH(aq) + H_2SO_4(aq) \longrightarrow Na_2SO_4(aq) + 2H_2O(l)$$

b. The driving force behind the reaction is the formation of a precipitate.

$$BaCl_2(aq) + K_2CO_3(aq) \longrightarrow BaCO_3 \downarrow + KCl(aq)$$
(unbalanced)

$$BaCl_2(aq) + K_2CO_3(aq) \longrightarrow BaCO_3 \downarrow + 2KCl(aq)$$

c. A gas is formed in this double-replacement reaction.

$$FeS(s) + HCl(aq) \longrightarrow H_2S \uparrow + FeCl_2(aq) \text{ (unbalanced)}$$

$$FeS(s) + 2HCl(aq) \longrightarrow H_2S \uparrow + FeCl_2(aq)$$

Remember to use ionic charges to write correct formulas for the products.

Problem

8. Write a balanced equation for each of these reactions.
   a. $H_2SO_4 + Al(OH)_3 \longrightarrow$
   b. $KOH + H_3PO_4 \longrightarrow$
   c. $SrBr_2 + (NH_4)_2CO_3 \longrightarrow$

—— Science, Technology, and Society ——

7·B Qualitative and Quantitative Analysis

Finding out what is in a chemical sample is the job of the analytical chemist. The analysis may determine what elements are in a particular compound, or what compounds are in a mixture. Perhaps only a particular grouping of atoms within a large molecule is of interest. If the results must show the exact amount of each substance present, the process is called *quantitative analysis*. If the results show only what those substances are, the process is called *qualitative analysis*.

Most high school chemistry courses include a qualitative analysis experiment. Generally, this involves identifying a compound, or part of a compound, by doing simple chemical tests. These usually include single-replacement and double-replacement reactions. The tests are called "wet chemical methods" because they involve mixing chemicals in ordinary glassware and drawing conclusions from the visible reactions. Wet chemical methods used in *quantitative* analysis require very accurate mass and volume measurements. After a substance has been isolated by a chemical reaction, it must be collected very carefully and measured exactly. Professional chemists use some wet chemical methods, but most analytical work today is done with instruments.

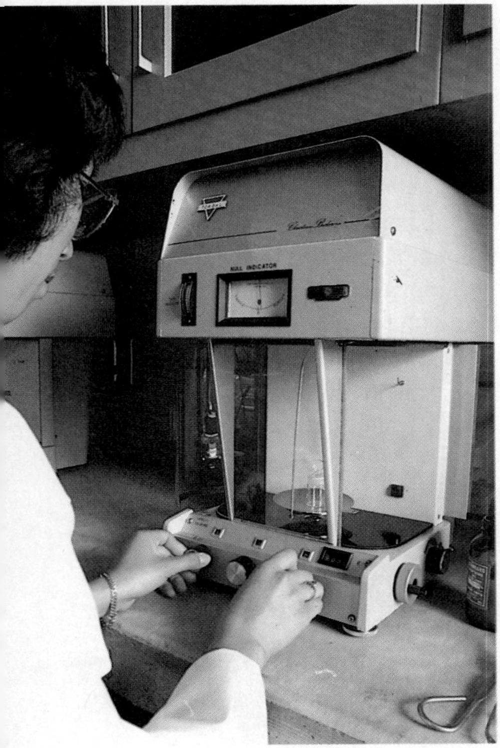

The analytical chemist must often work with an extremely small sample. If the mass of the sample is greater than 0.1 gram, the work is called macroanalysis. If the sample mass is between 0.01 and 0.1 gram, it is called semimicroanalysis. There are also categories of micro- and submicroanalysis. The last category deals with samples that weigh less than 1×10^{-6} gram! A balance used for submicroanalysis can detect 2×10^{-8} grams.

Many instrumental methods work by detecting the type of light or other radiant energy given off or absorbed by atoms or molecules. These patterns of energy, called spectra, identify a chemical as fingerprints identify a person. (See Chapter 11 for more information on spectra.) The mass spectrometer is used to identify fragments of molecules by their mass. This is often helpful in analyzing large organic molecules such as drugs. (See Section 4·D for a description of the mass spectrometer.) Many other techniques detect characteristic behavior that occurs when molecules or atoms are subjected to

Figure 7·15
This balance is used for microanalysis because it has a precision of 0.001 mg or 10^{-6} g.

various conditions. Some of these techniques are described in later chapters (see Sections 11·B and 16·A).

Crime laboratories, medical centers, academic and government research facilities, and chemical industries of all types employ analytical chemists. A forensic chemist is asked to analyze a paint chip to determine its lead content. A surgeon requests constant monitoring of the oxygen and carbon dioxide in a patient's blood. A professor wants to determine the products of a newly devised chemical reaction. A government water-quality chemist analyzes lake water for mercury. A pharmaceutical chemist must determine the exact molecular structure of a new drug. All these situations require the skills of an analytical chemist. Analytical chemists may also specialize in designing new instruments and methods of analysis.

A person can prepare for a career in analytical chemistry by obtaining a college degree in chemistry with emphasis on analytical methods and instrumentation. A knowledge of computers is also essential because they are used to operate many instruments and to process the results.

7·7 Combustion Reactions

■ A combustion reaction occurs when a substance is burned in oxygen.

In a **combustion reaction** *oxygen reacts with another substance, often producing energy in the form of heat and light.* Combustion reactions commonly involve hydrocarbons which are compounds of hydrogen and carbon. The complete combustion of a hydrocarbon produces the compounds carbon dioxide and water. Figure 7·16 shows the combustion of methane, the major component of natural gas. If sufficient oxygen is not available during the course of a reaction, combustion will be incomplete. Elemental carbon and poisonous carbon monoxide may be additional products. A combustion reaction might be viewed as a variation of a combination reaction. The oxygen combines with both the carbon and the hydrogen. The complete combustion of a hydrocarbon produces a large amount of energy (Section 8·7). For this reason, hydrocarbons such as methane (CH_4), propane (C_3H_8), butane (C_4H_{10}), and octane (C_8H_{18}) are important fuels.

Breathing carbon monoxide is dangerous to humans because it interferes with the oxygen transport by red blood cells.

Figure 7·16
Carbon dioxide and water are the common products of the *combustion* of methane and other compounds containing carbon and hydrogen.

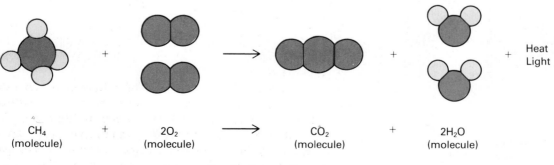

| CH_4 | + | $2O_2$ | ⟶ | CO_2 | + | $2H_2O$ | + | Heat Light |
| (molecule) | | (molecule) | | (molecule) | | (molecule) | | |

Combustion reaction

The general equation for combustion of a hydrocarbon is:

$$C_xH_y + O_2 \longrightarrow CO_2 + H_2O$$

Safety

Combustion reactions can be very rapid, even explosive. In pure oxygen, combustion takes place much faster than in air. Use extreme care when working around a source of oxygen.

Example 10

Write a balanced equation for the complete combustion of these fuels.
a. benzene, $C_6H_6(l)$ **b.** methyl alcohol, $CH_3OH(l)$

Solution

a. Oxygen is the other reactant in a combustion reaction. The products are CO_2 and H_2O.

$$C_6H_6(l) + O_2(g) \longrightarrow CO_2(g) + H_2O(g) \text{ (unbalanced)}$$

$$2C_6H_6(l) + 15O_2(g) \longrightarrow 12CO_2(g) + 6H_2O(g)$$

b. $2CH_3OH(l) + 3O_2(g) \longrightarrow 2CO_2(g) + 4H_2O(g)$

Problem

9. Write a balanced equation for the complete combustion of these compounds. **a.** glucose, $C_6H_{12}O_6$ **b.** octane, C_8H_{18}

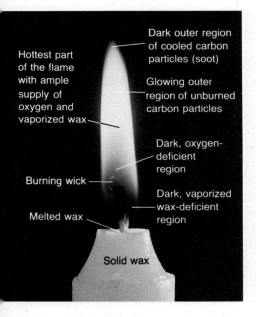

Hottest part of the flame with ample supply of oxygen and vaporized wax

Burning wick

Melted wax

Dark outer region of cooled carbon particles (soot)

Glowing outer region of unburned carbon particles

Dark, oxygen-deficient region

Dark, vaporized wax-deficient region

Solid wax

Figure 7·17
The different colors of a candle flame represent different stages in the combustion process.

Safety

When working with flames, be sure that there is no possibility of loose clothing or hair falling into the flame. Do not reach across a flame.

7·C What's in a Flame?

When you light a candle, you see a common combustion reaction. Wax vaporizes and reacts with oxygen in the air to produce heat, light, and combustion products. This reaction, however, takes place in different ways in different parts of the flame. A simple candle flame involves a series of complex reactions that occur in distinct regions.

When a lit match is held near the wick of an unlit candle, the wax begins to melt and vaporize. The heat of the match starts the combustion reaction of candle wax and oxygen. This reaction gives off heat, which keeps the reaction going even after the match has been taken away. Vaporized wax—the fuel—rises upward through the flame, while the air supply is drawn upward into the flame from below the burning wick. Thus the upper region of the flame has less oxygen than the area near the wick. This means that some of the vaporized wax cannot be completely burned in the upper part of the flame. Instead it remains as tiny carbon particles, which glow in the heat of the flame. These glowing particles give the flame a bright yellow color. Some of the particles may leave the flame as soot.

By contrast, combustion takes place more completely in the upper region of the flame where there is more oxygen. Here, in the hottest part of the flame, the reaction products are mostly carbon dioxide and water.

Unlike a candle flame, the flame of a Bunsen burner can be changed from one kind to another. A Bunsen burner is also different because it burns methane, a fuel that is gaseous to begin with. If the air vents of a Bunsen burner are completely shut, air can only mix with the gaseous fuel by diffusing into the flame. Under this condition the flame is like that of a candle and has the same bright yellow color. If

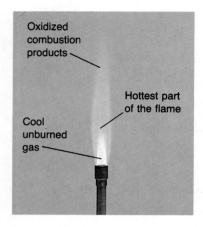

Oxidized combustion products

Hottest part of the flame

Cool unburned gas

Figure 7·18
A Bunsen burner flame has three distinct regions: a cool dark inner area of unburned gas, a hot middle cone where combustion takes place, and a cooler outer cone where the combustion products form carbon dioxide and water.

Recognizing reactions by type can be helpful in the prediction of the products of a chemical reaction.

the air vents are open, however, air is drawn in at the bottom of the burner tube. It mixes with the gas as it passes up the tube, and the mixture that comes out at the top of the burner has sufficient oxygen for nearly complete combustion. This is called a premixed flame.

The innermost dark cone of this flame is cold unburned methane and air. A match suspended in this area of the flame will not light, even though it is surrounded by the flame. In the bright blue region outside the innermost cone, the premixed fuel and oxygen are burning. This is the hottest part of the flame. Even here, however, there is not enough oxygen for all of the methane to be converted to carbon dioxide and water. Some of the fuel becomes carbon monoxide. This gas leaves the inner cone and enters the outermost area of the flame. Here it combines with air that diffuses into the flame and is burned to give carbon dioxide and water. This outer region is not as hot as the inner, premixed region of the flame.

7·8 Summary of the Types of Reactions

Many chemical reactions can be classified as one of five general types. To classify a reaction as to type, you must look at what happens as reactants are converted into products. In a *combination reaction,* two or more elements or compounds combine chemically to form a new compound. By contrast, a single reactant is the identifying characteristic of a *decomposition reaction.* In a typical decomposition reaction, a compound is decomposed into two or more elements and/or compounds.

When oxygen reacts with a compound composed of the elements carbon and hydrogen (and perhaps, oxygen), a *combustion reaction* takes place. If all the carbon in the product is present as carbon dioxide, the reaction is complete combustion. If carbon monoxide is formed, however, the reaction is incomplete combustion. Whether a combustion reaction is complete or incomplete depends upon the amount of oxygen present. It also depends on reaction conditions, such as temperature.

Single- and double-replacement reactions are the final two general types of reactions. In a *single-replacement reaction* an element replaces another element from a compound. A new compound and a new element are formed. The activity series of metals is used to determine whether a single-replacement reaction will occur. The element that is being displaced must be less active than the element that is doing the displacing for a single replacement reaction to occur.

In a *double-replacement reaction* two ionic compounds react by exchanging cations to form two different compounds. A double-replacement reaction usually takes place in aqueous solution. Double-replacement reactions are driven by the formation of a precipitate, a gaseous product, or water.

Keys to identifying chemical reactions by type and the means of predicting products of a reaction are shown in Table 7·3. Nevertheless, some reactions do not fall into any of these five general types. Another very im-

Table 7·3 Keys to Identifying Types of Chemical Reactions

General equation	Reactants	Probable reaction type	Probable products
$R + S \rightarrow RS$	Two elements	Combination	A single compound
	Two compounds, at least one a molecular compound	Combination	A single compound
$RS \rightarrow R + S$	A single binary compound	Decomposition	Two elements
	A single ternary compound	Decomposition	Two or more elements and/or compounds
$T + RS \rightarrow TS + R$	An element and a compound	Single-replacement	A different element and a new compound
$R^+S^- + T^+U^- \rightarrow R^+U^- + T^+S^-$	Two ionic compounds	Double-replacement	Two new compounds
$C_xH_y + \left(x + \dfrac{y}{4}\right)O_2 \longrightarrow xCO_2 + \dfrac{y}{2}H_2O$	Oxygen and a compound of C, H, (O)	Combustion	CO_2 and H_2O (with incomplete combustion, C and/or CO may be additional products)

portant type of chemical reaction is an oxidation-reduction, or redox reaction. This type of reaction is discussed in Chapters 20 and 21. The only way to determine the products of any reaction with certainty is to carry out the reaction in the laboratory.

Problem

10. After balancing each of these equations, identify them as to type.
   a. $Mg + H_2SO_4 \longrightarrow MgSO_4 + H_2$
   b. $Fe + O_2 \longrightarrow Fe_3O_4$
   c. $Pb(NO_3)_2 \longrightarrow PbO + NO_2 + O_2$
   d. $C_2H_6 + O_2 \longrightarrow CO_2 + H_2O$
   e. $Pb(NO_3)_2 + NaI \longrightarrow PbI_2 + NaNO_3$
   f. $(NH_4)_2SO_4 + NaOH \longrightarrow NH_3 + H_2O + Na_2SO_4$
   (Hint: There are two stepwise reactions in this equation.)

7·D The Chemistry of Firefighting

In North America alone, fire kills more than 20 000 people and damages millions of dollars in property each year. Controlling fires is an important application of chemistry.

Fire, or combustion, is the rapid reaction of a fuel with oxygen. The fuel is heated initially with a match, an electric current, intense light, or a bolt of lightning. Once heated, the fuel is converted into very reactive species, called combustion intermediates. These, in turn, can react with oxygen. Heat is liberated in the reaction, and this heat keeps the reaction going. A fire can be put out in three ways: (1) by cooling the fuel, (2) by cutting off the supply of oxygen, or (3) by trapping the combustion intermediates so they cannot react with oxygen. All three of these chemical principles come into play in firefighting.

Firefighters are called upon most often to fight fires involving a solid fuel—such as the wooden frame of a typical house. This kind of fire is known as a Class A fire.

Water is usually used to put out a Class A fire. It works against the fire in two ways. First, water can cool the fuel below its ignition point. Second, the heat of the fire changes the water to steam, which dilutes the oxygen in the air around the fire. When a Class A fire is caught in time, it can be put out with a pressurized water fire extinguisher. The water in the extinguisher may contain a dissolved salt to keep it from freezing. It may also contain a chemical that helps it "stick" to the burning fuel better. The water is pressurized by carbon dioxide.

Safety

Be sure that you know where the fire extinguishing equipment is in your laboratory and how it should be used.

Figure 7·19
Although controlled combustions can be useful, uncontrolled combustions can be destructive and dangerous.

Figure 7·20
Airplanes can be used to dump fire retardant chemicals on brush and forest fires. The powder also contains a nutrient to help vegetation grow in the burned area.

Figure 7·21
Dry chemical powder fire extinguishers can be used on many kinds of burning fuels, including solids, liquids, and those involving electrical circuits. Some can also be used on fires in which the fuel is a metal.

Water is not always the best fire extinguishing agent. In a Class B fire, where the burning material is a liquid or gas, a stream of water can sometimes spread the fire. Also, because of the salt dissolved in most pressurized water fire extinguishers, the water will conduct electricity. Thus water fire extinguishers are not used in a fire involving an energized electrical circuit, known as a Class C fire. Finally, water can actually be a source of oxygen for a Class D fire, in which a metal is burning.

There are several types of fire extinguishers that can be used on these other kinds of fires. Carbon dioxide extinguishers can be used on Class B and C fires. They cannot be used on Class D fires because carbon dioxide may react with the hot metal. Carbon dioxide extinguishes fires by cutting off the oxygen supply. Because carbon dioxide is more dense than air, it will blanket the burning fire, as long as there is not much wind in the area. Pressurized carbon dioxide extinguishers contain liquid carbon dioxide at a pressure of 50 to 60 atmospheres.

Dry chemical powder fire extinguishers contain one of several types of dry and finely powdered chemicals, pressurized by carbon dioxide. The simplest of these chemicals, sodium hydrogen carbonate, works by generating carbon dioxide when exposed to heat, which shuts off the oxygen supply to the fire.

$$2NaHCO_3(s) + \text{heat} \rightarrow Na_2CO_3(s) + H_2O(g) + CO_2(g)$$

Other dry chemical extinguishers work by combining with the combustion intermediates, thus preventing them from reacting with oxygen. Monoammonium phosphate (MAP) is the extinguishing agent commonly used in the portable fire extinguishers found in the home or car. This kind of extinguisher is suitable for Class A, B, and C fires.

7 Chemical Reactions

Chapter Review

Key Terms

activity series of metals	7·5
balanced equation	7·2
catalyst	7·1
chemical equation	7·1
coefficient	7·2
combination reaction	7·3
combustion reaction	7·7
decomposition reaction	7·4
double-replacement reaction	7·6
single-replacement reaction	7·5
skeleton equation	7·1

Chapter Summary

A chemical reaction can be concisely represented by a chemical equation. The substances that undergo a chemical change are the reactants. The new substances formed are the products. In accordance with the law of conservation of matter, a chemical equation must be balanced. In balancing an equation, coefficients are placed in front of the reactants and products so that the same number of atoms of each element are on each side of the equation.

Special symbols are written after formulas in equations to show a substance's state. The designations for a solid, liquid, or gas are (s), (l), and (g) respectively. A substance dissolved in water is designated (aq). If a catalyst is used to increase the speed of a chemical reaction, its formula is written above the arrow.

Many chemical equations can be classified as to type. In a combination reaction there is always a single product. The reactancts are two or more elements and/or compounds. In a decomposition reaction, a single compound is broken down into two or more simpler substances.

In a single-replacement reaction, the reactants and products are an element and a compound. The activity series of metals can be used to predict whether most single-replacement reactions will take place. A double-replacement reaction involves the exchange of cations (or anions) between two compounds. This reaction generally takes place between two ionic compounds in aqueous solution. One of the reactants in a combustion reaction is oxygen. The products of the complete combustion of a hydrocarbon are carbon dioxide and water.

Practice Questions and Problems

11. Write formulas and other symbols for these substances.   *7·1*
  a. metallic potassium
  b. sodium chloride dissolved in water
  c. heat supplied to a chemical reaction
  d. liquid mercury
  e. zinc chloride as a catalyst
  f. carbon dioxide gas

12. What is the purpose of a catalyst?   *7·1*

13. Write a sentence that describes each chemical reaction.   *7·1*
  a. $2K(s) + 2H_2O(l) \longrightarrow 2KOH(aq) + H_2(g)$
  b. $NaOH(aq) + HNO_3(aq) \longrightarrow H_2O(l) + NaNO_3(aq)$
  c. $FeO(s) + C(s) \xrightarrow{\Delta} Fe(s) + CO(g)$

14. Balance the following equations.   *7·2*
  a. $Pb(NO_3)_2 + K_2CrO_4 \longrightarrow PbCrO_4 + KNO_3$
  b. $MnO_2 + HCl \longrightarrow MnCl_2 + H_2O + Cl_2$
  c. $C_3H_6 + O_2 \longrightarrow CO_2 + H_2O$
  d. $Zn(OH)_2 + H_3PO_4 \longrightarrow Zn_3(PO_4)_2 + H_2O$
  e. $CO + Fe_2O_3 \longrightarrow Fe + CO_2$
  f. $CS_2 + Cl_2 \longrightarrow CCl_4 + S_2Cl_2$
  g. $CH_4 + Br_2 \longrightarrow CH_3Br + HBr$
  h. $Ba(CN)_2 + H_2SO_4 \longrightarrow BaSO_4 + HCN$

15. How is the law of conservation of matter related to the balancing of a chemical equation? *7·2*

16. What is a characteristic of every combination reaction? *7·3*

17. Complete and balance these combination reactions. *7·3*
 a. $Ca + Br_2 \longrightarrow$ **c.** $SO_2 + H_2O \longrightarrow$
 b. $Mg + N_2 \longrightarrow$ **d.** $Be + O_2 \longrightarrow$

18. What is a distinguishing feature of every decomposition reaction? *7·4*

19. Complete and balance these decomposition reactions. In some cases one of the decomposition products is given. *7·4*
 a. $H_2O_2 \xrightarrow{\text{MnO}_2} H_2O +$
 b. $Mg(ClO_3)_2 \longrightarrow MgCl_2 +$
 c. $HI \longrightarrow$
 d. $NaNO_3 \longrightarrow NaNO_2 +$

20. Complete the equations for these single-replacement reactions that take place in water solution. Balance each equation. If a reaction does not occur (use the activity series), write "No reaction." *7·5*
 a. $Al + CuSO_4 \longrightarrow$
 b. $Ca + H_2O \longrightarrow$
 c. $Zn + Pb(NO_3)_2 \longrightarrow$
 d. $Fe + H_2O \longrightarrow$
 e. $F_2 + KCl \longrightarrow$

21. What are the three "driving forces" for double-replacement reactions? *7·6*

22. Write the products for these double-replacement reactions. Then balance each equation. Assume that each reaction will occur. *7·6*
 a. $CdCl_2 + H_2S \longrightarrow$
 b. $NH_4C_2H_3O_2 + AgNO_3 \longrightarrow$
 c. $H_2C_2O_4 + NaOH \longrightarrow$
 d. $Fe(NO_3)_3 + LiOH \longrightarrow$
 e. $BaCl_2 + H_3PO_4 \longrightarrow$

23. Distinguish between complete and incomplete combustion. *7·7*

24. Write a balanced equation for the complete combustion of each of these compounds. *7·7*
 a. acetic acid, $HC_2H_3O_2$ **c.** glycerol, $C_3H_8O_3$
 b. decane, $C_{10}H_{22}$ **d.** sucrose, $C_{12}H_{22}O_{11}$

25. Balance each of these equations. Identify each as to type. *7·8*
 a. $HCl + Fe_2O_3 \longrightarrow FeCl_3 + H_2O$
 b. $Li + O_2 \longrightarrow Li_2O$

 c. $MgCO_3 \longrightarrow MgO + CO_2$
 d. $Fe_3O_4 + H_2 \longrightarrow Fe + H_2O$
 e. $C_6H_5OH + O_2 \longrightarrow CO_2 + H_2O$
 f. $KClO_4 \longrightarrow KCl + O_2$
 g. $AgNO_3 + H_2S \longrightarrow Ag_2S + HNO_3$
 h. $Cl_2 + KI \longrightarrow KCl + I_2$
 i. $FeCl_3 + NaOH \longrightarrow Fe(OH)_3 + NaCl$

Mastery Questions and Problems

26. Each of these equations is "balanced" but incorrect. Find the errors and correctly balance each equation.
 a. $2Mg + H_2SO_4 \longrightarrow Mg_2SO_4 + H_2$
 b. $H_2 + Cl_2 \longrightarrow H_2Cl_2$
 c. $NH_3 \longrightarrow N + H_3$
 d. $Cl_2 + NaI \longrightarrow NaCl_2 + I$
 e. $Na + O_2 \longrightarrow NaO_2$
 f. $MgCl + CaOH \longrightarrow MgOH + CaCl$

27. Write a balanced chemical equation for each of these reactions. Use the necessary symbols from Table 7·1 to completely describe the reaction.
 a. Bubbles of hydrogen gas and aqueous iron(III) chloride are produced when metallic iron is dropped into hydrochloric acid.
 b. Mercury metal is produced by heating a mixture of mercury(II) sulfide and calcium oxide. Additional products are calcium sulfide and calcium sulfate.
 c. Solid silver oxide can be heated to give silver and oxygen gas.
 d. Solid tetraphosphorus decoxide reacts with water to produce phosphoric acid.
 e. Bubbling chlorine gas through a solution of potassium iodide gives elemental iodine and a solution of potassium chloride.
 f. Iodine crystals react with chlorine gas to form solid iodine trichloride.

28. Write balanced chemical equations for these double-replacement reactions that occur in aqueous solution.
 a. Sodium hydroxide reacts with nitric acid.
 b. Solutions of potassium fluoride and calcium nitrate are mixed.
 c. Zinc sulfide is added to sulfuric acid.
 d. A solution of calcium iodide is poured into a solution of mercury(II) nitrate.

29. Write balanced chemical equations for each of these combination reactions.
  a. dichlorine heptoxide + water $\longrightarrow$
  b. hydrogen + bromine $\longrightarrow$
  c. sodium oxide + water $\longrightarrow$
  d. aluminum + chlorine $\longrightarrow$

30. Write balanced chemical equations for each of these single-replacement reactions that take place in water solution. Then balance each equation. Write "No reaction" if they do not occur.
  a. A piece of silver jewelry is dropped in hydrochloric acid.
  b. Bromine reacts with aqueous barium iodide.
  c. A piece of steel wool (iron) is placed in sulfuric acid.
  d. Mercury is poured into an aqueous solution of zinc nitrate.

31. Test tubes A and B contain water. A piece of sodium metal has been dropped in one test tube; a piece of magnesium metal in the other.
  a. Which tube contains the sodium metal?
  b. Write an equation for the reaction in the tube containing the sodium metal.
  c. What type of reaction is occurring in this tube?

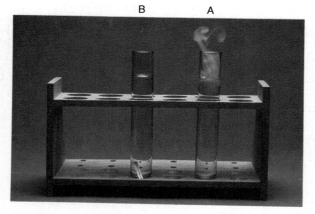

32. Write balanced equations for the incomplete combustion of each compound. Assume that the products are carbon monoxide and water.
  a. glycerol, $C_3H_8O_3$
  b. acetic acid, $HC_2H_3O_2$
  c. glucose, $C_6H_{12}O_6$
  d. acetylene, C_2H_2

33. Write balanced chemical equations for these decomposition reactions.
  a. Passing an electric current through melted crystals of magnesium chloride decomposes it into its elements.
  b. Silver carbonate decomposes into silver oxide and carbon dioxide when it is heated.
  c. Heating tin(IV) hydroxide gives tin(IV) oxide and water.
  d. Aluminum can be obtained from aluminum oxide with the addition of a large amount of electrical energy.

Critical Thinking Questions

34. Choose the term that best completes the second relationship.
  a. charcoal:ashes   reactant:_____
    (1) equation   (3) heat
    (2) reaction   (4) product
  b. tool:hammer   catalyst:_____
    (1) reaction   (3) sulfur
    (2) MnO_2 (4) combustion
  c. east:west   combination:_____
    (1) decomposition   (3) coefficient
    (2) reaction   (4) replacement

35. Why is no smoking permitted near an oxygen source? What would happen if a match were struck in an empty room filled with oxygen?

36. What type of fire extinguisher would you use to fight fires involving each of the following objects? Explain your answer.
  a. car engine   **c.** newspaper
  b. Christmas tree   **d.** airplane

Review Questions and Problems

37. Calculate the number of moles in each of the following substances.
  a. 34.8 g sodium hypochlorite
  b. 2.00×10^{21} molecules carbon monoxide
  c. 78.0 L nitrogen dioxide (at STP)
  d. 0.859 g magnesium ions

38. What mass of gold (density = 19.3 g/cm³) occupies the same volume as 2.20 mol of aluminum (density = 2.70 g/cm³)?

39. Many coffees and colas contain the stimulant caffeine. The percent composition of caffeine is 49.5% C, 5.20% H, 16.5% O, and 28.9% N. What is the molecular formula of caffeine if its molar mass is 194.1 g/mol?

40. The white solid calcium chloride, $CaCl_2$, is used as a drying agent. The maximum amount of water absorbed by different quantities of $CaCl_2$ is given in the table below.
  a. Complete the table.
  b. Plot the moles of water absorbed (y-axis) versus the moles of $CaCl_2$.
  c. Based on your graph, how many molecules of water does each formula unit of $CaCl_2$ absorb?

$CaCl_2$ (g)	$CaCl_2$ (mol)	H_2O (g)	H_2O (mol)
17.3	____	5.62	____
48.8	____	15.8	____
124	____	40.3	____
337	____	109	____

41. Name these compounds.
  a. $Sr(NO_3)_2$
  b. CdS
  c. $NaHCO_3$
  d. HF
  e. KCN
  f. $SnBr_4$
  g. B_2O_3
  h. $Al(OH)_3$
  i. $Fe_2(SO_4)_3$
  j. $LiClO_3$

Challenging Questions and Problems

42. Write balanced chemical equations for each of these reactions. Classify each as to type.
  a. Aluminum reacts with sulfuric acid to produce hydrogen and aluminum sulfate.
  b. Heating sulfuric acid produces water, oxygen, and sulfur dioxide.
  c. Pentane, C_5H_{12}, reacts with oxygen.
  d. Sodium iodide reacts with phosphoric acid to form hydrogen iodide and sodium phosphate.
  e. Potassium oxide reacts with water to form potassium hydroxide.

43. The mass of a proton and an electron are 1.67×10^{-24} g and 9.11×10^{-28} g respectively. What is the mass of a mole of protons and a mole of electrons? How many electrons are equal in mass to one proton?

44. Complete and balance these equations.
  a. $Bi(NO_3)_3 + H_2S \longrightarrow$
  b. $Br_2 + NaI \longrightarrow$
  c. $Al + Cl_2 \longrightarrow$
  d. $H_2O_2 \longrightarrow$
  e. $Ba(OH)_2 + HNO_3 \longrightarrow$
  f. $K + H_2O \longrightarrow$
  g. $C_2H_5OH + O_2 \longrightarrow$

Research Projects

1. Identify one or more everyday examples of each type of chemical equation.

2. Use a file to scrape the edge off two pennies, one minted before 1982, the other after. Place the pennies in a glass of vinegar. Explain what you observe.

3. What chemical reactions are commonly used in cooking?

4. What are plastic explosives? When and how are they used?

5. How does a miner's safety lamp work? Why was this an important invention?

6. Investigate single-replacement reactions in a gel such as agar.

7. How do matches work? What led to their invention and how did they change people's lives?

8. How are large buildings designed to control fires?

Readings and References

Beller, Joel. *So You Want to Do a Science Project!* New York: Arco, 1982.

Coulson, E. H., A. E. J. Trinder, and Aaron E. Klein. *Test Tubes and Beakers: Chemistry for Young Experimenters*. New York: Doubleday, 1971.

Henahan, John F. "FIRE." *Science 80* (January/February 1980), pp. 28–39.

McMurry, Linda O. *George Washington Carver: Scientist and Symbol*. New York: Oxford U Pr., 1981.

Moorman, Thomas. *How to Make Your Science Project Scientific*. New York: Atheneum, 1974.

Smith, Norman F. *How Fast Do Your Oysters Grow? Investigate and Discover through Science Projects*. New York: Messner, 1982.

8 Stoichiometry

Chapter Preview

8·1 Interpreting Chemical Equations

8·2 Mole–Mole Calculations

8·3 Mass–Mass Calculations

8·4 Other Stoichiometric Calculations

8·A Chemical Sleuthing

8·5 Limiting Reagent

8·6 Percent Yield

8·7 Energy Changes in Chemical Reactions

8·B Calorimetry

8·8 The Heat of Reaction

Chemicals are used in the manufacture of almost everything we buy, for example, soaps, foods, drugs, and paints. The processes used to manufacture these items must be carried out efficiently. Any waste will cost money. This is where balanced chemical equations come in. Equations are the recipes that tell chemists what amounts of reactants to mix and what amounts of products to expect.

Balanced equations allow us to calculate the quantities of reactants and products in a reaction. When you know the quantity of one substance, you can calculate the quantity of any other substance. By quantity, we usually mean the amount of a substance expressed in grams or moles. It could just as well be a quantity in liters, tons, or molecules. Calculations using balanced equations are called *stoichiometric* calculations. **Stoichiometry** *is the calculation of quantities in chemical equations*. For chemists it is a form of bookkeeping.

Greek: *stoikheion* = element; *metron* = to measure.

8·1 Interpreting Chemical Equations

Ammonia is widely used as a fertilizer. It is produced industrially by the reaction of hydrogen with nitrogen.

$$N_2(g) + 3H_2(g) \longrightarrow 2NH_3(g)$$

What kinds of information can be derived from this equation?

1. Particles. We might look at interacting particles. One molecule of nitrogen reacts with three molecules of hydrogen to produce two molecules of ammonia. Nitrogen and hydrogen will always react together to form ammonia in this $1:3:2$ ratio of molecules. If we could make 10

Figure 8·1
Without balance, a gymnast would fall. In a similar way, balanced chemical equations are essential for chemical calculations.

169

Figure 8·2
The balanced chemical equation for the formation of ammonia can be interpreted in several ways.

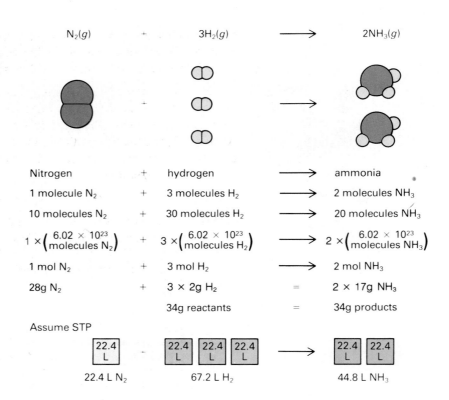

$$N_2(g) \quad + \quad 3H_2(g) \quad \longrightarrow \quad 2NH_3(g)$$

Nitrogen	+	hydrogen	$\longrightarrow$	ammonia
1 molecule N$_2$	+	3 molecules H$_2$	$\longrightarrow$	2 molecules NH$_3$
10 molecules N$_2$	+	30 molecules H$_2$	$\longrightarrow$	20 molecules NH$_3$
$1 \times \left(\begin{array}{c} 6.02 \times 10^{23} \\ \text{molecules N}_2 \end{array}\right)$	+	$3 \times \left(\begin{array}{c} 6.02 \times 10^{23} \\ \text{molecules H}_2 \end{array}\right)$	$\longrightarrow$	$2 \times \left(\begin{array}{c} 6.02 \times 10^{23} \\ \text{molecules NH}_3 \end{array}\right)$
1 mol N$_2$	+	3 mol H$_2$	$\longrightarrow$	2 mol NH$_3$
28g N$_2$	+	3 × 2g H$_2$	=	2 × 17g NH$_3$
		34g reactants	=	34g products

Assume STP

22.4 L N$_2$ 67.2 L H$_2$ 44.8 L NH$_3$

molecules of nitrogen react with 30 molecules of hydrogen, we would expect to get 20 molecules of ammonia (Figure 8·2). Of course, we cannot count out small numbers of molecules and allow them to react. We could, however, take Avogadro's number of nitrogen molecules and react them with three times Avogadro's number of hydrogen molecules. This would be the same 1 : 3 ratio of molecules of the reactants. This reaction would form two times Avogadro's number of ammonia molecules.

2. Moles. You know that Avogadro's number of representative particles is 1 mol of a substance. On the basis of the particle interpretation just given, this equation tells us the number of moles of reactants and products. One mole of nitrogen molecules reacts with three moles of hydrogen molecules. These reactants form two moles of ammonia molecules. This is the most important information that a chemical equation provides. *The coefficients of a balanced chemical equation tell us the relative number of moles of reactants and products in a chemical reaction.* From this the amounts of reactants and products can be calculated.

▪ The coefficients in a balanced chemical equation indicate the relative number of moles of reactants and products that react in a chemical reaction.

3. Mass. A balanced chemical equation must obey the law of conservation of mass. The mole interpretation supports this. The mass of 1 mol of nitrogen (28.0 g) plus the mass of 3 mol of hydrogen (6.0 g) does equal the mass of 2 mol of ammonia (34.0 g).

This interpretation reflects the law of definite proportions (Section 5·3).

4. Volume. If we assume standard temperature and pressure, this equation also tells us about the volumes of gases. Recall that 1 mol of any gas at STP occupies a volume of 22.4 L. It follows that 22.4 L of nitrogen reacts with 67.2 L (3 × 22.4) of hydrogen to form 44.8 L (2 × 22.4) of ammonia. Section 8·4 will show that at *any* constant temperature and pressure, nitrogen and hydrogen will react together in a 1 : 3 ratio of volume. Two volumes of ammonia will be formed.

Notice in Figure 8·2 that mass and the number of atoms are conserved in this chemical reaction. Mass and atoms will be conserved in every chemical reaction. The mass of the reactants equals the mass of the products. Representative particles, moles, and volumes of gases will not generally be conserved, although they may. Take, for example, the formation of hydrogen iodide.

$$H_2(g) + I_2(g) \longrightarrow 2HI(g)$$

In this reaction, moles, molecules, and volume all happen to be conserved. In the majority of reactions, however, they are not. In the formation of ammonia (Figure 8·2), moles, molecules, and volume of gas are not conserved. *Only mass and atoms are conserved in every chemical reaction.*

Example 1

Interpret this balanced chemical equation in terms of the interaction of relative quantities in the following four ways.
a. number of representative particles
b. number of moles
c. masses of reactants and products
d. volumes of gases at STP

$$2H_2S(g) + 3O_2(g) \longrightarrow 2SO_2(g) + 2H_2O(g)$$

Solution

a. Two molecules of H_2S react with three molecules of O_2 to form two molecules of SO_2 and two molecules of H_2O.
b. Two moles of H_2S react with 3 mol of O_2 to produce 2 mol of SO_2 and 2 mol of H_2O.
c. Apply the mole interpretation and use the gram molecular mass of the reactants and products.

$$2 \text{ mol } H_2S + 3 \text{ mol } O_2 \longrightarrow 2 \text{ mol } SO_2 + 2 \text{ mol } H_2O$$

$$(2 \text{ mol} \times 34.1 \text{ g/mol}) + (3 \text{ mol} \times 32.0 \text{ g/mol}) \longrightarrow$$

$$(2 \text{ mol} \times 64.1 \text{ g/mol}) + (2 \text{ mol} \times 18.0 \text{ g/mol})$$

$$68.2 \text{ g} + 96.0 \text{ g} \longrightarrow 128.2 \text{ g} + 36.0 \text{ g}$$

$$164.2 \text{ g} = 164.2 \text{ g}$$

d. $(2 \text{ mol } H_2S \times 22.4 \text{ L/mol}) + (3 \text{ mol } O_2 \times 22.4 \text{ L/mol}) \longrightarrow$

$(2 \text{ mol } SO_2 \times 22.4 \text{ L/mol}) + (2 \text{ mol } H_2O \times 22.4 \text{ L/mol})$

Therefore, 44.8 L of $H_2S(g)$ reacts with 67.2 L of $O_2(g)$ to form 44.8 L of $SO_2(g)$ and 44.8 L of $H_2O(g)$.

Figure 8·3
Hydrogen sulfide, H_2S, smells like rotten eggs. It escapes from the ground through fumaroles in volcanic areas.

Problem

1. After balancing each of these equations, interpret them in terms of relative quantities in four ways. (1) number of representative particles (2) number of moles (3) masses of reactants and products (4) volumes of gases at STP (where appropriate)
   a. $2CO(g) + O_2(g) \longrightarrow 2CO_2(g)$
   b. $2Na(s) + 2H_2O(l) \longrightarrow 2NaOH(aq) + H_2(g)$
   c. $2C_2H_2(g) + 2O_2(g) \longrightarrow 2CO_2(g) + 2H_2O(g)$

8·2 Mole–Mole Calculations

Return now to the balanced equation for the production of ammonia.

$$N_2(g) + 3H_2(g) \longrightarrow 2NH_3(g)$$

The most important interpretation of this equation is that 1 mol of nitrogen reacts with 3 mol of hydrogen to form 2 mol of ammonia. With this interpretation we can relate moles of reactants to moles of products. The coefficients from the balanced equation can be used to write conversion factors. Then the number of moles of a product can be calculated from a given number of moles of reactant. Using other mole-quantity relationships, we can introduce mass, volume, and particles into our calculations.

Example 2

How many moles of ammonia are produced when 0.60 mol of nitrogen reacts with hydrogen?

Solution

We will work from a known number of moles of N_2 to the unknown number of moles of NH_3: mol $N_2 \longrightarrow$ mol NH_3. From the balanced equation just given, we see that 1 mol of N_2 produces 2 mol of NH_3. This relationship can be written and used as a conversion factor relating N_2 and NH_3. The conversion factor is called a *mole ratio*.

$$\frac{1 \text{ mol } N_2}{2 \text{ mol } NH_3} \quad or \quad \frac{2 \text{ mol } NH_3}{1 \text{ mol } N_2}$$

To solve the problem, we start with our given quantity, 0.60 mol of N_2. Then we multiply it by the form of the mole ratio that allows the given unit to cancel.

$$0.60 \text{ mol } N_2 \times \frac{2 \text{ mol } NH_3}{1 \text{ mol } N_2} = 1.2 \text{ mol } NH_3$$

Note that mole ratios from balanced equations are considered to be exact. They do not enter into the determination of significant figures in the answer.

Figure 8·4
This plant produces ammonia by combining nitrogen with hydrogen. Ammonia is used in cleaning products, fertilizers, and in the manufacture of other chemicals.

ChemDirections

The hazardous by-products of chemical production have to be disposed of under stringent regulations. To find out how hazardous chemicals can be managed with the least harm done to humans and the environment, read **ChemDirections** *Hazardous Wastes*, page 658.

Issues in Chemistry

Do you send grass clippings, newspapers, and plastic and glass containers to the landfill? Landfills around the country are quickly filling up and appropriate sites for new landfills are lacking. What are some better ways to handle the disposal of these solid wastes?

Example 3

Calculate the number of moles of reactants required to make 7.24 mol of ammonia.   **a.** nitrogen **b.** hydrogen

$$N_2(g) + 3H_2(g) \longrightarrow 2NH_3(g)$$

Solution

The conversions are similar: **a.** mol $NH_3 \longrightarrow$ mol N_2; **b.** mol $NH_3 \longrightarrow$ mol H_2. From the balanced equation we can get mole ratios relating each of the reactants to the product.

$$\frac{1 \text{ mol } N_2}{2 \text{ mol } NH_3} \quad or \quad \frac{2 \text{ mol } NH_3}{1 \text{ mol } N_2} \quad and \quad \frac{3 \text{ mol } H_2}{2 \text{ mol } NH_3} \quad or \quad \frac{2 \text{ mol } NH_3}{3 \text{ mol } H_2}$$

In each case, start with the given amount of product. Multiply by the appropriate mole ratio so that the given unit cancels.

a. $7.24 \text{ mol } NH_3 \times \dfrac{1 \text{ mol } N_2}{2 \text{ mol } NH_3} = 3.62 \text{ mol } N_2$

b. $7.24 \text{ mol } NH_3 \times \dfrac{3 \text{ mol } H_2}{2 \text{ mol } NH_3} = 10.86 = 10.9 \text{ mol } H_2$

Each solution in these two examples follows the same pattern. The moles of the given quantity (G), whether reactant or product, are multiplied by a mole ratio with the given in the denominator.

$$\frac{b \text{ moles of } W}{a \text{ moles of } G}$$

In this mole ratio, W is the "wanted" quantity. The values of a and b are the coefficients from the balanced equation. A general solution for a mole–mole problem ($aG \rightarrow bW$) is given by the following equation.

$$\underset{\text{Given}}{\frac{x \text{ mol } G}{1}} \times \underset{\text{Mole ratio}}{\frac{b \text{ mol } W}{a \text{ mol } G}} = \underset{\text{Calculated}}{\frac{xb}{a} \text{ mol } W}$$

Problem

2. The formation of aluminum oxide from its constituent elements is represented by this equation.

$$4Al + 3O_2 \longrightarrow 2Al_2O_3$$

  a. Write the six ratios of moles that can be derived from this equation.

  b. How many moles of aluminum are needed to form 2.3 mol of Al_2O_3?

  c. How many moles of oxygen are required to react completely with 0.84 mol of Al?

  d. Calculate the number of moles of Al_2O_3 formed when 17.2 mol of O_2 reacts with aluminum.

8·3 Mass–Mass Calculations

No laboratory balance can measure substances directly in moles. Instead the amount of a substance is determined by measuring its mass in grams. Thus, we must also be able to calculate the mass of a reactant or product from a chemical equation. The mole interpretation of a balanced equation still holds. If the given quantity is measured in grams, the gram formula mass is used to convert the mass to moles. Then the mole ratio from the balanced equation is used to calculate the unknown. Similarly, if we need to know the mass of the unknown, we first find the number of moles of unknown. The gram formula mass is then used to calculate the mass. As in mole–mole calculations, the unknown can be either a reactant or a product.

Stoichiometric calculations in which quantities are measured in grams still use the mole ratio from the balanced chemical equation.

Example 4

Calculate the number of grams of NH_3 produced by the reaction of 5.40 g of hydrogen with nitrogen. The balanced equation is given.

$$N_2(g) + 3H_2(g) \longrightarrow 2NH_3(g)$$

Solution

We are given the mass, in grams, of hydrogen and asked to find the mass, in grams, of ammonia: g $H_2 \longrightarrow$ g NH_3. The coefficients in the equation tell us that 3 mol of H_2 can react with 1 mol of N_2 to produce 2 mol of NH_3. Thus we need two more steps to do the calculation: g $H_2 \longrightarrow$ mol $H_2 \longrightarrow$ mol $NH_3 \longrightarrow$ g NH_3.

First, change the given, 5.40 g of H_2, to moles by using the gram formula mass of hydrogen: g $H_2 \longrightarrow$ mol H_2.

$$5.40 \, \cancel{g \, H_2} \times \frac{1 \, mol \, H_2}{2.0 \, \cancel{g \, H_2}} = 2.70 \, mol \, H_2$$

The gram formula mass of a substance is used to convert moles to grams and grams to moles.

Now use the mole ratio from the equation to calculate the number of moles of NH_3: mol $H_2 \longrightarrow$ mol NH_3.

$$2.70 \text{ mol } H_2 \times \frac{2 \text{ mol } NH_3}{3 \text{ mol } H_2} = 1.80 \text{ mol } NH_3$$

Finally, convert 1.80 mol of NH_3 to grams: mol $NH_3 \longrightarrow$ g NH_3. The gram formula mass of NH_3 is 17.0 g/mol.

$$1.80 \text{ mol } NH_3 \times \frac{17.0 \text{ g } NH_3}{1 \text{ mol } NH_3} = 30.6 \text{ g } NH_3$$

The three steps just described can be combined into one "solution": g $H_2 \longrightarrow$ mol $H_2 \longrightarrow$ mol $NH_3 \longrightarrow$ g NH_3.

$$5.40 \text{ g } H_2 \times \underbrace{\frac{1 \text{ mol } H_2}{2.0 \text{ g } H_2}}_{} \times \underbrace{\frac{2 \text{ mol } NH_3}{3 \text{ mol } H_2}}_{} \times \underbrace{\frac{17.0 \text{ g } NH_3}{1 \text{ mol } NH_3}}_{} = 30.6 \text{ g } NH_3$$

| Given quantity | Change given unit to moles | Mole ratio from balanced equation | Change moles of "wanted" to grams |

In view of the law of conservation of matter, is it possible to make 30.6 g of NH_3 from only 5.40 g of H_2? Check the equation for the reaction. From it you will see that another reactant, nitrogen, is also involved.

Figure 8·5
Clouds of condensed ammonia hide the surface of Saturn.

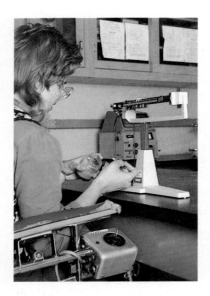

Figure 8·6
To determine the number of moles in a sample of a compound, first measure the mass of the sample. Then, use the gram formula mass to calculate the number of moles in that mass.

Figure 8·7
This general solution to a mass–mass problem shows how the gram formula mass is used to convert mass to moles and moles to mass. Conversion factors are shown in blue. The given substance (G) and the wanted substance (W) can be either reactants or products.

Example 5

How many grams of nitrogen are needed to produce the 30.6 g of NH_3 in the previous example? The equation is the same.

$$N_2(g) + 3H_2(g) \longrightarrow 2NH_3(g)$$

Solution

The balanced equation tells us that 1 mol of N_2 is needed to make 2 mol of NH_3. To use this fact, change the 30.6 g of NH_3 to moles. Then use the mole ratio to find moles of N_2. Finally, change moles of N_2 to grams of N_2. The process is: g $NH_3 \longrightarrow$ mol $NH_3 \longrightarrow$ mol $N_2 \longrightarrow$ g N_2.

$$30.6 \text{ g NH}_3 \times \frac{1 \text{ mol NH}_3}{17.0 \text{ g NH}_3} \times \frac{1 \text{ mol N}_2}{2 \text{ mol NH}_3} \times \frac{28.0 \text{ g N}_2}{1 \text{ mol N}_2} = 25.2 \text{ g N}_2$$

| Given quantity | Change given unit to moles | Mole ratio from balanced equation | Change moles of "wanted" to grams |

Our answer can be checked by using the law of conservation of matter and the answers from the last two examples.

$$\text{Mass of reactants} = \text{mass of products}$$

$$\text{Mass of N}_2 + \text{mass of H}_2 = \text{mass of NH}_3$$

$$25.2 \text{ g N}_2 + 5.40 \text{ g H}_2 = 30.6 \text{ g NH}_3$$

$$30.6 \text{ g} = 30.6 \text{ g}$$

Mass–mass stoichiometric problems can be solved in basically the same way as mole–mole problems. Figure 8·7 illustrates the steps.

1. The mass of the given quantity must be changed to moles by using the ratio, 1 mol G/gfm G: mass $G \longrightarrow$ mol G.
2. The moles of G are changed to moles of W (wanted unit) by using the mole ratio from the balanced equation: mol $G \longrightarrow$ mol W.
3. The moles of W are changed to grams of W by using the ratio, gfm W/1 mol W: mol $W \longrightarrow$ mass W.

From Figure 8·7 you can also see the steps for doing mole–mass and mass–mole stoichiometric calculations. For a mole–mass problem, the first conversion (from mass to moles) is skipped. For a mass–mole problem, the last conversion (from moles to mass) is not done.

$$aG \longrightarrow bW$$
(given quantity)   (wanted quantity)

$$\text{mass of } G \times \frac{1 \text{ mol } G}{\text{gfm } G} \longrightarrow \text{mol } G \times \frac{b \text{ mol } W}{a \text{ mol } G} \longrightarrow \text{mol } W \times \frac{\text{gfm } W}{\text{mol } W} \longrightarrow \text{mass of } W$$

Mass-mole conversion   Mole ratio from balanced equation   Mole-mass conversion

Problems

3. The combustion of acetylene gas is represented by this equation.

$$2C_2H_2(g) + 5O_2(g) \longrightarrow 4CO_2(g) + 2H_2O(g)$$

a. How many grams of oxygen are required to "burn" 13.0 g of C_2H_2?

b. How many grams of CO_2 and grams of H_2O are produced when 13.0 g of C_2H_2 reacts with the oxygen required to burn 13.0 g of C_2H_2?

c. Use the answers from **a** and **b** to show that this equation obeys the law of conservation of matter.

4. Acetylene gas, C_2H_2, is produced by adding water to calcium carbide, CaC_2.

$$CaC_2(s) + 2H_2O(l) \longrightarrow C_2H_2 \uparrow + Ca(OH)_2(aq)$$

a. How many grams of acetylene are produced by adding water to 5.00 g of CaC_2?

b. How many moles of CaC_2 are needed to react completely with 98.0 g of H_2O?

c. How many grams of $Ca(OH)_2$ are produced when 5.34 mol of C_2H_2 is produced?

8·4 Other Stoichiometric Calculations

Any measurement of matter related to the mole can be worked into stoichiometric calculations.

A balanced chemical equation shows us the relative number of moles of reactants and products. From this foundation, stoichiometric calculations can be expanded to include any unit of measurement that is related to the mole. The problems can include *mass–volume, volume–volume,* and *particle–mass* calculations. The given quantity can be expressed in numbers of representative particles, units of mass, or volumes of gases at STP. The given quantity is first converted to moles. Then the mole ratio from the balanced equation can be used to calculate the number of moles of the wanted substance. Once this unit has been determined, the moles can be converted to any other unit of measurement. These relationships are shown in Figure 8·8 and illustrated in the next two examples.

Figure 8·8
This diagram can help you solve various stoichiometry problems. Possible given quantities (G) are shown in green. Conversion factors are used to calculate the wanted quantities (W) shown in yellow.

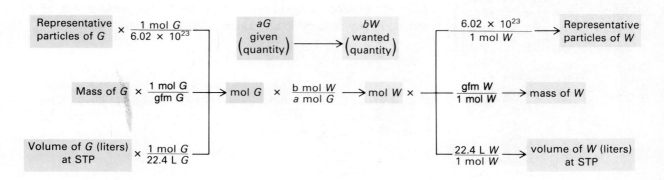

Example 6

How many molecules of oxygen are produced when 29.2 g of water is decomposed according to this balanced equation?

$$2H_2O(l) \xrightarrow{\text{electricity}} 2H_2(g) + O_2(g)$$

Solution

Figure 8·8 shows us the steps to follow.

Mass $H_2O \longrightarrow$ moles $H_2O \longrightarrow$ moles $O_2 \longrightarrow$ molecules O_2

$$29.2 \text{ g } H_2O \times \underset{\text{Change to moles}}{\frac{1 \text{ mol } H_2O}{18.0 \text{ g } H_2O}} \times \underset{\text{Mole ratio}}{\frac{1 \text{ mol } O_2}{2 \text{ mol } H_2O}} \times \underset{\text{Change to molecules}}{\frac{6.02 \times 10^{23} \text{ molecules } O_2}{1 \text{ mol } O_2}}$$

(Given quantity)

$$= 4.88 \times 10^{23} \text{ molecules } O_2$$

A mole of any substance contains 6.02×10^{23} representative particles.

Example 7

Assuming STP, how many liters of oxygen are needed to produce 19.8 L of SO_3 according to this balanced equation?

$$2SO_2(g) + O_2(g) \longrightarrow 2SO_3(g)$$

A mole of any gas at STP occupies a volume of 22.4 L.

Solution

Again using Figure 8·8 as a guide, we find the steps to follow.

Liters $SO_3 \longrightarrow$ moles $SO_3 \longrightarrow$ moles $O_2 \longrightarrow$ liters O_2

$$19.8 \text{ L } SO_3 \times \underset{\text{Change to moles}}{\frac{1 \text{ mol } SO_3}{22.4 \text{ L } SO_3}} \times \underset{\text{Mole ratio}}{\frac{1 \text{ mol } O_2}{2 \text{ mol } SO_3}} \times \underset{\text{Change to liters}}{\frac{22.4 \text{ L } O_2}{1 \text{ mol } O_2}} = 9.9 \text{ L } O_2$$

(Given quantity)

In the previous example, did you notice that the 22.4 L/mol factors canceled out? This will always be true in a volume–volume problem. *The coefficients in a balanced chemical equation tell you the relative number of moles. The coefficients also tell the relative volumes of interacting gases at the same temperature and pressure.* The volume can be expressed in any unit: cubic decimeters, milliliters, liters, and so on.

Figure 8·9
In Example 8 two moles of nitrogen dioxide, the brown gas shown below, form when 2 mol of nitrogen monoxide reacts with 1 mol of oxygen. Since the reactants and product are gases, we know that 2 L of nitrogen monoxide reacts with 1 L of oxygen to form 2 L of nitrogen dioxide.

Example 8

How many milliliters of nitrogen dioxide are produced when 3.4 mL of oxygen reacts with an excess of nitrogen monoxide? Assume conditions of STP.

$$2NO(g) + O_2(g) \longrightarrow 2NO_2(g)$$

Solution

The coefficients indicate the relative number of reacting milliliters of gases. The problem is solved in one step: mL $O_2 \longrightarrow$ mL NO_2.

$$3.4 \; \cancel{mL \; O_2} \times \underset{\text{Volume ratio}}{\frac{2 \; mL \; NO_2}{1 \; \cancel{mL \; O_2}}} = 6.8 \; mL \; NO_2$$

$\underset{\text{quantity}}{\text{Given}}$

Problems

5. Using Figure 8·8 as a guide, outline the steps necessary to convert these quantities. (G = given; W = wanted)
   a. volume of gas G (at STP) to mass of W
   b. representative particles of G to moles of W
   c. mass of G to representative particles of W

6. Tin(II) fluoride, formerly found in many toothpastes, is formed in this reaction.

$$Sn(s) + 2HF(g) \longrightarrow SnF_2(s) + H_2(g)$$

   a. How many grams of SnF_2 can be made by reacting 7.42×10^{24} molecules of HF with tin?
   b. How many liters of hydrogen gas (at STP) are produced by reacting 23.4 g of Sn with HF?
   c. How many liters of HF are needed to produce 14.2 L of H_2? (Assume STP.)
   d. How many molecules of H_2 are produced by the reaction of tin with 80.0 L of HF (at STP)?

Science, Technology, and Society

8·A Chemical Sleuthing

Does the blood on a murder victim's hand belong to the victim or to the murderer? Does the charred material from a warehouse fire show any signs that a chemical was used to spread the fire? Is the ink on a document really 20 years old, or was it forged last week? Answering questions like these is the job of the forensic chemist.

Forensic chemistry is involved with the collection and representation of evidence that may be used in a court of law. Most forensic laboratories handle evidence involved in criminal cases. Several hundred such "crime labs" exist in the United States and Canada. Other laboratories prepare evidence in cases concerning violations of federal or state regulations. These may involve the chemical compositions of pollutants, imported materials, or products sold to the public. In both types of laboratories, standard methods are used for the analysis and comparison of materials. Because the materials may

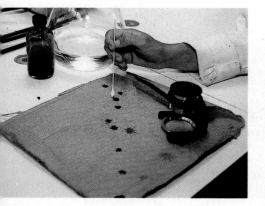

Figure 8·10
Evidence from a crime must be carefully collected, analyzed, and preserved.

Figure 8·11
Forensic chemists must have strong problem-solving skills, and a broad background in analytical chemistry.

be used in a court case, however, they must be collected, preserved, and analyzed according to very strict and well-documented methods.

In a crime laboratory body fluids and internal organs are commonly analyzed for poison, drugs, and alcohol. Similarly, analyses of pills, liquor, fibers, paint, glass, plastics, metals, and inks are performed. Blood, saliva, and hair can be typed. Charred documents can be restored, and fire debris can be analyzed. Many tests must be run on very tiny samples. Sometimes the tests must measure the exact amount of some substance in a sample, as in the analysis of pills for drugs. This requires the techniques of quantitative analysis. Often the tests only compare a material to some other sample, as in matching the paint on a crowbar to the paint on a door. These tests employ the methods of qualitative analysis. A forensic chemist must be an expert in both fields.

A career in forensic chemistry requires a degree in analytical chemistry and several years of on-the-job training. Skill in the use of all modern analytical instruments is very important. Specialized criminalistics courses include crime scene and firearms investigation, the identification of documents and fingerprints, and the collection and storage of evidence. Presenting evidence in court is an important part of the forensic chemist's job. Thus courses in public speaking and constitutional law are recommended. The chemist must also be able to explain complex procedures and results in simple language to the judge and jury. Lastly, a forensic chemist must have the personal strength to stand up to the threats of a criminal or the hostile cross-examination of a defense attorney.

8·5 Limiting Reagent

Most cooks follow a recipe when making a new dish. Before the actual mixing begins, sufficient quantities of all the ingredients must be available. For example it is impossible to make two loaves of bread with only one metric cup of flour. Similarly it is impossible for a chemist to make a certain amount of compound if there is an insufficient quantity of any of the reactants.

We have likened a balanced chemical equation to a chemist's recipe. This recipe can be interpreted on a microscopic scale (interacting particles) or on a macroscopic scale (interacting moles). For example, let us return to the now familiar equation for the preparation of ammonia.

$$3H_2(g) + N_2(g) \longrightarrow 2NH_3(g)$$

Two molecules (moles) of NH_3 will be produced when three molecules (moles) of H_2 are reacted with one molecule (mole) of N_2. What would happen, however, if three molecules (moles) of H_2 were reacted with two molecules (moles) of N_2? Would more than two molecules (moles) of NH_3 be formed? Figure 8·12 shows both the particle and the mole interpretations of this problem.

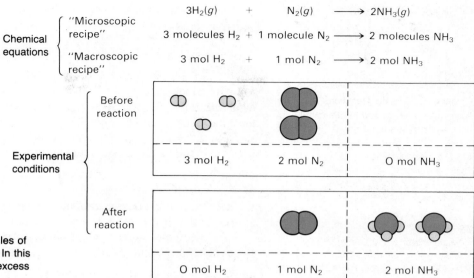

Chemical equations

"Microscopic recipe"

$$3H_2(g) \quad + \quad N_2(g) \quad \longrightarrow \quad 2NH_3(g)$$

3 molecules H_2 + 1 molecule N_2 $\longrightarrow$ 2 molecules NH_3

"Macroscopic recipe"

3 mol H_2 + 1 mol N_2 $\longrightarrow$ 2 mol NH_3

Experimental conditions

Before reaction

| 3 mol H_2 | 2 mol N_2 | O mol NH_3 |

After reaction

| O mol H_2 | 1 mol N_2 | 2 mol NH_3 |

Figure 8·12
The "recipe" calls for 3 molecules of H_2 for every 1 molecule of N_2. In this particular experiment N_2 is in excess and H_2 is the limiting reagent.

Before the reaction takes place, hydrogen and nitrogen are present in a $3:2$ molecule (mole) ratio. The reaction takes place according to the balanced equation. Three molecules (moles) of H_2 react with one molecule (mole) of N_2 to produce two molecules (moles) of NH_3. At this point all the hydrogen has been used up, and the reaction stops. In the reaction container one molecule (mole) of unreacted nitrogen is left in addition to the two molecules (moles) of NH_3.

In this reaction, only the hydrogen is completely used up. It is called the limiting reagent. As the name implies, *the* **limiting reagent** *limits or determines the amount of product that can be formed in a reaction.* The reaction occurs only until the limiting reagent is used up. By contrast, *the quantity of an* **excess reagent** *is more than enough to react with a limiting reagent.* In the present example, nitrogen is the excess reagent.

■ Whenever quantities of two or more reactants in a chemical reaction are given, you must identify the limiting reagent.

Figure 8·13
The amount of product is determined by the quantity of the limiting reagent. In this example, bread is the limiting reagent. No matter how much peanut butter and jelly you have, with eight slices of bread, you can make only four sandwiches.

Safety

Chlorine gas is toxic. Reactions involving toxic gases and fumes should only be done in a fume hood.

The limiting reagent is used up first in a chemical reaction.

Example 9

Sodium chloride can be prepared by the reaction of sodium metal with chlorine gas.

$$2Na(s) + Cl_2(g) \longrightarrow 2NaCl(s)$$

What will occur when 6.70 mol of Na reacts with 3.20 mol of Cl_2?
a. What is the limiting reagent?
b. How many moles of NaCl are produced?
c. How much of the excess reagent remains unreacted?

Solution

a. Start with the known amount of one of the reactants. Use the mole ratio from the balanced equation to calculate the required amount of the other reactant. Sodium is chosen arbitrarily here: mol Na $\longrightarrow$ mol Cl_2.

$$6.70 \text{ mol Na} \times \frac{1 \text{ mol } Cl_2}{2 \text{ mol Na}} = 3.35 \text{ mol } Cl_2$$

Given amount of sodium   Mole ratio   Required amount of chlorine

This calculation shows that 3.35 mol of Cl_2 is needed to react with 6.70 mol of Na. Because only 3.20 mol of Cl_2 is available, chlorine is the limiting reagent. The sodium, then, must be in excess.

b. The given amount of limiting reagent, 3.20 mol of Cl_2, is used to calculate the maximum amount of product that can be formed: mol Cl_2 $\longrightarrow$ mol NaCl.

$$3.20 \text{ mol } Cl_2 \times \frac{2 \text{ mol NaCl}}{1 \text{ mol } Cl_2} = 6.40 \text{ mol NaCl}$$

c. The amount of the excess reagent remaining is the difference between the given amount (6.70 mol of Na) and the amount of sodium needed to react with the limiting reagent. We need to calculate the sodium that is needed: mol Cl_2 $\longrightarrow$ mol Na.

$$3.20 \text{ mol } Cl_2 \times \frac{2 \text{ mol Na}}{1 \text{ mol } Cl_2} = 6.40 \text{ mol Na}$$

Limiting reagent   Mole ratio   Amount of excess reagent used up

The amount of remaining sodium can now be calculated.

$$6.70 \text{ mol} - 6.40 \text{ mol} = 0.30 \text{ mol Na in excess}$$

Problem

7. Two equations for the combustion of ethene, C_2H_4, are possible.

$$C_2H_4(g) + 3O_2(g) \longrightarrow 2CO_2(g) + 2H_2O(g)$$

(complete combustion)

and

$$C_2H_4(g) + 2O_2(g) \longrightarrow 2CO(g) + 2H_2O(g)$$

(incomplete combustion)

What will occur for each reaction if 2.70 mol of C_2H_4 is reacted with 6.30 mol of O_2?

a. Identify the limiting reagent.

b. Calculate the moles of water produced.

c. Calculate the moles of excess reagent remaining.

The given quantities of reactants may be expressed in units other than moles. In that case, the first step in the solution of the problem is to convert each reactant to moles. The limiting reagent is then identified as in the previous example. The amount of product is determined from the given amount of limiting reagent.

Figure 8·14
Black crystalline copper(II) sulfide is formed from copper and sulfur. The reaction will take place if a mixture of the reactants is heated.

Example 10

Copper reacts with sulfur to form copper(I) sulfide.

$$2Cu(s) + S(s) \longrightarrow Cu_2S(s)$$

What is the maximum number of grams of Cu_2S that can be formed when 80.0 g of Cu reacts with 25.0 g of S?

Solution

First, find the number of moles of each reactant: g Cu $\longrightarrow$ mol Cu, and g S $\longrightarrow$ mol S.

$$80.0 \text{ g Cu} \times \frac{1 \text{ mol Cu}}{63.5 \text{ g Cu}} = 1.26 \text{ mol Cu}$$

$$25.0 \text{ g S} \times \frac{1 \text{ mol S}}{32.1 \text{ g S}} = 0.779 \text{ mol S}$$

Next, use the balanced equation to calculate the number of moles of one reactant needed to react with the given amount of the other reactant: mol Cu $\longrightarrow$ mol S.

$$\underset{\substack{\text{Given} \\ \text{quantity}}}{1.26 \text{ mol Cu}} \times \underset{\substack{\text{Mole} \\ \text{ratio}}}{\frac{1 \text{ mol S}}{2 \text{ mol Cu}}} = \underset{\substack{\text{"Needed"} \\ \text{amount}}}{0.630 \text{ mol S}}$$

Compare the amount of sulfur needed, 0.630 mol of S, with the given amount, 0.779 mol of S. The sulfur is in excess. Therefore, in this problem, copper is the limiting reagent.

Finally, use the limiting reagent to calculate the maximum amount of Cu_2S formed: mol Cu $\longrightarrow$ mol Cu_2S $\longrightarrow$ g Cu_2S.

$$1.26 \text{ mol Cu} \times \frac{1 \text{ mol Cu}_2\text{S}}{2 \text{ mol Cu}} \times \frac{159.1 \text{ g Cu}_2\text{S}}{1 \text{ mol Cu}_2\text{S}} = 1.00 \times 10^2 \text{ g Cu}_2\text{S}$$

We could have begun this last step with the given quantity of copper, 80.0 g. In that case the first step in the solution of the problem would be to change to moles of copper, which we had previously done.

Figure 8·15
The productivity of a farm is measured in yield. Because growing conditions may vary from year to year, the actual yield often differs from the theoretical yield.

The theoretical yield is always a calculated number.

The percent yield is a measure of the efficiency of a reaction in changing reactants to products.

Problem

8. Hydrogen gas can be produced in the laboratory by the reaction of magnesium metal with hydrochloric acid.

$$Mg(s) + 2HCl(aq) \longrightarrow MgCl_2(aq) + H_2(g)$$

   a. How many grams of hydrogen can be produced when 4.00 g of HCl is added to 3.00 g of Mg?
   b. Assuming STP, what is the volume of this hydrogen?

8·6   Percent Yield

In theory, when a teacher gives an exam to the class, everyone should get a grade of 100%. For a variety of reasons (not all of which are the teacher's fault!) this is usually not the case. Instead, the performance of the class is usually spread over a range of grades.

In doing stoichiometric problems, we have assumed that things do not "go wrong" in chemical reactions. This assumption is as faulty as assuming that all students will score 100% on an exam. *When an equation is used to calculate the amount of product that will form during a reaction, then a value for the* **theoretical yield** *is obtained*. This is the maximum amount of product that could be formed from a given amount of reactant. By contrast, *the amount of product that forms when the reaction is carried out in the laboratory is called the* **actual yield.** It is often less than the theoretical yield. *The* **percent yield** *is the ratio of the actual yield to the theoretical yield*. It measures the efficiency of the reaction.

$$\text{Percent yield} = \frac{\text{actual yield}}{\text{theoretical yield}} \times 100\%$$

A percent yield should not normally be larger than 100%. Many factors cause percent yields to be less than 100%. Reactions do not always go to completion. Impure reactants and competing side reactions may cause other products to be formed. Usually some of the product is lost during purification. An actual yield is an experimental value. If you are asked to calculate a percent yield, the value of an actual yield must be given. For reactions in which percent yields have been determined, an "actual" yield can be calculated if the reaction conditions remain the same.

Example 11

Calcium carbonate can be decomposed by heating.

$$CaCO_3(s) \longrightarrow CaO(s) + CO_2(g)$$

What is the percent yield of this reaction if 24.8 g of $CaCO_3$ is heated to give 13.1 g of CaO?

Figure 8·16
The masses of reactants and products must be carefully measured when the percent yield of a reaction is determined. Here the mass of sodium bicarbonate, the reactant, is measured before it is heated. The mass of the product, sodium carbonate, is measured after the reaction is completed.

Solution

The actual yield is 13.1 g of CaO. The theoretical yield can be calculated: g $CaCO_3 \longrightarrow$ mol $CaCO_3 \longrightarrow$ mol CaO $\longrightarrow$ g CaO.

$$24.8 \text{ g CaCO}_3 \times \frac{1 \text{ mol CaCO}_3}{100.1 \text{ g CaCO}_3} \times \frac{1 \text{ mol CaO}}{1 \text{ mol CaCO}_3} \times \frac{56.1 \text{ g CaO}}{1 \text{ mol CaO}}$$

$$= 13.9 \text{ g CaO}$$

The percent yield is the ratio of the actual yield over the theoretical yield expressed as a percent.

$$\text{Percent yield} = \frac{\text{actual yield}}{\text{theoretical yield}} \times 100\%$$

$$= \frac{13.1 \text{ g CaO}}{13.9 \text{ g CaO}} \times 100\% = 94.2\%$$

Problem

9. What is the percent yield if 3.74 g of copper is produced when 1.87 g of aluminum is reacted with an excess of copper(II) sulfate?

$$2\text{Al}(s) + 3\text{CuSO}_4(aq) \longrightarrow \text{Al}_2(\text{SO}_4)_3(aq) + 3\text{Cu}(s)$$

8·7 Energy Changes in Chemical Reactions

■ Energy, most often in the form of heat, is either released or absorbed in a chemical reaction.

Greek: *exo* = out of;
 thermos = heat.

Energy changes occur whenever a chemical reaction takes place. Some of these energy changes involve heat. **Exothermic reactions** *release energy in the form of heat*. The combustion of coal is an exothermic reaction. Heat is one of the products.

$$\text{C}(s) + \text{O}_2(g) \longrightarrow \text{CO}_2(g) + 393.5 \text{ kJ}$$

Figure 8·17
This energy diagram is typical for exothermic reactions. The energy content of C and O_2 is higher than the energy content of CO_2. Therefore, energy is released when C burns to CO_2.

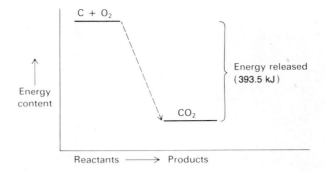

Greek: *endo* = within;
thermos = heat.

The energy released in the combustion of carbon must have a source. This is because the law of conservation of energy states that energy is neither created nor destroyed in ordinary chemical processes. The energy source is the chemical bonds that hold the oxygen molecules together and the atoms of carbon together. The energy stored in the carbon-oxygen bonds of carbon dioxide is less than the sum of the energy stored in the oxygen-oxygen bonds of oxygen and the carbon-carbon bonds of carbon. When carbon is burned, some of the energy stored in the bonds of the reactants is released as heat. The rest of the energy in the bonds of the reactants goes into the bonds of the products (Figure 8·17).

Not all reactions are exothermic. *In an* **endothermic reaction,** *energy is absorbed.* The production of calcium oxide (lime) from calcium carbonate (limestone) is an example of an endothermic process.

$$CaCO_3(s) \ + \ 176 \text{ kJ} \xrightarrow{\ 850° \ } CaO(s) \ + \ CO_2(g)$$

For each mole of calcium carbonate decomposed, 176 kJ of heat is absorbed. In this and other endothermic processes, the energy content of the products is higher than the energy content of the reactants (Figure 8·18). (The reason such reactions occur will be explained in Chapter 17.)

An equation that includes the amount of heat produced or absorbed by a reaction is a **thermochemical equation.** Standard conditions for a thermochemical equation are 25°C and 1 atm pressure. The heat term in a thermochemical equation can be treated like any other reactant or product as shown in the following example.

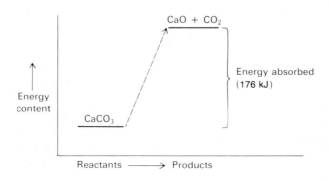

Figure 8·18
This energy diagram is typical for endothermic reactions. The energy content of $CaCO_3$ is lower than that of Ca and CO_2. Energy is absorbed when $CaCO_3$ is decomposed to Ca and CO_2.

Figure 8-19
Limestone (CaCO₃) is a common mineral (top). It is heated to produce the white powder, lime (CaO), which is used in mortar.

Example 12

Using the equation just given, calculate the energy required to decompose 5.20 mol of $CaCO_3(s)$. **a.** kilojoules **b.** kilocalories

Solution

a. The equation indicates that 176 kJ are needed to decompose 1 mol of $CaCO_3$. This can be used to write a conversion factor.

$$5.20 \text{ mol } CaCO_3 \times \frac{176 \text{ kJ}}{1 \text{ mol } CaCO_3} = 915 \text{ kJ}$$

b. The number of joules can be converted to calories given that 1.00 kcal = 4.18 kJ.

$$915 \text{ kJ} \times \frac{1.00 \text{ kcal}}{4.18 \text{ kJ}} = 219 \text{ kcal}$$

Problems

10. The air pollutant sulfur trioxide reacts with water in the atmosphere to produce sulfuric acid and heat.

$$SO_3(g) + H_2O(l) \longrightarrow H_2SO_4(aq) + 129.6 \text{ kJ}$$

How much heat is released when 583 g of $SO_3(g)$ reacts with water?
a. kilojoules **b.** kilocalories

11. Carbon dioxide can be decomposed into carbon monoxide and oxygen by the absorption of heat.

$$2CO_2(g) + 43.9 \text{ kJ} \longrightarrow 2CO(g) + O_2(g)$$

How many molecules of $CO_2(g)$ can be decomposed by the addition of 22.2 kJ of heat energy?

Figure 8·20
A foam cup calorimeter can be used to measure the heat given off by a reaction.

Science, Technology, and Society

8·B Calorimetry

Many chemists have been curious about the heat effects that accompany chemical reactions. In the early 1780s the French chemist Lavoisier (see Section 2·A) measured the heat given off by a guinea pig. He kept the animal in an enclosed container so that its body heat would melt ice. From the amount of ice melted, he calculated the heat produced by the animal.

The apparatus used to contain the guinea pig and measure its body heat was an ice calorimeter. It was a forerunner of a family of devices used to measure heat in chemical reactions. In all calorimeters a reaction is allowed to take place inside an insulated container. The heat given off by the reaction is transferred (with as little loss as possible) to some other material. The heat can then be calculated from the change in temperature.

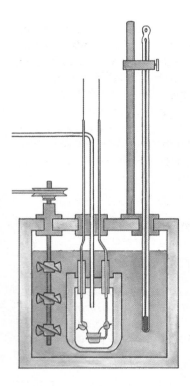

Figure 8·21
To start the combustion in a bomb calorimeter, oxygen is pumped into the reaction chamber containing the sample, and an electric charge is sent through the ignition wire. A thermometer measures the change in temperature of the water in the insulated water bath. This temperature is related to the heat given off by the reaction.

High school chemistry courses often include an experiment using a disposable foam cup as a calorimeter. A reaction involving aqueous solutions of chemicals is done in the cup. The initial temperature of the solutions and the final temperature of the mixture are measured. Using the change in temperature, the specific heat capacity of water, and the total volume of water involved in the reaction, the heat given off can be calculated (see Sections 2·12 and 2·13). There are some obvious sources of error in this type of experiment. Heat from the reaction will also go into heating the cup. Some will be lost to the environment because the cup is not closed. If reactants are not completely mixed, the temperature measurements will not be accurate. The mixture in the cup is not water but an aqueous solution. Its heat capacity will be different from that of pure water.

More sophisticated calorimeters are designed to overcome these problems. In most, the reaction takes place in a closed chamber surrounded by a material that absorbs heat. The heat of the reaction is transferred to all parts of the calorimeter, raising its temperature. The amount of heat that is given off is equal to the heat capacity of the calorimeter times the change in temperature. The heat capacity of the calorimeter is calculated by doing a reaction for which the heat produced is known. Thereafter, the calculated heat capacity can be used to measure the heat given off by any reaction.

The material in the reservoir may be water, other liquids, or a metal. The choice depends on what the highest temperature in the calorimeter will be. The reservoir is equipped with a thermometer to measure its change in temperature. If the reservoir contains a liquid, it also has a mechanical stirrer. The reservoir is surrounded by an insulating jacket to reduce heat loss to the environment. It may be made of an insulating material, like plastic foam.

A bomb calorimeter is used to measure the heat released by the combustion of a compound. A sample of the compound with a measured mass is placed in the inner chamber of the calorimeter. This chamber is filled with oxygen at high pressure, up to 30 atmospheres. A wire leads into the sample from outside of the calorimeter. When current is passed through this wire, it glows and ignites the sample which reacts with the oxygen. The heat from this reaction passes through the walls of the chamber into the water in the reservoir, which is constantly stirred. The temperature increase of the calorimeter is measured and used to calculate the heat given off by the compound.

8·8　The Heat of Reaction

At the start of any reaction each of the reactants has a heat content. **Enthalpy** *is the amount of heat that a substance has at a given temperature and pressure.* Enthalpy is symbolized by H. Unfortunately, the enthalpy of individual substances cannot be measured directly. Nevertheless, changes in enthalpy can be measured. A change in enthalpy is

An exothermic reaction has a $-\Delta H$.
An endothermic reaction has a $+\Delta H$.

Figure 8·22
The number of calories in a food sample is determined by measuring the heat of reaction of the sample in a calorimeter.

Fractional coefficients are needed here because we are dealing with the production of 1 mol of H₂O.

The heat of a reaction is calculated from the standard heats of formation of the reactants and products.

The (^{0}) in ΔH_f^0 indicates the use of standard conditions (25°C and 1 atm). The subscript ($_f$) indicates that the change in enthalpy is for the *formation* of the compound.

symbolized by ΔH (read "delta H"). Changes in enthalpy occur whenever heat is released or absorbed in chemical reactions. The standard conditions for measuring ΔH are 25°C and 1 atm. A reaction may be conducted at other conditions, say, at 50°C and 1 atm. In this situation, ΔH includes the heat changes involved in raising the temperature of the reactants from 25°C and lowering the temperature of the products back to that.

The heat that is released or absorbed during a chemical reaction is the **heat of reaction.** It is equivalent to ΔH, the change in enthalpy. For example, if 1 mol of carbon in its standard state is burned, 393 kJ of heat is released. The ΔH for this reaction is -393.5 kJ per mole.

$$C(s) + O_2(g) \longrightarrow CO_2(g) \qquad \Delta H = -393.5 \text{ kJ}$$

The sign of ΔH is always negative for an exothermic reaction. As Figure 8·17 shows, the products have *less* energy than the reactants. *In the special case in which one mole of a substance is completely burned, ΔH is called the* **heat of combustion.**

The sign of ΔH is always positive for an endothermic reaction. When calcium carbonate is decomposed, the products absorb 176 kJ of energy per mole of calcium carbonate decomposed. The heat content of the products is higher than that of the reactants by 176 kJ (Figure 8·18).

$$CaCO_3(s) \longrightarrow CaO(s) + CO_2(g) \qquad \Delta H = +176 \text{ kJ}$$

Note that the physical state of the reactants and products in a thermochemical reaction must be stated. (The symbols are g for *gas*, l for *liquid*, and s for *solid*.) Compare the following two equations.

$$H_2(g) + \tfrac{1}{2}O_2(g) \longrightarrow H_2O(g) \qquad \Delta H = -241.8 \text{ kJ}$$
$$H_2(g) + \tfrac{1}{2}O_2(g) \longrightarrow H_2O(l) \qquad \Delta H = -285.8 \text{ kJ}$$

In one case the product is a liquid. In the other case the product is a gas. The condensation of 1 mol of water vapor to 1 mol of liquid water at 25°C produces 44.0 kJ of heat.

Problem

12. If 44.0 kJ of heat is liberated when $H_2O(g) \longrightarrow H_2O(l)$, calculate the energy released in calories when 1 g of $H_2O(g)$ at 25°C is changed to 1 g of $H_2O(l)$ at the same temperature.

Enthalpy changes occur when a compound is formed from its elements. In enthalpy measurements the ΔH of a free element in its standard state is arbitrarily set to zero. For example, for the diatomic molecules— $H_2(g)$, $N_2(g)$, $O_2(g)$, $F_2(g)$, $Cl_2(g)$, $Br_2(l)$, and $I_2(s)$—the enthalpy is 0. Similarly, the enthalpy for carbon in the form of graphite, $C(s, \text{graphite})$ is also 0. *The ΔH for a reaction in which 1 mol of a compound is formed from its elements is the* **standard heat of formation,** ΔH_f^0, *of that compound.* Table 8·1 gives ΔH_f^0 for some common substances. The more negative the value of ΔH_f^0, the more stable the compounds.

Table 8-1 Standard Heats of Formation (ΔH_f^0) at 25°C and 1 atm

Substance	ΔH_f^0 (kJ/mol)
$Al_2O_3(s)$	-1676.0
$Br_2(g)$	30.91
$Br_2(l)$	0.0
$C(s, \text{diamond})$	1.9
$C(s, \text{graphite})$	0.0
$CH_4(g)$	-74.86
$CO(g)$	-110.5
$CO_2(g)$	-393.5
$CaCO_3(s)$	-1207.0
$CaO(s)$	-635.1
$Cl_2(g)$	0.0
$F_2(g)$	0.0
$Fe(s)$	0.0
$Fe_2O_3(s)$	-822.1
$H_2(g)$	0.0
$H_2O(g)$	-241.8
$H_2O(l)$	-285.8
$H_2O_2(l)$	-187.8
$HCl(g)$	-92.31
$H_2S(g)$	-20.1
$I_2(g)$	62.4
$I_2(s)$	0.0
$N_2(g)$	0.0
$NH_3(g)$	-46.19
$NO(g)$	90.37
$NO_2(g)$	33.85
$Na_2CO_3(s)$	-1131.1
$NaCl(s)$	-411.2
$O_2(g)$	0.0
$O_3(g)$	142.0
$P(s, \text{white})$	0.0
$P(s, \text{red})$	-18.4
$S(s, \text{rhombic})$	0.0
$S(s, \text{monoclinic})$	0.30
$SO_2(g)$	-296.8
$SO_3(g)$	-395.7

Standard heats of formation of compounds are useful for calculating heats of reaction under conditions of standard temperature and pressure. Because of the law of conservation of energy, the ΔH for a reaction is the difference between the standard enthalpies of formation of all reactants and all products.

$$\Delta H = \Delta H_f^0(\text{products}) - \Delta H_f^0(\text{reactants})$$

This is demonstrated in the following example.

Example 13

What is the change in enthalpy (ΔH) for the following reaction when all reactants and products are in the gaseous state?

$$\text{Sulfur dioxide} + \text{oxygen} \longrightarrow \text{sulfur trioxide}$$

Solution

First, write a balanced equation, including the physical states of all products and reactants.

$$2SO_2(g) + O_2(g) \longrightarrow 2SO_3(g)$$

Then, use Table 8·1 to find the standard heats of formation per mole.

$$\Delta H_f^0 \; O_2(g) = 0.0 \text{ kcal/mol (free element)}$$
$$\Delta H_f^0 \; SO_2(g) = -296.8 \text{ kJ/mol}$$
$$\Delta H_f^0 \; SO_3(g) = -395.7 \text{ kJ/mol}$$

Sum the ΔH_f^0 of the reactants while taking into account the number of moles of each.

$$2 \text{ mol } SO_2 \times \frac{-296.8 \text{ kJ}}{1 \text{ mol } SO_2} + 1 \text{ mol } O_2 \times \frac{0.0 \text{ kJ}}{1 \text{ mol } O_2} = -593.6 \text{ kJ}$$

Sum the ΔH_f^0 of the products in a similar way.

$$2 \text{ mol } SO_3 \times \frac{-395.7 \text{ kJ}}{1 \text{ mol } SO_3} = -791.4 \text{ kJ}$$

The change in enthalpy is equal to the heats of formation of the products minus the heats of formation of the reactants.

$$\Delta H = (-791.4 \text{ kJ}) - (-593.6 \text{ kJ}) = -197.8 \text{ kJ}$$

The reaction of oxygen and sulfur dioxide to give sulfur trioxide is exothermic.

Problem

13. Calculate the change in enthalpy for these reactions.
   a. $CH_4(g) + 2O_2(g) \longrightarrow CO_2(g) + 2H_2O(l)$
   b. $2CO(g) + O_2(g) \longrightarrow 2CO_2(g)$

8 Stoichiometry
Chapter Review

Key Terms

actual yield	8·6	limiting reagent	8·5
endothermic		percent yield	8·6
reaction	8·7	standard heat of	
enthalpy (H)	8·8	formation (ΔH_f^0)	8·8
excess reagent	8·5	stoichiometry	8·0
exothermic reaction	8·7	theoretical yield	8·6
heat of combustion	8·8	thermochemical	
heat of reaction	8·8	equation	8·7

Chapter Summary

A balanced chemical equation may be interpreted in several different ways. For example, the coefficients tell us the relative number of moles of reactants and products. Chemists use moles to do chemical arithmetic, or stoichiometry. All stoichiometric calculations involving chemical reactions begin with a balanced equation. Balanced equations are necessary because mass is conserved in every chemical reaction. The number and kinds of atoms in the reactants equal the number and kinds of atoms in the products.

Stoichiometric problems are solved using conversion factors derived from a balanced chemical equation. The conversion factor relates the moles of a given substance to the moles of the desired substance. Units such as mass, volume of gases (at STP), and particles are converted to moles when working stoichiometry problems.

Whenever quantities of two or more reactants are given in a stoichiometry problem, we must identify the limiting reagent. A limiting reagent is completely used up in a chemical reaction. The amount of limiting reagent determines the amount of product that is formed in a chemical reaction. If there is a single limiting reagent in a reaction, all the other reactants are in excess.

A theoretical yield is the maximum amount of product that can be obtained from a given amount of reactants in a chemical reaction. An actual yield is the amount of product obtained when the reaction is carried out in the laboratory. A ratio of the actual yield to the theoretical yield, expressed as a percent, is the percent yield of a reaction.

Energy changes are always associated with chemical reactions. The enthalpy change (ΔH) is a measure of the heat absorbed or released during a chemical reaction. A reaction in which heat is released (negative ΔH) is termed exothermic. In an endothermic reaction heat is absorbed, and ΔH is positive. The heat absorbed or evolved when one mole of a compound is formed from its elements is the standard heat of formation.

Practice Questions and Problems

14. Interpret each of these chemical equations in terms of interacting particles. *8·1*
  a. $2Zn(s) + O_2(g) \longrightarrow 2ZnO(s)$
  b. $2KClO_3(s) \longrightarrow 2KCl(s) + 3O_2(g)$
  c. $4NH_3(g) + 6NO(g) \longrightarrow 5N_2(g) + 6H_2O(g)$

15. Interpret each of the equations in Problem 14 in terms of interacting numbers of moles of reactants and products. *8·1*

16. Calculate and compare the mass of the reactants with the mass of the products for each of the equations in Problem 14. Show that each balanced equation obeys the law of conservation of mass. *8·1*

17. Carbon disulfide is an important industrial solvent. It is prepared by the reaction of coke with sulfur dioxide. *8·2*

$$5C + 2SO_2 \longrightarrow CS_2 + 4CO$$

  a. How many moles of CS_2 form when 6.30 mol of C reacts?

b. How many moles of carbon are needed to react with 7.24 mol of SO_2?

c. How many moles of carbon monoxide form at the same time that 0.762 mol of CS_2 forms?

d. How many moles of SO_2 are required to make 182 mol of CS_2?

18. Methanol, CH_3OH, is used in the production of many chemicals. Methanol is made by reacting carbon monoxide and hydrogen at high temperature and pressure. *8·3*

$$CO(g) + 2H_2(g) \longrightarrow CH_3OH(g)$$

a. How many moles of each reactant are needed to form 6.00×10^2 g of CH_3OH?

b. Calculate the number of grams of each reactant needed to produce 10.0 mol of CH_3OH.

c. How many grams of hydrogen are necessary to react with 5.74 mol of CO?

19. Oxygen is produced by the reaction of sodium peroxide and water. *8·3*

$$2Na_2O_2(s) + 2H_2O(l) \longrightarrow O_2(g) + 4NaOH(aq)$$

a. Calculate the mass of Na_2O_2 in grams needed to form 3.20 g of oxygen.

b. How many grams of NaOH are produced when 3.20 g of O_2 is formed?

c. When 0.48 g of Na_2O_2 is dropped in water, how many grams of O_2 are formed?

20. Lithium nitride reacts with water to form ammonia and aqueous lithium hydroxide. *8·4*

$$Li_3N(s) + 3H_2O(l) \longrightarrow NH_3(g) + 3LiOH(aq)$$

a. What mass of water is needed to react with 98.7 g of Li_3N?

b. When the above reaction takes place, how many molecules of NH_3 are produced?

c. Calculate the number of grams of Li_3N that must be added to an excess of water to produce 45.0 L of NH_3 (at STP).

21. How would you identify a limiting reagent in a chemical reaction? *8·5*

22. For each of these balanced equations, identify the limiting reagent for the given combination of reactants. *8·5*

a. $2H_2 + O_2 \longrightarrow 2H_2O$
   5.0 mol   2.3 mol

b. $4P + 5O_2 \longrightarrow P_4O_{10}$
   7.0 mol   8.0 mol

c. $2Al + 3Cl_2 \longrightarrow 2AlCl_3$
   5.4 mol   8.0 mol

d. $P_4O_{10} + 6H_2O \longrightarrow 4H_3PO_4$
   0.37 mol   2.50 mol

23. For each reaction in Problem 22, calculate the number of moles of product formed. *8·5*

24. For each reaction in Problem 22, calculate the number of moles of excess reagent remaining after the reaction. *8·5*

25. Acetylene, C_2H_2, will burn in the presence of oxygen. *8·6*

$$2C_2H_2(g) + 5O_2(g) \longrightarrow 4CO_2(g) + 2H_2O(g)$$

How many grams of water can be produced by the reaction of 4.80 mol of C_2H_2 with 14.8 mol of O_2?

26. What is a percent yield of a chemical reaction a measure of? *8·6*

27. When 84.8 g of iron(III) oxide reacts with an excess of carbon monoxide, then 57.8 g of iron is produced.

$$Fe_2O_3(s) + 3CO(g) \longrightarrow 2Fe(s) + 3CO_2(g)$$

What is the percent yield of this reaction? *8·6*

28. When 50.0 g of silicon dioxide is heated with an excess of carbon, 32.2 g of silicon carbide is produced.

$$SiO_2(s) + 3C(s) \longrightarrow SiC(s) + 2CO(g)$$

What is the percent yield of this reaction? *8·6*

29. The manufacture of compound F requires five separate chemical reactions. The initial reactant, compound A, is converted to compound B; compound B is converted to compound C, and so forth. The diagram below summarizes the step-wise manufacture of compound F, including the percent yield for each step. Provide the missing quantities or missing percent yields. Assume that the reactant and product in each step react in a one-to-one mole ratio. *8·6*

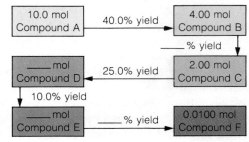

30. Distinguish between an exothermic and an endothermic reaction. *8·7*

31. The reaction of iron with carbon dioxide to form iron(III) oxide and carbon monoxide is an endothermic reaction. *8·7*

$$2Fe(s) + 3CO_2(g) + 26.3\ kJ \longrightarrow Fe_2O_3(s) + 3CO(g)$$

a. How many kilojoules of energy are needed for 7.3 mol of Fe to react with an excess of CO_2?
b. Is this a positive or negative enthalpy change?

32. The burning of magnesium in oxygen is a very exothermic reaction. *8·7*

$$2Mg(s) + O_2(g) \longrightarrow 2MgO(s) + 288\ kcal$$

a. How much heat, in kilocalories, is given off when 0.12 mol of Mg reacts with an excess of oxygen?
b. Express the answer in kilojoules of energy.

33. Why must the physical state of a substance be given in a thermochemical reaction? *8·7*

34. What is the enthalpy change (or heat of reaction) of a chemical reaction? *8·8*

35. What is the standard heat of formation of a compound? *8·8*

36. Which of these compounds has a negative standard heat of formation? *8·8*
a. $NO_2(g)$ **b.** $SO_2(g)$ **c.** $CaO(s)$ **d.** $H_2O_2(l)$

37. Calculate the change in enthalpy (ΔH) for these reactions. *8·8*
a. $I_2(s) \longrightarrow I_2(g)$
b. $3CO(g) + 2Fe_2O_3(s) \longrightarrow Fe(s) + 3CO_2(g)$
c. $2NO_2(g) \longrightarrow 2NO(g) + O_2(g)$

Mastery Questions and Problems

38. Zinc reacts with nitric acid to form zinc nitrate, ammonium nitrate, and water.

$$4Zn(s) + 10HNO_3(aq) \longrightarrow 4Zn(NO_3)_2(aq) + NH_4NO_3(aq) + 3H_2O(l)$$

a. How many atoms of zinc react with 5.96 g of HNO_3?
b. Calculate the number of grams of zinc that must react with an excess of HNO_3 to form 87.3 g of NH_4NO_3.

39. The formation of nitrogen monoxide from its elements is an endothermic process.

$$N_2(g) + O_2(g) + 43.2\ kcal \longrightarrow 2NO(g)$$

Given an unlimited amount of nitrogen and oxygen, how many grams of NO can be formed using 2.0×10^3 kcal of heat energy?

40. Hydrazine, N_2H_4, is used as a rocket fuel. It reacts with oxygen to form nitrogen and water.

$$N_2H_4(l) + O_2(g) \longrightarrow N_2(g) + 2H_2O(g)$$

a. How many liters of N_2 (at STP) form when 1.0 kg of N_2H_4 reacts with 1.0 kg O_2?
b. How many grams of the excess reagent remain after the reaction?

41. The pollutant sulfur dioxide can be removed from the emissions of an industrial plant by reaction with calcium carbonate and oxygen.

$$2CaCO_3(s) + 2SO_2(g) + O_2(g) \longrightarrow 2CaSO_4(s) + 2CO_2(g)$$

If this reaction proceeds with a 96.8% yield, how many kilograms of $CaSO_4$ are formed when 5.24 kg of SO_2 reacts with an excess of $CaCO_3$ and O_2?

42. The combustion of ethene, C_2H_4, liberates 1.39×10^3 kJ/mol.

$$C_2H_4(g) + 3O_2(g) \longrightarrow 2CO_2(g) + 2H_2O(l) + 1.39 \times 10^3\ kJ$$

Calculate the amount of heat liberated when 1.87 g of C_2H_4 reacts with excess oxygen.

43. Ammonium nitrate will decompose explosively at high temperatures to form nitrogen, oxygen, and water vapor.

$$2NH_4NO_3(s) \longrightarrow 2N_2(g) + 4H_2O(g) + O_2(g)$$

a. What is the total number of liters of gas formed when 45.6 g of NH_4NO_3 is decomposed? (Assume STP.)
b. Calculate the number of grams of water formed when 10.0 L of N_2 is produced in this reaction. (Assume STP.)

44. When 50.0 g of silicon dioxide is heated with an excess of carbon, 32.2 g of silicon carbide is produced.

$$SiO_2(s) + 3C(s) \longrightarrow SiC(s) + 2CO(g)$$

a. What is the percent yield of this reaction?
b. How many grams of CO gas are made?

Critical Thinking Questions

45. Choose the term that completes the second relationship.
- **a.** equation:coefficients   balance: _____
  - (1) moles
  - (3) weight
  - (2) standard masses
  - (4) atoms
- **b.** up:down   endothermic: _____
  - (1) combustion
  - (3) moles
  - (2) heat
  - (4) exothermic
- **c.** meter stick:length   calorimeter: _____
  - (1) heat
  - (3) product
  - (2) exothermic
  - (4) foam cup

46. Given a certain quantity of reactant, you calculate that a particular reaction should produce 55 g of a product. When you perform the reaction you find that you have produced 63 g of product. What is your percent yield? What could have caused a percent yield over 100%?

47. Heat of reaction, heat of combustion, and heat of formation are similar terms but have different meanings. For each term, write a chemical equation for a reaction in which that term best describes the change in enthalpy that occurs.

48. Standard heat of formation is measured at 1 atm and 25°C. Why would scientists choose these conditions instead of STP?

Review Questions and Problems

49. What is the molecular formula of oxalic acid, gram molecular mass 90 g/mol. Its percent composition is 71.1% O, 26.7% C, and 2.2% H.

50. What is the mass, in grams, of a molecule of benzene, C_6H_6?

51. How many electrons, protons, and neutrons are in an atom of each isotope?
- **a.** magnesium-26
- **c.** oxygen-18
- **b.** titanium-47
- **d.** tin-120

52. Write these formulas.
- **a.** manganese(II) chromate
- **b.** hydrobromic acid
- **c.** aluminum carbonate
- **d.** silicon dioxide
- **e.** potassium sulfide

53. How many grams of beryllium are in 255 g of the mineral beryl, $Be_3Al_2Si_6O_{18}$?

54. Write a balanced chemical equation for each of these reactions.
- **a.** When a mixture of aluminum and iron(II) oxide is heated metallic iron and aluminum oxide are produced.
- **b.** Sodium bicarbonate reacts with acetic acid to produce sodium acetate, water, and carbon dioxide.
- **c.** The complete combustion of isopropyl alcohol, C_3H_7OH, produces carbon dioxide and water vapor.
- **d.** Lead(II) nitrate when heated decomposes to form lead(II) oxide, nitrogen dioxide, and molecular oxygen.

Challenging Questions and Problems

55. A car gets 9.2 kilometers to a liter of gasoline. Assuming that gasoline is 100% octane, C_8H_{18} (which has a specific gravity of 0.69), how many liters of air (21% oxygen by volume at STP) will be required to burn the gasoline for a 1250-km trip? Assume complete combustion.

56. Ethyl alcohol, C_2H_5OH, can be produced by the fermentation of glucose, $C_6H_{12}O_6$. If it takes 5.0 hr to produce 8.0 kg of alcohol, how many days will it take to consume 1.0×10^3 kg of glucose? An enzyme (biological catalyst) is used to increase the rate of this reaction.

$$C_6H_{12}O_6 \xrightarrow{\text{enzyme}} C_2H_5OH + CO_2$$

57. A 435.0-g sample of $CaCO_3$ that is 95.0% pure gives 225 L of CO_2 when reacted with an excess of hydrochloric acid.

$$CaCO_3 + HCl \longrightarrow CaCl_2 + CO_2 + H_2O$$

What is the density (in grams/liter) of the CO_2?

58. Aspirin, $C_9H_8O_4$, is made by the reaction of salicylic acid, $C_7H_6O_3$, and acetic anhydride, $C_4H_6O_3$.

$$C_7H_6O_3 + C_4H_6O_3 \longrightarrow C_9H_8O_4 + HC_2H_3O_2$$

How many grams of each reactant should be used to produce 75.0 g of aspirin?

59. SO_3 can be produced in this two-step process.

$$FeS_2 + O_2 \longrightarrow Fe_2O_3 + SO_2$$

$$SO_2 + O_2 \longrightarrow SO_3$$

Assuming that all the FeS_2 reacts, how many grams of SO_3 are produced when 20.0 g of FeS_2 reacts with 16.0 g of O_2?

60. The white limestone cliffs of Dover, England, contain a large percentage of calcium carbonate, $CaCO_3$. A sample of limestone weighing 84.4 g reacts with an excess of hydrochloric acid to form calcium chloride.

$$CaCO_3 + 2HCl \longrightarrow CaCl_2 + H_2O + CO_2$$

The mass of calcium chloride formed is 81.8 g. What is the percent of calcium carbonate in the limestone?

Research Projects

1. Some reactions such as those between copper and sulfur are nonstoichiometric (or seemingly so). Study the conditions that influence the combinations of copper and sulfur that can be obtained.

2. Investigate the factors that affect the percent yield of a reaction.

3. What chemistry is used in the detection of fingerprints?

4. Find out more about the methods used by forensic chemists. Write a short story that includes a mystery which is solved by a forensic chemist.

5. Why was the development of Portland cement important? How is cement used to make concrete? What are the principal types of concrete and how are they used?

6. Compare the chemical composition of earth, the moon, and asteroids. How do scientists identify the source of meteorites?

7. Compile a list of common everyday chemical reactions that are exothermic and a list of reactions that are endothermic.

8. How are waste products turned into energy sources?

9. Discuss some exothermic reactions in which the heat given off is useful, and some in which it is not useful. Do the same for endothermic reactions and the heat absorbed.

Readings and References

Blassingame, Wyatt. *Science Catches the Criminal.* New York: Dodd, 1975.

Gerber, Samuel M., ed. *Chemistry and Crime: From Sherlock Holmes to Today's Courtroom.* Washington, DC: American Chemical Society, 1983.

O'Connor, Rod, and Charles Mickey. *Solving Problems in Chemistry: With Emphasis on Stoichiometry and Equilibrium.* New York: Harper & Row, 1974.

Smyth, Frank. *Cause of Death: The Study of Forensic Science.* New York: Van Nostrand Reinhold, 1980.

9 The States of Matter

Chapter Preview

9·1 Kinetic Theory and the Nature of Gases

9·2 Kinetic Energy and Temperature

9·3 Pressure

9·A Measuring Pressure

9·4 Avogadro's Hypothesis

9·5 The Nature of Liquids

9·6 Vaporization

9·7 The Boiling Point of a Liquid

9·B Liquid Crystals

9·8 The Nature of Solids

9·9 Phase Changes

9·10 Energy and Phase Changes

9·C Plasma: The Fourth State of Matter

We know that matter exists as solids, liquids, and gases. Liquids solidify, and gases liquefy when cooled. Why do changes in the state of matter occur with changes in temperature? We will find out in this chapter with the help of a model called the kinetic theory.

9·1 Kinetic Theory and the Nature of Gases

If you open a bottle of perfume, the aroma soon reaches the nose of anyone nearby. Obviously, the molecules of the fragrance have diffused from the bottle. This diffusion is simple evidence that molecules move.

The word *kinetic* means motion. *The* **kinetic theory** *says that the tiny particles in all forms of matter are in constant motion.* These particles may be atoms, ions, or molecules of gases, liquids, or solids. For the present let us concentrate on the kinetic theory as it applies to gases. We know that a gas is easily compressed and fills whatever container it is in. This behavior and many other properties of gases can be explained by the kinetic theory.

Here are the basic assumptions of the kinetic theory of gases.

1. *A gas is composed of particles, usually molecules or atoms.* These particles are considered to be small hard spheres with negligible volume and far from one another. Between the particles is empty space, absolute nothingness. No attractive or repulsive forces exist between the particles.

Greek: *kinetos* = to move.

The kinetic theory describes the motion of particles.

Figure 9·1
Life on planet Earth is dependent on water in all three states: as a solid, liquid and gas. Water vapor (a gas) is invisible, but it is an important component of air.

197

Figure 9·2
a Gas particles have random and chaotic movements. They are constantly colliding with one another and with the walls of the container.
b As a single gas particle moves through space it frequently changes direction due to collisions with other particles.

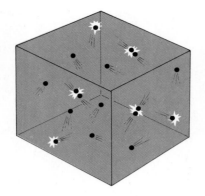 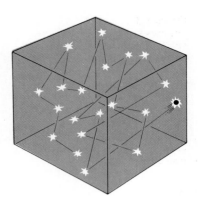

2. *The particles in a gas move rapidly in constant random motion.* They travel in straight lines and move independently of each other. The particles change direction when they rebound from collisions with one another or with other objects (Figure 9·2).

3. *All collisions are perfectly elastic.* Perfectly elastic means that energy is transferred from one particle to another during collisions, but the total kinetic energy remains constant.

Measurements indicate that the average speed of oxygen molecules in air at 20°C is about 1656 km/hr. We say "average speed" because some particles are moving faster and some slower. At such high speeds the molecules in a bottle of perfume opened in Washington, D.C. should reach Vancouver, B.C. in about three hours. They never get there, however, because their path of unimpeded travel is very short. The perfume molecules are constantly striking air molecules and rebounding in other directions. The aimless path they take is called a *random walk* (Figure 9·2).

The physical behavior of a gas depends on its volume, temperature, and pressure. Gases in sealed containers are *contained gases*. The volume of a contained gas is the volume of the container, because gases expand to completely fill their containers. Gases that are not in sealed containers are uncontained gases. Air is usually an *uncontained gas,* although it may be contained by sealing it in a closed vessel. *Temperature* and *pressure* are terms we often use rather loosely, but what do they really mean? We will find out in the next two sections.

> Temperature and pressure affect the volume of a gas.

9·2 Kinetic Energy and Temperature

Motion suggests energy. *The energy an object has because of its motion is kinetic energy.* Because gas particles are in motion, they have kinetic energy. When a gas is heated, the particles in the gas absorb thermal energy. Some of it goes into increasing the energy within the particles. The rest goes into increasing the motion of particles, that is, increasing their kinetic energy.

Figure 9·3
The curve on the left shows the kinetic energy distribution of a typical collection of molecules. Notice that most molecules have intermediate energies. The curve on the right shows the energy distribution of the same collection of molecules at a higher temperature. The average kinetic energy has increased. The range of energies of the molecules is also greater.

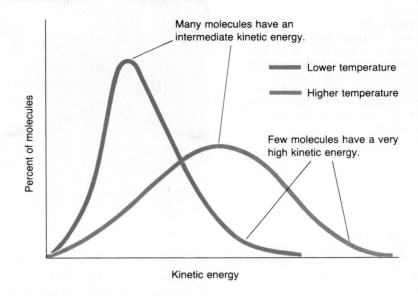

Many molecules have an intermediate kinetic energy.

Lower temperature
Higher temperature

Few molecules have a very high kinetic energy.

Percent of molecules

Kinetic energy

■ Temperature measures the average kinetic energy of particles in a substance.

In a collection of molecules there will be a wide range of kinetic energies from very low to very high. Most molecules will have speeds in between (Figure 9·3). As the temperature rises so does the kinetic energy. When discussing the kinetic energy of a substance, scientists use the average kinetic energy of the particles. *The average kinetic energy of the particles of a substance is proportional to the temperature of the substance.* We can measure how much the average kinetic energy has been increased by measuring the temperature increase. Temperature is a measure of the average kinetic energy of the particles in a substance. Note that it is not the total thermal energy the substance has absorbed.

Particles of all substances at the same temperature have the same average kinetic energy. An increase in the temperature of a substance signifies an increase in the average kinetic energy of the particles of the substance. A decrease in temperature signifies a decrease in the average kinetic energy. Figure 9·3 shows the distribution of kinetic energies at two different temperatures. Notice that at the higher temperature, there is a wider range of energies.

Problems

1. How is the average kinetic energy of water molecules affected when hot water is poured from a kettle into cups at the same temperature as the water?

2. Is the average kinetic energy of the particles in a block of ice at 0°C the same as or different from the average kinetic energy of the particles in a gas-filled dirigible at 0°C?

3. Does the average kinetic energy of the gas particles in an inflated life raft increase or decrease if the sun heats the life raft from 25°C to 37°C?

Figure 9·4
Many substances have no electrical resistance at temperatures close to absolute zero. This property is known as superconductivity. Superconducting circuits are used in large computers where ordinary circuits would generate too much heat.

 ChemDirections

Superconductivity is the lack of any resistance to an electric current. Only a few materials are superconductors, and then only at very low temperatures. Can you think of possible applications of superconductors? For more information, turn to **ChemDirections** *Superconductors*, page 656.

The temperature to which a substance can be raised has no theoretical upper limit. By contrast, however, there is a lower limit. At some low temperature, the particles of all substances should stop moving. The particles would have no kinetic energy at this temperature because they would have no motion. *The temperature at which the motion of particles ceases is known as absolute zero.* It falls at about $-273°C$. Taking a substance to a temperature *below* this lower limit is impossible. The Kelvin temperature scale is based on the idea of an absolute zero (Section 2·11). Zero degrees Kelvin (0 K) is $-273°$ on the Celsius scale. There are no minus numbers on the Kelvin scale because they are not needed. The Kelvin scale is extremely useful because it is a direct measure of the average kinetic energy of the particles of a substance. For example, the particles in a substance at 200 K have twice the average kinetic energy of the particles in a substance at 100 K.

Problem

4. By what factor does the average kinetic energy of the molecules of a gas in an aerosol container increase when the temperature is raised from 300 K (27°C) to 900 K (627°C)?

9·3 Pressure

Moving bodies exert forces when they collide with other bodies. Although a gas particle is a moving body, the force exerted by a single gas particle would not make a gnat hair twitch. It is not hard to imagine, however, that many simultaneous collisions would produce a measurable force on an object. **Gas pressure** *is the result of simultaneous collisions of billions upon billions of gas particles on an object.* When there are no gas particles present, they do not collide and cause pressure. *We call the resulting empty space a* **vacuum.**

Air exerts pressure because gravity holds air molecules in the earth's atmosphere. **Atmospheric pressure** *results from the collisions of air*

Figure 9·5
Barometers measure atmospheric pressure. This type of barometer is commonly used in laboratories. The average atmospheric pressure at sea level is 1 atm, or 760 mm Hg.

Collisions of gas particles with objects in their paths generate gas pressure.

1 atm = 760 mm Hg = 101.3 kPa.

Another unit of pressure, the *torr,* is named after Evangelista Torricelli (1608–1647), who invented the barometer. 1 torr = 1 mm Hg.

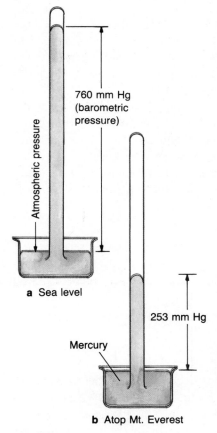

Figure 9·6
Normal atmospheric pressure pushing on a simple mercury barometer supports a column of mercury about 760 mm high. On top of Mt. Everest (at 9000-m altitude) the air exerts enough push to support a column of mercury only 253 mm high.

molecules with objects. It decreases with an increase in elevation. This is because the air layer around the earth thins out at high elevations.

Barometers *are commonly used to measure atmospheric pressure.* This pressure varies slightly, depending on the weather. *The SI unit of pressure is the* **pascal** (Pa). Atmospheric pressure at sea level is about 101.3 kilopascal (kPa). Two older units of pressure are millimeters of mercury and the atmosphere. **One millimeter of mercury** (1 mm Hg) *is the pressure needed to support a column of mercury 1 mm high.* This unit developed from the early use of mercury barometers. **One standard atmosphere** (1 atm) *is the pressure required to support 760 mm of mercury* (760 mm Hg) *in a mercury barometer at 25°C.* This is the average atmospheric pressure at sea level. Thus 1 atm equals 760 mm Hg. We will use atmospheres and mm Hg as the units of pressure. *Standard conditions when working with gases are a temperature of 0°C and a pressure of 1 atm.* This is standard temperature and pressure (STP).

Example 1

How many millimeters of mercury does a gas exert at 1.50 atm of pressure?

Solution

$$1.50 \text{ atm} \times \frac{760 \text{ mm Hg}}{1 \text{ atm}} = 1140 \text{ mm Hg} = 1.14 \times 10^3 \text{ mm Hg}$$

Problems

5. Convert 190 mm Hg to atmospheres of pressure.

6. The pressure at the top of Mount Everest is 253 mm Hg. Is this greater or less than 0.25 atm?

─── Science, Technology, and Society ───

9·A Measuring Pressure

The first barometers were straight glass tubes closed at one end. Each tube was filled with mercury and inverted so that the open end was below the surface of a dish of mercury. The level of mercury in the tube depended on the number of collisions of air molecules with the pool of mercury in the dish. At sea level these collisions are sufficient to support a 760 mm column of mercury in the tube. This pressure, 760 mm Hg, is called one standard atmosphere (1 atm).

Today many barometers do not contain mercury. They are called aneroid barometers. In these devices atmospheric pressure is related to the number of collisions of air molecules with a sensitive metal diaphragm. The diaphragm controls the movement of a pointer, which gives the pressure reading.

9·4 Avogadro's Hypothesis

The particles that make up different gases must have different sizes. For example, chlorine molecules have large numbers of electrons, protons, and neutrons. They must be bigger and occupy more volume than hydrogen molecules, which have only two protons and two electrons. Early scientists recognized this. Thus many reacted in disbelief in 1811 when they heard of **Avogadro's hypothesis:** *Equal volumes of gases at the same temperature and pressure contain equal numbers of particles.* It was as if Amedeo Avogadro were suggesting that two rooms of the same size could be filled by the same number of objects, no matter whether the objects were marbles or oranges.

What Avogadro had in mind is not so mysterious. It makes sense if the particles in a gas are very far apart with nothing but space in between (Figure 9·7). You can suspend ten marbles or ten oranges in a ballroom without decreasing the volume of the room very much. If we put the marbles or oranges into motion, they would occasionally collide with one another or with the walls of the room. On the average there would be large expanses of space between them. The same concept can be applied to gases. This was Avogadro's great insight, and it is easily demonstrated by experiment. At STP, 1 mol (6.02×10^{23} particles) of any gas, regardless of the size of the particles, occupies 22.4 L.

We can also understand Avogadro's hypothesis by thinking of the modern explanation for gas pressure. The same number of particles in a given volume at the same temperature should exert the same pressure. This is because the particles have the same average kinetic energy and are contained within an equal amount of space. *Whenever we have equal volumes of gases at the same temperature and pressure, the volumes must contain equal numbers of particles.*

A gas is mostly empty space.

Avogadro's hypothesis: At the same conditions of temperature and pressure, equal volumes of gases contain the same number of particles.

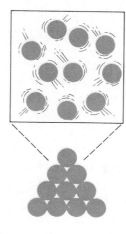
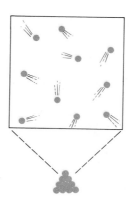

Figure 9·7
The volume of a container easily accommodates an equal number of rapidly moving large or small particles as long as the particles are not tightly packed. When the particles are tightly packed, large particles take up more space than small particles.

Example 2

Determine the volume, in liters, occupied by 0.030 mol of a gas at STP.

Solution

We know that 1 mol of a gas at STP has a volume of 22.4 L.

$$0.030 \text{ mol} \times \frac{22.4 \text{ L}}{1 \text{ mol}} = 0.67 \text{ L}$$

Example 3

How many oxygen molecules are in 3.36 L of oxygen gas at STP?

Solution

At STP, 22.4 L of oxygen contains 1 mol. One mole of oxygen contains 6.02×10^{23} oxygen molecules.

$$3.36 \text{ L } O_2 \times \frac{1 \text{ mol } O_2}{22.4 \text{ L } O_2} \times \frac{6.02 \times 10^{23} \text{ molecules } O_2}{1 \text{ mol } O_2}$$
$$= 9.03 \times 10^{22} \text{ molecules } O_2$$

Problems

7. How many moles is 11.2 L of hydrogen gas at STP? How many particles (hydrogen molecules) does it contain?

8. What is the volume occupied by 0.25 mol of a gas at STP?

9. Hydrogen reacts with oxygen to give water according to this familiar combustion reaction.

$$2H_2 + O_2 \longrightarrow 2H_2O$$

   What volume of oxygen (liters) at STP is required to combine with 1 mol of hydrogen?

Example 4

Determine the volume in liters occupied by 14.0 g of nitrogen gas at STP.

Solution

Nitrogen is a diatomic molecule (N_2). One mole of nitrogen gas has a mass of 28.0 g and occupies a volume of 22.4 L at STP.

$$14.0 \text{ g } N_2 \times \frac{1 \text{ mol } N_2}{28.0 \text{ g } N_2} \times \frac{22.4 \text{ L } N_2}{1 \text{ mol } N_2} = 11.2 \text{ L } N_2$$

1 mol of any gas at STP has 6.02×10^{23} particles and occupies 22.4 L.

Issues in Chemistry

At the mention of air pollution you might envision choking exhaust fumes streaming from a bus or a smokey haze hanging over an industrial complex. Unfortunately, air pollution inside your home, school, and workplace can be just as harmful as outdoor pollution. What are the sources and how can common indoor air pollutants be reduced or eliminated?

Problem

10. What is the volume of a container if it holds 8.8 g of carbon dioxide at STP?

9·5 The Nature of Liquids

Attractive forces tend to balance disruptive forces in liquids.

We have been able to explain many properties of gases by assuming that gas particles have motion and that there is no attraction between them. The particles that constitute liquids are also in motion. They are free to slide past one another. Hence both liquids and gases flow (Figure 9·9). Unlike gas particles, the particles of a liquid are held together by weak attractive forces. Most particles of a liquid do not have enough kinetic energy to overcome the attractive forces and escape. The forces also reduce the amount of space between the particles of the liquid (Figure 9·8). Thus liquids are much denser than gases. Increasing the pressure on a liquid has hardly any effect on its volume. For this reason, liquids and solids are known as the *condensed states of matter*. Many physical properties of liquids are determined by the interplay between the disruptive motions of particles and the attractive forces between them. Several of these properties are the vapor pressure, the heat of vaporization, and the boiling point of a liquid.

Figure 9·8
The particles in a liquid change places because of their motions. Most of the particles do not escape the liquid because of weak attractive forces between them.

Figure 9·9
Both liquids and gases flow. If a gas is denser than air, then it can be poured from one container to another. The liquid on the left is colored water. The gas on the right is bromine vapor. These pictures were taken in a fume hood because bromine is both toxic and corrosive.

9·6 Vaporization

Vaporization is the escape of molecules from the surface of a liquid.

The scientific term for the conversion of a liquid to a gas or vapor below its boiling point is **vaporization.** We all know that water in an open vessel goes into the air upon standing (Figure 9·10). *When vaporization of an uncontained liquid occurs, we commonly call it* **evaporation.**

Figure 9·10
a In an open container, water molecules evaporate from the liquid and escape from the container. Equilibrium cannot be established in this system.
b In a closed container, water molecules do not escape but collect as a vapor above the liquid. A dynamic equilibrium between the vapor and the liquid is established.

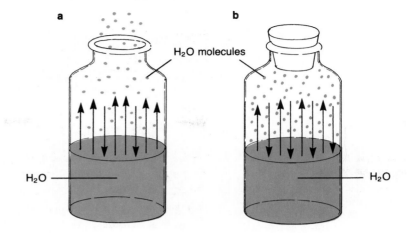

Safety

Never heat flammable liquids over an open flame. Use a steam bath or electric hot plate.

Figure 9·11
This terrarium is at equilibrium. The rate of evaporation equals the rate of condensation. Notice the condensation on the inner surface of the glass.

Molecules at the surface of the liquid break away and go into the gas phase. Only those liquid molecules that possess a certain minimum kinetic energy can leave the surface. Some escaping particles collide with air molecules and return to the liquid, but others escape completely.

A liquid evaporates faster when heated because the kinetic energy of its particles increases. This enables more particles to overcome the attractive forces keeping them in the liquid. Although it requires heat, evaporation itself is a cooling process. This is because the particles with the highest kinetic energy (highest temperature) escape first. The particles left in the liquid have a lower average temperature than the particles that have escaped. When water molecules in perspiration evaporate from the skin's surface, the remaining perspiration is cooler. Therefore the remaining perspiration cools us by absorbing more body heat.

The vaporization of a liquid in a closed container is somewhat different from evaporation. This is because no particles can escape into the atmosphere (Figure 9·10). When a partially filled container of liquid is sealed, some of the particles in the liquid vaporize. *As vaporized particles collide with the walls of a sealed container, they produce a **vapor pressure** above the liquid.* As the container stands, the number of particles entering the vapor increases. Eventually some of the particles will condense and return to the liquid state.

$$\text{Liquid} \; \underset{\text{condensation}}{\overset{\text{evaporation}}{\rightleftharpoons}} \; \text{vapor (gas)}$$

After a time, the number of liquid particles vaporizing equals the number of particles condensing. The container is now saturated with vapor. A *dynamic equilibrium* exists between the gas and the liquid.

$$\text{Rate of evaporation} = \text{rate of condensation}$$

At equilibrium, the particles in the system are still vaporizing and condensing. At this point, there is no net change in either number of particles. One sign that equilibrium is established is that the inner walls of the container "sweat." The liquid that once vaporized is now condensing.

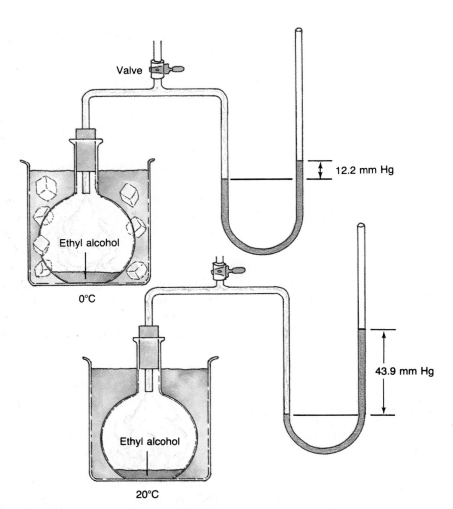

Figure 9·12
A manometer is similar to a barometer and is used to measure the pressure of a gas sample. This illustration shows how the vapor pressure of a liquid can be measured at two different temperatures. The height difference in the U-tube equals the vapor pressure at that temperature.

Valve

12.2 mm Hg

Ethyl alcohol

0°C

43.9 mm Hg

Ethyl alcohol

20°C

An increase in the temperature of a contained liquid increases the vapor pressure. This is because the particles in the warmed liquid have an increased kinetic energy. As a result, more of them will have the minimum kinetic energy necessary to escape the surface of the liquid. The particles escape the liquid and collide with the walls of the container. The vapor pressures of some common liquids at various temperatures are given in Table 9·1. Figure 9·12 illustrates one way of determining the vapor pressure of a liquid.

Safety

Solids also have vapor pressures. Keep all containers of chemicals (liquids and solids) tightly closed except when they are in use.

Table 9·1	Vapor Pressures (mm Hg) of Several Substances at Various Temperatures					
	0°C	20°C	40°C	60°C	80°C	100°C
Water	4.6	17.5	55.3	149.4	355.1	760.0
Ethyl alcohol	12.2	43.9	135.3	352.7	812.6	1693.3
Diethyl ether	185.3	442.2	921.1	1730.0	2993.6	4859.4

9·7 The Boiling Point of a Liquid

The rate of evaporation of a liquid in an open container increases with the addition of heat. This allows even larger numbers of particles at the liquid's surface to break the attractive forces keeping them in the liquid. The remaining particles in the liquid grow more and more agitated as they absorb thermal energy. Hence their average kinetic energy increases. When the liquid is finally heated to its boiling point, many of the particles in the liquid (not just those at its surface) have enough kinetic energy to vaporize. *The **boiling point** (bp) is the temperature at which the vapor pressure of the liquid is just equal to the external pressure.* Bubbles of vapor form throughout the liquid. The vapor pressure in these bubbles is equal to the atmospheric pressure. Vapor escapes to the atmosphere (Figure 9·13). This stage marks the onset of boiling.

The boiling point of a liquid at a pressure of 1 atm is the **normal boiling point** (Table 9·2). The boiling point of a liquid changes as the pressure changes. At lower external pressures, the boiling point decreases. This is because particles in the liquid need less kinetic energy to escape. The normal boiling point of water is 100°C. In Denver, however, water boils at 95°C. At 1600 m above sea level, Denver has an average

Evaporation is a process that occurs only at the surface of a liquid. Boiling occurs throughout a liquid.

At the normal boiling point, the vapor pressure of a liquid is 1 atm.

Figure 9·13
A liquid boils when the vapor pressure of particles in the bulk of the liquid equals the atmospheric pressure. **a** At 70°C and 760 mm Hg the atmospheric pressure at the surface of the liquid is greater than the vapor pressure of the liquid. Bubbles of vapor cannot form in the liquid. **b** At the boiling point the vapor pressure is equal to atmospheric pressure. Bubbles of vapor form inside the liquid, and the liquid boils. **c** At higher altitudes the atmospheric pressure is lower than at sea level. Thus the liquid boils at a lower temperature.

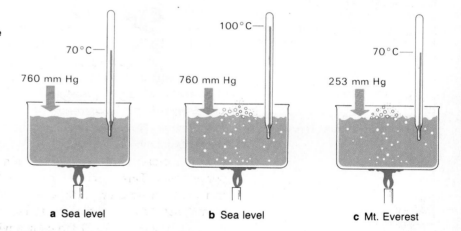

Table 9·2	The Normal Boiling Points of Several Common Substances	
	Name and formula	Boiling point (°C)
	Carbon disulfide (CS_2)	46.0
	Chloroform ($CHCl_3$)	61.7
	Ethanol (C_2H_6O)	78.5
	Octane (C_8H_{18})	126.0
	Water (H_2O)	100.0

Figure 9·14
The vapor pressure of a liquid increases with temperature. A liquid boils when its vapor pressure equals the external pressure. The intersection of a curve with the line 760 mm Hg pressure (1 atm) indicates the normal boiling point of that substance. At about what temperature would ethyl alcohol boil in Denver where the average atmospheric pressure is 640 mm Hg?

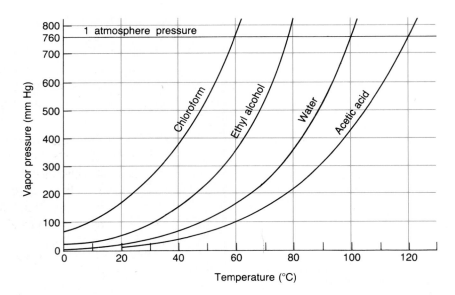

Recipes sometimes carry special instructions for people who live in cities high above sea level, such as Denver, Colorado. Those directions are needed because water boils at a lower temperature there, which means food takes longer to cook.

 Any liquid remains at a constant temperature while it is boiling.

atmospheric pressure of only 640 mm Hg. Similarly, at higher external pressures, a liquid's boiling point increases. This is because the particles in a liquid need more kinetic energy to escape. Pressure cookers reduce cooking time because at high pressure water boils at well above 100°C.

Odd as it may seem, boiling is a cooling process like evaporation. As in evaporation, the particles with the highest kinetic energy escape first when the liquid is at the boiling point. If no more heat is supplied, the temperature of the liquid will drop below its boiling point. If more heat is supplied, however, more particles acquire enough kinetic energy to escape. The result is a continual cooling effect on the remaining liquid. *Therefore the temperature of a boiling liquid never rises above its boiling point.* No matter how much heat is supplied, the liquid only boils faster. Eventually all the liquid boils away.

Figure 9·15
An autoclave is similar to a pressure cooker. It is commonly used to sterilize medical instruments. Autoclaves are usually set at 120°C and 1520 mm Hg (2 atm) for 15 minutes. The high pressure keeps the water at a high enough temperature to kill bacteria and their spores.

9·B Liquid Crystals

Not all substances change sharply from orderly solids to disorganized liquids when they melt. Some substances flow like liquids even though their molecules are still orderly. These materials are called liquid crystals.

The molecules in liquid crystals are shaped like rods. The three types of liquid crystals differ in the way the rodlike molecules are organized. In *nematic* substances, the rods remain parallel to each other as they move back and forth in the liquid. In *smectic* liquid crystals, the rods are not only parallel but also are arranged in layers. The layers can move, but individual molecules remain fixed within a layer. The molecules in one layer are parallel to those in another layer. In a *cholesteric* liquid crystal, the molecules are also in layers. Within a layer, the molecules are free to move back and forth, remaining parallel to each other, as they do in nematic crystals. In contrast to the smectic crystal, the molecules in one layer of a cholesteric crystal are not parallel to those in another layer.

When light strikes a smectic or cholesteric liquid crystal, it is reflected by the layers of molecules within the crystal. The color of light that is reflected depends on the distance between layers. The distance between layers changes as the temperature changes. Thus a temperature change in the crystal will show up as a color change. Liquid crystal thermometers can be placed on the skin to take body temperature or to locate a "hot spot," like an artery or a vein.

In a digital wristwatch or calculator, a nematic liquid crystal is contained between two layers of glass. The surface of the top layer of glass can be electrically charged in segments that can be grouped to create numbers. Under the influence of these charged segments, the molecules of the liquid crystal orient themselves in a way that prevents light from passing through. The numbers therefore appear black.

Liquid crystals flow like liquids but at certain temperatures may possess crystal-like ordered structures.

Figure 9·16
The molecules in liquid crystals are usually rod-like. **a** In nematic crystals the molecules are all parallel. **b** In smectic crystals the molecules are in layers as well as being parallel. **c** In cholesteric crystals the molecules are parallel within a layer but the layers are aligned as in a spiral.

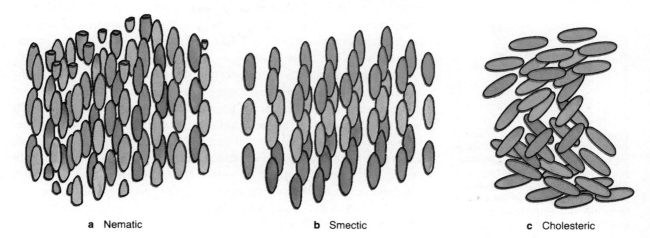

a Nematic   **b** Smectic   **c** Cholesteric

Figure 9·17
The regular shape of this sodium
chloride crystal (**a**) can be explained
by the arrangement of sodium ions
and chloride ions (**b**) within the crystal.

a

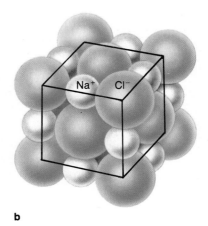

b

The particles in a solid vibrate
about fixed points.

ChemDirections

Ceramics are crystalline and glasses
are amorphous. To discover how the
physical properties of materials depend upon the arrangement of their
atoms read **ChemDirections** *Ceramics*, page 660.

9·8 The Nature of Solids

The motions in liquids are chaotic. In most solids, however, the particles
are packed against one another in a highly organized fashion. Rather than
sliding from place to place, they vibrate and rotate about fixed points
(Figure 9·18). Solids are dense and incompressible. Because of the fixed
position of their particles, solids do not flow to fit their containers.

When a solid is heated, its particles vibrate more rapidly. The kinetic energy of the particles increases. *The* **melting point** (mp) *is the
temperature at which the solid turns into a liquid.* At this temperature,
the disruptive vibrations of some of the particles are strong enough to
overcome the interactions that hold them in fixed positions. The solid begins to melt. The melting and freezing points of a substance are the
same. It is the temperature at which liquid and solid exist in equilibrium.

$$\text{Solid} \underset{\text{freezing}}{\overset{\text{melting}}{\rightleftharpoons}} \text{liquid}$$

In general, ionic solids have high melting points because they are
held together by relatively strong forces. Sodium chloride, an ionic compound, has a melting point of 801°C. By contrast, molecular solids have
relatively low melting points. For example, hydrogen chloride, a molecular compound, melts at $-112°C$. Not all solids melt. Wood, for example,
decomposes when heated.

Most solid substances are crystalline in nature. *In a* **crystal** *the
atoms, ions, or molecules are arranged in an orderly, repeating, three-
dimensional pattern.* This array is called the crystal lattice. All crystals
have a regular shape. The shape reflects the arrangement of the particles
within the solid. The angles at which the faces of a crystal intersect are
always the same for a given substance and are characteristic of that substance. Crystals are classified into seven crystal systems. They have characteristic shapes as shown in Figure 9·19. The seven crystal systems differ according to the angles between the faces and by how many of the
edges of the faces are equal.

Figure 9·18
The particles in a solid do not change
places but vibrate about fixed points.
A solid melts when its temperature is
raised to a point at which the vibrations of the particles become so intense that they disrupt the ordered
structure.

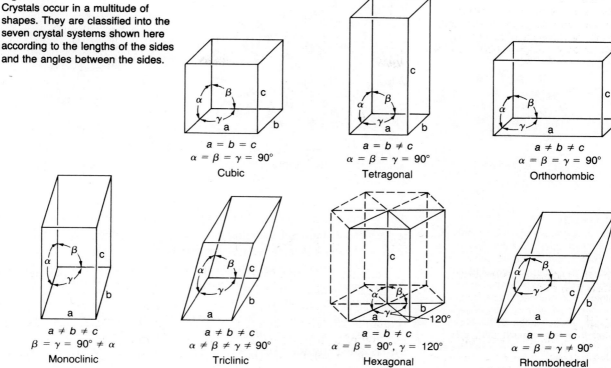

Figure 9·19
Crystals occur in a multitude of shapes. They are classified into the seven crystal systems shown here according to the lengths of the sides and the angles between the sides.

$a = b = c$
$\alpha = \beta = \gamma = 90°$
Cubic

$a = b \neq c$
$\alpha = \beta = \gamma = 90°$
Tetragonal

$a \neq b \neq c$
$\alpha = \beta = \gamma = 90°$
Orthorhombic

$a \neq b \neq c$
$\beta = \gamma = 90° \neq \alpha$
Monoclinic

$a \neq b \neq c$
$\alpha \neq \beta \neq \gamma \neq 90°$
Triclinic

$a = b \neq c$
$\alpha = \beta = 90°, \gamma = 120°$
Hexagonal

$a = b = c$
$\alpha = \beta = \gamma \neq 90°$
Rhombohedral

The shape of a crystal depends on the arrangement of the particles within it. *The smallest group of particles within a crystal that retains the geometric shape of the crystal is known as a* **unit cell.** A crystal lattice is a repeating array of any one of the fourteen known unit cells. The three unit cells that are part of the cubic crystal system are shown in Figure 9·20. Each crystal system has from one to four types of unit cells that are associated with that crystal system. The melting points of crystals are determined by how the atoms are bonded.

Some substances can exist in more than one type of solid state. A good example is the element carbon. Diamond is one crystalline form of carbon (Figure 9·21). It forms when carbon crystallizes under tremendous pressure (thousands of atmospheres). The other crystalline form of carbon is graphite. The "lead" in a pencil is actually graphite. In graphite

Figure 9·20
Three unit cells are part of the cubic crystal system. **a** In the simple cubic unit cell the atoms or ions are at the corners of an imaginary cube.
b In a face-centered cubic unit cell the atoms or ions are also in the center of each face of the imaginary cube. **c** In a body-centered cubic unit cell the atoms or ions are at the corners and in the center of an imaginary cube.

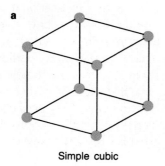

a

Simple cubic

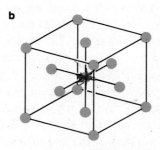

b

Face-centered cubic

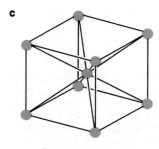

c

Body-centered cubic

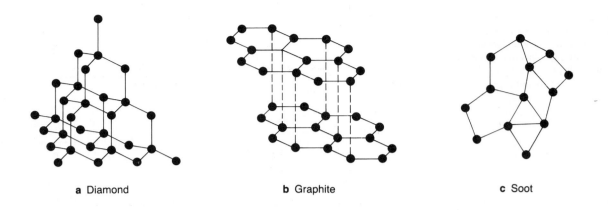

a Diamond   **b** Graphite   **c** Soot

Figure 9·21
The arrangement of carbon atoms affects the properties of the three forms of carbon. **a** In diamonds, atoms are packed in a compact symmetrical array. Each atom is bonded to four others. **b** In graphite the atoms are bonded together in sheets. Weak bonds between these sheets allow them to slide over each other. **c** In soot, the atoms are randomly bonded to one another.

 **ChemDirections**

New uses for glass include optical fibers and light-sensitive eyeglasses. These new technologies are described in **ChemDirections** *Glass*, page 661.

Safety

Glass retains heat well, and hot glass looks just like cool glass. To avoid burns, be careful when bending or fire-polishing glass in the laboratory.

the carbon atoms are packed more loosely than in diamond. The physical properties of diamond and graphite are quite different. Diamond has a high density and is very hard; graphite has a low density and is soft.

Not all solids are crystalline in form; some solids are amorphous. **Amorphous** *solids lack an ordered internal structure*. Rubber, plastic, and asphalt are amorphous solids. Soot, a third form of solid carbon is also amorphous. Its atoms are in a random arrangement (Figure 9·21).

Glasses are amorphous solids which are sometimes called **supercooled liquids.** They are transparent fusion products of inorganic substances that have cooled to a rigid state without crystallizing. The irregular internal structures of glasses are intermediate between those of a crystalline solid and a free-flowing liquid. Glasses do not melt at a definite temperature but gradually soften when heated.

When a crystalline solid is shattered, the fragments have the same surface angles as the original solid. By contrast, when an amorphous solid such as glass is shattered, the fragments have irregular angles and jagged edges.

Figure 9·22
Glass does not have a definite melting point because it is an amorphous solid.

Figure 9·23
As this phase diagram shows, the temperature of a substance remains constant while a phase change is taking place.

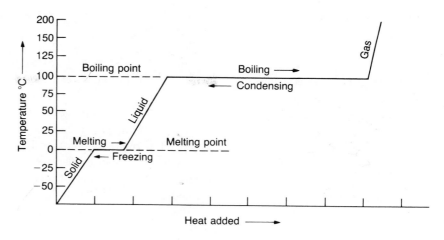

The temperature of a substance does not change while a phase change is taking place.

Figure 9·24
When solid iodine is heated, the crystals sublime, going directly from the solid to a gas. When the gas cools, solid crystals form without passing through the liquid state.

9·9 Phase Changes

A **phase change** *occurs whenever the physical state of a substance changes.* Melting, freezing, evaporation, and condensation involve phase changes. Imagine slowly warming a piece of ice, originally at $-50°C$ and normal atmospheric pressure (Figure 9·23). As heat energy is added, the temperature of the ice increases, but the ice does not change its physical appearance. At 0°C, however, if we continue to supply heat, the ice begins to melt and liquid water appears. The ice continues to melt, but the temperature stays at a constant 0°C until all the ice is gone. During this and other phase changes, adding or removing heat does not raise or lower the temperature of the system. It just changes the physical state of the substance. In the case of melting ice, the heat energy disrupts the forces that hold the ice crystal together.

If heating is continued after all the ice has melted, the temperature of the water rises steadily until it reaches 100°C. Then the water reaches its boiling point. Another phase change occurs as liquid water turns to vapor or steam. While vaporization (boiling) is taking place, the temperature again remains constant. The heat energy added at the boiling point does not change the temperature of the liquid, but it does cause the liquid to change to vapor. As long as liquid water is present, the temperature is at 100°C. Once all the water is vaporized (turned to steam), the addition of heat raises the temperature of the steam.

Most substances undergo the physical change from solid to liquid and from liquid to gas as the temperature is raised. One exception is iodine. This violet-black solid ordinarily changes into a purple vapor without passing through a liquid state (Figure 9·24). *The change of a substance from a solid to a gas or vapor without passing through the liquid state is called* **sublimation.** A violet-black solid unchanged in appearance and composition from the original solid is obtained when iodine vapor is cooled. If wet laundry is hung on a clothes line to dry on a very cold day, the water in the clothes quickly freezes to ice. After several hours, however, the clothes are dry although the ice never thaws. The ice goes to water vapor without passing through the liquid state.

9·10 Energy and Phase Changes

The term cal/(g × °C) can also be written cal/g°C.

Substances vary in the way in which they absorb heat. The specific heat of a substance (Section 2·13) is one measure of this property. The specific heat of liquid water is 1 cal/g°C. This means that 1 cal of heat raises the temperature of 1 g of water 1°C. Therefore 100 cal of heat is needed to raise the temperature of 1 g of water from 0°C to 100°C. By contrast, ice has a specific heat of 0.5 cal/g°C. Half a calorie of heat will raise the temperature of 1 g of ice 1°C. Thus the temperature of ice changes twice as much as the temperature of water for every calorie of heat absorbed. The specific heat of steam is 0.5 cal/g°C. The specific heat is indicated by the slope of the line on a heating curve (Figure 9·25).

■ The state of a substance determines the effect of heat on changing the temperature of that substance.

When a phase change occurs, the temperature of the substance remains constant. As heat is absorbed, the energy causes a change of state instead of a change of temperature. The heat of fusion describes the energy changes as a substance melts. *The heat required to melt one gram of a solid at its melting point is its* **heat of fusion.** For example, 80 cal of energy is needed to change 1 g of ice to 1 g of water at 0°C. Thus the heat of fusion for ice is 80 cal/g. The reverse of this process is solidification. *The* **heat of solidification** *is the amount of heat given up as one gram of liquid changes to a solid at the melting point.* Because melting and freezing are reverse processes, the magnitudes of the heat of fusion and heat of solidification are identical.

The heat of fusion and heat of solidification quantify the energy changes that take place when melting and freezing occur. The heat of vaporization and heat of condensation quantify the energy changes that take place when vaporization and condensation occur.

Similar energy changes take place when a liquid changes to a gas. *The number of calories required to change 1 g of a liquid to gas at the boiling point at atmospheric pressure is the liquid's* **heat of vaporization.** When a vapor condenses to a liquid, the opposite transfer of heat occurs. The heat of vaporization is passed to anything on which the vapor condenses. The temperature of the condensed liquid does not decrease until the heat of vaporization is lost. *The heat released when 1 g of a gas condenses to a liquid at the boiling point is the* **heat of condensation.**

Safety

Steam is invisible, but it can cause severe burns. Do not reach across a container of boiling water.

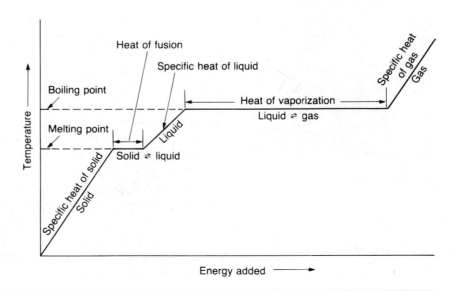

Figure 9·25
A phase diagram contains a wealth of information. Notice that the slope of the heating curve indicates the specific heat of the substance for a given state.

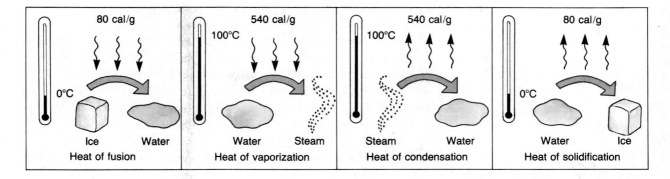

80 cal/g	540 cal/g	540 cal/g	80 cal/g
0°C Ice → Water	100°C Water → Steam	100°C Steam → Water	0°C Water → Ice
Heat of fusion	Heat of vaporization	Heat of condensation	Heat of solidification

Figure 9·26
The heat of fusion equals the heat of solidification. The heat of vaporization equals the heat of condensation.

Since vaporization and condensation are reverse processes, the heat of vaporization of a substance must be equal in magnitude to its heat of condensation. As you might expect, various liquids have different heats of vaporization and boiling points. This is because their intermolecular attractive forces vary.

Figure 9·27
The aurora borealis is caused by the presence of low density plasma in the upper atmosphere.

The fusion process and the problems faced in producing and containing hot plasma are described in Section 24·7.

======= Science, Technology, and Society =======

9·C Plasma: The Fourth State of Matter

What happens if you heat a gas to a very high temperature? First the kinetic energy of the molecules becomes great enough to separate molecules into atoms. Then at even higher temperatures, electrons are stripped off the gaseous atoms. The gas is ionized, becoming a mixture of electrons and positive ions. This new substance behaves like ordinary gases in some ways. It also has some very peculiar properties. It is really a fourth state of matter, called *plasma*.

A gas can be partially ionized at low temperatures by common events. Partial plasmas are created in fluorescent lights, neon signs, lightning bolts, and even in flames. In all of these examples, only a few of the atoms present are ionized at any moment. Even these will not remain ionized. A free electron will lose energy when it collides with anything, including other particles or the walls of its container. When it has lost enough energy, it can be recaptured by a gas ion. Because of this, plasmas in the atmosphere or those created in lights and flames are only weakly ionized.

A tremendous amount of energy is required to create highly ionized plasmas. Some materials can be mostly converted into plasmas at temperatures of 50 000 K–100 000 K. These are called "cold plasmas." In contrast, some plasmas require temperatures of 10 000 000 K–100 000 000 K and more. These "hot plasmas" are the material of the stars. Scientists are trying to create hot plasmas to produce energy by nuclear fusion, the same process that produces energy in our sun. No material can contain matter at these temperatures. Instead, scientists use a strong magnetic field to confine hot plasmas. One device used to do this is called a tokamak, in which the plasma is confined in a doughnut-shaped ring.

Even weakly ionized plasmas have unique properties that make them useful. The free electrons and ions in a plasma make it very corrosive. Metal surfaces can be made more receptive to certain kinds of coatings by etching them with a plasma. Many other applications make use of the electrical properties of plasmas. Because plasmas consist of charged particles, they conduct electricity. Because the charged particles are moving, plasmas generate a magnetic field. Like any conducting material, a plasma generates an electric current when it moves through an external magnetic field. Scientists hope to make use of these properties to develop plasma motors and electric generators.

Low-density plasma technology is the basis of a promising alternative source of electric power, called magnetohydrodynamic power generation, or MHD. The heat from burning a fossil fuel (coal, oil, or natural gas) is used to partially ionize a gas. This low-density plasma is passed through a strong magnetic field. An electric current is generated in the plasma. This is more efficient than the conventional way of producing electricity from burning coal. MHD research is only one of the challenging problems in the rapidly expanding field of plasma physics.

9 The States of Matter
Chapter Review

Key Terms

amorphous solid	9·8	kinetic theory	9·1	
atmospheric		melting point (mp)	9·8	
pressure	9·3	millimeter of mercury		
Avogadro's		(mm Hg)	9·3	
hypothesis	9·4	normal boiling		
barometer	9·3	point	9·7	
boiling point (bp)	9·7	pascal (Pa)	9·3	
crystal	9·8	phase change	9·9	
evaporation	9·6	standard atmosphere		
gas pressure	9·3	(atm)	9·3	
heat of		sublimation	9·9	
condensation	9·10	supercooled liquid	9·8	
heat of fusion	9·10	unit cell	9·8	
heat of		vapor pressure	9·6	
solidification	9·10	vaporization	9·6	
heat of				
vaporization	9·10			

Chapter Summary

The kinetic theory describes the motion of particles (atoms, ions, or molecules) in matter and the forces of attraction between them. The theory assumes that the volume occupied by a gas is mostly empty space and that the particles of a gas are far apart, move rapidly, and have chaotic motion. Avogadro knew that a gas is mostly empty space. He stated that equal volumes of gases, at the same temperature and pressure, contain equal numbers of particles. The pressure of a gas results from the countless collisions of the gas particles with an object. Barometers are used to measure atmospheric pressure.

Standard conditions when working with gases are a temperature of 0°C and a pressure of 1 atm. The temperature of a gas is directly proportional to the average kinetic energy of the particles. When the temperature of a gas is lowered, the particles slow down, and the gas condenses to a liquid. As the temperature is lowered further, the liquid solidifies. Solids are rigid and have fixed volumes. In a solid the movement of particles is restricted to vibrations about fixed points. Liquids have a fixed volume and characteristic vapor pressure. A liquid boils when its vapor pressure equals the external pressure. The normal boiling point of a liquid is the temperature at which the vapor pressure is equal to 1 atm.

Most substances change their physical state and melt or vaporize as the temperature is raised. They freeze or condense as the temperature is lowered. When the physical state of a substance changes, a phase change occurs. During a phase change the temperature of the system remains constant.

The particles in a solid vibrate more rapidly when heated. Solids melt when the vibrations are greater than the forces holding the particles together. Most solids are crystalline. The particles are arranged in a repeating three-dimensional pattern known as a crystal lattice. The smallest subunit of a crystal lattice is the unit cell. Each crystal belongs to one of seven crystal systems depending upon the shape of its unit cell. Some solids have a disordered array of particles and are amorphous.

Practice Questions and Problems

11. What is meant by the term "elastic collision"? *9·1*

12. Change 1656 km/hr to m/s.

13. How can you raise the average kinetic energy of the water molecules in a glass of water? *9·2*

14. What is significant about the temperature absolute zero? *9·2*

15. A cylinder of oxygen gas is cooled from 300 K (27°C) to 150 K (−123°C). By what factor does the average kinetic energy of the oxygen molecules in the cylinder decrease? *9·2*

16. What pressure in atmospheres does a gas exert at 380 mm Hg? *9·3*

17. How many millimeters of mercury pressure does a gas exert at 3.1 atm? *9·3*

18. Express 400 mm Hg in kilopascals. *9·3*

19. What does the abbreviation *STP* represent? *9·4*

20. Express standard temperature in Kelvin and standard pressure in millimeters of mercury. *9·4*

21. A gas has a volume of 2.24 L at STP. How many moles of gas is this? *9·4*

22. Explain why liquids and gases differ. *9·5*
  a. in physical state
  b. in compressibility

23. Why does evaporation lower the temperature of a liquid? *9·6*

24. Explain why increasing the temperature of a liquid increases its rate of evaporation. *9·6*

25. Would you expect an equilibrium vapor pressure to be reached above a liquid in an open container? Why? *9·6*

26. Describe the effect that increasing temperature has on the vapor pressure of a liquid. *9·6*

27. Explain why the boiling point of a liquid varies with atmospheric pressure. *9·7*

28. Use Figure 9·14 to determine the temperature at which water will boil in an open vessel when the atmospheric pressure is 400 mm Hg. *9·7*

29. Use Figure 9·14 to determine the boiling point of each of these liquids. *9·7*
  a. acetic acid at 200 mm Hg
  b. chloroform at 600 mm Hg
  c. ethyl alcohol at 400 mm Hg

30. At the top of Mt. Everest, water boils at only 69°C. Use Figure 9·14 to estimate the atmospheric pressure at the top of this mountain. *9·7*

31. Explain how boiling is a cooling process. *9·7*

32. Describe the terms *boiling point* and *normal boiling point*. *9·7*

33. Describe what happens when a solid is at its melting point. *9·8*

34. In general, molecular solids have lower melting points than ionic solids. Why? *9·8*

35. Any liquid stays at a constant temperature while it is boiling. Why? *9·9*

36. Describe the process of sublimation. What practical use can be made of this process? *9·9*

37. Define the terms. *9·10*
  a. heat of fusion
  b. heat of vaporization

Mastery Questions and Problems

38. Calculate the energy in calories that is required to change 15.0 g of ice at $-10°C$ to 15.0 g of water at $0°C$.

39. Describe vaporization, vapor pressure, and boiling point.

40. If liquid nitrogen is poured on a balloon, its volume decreases dramatically as shown in the photograph below. When no more liquid nitrogen is poured on the balloon, it "reinflates." Use kinetic theory to explain this sequence of events. The temperature of liquid nitrogen is $-196°C$.

41. What causes atmospheric pressure, and why is it much lower on the top of a mountain than at sea level?

42. What is the volume occupied by 0.30 mol of a gas at STP?

43. Why do we call the equilibrium that exists between a liquid and its vapor in a closed container a *dynamic equilibrium*?

44. The temperature of the gas in an aerosol container is $0°C$ (273 K). To what new temperature must the gas be raised to increase the average kinetic energy of the gas molecules by a factor of three?

45. If 3.20 kcal of energy is added to 1.00 kg of ice at $0°C$, how much water, at $0°C$, is produced? How much ice is left?

46. The amounts of energy required to change different quantities of liquid carbon tetrachloride, CCl_4, into vapor are given in the table below.
  a. Graph the data, using energy as the dependent variable.
  b. What is the slope of the line?

c. The heat of vaporization of CCl_4 is 53.8 cal/g. How does this value compare with the slope of the line?

Mass of $CCl_4(g)$	Energy (cal)
2.90	156
7.50	404
17.0	915
26.2	1410
39.8	2140
51.0	2740

47. Explain what happens at the melting point of a substance.

48. Calculate the energy in calories when 45.2 g of steam at $100°C$ condenses to water at the same temperature. What is this energy in joules?

49. Mt. McKinley (6194 m) in Alaska is the tallest peak in North America. The atmospheric pressure at the peak is 330 mm Hg. Find the boiling point of water there. Use Figure 9·14.

Critical Thinking Questions

50. Choose the term that best completes the second relationship.
  a. temperature:thermometer atmospheric pressure: _____
      (1) valve   (3) volume
      (2) barometer   (4) gauge
  b. education:knowledge vaporization: _____
      (1) evaporation   (3) gas
      (2) water   (4) condensation
  c. evolution:organisms kinetic theory: _____
      (1) particles   (3) heat
      (2) temperature   (4) pressure

51. What everyday evidence suggests that all matter is in constant motion?

52. Are perfectly elastic collisions possible between objects we can see?

53. How does perspiration help cool your body on a hot day?

54. Why do different compounds have different boiling points?

Review Questions and Problems

55. Balance these equations.
a. $V_2O_5 + H_2 \longrightarrow V_2O_3 + H_2O$
b. $(NH_4)_2Cr_2O_7 \longrightarrow Cr_2O_3 + N_2 + H_2O$
c. $NH_3 + O_2 \longrightarrow NO + H_2O$
d. $C_6H_{14} + O_2 \longrightarrow CO + H_2O$

56. Perchloric acid is formed by the reaction of water with dichlorine heptoxide.

$$Cl_2O_7 + H_2O \longrightarrow HClO_4$$

a. How many grams of Cl_2O_7 must be reacted with an excess of H_2O to form 56.2 g of $HClO_4$?
b. How many mL of water are needed to form 3.40 mol of $HClO_4$?

57. What is the density of krypton gas at STP?

58. How many moles is each substance?
a. 888 g of sulfur dioxide
b. 2.84×10^{22} molecules of ammonia
c. 0.47 L of carbon dioxide (at STP)

59. Hydrogen reacts with ethene (C_2H_4) to form ethane (C_2H_6).

$$C_2H_4 + H_2 \longrightarrow C_2H_6 + 137.2 \text{ kJ}$$

a. Is this an exothermic or endothermic reaction?
b. How many kJ of energy are produced when 66.2 g of C_2H_6 are made?
c. What is the limiting reagent when 40.0 g C_2H_4 reacts with 3.0 g H_2?

Challenging Questions and Problems

60. Calculate the energy required to change 42.6 g of ice at $-27.6°C$ to steam at $856°C$.
a. in calories   **b.** in kilocalories   **c.** in joules

61. If the volume of the container in which there is a liquid–vapor equilibrium is changed, the vapor pressure is not affected. Why?

62. An ice cube with a mass of 40.0 g melts in water originally at $25.0°C$.
a. How much heat does the ice cube absorb from the water when it melts? Report your answers in calories, kilocalories, and joules.
b. Calculate the number of grams of water that can be cooled to $0°C$ by the melting ice cube.

63. The ions in sodium chloride are arranged in a face-centered cubic pattern. Sketch a layer of the ions in a crystal of sodium chloride.

Research Projects

1. Measure the vapor pressure of several liquids.

2. How do scientists use cryogenics to study rates of reactions?

3. Investigate the effect of temperature on the rate of sublimation of moth balls.

4. How can crystal formation be used to determine the stucture of an unknown substance? How did Eilhardt Mitscherlich invent this technique?

5. Can you influence the crystalline form of ice by varying the conditions under which ice forms?

6. Compare the crystalline and amorphous structures of rocks and minerals using a hand lens or a low power microscope. How do the structures contribute to the properties of these substances?

7. How is glass made, and how do various glasses differ?

8. What are ceramics? In what ways have they traditionally been used? Describe some new applications that have been or are being developed.

9. How is X-ray crystallography used to study the structures of large molecules? What discoveries has this technique made possible?

10. How are plasmas currently used in industry? How are they important in microelectronic research?

Readings and References

Haines, Gail Kay. *Supercold/Superhot; Cryogenics and Controlled Thermonuclear Fusion*. New York: Watts, 1976.

Holden, Alan, and Phylis Morrison. *Crystals and Crystal Growing*. Cambridge, MA: MIT Press, 1982.

10 The Behavior of Gases

Chapter Preview

10·1 The Effect of Adding or Removing Gas
10·2 The Effect of Changing the Size of the Container
10·3 The Effect of Heating or Cooling a Gas
10·4 Real vs Ideal Gases
10·5 Dalton's Law of Partial Pressures
10·6 Boyle's Law for Pressure–Volume Changes
10·A Scuba Science
10·7 Charles' Law for Temperature–Volume Changes
10·8 Gay-Lussac's Law for Temperature–Pressure Changes
10·B The Liquefaction of Gases
10·9 The Combined Gas Law
10·10 The Ideal Gas Law
10·C Explosives and the Establishment of Nobel Prizes
10·11 Departures from the Gas Laws
10·12 Diffusion and Graham's Law

The atmosphere is too vast for us to have any control over atmospheric pressure. Contained gases are a different matter. We can add gas or remove it, shrink or expand the container, or heat or cool the gas. In this chapter we will examine the work of such eminent scientists as Robert Boyle, Jacques Charles, and Joseph Gay-Lussac. These scientists studied the effects of changes in the pressure, volume, and temperature of contained gases. From their results they proposed a set of relationships that together are known as the gas laws. Using the kinetic theory, however, we can often explain how gases will respond to a change of conditions without resorting to formal mathematical expressions. We will therefore begin this chapter with some examples of how simple kinetic theory is used to explain gas behavior. Our emphasis will be on gas pressure.

10·1 The Effect of Adding or Removing Gas

When we pump up a tire, the pressure inside it increases. The pressure exerted by an enclosed gas is caused by collisions of gas particles with the walls of the container. By adding gas we increase the number of gas particles. This increases the number of collisions and therefore the gas

Figure 10·1
Hot air balloonists take advantage of the behavior of gases as they take to the air.

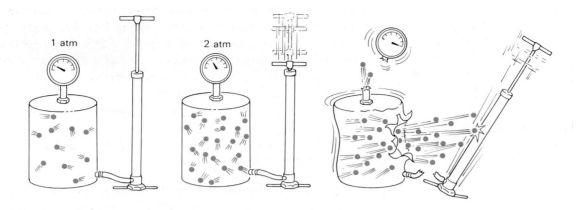

Figure 10·2
When gas is pumped into a closed rigid container, the pressure increases in proportion to the number of gas particles added. If the number of particles doubles, the pressure doubles. If the pressure exceeds the strength of the container, the container explodes.

pressure. Doubling the number of gas particles doubles the pressure if the temperature of the gas does not change (Figure 10·2). Tripling the number of gas particles triples the pressure and so forth. With a powerful pump and a strong container we can generate very high gas pressures. Once the pressure exceeds the strength of the container, however, the container will rupture. Overinflated balloons burst for this reason.

Problem

1. The manufacturer of an aerosol deodorant, wishes to produce a family-size package that will hold twice as much gas as the 150-mL regular size. The pressure in both containers must be kept the same. What will the volume of the larger package have to be?

The pressure exerted by a gas depends on the number of particles per unit volume.

Letting the air out of a tire or the gas out of a storage cylinder decreases the pressure in the container. Fewer particles are left. As you have probably already guessed, halving the number of particles in a given volume of gas decreases the pressure by one-half (Figure 10·3). When a sealed container of gas under pressure is opened, the gas always moves from the region of higher pressure to the region of lower pressure. This is

Figure 10·3
A gas under pressure in a sealed rigid container rushes into the atmosphere if the container is opened. The pressure in the container decreases in proportion to the number of gas particles released. If the number of particles decreases by one-half, the pressure also decreases by one-half.

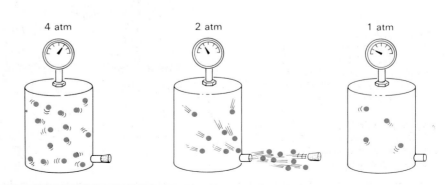

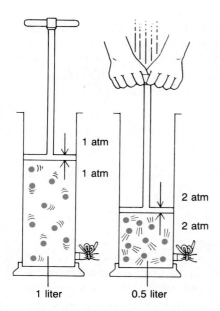

Figure 10·4
Doubling the force on a gas reduces the volume of the gas by one-half and doubles the pressure it exerts.

Gases cool when they expand and heat when they are compressed.

Safety

Always open a bottle of liquid slowly and cautiously to release any built-up pressure.

Figure 10·5
When a gas in a container is heated from 300 K (27°C) to 600 K (327°C), the kinetic energy of the gas particles doubles. The particles strike the sides of the container with a greater force at 600 K than they do at 300 K. The pressure exerted by the gas therefore doubles. The pressure buildup at high temperatures may cause the container to explode.

because there is more empty space for the gas particles to occupy. Gas particles increase their randomness by moving into this empty space, which is always a favorable process. Expansion continues until the gas pressures inside and outside the container are equal. Once the pressures are equal, the mixing of the gases in the container with gases in the surrounding atmosphere is only by diffusion.

10·2 The Effect of Changing the Size of the Container

We increase the pressure exerted by a contained gas when we reduce the size of the container. The more a gas is compressed, the greater the pressure it exerts on its container. Reducing the volume of the container by one-half has the same effect on pressure as doubling the quantity of gas while keeping the volume constant (Figure 10·4). Increasing the volume of the container has just the opposite effect. By doubling the volume we halve the gas pressure. There are now one-half as many gas particles in a given volume.

Gases cool when they expand and heat when they are compressed. When the gas is rapidly released from an aerosol can, the can becomes cooler. This is because the expanding gas absorbs some thermal energy from the container and its contents as it escapes. Compressing a gas always increases its temperature.

10·3 The Effect of Heating or Cooling a Gas

Raising the temperature of an enclosed gas increases the gas pressure. The kinetic energy of gas particles increases as they absorb thermal energy. Naturally, fast-moving particles bombard the walls of their container harder than slow-moving particles. The average kinetic energy of gas particles doubles with a doubling of the Kelvin temperature. Hence doubling the Kelvin temperature of an enclosed gas doubles the gas pressure (Figure 10·5).

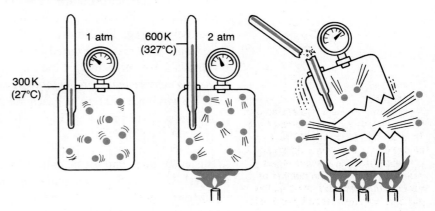

The pressure exerted by a gas depends on the average kinetic energy of the particles.

A heated gas sealed in a container generates enormous pressure. An aerosol can carelessly thrown on a fire is an explosion hazard. By contrast, as the temperature of an enclosed gas decreases, the particles move more slowly and strike the container walls with less force. By halving the Kelvin temperature of a gas in a rigid container, we decrease the gas pressure by one-half.

Problem

2. What happens to the volume of a balloon when it is taken outside on a cold winter day? Why?

10·4   Real vs Ideal Gases

An ideal gas follows the gas laws at all conditions of temperature and pressure.

In the discussions of the gas laws to follow, we will assume that we are dealing with ideal gases. An ideal gas is one that follows the gas laws at all conditions of pressure and temperature. An ideal gas is a substance that does not really exist. Kinetic theory assumes that the particles of an ideal gas have no volume and that the particles are not attracted to each other. There is no gas for which this is true. Nevertheless, under many conditions the behavior of a real gas is similar to that of an ideal gas.

Real gases can be liquefied and sometimes solidified by cooling them and applying pressure. Ideal gases cannot. For example when water vapor is cooled to less than 100°C, it condenses to liquid at atmospheric pressure. The behavior of other real gases is similar although lower temperatures and greater pressure may be required. Carbon dioxide goes from a gas to a solid without passing through the liquid state. The product is solid carbon dioxide, or "dry ice." Conversely, upon warming, solid carbon dioxide passes from the solid state to the gaseous state without passing through the liquid state. This is an example of sublimation.

Safety

Dry ice looks very much like ice, but it is much colder and can cause frostbite. Handle it only with tools and thick gloves.

10·5   Dalton's Law of Partial Pressures

Many gases, including air, are mixtures (Table 10·1). The particles in a gas at the same temperature have the same average kinetic energy. Gas pressure depends only on the number of gas particles in a given volume and their average kinetic energy. The kind of particle is unimportant. If we know the pressure exerted by each gas in a mixture, we can add the individual pressures and get the total gas pressure. *The contribution each gas in a mixture makes to the total pressure is the* **partial pressure** *exerted by that gas*. In a mixture of gases the total pressure is the sum of the partial pressures of the gases (Figure 10·6).

$$P_{total} = P_1 + P_2 + P_3$$

This equation is one mathematical form of **Dalton's law of partial pressures:** *At constant volume and temperature, the total pressure exerted by a mixture of gases is equal to the sum of the partial pressures.*

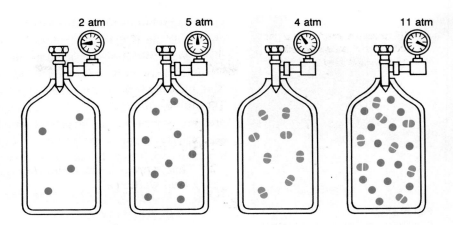

2 atm 5 atm 4 atm 11 atm

Figure 10·6
The sum of the pressures exerted by the gas in each container is the same as the total pressure exerted by a mixture of the gases in the same volume as long as the temperature stays the same. Dalton's law of partial pressures holds because each gas exerts its own pressure independent of the pressure exerted by other gases.

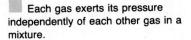

Each gas exerts its pressure independently of each other gas in a mixture.

Figure 10·7
Solid carbon dioxide is known as dry ice. It can injure the skin because it is much colder ($-78.5°C$) than frozen water. Why is dry ice not as messy to use as regular ice?

Example 1

Air contains oxygen, nitrogen, carbon dioxide, and trace amounts of other gases. What is the partial pressure of oxygen (P_{O_2}) at 1 atm of pressure if $P_{N_2} = 593.4$ mm Hg, $P_{CO_2} = 0.3$ mm Hg, and $P_{others} = 7.1$ mm Hg?

Solution

We know that 1 atm = 760 mm Hg = P_{total}

$$P_{total} = P_{O_2} + P_{N_2} + P_{CO_2} + P_{others}$$

$$P_{O_2} = P_{total} - (P_{N_2} + P_{CO_2} + P_{others})$$

$$= 760 \text{ mm Hg} - (593.4 \text{ mm Hg} + 0.3 \text{ mm Hg} + 7.1 \text{ mm Hg})$$

$$= 159.2 \text{ mm Hg}$$

Problem

3. Determine the total pressure of a gas mixture that contains oxygen, nitrogen, and helium if the partial pressures of the gases are $P_{O_2} = 150$ mm Hg, $P_{N_2} = 350$ mm Hg, and $P_{He} = 200$ mm Hg. Give your answer in millimeters of mercury and in atmospheres.

Table 10·1   Composition of Dry Air

Component	Volume (%)	Partial pressure (mm Hg)	(atm)
Nitrogen	78.08	593.4	0.781
Oxygen	20.95	159.2	0.209
Carbon dioxide	0.04	0.3	0.001
Argon and others	0.93	7.1	0.009
	100.00	760.0	1.000

Figure 10·8
Why must high altitude pilots have a separate oxygen supply available?

For a given mass of gas at constant temperature, the volume varies inversely with pressure.

Chemical Connections

Boyle's law helps explain how humans breathe. During inhalation, the rib cage expands and the diaphragm pulls downward, both causing an increase in the volume of the lungs. The increased volume results in decreased pressure compared to outside the lungs, causing air to rush inside.

The proportionate pressure exerted by each gas in a mixture does not change as the temperature, pressure, or volume changes. This fact has important implications for aviators and mountain climbers. When the total atmospheric pressure is reduced to 253 mm Hg (1/3 atm), as it is atop Mount Everest, the partial pressure of oxygen is reduced to 53 mm Hg. This is one-third the partial pressure of oxygen at 1 atm. This oxygen pressure is insufficient for respiration. Breathing air with a P_{O_2} of about 80 mm Hg supplies the minimum amount of oxygen. Airplanes and mountaineering expeditions must carry a supply of oxygen.

10·6 Boyle's Law for Pressure–Volume Changes

Consider the effect of pressure on the volume of a contained gas while the temperature is held constant. When the pressure goes up, the volume goes down. Similarly, when the pressure goes down, the volume goes up. In 1662 the British chemist Robert Boyle (1627–1691) proposed a law to describe this behavior of gases. **Boyle's law** *states that for a given mass of gas at constant temperature, the volume of the gas varies inversely with pressure.* In an inverse relationship, the product of the two quantities that change is always constant. This is shown in Figure 10·9.

In the figure, a volume of 1 L (V_1) is at a pressure of 1 atm (P_1). When the volume is increased to 2 L (V_2), the pressure decreases to 0.5 atm (P_2). Observe that the product $P_1 \times V_1$ (1 atm × 1 L = 1 L-atm) is the same as the product of $P_2 \times V_2$ (0.5 atm × 2 L = 1 L-atm). When the volume is decreased to 0.5 L, the gas pressure increases to 2 atm. Once again, the product of pressure times volume equals 1 L-atm. When an inverse relationship is graphed, the result is a curved line, as shown in Figure 10·10. The product of pressure and volume for a given mass of gas at any two sets of conditions is always constant at a given temperature. We can write Boyle's law as follows.

$$P_1 \times V_1 = P_2 \times V_2$$

Figure 10·9
Boyle's law and pressure–volume relationships are studied by using a cylindrical container fitted with a frictionless, weightless piston. When the pressure of a gas at constant temperature decreases (P_2), the volume increases (V_2). When the pressure increases (P_3), the volume decreases (V_3).

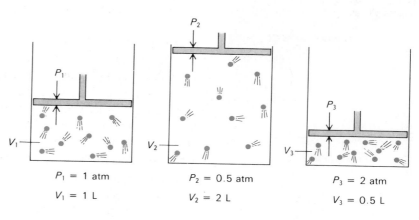

P_1 = 1 atm
V_1 = 1 L

P_2 = 0.5 atm
V_2 = 2 L

P_3 = 2 atm
V_3 = 0.5 L

$P_1 \times V_1 = P_2 \times V_2 = P_3 \times V_3$ = 1 L-atm

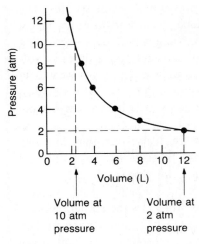

Figure 10·10
This graph illustrates Boyle's law. When the pressure of a gas decreases at constant temperature, the volume increases. At any point on this curve the product of pressure (P) times volume (V) is a constant: in this case, $P \times V = 24$ L-atm.

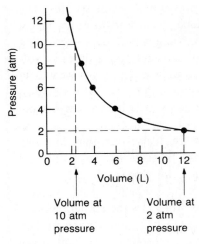

ChemDirections

The air in scuba tanks is under a lot of pressure; therefore the construction of the tanks must be very strong. Composite materials are now replacing metals as the preferred material for making scuba tanks. Find out why in **ChemDirections** *Composites*, page 662.

Example 2

A balloon is filled with 30 L of helium gas at 1 atm. What is the volume when the balloon rises to an altitude where the pressure is only 0.25 atm? (Assume that the temperature remains constant.)

Solution

We know that $P_1 = 1$ atm, $V_1 = 30$ L, and $P_2 = 0.25$ atm. We need to solve for V_2. Rearrange the expression for Boyle's law to obtain V_2.

$$P_1 \times V_1 = P_2 \times V_2 \qquad \text{Therefore:} \qquad V_2 = V_1 \times \frac{P_1}{P_2}$$

Substitute values for P_1, V_1, and P_2.

$$V_2 = 30 \text{ L} \times \frac{1 \text{ atm}}{0.25 \text{ atm}} = 120 \text{ L}$$

The balloon will expand to a volume of 120 L at 0.25 atm. This agrees with what kinetic theory predicts. At a constant temperature a decrease in pressure must correspond to an increase in volume.

Problem

4. The pressure on 2.50 L of anesthetic gas is changed from 760 mm Hg to 304 mm Hg. What will be the new volume if the temperature remains constant?

Science, Technology, and Society

10·A Scuba Science

In 1943 Jacques-Ives Cousteau and Emile Gagnan invented scuba: a Self-Contained Underwater Breathing Apparatus. No longer did divers need to be tethered to the surface by air hoses. The increased maneuverability and convenience gave rise to the sport of scuba diving. It also greatly increased the use of diving for scientific research. All divers must have a good understanding of the science involved in diving in order to dive safely.

At the water's surface, the pressure on your body due to the mass of air around you is 1 atmosphere. Under water, the pressure increases due to the added mass of the water. Every 10 meters of depth adds 1 atmosphere pressure. Thus the total pressure on your body at a depth of 10 m will be 2 atm, at 20 m 3 atm, and so on. At around 40 m the pressure on your chest would make it impossible for you to inflate your lungs to breathe. If the pressure in your lungs had also increased as you went down, however, you would be able to breathe normally. Scuba equipment provides air to the lungs at a pressure to match that of the underwater environment. This enables the diver to breathe comfortably.

A diver at 20 meters is under a pressure of 3 atmospheres. The scuba equipment is maintaining the same pressure in his or her lungs. The average lung capacity of a human being is 6–7 liters. According to Boyle's law, this amount of air will expand to three times its volume if the pressure is reduced to 1 atmosphere. If a diver ascends to the surface without exhaling steadily along the way, the air held in the lungs will expand as the pressure drops. The increase in volume can rupture the lungs.

An understanding of partial pressure is also important in scuba diving. When the pressure on a mixture of gases increases, the partial pressure of each of the gases increases proportionately. This means that in the compressed air that a diver breathes, every ingredient is at higher pressure. At depths of 30 meters the partial pressure of carbon dioxide in air is sufficient to poison a diver. Even oxygen can be toxic if the total gas pressure is 2 atmospheres. At depths over 10 meters the length of the dive becomes very important in preventing such toxic effects.

Decompression sickness, or "the bends," is explained by another important gas law. The solubility of a gas in a liquid is proportional to the pressure of the gas above the liquid. A diver is breathing air at higher than atmospheric pressure. Thus the nitrogen in the air will be more soluble in his or her blood. When the pressure drops as the diver ascends, the nitrogen again becomes less soluble. If the drop in pressure occurs too rapidly, the nitrogen will come out of the blood in the form of tiny bubbles. This effect is just like the bubbles that form in a carbonated drink when you open the cap and relieve the pressure in the bottle. If nitrogen bubbles form in the joints or muscles, they cause a great deal of pain. If they form in the spinal cord, brain, or lungs they can cause paralysis or death. Decompression sickness can be prevented by ascending from a dive slowly. If that is not possible, the diver can be brought to the surface rapidly and put in a recompression chamber. Here he or she is recompressed to a pressure of 6 atmospheres. Then the pressure is gradually reduced so that the nitrogen can be eliminated through the lungs.

Figure 10·11
Scuba divers must be careful not to stay down too long or to ascend too rapidly. The high partial pressures of carbon dioxide, nitrogen, and oxygen can become dangerous over time. Nitrogen bubbles can form in the blood if divers come up too quickly.

The volume of a fixed mass of gas at a constant pressure varies with the Kelvin temperature.

10·7 Charles' Law for Temperature–Volume Changes

In 1787 a French physicist, Jacques Charles (1746–1823), investigated the effect of temperature on the volume of a gas at constant pressure. In every experiment, he observed an increase in the volume of gas with an increase in temperature. Charles summarized his observations in a law. **Charles' law** *states that the volume of a fixed mass of gas is directly proportional to its Kelvin temperature if the pressure is kept constant.*

In a direct relationship the ratio of the two quantities that change is a constant. In Figure 10·12, for example, a 1-L sample of gas (V_1) is at a

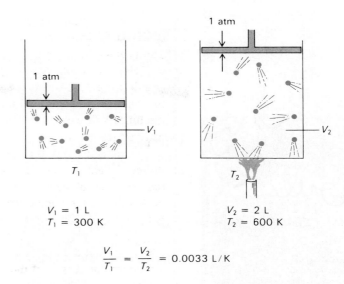

Figure 10·12
This apparatus illustrates Charles' law. When a gas at constant pressure is heated, the volume increases. When a gas at constant pressure is cooled, the volume decreases.

$V_1 = 1$ L
$T_1 = 300$ K

$V_2 = 2$ L
$T_2 = 600$ K

$$\frac{V_1}{T_1} = \frac{V_2}{T_2} = 0.0033 \text{ L/K}$$

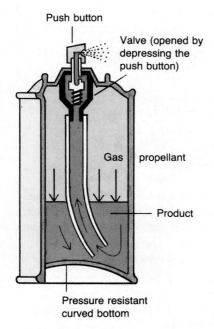

Push button

Valve (opened by depressing the push button)

Gas | propellant

Product

Pressure resistant curved bottom

Figure 10·13
The propellant in an aerosol container is a pressurized gas. When product is released, the propellant expands, cooling the container and the product.

temperature of 300 K (T_1). (*Temperature in gas law problems is always in Kelvin.*) When the temperature is increased to 600 K (T_2), the volume increases to 2 L (V_2). Observe that the ratio $V_1 \div T_1$ (1 L $\div$ 300 K = 0.0033 L/K) is equal to the ratio $V_2 \div T_2$ (2 L $\div$ 600 K = 0.0033 L/K). Thus, the ratio of volume to Kelvin temperature for a given mass of gas at two sets of conditions is constant. Charles' law is

$$\frac{V_1}{T_1} = \frac{V_2}{T_2}$$

as long as the pressure is held constant. The graph of a relationship that is a direct proportion is a straight line, as shown in Figure 10·14. This line would intersect the temperature axis at $-273°C$. That is, starting at $0°C$ the volume of a gas decreases by $1/273$ for every decrease of $1°C$. William Thompson (Lord Kelvin) realized that $-273°C$ is the temperature at which the average kinetic energy of gas particles would theoretically be zero. This is the basis of the Kelvin temperature scale in which 0 K corresponds to $-273.14°C$. Absolute zero has never been reached, but temperatures within one kelvin of 0 K have been produced.

Figure 10·14
This graph illustrates Charles' law. When the temperature of a gas increases, the volume increases if the pressure is held constant. At any point on this curve the ratio of volume (V) to temperature (T) is a constant: in this case, $V/T = 0.0033$ L/K.

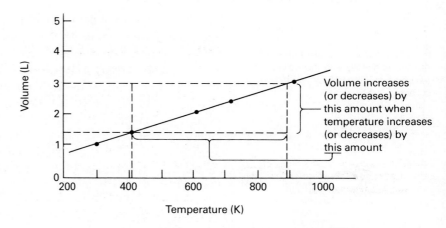

Volume increases (or decreases) by this amount when temperature increases (or decreases) by this amount

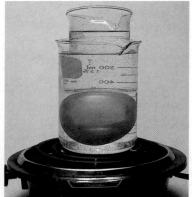

Figure 10·15
The volume of a balloon is less in a beaker of ice water than if the same balloon is in a beaker of hot water. Why does this happen?
The volume of air in the balloon increases with an increase in temperature.

The pressure of a fixed mass of gas varies directly with the Kelvin temperature if the volume is held constant.

Example 3

A balloon, inflated in an air-conditioned room at 27°C, has a volume of 4.0 L. It is heated to a temperature of 57°C. What is the new volume of the balloon if the pressure remains constant?

Solution

Temperature in all gas problems is expressed as kelvins. Therefore we must convert degrees Celsius to kelvins.

$$T_1 = 27°C + 273 = 300 \text{ K}$$

$$T_2 = 57°C + 273 = 330 \text{ K}$$

We know that $V_1 = 4.0$ L, $T_1 = 300$ K, and $T_2 = 330$ K. Therefore we use Charles' law to solve for V_2. First we rearrange the expression for Charles' law to obtain V_2.

$$\frac{V_1}{T_1} = \frac{V_2}{T_2} \qquad \text{Therefore: } V_2 = V_1 \times \frac{T_2}{T_1}$$

Then the values for T_1, V_1, and T_2 are substituted.

$$V_2 = \frac{4.0 \text{ L} \times 330 \,\cancel{K}}{300 \,\cancel{K}} = 4.4 \text{ L}$$

The balloon will expand from a volume of 4.0 L at 27°C to a volume of 4.4 L at 57°C. Using the kinetic theory, we would predict that the volume would increase with an increase in temperature (at constant pressure).

Problem

5. If a sample of gas occupies 6.8 L at 327°C, what will be its volume at 27°C if the pressure does not change?

10·8 Gay-Lussac's Law for Temperature–Pressure Changes

On a hot summer day the pressure in a car tire increases. This illustrates a relation discovered in 1802 by Joseph Gay-Lussac (1778–1850), a French chemist. **Gay-Lussac's law** *states that the pressure of a given mass of gas is directly proportional to the Kelvin temperature if the volume is kept constant.* If the Kelvin temperature is doubled, the gas pressure is doubled (Figure 10·16). Because Gay-Lussac's law involves direct proportions, the ratios $P_1 \div T_1$ and $P_2 \div T_2$ should be constant at constant volume. (Work out these ratios for the quantities shown in Fig. 10·16 to assure yourself this is true.) Thus, assuming the volume is held constant, we can write Gay-Lussac's law as follows.

$$\frac{P_1}{T_1} = \frac{P_2}{T_2}$$

Figure 10·16
This apparatus illustrates Gay-Lussac's law. When a gas at constant volume is heated, the pressure increases. When a gas at constant volume is cooled, the pressure decreases.

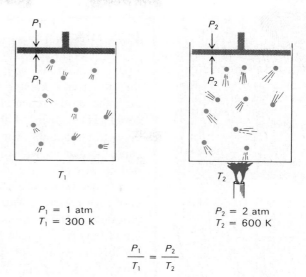

$P_1 = 1$ atm
$T_1 = 300$ K

$P_2 = 2$ atm
$T_2 = 600$ K

$$\frac{P_1}{T_1} = \frac{P_2}{T_2}$$

Safety

Never heat a closed container.

Example 4

The gas left in a used aerosol can is at a pressure of 1 atm at 27°C (room temperature). If this can is thrown onto a fire, what is the internal pressure of the gas when its temperature reaches 927°C?

Solution

First convert degrees Celsius to kelvins.

$$T_1 = 27°C + 273 = 300 \text{ K}$$

$$T_2 = 927°C + 273 = 1200 \text{ K}$$

We know that $P_1 = 1$ atm, $T_1 = 300$ K, and $T_2 = 1200$ K. Therefore we can solve Gay-Lussac's equation for P_2. We rearrange the expression to obtain P_2.

$$\frac{P_1}{T_1} = \frac{P_2}{T_2} \qquad \text{Therefore:} \qquad P_2 = \frac{P_1 \times T_2}{T_1}$$

Then we insert values for P_1, T_2, and T_1 into the equation.

$$P_2 = \frac{1 \text{ atm} \times 1200 \,\cancel{K}}{300 \,\cancel{K}} = 4 \text{ atm}$$

This is the pressure buildup in a can containing 1 atm of gas when it is heated from 27°C to 927°C. Once again, from the kinetic theory we would expect the increase in temperature of a gas at constant volume to give an increase in pressure.

Problem

6. A gas has a pressure of 50.0 mm Hg at 540 K. What will be the pressure at 200 K if the volume does not change?

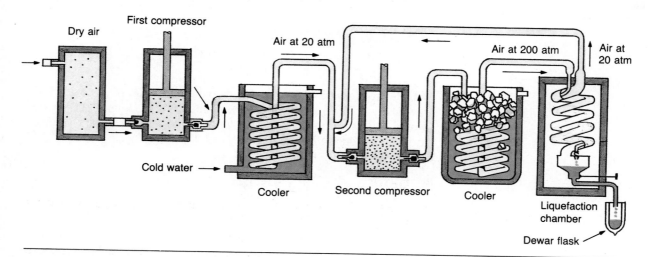

Issues in Chemistry

Since all risk associated with chemical manufacturing cannot be eliminated, would you be willing to stop using plastics, cosmetics, prescription drugs, synthetic fibers, and gasoline? If not, is there some level of risk that you would accept in order to permit the continued use of these products?

10·B The Liquefaction of Gases

Early chemists and physicists knew that solids turned to liquids and then to gases if enough heat was added. In the late eighteenth century, the great French chemist Lavoisier speculated that the reverse could also happen. If the earth's atmosphere could be made cold enough, he thought, it would turn to a liquid and then to a solid. Because he had no way of creating very low temperatures, Lavoisier had no way of testing this idea.

In the 1800s scientists were trying to liquefy gases by applying very high pressure. This work caused a good many spectacular explosions. In 1877 a French mining engineer, Louis Cailletet, was trying to liquefy acetylene gas. Unfortunately, his apparatus burst from the high pressure. Just as the accident took place, Cailletet saw a drop of liquid form. At the same time, the temperature of the acetylene gas dropped suddenly. Cailletet had witnessed adiabatic expansion, the cooling of a gas when it rapidly expands. He went on to use this phenomenon to bring gases to lower and lower temperatures. Eventually he was able to liquefy oxygen and nitrogen.

Today liquid oxygen, nitrogen, and argon are produced from liquid air. Hydrogen is produced as a gas by the electrolysis of water and then liquefied. Helium is liquefied after being separated from natural gas.

A gas can be liquefied in several ways. All of them depend on cooling the gas below its critical temperature. Above this temperature, no matter how much pressure is applied, the gas will not liquefy. The critical temperature of oxygen is 155 K, of nitrogen 126 K, of hydrogen 33 K, and of helium 5 K. Gases can be cooled to these low temperatures by first compressing them and then allowing them to expand rapidly. When a gas is compressed, it becomes warmer. If the heat is removed as the gas is compressed, however, it can be maintained at a constant temperature. Then if the gas is

Figure 10·17
Air is liquefied by compressing and cooling it. In the liquefaction chamber the highly compressed air is allowed to expand so that it cools sufficiently to become a liquid.

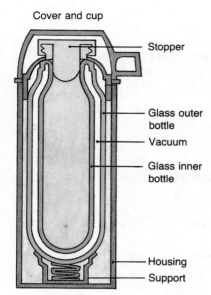

Cover and cup

Stopper

Glass outer bottle

Vacuum

Glass inner bottle

Housing

Support

Figure 10·18
A thermos bottle has basically the same construction as a Dewar flask. A vacuum is the best way to insulate against the transfer of heat.

The pressure-volume and temperature-volume laws are joined in the *combined gas law.*

ChemDirections

Large volumes of liquefied gases are transported annually on our highways in refrigerated vehicles. These materials are cryogens—extremely cold liquids—that cause immediate frostbite upon contact with the skin. Many of them are also highly flammable. Read **ChemDirections** *Trucking Chemicals,* page 659, to learn how to recognize which trucks carry these liquids.

allowed to expand rapidly, it will cool. This is the phenomenon that Cailletet observed. The gas can be repeatedly compressed and expanded until its temperature is below the critical temperature. It can then be liquefied by applying pressure.

A gas can also be cooled by surrounding it with a cold material. For example, helium can be cooled by putting it in a bath of liquid nitrogen. Liquid nitrogen boils at 77 K at room pressure. It can be made to boil at a lower temperature, however, by reducing the pressure in its container. A series of such baths can be used one after the other to cool another gas. When the temperature is low enough, the gas can be compressed to liquefy it.

The containers in which liquefied gases are stored and transported are called cryostats. They are designed to prevent heat from being transferred from the environment to the cold liquid. The most widely used cryostats are called Dewars after Sir James Dewar, who invented them in 1892. They are double-walled vessels with a vacuum between the walls, like the familiar thermos bottle used to carry hot or cold drinks. Cryostats are very light in comparison to pressure cylinders. A substance has a much smaller volume as a liquid than as a gas, even if the gas is under pressure. For these reasons, large amounts of gas can be stored and transported much more easily as a liquid than as a gas.

10·9 The Combined Gas Law

The three gas laws we have just discussed can be combined into a single expression called the **combined gas law.**

$$\frac{P_1 \times V_1}{T_1} = \frac{P_2 \times V_2}{T_2}$$

If you have been wondering how to remember the expressions for the other gas laws, it turns out that there is really no need. The other laws can be obtained from the combined gas law by holding one quantity (pressure, volume, or temperature) constant.

To illustrate, suppose we hold *temperature* constant ($T_1 = T_2$).

$$\frac{P_1 \times V_1}{T_1} = \frac{P_2 \times V_2}{T_2} \qquad \text{Therefore:} \qquad P_1 \times V_1 = \frac{P_2 \times V_2 \times \cancel{T_1}}{\cancel{T_2}}$$

We see that T_1 and T_2 cancel, and $P_1 V_1 = P_2 V_2$ (Boyle's law).

Problems

7. Show how Charles' law can be derived from the combined gas law.

8. Show how Gay-Lussac's law can be derived from the combined gas law.

Figure 10·19
When the temperature, pressure, and volume of a gas is measured, the ideal gas law allows the number of moles of the gas to be calculated. If the percent composition is known, the number of moles allows the molecular formula to be calculated.

Example 5

A cylinder of compressed oxygen gas has a volume of 30 L and 100 atm pressure at 27°C. The cylinder is cooled until the pressure is 5.0 atm. What is the new temperature of the gas in the cylinder?

Solution

First we convert degrees Celsius to kelvins: 27°C + 273 = 300 K.

We know that $T_1 = 300$ K, $P_1 = 100$ atm, and $P_2 = 5.0$ atm; V_1 and V_2 are both equal to 30 L. We can now rearrange the combined gas law to obtain T_2.

$$\frac{P_1 \times V_1}{T_1} = \frac{P_2 \times V_2}{T_2} \qquad \text{Therefore:} \qquad T_2 = \frac{P_2 \times V_2 \times T_1}{P_1 \times V_1}$$

Now we can substitute the known quantities into the expression for the combined gas law.

$$T_2 = \frac{5.0 \text{ atm} \times 300 \text{ K}}{100 \text{ atm}} = 15 \text{ K}$$

We then convert 15 K back to degrees Celsius.

$$\text{K} = {}^\circ\text{C} + 273 \qquad \text{Therefore:} \qquad \begin{aligned} {}^\circ\text{C} &= \text{K} - 273 \\ &= 15 - 273 \\ &= -258{}^\circ\text{C} \end{aligned}$$

Problem

9. A container with an initial volume of 1.0 L is occupied by a gas at a pressure of 1.5 atm at 25°C. By changing the volume, the pressure of the gas increases to 6.0 atm as the temperature is raised to 100°C. What is the new volume?

10·10 The Ideal Gas Law

Sometimes we wish to calculate the number of moles of a gas in a fixed volume at a known temperature and pressure. Such a calculation is possible if the combined gas law is modified. The modification may be understood by recognizing that the volume occupied by a gas at a specified temperature and pressure is directly proportional to the number of particles in the gas. The number of moles, n, of gas is also directly proportional to the number of particles. Hence moles must be directly proportional to volume as well. Therefore moles may be introduced into the combined gas law by placing n in the denominator on each side of the equation.

$$\frac{P_1 \times V_1}{T_1 \times n_1} = \frac{P_2 \times V_2}{T_2 \times n_2}$$

This equation says that $(P \times V)/(T \times n)$ is a constant for any ideal gas.

Table 10·2   Units and Numerical Values for the Gas Constant

Units of R	Numerical value	Units of P	Units of V
(L × atm)/(K × mol)	8.21×10^{-2}	atm	L
(L × mm Hg)/(K × mol)	62.4	mm Hg	L
(L × Pa)/(K × mol)	8.31×10^{3}	Pa	L

If we were able to evaluate the constant $(P \times V)/(T \times n)$, we could then calculate the number of moles of gas at any specified conditions of P, V, and T.

We have sufficient information to evaluate the constant because we know that 1 mol of every ideal gas occupies 22.4 L at STP. Inserting the values of P, V, T, and n into the right side of the equation, we obtain a constant.

$$\frac{P_1 \times V_1}{T_1 \times n_1} = \frac{1 \text{ atm} \times 22.4 \text{ L}}{273 \text{ K} \times 1 \text{ mol}} = 0.0821 \frac{\text{L} \times \text{atm}}{\text{K} \times \text{mol}}$$

The value of R, the ideal gas constant, depends on the units used.

*The **ideal gas constant** (R) is 0.0821 (L × atm)/(K × mol)*. As shown in Table 10·2, the numerical value of R depends on the units used. For example, R is 62.4 (L × mm Hg)/(K × mol), if 760 mm Hg is used for standard pressure. Rearranging the equation for R and dropping the subscripts we obtain the usual form of the **ideal gas law.**

$$\frac{P_1 \times V_1}{T_1 \times n_1} = R \quad \text{or} \quad P \times V = n \times R \times T$$

The ideal gas law describes the pressure, volume, temperature, or number of moles of a gas when any three of these four conditions are given.

An obvious advantage of the ideal gas law over the combined gas law is that it permits us to solve for the number of moles of a contained gas when P, V, and T are known.

Safety

Gas cylinders should always be securely strapped to the wall or to the cart on which they are transported. Each cylinder is very heavy. A cylinder could injure a person if it fell.

Example 6

A rigid steel cylinder with a volume of 20.0 L is filled with nitrogen gas to a final pressure of 200 atm at 27°C. How many moles of N_2 gas does the cylinder contain?

Solution

First we convert degrees Celsius to kelvins.

$$27°C + 273 = 300 \text{ K}$$

The conditions are $P = 200$ atm, $V = 20.0$ L, and $T = 300$ K. Because we want to find the number of moles of N_2 gas, we may rearrange the ideal gas law to obtain n.

$$n = \frac{P \times V}{R \times T}$$

Now we can substitute the known quantities into the equation.

$$n = \frac{200 \text{ atm} \times 20.0 \text{ L}}{0.0821 \dfrac{\text{L} \times \text{atm}}{\text{K} \times \text{mol}} \times 300 \text{ K}} = 162 \text{ mol}$$

Problem

10. When a rigid hollow sphere containing 680 L of helium gas is heated from 300 K to 600 K, the pressure of the gas increases to 18 atm. How many moles of helium does the sphere contain?

Example 7

A deep underground cavern contains 2.24×10^6 L of methane gas at a pressure of 15.0 atm and a temperature of 42°C. How many grams of methane does this natural gas deposit contain?

Solution

The problem calls for an answer in grams, but the ideal gas law permits us to obtain the solution only in moles. We must first find the number of moles, then convert to grams.

We know that $P = 15.0$ atm, $V = 2.24 \times 10^6$ L, and $T = 315$ K (42°C + 273). As in the previous example, we substitute the known quantities into the equation.

$$n = \frac{P \times V}{R \times T} \qquad n = \frac{15.0 \text{ atm} \times (2.24 \times 10^6 \text{ L})}{0.0821 \dfrac{\text{L} \times \text{atm}}{\text{K} \times \text{mol}} \times 315 \text{ K}}$$

The cavern contains 1.30×10^6 mol of methane.

One mole of methane has a mass of 16.0 g (C = 12 amu; 4 H = 4 amu). The number of grams of methane can be calculated using a mole $\longrightarrow$ mass conversion.

$$1.30 \times 10^6 \text{ mol CH}_4 \times \frac{16.0 \text{ g CH}_4}{1 \text{ mol CH}_4} = 20.8 \times 10^6$$

$$= 2.08 \times 10^7 \text{ g CH}_4$$

The cavern contains nearly 21 million grams of methane gas.

Problem

11. A child has a lung capacity of 2.2 L. How many grams of air do her lungs hold at a pressure of 1 atm and a normal body temperature of 37°C? Air is a mixture, but you may assume a "formula mass" of 29 g/mol for air because air is about 20% oxygen (gram formula mass of $O_2 = 32$) and 80% nitrogen (gram formula mass of $N_2 = 28$). Thus 0.20×32 g/mol + 0.80×28 g/mol = 29 g/mol.

Figure 10·20
Geologists use explosions to help map underground features. Explosives are also used in both the mining and road construction industries.

10·C Explosives and the Establishment of Nobel Prizes

An explosion is an extremely rapid combustion or decomposition reaction that produces gaseous products and heat. The most forceful explosions occur when liquid or solid reactants are instantaneously converted to gases. The volume of the gases will be much larger than that of the reactants. If the gases are confined, the resultant pressure will be great. Furthermore, the heat given off will increase this pressure. If the reaction is fast enough, however, the pressure will build up even if the gases are not confined. When the gas expands, the resulting release of pressure causes a shock wave. The kinetic energy of this "wave," the wind that follows it, and the heat from the reaction are the destructive force of the explosion.

An explosion reaction involves a fuel and an oxidizer. If the oxidizer is separate from the fuel, the reaction is called combustion. An example is the combustion of methane.

$$CH_4(g) + 2O_2(g) \longrightarrow CO_2(g) + 2H_2O(g) + \text{heat}$$

Such an explosion could occur if natural gas leaked into a building and was subsequently ignited by a spark or flame. If the fuel is its own oxidizer, a decomposition reaction takes place. An example of this type of explosive is nitroglycerine, $C_3H_5O_9N_3$. This compound is a thick, pale, oily liquid. It decomposes to give a variable mixture of products such as CO, CO_2, H_2O, and nitrogen oxides.

Some high explosives that are fuel and oxidizer in one, like nitroglycerine, are unstable. Even jarring them may be enough to cause detonation. Alfred Nobel (1833–1896) was very aware of this fact when his family started manufacturing nitroglycerine for use in the mining industry. In spite of their care, however, the Nobel family's factory exploded killing Alfred's brother.

At this point Nobel set to finding a way to make nitroglycerine less hazardous. He experimented on a barge in the middle of a lake to minimize the danger to others. At one point he found a cask of nitroglycerine that had leaked. Fortunately, the diatomaceous earth in which the cask was packed had absorbed the liquid. Nobel experimented with the mixture of nitroglycerine and diatomaceous earth. He found that it could be set off only by using a blasting cap. Moreover, it was equally explosive as the pure liquid. Nobel named his invention dynamite. It was soon one of the most widely used explosives because it could be handled relatively safely.

When Nobel died, he left his entire estate to establish a fund to provide annual prizes in five fields. He chose the fields of chemistry, physics, physiology and medicine, literature, and peace. Nobel Prizes are considered the highest honor a person can receive for achievements in these fields.

Safety

If you notice a strong odor of gas at home or in the laboratory, air out the room and have the gas lines checked for leaks before lighting any flame or using any electrical switch. (Methane is odorless, but compounds that do have odors are mixed with it so that it can be detected.)

Figure 10·21
Nobel Prize winners receive this medal and an award of over $100,000. The prestige of receiving the award is even greater than its monetary value.

10·11 Departures from the Gas Laws

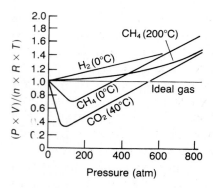

Figure 10·22
For an ideal gas, $(P \times V)/(n \times R \times T)$ always equals 1. By contrast, real gases deviate from ideality. What is the value of $(P \times V)/(n \times R \times T)$ for CO_2 at 40°C and 100 atm?

A gas that adheres to the gas laws at some conditions of temperature and pressure is said to exhibit ideal behavior. No gas behaves ideally, however, at all temperatures and pressures.

At any pressure the ideal gas law gives this ratio for an ideal gas: $(P \times V)/(n \times R \times T) = 1$. Figure 10·22 shows that this constant ratio plotted against pressure gives a horizontal line. The figure also shows that for several real gases at high pressures, $(P \times V)/(n \times R \times T)$ departs rather widely from the ideal. The ratio may be greater or less than one. The explanation for these departures from ideality is that two different factors are operating. These factors are attractions between molecules and the volume of gas molecules.

Simple kinetic theory assumes that gas particles are not attracted to each other and that the particles occupy a negligible volume. These assumptions are not strictly true. Gases and vapors could not be liquefied if there were no attractions between molecules. Of course, real gas particles must also occupy space. Furthermore, intermolecular forces tend to hold the particles in a gas together. This effectively reduces the distance between particles. The gas therefore occupies less volume than is assumed by the kinetic theory, and $(P \times V)/(n \times R \times T)$ tends to be less than one. At the same time, the molecules themselves occupy some volume and $(P \times V)/(n \times R \times T)$ tends to become greater than one. One or the other of these effects will usually dominate. In portions of the curves below the line, the intermolecular attractions dominate. In portions of the curves above the line, the volume of the molecules dominates. Which of these two effects dominates is determined by the temperature.

Compare the curves for CH_4 at 0°C and 200°C. At 0°C the methane molecules are moving relatively slowly. The attractions between the molecules are strong enough that part of the curve, at the lower pressures, is below the $(P \times V)/(n \times R \times T) = 1$ line. At higher pressures, the space between the molecules is reduced. The actual volume of the methane molecules becomes important, and the curve is above the $(P \times V)/(n \times R \times T) = 1$ line. Raising the temperature to 200°C increases the average kinetic energy of the methane molecules sufficiently to overcome weak attractive effects when the molecules collide. Thus $(P \times V)/(n \times R \times T)$ is near unity at lower pressures. The ratio increases to greater than one as the volume of the gas particles becomes important.

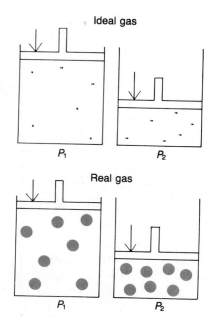

Figure 10·23
As pressure increases, the volume occupied by gas molecules is no longer negligible.

Problem

12. Small gas molecules such as N_2 and O_2 give the expected molar volumes of 22.41 L at STP. The nonideal behavior of gases is not limited, however, to extreme pressures and temperatures. The molar volumes of CH_4, CO_2, and NH_3 are respectively 22.37 L, 22.26 L and 22.06 L at the same conditions. Explain the reasons for these departures from ideality.

10·12 Diffusion and Graham's Law

Diffusion *is the tendency of molecules and ions to move toward areas of lower concentration until the concentration is uniform throughout the system.* For example, when a perfume bottle is opened, perfume molecules diffuse throughout a room.

Much of the early work on diffusion was done in the 1840s by the Scottish chemist Thomas Graham (1805–1869). Graham actually measured the rates of **effusion**, *which occurs as a gas escapes through a tiny hole in a container of gas.* Graham noticed that lighter gases effuse faster than heavier gases. From his observations, he proposed **Graham's law of effusion:** *The rate of effusion of a gas is inversely proportional to the square root of its formula mass.* Subsequently this relationship has also been shown to be true for the diffusion of gases. *The rate of diffusion of a gas is inversely proportional to the square root of its formula mass.*

Graham's law may be understood by an examination of the relationship of the mass and speed of a moving body to the force the body exerts when it strikes a stationary object. The mathematical expression that relates the mass (m) and the speed or velocity (v) of an object to its kinetic energy (KE) is $KE = \frac{1}{2} mv^2$. The kinetic energy influences the force the body is capable of exerting. Suppose a ball bearing with a mass of 2 g traveling at 5 m/s has just enough kinetic energy to shatter a pane of glass. A ball bearing with a mass of 1 g would need to travel at slightly more than 7 m/s to have the same kinetic energy. The lighter ball bearing would therefore have to move faster to shatter the same pane of glass.

There is an important principle here. *When two bodies of different mass have the same kinetic energy, the lighter body moves faster.* The particles in two gases at the same temperature have the same average kinetic energy. Thus a gas of low-formula mass should diffuse faster than a gas of high-formula mass at the same temperatures. It is easy to show this is true by comparing two balloons, one filled with helium and the other filled with air. Balloons filled with air stay inflated longer. This is because the main components of air, oxygen molecules and nitrogen molecules, move more slowly and therefore diffuse more slowly than helium atoms. Fast-moving helium atoms, with atomic masses of only 4 amu, rapidly diffuse through small pores in the balloon. The pores are large enough for both helium atoms and air molecules to pass through freely. Thus the rate of diffusion is related only to the particle's speed.

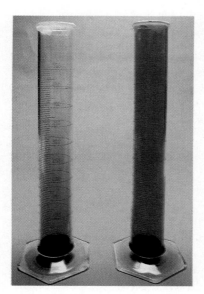

Figure 10·24
The diffusion of one substance through another is a relatively slow process. Here bromine is diffusing through air.

At a constant temperature, molecules with small mass diffuse more rapidly than molecules of large mass.

Figure 10·25
When ammonia (NH₃) and hydrochloric acid (HCl) are placed at either end of a tube, a white band of ammonium chloride (the reaction product) appears in the tube. At which end was the ammonia introduced, and at which end was hydrochloric acid introduced? *Hint:* Which molecule will diffuse faster?

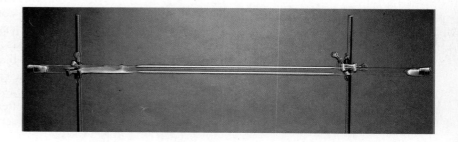

In a mathematical form Graham's law can be written as follows.

$$\frac{Rate_A}{Rate_B} = \frac{\sqrt{formula\ mass_B}}{\sqrt{formula\ mass_A}}$$

That is, the rates of effusion of two gases are inversely proportional to the square roots of their formula masses. Now let us compare the rates of effusion of nitrogen (formula mass, 28 amu) and helium (formula mass, 4 amu).

$$\frac{Rate_{He}}{Rate_{N_2}} = \frac{\sqrt{28}}{\sqrt{4}} = \frac{5.3}{2} = 2.7$$

We see that helium effuses nearly three times faster than nitrogen at the same temperature.

Problem

13. Which gas effuses faster: hydrogen or chlorine? How much faster?

10 The Behavior of Gases
Chapter Review

Key Terms

Chapter Summary

Gases expand to fill any volume available to them, and they diffuse from a region of high gas concentration to one of lower concentration. The smaller the formula mass of a gas, the greater its rate of diffusion (Graham's law). The collision of the particles in a gas with the walls of the container is gas pres-
sure. Gas pressure is measured in atmospheres or millimeters of mercury; 1 atm of pressure, at sea level, is 760 mm Hg. At a pressure of 1 atm and a temperature of 273 K (STP), the volume of 1 mol of any gas is 22.4 L. This is called the molar volume. Air is a mixture of gases, approximately 20% oxygen and 80% nitrogen. The total pressure in a mixture of gases is equal to the sum of the partial pressures of each gas present (Dalton's law).

The pressure and volume of a fixed mass of gas are inversely related. If one decreases, the other increases. This relationship is Boyle's law. The *pressure* of a fixed volume of gas is directly related to its Kelvin temperature (Gay-Lussac's law); the *volume* of a gas at constant pressure is directly related to its Kelvin temperature (Charles' law). These three separate gas laws can be written as a single expression called the combined gas law. Another expression is the ideal gas law, $PV = nRT$, that relates the moles of a gas to its pressure, temperature, and volume. The letter R is called the ideal gas constant. Its value depends on the units used.

Practice Questions and Problems

14. A metal cylinder contains 1 mol of nitrogen gas at STP. What will happen to the pressure if another mole of gas is added to the cylinder but the temperature and volume do not change? *10·1*

15. If a gas is compressed from 4 L to 1 L and the temperature remains constant, what happens to the pressure? *10·2*

16. A gas with a volume of 4 L is allowed to expand to a volume of 12 L. What happens to the pressure in the container if the temperature remains constant? *10·2*

17. Heating a contained gas at constant volume makes its pressure higher. Why? *10·3*

18. The gas in a container has a pressure of 3 atm at 27°C (300 K). What will the pressure be if the temperature is lowered to −173°C (100 K)? *10·3*

19. Describe an *ideal gas*. *10·4*

20. A gas mixture containing oxygen, nitrogen, and carbon dioxide has a pressure of 250 mm Hg. If $P_{O_2} = 50$ mm Hg and $P_{N_2} = 175$ mm Hg, what is P_{CO_2}? *10·5*

21. A gas with a volume of 4.0 L at a pressure of 0.90 atm is allowed to expand until the pressure drops to 0.20 atm. What is the new volume? *10·6*

22. A given mass of air has a volume of 6.0 L at 1 atm. What volume will it occupy at 190 mm Hg if the temperature does not change? *10·6*

23. Five liters of air at −50°C are warmed to 100°C. What is the new volume if the pressure remains constant? *10·7*

24. The pressure in an automobile tire is 2.0 atm at 27°C. At the end of a journey on a hot sunny day the pressure has risen to 2.2 atm. What is the temperature of the air in the tire? (Assume that the volume has not changed.) *10·8*

25. State the combined gas law. *10·9*

26. A 5.0-L air sample at a temperature of −50°C has a pressure of 800 mm Hg. What will be the new pressure if the temperature is raised to 100°C and the volume expands to 7.0 L? *10·9*

27. What volume will 12.0 g of oxygen gas (O_2) occupy at 25°C and a pressure of 0.520 atm? *10·10*

28. Calculate the number of liters occupied, at STP.
a. 2.5 mol N_2
b. 0.600 g H_2
c. 0.350 mol O_2 *10·10*

29. What pressure will be exerted by 0.450 mol of a gas at 25°C if it is contained in a vessel whose volume is 650 cm³? *10·10*

30. Determine the volume occupied by 0.582 mol of a gas at 15°C if the pressure is 622 mm Hg. *10·11*

31. No gas exhibits ideal behavior at all temperatures and pressures. Explain the meaning of this statement. *10·11*

32. Which gas effuses faster at the same temperature: molecular oxygen or atomic argon? *10·12*

33. Calculate the ratio of the velocity of helium atoms to neon atoms at the same temperature. *10·12*

34. Calculate the ratio of the velocity of helium atoms to fluorine molecules at the same temperature. *10·12*

Mastery Questions and Problems

35. If 4.50 g of methane gas (CH_4) is introduced into an evacuated 2.00-L container at 35°C, what is the pressure in the container, in atmospheres?

36. A gas with a volume of 300 mL at 150°C is heated until its volume is 600 mL. What is the new temperature of the gas if the pressure is unaltered?

37. At what conditions do real gases behave like ideal gases? Why?

38. A 5.00-L flask, at 25°C, contains 0.200 mol of Cl_2. What is the pressure in the flask?

39. A certain gas effuses four times as fast as oxygen, O_2. What is its molecular mass?

40. Calculate the volume of a gas in liters at 1 atm if its volume at 900 mm Hg is 1500 mL.

41. A 3.50-L gas sample at 20°C and a pressure of 650 mm Hg is allowed to expand to a volume of 8.00 L. The final pressure of the gas is 425 mm Hg. What is the final temperature of the gas in degrees Celsius?

42. Samples of neon and helium are placed in separate containers connected by a pinched rubber tube as shown below. When the clamp is removed from the rubber tube and the gases are allowed to mix, what is the partial pressure of each gas and the total pressure of the gas mixture? Assume that the volume of the rubber tubing connecting the containers is negligible.

Ne
5.00 L
634 mm Hg

He
3.00 L
522 mm Hg

43. A gas cylinder contains nitrogen gas at 10 atm pressure and a temperature of 20°C. The cylinder is left in the sun, and the temperature of the gas increases to 50°C. What is the pressure in the cylinder?

44. During an effusion experiment, it took 75 seconds for a certain number of moles of an unknown gas to pass through a tiny hole. Under the same conditions, the same number of moles of oxygen gas passed through the hole in 30 seconds. What is the molecular mass of the unknown gas?

Critical Thinking Questions

45. Choose the term that best completes the second relationship.
   a. ideal gas:real gas   fiction: _____
   (1) biography   (3) movie
   (2) novel   (4) nonfiction
   b. Charles' Law:temperature
   Boyle's Law: _____
   (1) pressure   (3) ideal mass
   (2) volume   (4) mass
   c. atm:mm Hg   °C: _____
   (1) mm   (2) L   (3) K   (4) mol

46. Gases will expand to fill a vacuum. Why do the earth's atmospheric gases not escape into the near-vacuum of space?

47. How does the vacuum used in thermos bottles and Dewar flasks prevent heat transfer?

48. What real gas comes closest to having the characteristics of an ideal gas? Why?

Review Questions and Problems

49. Calculate the gram formula mass of each of these substances.
   a. $Ca(C_2H_3O_2)_2$ **c.** $C_{12}H_{22}O_{11}$
   b. H_3PO_4 **d.** $Pb(NO_3)_2$

50. Name these compounds.
   a. $SnBr_2$ **c.** $Mg(OH)_2$
   b. $BaSO_4$ **d.** IF_5

51. An atom of lead-206 weighs 17.16 times as much as an atom of carbon-12.
   a. What is the atomic mass of this isotope of lead?
   b. How many protons, electrons, and neutrons are in this atom of lead?

52. How many kcal of heat are required to raise 40.0 g of water from $-12°C$ to $130°C$?

53. Write balanced equations for each of these chemical reactions.
   a. Calcium reacts with water to form calcium hydroxide and hydrogen gas.
   b. Tetraphosphorus decoxide reacts with water to form phosphoric acid.
   c. Mercury and oxygen are prepared by heating mercury(II) oxide.
   d. Aluminum hydroxide and hydrogen sulfide form when aluminum sulfide reacts with water.

54. Classify each of the reactions in Problem 53 as to type.

55. Calculate the molecular formula of each of these compounds.
   a. The empirical formula is C_2H_4O and the gmm = 88 g/mol.
   b. The empirical formula is CH and the gmm = 104 g/mol.
   c. The gmm = 90 g/mol. The percent composition is 26.7% C, 71.1% O and 2.2% H.

56. A piece of metal has a mass of 9.92 g and it measures 4.5 cm × 1.3 cm × 1.6 mm. What is the density of the metal?

57. Calculate the percent composition of isopropyl alcohol, C_3H_7OH.

58. Aluminum oxide is formed from its elements.

$$Al + O_2 \longrightarrow Al_2O_3$$

   a. Balance the equation.
   b. How many grams of each reactant is needed to form 583 g of Al_2O_3?

Challenging Questions and Problems

59. Oxygen is produced in the laboratory by heating potassium nitrate, KNO_3. Use the data table below, which gives the volume of oxygen produced at STP from varying quantities of KNO_3, to determine the mole ratio by which KNO_3 and O_2 react.

Mass of $KNO_3(g)$	Volume of $O_2(cL)$
0.84	9.3
1.36	15.1
2.77	30.7
4.82	53.5
6.96	77.3

60. Nitroglycerine has a density of 1.59 g/mL. It explodes to form several gases.

$$4C_3H_5O_9N_3 \longrightarrow 12CO_2 + O_2 + 6N_2 + 10H_2O$$

A sealed 1.00-mL container filled with nitroglycerine is detonated. Assuming standard temperature and assuming that the container would not break upon detonation, what is the pressure inside the container in atmospheres?

61. The following reaction takes place in a sealed 40.0-L container at a temperature of 120°C.

$$4NH_3(g) + 5O_2(g) \longrightarrow 4NO(g) + 6H_2O(g)$$

a. When 34.0 g NH_3 reacts with 96.0 g O_2, what is the partial pressure (in atm) of NO in the container?

b. What is the total pressure (in atm) in the container?

62. A mixture of acetylene, C_2H_2, and methane CH_4, occupied a certain volume at a total pressure of 126 torr. Upon burning the sample to form CO_2 and H_2O, the CO_2 was collected and its pressure found to be 192 torr in the same volume and at the same temperature as the original mixture. What percent of the original mixture was methane?

63. A 0.100-L container holds 3.0×10^{20} molecules of H_2 at 1.00 atm and 0°C.

a. If the volume of a hydrogen molecule is 6.7×10^{-24} mL, what percent of the volume of the gas is occupied by the its molecules?

b. If the pressure is increased to 1000 atm, the volume of the gas is 1×10^{-4} L. What fraction of the total volume do the hydrogen molecules now occupy?

Research Projects

1. How did early hot air balloonists such as the Montgolfier brothers influence the work of Charles and Gay-Lussac?

2. Carefully heat 10 mL of water to boiling in an aluminum soda can. Using tongs, quickly invert the can into a pan of cold water. Explain what happens.

3. Devise a method to determine the amount of gas in a bottle of soda. Compare different brands of soda.

4. How are liquefied gases used in scientific experiments?

5. How are modern submarines equipped to withstand high pressures underwater and to provide sufficient oxygen for the crew members?

6. Report on the research of one of the Nobel laureates in chemistry.

7. Trace the development of diving equipment.

8. Who invented the first "gas mask"? Report on the development of this type of equipment.

Readings and References

Morgan, Ralph A. *Collisions, Coalescence and Crystals: The States of Matter and Their Models.* New York: Methuen Educational (dist. by Harper & Row), 1973.

Morris, Daniel Luzon. "Cooking With Steam." *ChemMatters* (February 1987), pp. 17–19.

Nobel, Iris. *Contemporary Women Scientists of America.* New York: Messner, 1979.

Pierce, Conway, and R. Nelson Smith. *General Chemistry Workbook: How to Solve Chemistry Problems,* 5th ed. San Francisco: Freeman, 1975.

Walker, Mort. *Sport Diving, the Instructional Guide to Skin and Scuba.* Chicago: Regnery, 1977.

11 Electrons in Atoms

Chapter Preview

11·1 The Development of Atomic Models
11·2 The Quantum Mechanical Model of the Atom
11·3 Atomic Orbitals
11·A The Chemistry of Lighting
11·4 Electron Configurations
11·5 Exceptional Electron Configurations
11·B Flame Tests
11·6 Light and Atomic Spectra
11·C Spectroscopy: Identifying Atoms and Molecules
11·7 The Quantum Concept
11·8 Light as Particles: The Photoelectric Effect
11·9 An Explanation of Atomic Spectra
11·10 The Wave Motion of Matter and Quantum Mechanics
11·D Technical Illustration: Drawing the Unseen

Near the end of the nineteenth century, science was in turmoil. At that time, physicists began to probe the submicroscopic world of atoms. They found that ideas about matter and energy that had been relied on for over a century were inadequate. Chemists were confused about the reasons for the similarity in the chemistry of certain elements. When the fog lifted nearly 25 years later, the quantum mechanical model of the atom had emerged. The new model integrated the experimental discoveries and the theories of both chemists and physicists.

11·1 The Development of Atomic Models

For about 50 years past the time of John Dalton (1766–1844), the atom was considered to be a solid indestructible mass. Dalton's atomic theory was a great advance in explaining the nature of chemical reactions. It included, however, the idea of indivisible atoms. The discovery of subatomic particles shattered every theory people had about indivisible atoms.

The discoverer of the electron, Joseph John Thomson (1856–1940), realized that the accepted model of an indivisible atom did not take electrons and protons into account. Thomson therefore proposed

Figure 11·1
In this chapter you will learn how the brilliant colors of neon lights are due to the movements of electrons in atoms.

245

Figure 11·2
While he was still a student, Niels Bohr used his knowledge of light to develop a new atomic model.

■ Our understanding of the atom has been refined by successive atomic theories.

Figure 11·3
These drawings show how the model of the atom has changed as physicists have learned more about its structure. **a** In the nineteenth century, Thompson described the atom as a ball of electricity containing a number of electrons. **b** In the early twentieth century, Rutherford showed that most of an atom's weight is concentrated in a small positively charged area called the nucleus. **c** Later, Bohr proposed that electrons travel around the nucleus in definite orbits. **d** Modern physicists maintain that the atom has no defined shape and that electrons do not have precise orbits.

a revised model that was referred to as the "plum pudding atom" (Figure 11·3). It had negatively charged electrons (raisins) stuck into a lump of positively charged protons (the dough). Thomson's model did not include neutrons because they had not yet been discovered. The plum pudding model explained some electrical properties of atoms. It said nothing, however, about the number of protons, their arrangement in the atom, or the ease with which atoms are stripped of electrons to form ions.

Based on his discovery of the nucleus, Ernest Rutherford proposed the nuclear atom in which electrons surround a dense nucleus. He thought of the rest of the atom as empty space. Later experiments showed that the nuclei of most atoms are composed of protons and neutrons.

Oppositely charged particles attract each other. Thus one could argue that the negative electrons should be drawn into a positive nucleus. This would cause the atom to collapse. If the Rutherford model is correct, what prevents the electrons from falling into the nucleus?

In 1913 Niels Bohr (1885–1962), a young Danish physicist and a student of Rutherford's, came up with a new model. He proposed that electrons are arranged in concentric circular paths, or orbits, around the nucleus. This model was patterned after the motions of the planets around the sun and is often referred to as the planetary model. Bohr answered in a novel way the question of what prevents electrons from falling into the nucleus. He proposed that the electrons in a particular

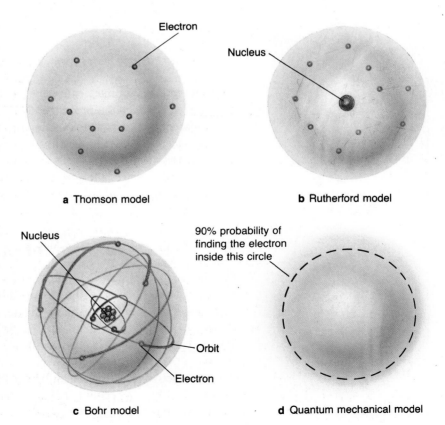

a Thomson model

b Rutherford model

c Bohr model

d Quantum mechanical model

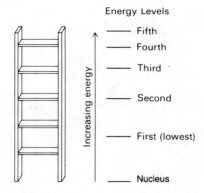

Energy Levels

Fifth

Fourth

Third

Second

First (lowest)

Nucleus

Increasing energy

Figure 11·4
The rungs of a ladder are analogous to the energy levels in an atom. The higher the energy level occupied by the electron, the more energetic it is and the further it usually is from the nucleus.

The quantum mechanical model of the atom describes the probability of finding an electron in a region of space around the nucleus.

path have a fixed energy. Thus they do not lose energy and fall into the nucleus. *The **energy level** of an electron is the region around the nucleus where it is likely to be moving.* The rungs of a ladder are analogous to the fixed energy levels of electrons. The lowest rung of the ladder corresponds to the lowest energy level (Figure 11·4). A person can climb up or down a ladder by going from rung to rung. Similarly, an electron can jump from one energy level to another. A person on a ladder cannot stand between the rungs. Similarly, the electrons in an atom cannot stop between energy levels. To move from one energy level to another, an electron must gain or lose just the right amount of energy.

The amount of energy gained or lost by every electron is not always the same. Unlike the rungs of a ladder, the energy levels in an atom are *not* equally spaced. *A **quantum** of energy is the amount of energy required to move an electron from its present energy level to the next higher one.* Thus the energies of electrons are said to be *quantized.* The term *quantum leap,* used to describe an abrupt change, comes from this concept. In general, the higher an electron is placed on the energy ladder, the larger is the diameter of the circular path or orbit of that electron.

11·2 The Quantum Mechanical Model of the Atom

In 1926 the Austrian physicist Erwin Schrödinger (1887–1961) took atomic models one step further. He used the new quantum theory to write and solve a mathematical equation describing the location and energy of an electron in a hydrogen atom. *The modern description of the electrons in atoms,* **the quantum mechanical model,** *derives from the mathematical solution to the Schrödinger equation.* Previous models were essentially physical models based on the extension of our knowledge of the motion of large objects. In contrast, the quantum mechanical model is primarily mathematical. It has few, if any, analogies in the visible world.

Like the Bohr model, the quantum mechanical model of the atom leads to quantized energy levels for an electron. Unlike the Bohr model, however, the quantum mechanical model does not define the exact path an electron takes around the nucleus. It is concerned with the likelihood of finding an electron in a certain position. This probability can be portrayed as a blurry cloud of negative charge. The cloud is most dense where the probability of finding the electron is large. It is less dense where the probability of finding the electron is small. Hence it is difficult to say where an electron cloud ends. There is at least a slight chance of finding the electron a considerable distance from the nucleus. A model is needed, however, to show where the electron probably is. By convention, a surface is drawn around the model of an electron cloud so that the electron is inside the surface 90% of the time. The surface is shaped so that the probability for finding the electron inside is the same for all points on the surface. The shape of the surface can now give us a useful picture of the shape of the cloud. Figure 11·3 illustrates this idea.

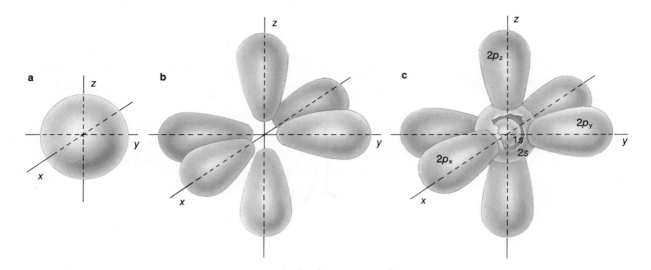

Do not confuse orbitals of the quantum mechanical model with orbits of the Bohr model.

An atomic orbital is identified by a principal quantum number and an energy sublevel. Together these identify the shape and location of the atomic orbital.

s = spherical orbital
p = dumbbell-shaped orbital

11·3 Atomic Orbitals

In the quantum mechanical model of the atom, the energy levels of electrons are designated by *principal quantum numbers* (n). These are assigned certain values: $n = 1, 2, 3, 4, 5, 6$, and so forth. Within each *principal energy level,* the complex mathematics of quantum mechanics gives us *several* electron cloud shapes. This occurs because the quantum mechanical model divides the energy levels into *energy sublevels.* Each sublevel corresponds to a different cloud shape. Each cloud shape is called an *atomic orbital.* (These are different from the orbits of the Bohr model.) *We may think of an* **atomic orbital** *as a region in space where there is a high probability of finding an electron.* Different atomic orbitals are denoted by letters. The *s orbitals* give spherical clouds, and *p orbitals* give dumbbell-shaped clouds. The shapes of *d orbitals* and *f orbitals* are far more complex.

The lowest principal energy level ($n = 1$) has only one sublevel, called $1s$. This orbital is spherical. The second principal energy level ($n = 2$) has two sublevels. The $2s$ orbital is spherical, and the $2p$ orbitals are dumbbell-shaped. The $2p$ sublevel is of higher energy than the $2s$. The $2p$ sublevel consists of three p orbitals of equal energy. The long axis of each dumbbell-shaped p orbital is perpendicular to the other two. It is convenient to label the axes $2p_x$, $2p_y$, and $2p_z$. Thus the second principal energy level has four orbitals: $2s$, $2p_x$, $2p_y$, and $2p_z$.

The third principal energy level ($n = 3$) has three sublevels, called $3s$, $3p$, and $3d$. The $3d$ sublevel consists of five d orbitals of equal energy. Thus the third principal energy level has nine orbitals (one $3s$, three $3p$, and five $3d$ orbitals). The fourth principal energy level ($n = 4$) has four sublevels, called $4s$, $4p$, $4d$, and $4f$. The $4f$ sublevel consists of seven f orbitals of equal energy. The fourth principal energy level, then, has 16 orbitals (one $4s$, three $4p$, five $4d$, and seven $4f$ orbitals). The shapes of the f orbitals are not shown. The sublevels and orbitals for each principal energy level are summarized in Table 11·1.

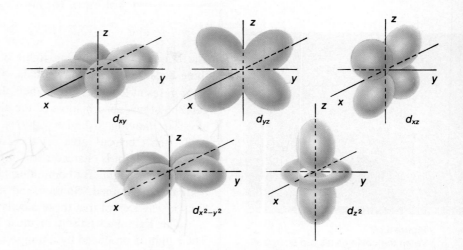

Figure 11·6
Four of the five *d* orbitals have the same shape but different orientations in space.

d_{xy} d_{yz} d_{xz}

$d_{x^2-y^2}$ d_{z^2}

The principal quantum number always equals the number of sublevels within that principal energy level. Furthermore, the maximum number of electrons that can occupy a given energy level is given by the formula, $2n^2$. Here n is the principal quantum number. The number of electrons allowed in each of the first four energy levels is as follows.

	Increasing energy (increasing distance from nucleus)			
Energy level n	1	2	3	4
Maximum number of electrons allowed	2	8	18	32

Problem

1. How many orbitals are in the following sublevels?
   a. 3*p* sublevel   d. 4*p* sublevel
   b. 2*s* sublevel   e. 3*d* sublevel
   c. 4*f* sublevel   f. third principal energy level

Table 11·1 Summary of Principal Energy Levels, Sublevels, and Orbitals		
Principal energy level	**Number of sublevels**	**Type of sublevel**
$n = 1$	1	1*s* (1 orbital)
$n = 2$	2	2*s* (1 orbital), 2*p* (3 orbitals)
$n = 3$	3	3*s* (1 orbital), 3*p* (3 orbitals) 3*d* (5 orbitals)
$n = 4$	4	4*s* (1 orbital, 4*p* (3 orbitals), 4*d* (5 orbitals), 4*f* (7 orbitals)

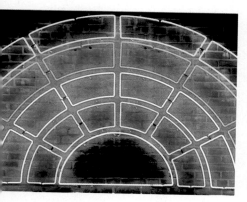

Figure 11·7
When the noble gases are energized they give off light of characteristic colors. Helium produces a yellowish light, and argon produces lavender. Krypton gives a whitish light, xenon gives blue, and neon shines orange-red.

Figure 11·8
Sodium vapor lamps produce a bright yellowish light. An electric current produces heat that vaporizes the sodium in the lamp. Atoms of sodium absorb energy and reradiate it as light. Sodium vapor lamps are used where energy efficiency and low maintenance are more important than realistic color.

11·A The Chemistry of Lighting

The famous, flashing neon lights of Times Square and Las Vegas shine because of electron orbitals. These lights contain neon gas or other gases at low pressure. When electric current passes through the gas, the atoms absorb some of the electrical energy by some of their electrons moving to higher orbitals. In dropping back to lower orbitals, the electrons give off the energy in the form of light. The color of the light is characteristic of the gas in the tube. Neon gives a bright orange light. By combining neon with other gases, other colors can be obtained. Sodium and mercury vapor lamps work in a similar way except that these metals must first be vaporized.

The light bulbs used in most homes are incandescent lamps. Their light is produced by a tungsten wire heated to white heat by an electric current. At such high temperatures the tungsten atoms give off white light much like sunlight. White light is really a combination of all the colors. This type of light is given off by hot solids and liquids and even by dense gases. Colors seen in incandescent light appear much as they do in natural light. The tungsten filament would rapidly burn up in the presence of oxygen. To prevent this an incandescent bulb is filled with argon or nitrogen. At such high temperatures the tungsten slowly vaporizes. Eventually the filament will break. Then we say the bulb is burned out. Even before the filament breaks, the vaporizing tungsten will condense on the inner surface of the bulb. This darkens the bulb and reduces the light given off.

Incandescent lighting is very inefficient at producing light from electricity. The heat that you feel when you hold your hand near a tungsten-filament bulb is energy that is lost as heat instead of as visible light. In fact, tungsten-filament lamps can be designed to radiate even more of their energy as heat. Fast-food restaurants keep food warm in the red glow of such infrared heat lamps.

Fluorescent lighting is much more efficient than incandescent lighting because no hot filament is involved. The long glass tube of a typical fluorescent bulb contains small amounts of mercury vapor and argon gas. The inside of the bulb is coated with a powder called a phosphor. When current passes through the bulb, electrons in the mercury atoms are excited to higher energy levels. In dropping back to the ground state, they emit ultraviolet (UV) light which is not visible. The UV light is absorbed by the electrons in the molecules of the phosphor coating of the tube. When the phosphor electrons return to their ground state, they emit visible white light that has more blue and less red than sunlight. Colors seen under this light seem "cold" in comparison to the same colors seen in natural light. A 40-watt fluorescent bulb produces as much light as a 150-watt incandescent bulb. It is no wonder that more than 75% of the lighting used in North America is fluorescent.

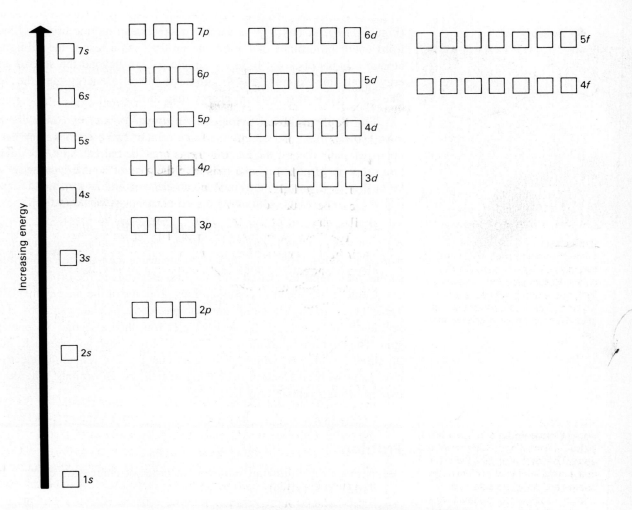

Figure 11·9
The energy levels of the various atomic orbitals are shown on this Aufbau diagram. Orbitals of greater energy are shown higher on the page.

The electron configurations of most elements can be written using three rules.

11·4 Electron Configurations

It is the nature of things to seek the lowest possible energy. High-energy systems are unstable. Unstable systems lose energy to become more stable. In the world of the atom, electrons and the nucleus interact to make the most stable arrangement possible. *The ways in which electrons are arranged around the nuclei of atoms are called* **electron configurations.**

Three rules govern the filling of atomic orbitals by electrons within the principal energy levels. These rules are the Aufbau principle, the Pauli exclusion principle, and Hund's rule. It is important to realize that these rules provide a way to obtain correct electron configurations for atoms. It is *not* the way that atoms are formed.

1. **The Aufbau principle:** *Electrons enter orbitals of lowest energy first.* The various orbitals within a sublevel of a principal energy level are always of equal energy. Further, within a principal energy level the *s* sublevel is always the lowest energy sublevel. Yet the range of energy levels within a principal energy level can overlap the energy levels of an adjacent principal energy level. This is shown on the Aufbau diagram

German: *Aufbau* = building up.

(Figure 11·9). A box (☐) is used to represent an atomic orbital. Electrons enter the orbitals of lowest energy first. As a result, the filling of atomic orbitals does not follow a simple pattern beyond the second energy level. For example, the $4s$ orbital is lower in energy than the $3d$. The $4f$ orbital is also lower in energy than the $5d$.

2. **The Pauli exclusion principle:** *An atomic orbital may describe at most two electrons.* To occupy the same orbital, two electrons must have opposite spins; that is, the electron spins must be paired. Spin is a quantum property of electrons and may be clockwise or counterclockwise. A vertical arrow (↑) is used to indicate an electron and its direction of spin (↑ or ↓). An orbital containing paired electrons is written as ↑↓.

3. **Hund's rule:** *When electrons occupy orbitals of equal energy, one electron enters each orbital until all the orbitals contain one electron with spins parallel* (↑ ↑ ↑). Second electrons then add to each orbital so that their spins are paired with the first electrons in the orbital.

Consider the electron configurations of atoms of the nine elements in Table 11·2. An oxygen atom, for example, contains eight electrons. The orbital of lowest energy, $1s$, gets one electron, then a second of opposite spin. The next orbital to fill is $2s$. Three electrons then go, one each, into the three $2p$ orbitals, which have equal energy. The remaining electron now pairs with an electron occupying one of the $2p$ orbitals. The other two $2p$ orbitals remain only half filled.

Problem

2. Arrange the following sublevels in order of decreasing energy: $2p$, $4s$, $3s$, $3d$, and $3p$.

Table 11·2 Electron Configurations for Some Selected Elements

Element	Orbital filling						Electron configuration
	$1s$	$2s$	$2p_x$	$2p_y$	$2p_z$	$3s$	
H	↑	☐	☐	☐	☐	☐	$1s^1$
He	↑↓	☐	☐	☐	☐	☐	$1s^2$
Li	↑↓	↑	☐	☐	☐	☐	$1s^2\,2s^1$
C	↑↓	↑↓	↑	↑	☐	☐	$1s^2\,2s^2\,2p^2$
N	↑↓	↑↓	↑	↑	↑	☐	$1s^2\,2s^2\,2p^3$
O	↑↓	↑↓	↑↓	↑	↑	☐	$1s^2\,2s^2\,2p^4$
F	↑↓	↑↓	↑↓	↑↓	↑	☐	$1s^2\,2s^2\,2p^5$
Ne	↑↓	↑↓	↑↓	↑↓	↑↓	☐	$1s^2\,2s^2\,2p^6$
Na	↑↓	↑↓	↑↓	↑↓	↑↓	↑	$1s^2\,2s^2\,2p^6\,3s^1$

A convenient shorthand method for showing the electron configuration of an atom can be used. This involves writing the energy level and the symbol for every sublevel occupied by an electron. A superscript indicates the number of electrons occupying that sublevel. For hydrogen, with one electron in a $1s$ orbital, the electron configuration is written $1s^1$. For helium, with two electrons in a $1s$ orbital, it is $1s^2$. For oxygen, with two electrons in a $1s$ orbital, two electrons in a $2s$ orbital, and four electrons in $2p$ orbitals, it is $1s^2 \, 2s^2 \, 2p^4$. Note that the sum of the superscripts equals the number of electrons in the atom.

Figure 11·10
Erwin Schrödinger developed and solved an equation which is the basis of the quantum mechanical model of the atom.

Example 1

Use Figure 11·9 to write the electron configuration of these atoms.
a. phosphorus   **b.** nickel.

Solution

Phosphorus has 15 electrons; nickel has 28 electrons. Using Figure 11·9, start placing electrons in the orbitals with the lowest energy ($1s$). Remember that there is a maximum of 2 electrons in each orbital. Electrons do not pair up in orbitals of equal energy until necessary.

a. phosphorus   **b.** nickel

P $1s^2 \, 2s^2 \, 2p^6 \, 3s^2 \, 3p^3$ Ni $1s^2 \, 2s^2 \, 2p^6 \, 3s^2 \, 3p^6 \, 3d^8 \, 4s^2$

When the configurations are written, the sublevels within the same principal energy level are written together. This is not always the same order as given on the Aufbau diagram.

Problems

3. Write electron configurations for atoms of the following elements. How many unpaired electrons do these atoms have?
a. boron   **b.** fluorine

4. How many electrons are in the highest occupied energy level of these atoms?   **a.** barium   **b.** sodium   **c.** aluminum   **d.** oxygen

11·5 Exceptional Electron Configurations

A few elements have electron configurations that do not follow the normal pattern.

We obtain correct electron configurations for the elements up to vanadium (atomic number 23) if we follow the Aufbau diagram for orbital filling. If we were to continue, we would assign chromium (atomic number 24) and copper (atomic number 29) the following configurations.

$$Cr\ 1s^2\ 2s^2\ 2p^6\ 3s^2\ 3p^6\ 3d^4\ 4s^2$$

$$Cu\ 1s^2\ 2s^2\ 2p^6\ 3s^2\ 3p^6\ 3d^9\ 4s^2$$

The correct electron configurations, confirmed experimentally, are shown below.

$$Cr\ 1s^2\ 2s^2\ 2p^6\ 3s^2\ 3p^6\ 3d^5\ 4s^1$$

$$Cu\ 1s^2\ 2s^2\ 2p^6\ 3s^2\ 3p^6\ 3d^{10}\ 4s^1$$

These arrangements give chromium a half-filled d sublevel and copper a filled d sublevel. Scientists have determined that filled energy sublevels are more stable than partially filled sublevels. Half-filled levels are not as stable as filled levels. Nevertheless, they are more stable than other configurations. With only one electron in the $4s$ sublevel, chromium atoms and copper atoms are more stable.

These unexpected results are worth knowing. It is more important, however, for you to achieve a good understanding of the general procedure for obtaining electron configurations.

a

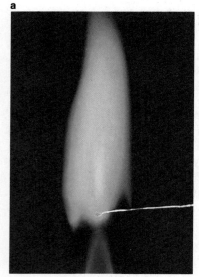

b

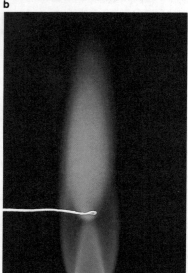

Figure 11·11
These two flame tests show the characteristic colors of barium (a) and strontium (b).

================ Science, Technology, and Society ================

11·B Flame Tests

Elements can be identified by the color of light they give off when their atoms absorb energy. Recognition of these colors is the basis for a simple analytical method called a flame test.

The metal or metal compound to be tested is dissolved in hydrochloric acid. This produces a solution of the metal chloride. A platinum or nichrome wire is dipped in the solution. It is then held in the hot, blue flame of a Bunsen burner. The metal atoms absorb energy from the heat of the flame. When they return to the ground state, they give off light of a characteristic color. Chemists learn to recognize the colors: pale green for barium, yellow for sodium, crimson for lithium, and so on. This simple test works for metals that form volatile chlorides.

The flame test was perfected by Robert Bunsen (1811–1899), a great German chemist of the nineteenth century. Bunsen was a university teacher and research chemist all his working life. He was famous for his emphasis on experimentation. Early in his career he lost an eye in a laboratory explosion. He also survived extensive work with arsenic compounds by developing an antidote for arsenic poisoning. He often invented or perfected the equipment that he needed, including batteries, calorimeters, and, of course, the Bunsen burner.

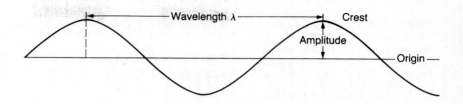

Figure 11·12
The amplitude of a wave is the height of the wave from the origin to the crest. The wavelength is the distance between crests.

11·6 Light and Atomic Spectra

The work that led to the development of the quantum mechanical model of the atom grew out of the study of light. Isaac Newton (1642–1727) had thought of light as consisting of particles. By the year 1900, however, the idea that light was a wave phenomenon was firmly ingrained among scientists.

In the wave model, light is considered to consist of electromagnetic waves that travel in a vacuum at a speed of 3.0×10^{10} cm/s. **Electromagnetic radiation** *includes radio waves, microwaves, visible light, infrared and ultraviolet light, X-rays, and gamma rays.* The parts of a wave at an instant of time are shown in Figure 11·12. Each wave cycle begins at the origin, then returns to the origin. *The* **amplitude** *of a wave is the height of the wave from the origin to the crest. The* **wavelength** (λ, the Greek letter lambda) *is the distance between the crests.* This can vary. In a traveling wave (Figure 11·13) **frequency** (ν, the Greek letter nu) *is the number of wave cycles to pass a given point per unit of time.* The frequency and wavelength of light are inversely related.

$$\nu = \frac{c}{\lambda}$$

The visible region is only a small portion of the electromagnetic spectrum.

■ Light and other electromagnetic radiation consist of waves of many lengths and frequencies.

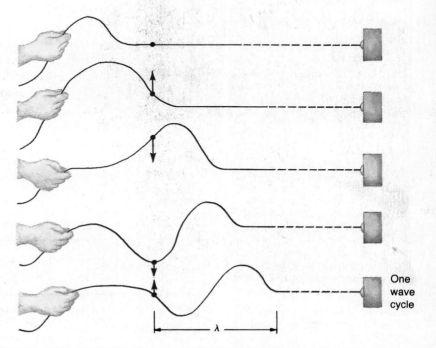

One wave cycle

Figure 11·13
The frequency of a wave is the number of wave cycles to pass a given point per unit time. This diagram shows the generation of one wave cycle.

Figure 11·14
The frequency and wavelength of light waves are inversely related. As the wavelength increases, the frequency decreases. The wavelength and frequency do not affect the amplitude.

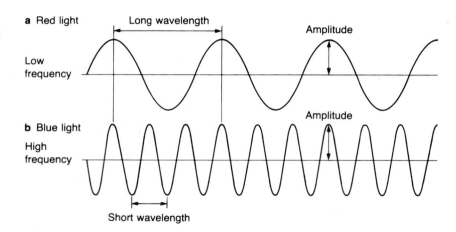

Long wavelength means low frequency. Short wavelength means high frequency.

Safety

Sunlight contains infrared and ultraviolet radiation as well as visible light. Both IR and UV radiation can be damaging to the eye. Do not look directly at the sun and be careful when working with lamps that produce these types of light.

 ChemDirections

Ozone in the upper atmosphere absorbs ultraviolet radiation from the sun. **ChemDirections** *Ozone Layer*, page 665, describes the consequences that may result from decreased ozone concentrations.

Figure 11·15
The electromagnetic spectrum consists of radiation over a broad band of wavelengths. The visible light portion is very small. It is in the 10^{-7} m wavelength range and 10^{15} hertz (s^{-1}) frequency range. What types of nonvisible radiation have wavelengths close to those of red light? To those of blue light?

As the wavelength of light increases, the frequency decreases (Figure 11·14). Hence the product of frequency and wavelength always equals a constant, c, the speed of light. The units of frequency are usually cycles per second. *In the SI, units of cycles per second are called* **hertz,** Hz. Frequency is expressed as reciprocal seconds (s^{-1}) because the term "cycles" is assumed as being understood.

Sunlight consists of light with a continuous range of wavelengths and frequencies. The wavelength and frequency of yellow light are different from those of red. *When sunlight is passed through a prism the light is separated into a* **spectrum** *of colors.* Each color blends into the next in order: red, orange, yellow, green, blue, indigo, and violet. In the visible spectrum, red light has the longest wavelength and the lowest frequency. Violet light has the shortest wavelength and highest frequency.

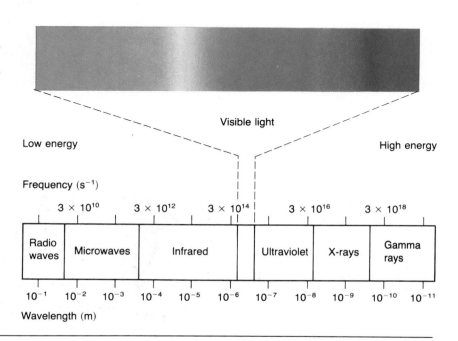

Example 2

Calculate the wavelength of the yellow light emitted by a sodium lamp if the frequency of the radiation is 5.10×10^{14} s^{-1}.

Solution

First rearrange the equation $\nu = c/\lambda$ to give $\lambda = c/\nu$. Substitute values for c and ν.

$$\lambda = \frac{3.00 \times 10^{10} \text{ cm/s}}{5.10 \times 10^{14} \text{ s}^{-1}}$$

$$= \frac{3.00 \times 10^{10} \text{ cm s}^{-1}}{5.10 \times 10^{14} \text{ s}^{-1}}$$

$$= 5.88 \times 10^{-5} \text{ cm}$$

Remember that s^{-1} equals 1/s.

Problem

5. What frequency is radiation whose wavelength is 5.00×10^{-6} cm? In what region of the electromagnetic spectrum is this radiation?

Chemical Connections

Dyes are substances that absorb light in certain parts of the spectrum, but not others. A red dye, for example, absorbs all colors of the spectrum except red, which it reflects back to the viewer.

Every element emits light if it is heated by passing an electric discharge through its gas or vapor. The atoms absorb energy, then lose the energy and emit it as light. *Passing the light emitted by an element through a prism gives the* **atomic emission spectrum** *of the element*. The emission spectra of elements are quite different from the spectrum of white light. White light gives a continuous spectrum. By contrast the atomic emission spectra consist of relatively few lines. Thus atomic emission spectra are line spectra or discontinuous spectra. Each line in an emission spectrum corresponds to one exact frequency of light being given off by the atom. Therefore each line corresponds to one exact amount of energy being given off. Figure 11·16 shows part of the emission spectrum of hydrogen.

The emission spectrum of each element is unique to that element. This makes these spectra extremely useful for the identification of unknown and sometimes otherwise inaccessible substances. Much of our knowledge of the composition of the universe, for example, comes from the atomic spectra of the stars, which are hot glowing bodies of gases.

Figure 11·16
The emission spectrum of an element consists of a series of bands which are characteristic for the element.
a The emission spectrum of atomic hydrogen consists of a few widely spaced bands. **b** Molecular hydrogen gives off light at more frequencies and thus has more bands in its spectrum.

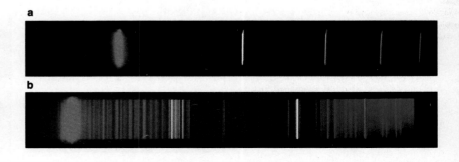

Problem

6. A hydrogen lamp emits several lines in the visible region of the spectrum. One of these lines has a wavelength of 6.56×10^{-5} cm. What is the color and frequency of this radiation?

11·C Spectroscopy: Identifying Atoms and Molecules

Spectroscopy is the process of producing and analyzing spectra. Robert Bunsen and Gustav Kirchhoff built the first spectroscope from two telescope eyepieces, a prism, and a Bunsen burner. With this apparatus they discovered a number of new elements including cesium and rubidium. They named these elements for the color of the spectra.

Spectroscopy is the chemist's most important means of identifying substances. Spectra can give both qualitative and quantitative information. The wavelengths detected tell the spectroscopist what is in the sample. The relative intensity of each wavelength tells how much of the substance is present.

A spectrum that is created by exciting atoms and then observing the light that they emit as they return to lower energy states is called an *emission spectrum*. In flame spectrophotometry, elements are excited by being burned in a flame. The ultraviolet or visible light emitted is then dispersed into its various wavelengths and detected.

A spectrum may also be created by determining the energy absorbed by an atom or molecule. Light or other energy is directed at a sample. The sample absorbs the quanta that it can. The remainder of the light passes through unchanged. The beam that comes out of the sample is compared to the beam that goes in, to determine which wavelengths have been absorbed. This is called an *absorption spectrum*. Atomic absorption spectroscopy is similar to flame spectrophotometry. The difference is that the light absorbed by the atoms is measured rather than the light emitted.

Spectral lines may be due to the emission or absorption of light. White light passed through the vapor of an element gives dark spectral lines because some of the wavelengths of light are absorbed by atoms of the element.

Figure 11·17
The emission spectra of the elements show different patterns. These spectra are for sodium (a), calcium (b), strontium (c), and barium (d).

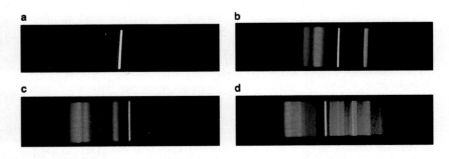

Figure 11·18
An absorption spectrometer can be used to measure the amount of a colored product in a chemical reaction. One wavelength of light is passed through the sample. The amount of light absorbed by the sample is measured and compared to a known standard to determine the amount of product.

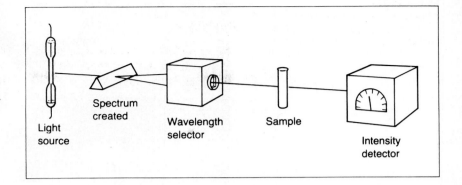

Like atoms, molecules also have quantized electron energy levels. In addition, they have energy levels due to their vibrations. In a molecule of water, for example, the hydrogen–oxygen bonds vibrate at particular frequencies. If the molecule absorbs energy, it can begin to vibrate at a different frequency, corresponding to a different energy. These vibrational energies are quantized, just like electron energies. Jumps from one vibrational level to another correspond to wavelengths in the infrared region. These wavelengths are very characteristic of the molecule. They depend on the atoms in the molecule, the forces in the molecular bonds, and the geometry of the molecule. These three characteristics uniquely define a particular molecule. Thus the vibrational spectrum also uniquely identifies that molecule. It is often called the "molecular fingerprint." The infrared spectra of thousands of compounds have been cataloged. Infrared absorption spectroscopy is the most common method of analysis used for covalently bonded molecules.

Many other types of spectroscopy have been developed. The source of the incident energy may be a lamp, an electric arc or spark, a flame, a laser, an X-ray tube, or a radioactive substance. The energy may change the energy level of electrons, bond vibrations or rotations, or even a nucleus. The resulting spectrum may be in any region of the electromagnetic spectrum. All these techniques are used to identify substances and to study chemical reactions.

 Issues in Chemistry
Dentists and physicians routinely use X-rays as a diagnostic tool. What are other practical applications of X-ray technology? Why is there a need to minimize exposure to X-rays?

Figure 11·19
Chemists use infrared spectroscopy to identify organic molecules. The tracing is characteristic for each compound. Each valley or peak in this infrared spectrum of ethene (C_2H_4) indicates bond stretching or bending in the molecule when it is subjected to infrared radiation.

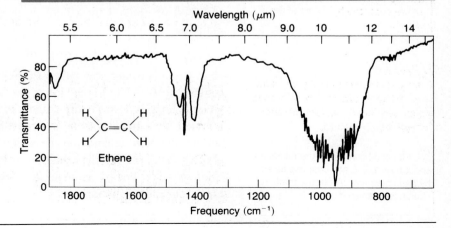

Figure 11·20
All objects change colors when they are heated. Although this looks like a continuous process, the atoms in the iron are giving off energy in small discrete units.

11·7 The Quantum Concept

The laws of classical physics hold that there is no limit to how small the energy gained or lost by an object may be. According to these laws, when a gaseous element or the vapor of an element is heated to glowing, the spectrum of the emitted light should be continuous. Thus classical physics is no help in explaining the emission spectra of atoms, which consist of lines. The seed of an idea that explained atomic spectra came in 1900 from the German physicist Max Planck (1858–1947).

The laws of classical physics are inadequate to explain atomic spectra.

Planck was trying to quantitatively describe why a body like a chunk of iron appears to change color as it is heated. First it appears black, then red, yellow, white, and blue as its temperature increases. Planck found he could explain this if he assumed that the energy of a body changes only in small discrete units. By analogy, a brickwall can be increased or decreased in size only by units of one or more bricks.

Problem

7. Explain the difference between the laws of classical physics and the quantum concept, for example, when describing the energy lost or gained by an object.

Planck showed mathematically that the amount of radiant energy, E, absorbed or emitted by a body is proportional to the frequency of the radiation.

$$E \propto \nu \quad \text{or} \quad E = h \times \nu$$

Here h is a constant now called **Planck's constant,** *which has a value of* 6.6262×10^{-34} J s (J is the joule, the SI unit of energy). The energy of a quantum equals $h \times \nu$. Any attempt to increase or decrease the energy of a system by a fraction of $h \times \nu$ must fail. The size of an emitted or absorbed quantum depends on the size of the energy change. A small energy change involves the emission or absorption of low-frequency radiation. A large energy change involves the emission or absorption of high-frequency radiation.

Planck showed that the emission spectra of atoms are best explained by assuming that electromagnetic radiation is quantized.

Planck's proposal of absorption or emission of quanta of energy was revolutionary. Everyday experience had led chemists to believe that there was no limitation to the smallness of permissible energy changes in a system. It appears, for example, that thermal energy may be continuously supplied to heat liquid water to any temperature between 0°C and 100°C. Actually, the water temperature increases by infinitesimally small steps. This occurs as individual molecules absorb quanta of energy. We are unable, however, to detect such small changes in temperature. Thus our everyday experience gives us no clue to the fact that energy is quantized.

11·8 Light as Particles: The Photoelectric Effect

In 1905 Albert Einstein, then a patent examiner in Zurich, Switzerland, returned to Newton's idea of particles of light. Einstein proposed that light could be described as quanta of energy that behave as particles. *Light quanta are called* **photons.** The energy of photons is quantized according to the familiar equation $E = h \times \nu$.

Figure 11·21
Albert Einstein is considered one of the greatest scientists of the twentieth century. His mathematical derivations helped usher in the atomic age. His work on the photoelectric effect resulted in the invention of the photoelectric cell, a device that made possible sound motion pictures, television, and other inventions.

The photoelectric effect occurs when Group 1A metals give off electrons when subjected to light of a given energy.

Example 3

Calculate the energy, in joules, of a quantum of radiant energy (the energy of a photon) whose frequency is 5.00×10^{15} s^{-1}.

Solution

Use the equation $E = h \times \nu$. Substitute values for h and ν to give E.

$$E = 6.62 \times 10^{-34} \text{ J s} \times 5.00 \times 10^{15} \text{ s}^{-1}$$
$$= 3.31 \times 10^{-18} \text{ J}$$

Problem

8. What is the energy of a photon of microwave radiation whose frequency is 3.20×10^{11} s^{-1}?

The dual wave-particle behavior of light was difficult for scientists trained in classical physics to accept. It was even more difficult to dispute because it provided an explanation for the previously mysterious photoelectric effect.

In the **photoelectric effect,** *electrons called photoelectrons are ejected by metals when light shines on them.* The alkali metals (Li, Na, K, Cs, and Rb) are particularly subject to the effect. Not just any light will do. Red light ($\nu = 4.3 \times 10^{14}$ s^{-1}–4.6×10^{14} s^{-1}) will not cause the ejection of photoelectrons from potassium no matter how intense the light. Yet even a very weak yellow light ($\nu = 5.1 \times 10^{14}$ s^{-1}–5.2×10^{14} s^{-1}) shining on potassium begins the effect.

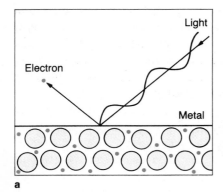
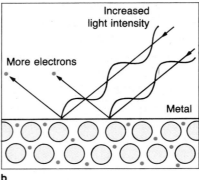
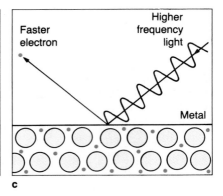

a

b

c

Figure 11·22
a When light strikes a metal surface, electrons are ejected. **b** If the threshold frequency has been reached, increasing the intensity only increases the number of electrons ejected. **c** If the frequency is increased, the ejected electrons will travel faster.

Figure 11·23
Photoelectric cells convert light energy into electrical energy.

 ChemDirections

To learn more about recent developments in solar cell technology, see **ChemDirections** *Solar Cells*, page 655.

The photoelectric effect could not be explained by classical physics, which had no quantum concept. Classical physics correctly viewed light as a form of energy. It assumed, however, that under weak light of any wavelength, an electron in a metal should eventually collect enough energy to be ejected. Obviously, the photoelectric effect presented a serious problem for the classical wave theory of light.

Einstein used his particle theory of light to explain the photoelectric effect. He recognized that there is a threshold value of energy below which the photoelectric effect does not occur. Because $E = h \times \nu$, all the photons in a beam of light of only one frequency (monochromatic light) have the same energy. If the frequency and therefore the energy of the light is too low, then no photoelectron will be ejected. It does not matter whether a single photon or a steady stream of low-energy photons strike an electron in the metal. Only if the frequency of light is above the threshold frequency will the photoelectric effect occur.

An analogous situation occurs with a table tennis ball that strikes a billiard ball. A table tennis ball is not energetic enough to budge the stationary billiard ball, no matter how many times a ball collides with it. By contrast, one golf ball moving at the same speed as the table tennis ball sets the target ball in motion. The golf ball is above the energy threshold. With the photoelectric effect, any excess energy of a photon beyond that needed to eject a photoelectron causes the ejected electron to travel faster. Increasing the intensity of light, however, only increases the number of photons striking the metal. Above the threshold frequency, increasing the intensity increases the number of electrons ejected. It does not, however, make them travel faster (Figure 11·22).

11·9 An Explanation of Atomic Spectra

Bohr's application of quantum theory to the energy levels of the electrons in atoms resulted in an explanation of the hydrogen spectrum. The lines observed in the spectrum are consistent with the idea that quantization limits the possible energies that an electron in a hydrogen atom can attain. Consider the lone electron of the hydrogen atom in *the lowest energy level, or* **ground state.** This energy level is designated by a quantum number, *n*. For the ground state, $n = 1$. Excitation of the electron

Bohr explained the spectrum of the hydrogen atom using the quantum concept.

raises it to an excited state so that $n = 2, 3, 4, 5$, or 6, and so forth. If the energy levels are quantized, it takes a quantum of energy, $h \times \nu$, to raise the electron from the ground state to an excited state. The same amount of energy is emitted as a photon when the electron drops from the excited state to the ground state. Only electrons in transition from higher to lower energy levels lose energy and emit light.

Figure 11·24 shows the explanation for the three groups of lines observed in the emission spectrum of hydrogen atoms. The positions of lines at the ultraviolet end of the hydrogen spectrum are the Lyman series. These match expected values for the emission due to transitions from higher energy levels to $n = 1$. The lines in the visible spectrum are the Balmer series. They are the result of transitions from higher energy levels to $n = 2$. Those in the infrared spectrum are the Paschen series. They correspond to transitions from higher energy levels to $n = 3$. Lines for the transitions from higher energy levels to $n = 4$ and $n = 5$ also exist but are not shown in the figure. Note that the spectral lines in each group become more closely spaced at increased values of n. This means that the energy differences between higher energy levels are smaller than those between lower levels. There is an upper limit to the frequency of emitted light for each set of lines. This is because a very excited electron completely escapes the atom.

■ The lines in the emission spectrum of hydrogen result from the transitions of electrons from higher to lower energy levels.

Bohr's theory of the atom was only partially satisfactory. It only explained the emission spectra of atoms and ions containing one electron. Moreover, it was of no help in understanding how atoms bond to form molecules. Eventually the Bohr model of the atom was displaced by a new and better model. The latter is based on the description of the motion of material objects as waves.

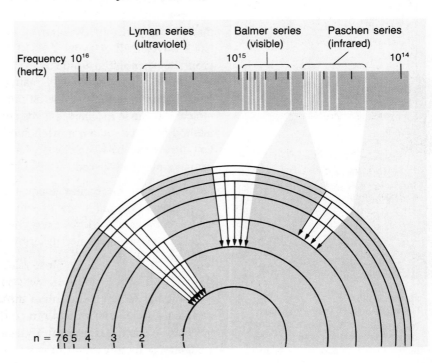

Figure 11·24
The three groups of lines in the hydrogen spectrum correspond to transitions from higher to lower energy levels. The Lyman series lines correspond to the transition to the $n = 1$ energy level. The Balmer series lines correspond to the transition to the $n = 2$ energy level. The Paschen series lines correspond to the transition to the $n = 3$ energy level.

Problem

9. What happens when a hydrogen atom absorbs a quantum of energy?

11·10 The Wave Motion of Matter and Quantum Mechanics

Such strange goings on! Energy absorbed or emitted in packages. Light behaving as waves *and* particles. Stranger things were yet to come. In 1924 Louis de Broglie, a French graduate student, asked an important question. Since light behaves as waves and particles, can particles of matter behave as waves? De Broglie derived an equation that described the wavelength, λ, of a moving particle.

$$\lambda = \frac{h}{mv}$$

In using this equation express the mass in kg and the velocity in m/s.

Here h is Planck's constant, m is the mass of the particle, and v is the velocity of the particle. From this equation it is easy to calculate the wavelength of a moving electron. With a mass of 9.11×10^{-28} g and moving at the speed of light, an electron has a wavelength of about 2×10^{-8} cm. This is about the diameter of an atom.

■ De Broglie's claim that all moving bodies have wavelike properties stimulated the development of quantum mechanics.

Indeed, **de Broglie's equation** *predicts that all matter exhibits wavelike motions.* Why then are we unaware of this wave motion? As with quanta, the answer is concerned with the size of the object in motion. Wavelengths of objects that are visible to the naked eye are too small to measure. A 200-g baseball moving at 30 m/s has a wavelength of approximately 10^{-32} cm. This is too small to detect by any experiment that we can perform. By contrast, an electron moving at the same speed has a wavelength of about 2×10^{-3} cm. This is easily measured by appropriate scientific instruments.

De Broglie's proposal of matter waves set the stage for an entirely new method of describing the motions of subatomic particles, atoms, and molecules. Since mechanics is the study of the motion of bodies, the new method is called quantum mechanics. Let us summarize the most important differences between classical mechanics and quantum mechanics that we have previously seen.

1. Classical mechanics adequately describes the motions of bodies much larger than the atoms of which they are composed. It appears that such a body gains or loses energy in any amount.

2. Quantum mechanics describes the motions of subatomic particles and atoms as waves. These particles gain or lose energy in packages called quanta.

Figure 11·25
DeBroglie said all matter exhibits wavelike motions. This phenomenon is more apparent in some forms of matter than in others.

Another feature of quantum mechanics that is not found in classical mechanics is the uncertainty principle. This was derived by the German physicist Werner Heisenberg (1901–1976) in 1927. The **Heisenberg uncertainty principle** *states that it is impossible to know exactly both the*

Figure 11·26
a The location of an electron can be determined only if it is struck by another particle such as a photon.
b After the impact, the electron velocity changes. Thus, it is impossible to know both the position and velocity of the electron at the same time.

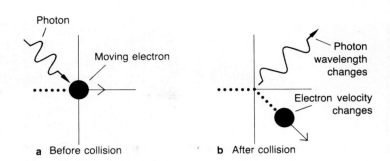

velocity and the position of a particle at the same time. As the measurement of velocity is made more accurately, the measurement of the position must become less accurate. Conversely, if the position of a moving particle is known accurately, the velocity is less well known.

The uncertainty principle is much more obvious with small bodies like electrons than with large objects like baseballs. The uncertainty in the position of a baseball traveling at 30 m/s is only about 10^{-21} cm, which is not measurable. The uncertainty in the position of an electron with a mass of 9.11×10^{-28} g moving at the same speed is nearly a billion centimeters! The quantum mechanical description by Schrödinger of the electrons in atoms shaped the concept of electron orbitals and configurations. It incorporates both the wave properties of the motions of bodies and the uncertainty principle.

———— Science, Technology, and Society ————

11·D Technical Illustration: Drawing the Unseen

If you glance through this textbook, you will see many illustrations that attempt to explain technical concepts. Often the objects pictured are things that have never been seen, like electron orbitals or the nucleus of an atom. Producing such pictures and diagrams is the job of the technical illustrator.

A technical illustrator must have artistic ability, a good background in drawing and design, and experience with various art media. General knowledge of the subject matter of the drawings also helps. Knowledge of chemistry, for example, is very helpful in illustrating a chemistry textbook. Most important is the ability to communicate with scientists, engineers, and technical writers to establish what is to be shown in the art.

Technical illustrators work on textbooks, scientific papers, operating manuals, product brochures, and many other types of publications. They may be employed by private industry, government agencies, or publishing concerns. They may also work on a free-lance basis. Technical illustration is an excellent field for a person who is interested in both art and science.

Figure 11·27
Technical illustrators work with scientists and engineers to help make their material clearer and more interesting.

11 Electrons in Atoms

Chapter Review

Key Terms

amplitude	$11\cdot6$	hertz (Hz)	$11\cdot6$
atomic emission		Hund's rule	$11\cdot4$
spectrum	$11\cdot6$	Pauli exclusion	
atomic orbital	$11\cdot3$	principle	$11\cdot4$
Aufbau principle	$11\cdot4$	photoelectric effect	$11\cdot8$
de Broglie's		photon	$11\cdot8$
equation	$11\cdot10$	Planck's constant	
electromagnetic		(h)	$11\cdot7$
radiation	$11\cdot6$	quantum	$11\cdot1$
electron		quantum mechanical	
configuration	$11\cdot4$	model	$11\cdot2$
energy level	$11\cdot1$	spectrum	$11\cdot6$
frequency (ν)	$11\cdot6$	wavelength (λ)	$11\cdot6$
ground state	$11\cdot9$		
Heisenberg uncertainty			
principle	$11\cdot10$		

Chapter Summary

Rutherford pictured the atom as a dense nucleus surrounded by electrons. Rutherford's model was refined by Niels Bohr. In the Bohr model of the atom, the electrons move in fixed circular paths around a dense, positively charged nucleus.

The energies of electrons in an atom are quantized. The quantum mechanical model is the modern description of the electrons in atoms. This model does not define the exact path of an electron. It does show the probability of finding an electron as a cloud of negative charge. Each cloud shape can be calculated from a mathematical expression called an atomic orbital.

The ways in which electrons are arranged around the nuclei of atoms are called electron configurations. Correct electron configurations for atoms may be written by using the Aufbau principle, the Pauli exclusion principle, and Hund's rule.

The Aufbau principle tells us the sequence in which the orbitals are filled. The Pauli exclusion principle states that a maximum of only two electrons can occupy each orbital. Hund's rule states that the electrons pair up only after each orbital in a sublevel is occupied by a single electron.

The concept of quantized electron energy levels in atoms grew out of the study of the interaction of light and matter. The line emission spectra of atoms are best explained if energy levels are quantized. The quantum concept developed in part from Planck's studies of light radiation from heated objects and Einstein's explanation of the photoelectric effect. De Broglie's proposal that all matter in motion has wavelike properties further stimulated the development of a new mathematical description of electron configuration. Erwin Schrödinger devised the most successful of these early quantum mechanical models. It also incorporates the Heisenberg uncertainty principle.

Practice Questions and Problems

10. Describe Rutherford's model of the atom and compare it with the model proposed by his student Niels Bohr. $11\cdot1$

11. How did Bohr answer the objection that an electron traveling in a circular orbit would radiate energy and fall into the nucleus? $11\cdot1$

12. The energies of electrons are said to be *quantized*. Explain. $11\cdot1$

13. In general terms, explain how the quantum mechanical model of the atom describes the electron structure of an atom. $11\cdot2$

14. Sketch $1s$, $2s$, and $2p$ orbitals using the same scale for each. $11\cdot3$

15. What are the three rules that govern the filling of atomic orbitals by electrons? $11\cdot4$

16. Write electron configurations for the elements that are identified only by these atomic numbers. *11·4*
a. 15 **b.** 12 **c.** 9 **d.** 18

17. What is meant by $3p^3$? *11·4*

18. Write complete electron configurations for each of the following kinds of atoms.
a. carbon   **c.** fluorine
b. argon   **d.** rubidium   *11·4*

19. Indicate which of these orbital designations is incorrect. *11·4*
a. $4s$ **b.** $3f$ **c.** $2d$ **d.** $3d$

20. How many electrons are in the second energy level of an atom of these elements?
a. chlorine **b.** phosphorus **c.** potassium *11·4*

21. What is the maximum number of electrons that can go into each of the following sublevels? *11·4*
a. $2s$ **c.** $4s$ **e.** $4p$ **g.** $4f$
b. $3p$ **d.** $3d$ **f.** $5s$ **h.** $5p$

22. Write electron configurations for atoms of these elements. *11·4*
a. selenium   **c.** nickel
b. vanadium   **d.** calcium

23. Why does one electron in a potassium atom go into the fourth energy level instead of squeezing into the third energy level along with the eight already there? *11·5*

24. List the colors of the visible spectrum in order of increasing wavelength. *11·6*

25. Arrange the following electromagnetic radiations in order of decreasing wavelength. *11·6*
a. infrared radiation from a heat lamp
b. ultraviolet light from the sun
c. dental X-rays
d. the signal from a short-wave radio station
e. green light

26. Use a diagram to illustrate the following terms.
a. wavelength **b.** amplitude **c.** wave cycle
11·6

27. What is the wavelength of radiation whose frequency is 1.50×10^{13} s^{-1}? Does this radiation have a longer or shorter wavelength than red light? *11·6*

28. Suppose that your favorite AM radio station broadcasts at a frequency of 1150 kHz. What is the wavelength in centimeters of the radiation from the station? *11·6*

29. How did Planck influence the development of modern atomic theory? *11·7*

30. Briefly describe the photoelectric effect and explain why it could not be explained by classical physics. *11·8*

31. What will happen if the following occur? *11·8*
a. Monochromatic light shining on the alkali metal cesium is just above the threshold frequency?
b. The intensity of light increases, but the frequency remains the same?
c. Monochromatic light of a shorter wavelength is used?

32. Explain the difference between a photon and a quantum. *11·8*

33. When white light is viewed through sodium vapor in a spectroscope, the spectrum is continuous except for a dark line at 589 nm. How can you explain this observation? *11·9*

34. What is the wavelength of a 2500-kg truck traveling at a rate of 75 km/hr? *11·10*

Mastery Questions and Problems

35. Give the symbol for the atom whose electron configuration corresponds to each of the following electron configurations.
a. $1s^2\,2s^2\,2p^6\,3s^2\,3p^6$
b. $1s^2\,2s^2\,2p^6\,3s^2\,3p^6\,3d^{10}\,4s^2\,4p^6\,4d^7\,5s^1$
c. $1s^2\,2s^2\,2p^6\,3s^2\,3p^6\,3d^{10}\,4s^2\,4p^6\,4d^{10}\,4f^7\,5s^2\,5p^6\,5d^1\,6s^2$

36. Write the electron configuration for a uranium atom. Calculate the total number of electrons in each level and state which levels are not full.

37. How many paired electrons are there in an atom of each of these elements?
a. helium   **c.** sodium
b. boron   **d.** oxygen

38. An atom of an element has two electrons in the first energy level and five electrons in the second energy level. Write the electron configuration and name the element. How many unpaired electrons does an atom of this element have?

39. What is the energy of a photon of green light whose frequency is 5.80×10^{14} s^{-1}?

40. A mercury lamp like that shown below emits radiation with a wavelength of 4.36×10^{-7} m.
  a. What is this wavelength in centimeters?
  b. In what region of the electromagnetic spectrum is this radiation?
  c. Calculate the frequency of this radiation.

41. Calculate the energy of a photon of red light whose wavelength is 6.45×10^{-5} cm. Compare your answer to that of the previous question and say whether red light is of higher or lower energy than green light.

42. Give the symbol and name of the elements whose atoms have these configurations.
  a. $1s^2\, 2s^2\, 2p^6\, 3s^1$
  b. $1s^2\, 2s^2\, 2p^3$
  c. $1s^2\, 2s^2\, 2p^6\, 3s^2\, 3p^2$
  d. $1s^2\, 2s^2\, 2p^4$
  e. $1s^2\, 2s^2\, 2p^6\, 3s^2\, 3p^6\, 4s^1$
  f. $1s^2\, 2s^2\, 2p^6\, 3s^2\, 3p^6\, 3d^2\, 4s^2$

Critical Thinking Questions

43. Choose the term that best completes the second relationship.
  a. orbital:energy level   apartment: _____
    (1) floor   (3) building
    (2) room   (4) stairway
  b. water:container   electrons: _____
    (1) frequency   (3) nuclei
    (2) sublevel   (4) orbitals
  c. electromagnetic radiation:visible light student body: _____
    (1) eleventh graders (3) teachers
    (2) principal   (4) textbooks

44. Traditional cooking methods make use of infrared radiation (heat). Microwave radiation cooks food faster. Could radio waves be used for cooking? Explain.

45. Think about the currently accepted models of the atom and of light. In what ways do these models seem strange to you? Why are these models not exact or definite?

Review Questions and Problems

46. A potassium atom has a diameter of about 0.406 nm. Express this in meters and micrometers. If the measurement is always the same, why are the two numbers different?

47. Calculate the percent composition of each compound.
  a. SiO_2
  b. $FeCl_3$
  c. H_2O
  d. H_2SO_4

48. Balance the following chemical equations.
  a. $KNO_3 + H_2SO_4 \longrightarrow K_2SO_4 + HNO_3$
  b. $Cu_2O + H_2 \longrightarrow Cu + H_2O$
  c. $NO + Br_2 \longrightarrow NOBr$
  d. $SnO_2 + CO \longrightarrow Sn + CO_2$

49. Calculate the volume of O_2 at STP required for the complete combustion of 5.00 L of acetylene (C_2H_2) at STP.

$$2C_2H_2(g) + 5O_2 \longrightarrow 4CO_2 + 2H_2O(l)$$

50. Write symbols for the following ions.
  a. iron(III)
  b. mercury(II)
  c. nitride
  d. bicarbonate
  e. oxide
  f. permanganate

51. Give the number of protons and electrons in each of the following.
  a. Cs
  b. Ag^+
  c. Cd^{2+}
  d. Se^{2-}

52. The temperature of a gas at STP is changed to 125°C at constant volume. Calculate the final pressure of the gas in atmospheres.

Challenging Questions and Problems

53. The average distance between Earth and Mars is about 2.08×10^8 km. How long does it take to transmit television pictures from the *Mariner* spacecraft to Earth from Mars?

54. The energy of a photon is related to its wavelength and its frequency.
  a. Complete the table below.
  b. Plot the energy of the photon (y-axis) versus the frequency (x-axis).
  c. Determine the slope of the line.
  d. What is the significance of this slope?

Energy of photon (J)	Wavelength (m)	Frequency (s^{-1})
3.45×10^{-21}	5.77×10^{-5}	___
2.92×10^{-20}	6.82×10^{-6}	___
6.29×10^{-20}	3.16×10^{-6}	___
1.13×10^{-19}	1.76×10^{-6}	___
1.46×10^{-19}	1.36×10^{-6}	___
3.11×10^{-19}	6.38×10^{-7}	___

55. In a photoelectric experiment a student shines light of greater than the threshold frequency upon the surface of a metal. She observes that after a long time the maximum energy of the ejected electrons begins to decrease. Can you explain why?

56. Bohr's atomic theory can be used to calculate the energy required to remove an electron from an orbit of a hydrogen atom or an ion containing only one electron. This is the *ionization energy* for that atom or ion. The formula for determining the ionization energy E is

$$E = \frac{Z^2 \times k}{n^2}$$

where Z is the atomic number, k is 1312 kJ per mole, and n is the energy level. What is the energy required to eject an electron from a hydrogen atom when the electron is in the ground state ($n = 1$)? In the second energy level? How much energy is required to eject a ground state electron from the species Li^{2+}?

57. Look up the electron configuration for rhodium. Why is it exceptional? Explain.

Research Projects

1. What are the chemical causes and effects of lightning?

2. What types of incandescent light bulbs are available? How is their usefulness dependent on the materials from which they are constructed?

3. How do microwaves compare to visual light waves? Describe some common applications of microwaves and the concerns about their use.

4. Construct a simple spectroscope. Use it to identify unknowns.

5. Compare absorption and emission spectroscopy. How is each technique used in the laboratory?

6. Is there a relationship between the color of an object and the amount of solar energy it absorbs? Add different colored food dyes to several beakers containing the same amount of water. Determine if the color affects how much energy the water absorbs when exposed to the sunlight.

7. How did Einstein's work on the photoelectric effect lead to the invention of sound motion pictures and television?

8. How do solar cells work? How are they being used? What are their limitations?

9. How is the laser used as a tool in chemistry? As a tool in medicine?

Readings and References

Born, Max. *My Life: Recollections of a Nobel Laureate*. New York: Scribner, 1978.

Holzman, David. "Superconductivity." *ChemMatters* (October 1987), pp. 18–21.

Pagels, Heinz R. *The Cosmic Code: Quantum Physics as the Language of Nature*. New York: Simon and Schuster, 1982.

Soloman, Burt. "Will Solar Sell?" *Science 84* (April 1984), pp. 70–76.

Townes, Charles H. "Harnessing Light." *Science 84* (November 1984), pp. 153–155.

Trefil, James S. *From Atoms to Quarks: An Introduction to the Strange World of Particle Physics*. New York: Scribner, 1980.

12 Chemical Periodicity

Chapter Preview

During the nineteenth century, chemists began to categorize the elements according to similarities in their physical and chemical properties. The end result of these studies was the modern periodic table. In this chapter you will learn how the periodic table is organized and its relationship to the atomic structure and properties of the elements.

12·1 The Development of the Periodic Table

Figure 12·1
These elements exhibit widely different chemical and physical properties. Keeping track of these properties would be quite difficult without the periodic table. It is our tool for understanding the periodicity of the elements.

About 70 elements had been described by the mid-1800s, but no common feature that would relate them had been found. Dmitri Mendeleev (1834–1907), a Russian chemist, had more success than most. Mendeleev listed the elements in several vertical columns in order of increasing atomic mass. He noticed a regular (periodic) recurrence of their physical and chemical properties. This led him to arrange the columns so that elements with the most similar properties were side by side. Mendeleev

Figure 12·2

Mendeleev designed his periodic table by grouping elements with similar properties in rows, and by ordering the rows according to the atomic masses.

```
                                        Ti = 50      Zr = 90      ? = 180.
                                        V = 51       Nb = 94      Ta = 182.
                                        Cr = 52      Mo = 96      W = 186.
                                        Mn = 55      Rh = 104,4   Pt = 197,4
                                        Fe = 56      Ru = 104,4   Ir = 198.
                                  Ni = Co = 59        Pl = 106,6   Os = 199.
    H = 1                               Cu = 63,4     Ag = 108     Hg = 200.
              Be = 9,4     Mg = 24      Zn = 65,2     Cd = 112
              B = 11       Al = 27,4    ? = 68        Ur = 116     Au = 197?
              C = 12       Si = 28      ? = 70        Su = 118
              N = 14       P = 31       As = 75       Sb = 122     Bi = 210
              O = 16       S = 32       Se = 79,4     Te = 128?
              F = 19       Cl = 35,5    Br = 80       I = 127
    Li = 7    Na = 23      K = 39       Rb = 85,4     Cs = 133     Tl = 204
                          Ca = 40      Sr = 87,6     Ba = 137     Pb = 207.
                          ? = 45       Ce = 92
                          ?Er = 56     La = 94
                          ?Yt = 60     Di = 95
                          ?In = 75,6   Th = 118?
```

The periodic table organizes the chemical elements according to their properties.

thus constructed the first periodic table. This is an arrangement of the elements according to similarities in their properties. Numerous blank spaces had to be left in the table because there were no known elements with the appropriate properties (Figure 12·2).

Mendeleev noted the properties of the elements adjacent to the blank spaces. Then he and others were able to predict the physical and chemical properties of the missing elements. Eventually these missing elements were discovered and found to have properties similar to those predicted.

In 1913 Henry Moseley (1887–1915), a young British physicist, determined the nuclear charge, also called the atomic number, of the atoms of the elements. Moseley arranged the elements in a table by order of atomic number. This is the way that periodic tables are arranged today.

--- Science, Technology, and Society ---

12·A Mendeleev and Moseley: Building on Others' Work

Dmitri Mendeleev's periodic table was the triumph of an outstanding career in chemistry. As a professor at the University of St. Petersburg, Mendeleev found no textbook that met the needs of his class. He decided to write a book that emphasized the similarities in the chemical properties of elements. Investigating the elements, Mendeleev wrote their names and characteristics on cards. Then he more or less played solitaire, arranging the elements into groups. Years of work led to his periodic law: Elements arranged according to their atomic masses present a clear periodicity of properties.

Figure 12·3
Mendeleev consolidated and interpreted the findings of many other chemists. He then used his periodic theory to predict the existence and properties of elements which had not yet been discovered.

Mendeleev's work was made possible by the Karlsruhe Conference held in Germany in 1860. This scientific meeting was held to standardize the atomic and molecular masses that were being used by chemists throughout the world. Once all masses were calculated by a common standard, Mendeleev was able to see the relationship among the elements. Using the experimental results of many chemists, Mendeleev formulated the periodic law.

This process of predicting a general theory from known details is called inductive reasoning. The opposite, predicting details from a known general theory, is called deductive reasoning. If general laws are known, the results can be reliably predicted. Inductive reasoning has more risk of error. It also has greater potential, however, for leading to a significant breakthrough. Mendeleev used inductive reasoning to build the periodic table. Then he used deductive reasoning to predict the properties of elements still unknown. When gallium was discovered in 1875, Mendeleev insisted that its discoverer recalculate its density. The experimental value did not agree with Mendeleev's prediction. To the astonishment of the scientific world, the recalculation proved that Mendeleev was right!

One scientist who built on Mendeleev's theory was Henry Moseley. If you look closely at the periodic table, you will find several places where the elements are not in order according to atomic mass. Moseley's work proved that the elements should be arranged by atomic number instead. Like Mendeleev, he predicted several missing elements which were later found. Unfortunately for science, Moseley was killed in World War I when he was only 28 years old. As a result of Moseley's death, the British government no longer assigns scientists to combat duty in time of war.

12·2 The Modern Periodic Table

The most commonly used form of the modern periodic table, sometimes called the long form, is shown in Figure 12·4 on the next two pages. In this table the elements are arranged in seven horizontal rows in order of increasing atomic number. Each element is identified by its symbol placed in a block. The atomic number of the element is shown above the symbol. The atomic mass and the name of the element are shown below the symbol. The circle with letter in the upper left corner indicates the physical state of the element at 25°C.

Horizontal rows = periods.
Vertical columns = groups or families.

The horizontal rows of the periodic table are called **periods.** There are seven periods. The vertical columns are called groups, or families. Each group is identified by a numeral and the letter A or B. Groups 1A through 7A and Group 0 make up the *representative elements*. They exhibit a wide variety of both physical and chemical properties. The elements in any group of the periodic table have similar physical and chemical properties. The properties of the elements in the periods change from group to group. The sequence of change is the same, however, in

Elements within a group have similar properties.

Figure 12-4 Periodic Table of the Elements

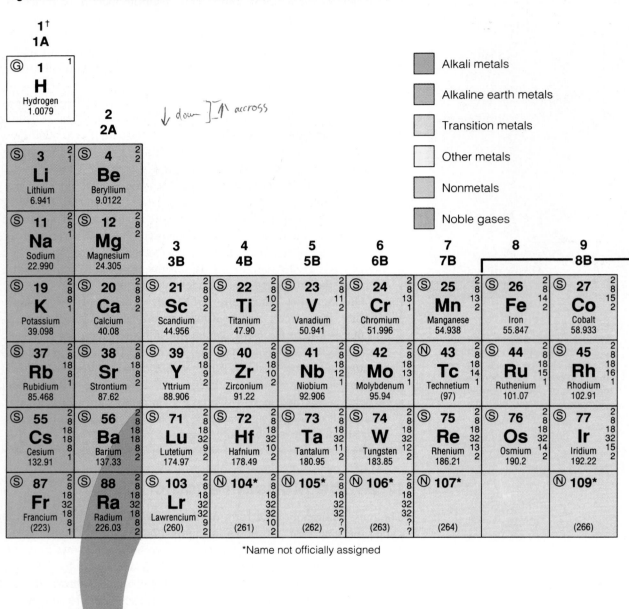

*Name not officially assigned

Lanthanide Series

Actinide Series

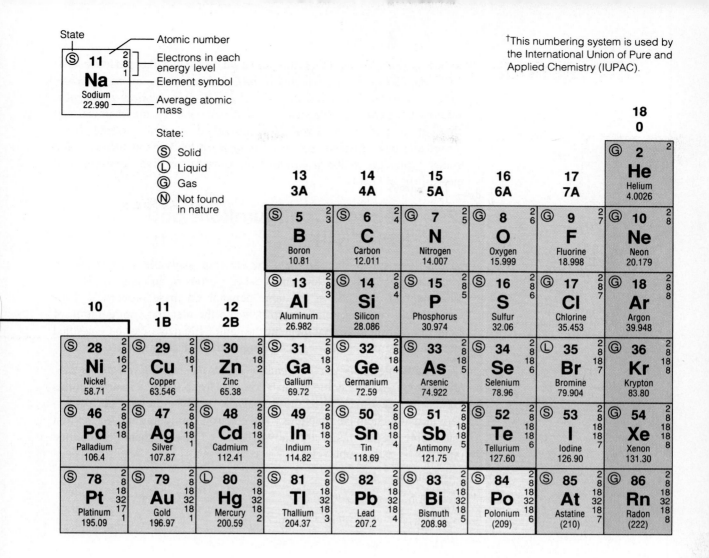

State — Atomic number
Electrons in each energy level
Element symbol
Average atomic mass

State:
Ⓢ Solid
Ⓛ Liquid
Ⓖ Gas
Ⓝ Not found in nature

†This numbering system is used by the International Union of Pure and Applied Chemistry (IUPAC).

all the periods. Because of this situation, we have the **periodic law:** *When the elements are arranged in order of increasing atomic number, there is a periodic pattern in their physical and chemical properties.* Without the periodic table, learning and remembering the chemical and physical properties of over 100 individual elements would be a formidable task. Instead, you need only learn the general behavior and trends within the major groups to have a useful working knowledge of most of the elements.

12·3 Electron Configurations and Periodicity

The properties of an element are related to the element's electron configuration.

Of the three subatomic particles, the electron plays the greatest part in determining the physical and chemical properties of an element. The arrangement of elements in the table depends on these properties. Thus there should be some relationship between the electron configuration of elements and their arrangement in the table. Elements can be classified into four different categories according to their electron configuration.

1. The noble gases *are elements in which the outermost s and p sublevels are filled.* The noble gases belong to Group 0. The elements in this group are sometimes called the *inert gases* because they do not participate in many chemical reactions. The electron configurations for the first four noble gas elements are shown below. These elements have filled outermost *s* and *p* sublevels.

Helium	$1s^2$
Neon	$1s^2\ 2s^2\ 2p^6$
Argon	$1s^2\ 2s^2\ 2p^6\ 3s^2\ 3p^6$
Krypton	$1s^2\ 2s^2\ 2p^6\ 3s^2\ 3p^6\ 3d^{10}\ 4s^2\ 4p^6$

2. The representative elements *are elements whose outermost s or p sublevels are only partially filled.* The representative elements are usually called the Group A elements. They may also include the noble gases. For any representative element the group number is equal to the number of electrons in the outermost occupied energy level. For example, the elements in Group 1A (lithium, sodium, potassium, rubidium, and cesium) have one electron in the outermost energy level.

Lithium	$1s^2\ 2s^1$
Sodium	$1s^2\ 2s^2\ 2p^6\ 3s^1$
Potassium	$1s^2\ 2s^2\ 2p^6\ 3s^2\ 3p^6\ 4s^1$

Carbon, silicon, and germanium, in Group 4A, have four electrons in the outermost energy level.

Carbon	$1s^2\ 2s^2\ 2p^2$
Silicon	$1s^2\ 2s^2\ 2p^6\ 3s^2\ 3p^2$
Germanium	$1s^2\ 2s^2\ 2p^6\ 3s^2\ 3p^6\ 3d^{10}\ 4s^2\ 4p^2$

3. The transition metals *are elements whose outermost s sublevel and the nearby d sublevel contain electrons.* The transition elements are called the Group B elements. They are characterized by having electrons added to the d orbitals.

4. The inner transition metals *are elements whose outermost s sublevel and the nearby f sublevel generally contain electrons.* The inner transition metals are characterized by the filling of f orbitals.

If we consider both the electron configurations and the positions of elements in the periodic table, another pattern emerges. The periodic table may be divided into sections, or blocks, that correspond to the sublevels that are filled with electrons (Figure 12·5).

The s block is the part of the periodic table that contains the elements with s^1 and s^2 electron configurations. It is composed of the elements in Groups 1A and 2A and the noble gas helium.

The p block is composed of elements in Groups 3A, 4A, 5A, 6A, 7A, and 0 with the exception of helium.

The transition metals belong to the d block, and the inner transition metals belong to the f block.

The electron configurations of elements can be determined by using the periodic table in Figure 12·5. Simply read the periodic table like a book from left to right and top to bottom until the element of interest is

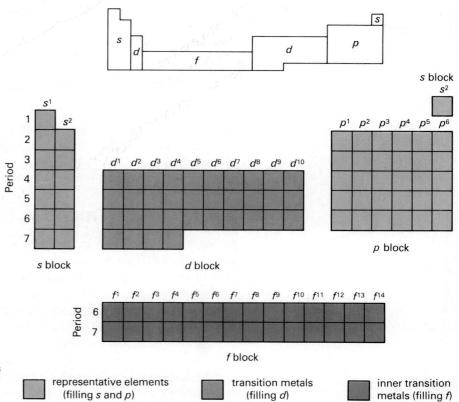

Figure 12·5
This block diagram identifies groups of elements according to the sublevels that are filled with electrons.

representative elements (filling s and p)   transition metals (filling d)   inner transition metals (filling f)

reached. Each period number on the periodic table corresponds to the principal energy level. The number of electrons in a partially filled sublevel is determined by counting over to the element, starting at the left side of the sublevel. For the transition elements, electrons are added to a d sublevel with a principal energy level that is one less than the period number. For the inner transition metals, the principal energy level is two less than the period number.

Example 1

Use the periodic table in Figure 12·5 to write the electron configuration of these elements. **a.** nitrogen **b.** nickel **c.** iodine

Solution

a. Nitrogen has seven electrons. From the periodic table in Figure 12·5 the first period is $1s^2$, and the second period is $2s^2 2p^3$. There are three electrons in the $2p$ sublevel because nitrogen is the third element in the $2p$ block.

b. Nickel has 28 electrons. From Figure 12·5 the first three periods are $1s^2 2s^2 2p^6 3s^2 3p^6$. Next is $4s^2$ and finally $3d^8$. Remember that the principal energy level number for the d block is always one less than the period number. The complete configuration is $1s^2 2s^2 2p^6 3s^2 3p^6 3d^8 4s^2$.

c. Iodine has 53 electrons. Using Figure 12·5 we find the electron configuration to be $1s^2 2s^2 2p^6 3s^2 3p^6 3d^{10} 4s^2 4p^6 4d^{10} 5s^2 5p^5$.

Problems

1. Use Figure 12·5 to write the electron configuration for these elements. **a.** boron **b.** magnesium **c.** vanadium **d.** strontium

2. Write the electron configuration of these elements. **a.** the inert gas in period 3 **b.** the element in Group 4A, period 4 **c.** the element in Group 2A, period 6

12·4   Periodic Trends in Atomic Size

We know from the quantum mechanical model that an atom does not have a sharply defined boundary to set the limit of its size. Therefore the radius of an atom cannot be measured directly. There are, however, several ways to estimate the relative sizes of atoms. The one we will use is the covalent atomic radius. *The **covalent atomic radius** is half of the distance between the nuclei of two atoms in a homonuclear diatomic molecule.* For example, the separation between the nuclei in the bromine molecule (Br_2) is 0.228 nm. Thus a value of 0.114 nm (0.228 ÷ 2) is assigned as the radius of the bromine atom. Figure 12·6 shows atomic radii for most of the representative elements.

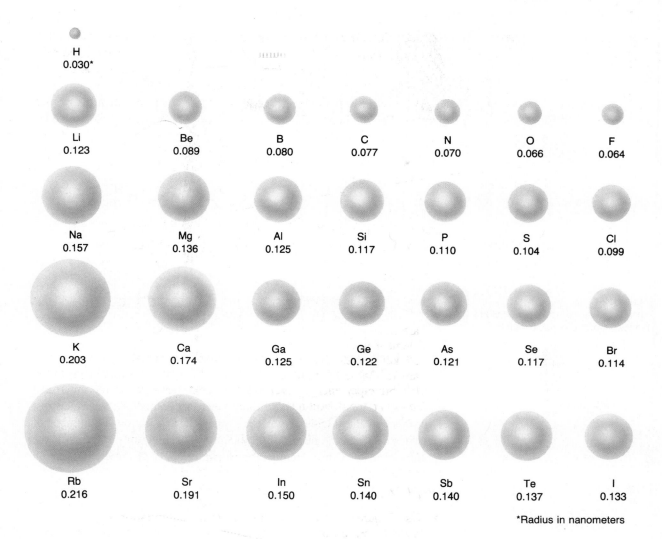

H 0.030*

Li 0.123 | Be 0.089 | B 0.080 | C 0.077 | N 0.070 | O 0.066 | F 0.064

Na 0.157 | Mg 0.136 | Al 0.125 | Si 0.117 | P 0.110 | S 0.104 | Cl 0.099

K 0.203 | Ca 0.174 | Ga 0.125 | Ge 0.122 | As 0.121 | Se 0.117 | Br 0.114

Rb 0.216 | Sr 0.191 | In 0.150 | Sn 0.140 | Sb 0.140 | Te 0.137 | I 0.133

*Radius in nanometers

Figure 12·6
The atomic radii of the representative elements are given here in nanometers. Note that the radii decrease as you read across, but they increase as you read down.

Atomic size decreases from left to right in a period and increases down a group.

Group trends. Atomic size generally increases as we move down a group of the periodic table. As we descend, electrons are added to successively higher principal energy levels, and the nuclear charge increases. The resulting increase in nuclear charge tends to decrease atomic size. The full nuclear charge, however, is shielded from the outer electrons that lie between them and the nucleus. As a result, the outermost orbital is larger as we move downward. The vertical columns in Figure 12·6 show how atomic size increases as we go down a group.

Periodic trends. Atomic size generally decreases as we move from left to right across a period. As we go across a period we remain in the same principal energy level. Each element has one proton and one electron more than the preceding element. The electrons are being added to the same principal energy level. The effect of the increasing nuclear charge on the outermost electrons is to pull them closer to the nucleus. Atomic size therefore decreases. If a plot is made of atomic radii versus

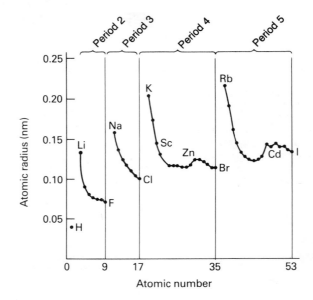

Figure 12·7
This plot of atomic radii versus atomic number shows a periodic variation.

atomic numbers, a periodic trend becomes obvious (Figure 12·7). The trend is less pronounced in periods where there are more electrons in the occupied principal energy levels between the nucleus and the outermost electrons. This is because these inner electrons help shield the outermost electrons and the nucleus from each other. In any period the number of electrons between the nucleus and the outermost electrons is the same for all the elements. Consequently, the *shielding effect* of these electrons on the nucleus is constant within a period.

Problem

3. Explain why fluorine has a smaller atomic radius than both oxygen and chlorine.

12·5   Periodic Trends in Ionization Energy

> The ionization energy is a measure of the ease with which an electron can be removed from a gaseous atom or ion.

When an atom gains or loses an electron it forms an ion. *The energy that is required to overcome the attraction of the nuclear charge and remove an electron from a gaseous atom is the* **ionization energy.** Removing one electron results in the formation of a positive ion with a 1+ charge.

$$\text{Na}(g) \longrightarrow \text{Na}^+(g) + e^-$$

The energy required to remove this first outermost electron is called the first ionization energy. To remove the outermost electron from the gaseous 1+ ion requires an amount of energy called the second ionization energy, and so forth. Table 12·1 gives the first three ionization energies of the first 20 elements.

The ionization energy is the energy required to remove a *single* electron from an atom. The amount of energy to do this is very small. A more realistic quantity is the amount of energy required to ionize a mole of atoms simultaneously. Therefore the unit used is kilojoules per mole.

Table 12·1   Ionization Energies of the First 20 Elements (in kilojoules per mole)

Symbol of element	Ionization Energy (kJ/mol)		
	First	Second	Third
H	1 312		
He (noble gas)	2 371	5 247	
Li	520	7 297	11 810
Be	900	1 757	14 840
B	800	2 430	3 659
C	1 086	2 352	4 619
N	1 402	2 857	4 577
O	1 314	3 391	5 301
F	1 681	3 375	6 045
Ne (noble gas)	2 080	3 963	6 276
Na	495.8	4 565	6 912
Mg	737.6	1 450	7 732
Al	577.4	1 816	2 744
Si	786.2	1 577	3 229
P	1 012	1 896	2 910
S	999.6	2 260	3 380
Cl	1 255	2 297	3 850
Ar (noble gas)	1 520	2 665	3 947
K	418.8	3 069	4 600
Ca	589.5	1 146	4 941

We can use the concept of ionization energy to explain how we were able to predict some of the ionic charges in Chapter 5. Look at the three Group 1A metals in Table 12·1. You will see a large increase in energy between the first and second ionization energies. It is relatively easy to remove one electron from a Group 1A metal to form an ion with a 1+ charge. It is very difficult, however, to remove an additional electron. For the three Group 2A metals, the large increase in ionization energy occurs between the second and third ionization energies. It is fairly easy to remove two electrons from a Group 2A metal (resulting in a 2+ ionic charge), but it is difficult to remove a third electron. We know that aluminum, in Group 3A, forms a 3+ ion. The large increase in ionization energy for aluminum occurs after the third electron is removed.

Groups trends.   The data in Table 12·1 show that, in general, the first ionization energy decreases as we move down a group of the periodic table. The size of the atoms is increasing as we descend. Thus the outermost electron is farther from the nucleus. It should be more easily removed and therefore have a lower ionization energy.

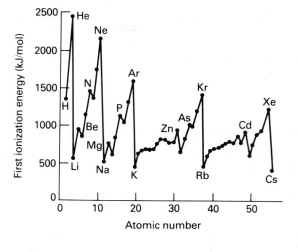

Figure 12·8
Here the first ionization energy is graphed versus the atomic number. Notice the ease with which Group 1A elements are ionized and the difficulty of ionizing noble gases.

The ionization energy increases with atomic number within a period and decreases with atomic number within a group.

Periodic trends. For the representative elements the first ionization energy generally increases as we move from left to right across a period. The nuclear charge is increasing, and the shielding effect is constant as we move across. A greater attraction of the nucleus for the electron therefore leads to the increase in ionization energy. The periodic trends of first ionization energies are shown in Figure 12·8.

12·6 Periodic Trends in Electron Affinity

The energy change that accompanies the addition of an electron to a gaseous atom is the **electron affinity** (EA). For example, energy is released when a fluorine atom gains an electron to become a negative ion.

$$F(g) + e^- \longrightarrow F^-(g)$$

Most elements release energy when they gain an electron. Most electron affinities are therefore negative.

A negative electron affinity means the gain of an electron is favorable because energy is released. Positive electron affinities are unfavorable.

Electron affinities (Table 12·2) give an indication of the relative ease by which atoms gain an electron. The halogens as a group have the highest electron affinities. The accepted convention is to assign a negative sign to an electron affinity when energy is released upon gain of an electron. A positive sign is assigned when energy is absorbed.

$$F(g) + e^- \longrightarrow F^-(g) + 328 \text{ kJ} \qquad EA = -328 \text{ kJ/mol}$$

$$Be(g) + e^- + 240 \text{ kJ} \longrightarrow Be^-(g) \qquad EA = +240 \text{ kJ/mol}$$

It is experimentally very difficult to get reliable electron affinity values. In contrast to ionization energies, the trends in electron affinity values are less clear. Referring to Table 12·2, we see that electron affinity generally increases as we move from left to right across a period. This is because atoms become smaller and the nuclear charge increases. As we move down a group, electron affinities generally decrease with increasing atomic size.

Table 12·2   Electron Affinities for the Representative Elements (in kilojoules per mole)*

1A	2A	3A	4A	5A	6A	7A
H −73						
Li −60	Be (+240)[†]	B −27	C −122	N +9	O −141	F −328
Na −53	Mg (+230)	Al −44	Si −134	P −72	S −200	Cl −348
K −48	Ca (+156)	Ga (−30)	Ge −120	As −77	Se −195	Br −325
Rb −47	Sr (+170)	In −30	Sn −121	Sb −101	Te −190	I −295
Cs −45	Ba (+52)	Tl −30	Pb −110	Bi −110	Po (−183)	At (−270)

*Negative values mean that the process $M + e^- \rightarrow M^-$ is exothermic, where M stands for any element.
[†]The values in parentheses are estimated.

Problem

4. In general, would you expect nonmetals to have larger electron affinities than metals? Why or why not?

12·7   Periodic Trends in Ionic Size

The atoms of metallic elements have low ionization energies. They form positive ions easily. By contrast, the atoms of nonmetallic elements readily form negative ions. How does the gain or the loss of electrons affect the size of the ion produced? Positive ions (cations) are always *smaller* than the neutral atoms from which they are formed. This is because the loss of outer shell electrons results in increased attraction by the nucleus for the fewer remaining electrons. The radius of the Na⁺ ion, 0.095 nm, is only about one-half that of the Na atom, 0.186 nm. In contrast, negative ions (anions) are always *larger* than the neutral atoms from which they are formed. This is because the effective nuclear attraction is less for an increased number of electrons. The additional electron also increases the repulsive forces between the electrons. The radius of the Cl⁻ ion, 0.181 nm, is about twice that of the Cl atom, 0.099 nm.

A periodic relationship among ionic radii of the elements is seen when the ions are arranged in the periodic table (Figure 12·9). Going from left to right, across a row, there is a gradual decrease in the size of positive ions. Then, beginning with Group 5, the negative ions (which are much larger in size) gradually decrease in size as you continue to move right. The atomic radius increases with both anions and cations as you go down each group.

Cations are smaller and anions are larger than the atoms from which they are produced.

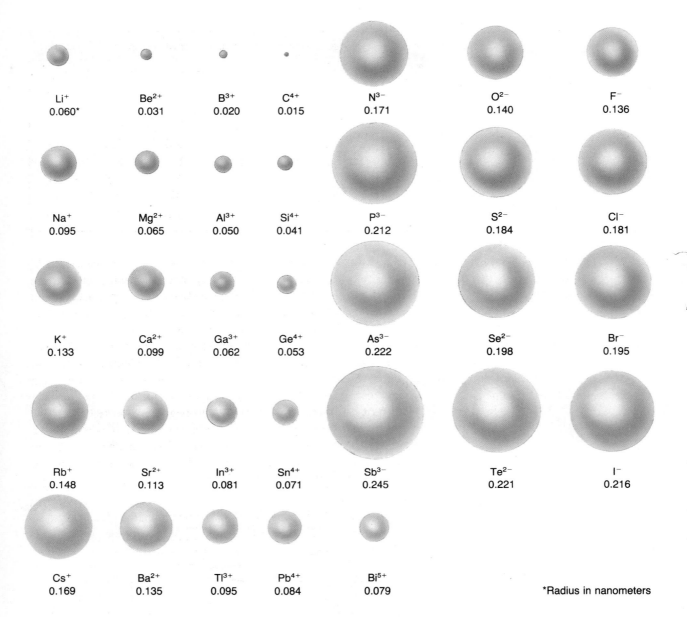

Li^+ 0.060*	Be^{2+} 0.031	B^{3+} 0.020	C^{4+} 0.015	N^{3-} 0.171	O^{2-} 0.140	F^- 0.136
Na^+ 0.095	Mg^{2+} 0.065	Al^{3+} 0.050	Si^{4+} 0.041	P^{3-} 0.212	S^{2-} 0.184	Cl^- 0.181
K^+ 0.133	Ca^{2+} 0.099	Ga^{3+} 0.062	Ge^{4+} 0.053	As^{3-} 0.222	Se^{2-} 0.198	Br^- 0.195
Rb^+ 0.148	Sr^{2+} 0.113	In^{3+} 0.081	Sn^{4+} 0.071	Sb^{3-} 0.245	Te^{2-} 0.221	I^- 0.216
Cs^+ 0.169	Ba^{2+} 0.135	Tl^{3+} 0.095	Pb^{4+} 0.084	Bi^{5+} 0.079		*Radius in nanometers

Figure 12·9
The ionic radii shown here demonstrate a periodic variation in size. Why are ionic radii different from atomic radii (Figure 12·6)?

12·8 Periodic Trends in Electronegativity

The **electronegativity** *of an element is the tendency for an atom to attract electrons to itself when it is chemically combined with another element.* Electronegativities have been calculated for the elements. They are expressed in arbitrary units on the *Pauling electronegativity scale*. This scale is based on a number of factors including the ionization energies and electron affinities of the atoms.

The electronegativities, arranged in the form of the periodic table, are presented in Table 12·3. Note that the noble gases are omitted because they do not form many compounds. Otherwise, each element is assigned an electronegativity number. Cesium, the least electronegative

Table 12·3   Electronegativity Values for Atoms of Selected Elements

H						
2.1						
Li	Be	B	C	N	O	F
1.0	1.5	2.0	2.5	3.0	3.5	4.0
Na	Mg	Al	Si	P	S	Cl
0.9	1.2	1.5	1.8	2.1	2.5	3.0
K	Ca	Ga	Ge	As	Se	Br
0.8	1.0	1.6	1.8	2.0	2.4	2.8

element is 0.7 and fluorine, the most electronegative element is 4.0. When fluorine is chemically bonded to any other element, it attracts the shared electrons, and it tends to form a negative ion. In contrast, cesium with the lowest electronegativity has the least tendency to attract electrons. It loses the electron tug-of-war and forms a positive ion.

As we go across a period from left to right, the electronegativity of the representative elements *increases*. The metallic elements at the far left of the periodic table have low electronegativities. By contrast, the nonmetallic elements at the far right (excluding the noble gases) have high electronegativities. Ordinarily, electronegativity decreases as we move down a given group. The trends in electronegativities among the transition metals are not so regular. As you will see in Chapters 13 and 14, electronegativity values help us to predict the type of bonding that can exist between atoms in compounds.

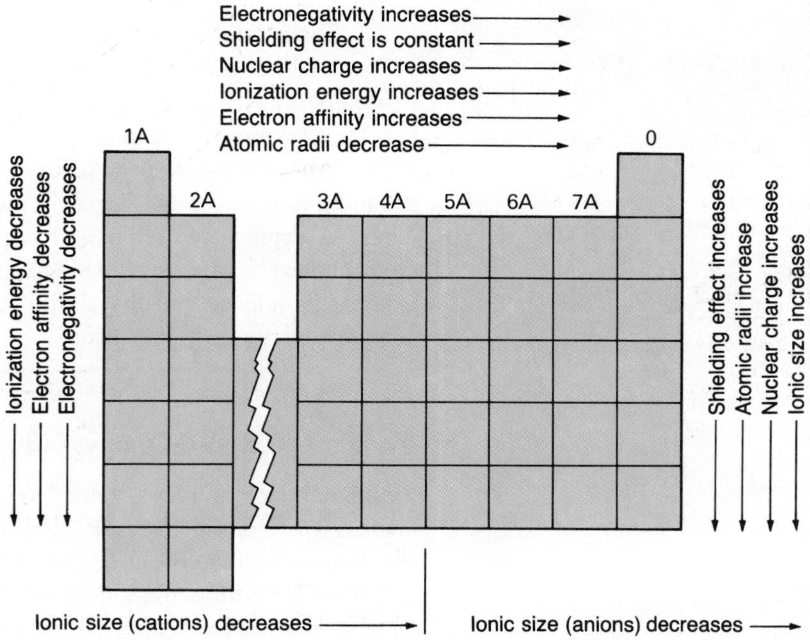

Electronegativity increases with atomic number within a period and decreases with atomic number within a group.

Figure 12·10
Periodic properties are those that vary as you move across and down the periodic table. These properties include the atomic radius, electron affinity, ionization energy, nuclear charge, shielding effect, and electronegativity of the elements.

We have explained five periodic trends by looking at variations in atomic structure. Figure 12·10 on the previous page summarizes the trends in atomic radii, ionization energy, electron affinity, ionic size, and electronegativity.

Problem

5. Which of these elements has a larger ionization energy?
   a. sodium or potassium   b. magnesium or phosphorus

Figure 12·11
Technical writers should have a broad science background and strong English skills. Reference books are important tools for their work.

■ The noble gases rarely form compounds.

=== Science, Technology, and Society ===

12·B Technical Writing: Communicating Clearly

The job of a technical writer is to communicate complex ideas clearly. Most technical writing explains something. The explanation must be clear and brief without sacrificing detail or accuracy.

Technical writers may write instruction manuals or contracts that involve scientific or technical issues. They may also write magazine features on science and technology, press releases, sales presentations, or articles for professional journals or books. Almost every company employs writers to help its scientists and engineers communicate their work. This may be for the public or for others in their profession. Large companies often have publication divisions that employ many technical writers, editors, and illustrators. The semiconductor and aerospace industries and the government are major employers in this field.

Some companies try to hire writers who have degrees in the area about which they will be writing. Other companies prefer to have writers with English or journalism degrees. Many colleges are now offering courses in technical writing in response to the high demand for able writers. A person looking for a job as a technical writer should have a college degree in one of these fields. Like other writers, a technical writer must have good writing skills, an extensive vocabulary, and a strong knowledge of grammar and spelling. An interest in science and the ability to communicate with scientists and engineers are very important.

12·9 The Noble Gases

Helium, neon, argon, krypton, xenon, and radon are all Group 0 elements. The name, *rare gases,* was originally used to describe these elements because they occur in the atmosphere in very small amounts. Early chemists also called these elements *inert gases* because they rarely combine with other elements. In 1962, however, a Canadian chemist named

Figure 12·12
Underwater living quarters, such as the one shown, commonly have an artificial atmosphere consisting primarily of oxygen and a noble gas.

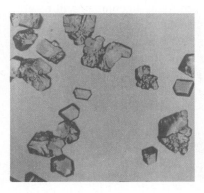

Figure 12·13
Xenon tetrafluoride is one of the few compounds that can be formed from a noble gas.

The alkali metals in Group 1A and the alkaline earth metals in Group 2A react with water to produce alkaline solutions.

Arabic: *al aqali* = the ashes.

Neil Bartlett prepared xenon tetrafluoride (XeF_4), a compound of xenon. Since that time, compounds of krypton and radon have also been prepared. Nevertheless, compared with all other elements, Group 0 elements are extremely unreactive. For this reason these elements are now called *noble gases*. This name emphasizes the tendency of these elements to exist as separate atoms rather than in combination with other atoms.

Despite their unreactivity, the noble gases have many uses. Helium is used to fill weather balloons. Although it is more dense than hydrogen, it is not explosive. Both helium and neon are used in artificial atmospheres such as those required in deep-sea diving. In a neon atmosphere, speech is less distorted than it is in a helium atmosphere. This is because the speed of sound is slower in neon than in helium. Neon, argon, krypton, and xenon are used to produce the inert atmospheres needed for photographic flashbulbs and aluminum welding.

12·10 The Alkali Metals and the Alkaline Earth Metals

The elements in Group 1A are the **alkali metals.** They have low densities, low melting points, and good electrical conductivity. They are soft enough to be cut with a knife. The freshly cut surface is shiny, but it quickly dulls on exposure to air. This is due to a rapid reaction with oxygen and moisture. These elements are not found in nature in the uncombined state. Many compounds of sodium and potassium were isolated from wood ash by early chemists. Each of the alkali metals reacts violently with cold water, producing hydrogen gas and a solution of the metal hydroxide (an alkali). The alkali metals react vigorously with water. Thus they should not be allowed to come into contact with your skin. In the stockroom they are usually stored under oil or kerosene to protect the metal from oxygen and moisture in the air.

a

b

Figure 12·14
a Freshly cut sodium is shiny, but it quickly reacts with oxygen in the air to become dull.
b Magnesium is present in asbestos, which is a fibrous form of the mineral serpentine. Asbestos was used as an insulation material until it was discovered that inhalation of its fibers causes lung cancer.

 **Issues in Chemistry**

Should the use of asbestos be eliminated or restricted?

Aluminum is the most common Group 3A element.

The elements in Group 2A are called **alkaline earth metals.** In their reaction with water they also produce alkaline solutions. They are extracted from the mineral ores that since early times have been called "earths." The alkaline earth metals are not found uncombined in nature, but they are less chemically reactive then the Group 1A metals. They need not be stored under oil. The alkaline earth metals are harder than the alkali metals. They have a gray-white luster but tarnish quickly in air with a thin oxide coating. The coating protects the metal, particularly beryllium and magnesium, from further oxidation. This allows alloys of these metals to be used as low-density structural materials.

12·11 The Aluminum Group

The elements in Group 3A include both metals and nonmetals. The non-metallic solid boron at the top of the group is followed by four metallic elements: aluminum, gallium, indium, and thallium.

Aluminum is a mechanically strong metal of low density that is especially corrosion-resistant. Like magnesium in Group 2A, it reacts

Figure 12·15
Aluminum is particularly useful in the construction of aircraft because it is lightweight and very strong. It also forms a protective coating that will not react with water.

Figure 12·16
Rubies are aluminum oxide in which a few of the aluminum ions are replaced by chromium ions.

rapidly with the oxygen in air. This reaction forms a thin, tough, protective coating of aluminum oxide. Aluminum, or its alloy with magnesium, is used widely as a lightweight structural material in aircraft production and in the manufacture of cookware. Because of its protective coating, aluminum does not react with water. Nevertheless, it will react rapidly with acids or bases, which dissolve the coat, liberating hydrogen.

Aluminum does not exist in the uncombined state in nature but is a major component of many rocks and minerals. Aluminum is commonly found as corundum (impure aluminum oxide). This very hard material is used as the abrasive in emery powder and grinding wheels. Gallium, indium, and thallium are quite rare and have few practical uses. One interesting use of gallium is in thermometers because of its extraordinarily wide liquid range. It has a melting point of 30°C and a boiling point of 1980°C.

12·12 The Carbon Group

 Group 4A, the carbon group, contains important nonmetals, metalloids, and metals.

ChemDirections

The conductivity of silicon can be changed dramatically by adding small amounts of other elements. For more information, turn to **ChemDirections** *Semiconductors,* page 654.

The elements in Group 4A continue the trend we saw in Group 3A. Carbon, at the top of the group, is a nonmetal. Silicon and germanium are metalloids with metallic and nonmetallic properties. Tin and lead are both metals.

Diamond and graphite are two forms of carbon. Diamond behaves like a typical nonmetal and is a nonconductor of electricity. Graphite has some of the properties of a metal and is a good conductor of electricity. Coke is a fairly pure form of carbon. Carbon-containing compounds, of which there are about 4 million, are called organic compounds. Organic compounds constitute the bulk of all living things.

Silicon, the second most abundant element of earth, occurs in nature in the combined state as sand and in rocks, soils, and clays. Silicon and germanium are semiconductors. Both substances are insulators at low temperatures. At high temperatures, however, they conduct electricity. These two elements when highly purified form the foundation of transistor technology. They are also used in photocells for solar power units.

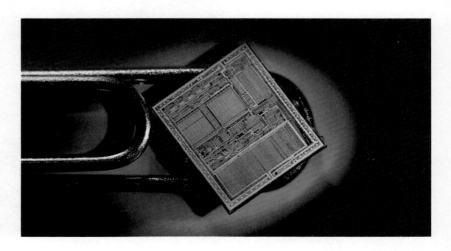

Figure 12·17
Semiconductors made of silicon are used in calculators and computers. Each chip may contain over a million transistors.

Safety

Many toxic substances can be absorbed through the skin. Avoid skin contact with chemicals whenever possible. Wash your hands thoroughly after handling chemicals, and never eat or drink in the laboratory.

Nitrogen is the most common Group 5A element.

Tin and lead are both typical metals. Tin is important in the manufacture of tinplate in which a thin coating of tin is applied to an iron can to prevent it from rusting. Plumber's solder is an alloy of tin and lead. Tetraethyl lead, an organic lead-containing compound, is the familiar antiknock additive in gasoline. It is highly toxic to humans. For this reason its use has been greatly decreased in recent years.

12·13 The Nitrogen Group

Another periodic trend, a change in physical state, is introduced with Group 5A. The first element, nitrogen, is a gas at room temperature. In descending order the next elements are phosphorus (a solid nonmetal) and arsenic and antimony, which are metalloids. The last element, bismuth, is a metal.

Nitrogen is essential to living organisms. Pairs of nitrogen-containing compounds called bases are the steps on the spiral staircase of double-stranded DNA (deoxyribonucleic acid). Proteins and enzymes are long chainlike molecules with carbon–nitrogen backbones. Although 80% of the air we breathe is nitrogen, we cannot use it to make these essential substances. Fortunately, bacteria that live in the root nodules of peas, beans, and other legumes can "fix" atmospheric nitrogen. Plants can use ammonia to synthesize proteins and other biologically important nitrogen-containing compounds.

Phosphorus is also essential to living organisms. It is present in the double strands of DNA, bones, and teeth. It is also part of ATP (adenosine triphosphate), which is the principal energy-storage molecule in living systems. Phosphorus occurs mainly in the form of phosphate rock. Pure phosphorus is prepared in a white form and a red form. The white form is very reactive. It is usually stored under water to prevent a reaction with oxygen from the air. Red phosphorus is a less active form used in the manufacture of matches.

Figure 12-18
The seeds of legumes are an excellent source of protein. Legumes do not depend on nitrogen in the soil.

Figure 12·19
White (or yellow) phosphorus is very reactive and must be stored in water. Red phosphorus is much more stable.

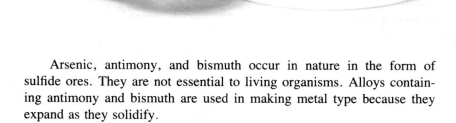

Safety

Even the friction of cutting can ignite white phosphorus. Make a point of remembering which chemicals are toxic or flammable. Handle all chemicals with care.

Arsenic, antimony, and bismuth occur in nature in the form of sulfide ores. They are not essential to living organisms. Alloys containing antimony and bismuth are used in making metal type because they expand as they solidify.

12·14 The Oxygen Group

The Group 6A elements are oxygen, sulfur, selenium, tellurium, and polonium. Oxygen is a gas and thereby continues the periodic trend started by nitrogen in the previous group. Sulfur is a nonmetal that occurs free in nature as a brittle yellow solid. Selenium and tellurium are both solids. They are borderline between metals and nonmetals. Polonium, the last element in the group, is a radioactive metal that occurs only in trace quantities in radium-containing ores.

Oxygen, the leading Group 6A element, is the most abundant element.

Oxygen is the most abundant element. It accounts for 20% by volume of the air we breathe, 60% by mass of the human body, and 50% by mass of the earth's crust. Most oxygen is combined in the silicate rocks of the earth's crust. It is produced by plants in photosynthesis. Oxygen gas is used in medicine, in the manufacture of steel (to remove impurities), and along with acetylene in oxyacetylene welding.

Sulfur occurs in the elemental state in large underground deposits. It is also a minor component of coal and petroleum. Sulfur is essential to living organisms where it is found mainly in disulfide bridges. These are the crosslinks that hold protein chains together. The major uses of sulfur are in the manufacture of sulfuric acid and in the vulcanization of rubber. Sulfuric acid is the most widely used industrial chemical.

Figure 12·20
Oxygen is present in combined form in silicate rocks and minerals. Those shown here (clockwise from the top) are feldspar ($KAlSi_3O_8$), talc ($Mg_3Si_4O_{10}(OH)_2$), biotite mica ($K_3AlSi_3O_{10}(OH)_2$), and jasper (primarily SiO_2).

Selenium is a semiconductor. It is a poor conductor of electricity in the dark, but its conductivity increases greatly in the light. Because of this property, selenium is used in photoelectric cells, in exposure meters for cameras, and in light-sensitive switches. The xerographic process of photocopying also depends on the photoconductivity of selenium.

Tellurium is one of the rarest elements. Its compounds are toxic, and the element itself plays no known role in living organisms.

12·15　The Halogens and Hydrogen

The **halogens** *are fluorine, chlorine, bromine, iodine, and astatine.* The halogens form a homogeneous family of nonmetals. The first two elements, fluorine and chlorine, are yellowish-green gases at room temperature and atmospheric pressure. They continue the periodic trend shown by nitrogen and oxygen. Bromine is a dark red liquid. Iodine is a purple-black crystalline solid with a metallic sheen. Tincture of iodine, a 3% solution of iodine in alcohol, was formerly widely used as an antiseptic. The last element, astatine, is an exceedingly rare radioactive solid that has not been well investigated.

The halogens do not exist in nature in the uncombined state, but their compounds are fairly abundant. These elements are named halogens because they are usually found as salts of the Group 1A or 2A metals. For example, the salts sodium chloride, sodium bromide, and sodium iodide are found in seawater and salt beds. Calcium fluoride is the mineral fluorspar. The free halogens are very reactive and must be handled with extreme caution. Nevertheless, compounds of fluorine, chlorine, and iodine are essential to our well-being and must be included in our diet. Fluorine, as fluoride ion, is beneficial in the formation and maintenance of healthy teeth. Chlorine, as chloride ion, is an important component of the blood and other body fluids. Iodine, as the iodide ion, is necessary to prevent goiter, an enlargement of the thyroid gland.

The halogens have many other uses in the home and in industry. A dilute solution of chlorine is used as a bleaching and disinfecting agent. Silver chloride and silver bromide are light-sensitive and are used to make photographic film. Fluorine is used in the manufacture of nonstick Teflon coatings that are applied to frying pans and other cookware.

Hydrogen is a group by itself. It is a reactive gas in that it forms an explosive mixture with oxygen. It also reacts violently with many other elements. Hydrogen is usually put at the top of Group 1A in the periodic table. It is not a metal, nor is it a good conductor of heat or electricity like the alkali metals. Like the alkali metals, however, hydrogen does react with the halogens. In some periodic tables hydrogen also appears at the top of Group 7A. This position has some validity because, like the halogens, hydrogen has one electron less than helium, the noble gas it precedes. Like the halogens, it reacts with the alkali metals. Thus hydrogen is unique.

12·16　The Transition Metals and Inner Transition Metals

The remainder of the elements are the transition metals and inner transition metals. Like the representative elements, most transition metals possess similar physical and chemical properties. They are therefore divided into groups. Starting with Group 3B on the left, they continue through 7B on the right. Group 7B is followed by three groups that together make up Group 8B. The last two groups are Groups 1B and 2B.

■ The halogens in Group 7A include two gases, a liquid, and two solids.

Greek: *halos* = salt; *gen* = born.

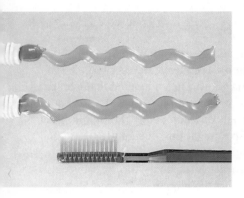

Figure 12·21
The fluoride ion is widely used in toothpastes to promote dental health.

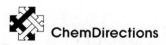

ChemDirections

Trace amounts of the transition metals iron and cobalt are essential for life. Learn about the importance of these and other elements in the proper functioning of our bodies in **ChemDirections** *Mineral Nutrients,* page 652.

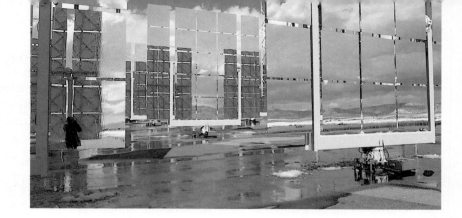

Figure 12·22
Because of its high luster, metallic silver is used to plate mirrors.

The transition metals are the Group B elements, and the inner transition metals are the lanthanides and actinides.

Similarly, there are 14 inner transition metals in the sixth period of the periodic table. These elements, from cerium through lutetium, are called the *lanthanides*. The pattern repeats in the seventh-period elements. The elements thorium through lawrencium are also inner transition metals and are called the *actinides*.

The transition and inner transition elements are typical metals. They have a metallic luster and are very good conductors of electricity and heat. Tungsten, a hard brittle solid with a melting point of 3400°C, is used in light-bulb filaments. At the other end of the scale is mercury, with a melting point of −38°C; it is used in making thermometers. The excellent reflective qualities of silver (the high luster) make it the ideal coating for mirrors. The production of copper wire in enormous quantities attests to the high electrical conductivity of copper. Steels with widely different characteristics are made by adding small amounts of cobalt, copper, chromium, nickel, or vanadium to iron. Our bodies also need transition metals to function normally. Iron is required in the production of hemoglobin. Cobalt is part of vitamin B_{12} molecules. Both zinc and copper are necessary components of many enzymes.

The transition and inner transition metals vary greatly in their chemical reactivity. The elements scandium, yttrium, and lanthanum are similar to the Group 1A and 2A metals. They are easily oxidized on exposure to air and react with water to liberate hydrogen. In contrast, platinum and

Figure 12·23
Chemistry is an important tool for the study of nutrition. It is used to determine the presence and availability of various nutrients. Some important transition element nutrients for humans are iron, zinc, copper, manganese, cobalt, chromium, and molybdenum.

gold are extremely unreactive and resist oxidation. Most compounds of the transition and inner transition metals are colored and show multiple-formula combinations with other elements.

Problem

6. Give the symbol and name for the element that occupies each of the designated positions in the periodic table.
   a. period 4, Group 4A
   b. period 3, Group 2A
   c. period 5, Group 5B
   d. period 3, Group 5A
   e. period 4, Group 4B
   f. period 2, Group 7A

Figure 12·24
Scientists speculate that the universe began with a big bang. Galaxies and stars are matter that has gathered because of gravitational attraction.

=== Science, Technology, and Society ===

12·C Before the Elements: The Big Bang

The orderly structure of the elements, each with its unique number of protons and electrons, leads us to wonder how they were formed. We know that in our sun, hydrogen nuclei are fused together to give helium nuclei. Physicists create new heavier elements by smashing small particles into heavy nuclei in accelerators. It is reasonable to think that all nuclei are built-up of lighter nuclei in this way. But what was in the universe even before hydrogen nuclei?

Physicists theorize that the universe began with an explosion of undescribable power, the Big Bang. At the moment of this explosion the temperature was many billions of degrees. All matter was in the form of quarks (see Section 4·C).

Within 10^{-43} second, strange new particles formed that could change quarks into leptons. At 10^{-4} s after the Big Bang, quarks formed neutrons, protons, and hundreds of other types of particles. By 10^{-2}s after the Big Bang, the temperature had cooled to 10^{11} K.

The lightest nuclei began to form just three minutes after the Big Bang. The temperature then was still at least 70 times the temperature of our sun. Matter was in the form of a plasma, a sea of positive nuclei and negative electrons. It probably took 500 000 years for electrons and nuclei to cool enough to combine to form atoms!

The data that lead to these theories come largely from experiments done with particle accelerators. When very high-energy collisions take place in an accelerator, energy is transformed into strange new particles. It is reasonable to think that such particles must have existed after the tremendous energy release of the Big Bang.

Scientists cannot test these ideas by creating another Big Bang. Furthermore, the strange particles and events that are predicted by the Big Bang theory would be extraordinarily rare under normal conditions. Statistically, however, they should occur sometime. Experiments are under way to try to detect free quarks, the decay of protons, or other occurrences that would support this theory.

12 Chemical Periodicity
Chapter Review

Key Terms

alkali metal	*12·10*	electronegativity	*12·8*
alkaline earth		halogen	*12·15*
metal	*12·10*	ionization energy	*12·5*
covalent atomic		noble gas	*12·3*
radius	*12·4*	period	*12·2*
electron affinity	*12·6*	periodic law	*12·2*

Chapter Summary

The periodic table organizes the elements into groups (vertical columns) and periods (horizontal rows) in order of increasing atomic number. Most of the elements in the periodic table are metals. The nonmetals are confined to a triangular area on the upper right-hand side of the table. Elements that have similar chemical properties are in the same group. The properties of any element are generally intermediate between those of its neighbors on either side in the same period and similar to those elements above and below it in the same group. The elements in Groups 1A through 7A are called the representative elements. They exhibit a wide range of both physical and chemical properties. The noble gases make up Group 0. They may be considered part of the representative elements. The elements in Groups 2A and 3A are interrupted in periods 4 and 5 by the transition metals and in periods 6 and 7 by the inner transition metals.

Elements that have similar properties also have similar electron configurations and are members of the same group. The atoms of the noble gas elements have their outermost s and p sublevels filled. The outermost s and p sublevels of the representative elements are only partially filled. The outermost s and nearby d sublevels of transition metals contain electrons. The outermost s and nearby f sublevels of inner transition metals contain electrons.

The regular changes in the electron configuration of the elements cause gradual changes in both the physical and the chemical properties of the elements within a group and within a period. Atomic radii generally decrease as we move from left to right in a given period of the periodic table because there is an increase in the nuclear charge while the number of inner electrons, and hence the shielding effect, remains constant. Atomic size generally increases within a given group because there are more occupied energy levels and an increased shielding effect. The ionization energy, the energy required to remove an electron from an atom, generally increases as we move from left to right across a period. It decreases as we move down a group. Electron affinity increases the ease with which an atom gains an electron. It increases as we move from left to right across a period and decreases as we move down a group. Electronegativity is the ability of a bonded atom to attract electrons to itself. It generally increases as we move from left to right across a period.

Practice Questions and Problems

7. What criterion did Mendeleev use in arranging his periodic table? What criterion was used in constructing the modern periodic table? *12·1*

8. What is the periodic law? *12·2*

9. How do the terms *group* and *period* relate to the periodic table? *12·2*

10. What are the representative elements, the transition elements, and the inner transition elements? *12·2*

11. Which of the following are representative elements: Na, Mg, Fe, Ni, Cl? *12·2*

12. How is an element's outer electron configuration related to its position in the periodic table? *12·3*

13. The graph below shows how many elements were discovered before 1750 and in each 50-year period since then. *12·1*

a. In which 50-year period were the most elements discovered?

b. How did Mendeleev's work contribute to the discovery of so many elements during this time?

c. What percent of the elements now known were discovered before 1750?

d. What characteristic do all the elements discovered since 1950 have in common?

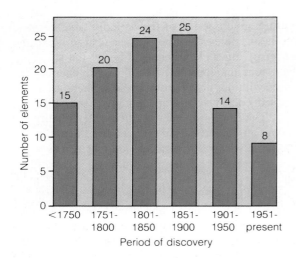

14. Use Figure 12·5 to write the electron configuration of these atoms. *12·3*
  a. fluorine **b.** zinc **c.** aluminum **d.** tin

15. What are the symbols for all the elements that have the following outer configurations? *12·3*
  a. s^1 **b.** s^2p^4 **c.** s^2d^{10}

16. Arrange these elements in order of decreasing atomic size: sulfur, chlorine, aluminum, and sodium. Does your arrangement demonstrate a periodic or a group trend? *12·4*

17. In general, would you expect metals or non-metals to have higher ionization energies? Why? *12·5*

18. Arrange the following elements in order of increasing ionization energy. *12·5*
  a. Be, Mg, Sr **b.** Bi, Cs, Ba **c.** Na, Al, S

19. Would you expect the noble gas neon to have a negative electron affinity? Explain your answer.
12·6

20. In each of the following pairs, which element is the most electronegative? *12·8*
  a. chlorine, fluorine **c.** magnesium, neon
  b. carbon, nitrogen **d.** arsenic, calcium

21. Why are alkali metals stored under kerosene or mineral oil? Why must they be kept dry? *12·10*

22. Why is aluminum so corrosion resistant? *12·11*

23. Name two forms of pure carbon. *12·12*

24. What are two forms of elemental phosphorus? How do they differ in reactivity with oxygen?
12·13

25. Name at least one industrial use of oxygen, sulfur, and selenium. *12·14*

26. Give colors and physical states at STP of chlorine, bromine, and iodine. *12·15*

27. Write a balanced equation for the explosive reaction of hydrogen with oxygen. *12·15*

28. Suggest reasons why copper, silver, and gold are valued for the manufacture of electronic devices. Why is copper preferred to silver and gold for home wiring? *12·6*

Mastery Questions and Problems

29. The Mg^{2+} and Na^+ ions each have ten electrons surrounding the nucleus. Which ion would you expect to have the smaller radius? Why?

30. Explain why it takes more energy to remove a $4s$ electron from zinc than calcium.

31. Give the name and symbol for the element found at each of the following locations in the periodic table.
  a. Group 1A, period 4
  b. Group 3A, period 3
  c. Group 6A, period 3
  d. Group 2A, period 6

32. Give the symbols for the elements in these groups.
  a. alkali metals
  b. alkaline earth metals
  c. halogens

33. Give the names, symbols, and electron configurations for the ten first-row transition metals.

Critical Thinking Questions

34. Choose the term that best completes the second relationship.
  a. sister:brother   oxygen: _____
      (1) hydrogen   (3) silicon
      (2) sulfur   (4) group 6A
  b. Mendeleev:atomic mass
    Moseley: _____
      (1) atomic number (3) atomic radius
      (2) periodicity   (4) neutrons
  c. magnesium:s orbital zinc: _____
      (1) s orbital   (3) d orbital
      (2) p orbital   (4) f orbital

35. Why did Mendeleev and later scientists classify the elements into groups?

36. Do you think there are more elements left to be discovered? Explain your answer.

Review Questions and Problems

37. A 2.00-L flask at 27°C contains 4.40 g of carbon dioxide and 2.00 g of nitrogen gas. What is the pressure, in atmospheres, of each of the two components?

38. Balance the following chemical equations.
  a. $Ag + S \longrightarrow Ag_2S$
  b. $Na_2SO_4 + Ba(OH)_2 \longrightarrow BaSO_4 + NaOH$
  c. $Zn + HNO_3 \longrightarrow Zn(NO_3)_2 + H_2$
  d. $H_2O + SO_2 + O_2 \longrightarrow H_2SO_4$

39. The smelting of iron ore consists of heating the ore with carbon.

$$2Fe_2O_3 + 3C \longrightarrow 4Fe + 3CO_2$$

What mass of iron can be obtained from 100 g of the ore?

40. Write chemical formulas for the following compounds.
  a. lithium sulfate
  b. zinc phosphate
  c. potassium permanganate
  d. strontium carbonate

41. If a gas sample at 25°C occupies a volume of 2.93 L, what will be the volume at 500°C if the pressure is unchanged?

Challenging Questions and Problems

42. The ions S^{2-}, Cl^-, K^+, Ca^{2+}, Sc^{3+} have the same *total* number of electrons as the noble gas argon. How would you expect the radii of these ions to vary? Would you expect to see the same variation in the series O^{2-}, F^-, Na^+, Mg^{2+}, and Al^{3+}, in which each ion has the same total number of electrons as the noble gas neon? Why or why not?

43. Using a handbook of chemistry, make a table for the Group 2A elements. Include densities, atomic masses, formulas of the chlorides and oxides, and first ionization potentials. Can you justify placing these elements in one group on the basis of these data?

Research Projects

1. How did the work of Jöns Berzelius lay the foundation for a periodic table of the elements? What other scientists were influenced by his work?

2. Devise a scheme to develop an activity series for some common transition metals. Use the reaction of tin, zinc, magnesium, copper, and iron with dilute hydrochloric acid.

3. What contributions did Glenn Seaborg make to the discovery of the actinide rare earth elements?

4. What are the current allowable levels of lead in gasoline? How has this changed in the last 20 years?

5. Report on selenium pollution and the use of selenium as a dietary supplement.

6. Design a method for detecting a toxic metal such as lead or selenium in the environment.

Readings and References

Asimov, Isaac. *Building Blocks of the Universe.* New York: Abelard-Schuman, 1974.

Ciparick, Joseph D. "Element X." *ChemMatters* (December 1987), pp. 8–9.

Sandage, Allan. "Inventing the Beginning." *Science 84* (November 1984), pp. 111–113.

13 Ionic Bonds

Chapter Preview

13·1 Valence Electrons
13·2 Stable Electron Configurations for Cations
13·3 Stable Electron Configurations for Anions
13·4 Ionic Compounds
13·A The Mixed Blessing of Curing Salts
13·5 Properties of Ionic Compounds
13·6 Metallic Bonds
13·B Alloys

In Chapters 11 and 12 you learned about the electron structure of atoms and the organization of the periodic table. This knowledge of atomic structure and periodicity will help you understand the chemical bonding that occurs between atoms. For example, the electron configurations of the sodium and chlorine atoms help explain why one atom of sodium combines with one atom of chlorine. It also explains why the formula unit for sodium chloride is NaCl and not Na_2Cl, $NaCl_2$, or Na_2Cl_2. Why does sodium chloride conduct an electric current when it is in the molten state? Why is sodium chloride a solid at room temperature with a melting point of 800°C whereas hydrogen chloride is a gas? Why do the metallic elements form positive ions whereas the nonmetallic elements form negative ions? Questions like these and many others will be answered in this chapter.

13·1 Valence Electrons

Valence is the capacity of an element to combine with another.

Latin: *valere* = to be strong.

Figure 13-1
Most ionic compounds are crystalline solids with high melting points. For example, calcium flouride, the mineral fluorite, melts at 1360°C.

Knowing electron configurations is important because the number of valence electrons largely determines the chemical properties of an element. **Valence electrons** *are the electrons in the highest occupied energy level of an element's atoms.* You may recall that when Mendeleev organized his periodic table, he did so with the properties of the elements in mind. Scientists later learned that all the elements in a particular group of the periodic table have the same number of valence electrons. For example, the elements in Group 1A (hydrogen, lithium, sodium, potassium, and so forth) all have one valence electron. For the representative elements an atom's number of valence electrons can be determined by looking up the group number of that element. Carbon and silicon, in Group 4A, have four valence electrons. Nitrogen and phosphorus, in Group 5A, have five valence electrons; and oxygen and sulfur, in Group 6A, have six. The noble gases (Group 0) are the one exception to this rule. Helium has two valence electrons, and all the others have eight.

Problem

1. How many valence electrons does each of the following atoms have?
   a. potassium **b.** carbon **c.** magnesium **d.** oxygen

Electron dot structures are also called *Lewis dot structures* in honor of Gilbert N. Lewis (1875–1946), an American chemist whose ideas concerning the nature of chemical bonds are very much a part of modern chemistry.

Valence electrons are usually the only electrons used in the formation of chemical bonds. Thus it is customary to show only the valence electrons in electron dot structures. **Electron dot structures** *depict valence electrons as dots. The inner electrons and the atomic nuclei are represented by the symbol for the element being considered.* Table 13·1 shows electron dot structures for atoms of some Group A elements.

Table 13·1		Electron Dot Structures of Some Group A Elements						
Period	1A	2A	3A	4A	5A	6A	7A	0
1	H·						H·	He:
2	Li·	·Be·	·B·	·C·	·N·	:O·	:F·	:Ne:
3	Na·	·Mg·	·Al·	·Si·	·P·	:S·	:Cl·	:Ar:
4	K·	·Ca·	·Ga·	·Ge·	·As·	:Se·	:Br·	:Kr:

Electron dot structures do not distinguish between *s* and *p* electrons.

13·2 Stable Electron Configurations for Cations

The atoms that make up elements are almost always found on earth as ions or in molecules. For example, sodium and potassium are present in your body as relatively unreactive cations, not as very reactive atoms. However noble gas atoms float about in the earth's atmosphere. The noble gases are of low chemical reactivity because they have relatively *stable electron configurations*. Why are elements other than noble gases found mainly as ions or in molecules?

Most stable means lowest energy.

In 1916 Gilbert Lewis provided an explanation for why atoms tend to form certain types of ions and molecules. He proposed the **octet rule:** *Atoms react by changing the number of their electrons so as to acquire the stable electron structure of a noble gas.* Recall that each noble gas,

Greek: *okto* = eight.

■ An outer structure of eight electrons is a stable configuration.

Safety

A modern-day chemist doing work similar to that of Gilbert Lewis would definitely wear safety goggles. Lewis was fortunate not to lose his eyesight.

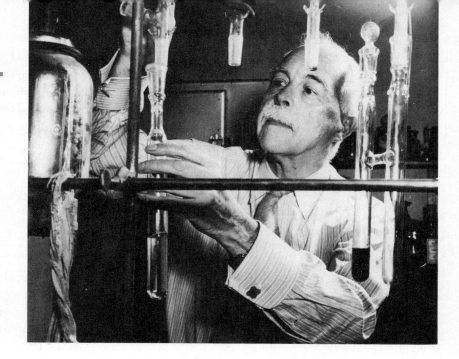

Figure 13·2
Gilbert Lewis proposed the octet rule and the electron dot method of showing valence electrons. He was also an important contributor to acid–base theory and thermodynamics.

Cation, a positive ion; anion, a negative ion.

The metallic elements tend to form positive ions, cations.

except helium, has eight electrons ($ns^2\,np^6$) in its highest energy level. The octet rule takes its name from this fact. Atoms of the metallic elements obey the octet rule by losing electrons. Atoms of some nonmetallic elements obey the rule by gaining electrons. *The loss of valence electrons from an atom produces a cation, or positively charged ion. The gain of valence electrons produces an anion, or negatively charged ion.*

You may recall from Section 5·2 that a *cation* is any atom or group of atoms with a positive charge. The most common cations are those produced by the loss of electrons from metal atoms. These atoms usually have up to three valence electrons that are easily removed. Sodium, in Group 1A of the periodic table, is typical. Sodium atoms have a total of 11 electrons, including 1 valence electron. When forming a compound, a sodium atom loses its 1 valence electron. It then has the same electron configuration as neon, a noble gas (Figure 13·3). The sodium *ion* has an octet (eight electrons) in its highest energy level. Because the number of

Figure 13·3
A sodium atom loses an electron to become a positively charged sodium ion. The sodium ion has an electron configuration like that of the noble gas neon. There are eight electrons (an octet) in the highest energy levels of Na^+ and Ne.

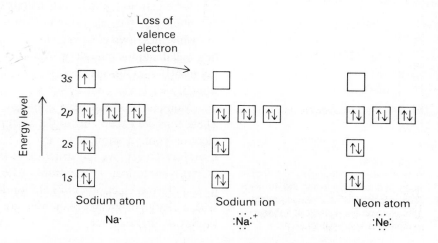

protons in the sodium nucleus is still 11, the lack of one unit of negative charge produces a sodium ion with a charge of $1+$. We can show the electron loss, or ionization, of the sodium atom by drawing the complete electron configuration of the atom and the ion formed.

$$\text{Na} \quad 1s^2\,2s^2\,2p^6\,3s^1$$

$$\text{Na}^+ \quad 1s^2\,\underbrace{2s^2\,2p^6}_{\text{octet}}$$

The electron configuration of the sodium ion is the same as that of the neon atom. Both have an octet of outer electrons.

$$\text{Ne} \quad 1s^2\,\underbrace{2s^2\,2p^6}_{\text{octet}}$$

The ionization can be shown more simply by using an electron dot structure for the atom.

$$\text{Na}\cdot \xrightarrow[\text{ionization}]{\substack{\text{loss of valence}\\ \text{electron}}} \text{Na}^+ \quad + \quad e^-$$

Sodium atom (electrically neutral, charge = 0)   Sodium ion (plus sign indicates one unit of positive charge)   Electron (minus sign indicates one unit of negative charge)

Magnesium (atomic number 12) belongs to Group 2A of the periodic table and therefore has two valence electrons. Magnesium atoms attain the electron configuration of neon by losing both valence electrons. The loss of the valence electrons produces a magnesium ion, a cation with twice the positive charge of a sodium ion.

$$\cdot\text{Mg}\cdot \longrightarrow \text{Mg}^{2+} + 2e^-$$

Magnesium atom   Magnesium ion

We have said that the cations of Group 1A elements always have a charge of $1+$. Similarly, the cations of Group 2A elements always have a charge of $2+$. We can now explain this constancy in terms of the loss of valence electrons by atoms of metals to attain the electron configuration of a noble gas. The elements in Group 2A all have two valence electrons. In losing these two electrons they form $2+$ cations. The charge of transition metal cations may vary. An atom of iron loses two electrons to form the iron(II), or ferrous, ion, Fe^{2+}. It loses three electrons in forming the iron(III), or ferric, ion, Fe^{3+}.

Many ions do not have noble-gas electron configurations ($ns^2\,np^6$). These ions are exceptions to the octet rule. Let us consider silver with the electron configuration of $1s^2\,2s^2\,2p^6\,3s^2\,3p^6\,3d^{10}\,4s^2\,4p^6\,4d^{10}\,5s^1$ as an example. To achieve the structure of the noble gas krypton, a silver atom would have to lose 11 electrons. Alternatively, it could gain 7 electrons to acquire the electron configuration of the noble gas xenon. Since ions with a charge of greater than three are uncommon, neither of these possibilities is likely.

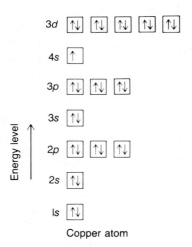

Figure 13·4
The electron configuration of a copper atom is shown. By losing the 4s electron, copper attains a pseudo noble gas electron configuration. It then becomes a copper(I) ion, Cu⁺.

Silver cannot acquire a noble-gas configuration. Yet its outer electron configuration will be $4s^2\ 4p^6\ 4d^{10}$ if it loses its $5s^1$ electron. This configuration, with 18 electrons in the outer energy level, is relatively stable. It is known as the *pseudo noble-gas* electron configuration. Silver forms a unipositive ion in this way. Other elements that behave similarly to silver are at the right of the transition metal series. Cu^+, Au^+, Cd^{2+}, and Hg^{2+} ions have pseudo noble-gas electron configurations.

Problem

2. Write electron configurations for the unipositive ions of copper and gold and the dipositive ions of cadmium and mercury.

13·3 Stable Electron Configurations for Anions

The nonmetallic elements tend to form negative ions with an octet of valence electrons.

An anion is an atom or a grouping of atoms with a negative charge. Atoms of nonmetallic elements attain stable electron configurations more easily by gaining electrons than by losing them. For example, chlorine belongs to Group 7A, the halogen family, of the periodic table. A gain of one electron converts a chlorine atom into a chloride ion. It is an anion with a single negative charge. Chlorine atoms therefore need one more valence electron to achieve the electron configuration of the nearest noble gas, argon.

$$Cl \quad 1s^2\ 2s^2\ 2p^6\ 3s^2\ 3p^5$$

$$Cl^- \quad 1s^2\ 2s^2\ 2p^6\ \underbrace{3s^2\ 3p^6}_{\text{octet}}$$

The chloride ion has eight electrons (an octet) in its highest energy level. It has the same electron configuration as the noble gas argon.

$$Ar \quad 1s^2\ 2s^2\ 2p^6\ 3s^2\ 3p^6$$

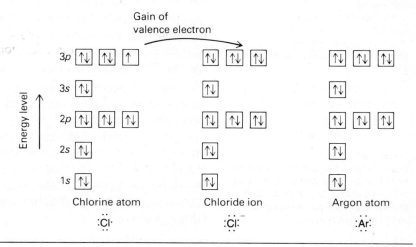

Figure 13·5
A chlorine atom gains an electron to become a negatively charged chloride ion. The chloride ion has an electron configuration like that of the noble gas argon. Each has an octet of electrons.

Electron dot structures can also be used to write an equation showing the formation of a chloride ion from a chlorine atom.

$$:\ddot{\underset{..}{Cl}}\cdot \ + \ e^- \ \xrightarrow[\text{(ionization)}]{\begin{array}{c}\text{gain of one}\\ \text{valence electron}\end{array}} \ :\ddot{\underset{..}{Cl}}:^-$$

Chlorine atom   Chloride ion (Cl⁻)

Atoms of chlorine and the other halogens gain electrons to form **halide ions.** An atom of each halogen has seven valence electrons. It needs to gain one electron to achieve the electron configuration of a noble gas. Therefore, as we have seen, all halide ions have a charge of $1-$: F^-, Cl^-, Br^-, and I^-.

Oxygen atoms each have six valence electrons. They attain the electron configuration of neon by gaining two electrons. The resulting oxide ions have charges of $2-$ and are written as O^{2-}.

$$:\ddot{O}\cdot \ + \ 2e^- \ \longrightarrow \ :\ddot{\underset{..}{O}}:^{2-}$$

Oxygen atom   Oxide ion

Problems

3. a. lose 2   **c.** lose 3
   b. gain 1   **d.** gain 2
4. a. S^{2-} **c.** F^-
   b. Na^+ **d.** Ba^{2+}

3. How many electrons will the following elements gain or lose in forming an ion?
   a. calcium (Ca)   **c.** aluminum (Al)
   b. fluorine (F)   **d.** oxygen (O)

4. What will be the formula of the ion formed when the following elements gain or lose valence electrons and attain noble gas configurations?
   a. sulfur (S)   **c.** fluorine (F)
   b. sodium (Na)   **d.** barium (Ba)

13·4 Ionic Compounds

Electrically charged objects can exert forces on each other even though they are not touching. Electrostatic forces result from the interaction of positive and negative charges.

Ionic compounds are composed of oppositely charged ions in fixed ratios.

Metals react with nonmetals to form ionic compounds.

Anions and cations have opposite charges. They attract one another by electrostatic forces. *The forces of attraction that bind oppositely charged ions together are called* **ionic bonds.** Ionic compounds are electrically neutral groups of ions joined by electrostatic forces. Most are also known as salts. In any sample of an ionic compound, the positive charges of the cations must equal the negative charges of the anions. We made use of this principle in using ionic charges to write formulas of ionic compounds in Chapter 5.

To learn about ionic bonds, consider the reaction between a sodium atom and a chlorine atom. In forming a compound, sodium has a single valence electron that it can easily lose. Chlorine has seven valence electrons and can easily gain one. When sodium and chlorine react to form a compound, they must do so in a one-to-one ratio. The sodium atom gives its one valence electron to a chlorine atom.

$$\text{Na}\cdot \longrightarrow \cdot \overset{..}{\underset{..}{\text{Cl}}}\colon \longrightarrow \text{Na}^+ \qquad \colon\overset{..}{\underset{..}{\text{Cl}}}\colon^- \qquad \textit{or } \text{NaCl}$$

$$1s^2 2s^2 2p^6 \boxed{3s^1} \quad 1s^2 2s^2 2p^6 3s^2 3p^5 \quad 1s^2 \underbrace{2s^2 2p^6}_{\text{octet}} \quad 1s^2 2s^2 2p^6 \underbrace{3s^2 3p^6}_{\text{octet}}$$

$$\text{Ne} \qquad\qquad \text{Ar}$$
$$1s^2 \underbrace{2s^2 2p^6}_{\text{octet}} \quad 1s^2 2s^2 2p^6 \underbrace{3s^2 3p^6}_{\text{octet}}$$

The chemical formula, NaCl, is a *formula unit*. It represents the smallest sample of an ionic compound that has the composition of the compound. The formula NaCl shows that one formula unit of sodium chloride contains one sodium ion and one chloride ion.

Aluminum and bromine react to form an ionic compound. Each aluminum atom has three valence electrons to lose. A bromine atom has seven valence electrons. It will easily gain an additional electron. When aluminum and bromine react, three bromine atoms react with each aluminum atom.

$$\text{Al} \quad + \quad \begin{matrix} \cdot\overset{..}{\underset{..}{\text{Br}}}\colon \\ \cdot\overset{..}{\underset{..}{\text{Br}}}\colon \\ \cdot\overset{..}{\underset{..}{\text{Br}}}\colon \end{matrix} \longrightarrow \quad \text{Al}^{3+} \quad \begin{matrix} \colon\overset{..}{\underset{..}{\text{Br}}}\colon^- \\ \colon\overset{..}{\underset{..}{\text{Br}}}\colon^- \\ \colon\overset{..}{\underset{..}{\text{Br}}}\colon^- \end{matrix} \quad \textit{or} \quad \text{AlBr}_3$$

The formula for aluminum bromide is therefore AlBr₃.

Figure 13·6
Aluminum and bromine combine to form aluminum bromide. Why do three bromine atoms combine with each aluminum atom?

Bromine, Br₂
Aluminum, Al
Aluminum bromide, AlBr₃

Example 1

Use electron dot formulas to predict the formulas of the ionic compounds formed from these elements.
a. potassium and oxygen
b. magnesium and nitrogen

Solution

Write down the correct electron dot formulas of each atom in the compound. Atoms of metals lose their valence electrons when forming an ionic compound. Atoms of nonmetals gain electrons to have an electron configuration like a noble gas (usually eight electrons). Enough atoms of each element must be used in the formula so that electrons lost equals electrons gained.

a. Start with the atoms: K· and ·Ö:

Oxygen needs two electrons. These electrons can come from two potassium atoms.

$$\begin{matrix} \text{K}\cdot \\ \text{K}\cdot \end{matrix} \quad + \quad \cdot\overset{..}{\text{O}}\colon \longrightarrow \quad \begin{matrix} \text{K}^+ \\ \text{K}^+ \end{matrix} \quad \colon\overset{..}{\underset{..}{\text{O}}}\colon^{2-}$$

Electrons *lost* now equals electrons *gained*. The formula of potassium oxide is K₂O.

b. Start with the atoms: $\dot{M}g$ and $\cdot\dot{N}:$

Nitrogen needs three electrons, but each magnesium atom can lose only two. Thus we need three magnesium atoms for every two nitrogen atoms.

$$
\begin{array}{ccccc}
\dot{M}g & & & Mg^{2+} & \\
\dot{M}g & + & \cdot\dot{N}: & \longrightarrow & Mg^{2+} & :\ddot{N}:^{3-} \\
 & & \cdot\dot{N}: & & Mg^{2+} & :\ddot{N}:^{3-} \\
\dot{M}g & & & Mg^{2+} &
\end{array}
$$

The formula of magnesium nitride is Mg_3N_2.

Problems

5. Use electron dot formulas to determine chemical formulas of the ionic compounds formed when the following elements combine.
   a. potassium and iodine   **c.** aluminum and oxygen
   b. calcium and sulfur   **d.** sodium and phosphorus

6. Name the compounds formed in Problem 5.

7. Use ionic charges to write chemical formulas of the compounds named in Problem 5.

Figure 13·7
The meat-packing industry was revolutionized when Lloyd Hall and others developed new types of curing salts.

Issues in Chemistry
What criteria should be used to determine if a food additive is safe?

━━━ Science, Technology, and Society ━━━

13·A   The Mixed Blessing of Curing Salts

Curing salts are ionic substances used to preserve meat. For centuries people have packed meat and fish in sodium chloride or soaked it in a salt solution to prevent bacterial growth. Many recipes for curing salts and pickling solutions used today are the work of Lloyd Hall. At one time he held more than 25 patents for food processing and packaging methods, which he developed as a research chemist.

One controversial substance used in meat processing is sodium nitrite. The nitrates and nitrites of sodium and potassium are used to help preserve ham, bacon, frankfurters, sausage, and other cured meats. These salts are used in small amounts along with sodium chloride. The nitrate ion is easily converted by enzymes or bacteria into nitrite. This ion prevents the growth of the bacteria that cause botulism, an extremely dangerous form of food poisoning. Nitrate and nitrite also enhance the flavor of meat and give it a pink color.

At high temperatures, nitrite ions react with amines, substances present in all meat. The product of this reaction is a group of chemicals called nitrosamines. These compounds have been found to cause cancer in every species of animal in which they have been tested. In addition, in high concentrations they can cause mutations and birth defects. Nitrosamines can form if meat containing nitrites is cooked at a high temperature, such as when bacon is fried.

Figure 13·8
Curing salts are used to prevent bacterial growth in meat.

An ionic compound is characterized by a high-melting point, conductivity of electricity in the molten state, existence in a crystalline form, and tendency to dissolve in water.

Figure 13·9
Cesium chloride and sodium chloride both form clear cubic crystals. Nevertheless their internal crystal structures differ. The unit cell of sodium chloride is face-centered cubic. By contrast the unit cell of cesium chloride is body-centered cubic. How many chlorine atoms surround each sodium atom? How many chlorine atoms surround each cesium atom?

No conclusive evidence exists that nitrites themselves cause cancer. They can react with the hemoglobin in blood, however, to reduce the capacity of blood to carry oxygen. This condition is particularly serious in babies and unborn children.

Cured meats are not the only source of nitrite we encounter. Saliva contains nitrite, and nitrate that is present naturally in food and water is converted to nitrite in the intestines. Thus it is advisable to control nitrite consumption wherever possible. Some experts feel, however, that a total ban on the use of nitrites in cured meats could cause an outbreak of botulism. The amount of nitrite that can be used in foods is restricted by government regulations. Scientists are also looking for ways to inhibit the formation of nitrosoamines.

13·5 Properties of Ionic Compounds

Most ionic compounds are crystalline solids at room temperature. *The component molecules, atoms, or ions of a crystal are arranged in repeating three-dimensional patterns.* The composition of a crystal of sodium chloride is typical. In solid NaCl each sodium ion is surrounded by six chloride ions and each chloride ion is surrounded by six sodium ions. In this arrangement each ion is attracted strongly to each of its neighbors and repulsions are minimized. The large attractive forces result in a very stable structure. This is reflected in the fact that NaCl and ionic compounds in general have very high melting temperatures.

The three dimensional arrangement of ions in NaCl is shown in Figure 13·9. Because each Na^+ ion is surrounded by six Cl^- ions, it has a coordination number of 6. *The* **coordination number** *gives the number of ions of opposite charge that surround each ion in a crystal.* Each Cl^- ion is surrounded by six Na^+ ions and also has a coordination number of 6.

Cesium chloride (CsCl) has a formula unit that is similar to NaCl. Both these compounds have cubic crystals but their internal crystal structures are different. Each Cs^+ ion is surrounded by eight Cl^- ions, and each Cl^- ion is surrounded by eight Cs^+ ions. Both the anion and cation in cesium chloride have a coordination number of 8.

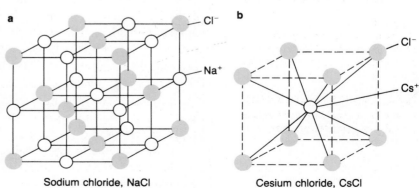

Sodium chloride, NaCl   Cesium chloride, CsCl

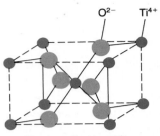

Titanium dioxide, TiO$_2$

Figure 13·10
The shape of a crystal depends on its internal structure. Crystals of the mineral rutile (titanium dioxide) are tetragonal.

Rutile is the crystalline form of titanium dioxide (TiO$_2$). In this compound the coordination number for the cation (Ti^{4+}) is 6. Each Ti^{4+} ion is surrounded by six O^{2-} ions. The coordination number of the anion (O^{2-}) is 3. Each O^{2-} ion is surrounded by three Ti^{4+} ions. The crystals are tetragonal instead of cubic (Figure 13·10). The internal structures of crystals are determined using a technique known as X-ray diffraction crystallography. Patterns are formed when X-rays pass through a crystal onto X-ray film. These patterns are used to calculate the positions of ions in the crystal (Figure 13·14).

Ionic compounds conduct an electric current when in the molten state. For example, when sodium chloride is melted (melting point 800°C) the orderly crystal structure breaks down. Each ion is then free to move throughout the molten mass. If a voltage is applied across this melt, cations will migrate to one electrode and anions will migrate to the other. This movement of ions means that there is a flow of electricity between the two electrodes. It is a characteristic property of ionic compounds that they conduct an electric current when melted. Ionic compounds also produce electrical conductivity if they can be dissolved in water. When such compounds dissolve, their ions are free to move about.

Figure 13·11
When sodium chloride melts, the sodium and chloride ions move freely through the molten salt. If a voltage is applied, the sodium ions will migrate to the negatively charged electrode (the cathode), and the chloride ions will migrate to the positively charged electrode (the anode).

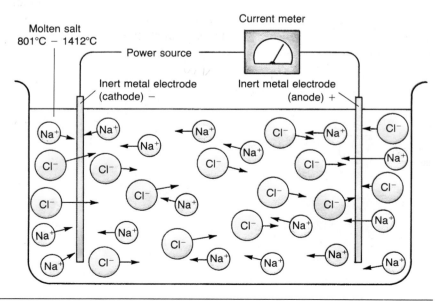

Metals are crystalline, ductile, malleable, and good conductors of electricity.

Safety

Metals are excellent conductors of heat. When heating glassware, position clamps or iron rings so that they will not be heated by the flame. Be careful not to burn yourself when disassembling the equipment.

Figure 13·13
a A metal rod can be forced through a narrow opening in a *die* to produce wire. **b** As this occurs, the metal changes shape but remains in one piece. **c** If an ionic crystal were forced through the die it would shatter. Why?

13·6 Metallic Bonds

Scientists believe that a piece of pure metal such as copper or iron consists not of metal atoms but of closely packed cations. The cations are surrounded by mobile valence electrons that are free to drift from one part of the metal to another. **Metallic bonds** *consist of the attraction of the free-floating valence electrons for the positively charged metal ions.* These are the forces of attraction that hold metals together.

Scientists have accounted for many physical properties of metals on the basis of this picture of metallic bonding. *Metals are good conductors of electrical current* (flow of electrons) *because as electrons enter one end of a bar of metal, an equal number leave the other end.* Metals are also *malleable* in that they can be hammered into different shapes (Figure 13·12). They are *ductile* in that they can be drawn into wires (Figure 13·13). The electron mobility theory says that the metal cations are insulated from one another by electrons. When a metal is subjected to pressure, the metal cations easily slide past one another like ball bearings immersed in oil. In contrast, if an ionic crystal is struck with a hammer, the blow tends to push ions of like charge into contact. They repel, and the crystal shatters (Figure 13·13).

b Metal

c Ionic crystal

a

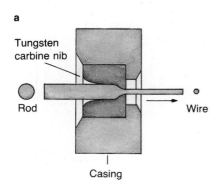

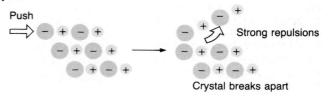

Figure 13·14
The techniques of X-ray diffraction and electron diffraction are used to determine the arrangement of atoms within a metal. This electron diffraction pattern is for stainless steel 306. Calculations based on the distances and angles within the pattern indicate that stainless steel has a body-centered cubic structure.

The atoms in metals are packed in several compact crystal forms:
1. body-centered cubic
2. face-centered cubic
3. hexagonal close-packed

ChemDirections

Ceramics are crystalline. To discover how the physical properties of materials depend upon the arrangement of their atoms read **ChemDirections** *Ceramics*, page 660.

Metals are among the simplest crystalline solids. Each metal contains just one kind of atom. Metal atoms are arranged in a very compact and orderly pattern, just as we might pack tennis balls in a box. Several arrangements give the closest packing possible for spheres of identical size, like metal atoms. These are body-centered cubic, face-centered cubic, and hexagonal close-packed (Figure 13·15).

In a *body-centered cubic* structure, every atom (except those on the surface) has 8 neighbors. The metallic elements sodium, potassium, iron, chromium, and tungsten crystallize in the body-centered cubic pattern.

In a *face-centered cubic* arrangement, every atom has 12 neighbors. Among the metals that form a face-centered cubic lattice are copper, silver, gold, aluminum, and lead.

In a *hexagonal close-packed* arrangement, every atom has 12 neighbors. The pattern is different, however, from the face-centered cubic. Metals that have this crystal structure include magnesium, zinc, and cadmium.

Figure 13·15
The three most common arrangements for atoms in metals are body-centered cubic (**a**) face-centered cubic (**b**) and hexagonal close-packed (**c**).

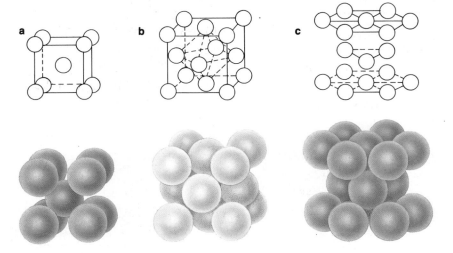

The properties of alloys are different from those of pure metals.

Figure 13·16
a The component metals of substitutional alloys have atoms of similar sizes.
b The component elements of an interstitial alloy have atoms of very different sizes.

Science, Technology, and Society

13·B Alloys

Solid solutions made by dissolving metals in other metals are called alloys. They are prepared by melting the metals together and cooling the mixture. If the atoms of the metals are about the same size, they can replace each other in the metal crystals. This type of mixture is called a substitutional alloy. Sterling silver, an alloy of copper and silver, is an example of a substitutional alloy.

Sometimes the atomic sizes of the metals in the alloy are quite different. Then the smaller atoms can fit into the interstices, the spaces between the larger atoms. This is called an interstitial alloy. In the various types of steel, for example, carbon atoms occupy the spaces between the iron atoms.

Alloys of mercury are called amalgams. They may be liquids or solids. Dental amalgam, used for filling cavities in teeth, is an alloy of mercury, silver, and zinc. It is a liquid when first prepared but hardens quickly.

The properties of alloys differ from those of their component metals. Sterling silver is harder than pure silver but still soft enough to be made into jewelry and tableware. Stainless steel is an alloy of iron, chromium, carbon, and nickel. This steel is stronger than iron and more resistant to corrosion. Table 13·2 gives the composition and use of some common alloys.

| a | | | Metal A | Metal B | Substitutional alloy | b | | | Metal A | Metal B | Interstitial alloy |

Table 13·2	Composition of Some Common Alloys	
Name	Composition (by mass)	Use
Sterling silver	Ag 92.5%; Cu 7.5%	Jewelry, tableware
Coinage silver	Ag 90%; Cu 10%	Medals, coins
Plumber's solder	Pb 67%; Sn 33%	Soldering joints
Pewter	Sn 85%; Cu 7.3%; Bi 6%; Sb 1.7%	Tableware
Brass	Cu 60%; Zn 39%; Sn 1%	Gears
Steel	Fe 99%; C 1%	Structural material
Cast iron	Fe 96%; C 4%	Castings
Stainless steel	Fe 80.6%; Cr 18%; C 0.4%; Ni 1%	Cutlery
Spring steel	Fe 98.6%; Cr 1%; C 0.4%	Cutting tools, springs
Surgical steel	Fe 67%; Cr 18%; Ni 12%; Mo 3%	Skeletal implants
18-Carat gold	Au 75%; Ag/Cu 25%	Gold jewelry
Duraluminum	Al 94.5%; Cu 5%; Mg 0.5%	Aircraft parts

13 Ionic Bonds
Chapter Review

Key Terms

coordination number	ionic bond	13·4
13·5	metallic bond	13·6
electron dot structure	octet rule	13·2
13·1	valence electron	13·1
halide ion 13·3	ion	

Chapter Summary

Atoms are held together in compounds by chemical bonds. Chemical bonds result from the sharing or transfer of valence electrons between pairs of atoms. Bonded atoms attain the stable electron configuration of a noble gas. The noble gases themselves exist as isolated atoms because that is their most stable condition. The transfer of one or more valence electrons between atoms produces positively and negatively charged ions: cations and anions.

The attraction between an anion and a cation is an ionic bond. A substance with ionic bonds is an ionic compound. Nearly all ionic compounds are crystalline solids at room temperature. They generally have high melting points. These solids consist of positive and negative ions packed in an orderly arrangement. The total positive charge is balanced by the total negative charge. Ionic compounds, also known as salts, are electrically neutral. Many ionic compounds contain polyatomic ions. The coordination number indicates the number of ions of opposite charge that surround each ion in a crystal.

Metals are like ionic compounds in some ways. They consist of positive metal ions packed together and surrounded by a sea of their valence electrons. This arrangement constitutes the metallic bond. The valence electrons are mobile and can travel from one end of a piece of metal to the other. This electron mobility accounts for the excellent electrical conductivity of metals and helps explain why metals are malleable and ductile. Metals are among the simplest crystalline solids. The metal atoms are commonly packed in a body-centered cubic, a face-centered cubic, or a hexagonal close-packed arrangement.

Practice Questions and Problems

8. Define the term *valence electrons*. $\qquad$ 13·1

9. Name the first four halogens. What group are they in, and how many valence electrons does each of them have? $\qquad$ 13·1

10. How many electrons does each of the following atoms have? What groups are they in? $\qquad$ 13·1
a. nitrogen $\qquad$ **c.** phosphorus
b. lithium $\qquad$ **d.** barium

11. Write electron dot structures for each of the following elements. $\qquad$ 13·1
a. Cl $\qquad$ **b.** S $\qquad$ **c.** Al $\qquad$ **d.** Li

12. The atoms of the noble gas elements are stable. Explain. $\qquad$ 13·2

13. How many electrons must be lost by each of the following atoms to attain a noble gas electron configuration? $\qquad$ 13·2
a. Ca $\qquad$ **b.** Al $\qquad$ **c.** Li $\qquad$ **d.** Ba

14. Write the formula for the ion you get when the following elements lose their valence electrons. 13·2
a. aluminum $\qquad$ **d.** potassium
b. lithium $\qquad$ **e.** calcium
c. barium $\qquad$ **f.** strontium

15. Write complete electron configurations for the following atoms and ions. For each group, comment on the results. $\qquad$ 13·2
a. Ar, K^+, Ca^{2+}
b. Ne, Na^+, Mg^{2+}, Al^{3+}

16. What are anions? How and why are anions produced? $\qquad$ 13·3

17. How many electrons must be gained by each of the following atoms to achieve a stable electron configuration? *13·3*
   a. N   **b.** S   **c.** Cl   **d.** P

18. Write the formula for the ion you get when each of the following elements gains electrons and attains a noble gas configuration. *13·3*
   a. Br   **b.** H   **c.** As   **d.** Se

19. Write electron configurations for the following and comment on the result. *13·3*
   a. N^{3-} **b.** O^{2-} **c.** F^- **d.** Ne

20. Why are ionic compounds electrically neutral? *13·4*

21. Which of the following pairs of elements are most likely to form ionic compounds? *13·4*
   a. chlorine and bromine
   b. potassium and helium
   c. lithium and fluorine
   d. iodine and sodium

22. Which pairs of elements will not form ionic compounds? *13·4*
   a. sulfur and oxygen
   b. sodium and calcium
   c. sodium and sulfur
   d. oxygen and chlorine

23. Write the correct chemical formula, the formula unit, for the following pairs of ions. *13·4*
   a. K^+, S^{2-} **c.** Na^+, SO_4^{2-}
   b. Ca^{2+}, O^{2-} **d.** Al^{3+}, PO_4^{3-}

24. Write the formula for the ions in the following compounds. *13·4*
   a. KCl   **c.** $MgBr_2$
   b. $BaSO_4$ **d.** Li_2CO_3

25. Write formulas for the following compounds.
   a. potassium nitrate *13·4*
   b. barium chloride
   c. magnesium sulfate
   d. lithium oxide
   e. ammonium carbonate
   f. calcium phosphate

26. In your own words describe a *crystal*. *13·5*

27. What determines the crystal structure of an ionic compound? *13·5*

28. Most ionic substances are brittle. Why? *13·5*

29. Why does molten $MgCl_2$ conduct an electric current although crystalline $MgCl_2$ does not? *13·5*

30. In your own words define a *metallic bond*. *13·6*

31. Describe what is meant by the terms *ductile* and *malleable*. *13·6*

32. Why is it possible to bend metals but not ionic crystals? *13·6*

33. Explain briefly why metals are good conductors of electricity. *13·6*

34. Name the three crystal arrangements that give close-packing of metal atoms. Give an example of a metal that crystallizes in each arrangement. *13·6*

Mastery Questions and Problems

35. Which of the following substances are most probably *not* ionic?
   a. H_2O **d.** CaS
   b. Na_2O **e.** NH_3
   c. CO_2 **f.** SO_2

36. Construct a table that shows the relationship among the group number, valence electrons lost or gained, and the formula of the cation or anion produced for the following metallic and nonmetallic elements. Na, Ca, Al, N, S, Br

37. Write electron dot formulas for the following atoms.
   a. C   **d.** F
   b. Be   **e.** Na
   c. O   **f.** P

38. Why does a cation have a positive charge?

39. Metallic cobalt crystallizes in a hexagonal close-packed structure. How many neighbors will a cobalt atom have?

40. Write electron configurations for the dipositive ions of these elements.
   a. Fe **b.** Co **c.** Ni

41. Why does an anion have a negative charge?

42. Some elements are found in nature mainly as their ions. Why?

43. Write electron configurations for these atoms and ions, and comment on the result.
   a. Ar   **c.** S^{2-}
   b. Cl^- **d.** P^{3-}

44. Show the relationship between the electron dot structure of an element and the location of the element in the periodic table.

45. Write electron configurations for the tripositive ions of these elements.
   a. chromium
   b. manganese
   c. iron

46. Explain how hexagonal close-packed, face-centered cubic, and body-centered cubic unit cells are different.

47. The spheres below represent the relative diameters of atoms or ions. Rearrange each of the following sequences so that the relative sizes of the particles in the sequence correspond to the increasing size of the particles in the illustration.
   a. oxygen atom, oxide ion, sulfur atom, sulfide ion
   b. sodium atom, sodium ion, potassium atom, potassium ion

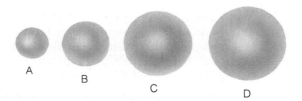

A   B   C   D

Critical Thinking Questions

48. Choose the term that best completes the second relationship.
   a. cow:horse   ionic bond: _____
      (1) metallic bond   (3) chemical bond
      (2) cation   (4) covalent bond
   b. Cl:Cl⁻   Mg: _____
      (1) Mg^{2+} (3) Al^{3+}
      (2) Mg^{2-} (4) Mn
   c. pipe:water   metal: _____
      (1) ions   (3) electricity
      (2) ionic bond (4) conductor

49. Compare the physical and chemical characteristics of metals and ionic compounds.

50. Which kind of metal alloy, interstitial or substitutional, is likely to have the greatest density?

Review Questions and Problems

51. Hydrogen and oxygen react to give water according to this equation.

$$2H_2 + O_2 \longrightarrow 2H_2O$$

How many liters of hydrogen at STP are needed to produce 0.50 mol of water?

52. Distinguish among gases, liquids, and solids with respect to shape, volume, relative density, and motion of particles.

53. What is the volume in liters occupied by 8.0 g of oxygen gas at STP?

54. If you raise the temperature of a gas, what actually happens to the gas particles?

55. A gas occupies 750 cm^3 at 27°C and 0.016 atm. Find its volume at STP.

56. Explain each of these observations on the basis of the kinetic theory and the forces of attraction that exist between the particles in matter.
   a. Water evaporates faster at 40°C than 20°C.
   b. A burn from steam at 100°C is worse than a burn from water at 100°C.
   c. A cap "pops" off a bottle of root beer that has been kept in the trunk of a car on a hot day.
   d. Diethyl ether, $C_4H_{10}O$, has a vapor pressure of 442 mm Hg at 20°C, whereas the vapor pressure of water at 20°C is only 17.5 mm Hg.
   e. A melting ice cube cools a glass of tea.
   f. Foods are dehydrated (water is removed) by using low pressures rather than high heat.

57. Use the gas laws and the kinetic theory to complete these statements. Unless otherwise stated, assume a constant amount of gas.
   a. As the volume of a gas increases at constant temperature, its pressure _____ .
   b. As the temperature of a gas increases and its pressure decreases, its volume _____ .
   c. At constant pressure, a decrease in the volume of a gas is caused by a/an _____ in its temperature.
   d. An increase in the volume and pressure of a gas is caused by a/an _____ in its temperature.
   e. At constant volume, a decrease in the temperature of a gas causes the pressure to _____ .
   f. If the volume of a gas is increased while its temperature is increased, the pressure will _____ .

58. A gas mixture contains 1.50 g of O_2 and 1.50 g of N_2. If the partial pressure of O_2 is 475 mm Hg, what is the total pressure of the mixture?

59. At what temperature will the average kinetic energy of a nitrogen molecule be twice the average kinetic energy of a nitrogen molecule at 25°C?

60. The reaction of zinc with an acid produces hydrogen gas. The table below gives the amount of hydrogen gas produced at STP when various amounts of zinc are reacted with an excess of hydrochloric acid.
  a. Calculate the moles of zinc used.
  b. Make a graph of the volume of hydrogen produced (y-axis) versus the moles of zinc reacting (x-axis).
  c. Determine the slope of the line. Include units on the slope.
  d. Based on the value of the slope, what can you surmise about the moles of hydrogen produced per mole of zinc that reacts?

Mass of Zn (g)	Moles of Zn	Volume of H_2(L)
0.960	_____	0.329
2.44	_____	0.835
4.18	_____	1.43
5.75	_____	1.96

Challenging Questions and Problems

61. The chemically similar alkali metal chlorides NaCl and CsCl have different crystal structures whereas the chemically different NaCl and MnS have the same crystal structures. Why?

62. Silver crystallizes in a face-centered cubic unit cell. A silver atom is at the edge of each lattice point. The length of the edge of the unit cell is 0.4086 nm. What is the atomic radius of silver?

63. Write electron dot structures for each of the following atoms.
  a. Pb
  b. Re
  c. Mo
  d. Rh
  e. Au
  f. Ag
  g. Rn
  h. Ho

Research Projects

1. Grow samples of crystals. Test the effects of varying temperature, varying concentrations, and impurities.

2. Devise a way to compare the space occupied by an ion and by an atom of the same element.

3. What is the historical significance of salt (NaCl)?

4. Trace the history of food preservation.

5. What are some common chemical food additives? Why are some considered harmful in high concentrations?

6. Report on studies linking nitrosamines to cancer.

7. Describe the techniques used by geochemists to analyze rock samples. How do other scientists use this information?

8. Are alloys mixtures or compounds? Devise experiments to test your answer.

9. Construct a model for each type of unit cell, and find a chemical example of each.

10. What alloys are used in structural materials? Why are they used?

Readings and References

Freydberg, Nicholas, and Willis A. Gortner. *The Food Additives Book*. New York: Bantam Books, 1982.

Hansen, James. "Metals that Remember." *Science 81* (June 1981), pp. 44–47.

Packard, Vernal S., Jr. *Processed Foods and the Consumer*. Minneapolis: University of Minnesota Press, 1976.

Sanders, Kevin. "Muscle Metal." *Discover* (October 1981), pp. 92–96.

Stine, William R. *Chemistry for the Consumer*. Boston: Allyn and Bacon, 1978.

Winter, Ruth. *A Consumer's Dictionary of Food Additives*. New York: Crown, 1978.

14 Covalent Bonds

Chapter Preview

14·1 Single Covalent Bonds
14·2 Double and Triple Covalent Bonds
14·3 Covalent Compounds
14·4 Coordinate Covalent Bonds
14·5 Resonance
14·6 Exceptions to the Octet Rule
14·7 Molecular Orbitals
14·A Molecular Models
14·8 VSEPR Theory

14·9 Hybrid Orbitals
14·10 Polar Bonds
14·11 Polar Molecules
14·12 Bond Dissociation Energies
14·B Johannes van der Waals: Using the Power of Reason
14·13 Intermolecular Attractions
14·C Bonding with Adhesives
14·14 Properties of Molecular Substances

The previous chapter showed how ions are formed from atoms by the gain or loss of electrons. Electrostatic attractions between ions result in ionic bonds. This chapter covers the sharing of electrons between atoms to form a different kind of bond, the covalent bond.

14·1 Single Covalent Bonds

Atoms share electrons in covalent bonds.

In the previous chapter the Lewis theory of bonding was applied to the formation of cations, anions, and ionic bonds. Another part of the Lewis theory holds that some atoms share electrons to attain noble-gas electron configurations. Covalent bonds are the result of electron-sharing between atoms. Atoms of hydrogen and the nonmetallic elements in Groups 4A, 5A, 6A, and 7A of the periodic table are particularly prone to form covalent bonds. This tendency is summarized in the octet rule for covalent bonding: *The sharing of electrons occurs when the atoms involved can thus acquire particularly stable electron configurations.* Often the configurations contain eight valence electrons (an octet). The notable exception is hydrogen. When hydrogen shares two electrons it acquires the stable electron configuration of helium.

Hydrogen is the simplest molecule. A hydrogen atom has a single valence electron. Pairs of hydrogen atoms share electrons to form simple

Figure 14·1
Chemists use molecular models to help explain the behavior of substances. This is a ball-and-stick model of isopropyl alcohol (rubbing alcohol).

317

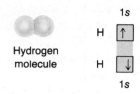

The quantum mechanical model of the atom was not yet developed when Lewis developed his theory of bonding. Thus Lewis did not use electron configurations to show the electrons that are involved in bonding. The electron configurations are included here as an alternate way of showing which electrons are involved in bonds.

diatomic molecules. Each hydrogen gets the stable electron configuration of helium, which has two valence electrons. *A **single covalent bond** is formed when a pair of electrons is shared between two atoms.* The chemical formula for the hydrogen molecule is therefore H_2.

$$H\cdot \quad + \quad \cdot H \quad \longrightarrow \quad H : H \quad \text{shared pair of electrons}$$

Hydrogen atom Hydrogen atom Hydrogen molecule

Sometimes it is helpful to show the pair of electrons in a covalent bond as a dash, as in H—H for hydrogen. Notations of this type are structural formulas. **Structural formulas** *are chemical formulas that show the arrangement of atoms in molecules and polyatomic ions.* The dashes between the atoms in structural formulas always indicate a pair of shared electrons. The dashes are never used to show ionic bonds.

The formula of hydrogen, H_2, brings up a difference between the chemical formulas of ionic and covalent compounds. The chemical formulas of ionic compounds are correctly described only as *formula units*. The chemical formulas of covalent compounds are correctly described only as *molecular formulas*. Ionic compounds do not have molecular formulas because they do not contain molecules. The formula unit of copper(II) oxide is CuO and not Cu_2O_2 because CuO represents the smallest electrically neutral sample that is representative of the composition of

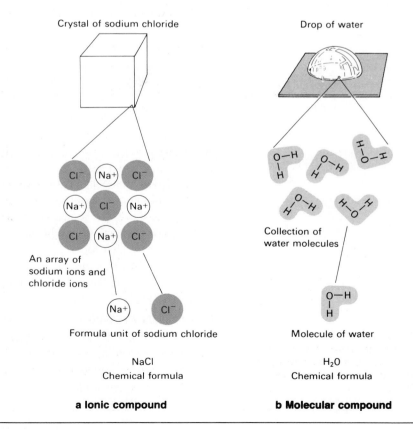

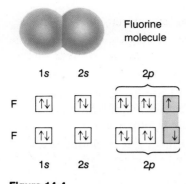

Fluorine molecule

1s   2s   2p

F $\uparrow\downarrow$ $\uparrow\downarrow$ $\uparrow\downarrow$ $\uparrow\downarrow$ $\uparrow$

F $\uparrow\downarrow$ $\uparrow\downarrow$ $\uparrow\downarrow$ $\uparrow\downarrow$ $\downarrow$

1s   2s   2p

Figure 14·4
The two bonding electrons in the fluorine molecule come from the $2p$ atomic orbitals of the fluorine atoms.

copper(II) oxide. In a similar way, only one pair of shoes, not two, is needed to show someone what a pair of shoes is. On the other hand, a single hydrogen molecule actually contains two hydrogen atoms bonded together. Writing the molecular formula of hydrogen as H would be similar to showing one shoe to someone who asks to see a pair. Sometimes the correct formulas of molecular compounds have subscripts that are multiples of the lowest whole number ratio, as in C_3H_6 or C_4H_{10}. Those of ionic compounds seldom do. Figure 14·3 illustrates the essential difference between ionic compounds and covalent compounds.

The halogens also form single covalent bonds in their diatomic molecules. Fluorine is one example, and the other halogens are similar. Because a fluorine atom has seven valence electrons, it needs one more to attain the electron configuration of neon. By electron-sharing, two fluorine atoms gain stability. By this process they form a single covalent bond.

$$:\!\ddot{F}\!\cdot \;+\; \cdot\!\ddot{F}\!: \;\longrightarrow\; :\!\ddot{F}\!:\!\ddot{F}\!: \quad or \quad :\!\ddot{F}\!-\!\ddot{F}\!:$$

Fluorine atom · Fluorine atom · Fluorine molecule

Notice that each fluorine nucleus shares only one pair of valence electrons. *The pairs of valence electrons that are not shared between atoms are called unshared pairs of electrons, or* **unshared pairs.** They are also called *lone pairs,* or *nonbonding pairs*.

Problem

1. Draw electron dot structures for these diatomic halogen molecules.
   a. chlorine **b.** bromine **c.** iodine

Some pairs of atoms share two or three pairs of electrons.

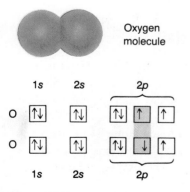

Oxygen molecule

1s   2s   2p

O $\uparrow\downarrow$ $\uparrow\downarrow$ $\uparrow\downarrow$ $\uparrow$ $\uparrow$

O $\uparrow\downarrow$ $\uparrow\downarrow$ $\uparrow\downarrow$ $\downarrow$ $\uparrow$

1s   2s   2p

Figure 14·5
The oxygen molecule is an exception to the octet rule. It forms only a single bond, and it has two unpaired electrons.

14·2   Double and Triple Covalent Bonds

Atoms can sometimes share more than one pair of electrons to attain stable, noble-gas electron configurations. **Double covalent bonds** *involve two shared pairs of electrons.* **Triple covalent bonds** *include three shared pairs of electrons.*

Oxygen, O_2, is an example of a molecule that should have a double covalent bond according to the octet rule. An oxygen atom, with six valence electrons, could share two of these electrons with another oxygen atom to form the double bond.

$$:\!\ddot{O}\!: \;+\; :\!\ddot{O}\!: \;\longrightarrow\; :\!\ddot{O}\!:\!:\!\ddot{O}\!: \quad or \quad :\!\ddot{O}\!=\!\ddot{O}\!:$$

Oxygen atom · Oxygen atom · Oxygen molecule

Experimental evidence indicates, however, that the electrons in the oxygen molecule are not all paired (see Section 14·6). Thus oxygen represents an exception to the octet rule.

The nitrogen molecule contains a triple covalent bond. Each nitrogen atom has five valence electrons. They each need three more electrons to

Table 14·1	The Diatomic Elements		
Name	Chemical formula	Structure	Properties and Uses
Fluorine	F_2	$:\ddot{F}-\ddot{F}:$	Greenish-yellow reactive toxic gas. Compounds of fluorine, a halogen, are added to drinking water and toothpaste to promote healthy teeth.
Chlorine	Cl_2	$:\ddot{Cl}-\ddot{Cl}:$	Greenish-yellow reactive toxic gas. Chlorine is a halogen used in household bleaching agents.
Bromine	Br_2	$:\ddot{Br}-\ddot{Br}:$	Dense red-brown liquid with pungent odor. Compounds of bromine are used in the preparation of photographic emulsions.
Iodine	I_2	$:\ddot{I}-\ddot{I}:$	Dense gray-black solid that produces purple vapors; a halogen. A solution of iodine in alcohol (tincture of iodine) is used as an antiseptic.
Hydrogen	H_2	$H-H$	Colorless, odorless, tasteless gas. Hydrogen is the lightest known element.
Nitrogen	N_2	$:N\equiv N:$	Colorless, odorless, tasteless gas. Air is 80% nitrogen by volume.
Oxygen	O_2	$:\ddot{O}-\ddot{O}:$	Colorless, odorless, tasteless gas that is vital for life. Air is 20% oxygen by volume.

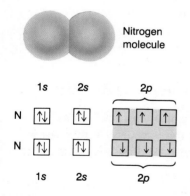

Figure 14·6
Three pairs of electrons are shared in a nitrogen molecule.

attain the electron configuration of neon. In the nitrogen molecule each nitrogen has one unshared pair of electrons.

$$:\dot{N}\cdot \;+\; \cdot\dot{N}: \longrightarrow :N::N: \quad or \quad :N\equiv N:$$

Nitrogen atom   Nitrogen atom   Nitrogen molecule

Up to this point, the examples of covalent bonds have all been in molecules of diatomic elements. Table 14·1 lists these elements and their properties.

14·3 Covalent Compounds

Electron dot formulas for molecules of compounds can be written in much the same way as for molecules of the diatomic elements. This will be shown with four examples: water, ammonia, methane, and carbon dioxide.

Water is a triatomic molecule. Two hydrogen atoms share electrons with one oxygen atom. The hydrogen and oxygen atoms attain stable noble-gas configurations by electron-sharing. The oxygen atom in water has two unshared pairs of valence electrons.

$$2H\cdot \;+\; :\dot{O}\cdot \longrightarrow :\ddot{O}:H \quad or \quad :\ddot{O}-H$$

Hydrogen atoms   Oxygen atom   Water molecule

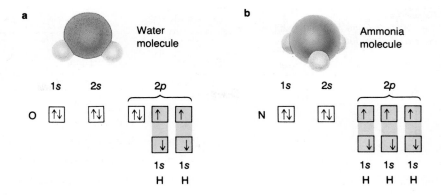

Figure 14·7
a In the water molecule two hydrogen atoms form single covalent bonds with an oxygen atom.
b In the ammonia molecule three hydrogen atoms form single covalent bonds with a nitrogen atom.

Covalent molecules share electrons, thus obtaining stable noble gas configurations.

Ammonia, a suffocating gas, is formed in a similar way. The ammonia molecule has one unshared pair of electrons.

$$3\text{H}\cdot \; + \; :\overset{\cdot}{\underset{\cdot}{\text{N}}}\cdot \; \longrightarrow \; :\overset{\cdot\cdot}{\underset{\text{H}}{\overset{\text{H}}{\text{N}}}}:\text{H} \quad or \quad :\text{N}\overset{\overset{\displaystyle\text{H}}{|}}{\underset{\underset{\displaystyle\text{H}}{|}}{}}\!\!\!-\text{H}$$

Hydrogen   Nitrogen   Ammonia molecule
atoms   atom

The carbon atom has four valence electrons and needs four more to attain a noble-gas configuration. The methane molecule contains four hydrogen atoms that each share one electron with carbon. In this way four identical carbon–hydrogen bonds are formed.

$$4\text{H}\cdot \; + \; \cdot\overset{\cdot}{\underset{\cdot}{\text{C}}}\cdot \; \longrightarrow \; \text{H}:\overset{\text{H}}{\underset{\text{H}}{\text{C}}}:\text{H} \quad or \quad \text{H}-\overset{\overset{\displaystyle\text{H}}{|}}{\underset{\underset{\displaystyle\text{H}}{|}}{\text{C}}}-\text{H}$$

Hydrogen   Carbon   Methane molecule
atoms   atom

When carbon bonds with other atoms it usually forms four bonds. You might not have predicted this from the electron configuration of atomic carbon.

$$\underset{1s^2}{\boxed{\uparrow\downarrow}} \quad \underset{2s^2}{\boxed{\uparrow\downarrow}} \quad \underbrace{\boxed{\uparrow}\;\boxed{\uparrow}\;\boxed{}}_{2p^2}$$

Any attempt to generate covalent C—H bonds for methane by combining the two $2p$ electrons of the carbon with two $1s$ electrons of hydrogen atoms would produce a molecule with the formula CH_2. Yet carbon tends to form four bonds to other atoms. This can be explained if a small amount of energy is used to promote one $2s$ electron to the vacant $2p$ orbital.

$$\underset{1s^2}{\boxed{\uparrow\downarrow}} \quad \underbrace{\boxed{\uparrow}\;\boxed{\uparrow}\;\boxed{\uparrow}\;\boxed{\uparrow}}_{2s \text{ and } 2p}$$

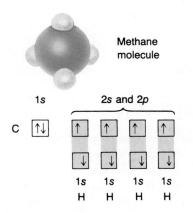

Figure 14·8
The methane molecule has four carbon-hydrogen bonds. In each bond the carbon and a hydrogen share a 1s electron from hydrogen and an electron from carbon.

The electron promotion provides four electrons of carbon capable of entering into covalent bonds with four hydrogen atoms (Figure 14·8).

Figure 14·9
The carbon dioxide molecule has two
carbon–oxygen double bonds.

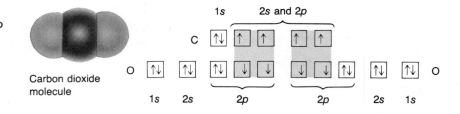

Carbon dioxide
molecule

The carbon dioxide molecule contains two oxygens that each share two electrons with carbon to form two carbon–oxygen double bonds.

$$:\ddot{O}. + .\ddot{C}. + .\ddot{O}: \longrightarrow :\ddot{O}::C::\ddot{O}: \quad or \quad :\ddot{O}=C=\ddot{O}:$$

Oxygen   Carbon   Oxygen   Carbon dioxide
atom   atom   atom   molecule

Example 1

Hydrogen chloride (HCl) is a diatomic molecule with a single covalent bond. Draw the electron dot structure for HCl.

Solution

To form a single covalent bond, hydrogen and chlorine atoms must share a pair of electrons. First, write electron dot structures for the two atoms. Then show the electron-sharing.

$$H· \quad + \quad ·\ddot{Cl}: \quad \longrightarrow \quad H:\ddot{Cl}:$$

Hydrogen   Chlorine   Hydrogen chloride
atom   atom   molecule

Through electron-sharing, hydrogen and chlorine atoms attain the electron configurations of the noble gases helium and argon.

Problem

2. Draw electron dot structures for the following covalent molecules which have only single covalent bonds.
   a. H_2S **b.** PH_3 **c.** ClF

14·4   Coordinate Covalent Bonds

When one atom contributes both bonding electrons in a covalent bond, a **coordinate covalent bond** *is formed.* An example is carbon monoxide. A carbon atom is four electrons short of the electron configuration of neon. An oxygen atom is two electrons short. Yet it is possible for both atoms to achieve stable electron configurations by coordinate covalent bonding. To see how, begin by making a double covalent bond between carbon and oxygen.

$$:\ddot{C}. + .\ddot{O}: \longrightarrow :C::\ddot{O}:$$

Carbon   Oxygen
atom   atom

In a coordinate covalent bond one atom contributes both bonding electrons.

Safety

Carbon monoxide is a colorless, odorless, toxic gas. It may be formed when fuels burn in a limited supply of oxygen. Do not burn a fuel in an enclosed area.

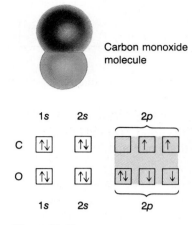

Carbon monoxide
molecule

	1s	2s	2p		
C	↑↓	↑↓		↑	↑
O	↑↓	↑↓	↑↓	↓	↓
	1s	2s	2p		

Figure 14·10
In a coordinate covalent compound such as is formed by carbon monoxide, one atom contributes both electrons of a bonding pair.

With the double bond in place, the oxygen atom has a stable configuration but the carbon atom does not. The dilemma is solved if the oxygen also donates one of its unshared pairs of electrons to make a coordinate covalent bond.

$$:C::O \longrightarrow :C::O:$$

Carbon monoxide
molecule

Coordinate covalent bonds are shown in structural formulas as arrows. They point from the atom donating the pair of electrons to the atom receiving them. The structural formula of carbon monoxide, with two covalent bonds and one coordinate covalent bond, is $C≡O$. Coordinate covalent bonds are often shown by arrows. It is important to realize though, that once formed, a coordinate covalent bond is like any other covalent bond. The only difference is the source of the bonding electrons. The polyatomic ammonium ion (NH_4^+) has a coordinate covalent bond. It is formed when a hydrogen ion is attracted to the unshared electron pair of an ammonia molecule.

Unshared electron pair

$$H^+ + :N:H \longrightarrow \left[H:N:H \right]^+ \quad or \quad H \leftarrow N^+ - H$$

Hydrogen   Ammonia   Ammonium
ion   molecule   ion (NH_4^+)
(proton)   (NH_3)

Most polyatomic cations and anions contain covalent and coordinate covalent bonds. Table 14·2 lists some common covalent compounds.

Example 2

The polyatomic hydronium ion (H_3O^+) contains a coordinate covalent bond. It forms when a hydrogen ion is attracted to an unshared electron pair of a water molecule. Write the electron dot structure for the hydronium ion.

Solution

$$H^+ + :O:H \longrightarrow \left[H:O:H \right]^+ \quad or \quad H \leftarrow O - H$$

Hydrogen   Water   Hydronium
ion   molecule   ion
(proton)   (H_2O)   (H_3O^+)

Many polyatomic ions have a negative charge. Because the atoms in polyatomic ions are covalently bonded, electron dot structures can be written for these ions. The negative charge of a polyatomic ion shows the number of electrons *in addition* to the valence electrons of the atoms

Figure 14·11
The polyatomic ammonium ion is an important component of fertilizer.

Table 14·2 Some Common Covalent Compounds

Name	Chemical formula	Structure	Properties and Uses
Carbon monoxide	CO		Colorless, highly toxic gas. It is a major air pollutant present in cigarette smoke and automobile exhaust.
Carbon dioxide	CO_2		Colorless unreactive gas. This normal component of the atmosphere is exhaled in the breath of animals and is essential for plant growth.
Water	H_2O		Colorless, odorless, tasteless liquid with melting point of 0°C and boiling point of 100°C. The human body is approximately 60% water.
Hydrogen peroxide	H_2O_2		Colorless, unstable liquid when pure. It is used as rocket fuel. A 3% solution is used as a bleach and antiseptic.
Sulfur dioxide	SO_2		Oxides of sulfur are produced in combustion of petroleum products and coal. They are major air pollutants in industrial areas. Oxides of sulfur can lead to respiratory problems.
Sulfur trioxide	SO_3		
Ammonia	NH_3		Colorless gas with pungent odor; extremely soluble in water. Household ammonia is a solution of ammonia in water.
Nitric oxide	NO		Oxides of nitrogen are major air pollutants produced by the combustion of fossil fuels in automobile engines. They irritate the eyes, throat, and lungs. Nitrogen dioxide, a dark brown gas, readily converts to colorless dinitrogen tetroxide.
Nitrogen dioxide	NO_2		
Dinitrogen tetroxide	N_2O_4		
Nitrous oxide	N_2O		Colorless sweet-smelling gas. It is used as an anesthetic commonly called laughing gas.
Hydrogen cyanide	HCN		Colorless toxic gas with the smell of almonds.
Hydrogen fluoride	HF		Four hydrogen halides, all extremely soluble in water. Hydrogen chloride, a colorless gas with pungent odor, readily dissolves in water to give a solution called hydrochloric acid.
Hydrogen chloride	HCl		
Hydrogen bromide	HBr		
Hydrogen iodide	HI		

present. Because a polyatomic ion is found as part of an ionic compound, these additional electrons are balanced by the positive charge of the cation of the compound.

The hydroxide ion is OH^-. The components of this ion are $H\cdot$, $\cdot\ddot{O}\cdot$, and an electron ($\cdot$). The additional electron is signified by the $1-$ charge. The stable electron dot formula is $[H\!:\!\ddot{O}\!:]^{1-}$.

Example 3

Draw the electron dot structure for sulfite, SO_3^{2-}, where sulfur is the central atom.

Solution

Start with the atoms and their valence electrons and the two "extra" electrons indicated by the charge.

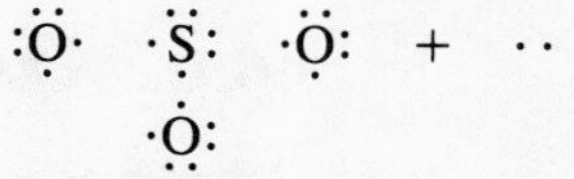

$$:\ddot{O}\cdot \quad \cdot\ddot{S}: \quad \cdot\ddot{O}: \quad + \quad \cdot\cdot$$
$$\cdot\ddot{O}:$$

Join two of the oxygens to sulfur by single covalent bonds.

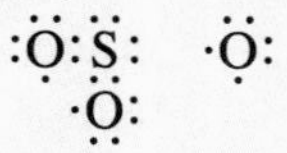

$$:\ddot{O}\!:\!\ddot{S}: \quad \cdot\ddot{O}:$$
$$\cdot\ddot{O}:$$

Join the remaining oxygen by a coordinate covalent bond, and add the "extra" two electrons.

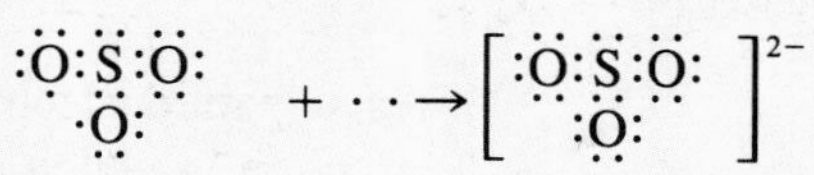

$$:\ddot{O}\!:\!\ddot{S}\!:\!\ddot{O}: \quad + \cdot\cdot \rightarrow \left[\, :\ddot{O}\!:\!\ddot{S}\!:\!\ddot{O}: \atop :\ddot{O}: \,\right]^{2-}$$

Problem

3. Draw the electron dot structures for sulfate, SO_4^{2-}, and carbonate, CO_3^{2-}. Sulfur and carbon are the central atoms, respectively.

14·5 Resonance

Consider the two electron dot structures for ozone shown below. Notice that the structure on the left can be converted to the one on the right by shifting electron pairs *without changing the positions of the oxygen atoms*.

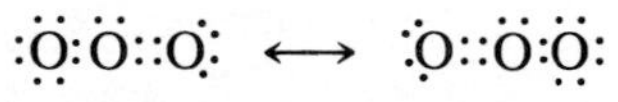

$$:\ddot{O}\!:\!\ddot{O}\!:\!:\!\ddot{O}: \quad \longleftrightarrow \quad :\ddot{O}\!:\!:\!\ddot{O}\!:\!\ddot{O}:$$

The structures suggest that the bonding in ozone consists of one coordinate covalent bond and one double covalent bond. Double covalent bonds are usually shorter than single covalent bonds. Experimental measurements show, however, that the two bonds in ozone are the same length. This result can be explained if the actual bonding in the ozone molecule is the average of the two electron dot structures.

A resonance structure cannot be adequately represented by a single electron dot diagram.

Figure 14·12
Resonance is similar to the mixing of two colors. The result is intermediate between the two extremes and different from each of them.

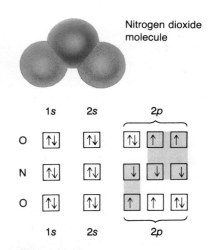

Nitrogen dioxide molecule

Figure 14·13
The nitrogen dioxide molecule is unusual in that it contains an unpaired electron.

Resonance *occurs when two or more equally valid electron dot structures can be written for a molecule.* How resonance is envisioned has changed somewhat with time. Earlier chemists imagined the electron pairs rapidly flipping back and forth, or resonating, between the various electron dot structures. Thus double-headed arrows are used to indicate that two or more structures are in resonance. A more modern interpretation is that electron pairs do not resonate. The actual resonance bonding is considered to be a hybrid, or mixture, of the extremes represented by the resonance forms. A mule is an example of a hybrid animal. The offspring of a donkey and a horse, a mule is a distinct creature with some characteristics of both parents. It is neither half-donkey nor half-horse, nor does it change continually between donkey and horse. Likewise, a resonance hybrid has some characteristics of its resonance forms, but it is a distinct species. The more resonance structures that can be drawn, the more stable is the ion or molecule.

14·6 Exceptions to the Octet Rule

Sometimes it is impossible to write electron dot structures that fulfill the octet rule. This happens whenever the total number of valence electrons in the species is an odd number. The NO_2 molecule, for example, contains a total of 17 valence electrons. Each oxygen contributes six electrons and the nitrogen contributes five. Two plausible resonance structures can be written for the NO_2 molecule.

$$\ddot{O} = \dot{N} - \ddot{O}\cdot \quad \longleftrightarrow \quad \cdot\ddot{O} - \dot{N} = \ddot{O}$$

An unpaired electron is present in each of these structures. No Lewis structure in which all of the atoms achieve an octet can be written.

Electrons may be considered as small, spinning, electric charges. These electric charges create magnetic fields, much as the current in an electric motor creates a magnetic field. Paired electrons can be thought of as having spins in opposite directions. The magnetic effect of paired electrons essentially cancels. Substances in which all of the electrons are paired are said to be *diamagnetic*. Diamagnetic substances are weakly repelled by an external magnetic field. In contrast, **paramagnetic** *substances show a relatively strong attraction to an external magnetic field.* These substances have molecules containing one or more unpaired electrons. Paramagnetism can be detected by measuring the mass of a substance in the absence and then in the presence of a magnetic field. The mass of the substance will appear to be greater in the magnetic field if the substance is paramagnetic (Figure 14·14).

Paramagnetism should not be confused with ferromagnetism. The latter is a much stronger attraction of iron, cobalt, and nickel for magnetic fields. The ions Fe^{2+}, Co^{2+}, and Ni^{2+} all have unpaired electrons. Large groups of ions are randomly dispersed throughout the metals. In a magnetic field these groups of ions line up in an orderly fashion with the field. This creates a strong magnetic attraction. The order remains even when the magnetic field is removed, leading to permanent magnetism.

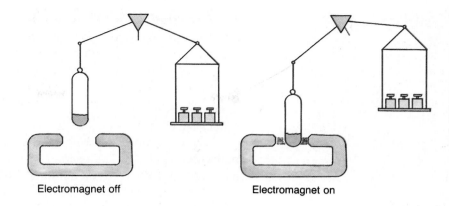

Figure 14·14
The mass of a paramagnetic substance appears to be greater in a magnetic field.

Electromagnet off   Electromagnet on

Some stable substances contain unpaired electrons or do not have an octet of electrons.

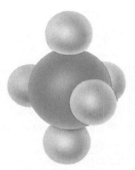

Phosphorus pentachloride

Sulfur hexafluoride

Figure 14·15
Phosphorus pentachloride and sulfur hexafluoride are exceptions to the octet rule.

A structure can be written for the oxygen molecule in which both oxygen atoms are surrounded by eight electrons. In this structure all the electrons are paired. We said in Section 14·2, however, that this structure is incorrect. The experimental evidence for this statement is that oxygen is paramagnetic. This property can be explained only if the oxygen molecule contains unpaired electrons. The measured distance between the oxygen atoms indicates that the oxygen molecule has some multiple bond character. This information suggests that oxygen is a resonance hybrid of these structures.

$$:\ddot{O}-\ddot{O}: \quad \longleftrightarrow \quad :\ddot{O}=\ddot{O}:$$

Several other molecules such as some compounds of boron also fail to follow the octet rule. The boron atom of boron trifluoride (BF_3), for example, is deficient by two electrons. Boron trifluoride readily reacts with ammonia to make the compound $BF_3 \cdot NH_3$. By doing so, the boron atom accepts the unshared electron pair from ammonia and completes the octet.

A few elements, especially phosphorus and sulfur, sometimes expand the octet to include ten or twelve electrons. Phosphorus trichloride (PCl_3) and phosphorus pentachloride (PCl_5) are stable compounds (Figure 14·15). In both compounds all of the chlorines are bonded to the phosphorus. Covalent bonding in PCl_3 follows the octet rule because all the atoms acquire eight electrons. An electron dot structure for PCl_5 can be written only if phosphorus has ten valence electrons. In sulfur hexafluoride (SF_6) the sulfur atom must have twelve valence electrons.

14·7 Molecular Orbitals

■ Molecular orbital theory is the logical extension to molecules of the quantum mechanical description of the atom.

There are other models for covalent bonding besides the one we have described. Just as there is a quantum mechanical description of the atom, there is also a quantum mechanical description of bonding. As discussed in Section 11·3, the quantum mechanical model describes the electrons in atoms by means of atomic orbitals. Similarly, quantum mechanics describes the electrons in molecules by *molecular orbitals*.

When two atoms combine, their atomic orbitals overlap to produce **molecular orbitals.** The overlap of two atomic orbitals produces two molecular orbitals. One is a **bonding orbital,** *a molecular orbital whose energy is lower than that of the atomic orbitals from which it is formed.* The other is an **antibonding orbital,** *a molecular orbital whose energy is higher than that of the atomic orbitals from which it is formed.*

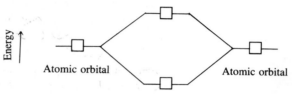

An atomic orbital belongs to a particular atom. A molecular orbital belongs to the molecule as a whole. The number of molecular orbitals is equal to the number of atomic orbitals which overlap.

Each atomic orbital describes at most two electrons. Thus an atomic orbital is half-filled if it contains one electron. It is filled if it contains two electrons. Similarly, two electrons are required to fill the energy level corresponding to a bonding or an antibonding molecular orbital.

These concepts can be used to explain bonding in the hydrogen molecule, H_2. The $1s$ atomic orbitals of two hydrogen atoms overlap in the formation of a hydrogen molecule. Two electrons, one from each hydrogen atom, are available. Since electrons seek the lowest energy, they fill the bonding molecular orbital energy level.

■ Do not confuse orbitals with electrons. An orbital, a region of space, has a predictable shape and energy regardless of whether it contains electrons.

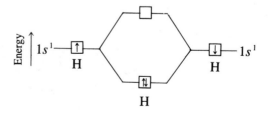

The energy of the electrons in the bonding orbital is lower than it is in separate hydrogen atoms. This makes the hydrogen molecule stable. Figure 14·16 shows the formation of the bonding and antibonding molecular orbitals in the hydrogen molecule.

Figure 14·16
The overlap of *s* orbitals produces sigma-bonding and sigma-antibonding molecular orbitals. In a bonding molecular orbital, the electron density between the nuclei is high. In an antibonding orbital the electron density between the nuclei is extremely low. In the hydrogen molecule the two electrons occupy the bonding orbital.

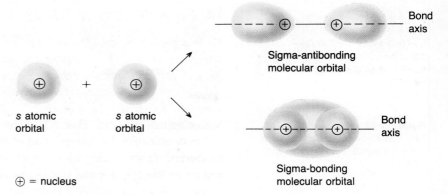

In the bonding molecular orbital there is a high probability of finding the electrons between the nuclei of the combining atoms. This orbital is symmetrical along the axis between the hydrogen atoms. *A* **sigma bond** *is formed when two atomic orbitals combine to form a molecular orbital that is symmetrical along the axis connecting two atomic nuclei.* The symbol for the Greek letter sigma is σ.

Bonding results from an imbalance between attractive and repulsive forces. Because they are oppositely charged, nuclei and electrons attract each other. Because they are similarly charged, nuclei repel other nuclei and electrons repel other electrons. In the hydrogen molecule the attraction of hydrogen nuclei to electrons tips the balance in favor of the attractive forces. In higher-energy or antibonding molecular orbitals, however, the electrons are *not* between the nuclei. Thus the balance favors the repulsive forces. Repulsive forces would occur if two helium atoms were to combine into a He_2 molecule. Each atom has two $1s$ electrons. Two of these electrons can go into the bonding energy level, but the other two must go into the antibonding level.

In this case the antibonding forces are larger than the bonding forces. Hence a He_2 molecule is unstable compared with two separate helium atoms, and helium exists only as atoms.

Atomic *p* orbitals also overlap to form molecular orbitals. For example, a fluorine atom has a half-filled $2p$ orbital. When two fluorine atoms combine, these orbitals overlap to make a filled bonding molecular orbital (Figure 14·17). The bonding molecular orbital shows a high probability of finding a pair of electrons between the positively charged nuclei of the two fluorines. The fluorine nuclei are attracted to this region of

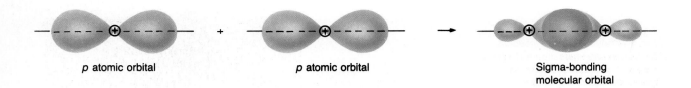

| p atomic orbital | p atomic orbital | Sigma-bonding molecular orbital |

Figure 14·17
Two p atomic orbitals can combine to form a sigma-bonding molecular orbital.

high-electron density. This attraction holds the atoms together in the fluorine molecule, F_2. The overlap of the $2p$ orbitals produces a symmetrical bonding molecular orbital when viewed along the F—F bond axis. Therefore the F—F bond is a sigma bond.

Problem

4. What is a sigma bond? Describe, with the aid of a diagram, how the overlap of two half-filled $1s$ orbitals produces a sigma bond.

In the fluorine molecule the p atomic orbitals overlap end-to-end. In some molecules these orbitals can also overlap side-by-side. The side-by-side overlap of atomic p orbitals produces pi molecular orbitals (Figure 14·18). A pi bond results when a pi molecular orbital is filled with two electrons. The symbol for the Greek letter pi is π. *In a* **pi bond,** *the bonding electrons are most likely to be found in sausage-shaped regions above and below the bond axis of the bonded atoms.* Orbital overlap in pi bonding is not as extensive as it is in sigma bonding. Therefore pi bonds tend to be weaker than sigma bonds. We will return to pi bonding when we discuss the shapes of molecules in Section 14·9.

Figure 14·18
The side-by-side overlap of two p atomic orbitals produces a pi-bonding molecular orbital. The *two* sausage-shaped regions in which a bonding electron pair is most likely to be found constitute *one* pi-bonding molecular orbital.

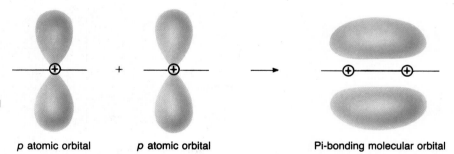

| p atomic orbital | p atomic orbital | Pi-bonding molecular orbital |

— Science, Technology, and Society —

14·A Molecular Models

In the sections ahead you will see that the properties of a substance depend on the shape of its molecules. Physical models of molecules help us to visualize them and better understand their behavior.

Ball-and-stick models represent atoms by means of balls of different colors. Yellow usually represents hydrogen, black carbon, and red oxygen. Bonds are represented by wooden sticks or metal springs that fit into holes drilled into the balls at angles of 109.5°. The

Figure 14·19
The ball-and-stick and space-filling models of methane show the advantages and disadvantages of each model. **a** The bond angles for methane are most apparent from the ball-and-stick model. **b** The relative sizes of the atoms are more apparent from the space-filling models. **c** Computers can generate 3-dimensional models of large molecules.

a Ball-and-stick model

b Space filling model

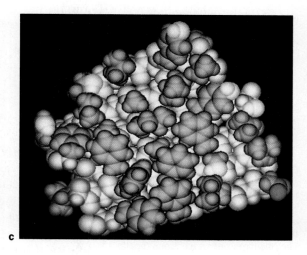

c

number of holes in each ball is the same as the number of bonds that atom makes. Hydrogen has one, oxygen two, and carbon four. Unshared pairs are not shown. Ball-and-stick models show very clearly the bond angles in a molecule. They do not, however, accurately represent the relative size of the atoms or the bond lengths.

Space-filling models are made of plastic balls. The size of each ball accurately represents the relative size of the atom. Bond lengths are also to scale. These models do not give the erroneous impression of empty space between the atoms of a molecule. They do make it more difficult, however, to see bond angles.

Organic chemists and biochemists use computers to create models of molecules composed of hundreds of atoms. Many biochemical reactions depend on a compound having the proper shape as well as the proper formula. Using a special viewing apparatus, chemists are able to see three-dimensional representations of complex molecules on the computer monitor. It is possible to turn the molecule over to see the other side, to shrink it, or to cut it in pieces on the monitor.

14·8 VSEPR Theory

A photograph or sketch may fail to do justice to a person's appearance. Similarly, electron dot structures and structural formulas fail to reflect the three-dimensional shapes of molecules. For example, the electron dot structure and structural formula of methane (CH_4) show the molecule in only two dimensions.

Methane
(electron dot
structure)

Methane
(structural
formula)

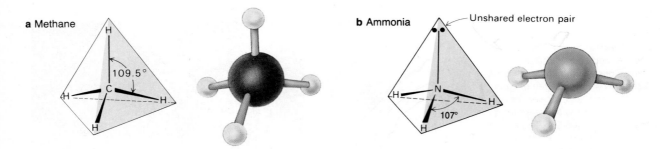

a Methane

109.5°

b Ammonia — Unshared electron pair

107°

Figure 14·20
a Methane is a good example of a tetrahedral molecule. The hydrogens in methane are at the four corners of a regular tetrahedron.
b An ammonia molecule is pyramidal. The unshared pair of electrons repels the bonding pairs causing the bond angle to be about 107°.

Molecules and ions have three-dimensional shapes.

In reality, methane molecules exist in three dimensions. The hydrogens in the methane molecule are at the four corners of a geometric solid, the regular tetrahedron. In this arrangement, all the H—C—H angles are *109.5°, the* **tetrahedral angle** (Figure 14·20). The *valence-s*hell *e*lectron-*p*air *r*epulsion theory, or VSEPR theory, explains this shape. **VSEPR theory** *states that because electron pairs repel, molecules adjust their shapes so that the valence-electron pairs are as far apart as possible.* The methane molecule has four bonding electron pairs and no unshared pairs. The bond pairs are farthest apart when the angle between the central carbon and its attached hydrogens is 109.5°. This is the H—C—H bond angle found by experiment. Any other arrangement tends to bring two bonding pairs of electrons closer together.

Unshared pairs of electrons are important when we are trying to predict the shapes of molecules. The nitrogen in ammonia (NH_3) is surrounded by four pairs of valence electrons, but one of these is an unshared pair (Figure 14·20). No bonding atom is vying for the unshared electrons. Thus they are held closer to the nitrogen than are the bonding pairs. The unshared pair strongly repels the bonding pairs, pushing them closer together. The experimentally measured H—N—H bond angle is only 107°. The shape of the ammonia molecule is *pyramidal*.

In a water molecule, oxygen forms single covalent bonds with two hydrogen atoms. The two bonding pairs and two unshared pairs of electrons form a tetrahedral arrangement around the central oxygen (Figure 14·21). Thus the water molecule is planar (flat) but *bent*. With two unshared pairs repelling the bonding pairs, the H—O—H bond angle is compressed. The experimentally measured bond angle is about 105°.

The carbon in CO_2 has no unshared pairs. The double bonds joining the oxygens to the carbon are farthest apart when the O=C=O bond angle is 180° (Figure 14·21). Thus CO_2 is a *linear* molecule. Figure 14·22 shows some common molecular shapes.

Figure 14·21
a The water molecule is bent because the bonding electrons are repelled by the unshared pairs.
b By contrast, the carbon dioxide molecule is linear. It has no unshared pairs of electrons.

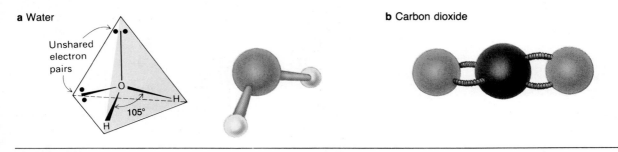

a Water

Unshared electron pairs

105°

b Carbon dioxide

Figure 14·22
Some commonly found shapes of molecules are linear triatomic, trigonal planar, bent triatomic, pyramidal, tetrahedral, and trigonal bipyramidal.

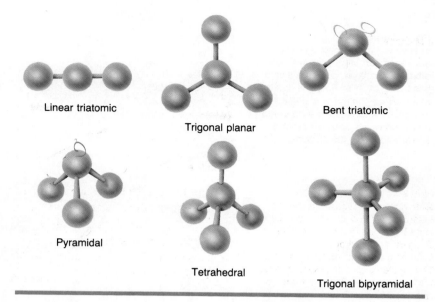

Linear triatomic

Trigonal planar

Bent triatomic

Pyramidal

Tetrahedral

Trigonal bipyramidal

Problem

5. The BF_3 molecule is planar. The attachment of a fluorine ion to the boron in BF_3 through a coordinate covalent bond creates the BF_4^- ion. What is the geometric shape of this ion?

14·9 Hybrid Orbitals

Electron dot structures and the combination of unpaired electrons in boxes representing orbitals are two ways of describing covalent bonding. The VSEPR theory does a good job of describing molecular shapes. Another way to describe molecules that is informative of *both* bonding and shape is orbital hybridization. *With* **hybridization** *several atomic orbitals mix to form the same number of equivalent hybrid orbitals*.

Orbital hybridization can be used to describe the methane molecule. One $2s$ orbital and three $2p$ orbitals of a carbon atom mix to form four

The number of hybrid orbitals is always the same as the number of orbitals mixed.

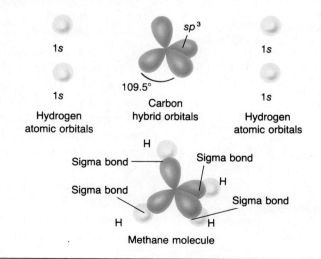

Figure 14·23
In methane, each of the four sp^3 hybrid orbitals of carbon overlap with a $1s$ orbital of hydrogen. The result of each overlap is a sigma bond.

Figure 14·24
In ethene, two sp^2 hybrid orbitals from each carbon overlap with a $1s$ orbital of hydrogen to form a sigma bond. The other sp^2 orbitals overlap to form a carbon–carbon sigma bond. The p atomic orbitals overlap to form a pi bond which occupies regions above and below the carbons.

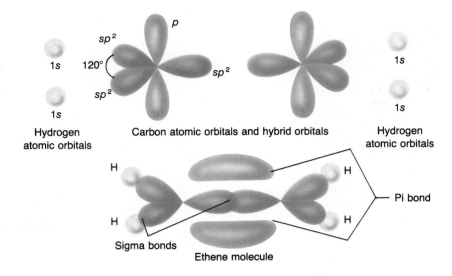

Hydrogen atomic orbitals

Carbon atomic orbitals and hybrid orbitals

Hydrogen atomic orbitals

Sigma bonds

Ethene molecule

Pi bond

sp^3 hybrid orbitals. These are at the tetrahedral angle of 109.5°. The four sp^3 orbitals of carbon overlap with the $1s$ orbitals of the four hydrogen atoms (Figure 14·23). The sp^3 orbitals extend farther into space than either s or p orbitals. Thus the overlap with a hydrogen $1s$ orbital can be greater. The eight available valence electrons fill the molecules' orbitals to form four C — H sigma bonds. The greater overlap results in an unusually strong covalent bond.

Hybridization is also useful in describing double covalent bonds. Ethene is a relatively simple molecule. It has one carbon–carbon double bond and four carbon–hydrogen single bonds.

$$\underset{\text{Ethene}}{\overset{\displaystyle H \qquad H}{\underset{\displaystyle H \qquad H}{C=C}}}$$

Experimental evidence indicates that the H—C—H bond angles are 120°. In ethene, sp^2 hybrid orbitals form from the combination of one $2s$ and two $2p$ atomic orbitals of carbon (Figure 14·24). Each hybrid orbital is separated from the other two by 120°. Two sp^2 hybrid orbitals of each carbon form sigma-bonding molecular orbitals with the four available hydrogen $1s$ orbitals. The third sp^2 orbitals of the two carbons overlap to form a carbon–carbon sigma–bonding orbital. The nonhybridized $2p$ (carbon) orbitals overlap side-by-side to form a pi-bonding orbital. A total of twelve electrons fill the six bonding orbitals. (The two carbons each contribute four electrons and the four hydrogens each contribute one.) Thus the ethene molecule is held together by five sigma bonds and one pi bond.

Both sigma and pi bonds are two-electron covalent bonds. These bonds are drawn alike in structural formulas. Pi bonds are weaker than sigma bonds. In chemical reactions of carbon–carbon double bonds, the pi bond is broken in preference to the sigma bond.

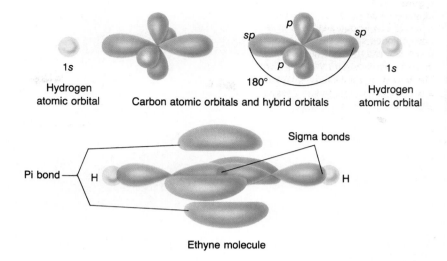

Figure 14·25
In ethyne, one *sp* hybrid orbital from each carbon overlaps with a 1*s* orbital of hydrogen to form a sigma bond. The other *sp* hybrid orbital of each carbon overlaps to form a carbon–carbon sigma bond. The two *p* atomic orbitals from each carbon overlap to form two pi bonds.

Hydrogen atomic orbital

Carbon atomic orbitals and hybrid orbitals

Hydrogen atomic orbital

Sigma bonds

Pi bond

H   H

Ethyne molecule

A third type of covalent bond is a triple bond such as is found in ethyne, C_2H_2. Another name for this compound is acetylene.

$$H-C \equiv C-H$$

As with other molecules, the hybrid orbital description of ethyne is guided by the properties of the molecule. Ethyne is a linear molecule. The best hybrid orbital description is obtained if a 2*s* atomic orbital of carbon mixes with only one of the three 2*p* atomic orbitals. The result is two *sp* hybrid orbitals for each carbon (Figure 14·25). A carbon–carbon sigma–bonding molecular orbital forms from the overlap of one *sp* orbital from each carbon. The other *sp* orbital of each carbon overlaps with the 1*s* orbital of each hydrogen. Sigma-bonding molecular orbitals are formed. The remaining pair of *p* atomic orbitals on each carbon overlap side-by-side. They form two pi-bonding molecular orbitals. These surround the central carbons. The ten available electrons completely fill the five bonding molecular orbitals. Thus the bonding of ethyne consists of three sigma bonds and two pi bonds.

14·10   Polar Bonds

Covalent bonds are formed by electron-sharing between atoms. Not all covalent bonds are the same. The character of these bonds in a given molecule depends on the kind and number of atoms joined together. These features in turn determine the properties of the molecules.

The bonding pairs of electrons in covalent bonds are pulled, as in a tug of war, between the nuclei of the atoms sharing the electrons. *When the atoms in a bond are the same, the bonding electrons are shared equally, and the bond is a* **nonpolar covalent bond.** Hydrogen (H_2), oxygen (O_2), and nitrogen (N_2) have nonpolar covalent bonds. *When two different atoms are joined by a covalent bond, and the bonding electrons are shared unequally, the bond is a* **polar covalent bond,** *or simply a* **polar bond.** The atom with stronger electron attraction (the more

Figure 14·26
The nuclei of atoms pull the bonding electrons as in a tug-of-war between two children.

Figure 14·27
This electron cloud picture for hydrogen chloride shows that the chlorine atom attracts more of the electron cloud than does the hydrogen atom. Chlorine is more electronegative than hydrogen.

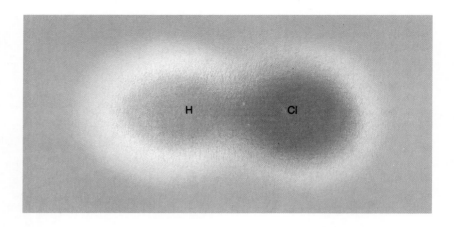

Covalent bonds joining unlike atoms are often polar.

electronegative atom) in a polar bond acquires a slightly negative charge. The less electronegative atom acquires a slightly positive charge. Table 12·3 gives electronegativities of some common elements. The higher the electronegativity value, the greater is the ability of an atom to attract electrons to itself.

Consider the hydrogen chloride molecule (HCl). Hydrogen has an electronegativity value of 2.1, and chlorine has an electronegativity value of 3.0. The covalent bond in hydrogen chloride is polar. The chlorine atom acquires a slightly negative charge. The hydrogen atom acquires a slightly positive charge. The Greek letter delta (δ) is used to show that atoms involved in the covalent bond acquire only partial charges, much less than $1+$ or $1-$.

$$\overset{\delta+}{\text{H}}\!-\!\overset{\delta-}{\text{Cl}}$$

The minus sign shows that chlorine has acquired a slightly negative charge. The plus sign shows that hydrogen has acquired a slightly positive charge. The polarity of the bond may also be represented with an arrow pointing to the more electronegative atom.

$$\overset{\longmapsto}{\text{H}\!-\!\text{Cl}}$$

The O—H bonds in the water molecule are also polar. The very electronegative oxygen pulls the bonding electrons away from hydrogen.

Table 14·3 Electronegativity Differences and Bond Types		
Electronegativity difference (approx.)	Type of bond	Example
0.0–0.4	Covalent (nonpolar)	H—H (0.0)
0.4–1.0	Covalent (moderately polar)	$\overset{\delta+}{\text{H}}\!-\!\overset{\delta-}{\text{Cl}}$ (0.9)
1.0–2.0	Covalent (very polar)	$\overset{\delta+}{\text{H}}\!-\!\overset{\delta-}{\text{F}}$ (1.9)
$\geq$2.0	Ionic	Na^+Cl^- (2.1)

The oxygen acquires a slightly negative charge. The hydrogens acquire a slightly positive charge.

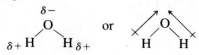

Electronegativities also indicate the type of bond that two atoms will tend to form. If the electronegativity difference between two atoms is greater than 2.0, a bond is ionic. If the electronegativity difference is less than 2.0 but greater than 0.4, a bond is covalent and polar. If the difference is less than 0.4, the bond is nonpolar covalent (Table 14·3).

Example 4

What type of bond (polar covalent, nonpolar covalent, or ionic) will form between atoms of the following pairs of elements?
a. N and H   **b.** F and F   **c.** Ca and O   **d.** Al and Cl

Solution

The types of bonds formed depend on the electronegativity differences between the bonding elements. The rules are given in Table 14·3.

Elements (electronegativities)	Electronegativity difference	Type of bond
a. N (3.0), H (2.1)	0.9	Moderately polar covalent
b. F (4.0), F (4.0)	0.0	Nonpolar covalent
c. Ca (1.0), O (3.5)	2.5	Ionic
d. Al (1.5), Cl (3.0)	1.5	Very polar covalent

Problems

6. Identify the bonds between atoms of the following pairs of elements as ionic, nonpolar covalent, or polar covalent.
   a. H and Br   **b.** K and Cl   **c.** C and O   **d.** Cl and F
   e. Li and O   **f.** Br and Br
7. Which covalent bond is the most polar?   **a.** H—Cl   **b.** H—Br
   c. H—S   **d.** H—C   **e.** F—F

14·11   Polar Molecules

A polar molecule always contains polar bonds, but some molecules with polar bonds are nonpolar.

The presence of a polar bond in a molecule often makes the entire molecule polar. *In a* **polar molecule** *one end of the molecule is slightly negative, and one end is slightly positive.* For example, in the hydrogen chloride molecule the partial charges on the hydrogen and chlorine atoms are electrically charged regions, or poles. *A molecule that has two poles is called a dipolar molecule, or* **dipole.** The hydrogen chloride molecule is a dipole.

Figure 14·28
When polar molecules are placed in an electric field, the negative ends of the molecules orient toward the positively charged plate and the positive ends of the molecules orient toward the negatively charged plate. What would happen if carbon dioxide molecules were placed in this field? Why?

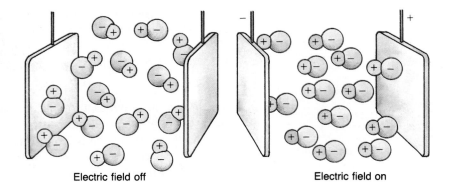

Electric field off   Electric field on

The effect of polar bonds on the polarity of an entire molecule depends on the shape of the molecule and the orientation of the polar bonds. A carbon dioxide molecule, for example, has two polar bonds and is linear.

$$\overset{\leftarrow +}{O} = \overset{+ \rightarrow}{C} = \overset{}{O}$$

The carbon and oxygens lie along the same axis. Therefore the bond polarities cancel because they are in opposite directions. Carbon dioxide is a nonpolar molecule.

The water molecule has two polar bonds, but the molecule is bent. Hence the bond polarities do not cancel, and a water molecule is polar.

14·12   Bond Dissociation Energies

A large quantity of heat is liberated when hydrogen atoms combine to form hydrogen molecules. This is evidence that the product is more stable than the reactants. Indeed, the covalent bond in the hydrogen molecule is so strong that 435 kJ of energy is required to dissociate 1 mol of hydrogen molecules to hydrogen atoms. *The energy required to break a single bond is known as the* **bond dissociation energy.** Hydrogen molecules have a bond dissociation energy of 435 kJ per mole.

$$H—H + 435 \text{ kJ} \longrightarrow H \cdot + \cdot H$$

■ The bond dissociation energy of a diatomic molecule is the energy required to dissociate the bond to atoms, *not* ions.

The carbon–carbon single covalent bond has a bond dissociation energy of about 347 kJ. The ability of carbon to form strong carbon–carbon bonds helps explain the stability of carbon compounds. Table 14·4 gives bond dissociation energies of several representative covalent bonds. Compounds with only C—C and C—H single covalent bonds are quite unreactive chemically. In part this is because of the high dissociation energies of these bonds.

Problem

8. How many kilojoules would be required to dissociate all the C—H single bonds in 0.1 mol of methane? Assume that the bond dissociation energy is the same for each bond.

Bond lengths are usually given in picometers (pm). 1 pm = 10^{-12} m and 1 pm = 10^{-3} nm.

Safety

Bromine is toxic and very corrosive. It has a high vapor pressure. When working with bromine solutions, avoid breathing the fumes, and keep the container tightly closed.

Table 14·4 Bond Dissociation Energies and Bond Lengths for Covalent Bonds

Bond	Bond energy (kJ/mol)	Bond length (pm)
H—H	435	74
C—H	393	109
C—O	356	143
C=O	736	121
C≡O	1074	113
C—C	347	154
C=C	657	133
C≡C	908	121
C—N	305	147
S—S	259	208
Cl—Cl	243	199
N—N	209	140
Br—Br	192	228
I—I	151	267
O—O	142	132
O—H	464	96

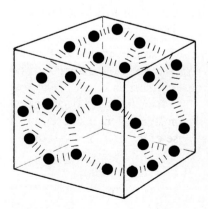

Figure 14·29
Van der Waals said that gases do not obey ideal behavior at low temperatures and high pressures because the gas molecules are weakly attracted to each other.

— Science, Technology, and Society —

14·B Johannes van der Waals: Using the Power of Reason

The lifelong work of Johannes van der Waals was the study of the behavior of gases and liquids. Van der Waals observed that at higher pressures or lower temperatures gases did not really obey the ideal gas equations. He reasoned that the assumptions involved in the ideal gas laws must be incorrect.

The ideal gas laws assume that the molecules of a gas are not attracted to each other. Van der Waals noted that at low temperatures and high pressures the measured pressure is less than the calculated ideal pressure. He reasoned that under these conditions gas molecules must be weakly attracted to each other. The pressure that the molecules exert on the walls of the container is reduced because they are "held back" by the molecules around them.

By such reasoning van der Waals was able to calculate correction factors for the erroneous assumptions of the ideal gas equation. For this work he was awarded the Nobel Prize in physics in 1910. Two types of intermolecular attractions are named in his honor because his work dealt with the weak forces of attraction between molecules.

14·13 Intermolecular Attractions

Intermolecular forces determine some of the properties of substances.

Van der Waals forces are sometimes called fluctuating dipoles.

In addition to covalent bonds in molecules, there are attractions between molecules, or intermolecular attractions. These attractions are weaker than either an ionic or covalent bond. Nevertheless, do not underestimate the power of these forces. Among other things, they are responsible for whether a molecular compound is a gas, liquid, or solid.

The weakest attractions between molecules are collectively called **van der Waals forces.** They are named after the Dutch chemist Johannes van der Waals (1837–1923). Two major van der Waals forces are dispersion forces and dipole interactions.

Dispersion forces, *the weakest of all molecular interactions, are thought to be caused by the motion of electrons.* Generally speaking, the strength of dispersion forces increases as the number of electrons in a molecule increases. The halogens are an example of molecules whose major attraction for one another is caused by dispersion forces. Fluorine and chlorine, with relatively few electrons, are gases at STP. The larger number of electrons in bromine generate larger dispersion forces. Bromine molecules are, therefore, sufficiently attracted to each other to make bromine a liquid at STP. Iodine, with a still larger number of electrons, is a solid at STP.

Dipole interactions *occur when polar molecules are attracted to one another* (Figure 14·30). Electrostatic attractions occur between the oppositely charged regions of dipolar molecules. Dipole interactions are similar to but much weaker than ionic bonds. Dipole interactions in water, for example, result in a weak attraction of water molecules for one another. Each O—H bond in the water molecule is highly polar. The oxygens in water acquire a slightly negative charge because of oxygen's greater electronegativity. The hydrogens in water acquire a slightly positive charge. Polar molecules attract one another. Hence the positive region of one water molecule attracts the negative region of another. This dipolar attraction is weak, however, when compared with the strength of hydrogen bonds.

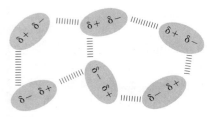

Figure 14·30
Polar molecules are attracted to one another by van der Waals forces called dipole interactions.

Figure 14·31
Liquids tend to form drops because of intermolecular attractions. Molecular substances with weak intermolecular attractions are gases at room temperature. Molecular substances with very strong intermolecular attractions are solids.

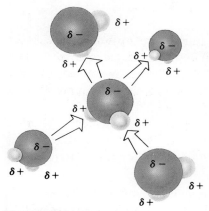

Figure 14·32
Hydrogen-bonding between water molecules accounts for many properties of water. The hydrogen bond has about 5% the strength of an average covalent bond.

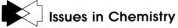 **Issues in Chemistry**

What do some adhesives used in plywood, permanent press clothing, paints, varnishes, and foam insulation have in common? They may be sources of formaldehyde vapor, an irritating substance that elicits an allergic response from many people. What is formaldehyde and how can you limit your exposure to it?

Hydrogen bonds *are attractive forces in which a hydrogen that is covalently bonded to a very electronegative atom is also weakly bonded to an unshared electron pair of an electronegative atom in the same molecule or in a nearby molecule.* Hydrogen-bonding always involves hydrogen. It is the only chemically reactive element whose valence electrons are not shielded from the nucleus by a layer of underlying electrons. A very polar covalent bond is formed when hydrogen bonds to an electronegative atom like oxygen, nitrogen, or fluorine. This leaves the hydrogen nucleus quite electron-deficient. The hydrogen makes up for its deficiency by sharing a nonbonding electron pair on a nearby electronegative atom. The resulting hydrogen bond has about 5% of the strength of an average covalent bond. Hydrogen bonds are the strongest of the intermolecular forces. They are extremely important in determining the properties of water and biological molecules like proteins. Figure 14·32 shows hydrogen-bonding between water molecules.

Problem

9. Depict the hydrogen-bonding between two ammonia molecules and between one ammonia molecule and one water molecule.

=== Science, Technology, and Society ===

14·C Bonding with Adhesives

An adhesive is a substance that can hold materials together at their surface. The earliest adhesives were natural substances like flour paste and egg white. Modern adhesives are mostly synthetic. Like their ancestors, however, they bond materials together by maximizing the intermolecular attractions.

If two perfectly smooth, clean surfaces are pressed together in a vacuum, they will stick together. If no air molecules are in the way, atoms or molecules in one surface will be attracted to those in the other surface. If the surfaces are very smooth, many atoms will be in contact. Intermolecular attractions will hold the surfaces together.

Ordinary surfaces, however, are not smooth. They are full of microscopic ridges and valleys. The contact area between such surfaces is actually quite small, and there is little adhesion. We can make two ordinary surfaces stick together, however, by coating each surface with a liquid that can fill up all the ridges and valleys. Many atoms will be in contact. Thus the adhesion between the liquid and each of the two solid surfaces will be very strong. If the liquid itself holds together, a strong bond will form between the two surfaces. To be a good adhesive, then, a liquid must spread well, wetting all the nooks and crannies of a surface. This means that its molecules must be attracted to those in the surface to be glued. For example, it should be polar if the material to be glued is polar. It should also have strong intermolecular forces or molecular bonds.

Figure 14·33
Glue works by filling in all the nooks and crannies. It provides a large area of contact between the two surfaces.

Glues and adhesives are generally polymers. These are very long molecules that are held together by strong covalent bonds. In addition, these molecules often have branches that are polar. Intermolecular attractions between the branches bind the molecules to other similar molecules. This makes these polymers very good adhesives.

Epoxy resins are a group of very strong synthetic adhesives. An epoxy prepolymer combines chemically with a curing agent to form an interlocking polymer (Figure 14·34). The end result of this kind of chemical reaction is that the glue layer is essentially one big molecule. The polar $-OH$ groups in the final polymer are strongly attracted to the surface to which the glue is applied. The attractions between the glue and the surface are extremely strong. So are the attractions within the glue itself. If an object glued with epoxy is subjected to stress, the object will break before the glued surface will!

Aircraft, cars, trucks, and boats are partially held together with epoxy adhesives. For example, the rear one-third of the wing surfaces of jet aircraft are glued together. The joints are strong and add little mass to the plane in comparison to welding or fastening with bolts. Approximately 500 kg of adhesive may be used in one plane.

Safety

The solvents used in many adhesives are toxic. Inhaling the fumes is dangerous. Work in a well-ventilated area.

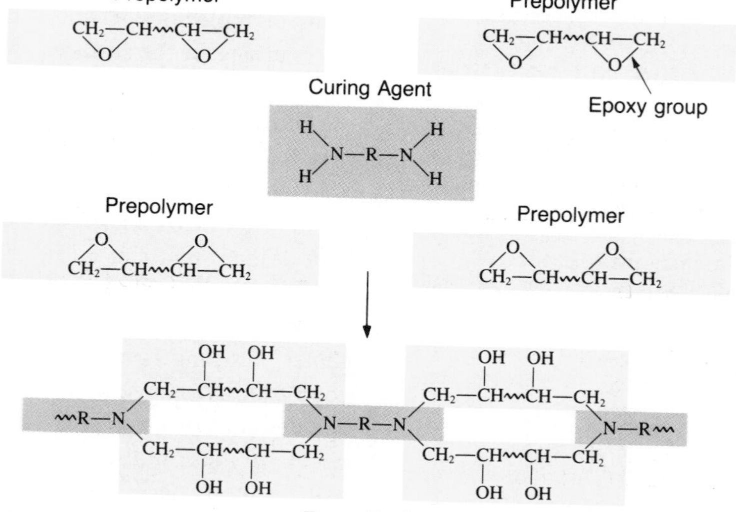

Figure 14·34
Epoxy resins are particularly strong adhesives. An epoxy prepolymer reacts with a curing agent to form essentially one large molecule with thousands of units. In this example, a curing agent molecule links four epoxy prepolymer molecules together to form an interlocking epoxy resin polymer chain.

342 Chapter 14 Covalent Bonds

14·14 Properties of Molecular Substances

Properties of molecular compounds are diverse.

The physical properties of a compound depend on the type of bonding it displays. Some covalent compounds are gases, some are liquids, and some are solids at room temperature. We have already touched on this point by noting that most ionic compounds are crystalline solids. Table 14·5 summarizes some of the differences between ionic and molecular substances. A great variety of physical properties occurs among covalent compounds. This is in large part because of intermolecular attractions.

The melting and boiling points of most molecular compounds are low compared with those of ionic compounds. A few molecular solids resist melting, however, except in the range of 1000°C to 3000°C. Some decompose without melting at all. Most of these *very stable substances are **network solids** (or network crystals) in which all of the atoms are covalently bonded to each other*. Diamond is a network solid because each carbon in a diamond is covalently bonded to four other carbons. Diamond-cutting requires breaking a multitude of these bonds. Diamond does not melt, but vaporizes to a gas at 3500°C and above. Silicon carbide, with the formula unit SiC and a melting point of about 2700°C, is also a network solid. Silicon carbide is so hard that it is used in grindstones and as an abrasive. The molecular structure of silicon carbide is like diamond except that every other carbon is replaced by silicon. Samples of diamond or silicon carbide can be thought of as single molecules. The melting of silicon carbide requires that covalent bonds be broken. To melt other molecular solids we need only to break the weak attractions between molecules.

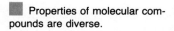

ChemDirections

High-temperature superconductors are an unusual group of covalently bonded compounds. At temperatures well above absolute zero, they offer no resistance to the flow of electrons through them. The search for materials that are superconducting at room temperatures has scientists busy because of the almost limitless applications for these compounds. For more information about these interesting materials, read **ChemDirections** *Superconductors*, page 656.

Figure 14-35
Ionic compounds have high melting points. Each ion in the crystal lattice is held firmly in place by electrostatic forces of attraction. By contrast, molecular crystals generally have low melting points. Even strong intermolecular bonds are much weaker than ionic bonds.

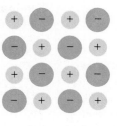

Ionic crystal

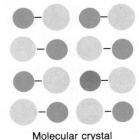

Molecular crystal

Table 14·5 Characteristics of Ionic and Covalent Compounds		
Characteristic	Ionic compound	Covalent compound
Representative unit	Formula unit	Molecule
Bond formation	Transfer of one or more electrons between atoms	Sharing of electron pairs between atoms
Type of elements	Metallic and nonmetallic	Nonmetallic
Physical state	Solid	Solid, liquid, or gas
Melting point	High (usually above 300°C)	Low (usually below 300°C)
Solubility in water	Usually high	High to low
Electrical conductivity of aqueous solution	Good conductor	Poor to nonconducting

14 Covalent Bonds
Chapter Review

Key Terms

antibonding orbital *14·7*
• bonding dissociation
   energy *14·12*
bonding orbital *14·7*
coordinate covalent
   bond *14·4*
• dipole *14·11*
dipole interaction *14·13*
dispersion force *14·13*
• double covalent
   bond *14·2*
• hybridization *14·9*
• hydrogen bond *14·13*
molecular orbital *14·7*
• network solid *14·14*
• nonpolar covalent
   bond *14·10*
paramagnetic *14·6*

• pi bond (π bond) *14·7*
polar bond *14·10*
polar covalent
   bond *14·10*
• polar molecule *14·11*
resonance *14·5*
sigma bond
   (σ bond) *14·7*
• single covalent
   bond *14·1*
• structural formula *14·1*
• tetrahedral angle *14·8*
• triple covalent
   bond *14·2*
• unshared pair *14·1*
van der Waals
   force *14·13*
• VSEPR theory *14·8*

affected by bond polarity and the electronegativity of the joined atoms. The attractions between opposite poles of polar molecules constitute dipole interactions. The dipole interaction is one of several weak attractions between molecules. Another weak attractive force is the hydrogen bond. These weak forces determine whether a covalent compound will be a solid, liquid, or gas.

As a general rule, molecules adjust their three-dimensional shapes so that the valence-electron pairs around a central atom are as far apart as possible. This is the guiding principle in the valence-shell electron-pair repulsion, or VSEPR, theory of molecular geometries. Molecular orbital theory is a logical extension of the quantum mechanical description of the electron structure of atoms. Covalent bonding is described in terms of sigma and pi bonds. In some instances molecular geometry is adequately described by simple overlap of atomic orbitals. In others a description of molecular shape that better fits experimental results is obtained from hybridized atomic orbitals.

Chapter Summary

When atoms share electrons to gain the stable electron configuration of a noble gas, they form covalent bonds. A shared pair of valence electrons constitutes a single covalent bond. Sometimes two or three pairs of electrons may be shared to give double or triple covalent bonds. In some cases only one of the atoms in a bond provides the pair of bonding electrons; this is a coordinate covalent bond.

When like atoms are joined by a covalent bond, the bonding electrons are shared equally, and the bond is nonpolar. When the atoms in a bond are not the same, the bonding electrons are shared unequally, and the bond is polar. The degree of polarity of a bond between any two atoms is determined by consulting a table of electronegativities. Some molecules are polar because they contain polar covalent bonds. Bond dissociation energies are

Practice Questions and Problems

10. Explain why helium is monatomic but hydrogen exists as diatomic molecules. *14·1*

11. The following covalent molecules have only single covalent bonds. Draw an electron dot structure for each. *14·1*
  a. H₂O **b.** H₂O₂ **c.** PCl₃ **d.** NH₃

12. Classify the following compounds as ionic or covalent. *14·1*
  a. MgCl₂ **b.** Na₂S **d.** H₂S **e.** H₂O

13. Explain why atoms form chemical bonds and describe the difference between an ionic and a covalent bond. *14·1*

14. How many electrons are shared by two atoms in a double covalent bond? In a triple covalent bond? *14·2*

15. Write plausible electron dot structures for the following substances. Each substance contains only single covalent bonds. *14·3*
   a. I_2 **b.** OF_2 **c.** H_2S **d.** NI_3

16. Draw the electron dot structures of these molecules. *14·3*
   a. F_2 **b.** HCl   **c.** HCCH   **d.** HCN

17. Explain why compounds containing C—N and C—O single bonds can form coordinate covalent bonds with H^+ but compounds containing only C—H and C—C single bonds cannot. *14·4*

18. Draw all resonance forms for sulfur dioxide, SO_2. The oxygens in SO_2 are attached to the sulfur. *14·5*

19. Draw resonance structures for the carbonate ion CO_3^{2-}. Each oxygen is attached to the carbon. *14·5*

20. Using electron dot structures draw at least two resonance structures for the nitrite ion, NO_2^-. The oxygens in NO_2^- are attached to the nitrogen. *14·5*

21. Predict whether the following species are diamagnetic or paramagnetic. *14·6*
   a. BF_3 **b.** O_2^- **c.** NO_2 **d.** F_2

22. Can you suggest a reason why phosphorus and sulfur expand octets in many of their compounds but nitrogen and oxygen never do? *14·6*

23. Which of the following species would you predict to be diamagnetic? Paramagnetic? *14·6*
   a. NO_3^- **b.** OH^- **c.** HO_2 **d.** SO_3

24. Draw molecular orbital diagrams for the possible diatomic molecule Li_2. Would you expect Li_2 to exist as a stable molecule? *14·7*

25. What is the total number of sigma bonds and pi bonds in each of the following molecules? *14·7*

   a. $H—C\equiv N$ **b.** $H—\overset{\displaystyle H}{\underset{\displaystyle H}{\overset{|}{\underset{|}{C}}}}—N=C=O$

26. Explain how the VSEPR theory can be used to predict bond angles in these molecules. *14·8*
   a. methane   **b.** ammonia   **c.** water

27. Use VSEPR theory to predict the shapes of the following species. *14·8*
   a. CO_2 **d.** SCl_2
   b. $SiCl_4$ **e.** CO
   c. SO_3 **f.** I_3^-

28. The molecule CO_2 has two carbon–oxygen double bonds. Draw the bonding in the CO_2 molecule using hybridized orbitals for carbon and oxygen. *14·9*

29. What type of hybrid orbitals are involved in bonding of each atom in the following molecules? *14·9*
   a. CO   **c.** $H_2C=O$
   b. H_2O **d.** $N\equiv C—C\equiv N$

30. The bonds between the following pairs of elements are covalent. Arrange them according to polarity, naming the most polar bond first. *14·10*
   a. H—Cl   **d.** H—O
   b. H—C   **e.** H—H
   c. H—F   **f.** S—Cl

31. Arrange the following bonds in order of increasing ionic character. *14·10*
   a. Cl—F   **d.** C—H
   b. N—N   **e.** S—O
   c. K—O   **f.** Li—F

32. Draw the electron dot structure for each of the following molecules. Then identify polar covalent bonds by assigning $\delta+$ and $\delta-$ to the appropriate atoms. *14·11*
   a. HOOH   **b.** BrCl   **c.** HBr   **d.** H_2O

33. Based on the information about molecular shapes in Section 14·8, which of these molecules would you expect to be polar? *14·11*
   a. SO_2 **b.** H_2S **c.** CO_2 **d.** BF_3

34. Explain what is meant by the term *bond dissociation energy*. *14·12*

35. Assume that the total bond energy in a molecule is the sum of the individual bond energies. Calculate the total bond energy in a mole of ethyne, C_2H_2. (*Hint*: Write the electron dot structure to determine the kinds of bonds. Then refer to Table 14·4). *14·12*

36. Which compound in each pair exhibits the stronger intermolecular hydrogen-bonding? *14·13*
   a. H_2S and H_2O **c.** HBr and HCl
   b. HCl and HF   **d.** NH_3 and H_2O

37. What is a hydrogen bond? *14·13*

38. Why do compounds with strong intermolecular attractive forces have higher boiling points than compounds with weak intermolecular attractive forces? *14·14*

Mastery Questions and Problems

39. Devise a hybridization scheme and predict the shape of PCl_3 based on this scheme.

40. Write two electron dot structures for the molecule N_2O. Predict the magnetic properties and shape of this molecule.

41. The chlorines and oxygen in thionyl chloride, $SOCl_2$, are bonded directly to the sulfur. Write an acceptable electron dot structure for thionyl chloride.

42. Explain why each of the following electron dot structures is incorrect. Replace each structure with one that is more acceptable.

a. $[:C::N:]^-$

b. $:\ddot{F}:P::\ddot{F}:$ $:\ddot{F}:$

c. $H:\ddot{C}::O$ $:\ddot{C}l:$

d. $:\ddot{F}:$ $:\ddot{F}:B:\ddot{F}:$

43. Use VSEPR theory to predict the geometry of each of the following.
a. $SeCl_4(g)$ **b.** CO_3^{2-} **c.** $CCl_4(l)$ **d.** $SnCl_2(g)$

44. The graph below shows how the percent ionic character of a single bond varies according to the difference in electronegativity between the two elements forming the bond. Answer the following questions using this graph and Table 12·3 on page 285.
 a. What is the relationship between the percent ionic character of single bonds and the electronegativity difference between their elements?
 b. What electronegativity difference will result in a bond with a 50% ionic character?
 c. Estimate the percent ionic character of the bonds formed between (1) lithium and oxygen, (2) nitrogen and oxygen, (3) magnesium and chlorine, (4) nitrogen and fluorine.

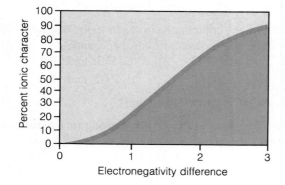

45. Describe differences between the members of each pair of terms.
 a. sigma bonds and pi bonds
 b. polar bonds and polar molecules
 c. diamagnetic and paramagnetic molecules
 d. sp^2 and sp^3 hybrid orbitals

46. Using bond dissociation energies estimate ΔH for the following reaction.

$$CO(g) + 2H_2(g) \longrightarrow CH_3OH(g)$$

47. Give the angles between the orbitals of each of the following hybrids.
 a. sp^3 hybrids **b.** sp^2 hybrids **c.** sp hybrids

48. Describe the difference between a bonding molecular orbital and an antibonding molecular orbital. How do their energies compare?

Critical Thinking Questions

49. Choose the term that best completes the second relationship.
 a. house:lumber   bond: _____
  (1) atoms   (3) valence electrons
  (2) molecules   (4) octet rule
 b. globe:earth
  electron dot formula: _____
  (1) atoms   (3) valence electrons
  (2) molecules   (4) octet rule

50. Make a list of the elements found in Table on page 324. What do the elements that form covalent bonds have in common?

51. Is there a clear difference between a *very* polar covalent bond and an ionic bond? If so, what criteria would you use to tell the difference?

Review Questions and Problems

52. Which of the following ions have the same number of electrons as a noble gas?
 a. Al^{3+} **c.** Br^-
 b. O^{2-} **d.** N^{3-}

53. A solution containing 0.10 mol of $BaCl_2$ is mixed with a solution containing 0.20 mol of Na_2SO_4 to give a precipitate of $BaSO_4$. What is the maximum yield in moles of $BaSO_4$? In grams?

54. Write correct electron configurations for atoms of the following elements.
- **a.** sodium
- **b.** sulfur
- **c.** phosphorus
- **d.** nitrogen

55. Calculate the gram formula mass of each of the following substances.
- **a.** $CaSO_4$
- **b.** H_2SO_4
- **c.** NI_3
- **d.** $FeCl_3$

56. Give the number of representative particles in the following molar quantities.
- **a.** 2.00 centimole
- **b.** 1.54 millimole
- **c.** 8.73 nanomole
- **d.** 3.00 micromole

57. Name and give the symbol for the element in the following position in the periodic table.
- **a.** Group 7B, period 4
- **b.** Group 3A, period 5
- **c.** Group 1A, period 7
- **d.** Group 6A, period 6

Challenging Questions and Problems

58. The electron structure and geometry of the methane molecule, CH_4, can be described by a variety of models. These include the electron dot structure, simple overlap of atomic orbitals, and orbital hybridization of carbon. Write the electron dot structure of CH_4. Sketch two molecular orbital pictures of the CH_4 molecule. For your first sketch assume that one of the paired $2s^2$ electrons of carbon has been "promoted" to the empty $2p$ orbital. Overlap each half-filled atomic orbital of carbon to a half-filled $1s$ orbital of hydrogen. What is the predicted geometry of the CH_4 molecule using this simple overlap method? In your second sketch assume hybridization of the $2s$ and $2p$ orbitals of carbon. Now what geometry would you predict for CH_4? Which picture is preferable based on the facts that all H—C—H bond angles in CH_4 are 109.5° and all C—H bond distances are identical?

59. There are some compounds in which one atom has more electrons than the corresponding noble gas. Examples are PCl_5, SF_6, and IF_7. Write the electron configuration of P, S, and I, and then draw the electron dot structures for these compounds. Looking at the outer shell configuration of P, S, and I, can you develop an orbital hybridization scheme to explain the existence of these compounds?

Research Projects

1. Make glue by gently heating a mixture of skim milk and vinegar. Try different quantities of reactants and techniques of purification to produce the best glue. Baking soda can be used to remove any excess acetic acid.

2. Devise a method to measure the polarity of some covalent bonds.

3. Relate the melting points of some covalently bonded compounds to the degree of polarity of the compounds.

4. What contributions did Jacobus van't Hoff make to the understanding of the three-dimensional structures of molecules? What other property of organic molecules did his research help explain?

5. Make models of a family of compounds such as H_2O, H_2Se, and H_2Te and relate the bond angles to the predictions of VSEPR theory.

6. How did Christian Schönbein discover ozone? What role does ozone play in the atmosphere? How is it affected by fluorocarbons?

7. How did Linus Pauling use wave theory to describe chemical bonds?

8. Diagram a simple protein molecule. What types of bonds and attractions contribute to the structure of the molecule?

9. Trace the development of adhesives.

Readings and References

Gordon, J. E. *The New Science of Strong Materials or Why You Don't Fall Through the Floor,* 2nd ed. Princeton, NJ: Princeton University Press, 1984.

Morgan, Ralph A. *Collisions, Coalescence and Crystals: The States of Matter and Their Models.* New York: Methuen Educational (dist. by Harper & Row), 1973.

Vergara, William C. *Science in Everyday Life.* New York: Harper & Row, 1980.

Wood, Clair G. "Dissolving Plastic." *ChemMatters* (October 1987), pp. 12–14.

15 Water and Aqueous Systems

Chapter Preview

15·1 The Water Molecule
15·2 Surface Properties of Water
15·3 The Heat Capacity of Water
15·4 The Vaporization of Water
15·5 Ice
15·6 Aqueous Solutions

15·7 Solvation
15·A When the Water is Hard
15·8 Water of Hydration
15·9 Electrolytes and Nonelectrolytes
15·10 Suspensions and Colloids
15·B Ellen Richards: Pioneer in Water Treatment

Figure 15·1
The unique chemical properties of water make it well suited to be the basis of life on our planet.

Water has a unique story because it is a unique compound. Liquid water covers about three-quarters of the surface of the earth, providing us with oceans, lakes, and rivers. Immense aquifers store water deep underground. Ice dominates the vast polar regions of the globe. It appears as icebergs in the oceans, and it whitens the temperate zones as snow in winter. Water vapor from the evaporation of surface water and from steam spouted from geysers and volcanoes is ever-present in the earth's atmosphere. Water is the foundation of everything that lives. Without it neither plant nor animal life as we know it could exist.

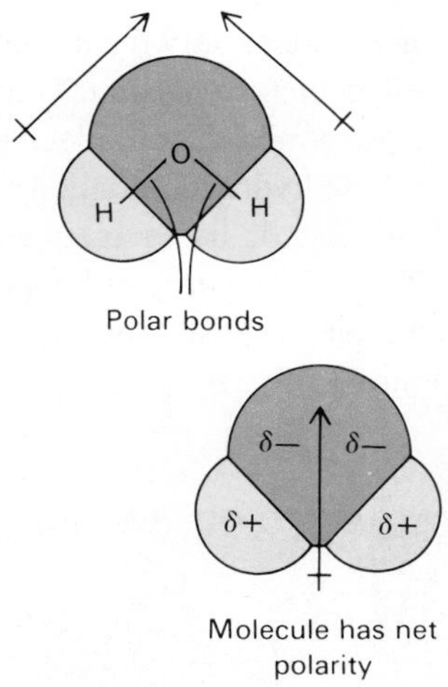

Polar bonds

Molecule has net polarity

Figure 15·2
The water molecule is bent. The bond polarities are equal, but the two dipoles do not cancel each other. There is a net polarity, and the molecule as a whole is polar.

15·1 The Water Molecule

Water is a simple triatomic molecule. Each O—H covalent bond in the water molecule is highly polar. Because of its greater electronegativity, oxygen attracts the electron pair of the covalent O—H bond and acquires a slightly negative charge. The less electronegative hydrogens acquire a slightly positive charge. The water molecule has a H—O—H bond angle of 105°. Because of the water molecule's bent shape, the two O—H bond polarities do not cancel, and the water molecule as a whole is polar. The region around oxygen is slightly negatively charged. The region around the hydrogens is slightly positively charged.

One property of polar molecules is that they attract one another. This leads to a dipolar interaction between the positive region of one water molecule and the negative region of another. This attraction is very weak, however, when compared with the strength of intermolecular

349

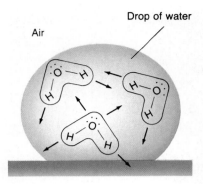

Air

Drop of water

Figure 15·3
Water molecules at the surface experience an uneven attraction. Because they cannot hydrogen-bond with air molecules, they tend to be drawn into the body of the liquid, producing surface tension.

■ Water molecules are polar, and they participate in hydrogen-bonding.

■ Water has unusual surface properties because of hydrogen-bonding.

hydrogen bonds. In water the molecules change position by sliding over one another. Because of hydrogen-bonding, however, most do not have enough kinetic energy to escape. *Hydrogen bonds* are the cause of many unique and important properties of water. They are responsible for its high surface tension, low vapor pressure, high specific heat, high heat of vaporization, and high-boiling point.

15·2   Surface Properties of Water

You have probably seen a glass so filled with water that the water surface bulges above the rim. You also may have noticed that water forms nearly spherical droplets at the end of a medicine dropper or when sprayed on a greasy surface. A needle floats if placed upon the surface of water, but it sinks immediately if it breaks through the surface. Apparently the surface of water acts like a skin. This phenomenon, called *surface tension,* is explained by the ability of water to form hydrogen bonds. The molecules within the liquid are completely surrounded by, and hydrogen-bonded to, adjacent water molecules. At the surface of the liquid, however, water molecules experience an uneven attraction. They are hydrogen-bonded on only one side. They are not attracted to the air because they cannot form hydrogen bonds with air molecules (Figure 15·3). As a result, the

Figure 15·4
Surface tension tends to hold drops of liquids in a spherical shape. At the same time the force of gravity tends to flatten the drops. The higher the surface tension, the more spherical the drop of liquid. All of the drops shown are of equal volume. From left to right they are mercury, water, and water with detergent.

Figure 15·5
Surface tension makes it possible for some insects such as this water strider to "walk" on water. The force of gravity pulling the insect downward is less than the forces of attraction between the water molecules.

surface water molecules tend to be drawn into the body of the liquid. *This inward force or pull which tends to minimize the surface area of a liquid is* **surface tension.**

A sphere has the smallest surface area for a given volume. The surface tension of a liquid tends to hold a drop of liquid in a spherical shape. A liquid that has strong intermolecular attractions has a high-surface tension. The higher the surface tension, the more nearly spherical the drop of that particular liquid (Figure 15·4).

All liquids have a surface tension, but water's surface tension is higher than most. The surface tension of water may be decreased by adding a *wetting agent* such as soap or detergent. Soaps and detergents are **surfactants** (*surf*ace *act*ive *a*gents). When a detergent is added to beads of water on a greasy surface, the detergent molecules interfere with the hydrogen-bonding between water molecules. As a result, surface tension is reduced. The beads collapse, and the water spreads out.

Hydrogen-bonding between water molecules also explains water's unusually low vapor pressure. As was covered in Section 9·6, the *vapor pressure* of a liquid is caused by the tendency of the molecules at the surface to escape. Because hydrogen bonds hold water molecules together, the tendency of these molecules to escape is low. If it were not, all the lakes and oceans, with their large surface areas, would rapidly evaporate!

The surface tension is strong for liquids having strong intermolecular attractions.

The low vapor pressure of water is the result of strong intermolecular forces of attraction between water molecules.

15·3   The Heat Capacity of Water

You may recall from Section 2·12 that it takes 1 cal of heat energy to raise the temperature of 1 g of water 1°C. The specific heat or heat capacity of water is nearly constant at 1 cal/g°C at temperatures between 0°C and 100°C. The specific heat of water is higher than that of many other substances. Iron has a specific heat of only 0.107 cal/g°C. For the same increase in temperature, iron requires about one-tenth as much heat as does an equal mass of water. On a sunny day a puddle of water remains relatively cool. Meanwhile a nearby manhole cover gets hot enough, so it is said, to fry an egg.

The units for specific heat, cal/(g × °C), can also be written cal/g°C.

Figure 15·6
People flock to the beach in hot weather because the shore is cooler than inland areas. Can you explain this phenomenon in terms of water's heat capacity?

Water has a higher specific heat than many other substances.

Water's high heat capacity helps moderate daily air temperatures around large bodies of water. On a warm day, water absorbs heat from its warmer environment, lowering the air temperature. On a cool night the transfer of heat is from the water to its cooler environment, raising the air temperature. Thus water is a storage medium for solar energy.

15·4 The Vaporization of Water

Because of hydrogen-bonding, water absorbs a large amount of heat before it vaporizes. The heat of vaporization is the amount of energy needed to convert one gram of a substance from a liquid to a gas at the boiling point. It takes 540 cal to convert 1 g of water at 100°C to 1 g of steam at 100°C. The extensive network of hydrogen bonds holds the molecules in water tightly together. The heat of vaporization of liquid ammonia, which is less hydrogen-bonded than water, is 327 cal/g. Liquid methane, which has no hydrogen-bonding, has a heat of vaporization of only 122 cal/g.

Water has the highest heat of vaporization of any substance that is a liquid at room temperature.

The reverse of vaporization is condensation. When 1 g of steam at 100°C condenses to 1 g of water at 100°C, 540 cal of heat is liberated. The heat of condensation of water is, therefore, 540 cal/g. As expected, this heat is numerically equal to the heat of vaporization. You can get a severe burn if you allow steam to condense on your hand. Not only is the temperature of steam 100°C, but the additional heat of condensation is absorbed by your skin.

The condensation of water is also important to geographical temperatures. Temperatures in the tropics would be much higher if water did not absorb heat while evaporating from the surface of the surrounding oceans. Temperatures in the polar regions would be much colder if water vapor did not release its heat while condensing out of the air.

Compounds of low formula mass are usually gases or low-boiling liquids at normal atmospheric pressure (Table 15·1). Ammonia is a typical

Table 15·1   Melting Points and Boiling Points of Some Substances with Low-Molecular Mass

Name of substance	Formula	Formula mass (amu)	Melting point (°C)	Boiling point (°C)
Methane	CH_4	16	-183	-164
Ammonia	NH_3	17	-77.7	-33.3
Water	H_2O	18	0	100
Neon	Ne	20	-249	-246
Methanol	CH_3OH	32	-93.9	64.9
Hydrogen sulfide	H_2S	34	-85.5	-60.7

example. Ammonia has a formula mass of 17 amu and boils at $-33°C$. Yet water, with a formula mass of 18 amu, is an exception. It has a boiling point of 100°C. The reason is again hydrogen-bonding. Hydrogen-bonding is more extensive in water than in ammonia. It takes much more heat to disrupt the attractions between water molecules than between ammonia molecules. If the hydrogen-bonding in water were as weak as it is in ammonia, water would be a gas at the usual temperatures found on earth.

Problem

1. Why does ammonia have a much higher boiling point ($-33°C$) than methane ($-164°C$) although their formula weights are almost the same?

15·5   Ice

Table 15·2   Density of Water and Ice

Temperature (°C)	Density (g/cm³)
100 (water)	0.9584
50	0.9881
25	0.9971
10	0.9997
4	1.0000*
0 (water)	0.9998
0 (ice)	0.9168

*Most dense.

As a typical liquid is cooled, it contracts slightly. Its density increases because its mass stays constant. (Recall that density = mass/volume.) If the cooling continues, the liquid eventually solidifies. Because the density of a typical solid is greater than that of the liquid, the solid sinks in its own liquid. (Lead shot sinks in molten lead.) As water is cooled, it first behaves like a typical liquid. Its density gradually increases until the temperature reaches 4°C. Below 4°C, the density of water starts to decrease (Table 15·2). At 0°C, ice begins to form. Ice has about 10% greater volume and therefore a lower density than water. As a result, it floats on water. This is certainly unusual behavior for a solid. Ice is one of very few solids that floats in its own liquid.

Why does ice behave so differently? The structure of ice is a very regular open framework of water molecules arranged like a honeycomb (Figure 15·7). Hydrogen-bonding holds the water molecules in place. At low temperatures they do not have sufficient kinetic energy to break out

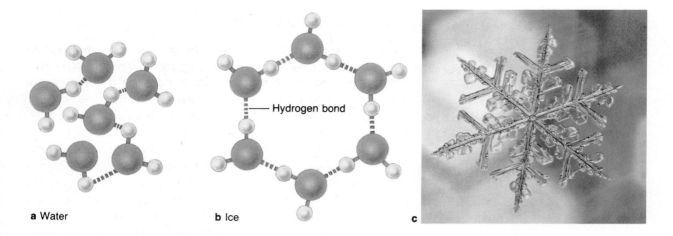

a Water

b Ice

— Hydrogen bond

c

Figure 15·7
a The molecules in liquid water are hydrogen bonded but free to move about.
b Extensive hydrogen bonding in ice causes the water molecules to be held further apart and in a more ordered arrangement than in liquid water.
c The hexagonal symmetry of snowflakes results from the hydrogen bonding of water molecules in ice.

Ice, like water, has unusual properties because of hydrogen-bonding.

of the rigid framework. When ice melts, the framework collapses. Then the water molecules pack together, and the water becomes more dense.

The fact that ice floats has important consequences for living organisms. The ice on a pond acts as an insulator for the water beneath, preventing it from freezing except under extreme conditions. Since the water at the bottom of a frozen pond is warmer than 0°C, aquatic life is able to survive. If ice were more dense than water, all bodies of water would tend to freeze solid during the winter months. Many types of aquatic life would be destroyed.

Water molecules require a considerable amount of kinetic energy to return to a liquid state from a frozen state. Ice melts at 0°C, a high temperature for such a small molecule (Table 15·1). The heat absorbed when 1 g of water changes from a solid to a liquid is 80 cal/g. This amount of energy could also raise the temperature of 1 g of water from 0°C to 80°C!

Figure 15·8
Ice packs are used to decrease swelling and pain from an injury. Water absorbs a large amount of heat as it changes from a solid to a liquid at 0°C.

Water is an excellent solvent.

Figure 15·9
The small size of solute particles allows them to pass through filter paper.

Safety

Many solutions are odor-free and have the appearance of water. Always label prepared solutions. Never taste a liquid to determine its identity.

15·6   Aqueous Solutions

Chemically pure water never exists in nature because water dissolves so many substances. Tap water contains varying amounts of dissolved minerals and gases. *Water samples containing dissolved substances are* **aqueous solutions.** In a solution, *the dissolving medium is the* **solvent,** and *the dissolved particles are the* **solute.** If sodium chloride is dissolved in water, water is the solvent and sodium chloride is the solute.

Solutions are homogeneous and stable. For example, sodium chloride does not settle out when its solutions are allowed to stand as long as other conditions such as temperature remain constant. Solute particles can be either ionic or molecular. Their average diameters are usually less than 1.0 nm (10^{-9} m). If a solution is filtered, both the solute and the solvent pass through the filter. Solvents and solutes may be gases, liquids, or solids. Table 1·3 lists some common types of solutions.

Substances that dissolve most readily in water include ionic compounds and polar covalent molecules. Nonpolar covalent molecules like methane and like those in grease and gasoline do not dissolve in water. Nevertheless, grease will dissolve in gasoline. Why the difference? To answer this question the structures of the solvent and the solute and the attractions between them must be discussed.

15·7   Solvation

Water molecules are in continuous motion because of their kinetic energy. When a crystal of sodium chloride is placed in water, the water molecules collide with it. The solvent molecules (H_2O) attract the solute (NaCl) ions. As individual sodium ions and chloride ions break away from the crystal, the sodium chloride dissolves (Figure 15·10). **Solvation** *occurs when a solute dissolves.* The negatively and positively charged ions become solvated, that is, surrounded by solvent molecules.

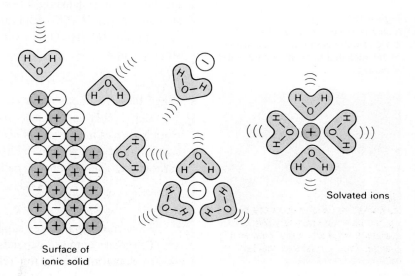

Solvated ions

Surface of ionic solid

Figure 15·10
When an ionic solid dissolves, the ions become solvated and are surrounded by solvent molecules.

In some ionic compounds the attractive forces within the crystals are stronger than the attractive forces exerted by water. These compounds cannot be solvated and are therefore insoluble. Barium sulfate ($BaSO_4$) and calcium carbonate ($CaCO_3$) are examples of nearly insoluble ionic compounds.

What about dissolving grease in gasoline? Both grease and gasoline are composed of nonpolar molecules. They mix to form a solution, but not because the solute and solvent are favorably attracted. Rather, it is because there are no repulsive forces between the two. As a rule, polar solvents dissolve ionic and polar compounds. Similarly, nonpolar solvents dissolve nonpolar compounds. This relationship can be summed up in the expression "like dissolves like."

Polar solvents such as water dissolve polar compounds, whereas nonpolar solvents dissolve nonpolar compounds.

Problem

2. Which of the following substances dissolve in water? Give reasons for your choice. **a.** HCl **b.** NaI **c.** NH_3 **d.** $MgSO_4$ **e.** CH_4 **f.** $CaCO_3$ **g.** gasoline

ChemDirections

Rainwater seeping through an abandoned hazardous waste site can dissolve or suspend substances that are dangerous to human health and the environment. These harmful substances may eventually appear in your drinking water. To discover what the government is doing to prevent this from happening read **ChemDirections** *Hazardous Wastes*, page 658.

Figure 15·11
Industrial boiler pipes can become clogged with calcium and magnesium carbonates that form when hard water is heated.

— Science, Technology, and Society —

15·A When the Water Is Hard

Have you ever found it difficult to get soap to lather? If so, then you may have been trying to wash in "hard" water. Hard water contains Ca^{2+}, Mg^{2+} and ions from dissolved minerals. Ground water is hard in many areas because of certain minerals in the ground that react with the carbon dioxide dissolved in the water.

When soap is added to hard water, undesirable precipitates form that hamper the cleansing action of soap. Soap is composed of sodium stearate ($NaC_{18}H_{35}O_2$). Calcium or magnesium ions combine with stearate ions ($C_{18}H_{35}O_2$) to form a gray scum of calcium or magnesium stearate.

$$Ca^{2+} + 2C_{18}H_{35}O_2^- \longrightarrow Ca(C_{18}H_{35}O_2)_2 \downarrow$$

When these reactions occur, it is hard to get soap to lather.

Hard water is not just troublesome when washing hands and clothes. When hard water is heated in industrial boilers, the dissolved calcium and magnesium hydrogen carbonates decompose to form carbonates. The carbonates, called "scale," can clog boilers and pipes.

$$Ca^{2+} + 2HCO_3^- \xrightarrow{heat} CaCO_3 \downarrow + H_2O + CO_2$$

Most municipal water supplies are treated to "soften" hard water. The treatment used depends on the types of anions present in the water. Sometimes only the relatively unstable hydrogen carbonate

Many methods for softening water replace Ca^{2+} and Mg^{2+} ions with Na^+ ions. People who must limit their sodium intake for medical reasons must be aware that softened water often contains elevated levels of Na^+ ions.

ion is present. Then the water can be softened by simply boiling the water. This precipitates the calcium and magnesium carbonates and removes them from the water supply.

Often more stable anions are present in the water. Then less stable anions must be added to force the calcium and magnesium ions to precipitate. For example, sodium carbonate (Na_2CO_3) can be added. Then the calcium and magnesium ions precipitate as calcium and magnesium carbonate.

$$Ca^{2+} + Na_2CO_3 \longrightarrow CaCO_3 + 2Na^+$$

Sodium phosphate (Na_3PO_4) added to the water causes the calcium and magnesium to precipitate as phosphates. Free sodium ions do not interfere with the action of soap.

$$3Ca^{2+} + 2Na_3PO_4 \longrightarrow Ca_3(PO_4)_2 + 6Na^+$$

Sometimes ion exchange is used to soften hard water. An ion exchanger can be made from natural minerals called "zeolites." Aluminum and silicate groups form a lattice in the zeolite in which free sodium ions are distributed. When hard water is passed through a zeolite, the sodium ions dissolve in the water, and the calcium and magnesium ions are caught in the zeolite. Most household water softeners use a synthetic zeolite that reacts more rapidly than natural zeolites.

Chemists often use a cation-exchange resin to remove positively charged ions like magnesium and calcium from water. The resin contains positively charged hydronium ions (H_3O^+). Like the sodium ions in zeolite, hydronium ions change places with the calcium and magnesium ions. Water can also be passed through an anion-exchanger. In that case hydroxide (OH^-) ions are exchanged for the negatively charged ions in the water. Water that is passed through a cation-exchanger and an anion-exchanger is said to be "deionized."

Figure 15·12
Ion exchange resins are used to exchange positive and negative ions in water. This is particularly important for laboratory experiments in which high concentrations of ions in the water may interfere with chemical reactions.

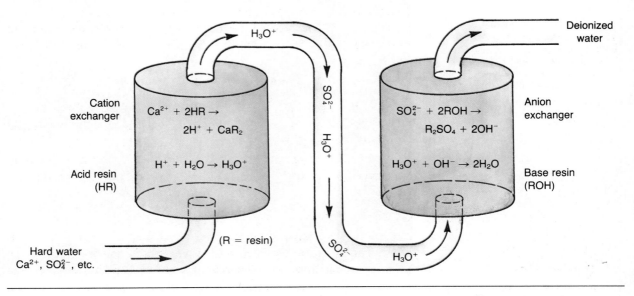

15·8 Water of Hydration

Water molecules are an integral part of the crystal structure of many substances. *The water in a crystal is called the* **water of hydration,** *or water of crystallization.* When an aqueous solution of copper(II) sulfate is allowed to evaporate, crystals of copper(II) sulfate pentahydrate are deposited. The chemical formula for this compound is $CuSO_4 \cdot 5H_2O$. A dot is used in the formula of a hydrate to connect the formula of the compound and the number of water molecules per formula unit. Crystals of copper sulfate pentahydrate always contain five molecules of water for every copper and sulfate ion pair. The deep blue crystals are dry to the touch. They are unchanged in composition or appearance in normally moist air. When heated or exposed to dry air, however, the crystals lose their water of hydration. They crumble to a white anhydrous powder whose formula is $CuSO_4$. If anhydrous copper sulfate is treated with water, the pentahydrate is regenerated.

$$CuSO_4 \cdot 5H_2O \underset{-heat}{\overset{+heat}{\rightleftharpoons}} CuSO_4 + 5H_2O$$

Table 15·3 lists some familiar hydrates. They all contain a fixed quantity of water and have a definite composition.

The forces holding the water molecules in hydrates are not very strong. This is shown by the ease with which water is lost and regained. Because the water molecules are held by weak forces, hydrates often have an appreciable vapor pressure. *If a hydrate has a vapor pressure higher than that of the water vapor in air, the hydrate will* **effloresce** *by losing the water of hydration.* For example, $CuSO_4 \cdot 5H_2O$ has a vapor

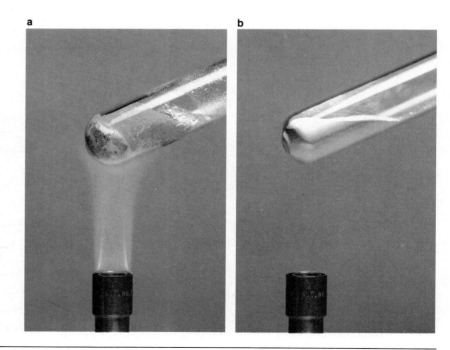

Figure 15·13
When hydrated copper sulfate **(a)** is heated, the water is driven off leaving anhydrous copper sulfate **(b).**

Figure 15·14
Dry sodium hydroxide pellets exposed to air (a) will, over time, remove water from the air (b). Eventually they will form a solution.

a

b

pressure of about 7.8 mm Hg at room temperature. The average pressure of water vapor at room temperature is usually about 10 mm Hg. This hydrate is stable until the humidity decreases. When the vapor pressure drops below 7.8 mm Hg the hydrate will effloresce.

Some hydrated salts, with a low vapor pressure, remove water from moist air to form higher hydrates.

$$CaCl_2 \cdot H_2O \xrightarrow{\text{moist air}} CaCl_2 \cdot 2H_2O$$

Salts and other compounds that remove moisture from air are said to be **hygroscopic.** Calcium chloride monohydrate is hygroscopic as are tetraphosphorus decoxide, sodium hydroxide, and concentrated sulfuric acid. *Hygroscopic substances are used as drying agents*, or **desiccants.** Some compounds are so hygroscopic that they become wet when exposed to normally moist air. These **deliquescent** *compounds remove sufficient water from the air to completely dissolve and form solutions*. Deliquescence does not necessarily involve hydrate formation. It occurs when the solution formed has a lower vapor pressure than that of the water in the air.

Table 15·3	Some Common Hydrates	
Formula	Chemical name	Common name
$MgSO_4 \cdot 7H_2O$	Magnesium sulfate heptahydrate	Epsom salts
$Ba(OH)_2 \cdot 8H_2O$	Barium hydroxide octahydrate	
$CaCl_2 \cdot 2H_2O$	Calcium chloride dihydrate	
$CuSO_4 \cdot 5H_2O$	Copper(II) sulfate pentahydrate	Blue vitriol
$Na_2SO_4 \cdot 10H_2O$	Sodium sulfate decahydrate	Glauber's salt
$KAl(SO_4)_2 \cdot 12H_2O$	Potassium aluminum sulfate dodecahydrate	Alum
$Na_2B_4O_7 \cdot 10H_2O$	Sodium tetraborate decahydrate	Borax
$FeSO_4 \cdot 7H_2O$	Iron(II) sulfate heptahydrate	Green vitriol
$H_2SO_4 \cdot H_2O$	Sulfuric acid hydrate (m.p. 8.6°C)	

Example 1

Calculate the percent by mass of water in washing soda ($Na_2CO_3 \cdot 10H_2O$).

Solution

The gram formula mass of $Na_2CO_3 \cdot 10H_2O$ is 286 g. The mass of water per gram formula mass of $Na_2CO_3 \cdot 10H_2O$ is $10 \times 18\text{ g} = 180\text{ g}$.

$$\text{Percent } H_2O = \frac{\text{mass of water}}{\text{mass of hydrate}} \times 100\%$$

$$= \frac{180 \not{g}}{286 \not{g}} \times 100\% = 62.9\%$$

Problem

3. What is the percent by mass of water in $CuSO_4 \cdot 5H_2O$?

15·9 Electrolytes and Nonelectrolytes

Safety

Even dilute aqueous solutions like tap water can conduct electricity. Do not use electrical equipment on a wet lab bench. Do not handle this equipment with wet hands or while standing on a wet floor.

Compounds that conduct an electric current in aqueous solution or the molten state are **electrolytes** (Figure 15·15). All ionic compounds are electrolytes. Sodium chloride, copper(II) sulfate, and sodium hydroxide are typical water-soluble electrolytes. They conduct electricity in solution and in the molten state. Barium sulfate is an example of an ionic compound that conducts electricity in the molten state but not in aqueous solution because it is insoluble.

All ionic compounds are electrolytes. Many water-soluble molecular compounds are nonelectrolytes.

Compounds that do not conduct an electric current in aqueous solution or the molten state are **nonelectrolytes.** Many molecular compounds are nonelectrolytes because they are nonionic. Most compounds of carbon such as cane sugar and rubbing alcohol are nonelectrolytes.

Some very polar molecular compounds are not electrolytes in the pure state, but they become electrolytes when they dissolve in water.

Table 15·4 Some Examples of Strong Electrolytes, Weak Electrolytes, and Nonelectrolytes

Strong electrolyte	Weak electrolyte	Nonelectrolyte
Acids (inorganic)	Heavy metal halides	Most organic compounds
HCl, HBr, HI, HNO_3, H_2SO_4, $HClO_4$	$HgCl_2$, $PbCl_2$	Sucrose (cane sugar)
Bases (inorganic)	Ammonia	Glycerol
$NaOH$, KOH	Organic acids and bases	
Soluble salts	acetic acid, aniline	
KCl, $MgSO_4$, $KClO_3$, $CaCl_2$	H_2O (very weak)	

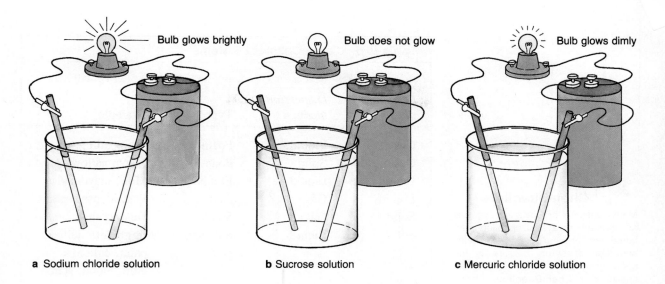

| **a** Sodium chloride solution | **b** Sucrose solution | **c** Mercuric chloride solution |
| Bulb glows brightly | Bulb does not glow | Bulb glows dimly |

Figure 15·15
The presence of an electrolyte in solution can be determined by a conductivity test. **a** Sodium chloride is a strong electrolyte. **b** Sucrose is a nonelectrolyte. **c** Mercuric chloride is a weak electrolyte.

This is because they ionize in solution. For example, neither ammonia (NH_3) nor hydrogen chloride (HCl) is an electrolyte. Yet an aqueous solution of ammonia conducts electricity. This is because ammonium ions (NH_4^+) and hydroxide ions (OH^-) are formed when ammonia dissolves. Similarly, hydrogen chloride produces hydronium ions (H_3O^+) and chloride ions (Cl^-). A hydrogen chloride solution is therefore an electrolyte.

$$NH_3 + H_2O \longrightarrow NH_4^+ + OH^-$$
$$HCl + H_2O \longrightarrow H_3O^+ + Cl^-$$

Not all electrolytes conduct an electric current to the same degree (Figure 15·15). In a simple conductivity test, a bulb glows dimly with the electrodes in a mercuric chloride solution. By contrast, the bulb glows brightly for a sodium chloride solution. Mercuric chloride is a weak electrolyte. *When a* **weak electrolyte** *is in solution, only a fraction of the solute exists as ions.* In a solution of mercuric chloride, most of the solute exists as unionized $HgCl_2$. In contrast, *when a* **strong electrolyte** *is dissolved, a large portion of the solute exists as ions.* Sodium chloride is a strong electrolyte. All the dissolved sodium chloride in a solution exists as Na^+ and Cl^- ions. Table 15·4 lists some common electrolytes and nonelectrolytes.

15·10 Suspensions and Colloids

Figure 15·16
In a suspension, the particles are not dissolved. Rather, they are suspended in the liquid medium. Suspended particles can be removed by filtration.

The emphasis of the chapter now shifts slightly to mixtures of water with substances that do not form true solutions. Two such mixtures are suspensions and colloids. **Suspensions** *are mixtures from which some of the particles will settle slowly upon standing.* A piece of clay shaken with water forms a suspension. The clay particles become suspended in the water, but they start to settle when shaking stops. A suspension differs from a solution because the component particles are much larger. The particles in a typical suspension have an average diameter greater than 100 nm. By contrast, in a solution the particle size is usually about 1 nm.

Table 15·5 Some Colloidal Systems

System			
Dispersed phase	*Dispersion medium*	Type	Example
Gas	Liquid	Foam	Whipped cream
Gas	Solid	Foam	Marshmallow
Liquid	Liquid	Emulsion	Milk, mayonnaise
Liquid	Gas	Aerosol	Fog, aerosol sprays
Solid	Gas	Smoke	Dust in air
Solid	Liquid	Sols and gels	Egg white, jellies, paint, colloidal gold, blood, starch in water

ChemDirections

Many consumer products are available as aerosol sprays. Chlorofluorocarbons were once used as propellants, and their effect upon ozone in the upper atmosphere led to the problem described in **ChemDirections** *Ozone Layer*, page 665.

Suspensions are unstable. The particles in a suspension settle out unless it is shaken or stirred.

The word *colloid* was coined by the English scientist Thomas Graham, in 1861. Greek: *kolla* = glue.

Chemical Connections

Colloids are widely used in the manufacture of cosmetics. For example, hairspray is an aerosol consisting of a resin suspended in a gas. Cold cream is an emulsion of rose water and other ingredients in beeswax and almond oil.

Suspensions are heterogeneous because at least two substances can be clearly identified. In our example, these were clay and water. If muddy water is filtered, the suspended clay particles are trapped by the filter and clear water passes through.

Colloids *are mixtures containing particles that are intermediate in size between those of suspensions and true solutions.* Thus the particles range in size from 1 nm to 100 nm. The particles are the *dispersed phase*. They are spread throughout the *dispersion medium*. The first colloids that were identified as such were glues. Many other materials such as blood, paint, aerosol sprays, and smoke are also colloids (Table 15·5).

The properties of colloids differ from those of solutions and suspensions. Many colloids are cloudy or milky in appearance but look clear when they are very dilute. The particles in a colloid cannot be retained by filter paper and do not settle out with time.

Colloidal particles exhibit the **Tyndall effect,** *which is the scattering of visible light in all directions.* We see a beam of light passed through a colloid just as we see a sunbeam in a dusty room. Suspensions also exhibit the Tyndall effect, but solutions never do (Figure 15·17).

Figure 15·17
Light is reflected in all directions from particles in colloids or in suspensions. This effect, called the Tyndall effect, is not observed with true solutions.

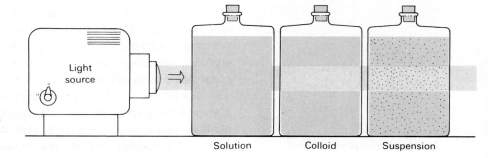

Light source | Solution | Colloid | Suspension

Table 15·6 Properties of Solutions, Colloids, and Suspensions

Property	System		
	Solution	*Colloid*	*Suspension*
Particle type	Ions, atoms, small molecules	Large molecules or particles	Large particles or aggregates
Particle size (approximate)	0.1–1 nm	1–100 nm	100 nm and larger
Effect of light	No scattering	Gives Tyndall effect	Gives Tyndall effect
Effect of gravity	Stable, does not separate	Stable, does not separate	Unstable, sediment forms
Filtration	Particles not retained on filter	Particles not retained on filter	Particles retained on filter
Uniformity	Homogeneous	Borderline	Heterogeneous

Figure 15·18
Many salad dressings are liquid-in-liquid colloids or emulsions. When vinegar and oil are shaken, a temporary emulsion forms. If an emulsifying agent such as egg yolk is added, the result is mayonnaise, a stable emulsion of vinegar and oil.

Flashes of light (scintillations) are seen when colloids are studied under a microscope. Colloids scintillate because the particles reflecting and scattering the light move erratically. *The chaotic movement of colloidal particles is called* **Brownian motion.** It was first observed by the Scottish botanist Robert Brown (1773–1858). Brownian motion is caused by the water molecules of the medium colliding with the small, dispersed colloidal particles. The buffeting action exerts such a force on the particles that they cannot settle.

Colloidal particles may also absorb charged particles from the surrounding medium onto their surface. Whether or not this occurs depends on the characteristics of the particles. Some colloidal particles can absorb positively charged particles and become positively charged. Some colloidal particles can absorb negatively charged particles and become negatively charged. All the particles in a system will have the same charge. The repulsion of the like-charged particles prevents them from forming aggregates. This keeps them dispersed throughout the medium. Charged colloidal particles form aggregates and precipitate from the dispersion when ions of opposite charge are added. Table 15·6 summarizes the properties of solutions, colloids, and suspensions.

Emulsions *are colloidal dispersions of liquids in liquids*. An emulsifying agent is essential for the formation of an emulsion and for maintaining its stability. For example, oils and greases are not soluble in water. They readily form a colloidal dispersion, however, if soap or detergent is added to the water. Soap and detergents are emulsifying agents. One end of the large soap molecule is polar and is attracted to water molecules. The other end of the soap molecule is nonpolar. It is soluble in oil or grease. Soaps and other emulsifying agents thus allow the formation of tiny stable droplets of one liquid in another liquid.

a

b

Figure 15·19
Two aspects of waste water treatment are shown. In a settling tank (a) suspended solids sink to the bottom and are periodically removed as sludge. Air can be added in an aeration tank (b) to allow bacteria to break down organic materials.

 Issues in Chemistry

Where and how is the waste water from your school treated? Is the treatment plant capable of removing or neutralizing the substances you pour down the drain in the lab?

15·B Ellen Richards: Pioneer in Water Treatment

Ellen Richards was America's first formally educated female chemist. In 1873, she received a B.S. degree in chemistry from the Massachusetts Institute of Technology (M. I. T.). She soon decided to devote her time to applying chemical knowledge to the public health problems of the day. Interested in a wide variety of problems such as nutrition and the environment, she focussed her attention on water quality and sanitation. She helped establish a department of sanitary chemistry at M. I. T. and remained an instructor in this department until her retirement in 1911.

Richards was concerned that drinking water was often contaminated by sewage and other harmful substances. She became a strong advocate for establishing public health standards. In one study she analyzed over 40 000 water samples. Richards also initiated significant advances in water treatment. She helped, for instance, to develop the use of straw filters for deodorizing waste water. When water passed through the straw, bacteria present in the water would collect on the straw. These bacteria captured the nitrogen compounds present in the water and removed the source of the odors.

During her career, Richards trained many men and women in sanitary chemistry. These people helped make water and sewage treatment commonplace by the middle of the twentieth century.

At a modern water treatment center, water is allowed to stand to allow silt and other fine particles to settle out. Instead of Richards's straw, a light, highly absorbent material called "floc" is added to the settling pond. Colloidal particles and bacteria become attached to the floc on the surface of the water and are removed. Next, the water is filtered through sand, pulverized coal, or a net of fibrous material. This removes additional suspended solids and bacteria. Chlorine may be added to kill any organisms not removed by filtration. Other possible steps include filtration through activated charcoal, aeration, and removal of excess calcium and magnesium ions (see Section 15·A).

Modern sewage treatment is a related process also influenced by Richards's work. It consists of up to three stages. In primary treatment, solids are separated from liquids and are sent to a landfill. In secondary treatment, liquid waste is treated with special bacteria to remove organic substances and nitrogen. This is a step that Richards initiated. Waste water may be released into a body of water after this step, or it may be first sent through an activated carbon filter.

Sanitary engineers help design and run water treatment and sewage treatment facilities. A background in chemistry allows the engineer to determine how to process the water and remove contaminants. Engineering skills are used in designing, constructing, and maintaining the equipment to carry out the processes.

Key Terms

aqueous solution	15·6	solute	15·6
Brownian motion	15·10	solvation	15·7
colloid	15·10	solvent	15·6
deliquescent	15·8	strong electrolyte	15·9
desiccant	15·8	surface tension	15·2
effloresce	15·8	surfactant	15·2
electrolyte	15·9	suspension	15·10
emulsion	15·10	Tyndall effect	15·10
hygroscopic	15·8	water of hydration	15·8
nonelectrolyte	15·9	weak electrolyte	15·9

Chapter Summary

Hydrogen-bonding between polar water molecules explains the high boiling point (100°C) and freezing point (0°C) of water, and it explains why ice floats. Hydrogen-bonding is also responsible for the high surface tension of water. Liquids tend to minimize their surface area and form spherical droplets because of their surface tension. Water is a polar liquid and an excellent solvent for many substances. Aqueous solutions are homogeneous mixtures of ions or molecules in water. The solubility of a solute depends on solute–solvent interactions. A good rule to remember is that "like dissolves like."

Substances that are in solution as ions are electrolytes. A solute that is completely ionized in solution is a strong electrolyte. A solute that is only partially ionized is a weak electrolyte. A solution of an electrolyte will conduct an electric current, whereas a solution of a nonelectrolyte is nonconducting.

True solutions are distinguished from colloidal dispersions and suspensions on the basis of particle size. Particles with an average diameter greater than 100 nm can be kept in suspension if the fluid (water) is kept agitated. Gravity or filtration, however, will separate the suspended particles from the liquid. The particles in a colloidal dispersion range in size from 1 nm to 100 nm. In general, they do not settle under gravity, and they pass through ordinary filter paper unchanged. Colloids are good at scattering light, as are suspensions, as evidenced by the Tyndall effect. Colloidal dispersions also exhibit Brownian motion. The particles in solutions are ions and small molecules. They cannot be trapped by filter paper, nor do they exhibit the Tyndall effect.

Many crystals are hydrates; they contain water of hydration. If water of hydration is lost from a hydrate when it is exposed to air, the process is called efflorescence. Some substances take up water from moist air and may eventually form a solution. They are deliquescent.

Practice Questions and Problems

4. Explain why water molecules are polar. *15·1*

5. Name four physical properties of water. *15·1*

6. What is meant by hydrogen-bonding? *15·1*

7. What is surface tension? Why do the particles at the surface of a liquid behave differently from those in the bulk of the liquid? *15·2*

8. Describe some observable effects that can be produced by the surface tension of a liquid. *15·2*

9. Explain how hydrogen-bonding is responsible for the high surface tension of water. *15·2*

10. What is a surfactant? Explain how it works. *15·2*

11. Define the term *vapor pressure* and explain why the vapor pressure of water is unusually low. *15·2*

12. How many calories are required to heat 256 g of water from 20°C to 99°C? How many calories are required to heat the same mass of iron through the same range of temperature? *15·3*

13. How many calories are liberated when 24 g of steam at 100°C condenses to 24 g of water at 100°C? *15·4*

14. Explain why water has a relatively high boiling point. *15·4*

15. Distinguish between the structures of water and ice. Then explain why ice floats on water. *15·5*

16. What would be some of the consequences if ice were denser than water? *15·5*

17. How much energy is required to change 47.6 g of ice at 0°C to 47.6 g of water at the same temperature? **a.** kilojoules **b.** kilocalories *15·5*

18. Why is water an excellent solvent for most ionic compounds and polar covalent molecules but not for nonpolar compounds? *15·6*

19. Define the terms *solution* and *aqueous solution*. *15·6*

20. Give a familiar example of solutions of each of these types. *15·6*
 a. gas in liquid **c.** solid in liquid
 b. liquid in liquid **d.** gas in gas

21. Suppose an aqueous solution contains both sugar and salt. Can you separate either of these solutes from the water by filtration? Explain. *15·6*

22. What is the significance of the statement "Like dissolves like"? What does "like" refer to? *15·6*

23. Describe how an ionic compound dissolves in water. *15·7*

24. Which of the following substances dissolve in water? Why? *15·7*
 a. CH_4 **c.** He **e.** cane sugar
 b. KCl **d.** $MgSO_4$ **f.** $NaHCO_3$

25. Define water of hydration. *15·8*

26. Name each of these hydrates. *15·8*
 a. $SnCl_4 \cdot 5H_2O$ **c.** $BaBr_2 \cdot 4H_2O$
 b. $FeSO_4 \cdot 7H_2O$ **d.** $FePO_4 \cdot 4H_2O$

27. Calcium chloride forms a hexahydrate. Write the equation for the formation of this hydrate from the anhydrous salt. *15·8*

28. Epsom salt ($MgSO_4 \cdot 7H_2O$) changes to the monohydrate form at 150°C. Write an equation for this change. *15·8*

29. Distinguish between an electrolyte and a nonelectrolyte. *15·9*

30. Contrast the characteristics of strong and weak electrolytes. *15·9*

31. Why does molten sodium chloride conduct electricity? *15·9*

32. What is the main distinction between an aqueous solution of a strong electrolyte and an aqueous solution of a weak electrolyte? *15·9*

33. Define the terms *suspension* and *colloid*. *15·10*

34. What is the basis for distinguishing among solutions, colloids, and suspensions? *15·10*

35. Explain the Tyndall effect. *15·10*

36. Solutions do not give the Tyndall effect. Why? *15·10*

37. What makes a colloidal dispersion stable? *15·10*

38. Define Brownian motion. *15·10*

39. Describe the term *emulsion* and tell how emulsions are stabilized. *15·10*

40. Mayonnaise is an example of what kind of colloidal dispersion? *15·10*

Mastery Questions and Problems

41. Water has its maximum density at 4°C. Explain why this is so, and discuss the consequences of this fact.

42. From your knowledge of intermolecular forces arrange these liquids in order of increasing surface tension: water, H_2O; hexane, C_6H_{14}; ethanol, C_2H_5OH; chloroform, $CHCl_3$.

43. Name these hydrates and determine the percent by mass of water in each.
 a. $Na_2B_4O_7 \cdot 10H_2O$ **c.** $MgSO_4 \cdot 7H_2O$
 b. $Na_2CO_3 \cdot H_2O$ **d.** $(CaSO_4)_2 \cdot H_2O$

44. If 5.0 g of steam, $H_2O(g)$, at 100°C condenses to 5.0 g of water, $H_2O(l)$, at 50°C, how much heat is liberated? **a.** calories **b.** kilocalories

45. Explain why ions become solvated in aqueous solution.

46. A block of ice at 0°C has a mass of 176.0 g. How much heat must be added to change 25% of this mass of ice to water at 0°C? (Express your answer in calories, kilocalories, joules, and kilojoules.) What is the mass of ice remaining?

47. You are given a solution containing either sugar or salt dissolved in water.
  a. Can you tell which it is by visual inspection?
  b. Give two ways by which you could easily tell which it is.

48. Water is a polar solvent; gasoline is a nonpolar solvent. Decide which of the following compounds are more likely to dissolve in water and which are more likely to dissolve in gasoline.
  a. sugar ($C_{12}H_{22}O_{11}$)
  b. chloroform ($CHCl_3$)
  c. Na_2SO_4
  d. methane (CH_4)
  e. KCl
  f. $BaSO_4$

49. Match each of the descriptions with the following terms. A description may apply to more than one term.
  1. true solution
  2. colloidal dispersion
  3. suspension

  a. Does not settle out on standing: _____
  b. Heterogeneous mixture: _____
  c. Particle size less than 1.0 nm: _____
  d. Particles can be filtered out: _____
  e. Gives Tyndall effect: _____
  f. Particles are invisible to the naked eye: _____
  g. Milk: _____
  h. Salt water: _____
  i. Jelly: _____

50. Explain which unique properties of water are responsible for these occurrences.
  a. Water in tiny cracks in rocks helps break up the rocks when it freezes.
  b. Water beads up on a newly waxed car.
  c. As you exercise and your body temperature increases, your body produces sweat.
  d. Grape vines are damaged at temperatures below 28°F. When severe frost is predicted, grape growers spray a mist of water on their vines.
  e. An efficient way of heating a large building is to generate steam in a boiler and circulate it through pipes to radiators throughout the entire building.

51. Explain why ethyl alcohol (C_2H_6O) will dissolve in both gasoline and water.

52. A 25.0 m × 10.0 m swimming pool is filled with fresh water to a depth of 1.7 m. The water temperature is initially at 25°C. How much heat must be removed from the water to change it all to ice at 0°C? Express your answer in kilocalories and kilojoules.

53. Are all liquids soluble in each other? Explain your answer.

54. Write equations to show how these substances ionize or dissociate in water.
  a. NH_4Cl
  b. $Cu(NO_3)_2$
  c. $HC_2H_3O_2$
  d. $HgCl_2$

55. The graph below shows the variation in the density of water over the temperature range 0°C to 20°C.
  a. What is the maximum density of water?
  b. At what temperature does the maximum density of water occur?
  c. Would it be correct to extend the smooth curve of the graph back below 0°C?

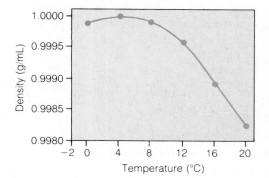

Critical Thinking Questions

56. Choose the term that best completes the second relationship.
  a. plant:green   water molecule: _____
    (1) polar   (3) frozen
    (2) ionic   (4) vapor
  b. east:west   condensation: _____
    (1) boiling   (3) vaporization
    (2) vapor pressure   (4) freezing
  c. colloid:emulsion   cat: _____
    (1) dog   (3) lion
    (2) Siamese   (4) fox

57. When the humidity is low and the temperature high, human beings must take in large quantities of water or face serious dehydration. Why do you think water is so important for the proper functioning of your body?

58. Describe as specifically as possible what would happen if a nonpolar molecular liquid were added to water. What would be formed if you shook this mixture?

Review Questions and Problems

59. A cylindrical vessel, 28.0 cm in height and 3.00 cm in diameter, is filled with water at 50°C. The density of water is 0.988 g/cm^3 at this temperature. What is the mass of water in the vessel?
 a. grams   **b.** milligrams   **c.** kilograms

60. A 1-L sample of steam at 100°C and 1 atm pressure is changed to 200°C at constant volume. Calculate the final pressure of the steam in atm.

61. What effect does a change in pressure on a water surface have on the boiling point of the water?
 a. an increase
 b. a decrease

62. The decomposition of hydrogen peroxide is given by this equation.

$$2H_2O_2(l) \longrightarrow 2H_2O(l) + O_2(g)$$

Calculate the mass of water in grams and the volume of oxygen at STP when 2.00×10^{-3} mol of hydrogen peroxide is decomposed.

63. How many grams each of hydrogen and oxygen are required to produce 4.50 mol of water?

64. Calculate the mass of water produced in the complete combustion of 8.00 L of propane (C_3H_8) at STP.

$$C_3H_8 + O_2 \longrightarrow CO_2 + H_2O$$

65. Acetaldehyde, C_2H_4O, is produced commercially by the reaction of acetylene with water.

$$C_2H_2 + H_2O \longrightarrow C_2H_4O$$

How many grams of C_2H_4O can be produced from 260 g H_2O, assuming sufficient C_2H_2 is present?

66. Hydrogen reacts with oxygen to form water.

$$2H_2 + O_2 \longrightarrow 2H_2O$$

How many moles of oxygen are required to produce 10.8 g H_2O? How many liters of oxygen is this at STP?

67. Balance these equations.
 a. $CO_2 + H_2O \longrightarrow C_6H_{12}O_6 + O_2$
 b. $Na + H_2O \longrightarrow Na^+ + OH^- + H_2$
 c. $C_6H_6 + O_2 \longrightarrow CO_2 + H_2O$
 d. $NaHCO_3 \longrightarrow Na_2CO_3 + H_2O + CO_2$

68. Give the number of molecules in 4.5×10^{-3} g of H_2O.

69. The normal boiling point of a substance depends on both the molecular mass and the intermolecular interactions. Considering these, arrange the following in order of decreasing boiling points and explain your answers.
 a. HBr, HCl, H_2O
 b. NH_3, H_2O, CH_4

70. If 20.0 g steam, $H_2O(g)$, occupies 6.25 L at 120°C, what is the pressure in atmospheres?

71. Calculate the percent composition.
 a. H_2O
 b. H_2S
 c. H_2Se
 d. H_2Te

72. What is the mass, in milligrams and in kilograms, of 2.7×10^3 mol H_2O?

73. Write the correct electron configuration for the oxide ion. Which noble gas has the same electron configuration?

74. When a proton is attracted to the unshared electron pair of a water molecule, the polyatomic hydronium ion (H_3O^+) is formed. Write electron dot structures to show the formation of this ion.

Challenging Questions and Problems

75. Deuterium oxide is often described as "heavy water". Why? List the differences in physical and chemical properties you would expect to exist between water and deuterium oxide. Give some uses for deuterium oxide.

76. What relationships exist between the following volumes?
 a. 1 g of ice at 0°C and 1 g of water at 0°C
 b. 1 g of water at 100°C and 1 g of steam at 100°C

77. A mixture of 40 cm³ of oxygen gas and 60 cm³ of hydrogen gas, at STP, is ignited. Determine the following.
  a. which gas is the limiting reagent
  b. the mass of water produced
  c. which gas remains after reaction
  d. the volume of the remaining gas

78. An iceberg with a mass of 3.50×10^6 kg is floating in fresh water at a temperature of 0°C. Calculate the mass and volume of the iceberg above the water level. How would this mass and volume change if the temperature of the water was raised to 4°C? (Use Table 15·2 and assume no melting of ice occurs.)

79. When ethyl alcohol (C_2H_6O) dissolves in water, the volume of the final solution is less than the separate volumes of the water and alcohol added together. Can you explain this result? Do you think it might be possible to mix two different liquids and get a mixture volume that is larger than the sum of the volume of the two components? Explain.

80. When an aqueous solution of sodium chloride starts to freeze, the ice that forms does not contain ions of the salt. Why?

81. Why is the midday sky blue and the early evening western sky often an orange or red color? (Hint: There are colloidal-size particles in the atmosphere.)

Research Projects

1. How do solutes affect the surface tension of water? Design an experiment to find the answer.

2. How does the volume of a solution compare to the volumes of the solute and the solvent? Give a few examples.

3. Compare the conductivities of solutions to various ions present.

4. What were Thomas Graham's contributions to colloid chemistry?

5. How did John Tyndall use his work on light scattering in solutions to describe why the sky is blue?

6. How does a household tap water purifier work?

7. How important is tertiary wastewater treatment? Why is it used in some areas but not in other areas?

8. How can wastewater be reused? How can sewer sludge be processed and reused?

9. What are the components of fabric softeners? How do they work?

10. What are the laws regulating water pollution in your area? What federal, state, and local agencies are responsible for enforcing the regulations?

11. Plaster of Paris incorporates water into its crystalline structure as it hardens. Devise an experiment to determine the number of water molecules taken up per formula unit of $CaSO_4$.

12. Test the effect of different soaps and detergents on the surface tension of water.

Readings and References

Branley, Franklyn M. *Water for the World*. New York: Harper & Row, 1982.

Goldin, Augusta. *Water: Too Much, Too Little, Too Polluted?* San Diego, CA: Harcourt Brace Jovanovich, 1983.

Hansen, James. "The Delicate Architecture of Cement." *Science 82* (December 1982), pp. 48–55.

Lightman, Alan. "Snow." *Science 84* (December 1984), pp. 22–26.

Maranto, Gina. "The Creeping Poison Underground." *Discover* (March 1985), pp. 83–85.

Pringle, Laurence. *Water: The Next Great Resource Battle*. New York: Macmillan, 1982.

Smith, Wesley D. "Skin Deep." *ChemMatters* (December 1987), pp. 4–7.

16 Properties of Solutions

Chapter Preview

16·1 Solution Formation
16·2 Solubility
16·3 Factors Affecting Solubility
16·A Chromatography
16·4 Molarity
16·5 Making Dilutions
16·B Nutrition: The Birth of a Science
16·6 Percent Solutions

16·7 Colligative Properties of Solutions
16·8 Molality and Mole Fraction
16·9 Calculating Boiling and Freezing Point Changes
16·10 Molecular Mass Determination

Chemical Connections

Blood is a solution that contains dissolved solids, dissolved gases, and suspended solids. The clear, yellowish solution part of blood is plasma. The suspended solids are red and white blood cells and platelets.

Solutions are homogeneous mixtures that can be grouped according to their physical state. Solid, liquid, and gaseous solutions exist. Certain metal alloys are solid solutions, air is a gaseous solution, and ocean water is a liquid solution. Chemists usually work with liquid solutions. This chapter begins with discussions of solution formation and the concept of solubility. Various ways of expressing the concentrations of solutions, both qualitatively and quantitatively, will then be examined. Colligative properties of solutions are very important. They depend on the number of solute particles dissolved in a given volume of solvent. The colligative properties that will be covered are vapor pressure lowering, boiling point elevation, and freezing point depression.

16·1 Solution Formation

The rate at which a solute dissolves is influenced by agitation, temperature, and particle size.

Figure 16·1
Rock formations can be created and eroded due to the properties of solutions.

The nature of the solvent and solute affects *whether* a substance will dissolve. Several other factors determine *how fast* a soluble substance dissolves. If sodium chloride crystals are placed in a flask containing water, the crystals will eventually dissolve. If the flask and its contents are shaken, the crystals disappear more quickly than if the flask is not shaken. Agitation makes the solute dissolve more rapidly because it brings fresh solvent into contact with the solute. This is why most people stir coffee after they have added sugar. It is important to realize, however, that agitation affects only the *rate* at which a solute dissolves. It cannot influence the *amount* of solute that dissolves. An insoluble substance remains undissolved no matter how much the system is agitated.

Figure 16·2
Heat and agitation increase the rate at which a solute dissolves. Sugar dissolves faster in hot tea than it does in iced tea. Stirring also makes the sugar dissolve faster.

For most substances there is a limit to how much solute will dissolve in a given volume of solvent.

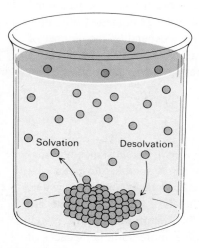

Figure 16·3
A state of dynamic equilibrium exists in a saturated solution. At these conditions the rate of solvation (dissolving) equals the rate of desolvation (crystallization).

Temperature also influences the rate at which solutes dissolve. A salt crystal dissolves much more rapidly in hot water than in cold water. The kinetic energy of the water molecules is greater at the higher temperature. This leads to an increased frequency and force of the collisions of water molecules with the crystal surfaces.

A third factor that determines the rate of dissolution of a solute is its particle size. A powder dissolves more rapidly than a single crystal. This is because a greater surface area of ions is exposed to the colliding water molecules. Solvation is a surface phenomenon.

16·2   Solubility

When 36 g of sodium chloride is added to 100 g of water at 25°C, all the salt dissolves. Yet if we add one more gram of salt and stir, no matter how vigorously or how long, only 0.2 g of the last portion goes into solution. Why? After all, the kinetic theory says that water molecules are in continuous motion. They must continue to bombard the excess solid, removing and solvating the ions. Seemingly, all the sodium chloride should eventually disappear, but that does not happen. What does take place is an exchange. New particles from the solid are solvated and enter into solution. At the same time an equal number of particles come out of solution. They become desolvated and are deposited as a solid. The undissolved crystals of sodium chloride change their shape over time, but their mass remains constant.

The particles move from the solid to the solvated state and back to the solid again. Yet there is no net change in the overall system. A state of dynamic equilibrium exists between the solution and the undissolved solute, provided that the temperature remains constant (Figure 16·3). The sodium chloride solution is saturated. *A **saturated solution** contains the maximum amount of solute for a given amount of solvent at a constant temperature.* For example, 36.2 g of sodium chloride in 100 g of water is a saturated solution at 25°C. *The **solubility** of a substance is the amount of substance that dissolves in a given quantity of a solvent at a given temperature to produce a saturated solution.* Solubility is usually expressed in grams of solute per 100 g of solvent. *A solution that contains less solute than a saturated solution is **unsaturated.*** The solubilities of some substances are given in Table 16·1.

Problem

1. What weight of NaCl can be dissolved in 750 g of water at 25°C?

*Two liquids are said to be **miscible** if they dissolve in each other.* For example, water and ethanol are infinitely soluble in one another. Any amount of ethanol will dissolve in a given volume of water and *vice versa.* Such a pair of liquids is said to be completely miscible. Liquids

Figure 16·4
Which of these two beakers contains a saturated solution?

a

b

that are slightly soluble in one another, for example, water and diethyl ether, are partially miscible. *Liquids that are insoluble in one another are* **immiscible.** Gasoline and water are immiscible.

16·3 Factors Affecting Solubility

At 25°C the solubility of sodium chloride in water is 36.2 g per 100 g of water. When the temperature is raised to 100°C, the solubility increases to 39.2 g of NaCl per 100 g of water. As shown in Figure 16·5, the solubility of most substances increases as the temperature of the solvent is increased. For some substances, however, the reverse occurs. The solubility of sodium sulfate in water drops from 50 g per 100 g at 40°C to 41 g per 100 g at 100°C. Table 16·1 lists the solubilities of some common substances at various temperatures.

Remember that 100 g of H_2O = 100 mL of H_2O at 4°C. This is not true for other liquids.

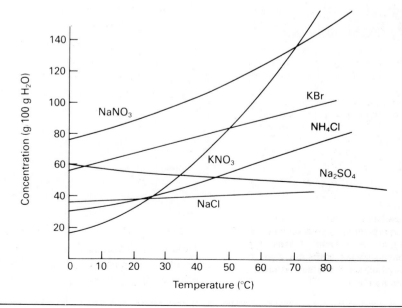

Figure 16·5
Changing the temperature may affect the solubility of a substance. Notice that increasing the temperature greatly increases the solubility of KNO_3 but decreases the solubility of Na_2SO_4.

Table 16·1 Solubilities of Some Substances in Water at Various Temperatures

Substance	Formula	Solubility (g/100 g of H_2O)			
		0°C	20°C	50°C	100°C
Barium sulfate	$BaSO_4$	0.00019	0.00025	0.00034	—
Lead(II) chloride	$PbCl_2$	0.60	0.99	1.70	—
Lithium carbonate	Li_2CO_3	1.5	1.3	1.1	0.70
Potassium chlorate	$KClO_3$	4.0	7.4	19.3	56.0
Potassium chloride	KCl	27.6	34.0	42.6	57.6
Sodium chloride	$NaCl$	35.7	36.0	37.0	39.2
Sodium nitrate	$NaNO_3$	74	88.0	114.0	182
Sodium sulfate	Na_2SO_4	4.76	62	50.0	41.0
Silver nitrate	$AgNO_3$	122	222.0	455.0	733
Lithium bromide	$LiBr$	143.0	166	203	266.0
Cane sugar	$C_{12}H_{22}O_{11}$	179	203.9	260.4	487
Hydrogen*	H_2	0.00019	0.00016	0.00013	0.0
Oxygen*	O_2	0.0070	0.0043	0.0026	0.0
Carbon dioxide*	CO_2	0.335	0.169	0.076	0.0

*Gas at 760 mm Hg total pressure

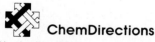
For most substances, solubility increases as temperature increases. For gases, solubility increases as temperature decreases.

The solubilities of gases are greater in cold water than in hot water. Bubbles form in water when it is heated as the dissolved gases escape from solution. The components of air all become less soluble in water as the temperature of the solution rises (Figure 16·6). When an industrial plant takes cool water from a lake and dumps hot water back into the lake, the temperature of the lake water rises. This is known as thermal

ChemDirections

Newly discovered organic compounds are capable of dissolving large amounts of oxygen. These substances show promise as blood substitutes in some medical applications. Look at **ChemDirections** *Blood Substitutes*, page 666, for further information on this topic.

Figure 16·6
Gases have different solubilities in water at different temperatures. Generally, as the temperature increases, the solubilities of gases decrease. Notice that nitrogen, oxygen, and nitric oxide are insoluble at 100°C.

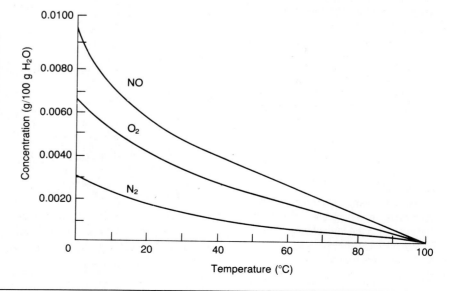

Figure 16·7
Warm but otherwise clean water can kill fish and other aquatic animals that depend on high concentrations of dissolved oxygen.

pollution. The temperature increase lowers the concentration of dissolved oxygen in the lake water. Aquatic animal life can be severely affected by the lack of oxygen.

The solubility of a gas increases as the partial pressure of the gas *above* the solution is increased. Carbonated beverages are a good example of this principle. These drinks contain large amounts of carbon dioxide (CO_2) dissolved in water. It is the dissolved CO_2 that makes your mouth tingle. The drinks are bottled under a high pressure of CO_2 gas. This forces large amounts of CO_2 into solution. When a carbonated beverage bottle is opened, the partial pressure of CO_2 above the liquid decreases, and the concentration of dissolved CO_2 decreases. Bubbles of CO_2 form in the liquid and escape from the open bottle. If the bottle is left open, the drink becomes flat as the solution loses its CO_2. **Henry's law** *states that at a given temperature the solubility of a gas in a liquid (S) is directly proportional to the pressure of the gas above the liquid (P).* In other words, as the pressure of the gas above the liquid increases, the solubility of the gas increases. Similarly, as the pressure of the gas decreases, the solubility of the gas decreases. This relationship can be written in mathematical form.

$$\frac{S_1}{P_1} = \frac{S_2}{P_2}$$

Chemical Connections

The quantity and kind of gases dissolved in aquarium water affect the health of the fish. When an aquarium is first filled, for example, the gaseous chlorine used to purify the tap water must be allowed to escape before the fish are added.

 Issues in Chemistry

The average temperature of a lake increases when its water is used as a coolant in a power plant. As the name implies, the resulting "thermal pollution" of the lake can have some negative effects. Can you think of some positive outcomes of using lake water as a power plant coolant?

Example 1

If the solubility of a gas in water is 0.77 g/L at 3.5 atm of pressure, what is its solubility, in grams per liter at 1.0 atm of pressure? (The temperature is held constant at 25°C.)

Solution

Use the same general method that was used for solving gas law problems in Chapter 10. The knowns are: $P_1 = 3.5$ atm, $S_1 = 0.77$ g/L, and $P_2 = 1.0$ atm. To find S_2, begin by rearranging Henry's law to give S_2.

$$S_2 = \frac{S_1 \times P_2}{P_1}$$

Then substitute values for the other variables in the equation.

$$S_2 = \frac{0.77 \text{ g/L} \times 1.0 \text{ atm}}{3.5 \text{ atm}} = 0.22 \text{ g/L}$$

When the pressure is reduced to 1.0 atm, the solubility of the gas decreases to 0.22 g/L.

Problem

2. A gas has a solubility in water at 0°C of 3.6 g/L at a pressure of 1.0 atm. What pressure is required to produce an aqueous solution containing 9.5 g/L of the same gas at 0°C?

Figure 16·8
Mineral deposits form around the edges of this hot spring because the hot water is saturated with minerals. As the water cools at the surface, the minerals crystallize because they are less soluble at a lower temperature.

(a) Supersaturated solution

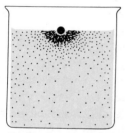

(b) Seed crystal added

(c) Excess solute crystallizes on seed

Figure 16·9
When a seed crystal **(b)** is added to a supersaturated solution **(a)**, the excess solute crystallizes **(c)**.

When the temperature of a saturated solution is raised, the excess solid will usually dissolve. If the system then cools slowly and undisturbed to its original temperature, the excess solute does not always immediately crystallize. *A solution which contains more solute than it can theoretically hold at a given temperature is a* **supersaturated solution.** In such a solution, a dynamic equilibrium cannot exist between the dissolved solute and the undissolved solid because there is no undissolved solid. Crystallization in a supersaturated solution can be initiated if a very small crystal, called a seed crystal, of the solute is added (Figure 16·9). Crystallization can also occur on a rough surface such as the inside of the container if it is scratched.

The rate at which excess solute deposits upon the surface of a seed crystal can be impressively rapid. Scientific rainmaking is done by seeding clouds, which are made of air supersaturated with water vapor. Tiny silver iodide (AgI) crystals are dusted on a cloud. Water molecules that are attracted to the ionic particles come together and form droplets that act as seeds for other water molecules. The water droplets grow and eventually fall as rain when they are large enough.

Science, Technology, and Society

16·A Chromatography

Chemists often need to separate the components of a solution to identify them. One of the most powerful techniques available for this purpose is chromatography. It is based on the fact that substances differ in their affinities for solvents and other media.

Paper chromatography is a relatively easy procedure to perform in the laboratory. It can be used to separate a mixture of colored substances. A small drop of the sample mixture is placed near the bottom edge of a piece of filter paper and is allowed to dry. The filter paper is placed in a shallow container so that the bottom edge just touches the chosen solvent. Capillary action causes the solvent to travel up the filter paper. Substances with little affinity for the paper

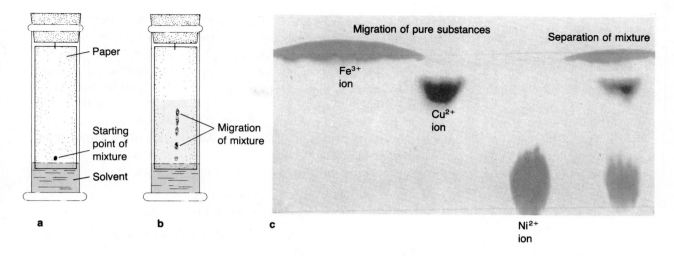

Migration of pure substances

Separation of mixture

Fe^{3+} ion

Cu^{2+} ion

Ni^{2+} ion

a   b   c

Figure 16·10
Paper chromatography can be used to detect the various ions. **a** The mixture to be identified is placed on paper which is rolled and put into a container of solvent. **b** The components of the mixture become separated as the solvent moves up the paper. **c** Three pure samples and one mixture were placed at the bottom of this chromatograph. In this case the mixture contained all three of the pure substances.

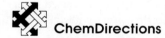

ChemDirections

Chromotography, together with mass spectrometry, is used in the analysis of small amounts of unknown materials, including unlawful substances. To find out more on this topic read **ChemDirections** *Drug Testing*, page 653.

Molarity is the most important unit of concentration in chemistry.

migrate up the strip almost as fast as the solvent. Components with greater affinity for the paper migrate at slower rates. The further the solvent migrates, the more the component substances of the sample mixture become separated. By changing the solvent, the degree of migration and separation can be varied.

Several other types of chromatography are also available. In some of these systems the liquid solvent, also called the mobile phase, is replaced by a gas solvent. The filter paper, the stationary phase, can be substituted by a liquid, a colloid, or more appropriate type of solid. The type of chromatographic system used depends on the substances that must be separated.

Commercial laboratories and research facilities make great use of chromatography. For example, the food industry uses a gas chromatograph to isolate and identify compounds responsible for food aromas. This is a first step in the process of producing flavorings that imitate natural flavors. Chromatography is used in many areas of chemical, medical, and biological research.

16·4 Molarity

So far this chapter has focused mainly on whether a particular substance does or does not dissolve in a particular solvent. This section focuses on ways of expressing the concentrations of solutions. *The* **concentration** *of a solution is a measure of the amount of solute that is dissolved in a given quantity of solvent.* The terms *concentrated* and *dilute* are a qualitative description of the amount of a solute in solution. *A* **dilute solution** *contains only a small amount of solute.* By contrast, *a* **concentrated solution** *contains a large amount of solute.* An aqueous solution of sodium chloride containing 1 g of NaCl per 100 g of H_2O might be described as dilute compared with another solution containing 30 g per 100 g. The first solution might be described as concentrated if it were compared with another solution containing only 1×10^{-2} g of NaCl per

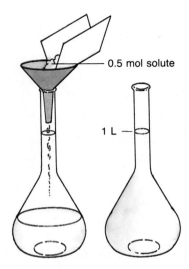

0.5 mol solute

1 L

Figure 16·11
To make a 0.5 mol solution, add 0.5 mol of solute to a 1-L volumetric flask half filled with water. Swirl the flask carefully to make the solute dissolve. Then fill the flask with water exactly to the 1 L mark.

Note that molarity is per liter of *solution,* not of *solvent.*

100 g of H_2O. For this reason, describing a solution as concentrated or dilute does not really express the concentration of the solution.

Just as the most important unit of mass in chemistry is the mole, the most important unit of concentration is molarity (abbreviated M). **Molarity** *is the number of moles of a solute dissolved in 1 L of solution.* Note that the volume is the volume of the *total solution,* not the volume of the solvent. If 0.5 mol of a solute is dissolved in water and diluted to a volume of exactly 1 L, a 0.5 molar (0.5 mol/L) or $0.5M$ solution is created. To determine the molarity of any solution, the number of moles in 1 L of the solution can be calculated using the following equation.

$$\text{Molarity } (M) = \frac{\text{number of moles of solute}}{\text{number of liters of solution}}$$

For example, if water is added to 2 mol of glucose to give 5 L of solution, the molarity is $0.4M$. To get this answer, divide the number of moles by the volume in liters.

$$\frac{2 \text{ mol of glucose}}{5 \text{ L of solution}} = 0.4 \text{ mol/L}$$

The solution is 0.4 molar ($0.4M$) in glucose.

Example 2

A saline solution contains 0.90 g of NaCl per 100 mL of solution. What is its molarity?

Solution

Molarity is moles per liter. The concentration must be converted from grams per 100 mL to moles per liter. To do this, start with the gram formula mass of sodium chloride.

$$\text{gfm NaCl} = 23.0 \text{ g} + 35.5 \text{ g} = 58.5 \text{ g}$$

Use a conversion factor based on the gram formula mass to calculate the moles of NaCl in 100 mL of saline.

$$\text{Moles} = 0.90 \text{ g} \times \frac{1.0 \text{ mol}}{58.5 \text{ g}} = 0.015 \text{ mol}$$

Now convert mol per 100 mL to mol per L (100 mL = 0.1 L).

$$\text{Molarity} = \frac{\text{moles}}{\text{liter}} = \frac{0.015 \text{ mol}}{0.1 \text{ L}} = 0.15M \text{ solution}$$

Problems

3. A salt solution has a volume of 250 mL and contains 0.70 mol of NaCl. What is the molarity of the solution?
4. An aqueous solution has a volume of 2.0 L and contains 36.0 g of glucose. If the gram formula mass of glucose is 180 g, what is the molarity of the solution?

$CuSO_4 \cdot 5H_2O$

NaCl

$KMnO_4$

1 mole

1 mole

1 mole

Figure 16·12
A 1 molar solution contains 1 mole of solute per liter of solution. One mole of each substance is shown: 58.5 g sodium chloride, 158 g potassium permanganate, 250 g copper sulfate pentahydrate.

Diluting a solution does not change the amount of solute.

Safety

To dilute an acid, always add the acid, with stirring, to the required amount of water. Never add water to the acid. The heat generated may splash the solution out of the container.

Example 3

How many moles of solute are present in 1.5 L of $0.20M$ Na_2SO_4?

Solution

Rearrange the expression for molarity to give moles of solute.

$$\text{Liters of solution} \times \text{molarity } (M) = \text{moles of solute}$$

Substitute in the known information, stating the molarity in mol/L.

$$1.5 \; \cancel{L} \times \frac{0.20 \; \text{mol}}{1.0 \; \cancel{L}} = 0.30 \; \text{mol}$$

To determine the weight of solute, multiply the number of moles of Na_2SO_4 by a conversion factor based on the gram formula mass.

$$0.30 \; \cancel{\text{mol}} \times \frac{142 \; \text{g}}{1.0 \; \cancel{\text{mol}}} = 42.6 \; \text{g} = 43 \; \text{g (to two significant figures)}$$

Problem

5. How many moles of solute are in 250 mL of $2.0M$ $CaCl_2$? How many grams of $CaCl_2$ is this?

16·5 Making Dilutions

Solutions of known molarity are usually available in the laboratory. Often other dilute solutions of known concentrations are required. A solution can be made less concentrated by diluting with solvent. *The number of moles of solute does not change when a solution is diluted.*

$$\begin{array}{c} \text{Number of moles of solute} \\ \textit{before} \text{ dilution} \end{array} = \begin{array}{c} \text{number of moles of solute} \\ \textit{after} \text{ dilution} \end{array}$$

The definition of molarity can be rearranged to show the relationship between molarity and volume.

$$\text{Molarity } (M) = \frac{\text{number of moles of solute}}{\text{number of liters of solution}}$$

$$\text{Moles of solute} = \text{molarity } (M) \times \text{liters of solution } (V)$$

Changing the amount of solvent does not change the amount of solute. The total number of moles remains constant upon dilution. If a solution is diluted from V_1 to V_2, the molarity of that solution changes according to this equation.

$$M_1 \times V_1 = M_2 \times V_2$$

In this equation M_1 and V_1 are the initial solution's molarity and volume, and M_2 and V_2 are the final solution's molarity and volume. Volume was expressed in liters to get the equation, but the volumes can be in liters or milliliters as long as the same units are used for both volumes.

Figure 16·13
To prepare 100 mL of 0.40M MgSO$_4$ from a stock solution of 2.0M MgSO$_4$, transfer 20 mL of the stock solution to a 100-mL volumetric flask. Carefully add water to make 100 mL of solution.

Accurate volume-measuring devices include the volumetric pipet, the buret, and the volumetric flask.

Chemists have a variety of volume-measuring devices available to them. Their choice of what to use depends on how accurate the concentration of the solution they are making must be. The following example shows how a dilute solution is prepared.

Example 4

How would you prepare 100 mL of 0.40M MgSO$_4$ from a stock solution of 2.0M MgSO$_4$?

Solution

Use the above equation to calculate the volume of 2.0M MgSO$_4$ and then describe the procedure. The knowns are: $M_1 = 2.0M$, $M_2 = 0.40M$, and $V_2 = 100$ mL. V_1 is the unknown. Rearrange the equation to give V_1.

$$M_1 \times V_1 = M_2 \times V_2 \text{ therefore } V_1 = \frac{M_2 \times V_2}{M_1}$$

Now substitute values into the equation.

$$V_1 = \frac{0.40M \times 100 \text{ mL}}{2.0M} = 20 \text{ mL}$$

Suppose the dilute solution must be very accurate. Measure 20 mL of the 2.0M MgSO$_4$ with a 20-mL volumetric pipet and transfer the solution to a 100-mL volumetric flask. (A buret can also be used to measure volumes accurately.) Next add distilled water to the flask up to the etched line. This dilutes the contents to exactly 100 mL. Then shake the flask. This produces 100 mL of 0.40M MgSO$_4$.

Safety

Never pipet a solution by mouth. Use a suction bulb.

Problem

6. You need 250 mL of 0.20M NaCl, but the only supply of sodium chloride you have is a solution of 1.0M NaCl. How do you prepare the required solution? Assume that you have the appropriate volume-measuring devices on hand.

Figure 16·14
Mary Swartz Rose used her knowledge of chemistry to make substantial contributions to the field of nutrition.

16·B Nutrition: The Birth of a Science

When Mary Swartz Rose (1874–1941) enrolled at Columbia University in 1905, no curriculum was available in the study of nutrition. Her desire to study the chemistry of food and nutrition led her to Yale University. She obtained a Ph.D. there in physiological chemistry. She returned to Columbia and was largely responsible for establishing a Department of Nutrition.

Rose was a brilliant and talented educator. Her department became a leading university center for the study of nutrition. She taught for 31 years and published over 40 scholarly papers on such topics as energy metabolism and the nutritive value of food proteins. She served on many national and international advisory councils concerned with diet and nutrition.

Nutrition is the study of the processes by which living organisms absorb and utilize food substances. Nutrients are transferred to the cells of the body by the circulatory system. Solution chemistry is a major aspect of nutritional science because body nutrients are more easily studied in the blood than in the body cells and tissues.

The world population is increasing more rapidly than the food supply. The agricultural and nutritional sciences are vital for our future. Nutritional chemists are helping to solve food supply problems by finding which foods provide adequate amounts of nutrients.

16·6 Percent Solutions

The two common ways of representing percent concentrations are volume/volume (v/v) percent and mass/volume (m/v) percent.

If both the solute and the solvent are liquids, a convenient way to make a solution is by measuring volumes. The concentration of the solute can then be expressed as a percent of the solution by volume. Percent means "parts per hundred." If 20 mL of rubbing alcohol is diluted with water to a total volume of 100 mL, the final solution is 20% alcohol by volume. It can be written as 20 percent (volume/volume), or 20% (v/v).

$$\text{Percent by volume} = \frac{\text{volume of solute}}{\text{volume of solution}} \times 100\%$$

Example 5

What is the percent by volume of ethanol (ethyl alcohol, C_2H_6O) in the final solution when 75 mL of ethanol is diluted to a volume of 250 mL with water?

Solution

$$\text{Percent by volume} = \frac{75 \ \text{mL}}{250 \ \text{mL}} \times 100\% = 0.30 \times 100\%$$
$$= 30\% \ \text{ethanol (v/v)}$$

Figure 16-15
Many product labels give the amounts of their ingredients by percent. Without units, however, we cannot be certain how the percent composition was determined.

Problem

7. If 10 mL of acetic acid ($HC_2H_3O_2$) is diluted with water to a total solution volume of 200 mL, what is the percent by volume of acetic acid in the solution?

A commonly used relationship for solutions of solids dissolved in liquids is percent (mass/volume). It is usually convenient to weigh the solute in grams and to measure the volume of the resulting solution in milliliters. Percent (mass/volume) is the number of grams of solute per 100 mL of solution. A solution containing 7 g of sodium chloride in 100 mL of solution is 7 percent (mass/volume) or 7% (m/v).

$$\text{Percent (mass/volume)} = \frac{\text{mass of solute (g)}}{\text{solution volume (mL)}} \times 100\%$$

Example 6

How many grams of glucose ($C_6H_{12}O_6$) would you need to prepare 2.0 L of 2.0% glucose (m/v) solution?

Solution

A 2.0% glucose (m/v) solution means that 100 mL of the solution contains 2.0 g of glucose. Use a conversion factor to determine the number of grams per liter.

$$\frac{2.0 \text{ g}}{100 \text{ mL}} \times \frac{1000 \text{ mL}}{1.0 \text{ L}} = 20 \text{ g/L}$$

Multiply the number of grams of solute per liter by the number of liters required.

$$2.0 \text{ L} \times \frac{20 \text{ g}}{1 \text{ L}} = 40 \text{ g}$$

The required mass of glucose is 40 g.

Problems

8. How many grams of magnesium sulfate are required to make 250 mL of a 1.6% $MgSO_4$ (m/v) solution?

9. A solution contains 2.7 g of $CuSO_4$ in 75 mL of solution. What is the percent (mass/volume) of the solution?

Percent composition can often be misleading. When a label says a product contains 5% glucose, what does it mean? It is probably specifying percent (mass/volume), but we cannot be certain unless the units are given. When using percentages to express concentration, be sure to state the units.

16·7 Colligative Properties of Solutions

Colligative properties depend on the number of particles in solution. They include: vapor pressure lowering, boiling point elevation, and freezing point depression.

The physical properties of a solution are different from those of the pure solvent. Some of these differences are due to the mere presence of solute particles in the solution. **Colligative properties** *depend on the number of particles dissolved in a given mass of solvent.* They do not depend on the chemical nature of the solute or the solvent. Three important colligative properties of solutions are vapor pressure lowering, boiling point elevation, and freezing point depression.

Vapor pressure is the pressure exerted by a vapor that is in dynamic equilibrium with its liquid in a closed system. A solution that contains a nonvolatile solute always has a lower vapor pressure than the pure solvent. A nonvolatile solute is one that does not vaporize. Glucose, a molecular compound, and sodium chloride, an ionic compound, are examples of nonvolatile solutes. Both glucose and sodium chloride would lower the vapor pressure of a pure solvent. Consider an aqueous sodium chloride solution as an example. Sodium ions and chloride ions are dispersed throughout the liquid. Both within the liquid and at the surface, the ions are surrounded by shells of water of solvation. The formation of these shells occupies much of the surface water. This reduces the number of solvent molecules that have enough kinetic energy to escape as vapor (Figure 16·16). As a result, the solution has a lower vapor pressure than the pure solvent.

The decrease in the vapor pressure is proportional *to the number of particles the solute makes in solution.* A solute that dissociates into several particles, like sodium chloride, has a greater effect on the vapor pressure than the same concentration of a nondissociating solute like glucose. For example, the vapor pressure lowering caused by 0.1 mol of sodium chloride in 1000 g of water is twice the vapor pressure lowering of 0.1 mol of glucose in 1000 g of water. Each formula unit of the ionic compound sodium chloride produces two particles in solution: a sodium ion and a chloride ion. When glucose dissolves, each molecule produces only one particle, the molecule itself. Similarly, 0.1 mol of $CaCl_2$ in 1000 mL of water produces three times the vapor pressure lowering of a 0.1 mol glucose solution. Each formula unit of calcium chloride produces three particles in solution.

An ionic compound dissociates into several particles in solution; a molecular compound does not dissociate in solution.

Figure 16·16
The vapor pressure of a solution of a nonvolatile solute is lower than the vapor pressure of a pure solvent. **a** Equilibrium is established between the liquid and the vapor in the pure solvent. **b** The equilibrium is disrupted when solute is added. Solvent particles form shells around the solute particles. This reduces the number of free solvent particles able to escape the liquid. Equilibrium is eventually reestablished at a lower vapor pressure.

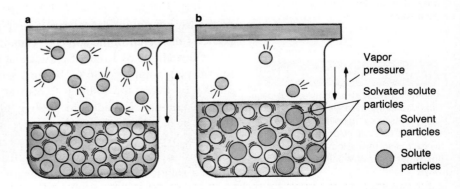

Figure 16·17
Temperatures below 0°C are needed to make ice cream. Thus rock salt is added to ice in a hand crank ice cream maker. A freezing point depression occurs and the temperature of the ice-water mixture decreases a few degrees below 0°C.

The boiling point of a solution is *higher* than the boiling point of the pure solvent.

The boiling point of a substance is the temperature at which the vapor pressure of the liquid phase equals the atmospheric pressure. Adding a nonvolatile solute to a liquid solvent decreases the vapor pressure of the solvent. Because of the decrease in vapor pressure, additional kinetic energy must be added to raise the vapor pressure of the liquid phase to atmospheric pressure. Thus the boiling point of the solution is higher than the boiling point of the pure solvent. *The **boiling point elevation** is the difference in temperature between the boiling points of a solution and of the pure solvent.*

Another way to think about the reason for a boiling point elevation is in terms of particles. Attractive forces exist between the solvent and solute particles. Additional kinetic energy must be added for solvent particles to overcome the attractive forces that keep them in the liquid phase. The magnitude of the boiling point elevation is proportional to the number of solute particles dissolved in the solvent. For example, the boiling point of water increases by 0.512°C for every mole of *particles* that the solute forms when dissolved in 1000 g of water. Thus boiling point elevation is a colligative property.

When a substance freezes, the particles in the solid take on an orderly pattern. The presence of a solute in water disrupts this pattern due to the shells of water of hydration. As a result, more kinetic energy must be withdrawn from a solution than from pure solvent for it to solidify. *The **freezing point depression** is the difference in temperature between the freezing points of a solution and of the pure solvent.* The magnitude of the freezing point depression is proportional to the number of solute particles dissolved in the solvent. Thus it is a colligative property.

The freezing point of a solution is *lower* than the freezing point of the pure solvent.

The addition of 1 mol of particles of solute to 1000 g of water lowers the freezing point by 1.86°C. When 1 mol (180 g) of glucose is added to 1000 g of water, the solution freezes at $-1.86°C$. If 1 mol (58.5 g) of sodium chloride is added to 1000 g of water, however, the solution freezes at $-3.72°C$. This is because 1 mol of NaCl produces 2 mol of particles. We take advantage of the freezing point depression of aqueous solutions to melt ice on sidewalks. Ice sprinkled with salt melts and forms a solution with a lower freezing point than that of pure water.

Chemical Connections

An antifreeze, such as ethylene glycol, is added to the water in a car's radiator to lower its freezing point and prevent it from freezing on cold days. The antifreeze also raises the boiling point of the water to prevent boiling over on hot days.

16·8 Molality and Mole Fraction

The colligative properties depend only on the ratio of the number of solute particles to solvent particles. Two convenient ways of expressing this ratio are in molality units and in mole fractions. **Molality** *(m) is the number of moles of solute dissolved in 1 kilogram (1000 g) of solvent.* Molality is also known as molal concentration.

$$\text{Molality} = \frac{\text{moles of solute}}{\text{kilogram of solvent}} = \frac{\text{moles of solute}}{1000 \text{ g of solvent}}$$

Molality, *m*, is moles of solute per kilogram of *solvent;* molarity, *M,* is moles of solute per liter of *solution.*

Note that *molality* is not the same as *molarity.* Molality refers to moles per kilogram of *solvent* rather than per liter of *solution.*

A solution that is 1.00 molal in glucose can be prepared by adding 1000 g of water to 1.00 mol (180 g) of glucose. A solution prepared by dissolving 0.500 mol (29.25 g) of sodium chloride in 1.000 kg (1000 g) of water is 0.500 molal (0.500 *m*) in NaCl.

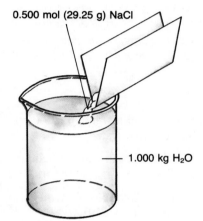

0.500 mol (29.25 g) NaCl

1.000 kg H₂O

Figure 16·18
To make a 0.500 *m* solution of NaCl use a balance to measure 1.000 kg of water, and add 29.25 g of NaCl to it.

Example 7

How many grams of potassium iodide must be dissolved in 500 g of water to produce a 0.060 molal KI solution?

Solution

According to the definition of *molal,* the final solution must contain 0.060 mol KI per 1000 g H₂O. Use this as a conversion factor.

$$500 \text{ g } H_2O \times \frac{0.060 \text{ mol KI}}{1000 \text{ g } H_2O} = 0.030 \text{ mol KI}$$

Then convert 0.030 mol of KI to grams.

$$\text{gfm KI} = 39.1 + 126.9 = 166.0 \text{ g}$$

$$0.030 \text{ mol KI} \times \frac{166.0 \text{ g KI}}{1 \text{ mol KI}} = 5.0 \text{ g KI}$$

Dissolve 5.0 g KI in 500 g of H₂O to get 0.060 molal KI.

Problem

10. Calculate the molality of a solution prepared by dissolving 10.0 g of NaCl in 600 g of water.

The ratio of the moles of solute in solution to the total number of moles of both solvent and solute is the **mole fraction** *of that solute.* In a solution containing n_A moles of solute and n_B moles of solvent, the mole fraction of solute, X_A, and mole fraction of the solvent, X_B, can be expressed mathematically.

$$X_A = \frac{n_A}{n_A + n_B} \qquad\qquad X_B = \frac{n_B}{n_A + n_B}$$

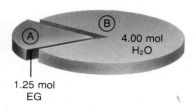

1.25 mol
EG

Total moles = Ⓐ + Ⓑ = 5.25 moles

Mole fraction EG $= \dfrac{Ⓐ}{Ⓐ + Ⓑ} = \dfrac{1.25}{5.25}$

Mole fraction H$_2$O $= \dfrac{Ⓑ}{Ⓐ + Ⓑ} = \dfrac{4.00}{5.25}$

Figure 16-19
The mole fraction is the ratio of the number of moles of one substance to the total number of moles of all substances in the solution.

Example 8

Compute the mole fraction of each component in a solution of 1.25 mol ethylene glycol (EG) and 4.00 mol water.

Solution

The mole fraction of ethylene glycol, X_{EG}, in the solution is the number of moles of ethylene glycol divided by the total number of moles in solution.

$$X_{EG} = \frac{n_{EG}}{n_{EG} + n_{H_2O}} = \frac{1.25 \text{ mol}}{1.25 \text{ mol} + 4.00 \text{ mol}} = 0.238$$

The mole fraction of water, X_{H_2O}, is the number of moles of water divided by the total number of moles in solution.

$$X_{H_2O} = \frac{n_{H_2O}}{n_{EG} + n_{H_2O}} = \frac{4.00 \text{ mol}}{1.25 \text{ mol} + 4.00 \text{ mol}} = 0.762$$

$$\text{Total} = 1.000$$

Note that mole fraction is a dimensionless quantity. *The sum of the mole fractions of all the components in a solution must equal unity.*

Problems

11. What is the mole fraction of each component in a solution made by mixing 300 g of ethanol (C$_2$H$_5$OH) and 500 g of water?

12. A solution is labeled 0.150 molal NaCl. What are the mole fractions of the solute and solvent in this solution?

13. Describe how you would make an aqueous solution of methanol, CH$_3$OH, in which the mole fraction of methanol is 0.40.

16·9 Calculating Boiling and Freezing Point Changes

Elevations of boiling points and depressions of freezing points are usually quite small. They can be measured accurately only with thermometers that can measure temperature changes as small as 0.001°C.

The boiling point of a solvent is raised by the addition of a nonvolatile solute. The magnitude of the boiling point elevation is directly proportional to the molal concentration when the solute is molecular (not ionic).

$$\Delta T_b \propto m$$

The change in the boiling temperature, ΔT_b, is the elevation of the boiling point of the solvent. It is the boiling point of the solution minus the boiling point of the solvent. The term m is the molal concentration of the solution. Adding a proportionality constant makes an equality.

$$\Delta T_b = K_b\, m$$

Boiling and freezing point changes are proportional to the number of particles in a solution.

386   Chapter 16   Properties of Solutions

Figure 16·20
The addition of a nonvolatile solute to water lowers its vapor pressure. This results in an increase in the normal boiling point of water and a decrease in the normal freezing point of water.

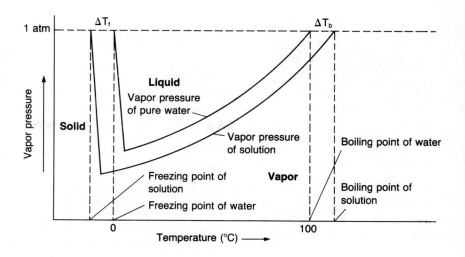

Table 16·2 K_b Values for Some Common Solvents

Solvent	$K_b(°C/m)$
Water	0.512
Ethanol	1.19
Benzene	2.53
Cyclohexane	2.79
Acetic acid	3.07
Phenol	3.56
Nitrobenzene	5.24
Camphor	5.95

The constant, K_b, the **molal boiling point elevation constant,** *is equal to the change in boiling point for a 1 molal solution of a nonvolatile molecular solute.* Values for the constant K_b for water and some other solvents are given in Table 16·2. The units of K_b are $°C/m$.

Example 9

What is the boiling point of a solution that contains 1.20 mol of sodium chloride in 800 g of water?

Solution

First determine the molality of the solution.

$$\frac{\text{mol NaCl}}{1000 \text{ g H}_2\text{O}} = \frac{1.20 \text{ mol NaCl}}{800 \text{ g}} \times \frac{1000 \text{ g}}{1 \text{ kg}} = 1.50 \ m$$

Each formula unit of NaCl is ionized into two particles.

$$\text{NaCl}(s) \longrightarrow \text{Na}^+(aq) + \text{Cl}^-(aq)$$

The molality of *total* particles is $2 \times 1.50 \ m = 3.00 \ m$. Now determine the boiling point elevation. From Table 16·2, K_b for H_2O is $0.512°C/m$.

$$\Delta T_b = K_b \times m$$
$$= 0.512°C/m \times 3.00 \ m$$
$$= 1.54°C$$

The boiling point of this solution is $100°C + 1.54°C = 101.54°C$.

Problem

14. What is the boiling point of a solution that contains 1.25 mol of $CaCl_2$ in 1400 g of water?

Table 16·3 K_f Values for Some Common Solvents	
Solvent	$K_f(°C/m)$
Water	1.86
Acetic acid	3.90
Benzene	5.12
Nitrobenzene	7.00
Phenol	7.40
Cyclohexane	20.2
Camphor	37.7

The freezing point depression of a solution can also be calculated.

$$\Delta T_f = K_f \times m$$

The change in the freezing temperature, ΔT_f, represents the freezing point depression. The molality is m. The **molal freezing point depression constant**, K_f, *is equal to the change in the freezing point for a 1 molal solution of a nonvolatile molecular solute* (Table 16·3).

Problems

15. What is the freezing point of these solutions?
  a. 1.40 mol of Na_2SO_4 in 1750 g of H_2O
  b. 0.60 mol of $MgSO_4$ in 1300 g of H_2O

16. One mole of a compound of iron and chlorine is dissolved in 1 kg of water. The boiling point of this aqueous solution is 102.05°C; the freezing point is −7.44°C. What is the formula of this compound?

16·10 Molecular Mass Determination

The changes in boiling and freezing points can be used to determine the molecular mass of a substance. A known amount of solute is added to a known amount of solvent, and the change in boiling or freezing point is measured. A solvent with a known molal freezing point depression constant (K_f) or molal boiling point elevation constant (K_b) must be used. Then the molecular mass for the solute can be calculated.

The molecular mass of a non-volatile solute may be determined by using the boiling point elevation or freezing point depression.

Example 10

A solution of 7.50 g of a nonvolatile compound in 22.60 g of water boils at 100.78°C at 760 mm Hg. What is the molecular mass of the solute? (Assume the solute exists as molecules and not ions.)

Solution

First use the boiling point elevation to calculate the molality of the solution. The observed boiling point elevation is:
100.78°C − 100.00°C = 0.78°C = ΔT_b.

$$\Delta T_b = K_b \times m \qquad \text{Thus} \qquad m = \frac{\Delta T_b}{K_b}$$

The K_b for water is $0.512 \dfrac{°C \times kg \text{ of water}}{mol \text{ of solute}}$

$$m = \frac{0.78°C}{0.512 \dfrac{°C \times kg \text{ of water}}{mol \text{ of solute}}}$$

$$m = 1.5 \frac{mol \text{ of solute}}{kg \text{ of water}}$$

Now calculate moles of solute in solution.

$$1.5 \frac{\text{mol of solute}}{\text{kg of water}} \times 22.6 \text{ g of water} \times \frac{1 \text{ kg}}{1000 \text{ g}} = 0.034 \text{ mol solute}$$

Finally use the number of moles of solute and its mass to determine the molecular mass of the solute.

$$\text{Molecular mass of solute} = \frac{\text{mass of solute}}{\text{moles of solute}}$$

$$= \frac{7.50 \text{ g}}{0.0344 \text{ mol}}$$

$$= 2.2 \times 10^2 \text{ g/mol}$$

Due to the difficulty of obtaining precise temperature measurements, most molecular mass determinations using boiling point elevation or freezing point depression are accurate to only two or three significant digits.

Problem

17. The freezing point for water is lowered to $-0.390°C$ when 3.90 g of a nonvolatile molecular solute is dissolved in 475 g of water. Calculate the molecular mass of the solute.

16 Properties of Solutions
Chapter Review

Key Terms

boiling point elevation	16·7	molal freezing point depression constant (K_f)	16·9
colligative property	16·7	molality (m)	16·8
concentrated solution	16·4	molarity (M)	16·4
concentration	16·4	mole fraction	16·8
dilute solution	16·4	saturated solution	16·2
freezing point depression	16·7	solubility	16·2
Henry's law	16·3	supersaturated solution	16·3
immiscible	16·2	unsaturated	16·2
miscible	16·2		
molal boiling point elevation constant (K_b)	16·9		

Chapter Summary

Changes in temperature and pressure of a system affect the solubility of a solute. The extent to which a gas dissolves in a liquid is proportional to the pressure of the gas in accordance with Henry's law. The solubility of a gas decreases with increasing temperature. Polar liquids tend to be soluble (miscible) in water whereas nonpolar liquids tend to be insoluble (immiscible). Generally the solubility of a solid in water increases with increasing temperature, but there are exceptions. The rate at which a solute dissolves is influenced by a number of factors, including the temperature of the solvent and the particle size of the solute.

The relative amounts of solute and solvent in a solution are best described by using quantitative units. include molar concentration, percent composition, molal concentration, and mole fraction.

The effects in solution of a nonvolatile solute on the properties of the solvent are called colligative properties. Colligative properties include freezing point and vapor pressure lowering and boiling point elevation. In each case, the magnitude of the effect is directly proportional to the number of solute molecules or ions present and is independent of the type of particle. Each solvent has a characteristic molal boiling point elevation constant and molal freezing point depression constant. The colligative properties are useful for demonstrating dissociation of solutes in solution and in determining molecular masses.

Practice Questions and Problems

18. Name and distinguish between the two components of a solution. *16·1*

19. Explain why the dissolved component does not settle out of a solution. *16·1*

20. Name three factors that influence the rate at which a solute dissolves in a solvent. *16·1*

21. Define the terms solubility, saturated solution, and unsaturated solution. *16·2*

22. What mass of $AgNO_3$ can be dissolved in 250 g of water at 20°C? *16·2*

23. If a clear saturated solution of sodium nitrate is cooled, what change(s) might you observe? *16·2*

24. Explain the terms *miscible* and *immiscible*. *16·2*

25. State Henry's law. *16·3*

26. What is the effect of pressure on the solubility of gases in liquids? *16·3*

27. The solubility of methane, the major component of natural gas, in water at 20°C and 1.00 atm pressure is 0.026 g/L. If the temperature remains constant, what will be the solubility of this gas be at the following pressures? *16·3*
   a. 0.60 atm   **b.** 1.80 atm

28. Calculate the molarity of each of the following solutions. *16·4*
   a. 1.0 mol of KCl in 750 mL of solution
   b. 0.50 mol of $MgCl_2$ in 1.5 L of solution
   c. 400 g of $CuSO_4$ in 4.00 L of solution
   d. 0.060 mol of $NaHCO_3$ in 1500 mL of solution

29. Define the term *molarity*. *16·4*

30. Calculate the number of moles and the number of grams of solute in each solution. *16·4*
   a. 1.0 L of 0.50M NaCl
   b. 500 mL of 2.0M KNO_3
   c. 250 mL of 0.10M $CaCl_2$
   d. 2.0 L of 0.30M Na_2SO_4

31. You have the following stock solutions available: 2.0M NaCl, 4.0M KNO_3, and 0.50M $MgSO_4$. Calculate the volumes you must dilute to make the following solutions. *16·5*
   a. 500 mL of 0.50M NaCl
   b. 2.0 L of 0.20M $MgSO_4$
   c. 50 mL of 0.20M KNO_3

32. Calculate the number of grams of solute required to make the following: *16·6*
   a. 2.5 L of normal saline solution (0.90% NaCl (m/v))
   b. 50 mL of 4.0% $MgCl_2$ (m/v)
   c. 250 mL of 0.10% $MgSO_4$ (m/v)

33. What is the concentration in percent (m/v) of the following solutions? *16·6*
   a. 20 g KCl in 600 mL of solution
   b. 32 g $NaNO_3$ in 2.0 L of solution
   c. 75 g K_2SO_4 in 1500 mL of solution

34. A solution with a nonvolatile solute has a lower vapor pressure than the pure solvent at the same temperature. Explain why. *16·7*

35. What is the boiling point of each solution? *16·7*
   a. 0.50 mol of glucose in 1000 g of H_2O
   b. 1.50 mol of NaCl in 1000 g of H_2O

36. How much NaCl would have to be dissolved in 1000 g of water to raise the boiling point 2.00°C? *16·7*

37. Distinguish between a lM solution and a l m solution. *16·8*

38. A solution is prepared by dissolving 2.50 g of K_2CrO_4 in 23.2 g of water. Calculate the molality of K_2CrO_4. *16·8*

39. How many kilograms of water must be added to 9.0 g of oxalic acid, $H_2C_2O_4$, to prepare a 0.025 m solution? *16·8*

40. A solution contains 50.0 g of carbon tetrachloride (CCl_4) and 50.0 g of chloroform ($CHCl_3$). Calculate the mole fraction of each component in the solution. *16·8*

41. Determine the freezing points of these 0.10 m aqueous solutions. *16·9*
  a. K_2SO_4 **b.** $CsNO_3$ **c.** $Al(NO_3)_3$

42. Estimate the freezing point of a solution of 12.0 g of carbon tetrachloride dissolved in 750 g of benzene (which has a freezing point of 5.48°C). *16·9*

43. A solution containing 16.9 g of a nonvolatile molecular compound in 250 g of water has a freezing point of −0.744°C. What is the molecular mass of the substance? *16·10*

Mastery Questions and Problems

44. Describe how you would prepare an aqueous solution of acetone, CH_3COCH_3, in which the mole fraction of acetone is 0.25.

45. You are asked to make 1.50 L of $0.250M$ HNO_3 by diluting concentrated $16.0M$ HNO_3. What volume of the concentrated acid will be required to make the dilution?

46. You are given a clear water solution containing KNO_3. How would you determine experimentally whether the solution is unsaturated, saturated, or supersaturated?

47. A mixture of glycerol and water is used as antifreeze in automobile engines. The freezing point and specific gravity of the glycerol/water mixture varies with the percent by mass of glycerol in the mixture. On the graph below, point A represents 20% glycerol by mass; point B 40%; and point C 60%.
  a. What is the specific gravity of the antifreeze mixture that freezes at −24°C?
  b. What is the freezing point of a mixture that has a specific gravity of 1.06?
  c. Estimate the freezing point of a mixture that is 30% by mass glycerol.

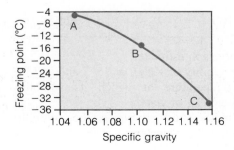

48. Calculate the freezing point and the boiling point of a solution that contains 15.0 g of urea, CH_4N_2O, in 250 g of water. Urea is a covalently bonded compound.

49. Calculate the mole fractions in a solution that is made of 25.0 g of ethyl alcohol, C_2H_5OH, and 40.0 g of water.

50. Estimate the freezing point of an aqueous solution of 20.0 g of glucose, $C_6H_{12}O_6$, dissolved in 500.0 g of water.

51. Plot a graph of solubility versus temperature for the three gases listed in Table 16·1.

52. The solubility of KCl in water is 34.0 g of $KCl/100$ g of H_2O at 20°C. A warm solution containing 50.0 g of KCl in 130 g of H_2O is cooled to 20°C.
  a. How many grams of KCl remain dissolved?
  b. How many grams came out of solution?

53. How many moles of ions are present when 0.10 mol of the following compounds are dissolved in water?
  a. K_2SO_4
  b. $Fe(NO_3)_3$
  c. $Al_2(SO_4)_3$
  d. $NiSO_4$

54. Complete the following table for aqueous solutions of glucose, $C_6H_{12}O_6$.

Mass Solute	Moles Solute	Volume of Solution	Molarity
12.5g	—	219 mL	—
—	1.08	—	0.519
—	—	1.62 L	1.08

55. A solution contains 26.5 g of NaCl in 75.0 g of H_2O at 20°C. Determine whether the solution is unsaturated, saturated, or supersaturated. (The solubility of NaCl at 20°C is 36.0 g/100 g H_2O.)

56. An aqueous solution freezes at −2.47°C. What is its boiling point?

57. Hydrogen peroxide is often sold commercially as a 3.0% (m/v) aqueous solution.
  a. If you buy a 250-mL bottle of 3.0% (m/v) H_2O_2, how many grams of hydrogen peroxide have you purchased?
  b. What is the molarity of this solution?

58. How many grams of $NaNO_3$ will precipitate if a saturated solution of $NaNO_3$ in 200 g of H_2O at 50°C is cooled to 20°C?

59. What is the molecular mass of a nondissociating compound if 5.76 g of the compound in 750 g of benzene gives a freezing point depression of 0.460°C?

60. The molality of an aqueous solution of sugar, $C_{12}H_{22}O_{11}$, is 1.62 m. Calculate the mole fractions of sugar and water.

61. Calculate the freezing and boiling point changes for a solution containing 12.0 g of naphthalene, $C_{10}H_8$, in 50.0 g of benzene.

62. The solubility of sodium hydrogen carbonate, $NaHCO_3$, in water at 20°C is 9.6 g/100 g of H_2O. What is the mole fraction of $NaHCO_3$ in a saturated solution? What is the molality of the solution?

Critical Thinking Questions

63. Choose the term that best completes the second relationship.
   a. tree:trunk   solution:_____
      (1) solubility   (3) water
      (2) solvent   (4) solute
   b. volume:liter   concentration:_____
      (1) moles   (3) solution
      (2) volume   (4) molarity
   c. molarity:L of solution   molality:_____
      (1) kg of solvent   (3) concentration
      (2) L of solvent   (4) moles of solute

64. Why is molarity measured in moles per liter of solution instead of grams per liter?

65. Why might calcium chloride spread on icy roads be more effective at melting ice than sodium chloride?

Review Questions and Problems

66. The solubility of hydrogen chloride gas in the polar solvent water is much greater than its solubility in the nonpolar solvent benzene. Why?

67. Discuss the effect that hydrogen bonding has upon the boiling points of compounds with low formula mass.

68. When soap is shaken with water do you get a solution, a suspension, or a colloid? Explain.

69. Explain the following terms as applied to solutions.
   a. solute   **e.** supersaturated
   b. solvent   **f.** nonelectrolyte
   c. saturated   **g.** weak electrolyte
   d. unsaturated   **h.** strong electrolyte

70. Write electron dot structures for each of the following atoms.
   a. I   **d.** Cs
   b. Sr   **e.** Sb
   c. Te   **f.** Xe

71. Indicate by simple equations how the following substances ionize or dissociate in water.
   a. NH_4Cl **d.** $HC_2H_3O_2$
   b. $Cu(NO_3)_2$ **e.** Na_2SO_4
   c. HNO_3 **f.** $HgCl_2$

72. What relationship exists between surface tension and intermolecular attractions in a liquid?

Challenging Questions and Problems

73. One way in which the solubility of a compound can be expressed is by moles of compound that will dissolve in 1 kg of water. Solubility depends on temperature. Plot a graph of the solubility of potassium nitrate, KNO_3, from the following data.

Temperature, °C	Solubility, mol/kg
0	1.61
20	2.80
40	5.78
60	11.20
80	16.76
100	24.50

From your graph estimate:
   a. the solubility of KNO_3 at 76°C and at 33°C.
   b. the temperature at which its solubility is 17.6 mol/kg of water.
   c. the temperature at which the solubility is 4.24 mol/kg of water.

74. An excess of zinc added to 800 mL of a hydrochloric acid solution evolves 1.21 L of hydrogen gas measured over water at 21°C and 747.5 mm Hg. What is the molarity of the acid? The vapor pressure of water at 21°C is 18.6 mm Hg.

75. How many milliliters of $2.50M$ HNO_3 contain enough nitric acid to dissolve an old copper penny with a mass of 3.94 g?

$$3Cu + 8HNO_3 \longrightarrow 3Cu(NO_3)_2 + 2NO + 4H_2O$$

76. A 250-mL sample of Na_2SO_4 is reacted with an excess of $BaCl_2$. If 5.28 g of $BaSO_4$ is precipitated, what is the molarity of the Na_2SO_4 solution?

77. The table below lists the most abundant ions in sea water and their molar concentrations.

Ion	Molarity
Chloride	0.568
Sodium	0.482
Magnesium	0.057
Sulfate	0.028
Calcium	0.011
Potassium	0.010
Hydrogen carbonate	0.002

Calculate the mass, in grams, of each component contained in 5.00 L of sea water. The density of sea water is 1.024 g/mL.

Research Projects

1. Describe the chemical composition of sea water. Devise a method for extracting a valuable element or compound.

2. Trace the development of chromatography starting with the work of Mikhail Tswett.

3. Compare two types of chromatography such as high-performance liquid chromatography and gas chromatography.

4. Make a diagram of how ground water becomes contaminated. How pure is the ground water in your area? What steps have been taken to safeguard its purity?

5. What techniques are used to measure levels of trace nutrients in foods?

6. What are the major sources of water pollution? What steps have been taken to control these sources?

7. What techniques are used to measure the levels of contaminants in drinking water?

8. What is the role of oxygen in water pollution, and what effects do temperature inversions have on water pollution?

9. What effects do combinations of solutes such as KCl and NaCl have on the colligative properties of solutions?

10. Compare the effect of several solutes on the vapor pressure of water.

11. Is the rate of dissolving of an ionic compound affected by the presence of another dissolved solute? Devise an experiment using sodium chloride and sugar to answer this question.

12. White chalk can be used as the stationary phase in the chromatography of ink from a black marker pen. Compare the separation achieved with chalk with that achieved using a paper towel.

Readings and References

Blaustein, Elliot H. *Your Environment and You: Understanding the Pollution Problems*. Dobbs Ferry, NY: Oceana, 1974.

Cagliotti, Luciano, and trans. by Mirella Giaconi. *The Two Faces of Chemistry*. Cambridge, MA: MIT Press, 1983.

Eagles, Juanita Archibald, Orrea Florence Pye, and Clara Mae Taylor. *Mary Swartz Rose, 1874–1941: Pioneer in Nutrition*. New York: Teachers College Press, 1979.

Epstein, Samuel S., Lester O. Brown, and Carl Pope. *Hazardous Waste in America*. San Francisco: Sierra Club, 1982.

McClure, Mike W. "Dog Gone." *ChemMatters* (February 1986), pp. 14–16.

Stark, Norman. *The Formula Book*. New York: Avon Books, 1979.

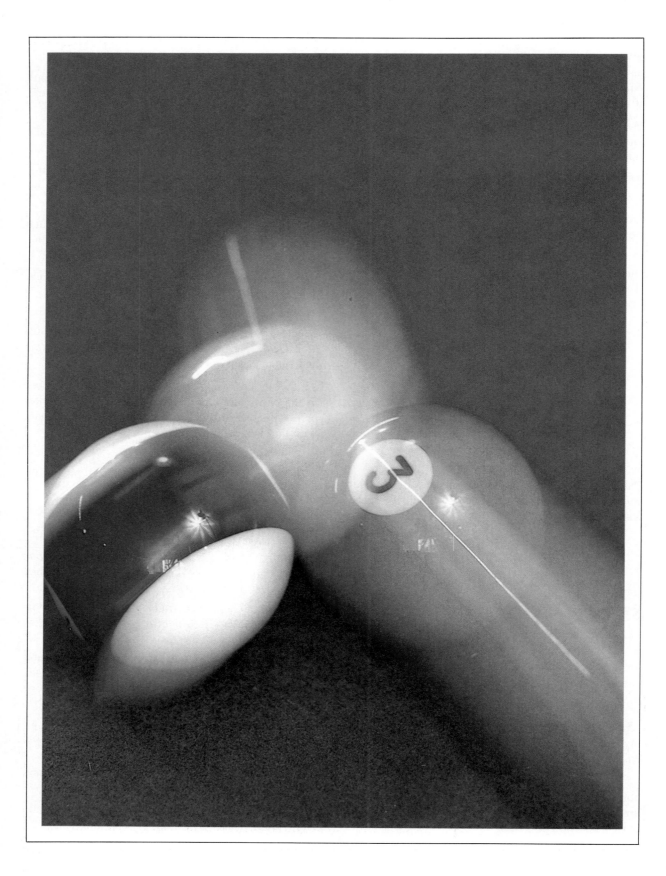

17 Reaction Rates and Equilibrium

Chapter Preview

17·1 Collision Theory
17·2 Reaction Rates
17·A The Contact Process
17·3 Rate Laws
17·4 Reaction Mechanisms
17·5 Entropy
17·6 Entropy Calculations
17·7 Spontaneous Reactions

17·8 Free Energy
17·9 Free Energy Calculations
17·10 Reversible Reactions
17·B Photochemical Smog
17·11 Equilibrium Constants
17·12 Le Châtelier's Principle
17·C Mining the Riches of the Atmosphere

Three questions can be asked about any single chemical reaction. Will the reaction go? How far will it go? How fast will it go? Answers to all these questions depend on the energy available in a chemical system and the energy required for a reaction to take place.

17·1 Collision Theory

How fast will a chemical reaction go? *The **reaction rate** is the number of atoms, ions, or molecules that react in a given time to form products.* Reactions proceed at different rates. The conversion of sodium metal to sodium hydroxide occurs rapidly at room temperature. The conversion of iron to iron(III) oxide (rust) occurs more slowly. Collision theory provides an explanation for why reactions proceed at different rates.

Most chemical reactions involve the transfer of atoms from one molecule to another. Thus contact between the reactants is very important. For any reaction involving two or more reactants, the reacting particles must collide. The more often the particles collide, the faster the reaction should go. Chemical reactions also involve chemical bond breaking and bond making. These processes require energy. The colliding particles must have sufficient energy for bond breaking and bond making to occur. Otherwise the particles collide without reacting. *The minimum energy colliding particles must have in order to react is the **activation energy.*** In a sense, activation energy is a barrier that the reactants must cross to be converted to products (Figure 17·2).

During a reaction, particles which are neither reactants nor products form momentarily. *An **activated complex** is the arrangement of atoms at*

■ In order to react, reactants must have sufficient energy when they collide.

Figure 17·1
Chemical reactions are the result of collisions between atoms, ions, and molecules. If a collision is not energetic enough, however, the particles will "bounce" off each other like billiard balls, without reacting.

395

Figure 17·2
The activation energy barrier must be crossed before reactants are converted to products. The activated complex is a temporary arrangement of particles that has sufficient energy to become either reactants or products.

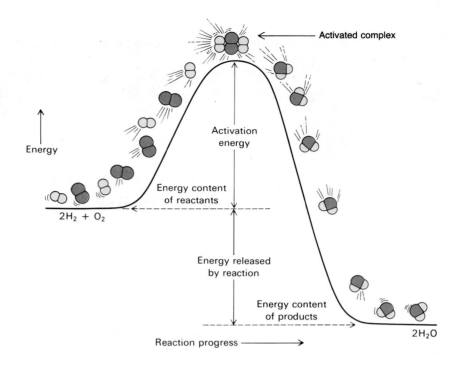

the peak of the activation energy barrier. This group of atoms is on its way to becoming an ionic or molecular product. The lifetime of an activated complex is only about 10^{-13} seconds. The very unstable activated complex is equally likely to return to the reactants as it is to go to the products. For this reason, *the activated complex is sometimes called the* **transition state.** Studies of transition states are conducted to learn what bonds are broken and formed during a given reaction.

Collision theory explains why some exothermic reactions do not occur at room temperature. The reaction of carbon and oxygen is exothermic, but it has a high activation energy. At room temperature the collisions of oxygen and carbon molecules are not energetic enough to break O—O and C—C bonds. Virtually none of the collisions are energetic enough to surmount the activation energy barrier. Thus the reaction rate of carbon with oxygen at room temperature is essentially zero.

17·2 Reaction Rates

The rates of chemical reactions can be affected in several ways.

The rate of reaction depends on the presence of catalysts and on the temperature, concentration, and particle size of the reactants.

Effect of Temperature. Usually, raising the temperature speeds up a reaction. Lowering the temperature usually slows reactions down. At higher temperatures the motions of the reactant particles are more chaotic and more energetic than at lower temperatures. The frequency of high energy collisions between reactants increases. More colliding molecules become energetic enough to slip over the activation energy barrier to become products.

Figure 17·3
A large amount of energy must be added to initiate the reaction of charcoal with oxygen even though the reaction is exothermic. Once the reaction is going it liberates enough energy to continue to drive the reactants to overcome the activation energy barrier.

As mentioned, coal does not burn at a measurable rate at room temperature. If it is given an "energy prod" in the form of heat, however, the result is dramatic. When a flame touches coal, the reactants (carbon and oxygen) collide with greater frequency and higher energy. Some of these collisions are at a high enough energy that the product (carbon dioxide) forms. The heat released by the reaction then supplies enough energy to get more carbon and oxygen over the activation energy barrier. When the flame is removed, the reaction continues. The coal burns as long as a supply of coal and oxygen is available.

Problem

1. Refrigerated food stays fresh for long periods. The same food stored at room temperature quickly spoils. Why?

Effect of Concentration. The number of reacting particles in a given volume affects the rate at which reactions occur. Cramming more particles into a fixed volume increases the collision frequency. Therefore increasing the concentration of reactants increases the reaction rate. A lighted splint vividly illustrates the effect. A splint glows in air, which is 20% oxygen. If a glowing splint is plunged into pure oxygen, however, it immediately bursts into flame (Figure 17·4). This is why smoking is forbidden in hospital areas where oxygen is being used.

Effect of Particle Size. The total surface area of a solid or liquid reactant has an important effect on the reaction rate. The smaller the particle size, the larger the surface area for a given mass of particles. An increase in surface area increases the collision frequency, and the reaction rate. As you probably know, a bundle of kindling burns faster than a log of equal mass.

Usually the best way to increase the surface area of solid reactants is to dissolve them. This separates the particles. For example, a marble on the floor is more accessible than a marble in the middle of a box of marbles. Homogeneous mixtures of reactants usually react more rapidly than

Figure 17·4
Why will a smoldering splint burst into flame if it is placed into a bottle of pure oxygen?

Figure 17·5
In 1932, this explosion destroyed a Chicago grain elevator. The small size of the reactant particles (grain dust) and the mixture of the grain dust and air caused the reaction to be explosive.

Safety

Use extreme care when mixing chemicals that react rapidly. A large amount of heat may be generated in a short time.

Chemical Connections

The catalytic convertor in your car contains pellets coated with a platinum-based catalyst. When exhaust gases and air pass over the pellets, pollutants such as hydrocarbons, carbon monoxide, and oxides of nitrogen are converted to carbon dioxide, water and nitrogen.

heterogeneous mixtures. In many reactions, however, it is impossible to obtain homogeneous mixtures, or solutions, of the reactants. *Reactions carried out with heterogeneous mixtures of reactants are called* **heterogeneous reactions.** When forced to conduct heterogeneous reactions, chemists often grind solids to a fine powder first. As coal miners know, coal dust suspended in the air presents an explosion hazard. By contrast, the large chunks of coal on the mine floor are no hazard at all.

Effect of Catalysts.   An increase in temperature is not always the best way to increase the rate of a reaction. A catalyst, if one can be found, is often better. A catalyst is a substance that increases the rate of a reaction without being used up itself. Catalysts provide reactants with a reaction path of lower activation energy than they could normally take (Figure 17·6). Presented with this lower-energy pathway, a larger fraction of reactants at a given temperature can form products. For instance, the

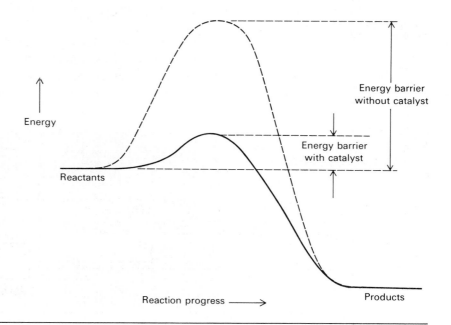

Figure 17·6
A catalyst increases the rate of a reaction by lowering the activation energy barrier.

398   Chapter 17   Reaction Rates and Equilibrium

Safety

Finely divided metals such as zinc, aluminum, and magnesium powders are very flammable because of their large surface area. Handle with care.

reaction of hydrogen and oxygen at room temperature is negligible. With a trace of catalyst, however, the reaction is rapid. Finely divided platinum catalyzes this reaction.

$$2H_2(g) + O_2(g) \xrightarrow{Pt} 2H_2O(l)$$

Catalysts are not used up in a reaction. Neither do they appear as reactants or products in the equation for the reaction. Catalysts called *enzymes* increase the rates of biological reactions. Our body temperature is held at 37°C and cannot be raised or lowered significantly without imperiling our lives. Yet without catalysts, few reactions in the body would proceed fast enough at this temperature. We could not live without enzymes to catalyze these reactions.

A substance that interferes with catalysis is an **inhibitor.** Some inhibitor molecules interfere with catalysis by reacting with ("poisoning") a catalyst. This reduces the amount of catalyst available.

Problem

2. Name two factors that influence the rate of a chemical reaction. How are the effects of these factors explained by the collision theory?

Science, Technology, and Society

17·A The Contact Process

Sulfuric acid (H_2SO_4) was first prepared on a commerical scale in the middle of the eighteenth century. Since that time, it has become one of the most widely used compounds in the chemical industry.

Today, sulfuric acid is produced mainly by the contact process. First, sulfur is burned in air to produce sulfur dioxide (SO_2). Then the sulfur dioxide is oxidized to sulfur trioxide (SO_3) in the presence of a catalyst.

Figure 17·7
This simplified diagram of the contact process shows all the major steps but does not show important secondary processes such as the heating, cooling, cleaning, and purifying of the chemicals involved.

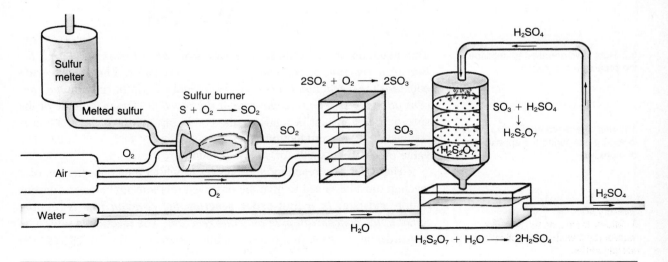

$$2SO_2(g) + O_2(g) \xrightarrow{V_2O_5} 2SO_3(g)$$

The catalyst used for this reaction is vanadium oxide (V_2O_5). The entire process is called the contact process because this key reaction takes place when the reactants are in contact with the surface of the solid catalyst.

The sulfur trioxide is then dissolved in concentrated sulfuric acid to form pyrosulfuric acid ($H_2S_2O_7$).

$$SO_3(g) + H_2SO_4(l) \longrightarrow H_2S_2O_7(l)$$

The pyrosulfuric acid is then added to water to produce commercial sulfuric acid.

$$H_2S_2O_7(l) + H_2O(l) \longrightarrow 2H_2SO_4(l)$$

Commercial sulfuric acid is a dense, colorless, oily liquid. It is normally 98% H_2SO_4.

17·3 Rate Laws

The rate of a reaction depends in part on the concentrations of the reactants. For a reaction of some reactant A going to a product B, a simple equation may be written.

$$A \longrightarrow B$$

The rate at which A is transformed to B is the change in concentration of A (ΔA) with time. Because the quantity of A is decreasing, the rate is given a minus sign.

$$\text{Rate} = -\frac{\Delta A}{\Delta t}$$

The rate of disappearance of A is proportional to its molar concentration.

$$-\frac{\Delta A}{\Delta t} \propto [A]$$

The proportionality can be expressed as a constant multiplied by [A].

$$\text{Rate} = -\frac{\Delta A}{\Delta t} = k \times [A]$$

Rate laws are used to describe the progress of a reaction.

This equation is an example of a **rate law,** *an expression relating the rate of a reaction to the concentration of reactants. The* **specific rate constant** *for a reaction (k) is a proportionality constant relating the concentrations of reactants to the rate of the reaction.* The magnitude of the specific rate constant depends on the conditions at which the reaction is conducted. If the production of B from A is very fast, k is large. The value of k will be small for a slow reaction.

The rate of a reaction can also be followed by measuring the appearance of a product.

The order of a reaction is the power to which the concentration of a reactant must be raised to give the observed relationship between concentration and rate. *In a* **first-order reaction** *the reaction rate is proportional to the concentration of only one reactant.* The reaction is said to be first-order in A. As a first-order reaction progresses, the rate of reaction

Figure 17·8
The rate of a first-order reaction decreases in direct proportion to the concentration of one reactant. The curved line shows the decrease in concentration of reactant A with time. The short colored lines show the rate at three points in the reaction.

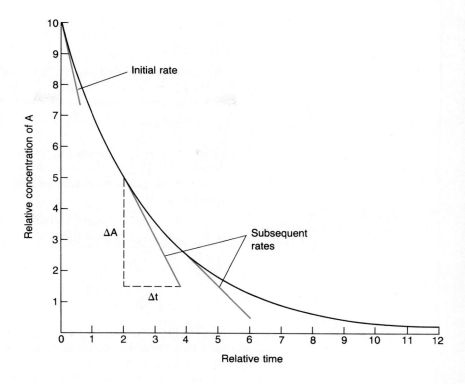

decreases (Figure 17·8). This is because the concentration of reactant decreases. On a graph, the rate at any point $(-\Delta A/\Delta t)$ equals the slope of the tangent to the curve at that point. For a first-order reaction, a halving of [A] results in a halving of the reaction rate.

Problems

3. Show that the unit of k for a first-order reaction is the reciprocal second (s^{-1}) or any other reciprocal unit of time.

4. A first-order reaction initially proceeds at a rate of 0.5 mol/(L × s). What will be the rate when half of the starting material remains? When one-fourth of the starting material remains?

In some kinds of reactions, such as double-replacement, two substances react to give products. The coefficients in the equation for such a reaction are represented by lower case letters.

$$a\text{A} + b\text{B} \xrightarrow{k} c\text{C} + d\text{D}$$

For a one-step reaction of A with B the rate of reaction is dependent on the concentration of both A and B.

$$\text{Rate} = k[\text{A}]^a[\text{B}]^b$$

The order of reaction in each reactant is the value of the exponent associated with that reactant. The overall order of the reaction is the sum of the exponents for each reactant.

Problem

5. Show that the unit of k for a second-order reaction can be expressed in $L/(mol \times s)$.

Example 1

The rate law for the one step reaction $aA \longrightarrow B$ is of the form: Rate $= k[A]^a$. From the data in the following table, find the kinetic order of the reaction with respect to A, and the overall order of the reaction.

The unit $mol/(L \times s)$ can also be written as mol/L-s.

Initial concentration of A(mol/L)	Initial rate (mol/L-s)
0.05	3×10^{-4}
0.10	12×10^{-4}
0.20	48×10^{-4}

Solution

If a reaction is first-order in A, then $a = 1$. In that case the initial rate would be directly proportional to the initial [A]. The reaction cannot be first-order in A because a doubling of A causes the rate to increase by four times. This suggests that $a = 2$, since $2^2 = 4$. Increasing [A] by four times should, therefore, increase the rate sixteen-fold ($4^2 = 16$). When the initial concentration of A is increased from 0.05 mol/L to 0.20 mol/L, the initial rate increases from 3×10^{-4} mol/L-s to 48×10^{-4} mol/L-s, a sixteen-fold increase. The reaction is second-order in A and second-order overall.

Problem

6. Ammonium ions and nitrite ions react in water to form nitrogen gas.

$$NH_4^+(aq) + NO_2^-(aq) \longrightarrow N_2(g) + 2H_2O(l)$$

From the following data, decide the kinetic order of the reaction with respect to NH_4^+ and NO_2^-, and the overall order of the reaction.

Initial concentration of NO_2^- (mol/L)	Initial concentration of NH_4^+ (mol/L)	Initial rate (mol/L-s)
0.0100	0.200	5.4×10^{-7}
0.0200	0.200	10.8×10^{-7}
0.0400	0.200	21.5×10^{-7}
0.0600	0.200	32.3×10^{-7}
0.200	0.0202	10.8×10^{-7}
0.200	0.0404	21.6×10^{-7}
0.200	0.0606	32.4×10^{-7}
0.200	0.0808	43.3×10^{-7}

17·4 Reaction Mechanisms

In an **elementary reaction** *reactants are converted to products in a single step.* Such a reaction has only one intervening activated complex. Most chemical reactions consist of a number of elementary reactions. *A* **reaction mechanism** *includes all of the elementary reactions of a complex reaction.* The reaction progress curve for a complex reaction consists of a number of hills and valleys. The hills correspond to the energy levels of the activated complexes. The valleys correspond to the energy levels of the intermediate products (Figure 17·9).

An **intermediate** *is a product of a reaction that immediately becomes a reactant of another reaction.* Intermediates have a significant lifetime compared with an activated complex. They have real ionic or molecular structures and some stability. They are reactive enough, however, to react further to eventually give the final product of the reaction.

Intermediates do not appear in the chemical equation for a reaction. For example, the decomposition of nitrous oxide (N_2O) is believed to occur in two elementary steps.

$$N_2O \longrightarrow N_2 + O$$
$$N_2O + O \longrightarrow N_2 + O_2$$

$$\overline{}$$

$$2N_2O \longrightarrow 2N_2 + O_2$$

Notice that oxygen atoms are intermediates in this reaction. They disappear when the individual reactants are summed to give the final chemical equation. Thus the overall chemical equation for a complex reaction gives no information about the reaction mechanism.

■ Most chemical reactions consist of more than one elementary step.

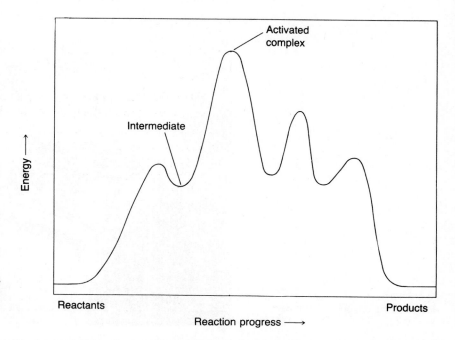

Figure 17·9
A reaction progress curve shows an activation energy peak for each elementary reaction. Valleys indicate the formation of intermediates. How many elementary reactions are part of this reaction? How many intermediates are formed?

17·5 Entropy

Energy changes accompany all chemical and physical processes. For example, the combustion of carbon is exothermic. The heat released is 393.5 kJ for each mole of carbon (graphite) burned.

$$C(s, \text{graphite}) + O_2(g) \longrightarrow CO_2(g) + 393.5 \text{ kJ/mol}$$

The heat content of the products is lower than that of the reactants in their stable states at 25°C and 1 atm.

Chemical systems tend to achieve the lowest possible energy. Thus we might expect that only exothermic reactions would occur. This reasonable assumption does not always work. The melting of ice, a physical process, is endothermic. One mole of ice at 25°C absorbs 6.0 kJ of heat from its surroundings as it melts from a solid to a liquid.

$$H_2O(s) + 6.0 \text{ kJ/mol} \longrightarrow H_2O(l)$$

Although the liquid is at a higher energy level than the solid, ice still melts. The change in heat content is obviously not the only factor in determining whether a reaction will proceed. Another factor, called entropy, is also involved.

The idea that physical and chemical systems attain the lowest possible energy has a corollary. *The **law of disorder** states that things move spontaneously in the direction of maximum chaos or disorder.* A handful of marbles is ordered in the sense that all the marbles are collected in one place. It is highly improbable that when permitted to fall, the marbles will end in the same neat arrangement. Instead, the marbles will scatter on the ground. The marbles will obey the law of disorder. *The measure of the disorder of a system is its* **entropy.** Scattered marbles have a higher entropy than gathered marbles.

Entropy can be described as the randomness of a system.

Figure 17·10
How does the entropy of your room vary over a week?

Figure 17·11
Entropy is a measure of the disorder of a system. **a** The entropy of a gas is greater than the entropy of a liquid or solid. **b** Entropy increases when a substance is divided into parts. **c** Entropy increases with an increase in temperature.

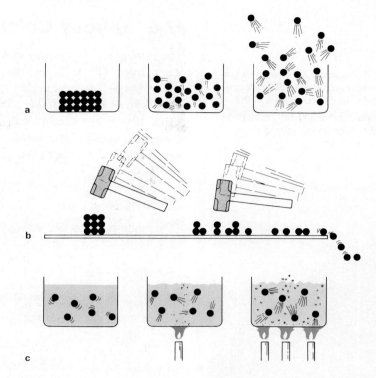

The law of disorder also operates at the level of atoms and molecules. On an atomic and molecular scale, the following concepts can be applied.

1. For a given substance, the entropy of a gas is greater than the entropy of a liquid or a solid. Similarly, the entropy of a liquid is usually greater than that of a solid. Entropy therefore increases in reactions in which solid reactants give liquid or gaseous products. It also increases when liquid reactants give gaseous products.

2. Entropy increases when a substance is divided into parts. Entropy therefore increases in chemical reactions in which the total number of product molecules is greater than the total number of reactant molecules.

3. Entropy increases with a temperature increase because the motions of molecules become more chaotic. Molecules are more disordered at high temperatures than at low temperatures. Figure 17·11 illustrates these concepts.

Problem

7. Which of the following systems has the higher entropy?
   a. a new pack of playing cards or playing cards in use?
   b. a sugar cube dissolved in water or a cube of sugar?
   c. a cup of water at room temperature or a cup of water at 50°C?
   d. a 1-g salt crystal or 1 g of powdered salt?

17·6 Entropy Calculations

The symbol for the entropy of a substance is S, with units of J/K. When the entropy (S) is given per mole of substance it has the unit of J/(K × mol). This unit is also written as J/K-mol. *The **standard entropy** of a substance in its stable state at 25°C and 1 atm is designated S^0.* Table 17·1 gives standard entropies for some common substances. The theoretical entropy of a perfect crystal at 0 K is zero. Other substances have positive entropies, even at absolute zero.

■ The quantity S^0 provides a way to compare entropies of different substances or different states of the same substance at a standard set of conditions.

Figure 17·12
What characteristics of an explosion would lead you to believe that an increase in entropy is occurring? (Hint: see Figure 17·11).

Example 2

Calculate the standard entropy change, ΔS^0, that occurs when 1 mol $H_2O(g)$ at 25°C is condensed to 1 mol $H_2O(l)$ at the same temperature.

Solution

The phase change can be written as a reaction.

$$H_2O(g) \longrightarrow H_2O(l)$$

From Table 17·1 find S^0 for $H_2O(g)$ and $H_2O(l)$.

$$S^0 \text{ for } H_2O(g) = 188.7 \text{ J/K-mol}$$

$$S^0 \text{ for } H_2O(l) = 69.94 \text{ J/K-mol}$$

The change in entropy ΔS^0, can be calculated by subtraction.

$$\Delta S^0 = S^0(\text{products}) - S^0(\text{reactants})$$

Only 1 mol of each reactant and product is involved.

$$\Delta S^0 = 69.94 \text{ J/K-mol} - 188.7 \text{ J/K-mol} = -118.8 \text{ J/K-mol}$$

The condensation of 1 mol of water vapor to liquid water at 25°C results in a large decrease in entropy.

Example 3

What is the standard change in entropy for the following reaction when all reactants and products are in the specified states?

$$NO(g) + O_2(g) \longrightarrow NO_2(g)$$

Solution

First, write the balanced equation including all the physical states.

$$2NO(g) + O_2(g) \longrightarrow 2NO_2(g)$$

Then use Table 17·1 to find the standard entropies.

$$S^0 \text{ for } NO(g) = 210.6 \text{ J/K-mol}$$

$$S^0 \text{ for } O_2(g) = 205.0 \text{ J/K-mol}$$

$$S^0 \text{ for } NO_2(g) = 240.5 \text{ J/K-mol}$$

Sum the S^0 of the reactants, taking into account the number of moles.

$$S^0 \text{ for NO} = 2 \text{ mol NO} \times 210.6 \text{ J/K-mol NO} = 421.2 \text{ J/K}$$

$$S^0 \text{ for O}_2 = 1 \text{ mol O}_2 \times 205.0 \text{ J/K-mol O}_2 = 205.0 \text{ J/K}$$

$$S^0 \text{(reactants)} = \overline{626.2 \text{ J/K}}$$

Calculate the S^0 of the products.

$$S^0 \text{ for NO}_2 = 2 \text{ mol NO}_2 \times 240.5 \text{ J/K-mol NO}_2 = 481.0 \text{ J/K}$$

Calculate the change in entropy ΔS^0 by subtraction.

$$\Delta S^0 = S^0 \text{(products)} - S^0 \text{(reactants)}$$

$$\Delta S^0 = (481.0 \text{ J/K}) - (626.2 \text{ J/K}) = -145.2 \text{ J/K}$$

The entropy declines because a total of *three* reactant molecules are reorganized to only *two* product molecules.

Problem

8. Calculate the change in entropy for these reactions.
   a. $CaCO_3(s) \longrightarrow CaO(s) + CO_2(g)$
   b. $2H_2(g) + O_2(g) \longrightarrow 2H_2O(l)$
   c. $H_2(g) + Cl_2(g) \longrightarrow 2HCl(g)$

Table 17·1   Standard Entropies (S^0) at 25°C and 1 atm

Substance	S^0(J/K-mol)	Substance	S^0(J/K-mol)
$Al_2O_3(s)$	50.99	$HCl(g)$	186.7
$Br_2(g)$	245.3	$H_2S(g)$	205.6
$Br_2(l)$	152.3	$I_2(g)$	260.6
$C(s, \text{diamond})$	2.439	$I_2(s)$	117
$C(s, \text{graphite})$	5.694	$N_2(g)$	191.5
$CH_4(g)$	186.2	$NH_3(g)$	192.5
$CO(g)$	197.9	$NO(g)$	210.6
$CO_2(g)$	213.6	$NO_2(g)$	240.5
$CaCO_3(s)$	88.7	$Na_2CO_3(s)$	136
$CaO(s)$	39.75	$NaCl(s)$	72.4
$Cl_2(g)$	223.0	$O_2(g)$	205.0
$F_2(g)$	203	$O_3(g)$	238
$Fe(s)$	27.2	$P(s, \text{white})$	44.4
$Fe_2O_3(s)$	90.0	$P(s, \text{red})$	29
$H_2(g)$	130.6	$S(s, \text{rhombic})$	31.9
$H_2O(g)$	188.7	$S(s, \text{monoclinic})$	32.6
$H_2O(l)$	69.94	$SO_2(g)$	248.5
$H_2O_2(l)$	92	$SO_3(g)$	256.2

17·7 Spontaneous Reactions

A balanced equation can be written for any reaction. That does not mean, however, that the reaction will actually occur. For chemical reactions, the most important question is: Under specified conditions, will the reaction go? **Spontaneous reactions** *are reactions that are known to produce the written products.* The reaction of sodium with water to produce sodium hydroxide is a spontaneous reaction at room temperature. **Nonspontaneous reactions** *do not give products under the specified conditions.* The terms *spontaneous* and *nonspontaneous* do not refer to how fast reactants go to products. The terms refer to whether the reactants go to products at all. Some reactions that are spontaneous go so slowly that they appear to be nonspontaneous. This section presents the two factors that together determine whether a reaction is spontaneous or not. These factors are entropy and the heat released or absorbed by the reaction.

On a molecular level, a reaction tends toward spontaneity whenever the reaction is exothermic. A reaction will also tend toward spontaneity if the entropy of the products is greater than the entropy of the reactants. In a given reaction, therefore, the heat changes and the entropy changes may both act in a favorable direction. They may both act in an unfavorable direction, or they may act in opposition to each other.

The combustion of carbon has both the heat changes and entropy changes in its favor. The reaction is exothermic. The entropy also increases as *solid* carbon, is converted to *gaseous* carbon dioxide. Hence the reaction is spontaneous.

The heat changes and entropy changes work in opposition when ice melts. The process is endothermic but still spontaneous above 0°C. The absorption of heat is more than offset by a favorable entropy change. The increase in entropy is large as *solid* water (ice) goes to *liquid* water.

Another chemical reaction shows the effect of temperature on the spontaneity of a reaction. Solid calcium carbonate decomposes to give calcium oxide and carbon dioxide. In this reaction the entropy increases

■ Heat and entropy changes together determine whether reactants will give products.

Table 17·2 How Changes in Heat Content and Entropy Affect Reaction Spontaneity		
Heat content	Entropy	Spontaneous reaction?
Decreases (exothermic)	Increases (more disorder in products than in reactants)	Yes
Increases (endothermic)	Increases	Depends on whether unfavorable heat content change is offset by favorable entropy change
Decreases (exothermic)	Decreases (less disorder in products than in reactants)	Depends on whether unfavorable entropy change is offset by favorable heat content change
Increases (endothermic)	Decreases	No

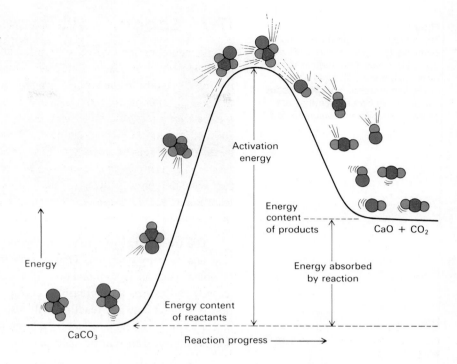

Some nonspontaneous reactions can be made to be spontaneous if the conditions are changed. For instance, water does not spontaneously break down into hydrogen and oxygen. With the application of electric energy, it will decompose.

All spontaneous reactions release free energy.

because one of the products formed from the *solid* reactant is a *gas*. The entropy increase is not great enough, however, for the reaction to be spontaneous at ordinary temperatures. This is because the reaction is endothermic. The heat content of the reactants is lower than that of the products. The effect of an entropy increase is magnified as the temperature is raised. At temperatures above 850°C the favorable entropy-temperature term outweighs the unfavorable energy term and the reaction is spontaneous.

Table 17·2 summarizes these ideas on how entropy and the heat absorbed or given off affect the spontaneity of chemical reactions.

17·8 Free Energy

When a reaction occurs some energy known as the **free energy** *of the system becomes available to do work.* Free energy is only part of the total energy of the system. *Spontaneous reactions release free energy and are said to be* **exergonic.** As a spontaneous reaction occurs, free energy is released that can be used to cause chemical or physical changes. *Nonspontaneous reactions absorb free energy and are said to be* **endergonic.**

A reaction that is nonspontaneous under one set of conditions may become spontaneous under other conditions. Changing the temperature or pressure may change whether or not a reaction is spontaneous. A nonspontaneous reaction can also be linked to a spontaneous reaction. If the spontaneous reaction produces a large release of free energy, it can be used to drive the nonspontaneous reaction. The free energy transfer in coupled reactions causes many otherwise nonspontaneous reactions to occur. This occurs frequently in living systems.

Figure 17·14
The entropy of the products of photosynthesis (complex sugars such as $C_6H_{12}O_6$) is less than that of the reactants (CO_2 and H_2O). Light energy drives this process. Nonspontaneous reactions can occur when they are coupled with spontaneous reactions.

The free energy of a reaction depends on the size and direction of the heat and entropy changes. That is why these changes determine whether or not a reaction is spontaneous. If a reaction is exothermic (heat is released) and causes an increase in entropy (more disorder), then the reaction is spontaneous. A reaction will also be spontaneous if a decrease in entropy (more order) is offset by a large release of heat. Similarly, an endothermic reaction will be spontaneous if an entropy increase offsets the

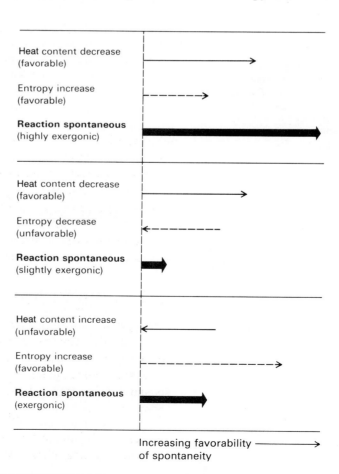

Figure 17·15
Free energy is released in spontaneous reactions. The reaction may be exothermic, the entropy of the system may increase, or both may occur.

Figure 17·16

A reaction may be nonspontaneous because the reaction is endothermic or the entropy of the system decreases, or both may occur.

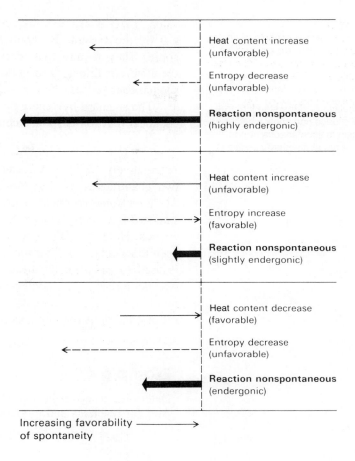

Heat content increase (unfavorable)

Entropy decrease (unfavorable)

Reaction nonspontaneous (highly endergonic)

Heat content increase (unfavorable)

Entropy increase (favorable)

Reaction nonspontaneous (slightly endergonic)

Heat content decrease (favorable)

Entropy decrease (unfavorable)

Reaction nonspontaneous (endergonic)

Increasing favorability of spontaneity ⟶

heat absorbed. Figure 17·15 summarizes the relationship between heat, entropy, and free energy changes for spontaneous reactions.

In nonspontaneous reactions the free energy of the system increases. Energy must be absorbed for the reaction to occur. The reaction may involve a large decrease in entropy (increased order) and/or the reaction may be highly endothermic. In either case the other changes are not favorable enough to cause the reaction to be spontaneous. Figure 17·16 summarizes these relationships.

17·9 Free Energy Calculations

In every spontaneous process, some energy becomes available to do useful work. *The **change in Gibbs free energy**, ΔG, is the maximum amount of energy that can be coupled to another process to do useful work*. The change in Gibbs free energy is related to the change in entropy ΔS and the change in enthalpy ΔH of the system.

$$\Delta G = \Delta H - T\Delta S$$

Temperature (T) in the equation is stated in Kelvin. All spontaneous processes release free energy; therefore, they are exergonic. The numerical value of ΔG is negative in spontaneous processes because free energy is lost by the system. By contrast, nonspontaneous processes are endergonic. They require that work be expended to make them go at the specified conditions. The numerical value of ΔG is positive for a nonspontaneous process. In that case free energy is absorbed by the system.

The standard free energy change ΔG^0 is used to determine whether a chemical reaction is spontaneous when ΔH^0 and ΔS^0 are known.

$$\Delta G^0 = \Delta H^0 - T\Delta S^0$$

■ Calculations of the standard free energy change can show whether a reaction is spontaneous or non-spontaneous at standard conditions.

Alternatively, ΔH^0 or ΔS^0 can be determined for a reaction if the other two quantities in the equation are known. Table 17·3 gives values of ΔG^0, for some common reactions. Table 17·4 lists values of ΔG_f^0, the standard free energy change for the formation of substances from their elements. Note that $\Delta G_f^0 = 0$ for elemental substances. The ΔG_f^0 may be used to calculate ΔG^0 for a reaction. This is similar to the way that ΔH_f^0 is used to calculate the heats of reaction. If ΔG^0 is calculated to see whether a reaction is spontaneous, the result only applies to reactants and products in their standard states. A reaction that is nonspontaneous at one set of conditions may be spontaneous under other conditions.

Example 4

Determine whether the following reaction is spontaneous by using the ΔG_f^0 for reactants and products.

$$C(s, \text{ graphite}) + O_2(g) \longrightarrow CO_2(g)$$

Solution

Find the ΔG_f^0 of all reactants and products in Table 17·4.

$$\Delta G_f^0 \text{ for } C(s, \text{ graphite}) = 0.0 \text{ kJ/mol}$$

$$\Delta G_f^0 \text{ for } O_2(g) = 0.0 \text{ kJ/mol}$$

$$\Delta G_f^0 \text{ for } CO_2(g) = -394.4 \text{ kJ/mol}$$

Sum the ΔG_f^0 of the reactants and sum the ΔG_f^0 of the products.

$$\Delta G_f^0 \text{ (reactants)} = 0.0 \text{ kJ/mol}$$

$$\Delta G_f^0 \text{ (products)} = -394.4 \text{ kJ/mol}$$

The ΔG^0 for the reaction is the difference between the ΔG_f^0 of the products and the ΔG_f^0 of the reactants.

$$\Delta G^0 = \Delta G_f^0 \text{ (products)} - \Delta G_f^0 \text{ (reactants)}$$

$$\Delta G^0 = -394.4 \text{ kJ/mol} - 0.0 \text{ kJ/mol}$$

$$= -394.4 \text{ kJ/mol}$$

The reaction is spontaneous. The spontaneity of the reaction can also be determined by using the changes in enthalpy and the entropies for the reactants and products, as shown in the next example.

Figure 17·17
Fireworks displays are the result of exergonic reactions. A large quantity of free energy is released.

Example 5

Using values for ΔH_f^0 and S^0, determine whether the reaction in Example 4 is spontaneous

Solution

First find the ΔH_f^0 and S^0 of all the reactants and products in Tables 8·1 and 17·1.

Substance	ΔH_f^0(kJ/mol)	S^0(J/(K-mol))
C(s, graphite)	0.0	5.69
$O_2(g)$	0.0	205
$CO_2(g)$	−393.5	214

The S^0 values of Table 17·1 are given in J/K-mol and must be converted to kJ/K-mol to match the units of ΔH_f^0.

$$S^0 \text{ for C}(s, \text{graphite}) = 5.7 \frac{J}{\text{K-mol}} \times \frac{1 \text{ kJ}}{1000 \text{ J}} = 0.0057 \text{ kJ/K-mol}$$

$$S^0 \text{ for O}_2(g) = 205 \frac{J}{\text{K-mol}} \times \frac{1 \text{ kJ}}{1000 \text{ J}} = 0.205 \text{ kJ/K-mol}$$

$$S^0 \text{ for CO}_2(g) = 214 \frac{J}{\text{K-mol}} \times \frac{1 \text{ kJ}}{1000 \text{ J}} = 0.214 \text{ kJ/K-mol}$$

The standard change in entropy ΔS^0 for the reaction is calculated from the standard entropies of the reactants and products.

$$\Delta S^0 = S^0(\text{products}) - S^0(\text{reactants})$$

$$\Delta S^0 = 0.214 \frac{\text{kJ}}{\text{K-mol}} - \left(0.0057 \frac{\text{kJ}}{\text{K-mol}} + 0.205 \frac{\text{kJ}}{\text{K-mol}} \right)$$

$$= 0.003 \text{ kJ/K-mol}$$

The standard enthalpy of the reaction is calculated in a similar way.

$$\Delta H^0 = \Delta H_f^0(\text{products}) - \Delta H_f^0(\text{reactants})$$

$$\Delta H^0 = -393.5 \text{ kJ/mol} - (0.0 \text{ kJ/mol} + 0.0 \text{ kJ/mol}) = -393.5 \text{ kJ/mol}$$

The value of ΔG^0 is calculated from the changes in enthalpy and in the changes in entropy.

$$\Delta G^0 = \Delta H^0 - T\Delta S^0$$

Kelvin temperature $T = 273.15 + 25°C = 298.15 \text{ K}$

$$\Delta G^0 = -393.5 \frac{\text{kJ}}{\text{mol}} - \left(298.15 \text{ K} \times 0.003 \frac{\text{kJ}}{\text{K-mol}} \right) = -394.4 \frac{\text{kJ}}{\text{mol}}$$

The reaction is spontaneous due to the large decline in ΔH^0 and due to the increase in entropy when solid graphite is converted to gaseous CO_2.

Problems

9. Determine whether the following reactions are spontaneous. Assume all substances are in their stable states at 25°C and 1 atm.
   a. $2Na(s) + Cl_2(g) \longrightarrow 2NaCl(s)$
   b. $4Al(s) + 3O_2(g) \longrightarrow 2Al_2O_3(s)$

10. When gaseous nitrogen and hydrogen are converted to gaseous ammonia, then ΔG^0 equals -16.64 kJ/mol.

$$3H_2(g) + N_2(g) \longrightarrow 2NH_3(g)$$

Using the S^0 values of Table 17·1, calculate the ΔH_f^0 for the formation of ammonia.

Table 17·3   Free Energy Values (ΔG^0) for Some Spontaneous Processes at 25°C

Reaction		Free energy	
		Kilocalories	Kilojoules
$H_2(g) + Cl_2(g) \longrightarrow 2HCl(g)$ Hydrogen Chlorine Hydrogen chloride		-45.6	-191
$S(s) + O_2(g) \longrightarrow SO_2(g)$ Sulfur Oxygen Sulfur dioxide		-71.7	-300
$2N_2O_5(s) \longrightarrow 4NO_2(g) + O_2(g)$ Dinitrogen Nitrogen Oxygen pentoxide dioxide		-7.2	-30
$C_6H_{12}O_6(s) + 6O_2(g) \longrightarrow 6CO_2(g) + 6H_2O(l)$ Glucose Oxygen Carbon Water dioxide		-686.0	-2868

Table 17·4   Standard Gibbs Free Energies of Formation (ΔG_f^0) at 25°C and 1 atm.

Substance	ΔG_f^0(kJ/mol)	Substance	ΔG_f^0(kJ/mol)	Substance	ΔG_f^0(kJ/mol)
$Al_2O_3(s)$	-1576.4	$Fe(s)$	0.0	$NO(g)$	86.69
$Br_2(g)$	3.14	$Fe_2O_3(s)$	-741.0	$NO_2(g)$	51.84
$Br_2(l)$	0.0	$H_2(g)$	0.0	$Na_2CO_3(s)$	-1048
$C(s, diamond)$	2.866	$H_2O(g)$	-288.6	$NaCl(s)$	-384.03
$C(s, graphite)$	0.0	$H_2O(l)$	-237.2	$O_2(g)$	0.0
$CH_4(g)$	-50.79	$H_2O_2(l)$	-114.0	$O_3(g)$	163.4
$CO(g)$	-137.3	$HCl(g)$	-95.27	$P(s, white)$	0.0
$CO_2(g)$	-394.4	$H_2S(g)$	-33.02	$P(s, red)$	-14
$CaCO_3(s)$	-1127.7	$I_2(g)$	19.4	$S(s, rhombic)$	0.0
$CaO(s)$	-604.2	$I_2(s)$	0.0	$S(s, monoclinic)$	0.096
$Cl_2(g)$	0.0	$N_2(g)$	0.0	$SO_2(g)$	-300.4
$F_2(g)$	0.0	$NH_3(g)$	-16.64	$SO_3(g)$	-370.4

17·10 Reversible Reactions

How far will a reaction go? So far we have assumed that spontaneous reactions go completely to products as written. In some reactions, however, products are favored only slightly over reactants or vice versa. Such reactions are reversible. *In* **reversible reactions** *the conversion of reactants into products and the conversion of products into reactants occur simultaneously.* The conversion of reactants to products is called the forward reaction. The conversion of products to reactants is called the reverse reaction. One example of a reversible reaction is the reaction of sulfur dioxide with oxygen to give sulfur trioxide. This is one step in the process of making sulfuric acid. The double arrows show the reaction is reversible.

$$2SO_2(g) + O_2(g) \rightleftharpoons 2SO_3(g)$$

Sulfur dioxide   Oxygen   Sulfur trioxide

A spontaneous chemical reaction begins when the reactants are mixed. In a reversible reaction the rate of the reverse process is zero at first. At that point there are no products to go back to reactants. As the concentrations of products build up, however, some products revert to reactants by the reverse reaction. As the reactants are used up, their concentrations decrease. Then the forward reaction slows down. As the products build up, their concentrations increase. Then the reverse reaction speeds up. Eventually, the products are going to reactants at the same rate as reactants are going to products. *The reaction has reached* **chemical equilibrium** *when the forward and reverse reactions are taking place at the same rate.* At chemical equilibrium there is no net

When chemical equilibrium has been achieved, the rates of the forward and reverse reactions are the same.

Figure 17·18
A reversible reaction achieves equilibrium regardless of the direction from which it is approached. Molecules of SO_2 and O_2 react to give SO_3; molecules of SO_3 decompose to give SO_2 and O_2. At equilibrium, all three types of molecules are present in the mixture.

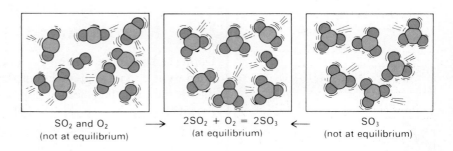

SO_2 and O_2 (not at equilibrium) → $2SO_2 + O_2 = 2SO_3$ (at equilibrium) ← SO_3 (not at equilibrium)

Figure 17·19
These graphs show the variation with time of the concentration of O_2, SO_2, and SO_3 in two situations at 1000 K. **a** Initially twice as much SO_2 as O_2 is present. At equilibrium, a mixture of all three gases is obtained. **b** Initially only SO_3 is present. At equilibrium the amounts of O_2, SO_2, and SO_3 are the same as in **a**.

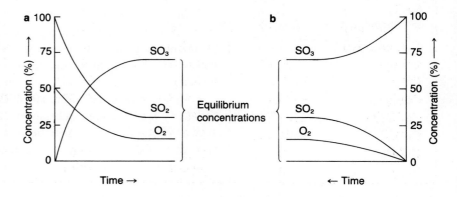

change in the actual amounts of reactants and products in the system (Figure 17·18). Chemical equilibrium is similar to the dynamic equilibrium of a saturated solution (Section 16·2).

The **equilibrium position** *of a reaction is given by the relative concentrations of reactants and products at equilibrium.* The equilibrium position indicates whether the reactants or the products are favored in a reversible reaction. For example, if A reacts to give B and the equilibrium mixture contains 1% of A and 99% of B, then the *products* are favored.

$$A \rightleftharpoons B$$
$$1\% \qquad 99\%$$

If at equilibrium the mixture contains 99% of A and 1% of B, then the *reactants* are favored.

$$A \rightleftharpoons B$$
$$99\% \qquad 1\%$$

The longer arrow indicates the favored direction of the reaction. Reactions that are exergonic give larger amounts of products than reactants at equilibrium. Reactions that are endergonic give larger amounts of reactants than products at equilibrium.

Catalysts speed up the forward and backward reactions equally. This is because the activation energy is reduced by the same amount for both the forward and reverse directions. Therefore catalysts do not affect the amounts of reactants and products present at equilibrium. They simply decrease the time it takes for equilibrium to be established.

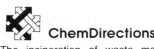

ChemDirections

The incineration of waste materials contributes to the problems of air pollution. **ChemDirections** *Plastics Recycling,* page 663, explains why it is better to recycle rather than incinerate discarded plastic items.

═══ Science, Technology, and Society ═══

17·B Photochemical Smog

During certain times of the year, citizens of cities such as Los Angeles often wake up to disturbing but predictable news on the radio or TV: the air they will breathe that day is dangerous to their health. The brown haze over their city, called photochemical smog, is an example of one of several serious global air pollution problems.

The word "smog" was coined in England in the 1950s to describe the combination of smoke and fog that dirtied the air above London. London's smog was due primarily to the presence of sulfur dioxide (SO_2) from the burning of high-sulfur coal. In contrast, the photochemical smog hanging over most cities today is produced mainly by the action of sunlight on the compounds present in automobile exhaust.

Photochemical reactions are initiated when light energy is absorbed by light-sensitive substances. Some photochemical reactions, such as photosynthesis, create beneficial products. The production of photochemical smog, however, is undesirable and harmful.

The major pollutants in automobile exhaust are nitrogen oxide (NO), carbon monoxide (CO), and unburned hydrocarbons from the

Figure 17·20
Photochemical smog builds up when the air is still over a city or other pollution source. Persistent smog can pose a health danger.

▶ Issues in Chemistry

Should the government do more to control air pollution resulting from auto exhausts? Should the cost be borne by individual drivers or by society as a whole? Is there anything that you can do to help alleviate this problem?

incomplete combustion of gasoline. The nitrogen oxide is produced by the direct combination of oxygen and nitrogen in the cylinders of internal combustion engines.

$$N_2(g) + O_2(g) \longrightarrow 2NO(g)$$

The process leading to the formation of photochemical smog begins when the nitric oxide from automobile exhaust is oxidized to nitrogen dioxide in the atmosphere.

$$2NO(g) + O_2(g) \longrightarrow 2NO_2(g)$$

Sunlight then causes the photochemical decomposition of nitrogen dioxide into nitric oxide and atomic oxygen.

$$NO_2(g) + \text{light} \longrightarrow NO(g) + O(g)$$

Atomic oxygen is very reactive and initiates further reactions. One of these reactions leads to the production of ozone.

$$O(g) + O_2(g) \longrightarrow O_3(g)$$

Ozone, nitrogen oxides, and unburned hydrocarbons can then combine in a complex series of photochemical reactions. Among the products of these reactions are peroxyacyl nitrates or PANs. PANs have the general formula $C_x H_y O_3 NO_2$.

Nitrogen dioxide, ozone, PANs, and carbon monoxide are all extremely toxic. At low concentrations, they irritate the eyes and mucous membranes of the nose and throat and may cause difficulty in breathing. Ozone damages the leaves of plants. The carbon monoxide from car exhaust contributes to the unhealthy effects of these photochemically produced compounds.

To reduce smog, the production of pollutants that contribute to its formation must be reduced. Catalytic convertors have been developed for this purpose. They reduce the levels of the two major pollutants–nitrogen oxides and unburned hydrocarbons. The hydrocarbons and carbon monoxide are oxidized to carbon dioxide and water; the nitrogen oxides are reduced to nitrogen. So far, however, the widespread use of catalytic convertors has not prevented the smog problem from growing worse in many areas.

17·11 Equilibrium Constants

Chemists seldom use percentages to indicate the position of equilibrium. Rather, they use equilibrium constants. These numbers relate the amounts of reactants to products at equilibrium. Consider the hypothetical reaction in which a mol of reactant A and b mol of reactant B react to give c mol of product C and d mol of product D at equilibrium.

$$a\text{A} + b\text{B} \rightleftharpoons c\text{C} + d\text{D}$$

Because the reaction is at equilibrium, there is no *net change* in the amounts of A, B, C, or D at any given instant. *The* **equilibrium constant** K_{eq} *is the ratio of product concentrations to reactant concentrations, with each concentration raised to a power given by the number of moles of that substance in the balanced chemical equation.*

$$K_{eq} = \frac{[\text{C}]^c \times [\text{D}]^d}{[\text{A}]^a \times [\text{B}]^b}$$

The brackets indicate that the amounts of substances are in moles per liter. The value of K_{eq} for a reaction depends on the temperature. If the temperature changes, the value of K_{eq} also changes.

Safety

Example 6

Dinitrogen tetroxide (N_2O_4) a colorless gas, and nitrogen dioxide (NO_2), a dark brown gas, exist in equilibrium with each other.

$$N_2O_4(g) \rightleftharpoons 2NO_2(g)$$

A liter of a gas mixture at 100°C at equilibrium contains 0.0045 mol of dinitrogen tetroxide and 0.030 mol of nitrogen dioxide.

a. Write the expression for the equilibrium constant.

b. Calculate the equilibrium constant, K_{eq}, for the reaction.

Solution

a. At equilibrium, there is no net change in the amount of N_2O_4 or NO_2 at any given instant. Nitrogen dioxide is the only product of the reaction. It has a coefficient of 2. According to the rule for writing the equilibrium constant, the concentration of nitrogen dioxide, raised to the second power, is in the numerator. The reactant is dinitrogen tetroxide. Its coefficient is 1. Thus the concentration of this substance, raised to the first power, is in the denominator.

$$K_{eq} = \frac{[NO_2]^2}{[N_2O_4]}$$

b. Substitute the given data into the equilibrium constant expression.

$$K_{eq} = \frac{[NO_2]^2}{[N_2O_4]} = \frac{(0.030 \text{ mol/L})^2}{0.0045 \text{ mol/L}} = \frac{0.030 \text{ mol/L} \times 0.030 \text{ mol/L}}{0.0045 \text{ mol/L}}$$

$$= 0.20 \text{ mol/L}$$

Figure 17·21
Dinitrogen tetroxide is a colorless gas, whereas nitrogen dioxide is brown. How does temperature affect the equilibrium of a mixture of these gases?

Example 7

One mole of hydrogen gas and 1 mol of iodine are sealed in a 1-L flask and allowed to react at 450°C. At equilibrium 1.56 mol of hydrogen iodide is present. Calculate K_{eq} for the reaction.

$$H_2(g) + I_2(g) \rightleftharpoons 2HI(g)$$

Solution

The balanced equation indicates that 1.00 mol of hydrogen and 1.00 mol of iodine form 2.00 mol of hydrogen iodide. To make 1.56 mol of hydrogen iodide therefore consumes 0.78 mol of hydrogen and 0.78 mol of iodine. First calculate how much of the H_2 and I_2 are left in the flask at equilibrium.

$$\text{mol } H_2 = \text{mol } I_2 = 1.00 \text{ mol} - 0.78 \text{ mol}$$
$$= 0.22 \text{ mol}$$

These concentrations can be substituted in the equation for K_{eq}.

$$K_{eq} = \frac{[HI]^2}{[H_2] \times [I_2]}$$

$$= \frac{(1.56 \text{ mol})^2}{0.22 \text{ mol} \times 0.22 \text{ mol}}$$

$$= \frac{1.56 \text{ mol} \times 1.56 \text{ mol}}{0.22 \text{ mol} \times 0.22 \text{ mol}}$$

$$= 50.3 = 50 \text{ (to two significant figures)}$$

Problems

11. Give the equilibrium-constant expression for the formation of ammonia from hydrogen and nitrogen.

$$3H_2(g) + N_2(g) \rightleftharpoons 2NH_3(g)$$

12. Analysis of an equilibrium mixture of nitrogen, hydrogen, and ammonia contained in a 1-L flask at 300°C gives the following results: hydrogen 0.15 mol; nitrogen 0.25 mol; ammonia 0.10 mol. Calculate K_{eq} for the reaction.

$$3H_2(g) + N_2(g) \rightleftharpoons 2NH_3(g)$$

Example 8

Bromine chloride, BrCl, decomposes to form chlorine and bromine.

$$2BrCl(g) \rightleftharpoons Cl_2(g) + Br_2(g)$$

At a certain temperature the equilibrium constant for the reaction is 11.1, and the equilibrium mixture contains 4.00 mol Cl_2. How many moles of Br_2 and BrCl are present in the equilibrium mixture?

Solution

$$2BrCl(g) \rightleftharpoons Cl_2(g) + Br_2(g)$$

According to the equation, when BrCl breaks down, equal moles of Cl_2 and Br_2 are formed. The measured equilibrium concentration of Cl_2 was given as 4.00 mol. Thus there must be 4.00 mol Br_2 in the equilibrium mixture. Substitute the known values in the equilibrium expression for the reaction. (K_{eq} was given as 11.1.)

$$K_{eq} = \frac{[Cl_2] \times [Br_2]}{[BrCl]^2} \qquad 11.1 = \frac{(4.00 \text{ mol}) \times (4.00 \text{ mol})}{[BrCl]^2}$$

Now solve for $[BrCl]$.

$$[BrCl]^2 = \frac{(4.00 \text{ mol}) \times (4.00 \text{ mol})}{11.1} = \frac{16.0 \text{ mol}^2}{11.1} = 1.44 \text{ mol}^2$$

$$[BrCl] = \sqrt{1.44 \text{ mol}^2} = 1.20 \text{ mol}$$

Problems

13. Write the equilibrium-constant expression for the decomposition of hydrogen iodide to hydrogen and iodine.

$$2HI(g) \rightleftharpoons H_2(g) + I_2(g)$$

14. The decomposition of hydrogen iodide at 450°C produces an equilibrium mixture that contains 0.50 mol of hydrogen. The equilibrium constant is 0.020 for the reaction. How many moles of iodine and hydrogen iodide are present in the equilibrium mixture?

$$2HI(g) \rightleftharpoons H_2(g) + I_2(g)$$

Equilibrium constants are valuable chemical information. Among other things, they show whether products or reactants are favored at equilibrium. An equilibrium constant is always written as a ratio of products to reactants. Thus a value of K_{eq} greater than 1 means that products are favored over reactants. Conversely, a value of K_{eq} less than 1 means that reactants are favored over products.

$$K_{eq} > 1 \qquad \text{products favored at equilibrium}$$
$$K_{eq} < 1 \qquad \text{reactants favored at equilibrium}$$

All spontaneous or exergonic reactions have $K_{eq} > 1$. All nonspontaneous or endergonic reactions have $K_{eq} < 1$.

Problem

15. A chemist has determined the equilibrium constants for several reactions. In which of these reactions are the products favored over the reactants?

a. $K_{eq} = 1 \times 10^2$ **c.** $K_{eq} = 3.5$
b. $K_{eq} = 0.003$ **d.** $K_{eq} = 6 \times 10^{-4}$

17·12　Le Châtelier's Principle

A delicate balance exists between reactants and products in a system at equilibrium. The application of stress to the system disrupts this balance. The French chemist Henri Le Châtelier (1850–1936) studied the changes in a system that result from changed conditions. He proposed **Le Châtelier's principle:** *If a stress is applied to a system in a dynamic equilibrium, the system changes to relieve the stress.* Three types of stress are easily applied to chemical systems: changes in concentration of reactants and products, changes in temperature, and changes in pressure.

Changes in Concentration.　　Changing the amount of any reactant or product in a system at equilibrium disturbs the equilibrium. The system changes to minimize the original change. When carbon dioxide dissolves in water, for example, it reacts with the water to a slight extent to form a substance known as carbonic acid. The equilibrium constant for this process is small. At equilibrium the amount of carbonic acid is 1%.

$$CO_2 + H_2O \rightleftharpoons H_2CO_3$$
$$99\% \qquad\qquad 1\%$$

If more carbon dioxide is added, the equilibrium will be disturbed. Suppose enough carbon dioxide is added to shift the ratio of carbonic acid to carbon dioxide (H_2CO_3/CO_2) from 1 : 99 to 0.5 : 99.5. As the carbon dioxide is added, it reacts with water to form more carbonic acid. This shifts the ratio back toward the original 1 : 99. In other words: *Adding a reactant always pushes a reversible reaction in the direction of products.* If carbon dioxide is removed, the H_2CO_3/CO_2 ratio increases to greater than 1 : 99. Then carbonic acid decomposes as the system readjusts itself toward a H_2CO_3/CO_2 ratio of 1 : 99. *Removing a reactant always pulls a reversible reaction in the direction of reactants.*

Our bodies use the removal of reactants to keep carbonic acid at low, safe levels in the blood. Blood contains dissolved carbonic acid in equilibrium with carbon dioxide and water. When we exhale carbon dioxide, the equilibrium shifts toward carbon dioxide and water. This reduces the amount of carbonic acid. The same principle applies to adding or removing products. *When a product is added to a system at equilibrium, the reaction shifts in the direction of the formation of reactants. When a product is removed a reaction shifts in the direction of formation of products.*

Removal of products is a trick often used by chemists to increase the yield of a desired product. As products are removed from a reaction mixture, the system continually changes to restore equilibrium with the products. Because the products are being removed as fast as they are formed, however, the reaction never builds up enough of them to establish an equilibrium. (If you have spent some time around henhouses, you know that a hen lays a clutch of eggs and then proceeds to hatch them. Only if you take the eggs away will she produce more eggs.) Similarly, the reactants continue to react to give products until they are completely used up.

When environmental factors such as the temperature or pressure of a system in dynamic equilibrium are changed, the position of equilibrium also changes.

Figure 17·22
The rapid exhalation of CO_2 following exercise helps reestablish the body's H_2CO_3/CO_2 equilibrium. This keeps the pH of the blood within a safe range.

Changes in Temperature. Increasing the temperature causes the equilibrium position of a reaction to shift in the direction that absorbs heat. For example, the production of SO_3 is an exothermic reaction.

$$2SO_2(g) + O_2(g) \rightleftharpoons 2SO_3(g) + \text{heat}$$

In this reaction, heat can be considered to be a product, just like SO_3. Heating the reaction mixture at equilibrium pushes the equilibrium position to the left. This favors the reactants. As a result, the product yield decreases. Cooling pulls the equilibrium to the right, and the product yield increases. Notice that the ratio of products to reactants changes with temperature. A change in temperature produces a change in K_{eq}.

Changes in Pressure. A change in pressure will affect only an equilibrium with an unequal number of moles of *gaseous* reactants and products. For example, an equilibrium is established between ammonia gas and the gaseous elements from which it is formed.

$$3H_2(g) + N_2(g) \rightleftharpoons 2NH_3(g)$$

A pressure increase causes a stress that can be relieved by a shift in the equilibrium to the right. This is because a decrease of pressure results when 4 mol of gas reacts to form 2 mol of gas.

The equilibrium for this reaction can be made to shift to the left (favoring reactants) by decreasing the pressure. When the reaction shifts left, the pressure is increased. This is because 4 mol of gas is produced for each 2 mol of NH_3 that decomposes.

> An increase of pressure on a gaseous system always favors a shift toward the side with the lowest total number of gaseous particles.

Example 9

What effect do the following changes have on the position of equilibrium for this reversible reaction?

$$PCl_5(g) + \text{heat} \rightleftharpoons PCl_3(g) + Cl_2(g)$$

a. addition of Cl_2
b. increase in pressure
c. removal of heat
d. removal of PCl_3 as it is formed.

Solution

a. The stress of adding Cl_2 is relieved if the equilibrium shifts to the left ($\longleftarrow$), forming more PCl_5.
b. There are 2 mol of gaseous products and 1 mol of gaseous reactant. The stress of an increase in pressure is relieved if the equilibrium shifts to the left ($\longleftarrow$). A decrease in the number of moles of gaseous substances gives a decrease in pressure.
c. Removal of heat causes a shift to the left ($\longleftarrow$) because the reverse reaction is heat-producing.
d. Removal of PCl_3 causes a shift to the right ($\longrightarrow$) to produce more PCl_3.

Problem

16. How is the equilibrium position of this reaction affected by the following changes?

$$C(s) + H_2O(g) + heat \rightleftharpoons CO(g) + H_2(g)$$

   a. lowering the temperature
   b. increasing the pressure
   c. removing H_2 from the equilibrium mixture

Science, Technology, and Society

17·C Mining the Riches of the Atmosphere

At the turn of the century, more and more land was being used as farmland in order to increase food production. Most of the best land had already been used. Much of the new cropland had marginal, nutrient-poor soil that required heavy fertilization. Fertilizers were obtained from nitrate deposits found only in scattered areas around the world. The most important element found in these fertilizers was nitrogen. It is essential for plant growth. The large demand for nitrogen-containing compounds was beginning to outstrip the supply.

The atmosphere contains huge quantities of nitrogen gas. (Air is 78% nitrogen by volume.) Unfortunately, this molecular form of nitrogen (N_2) is extremely unreactive. The German physical chemist Fritz Haber (1868–1934) was interested in the problem of synthesizing nitrogen compounds. Between 1907 and 1909, he successfully developed a laboratory procedure for making ammonia (NH_3) from nitrogen and hydrogen.

$$N_2 + 3H_2 \rightleftharpoons 2NH_3$$

The industrial production technique for this process was subsequently devised by Carl Bosch. It uses extremely high pressures and temperatures. An equally important factor, however, is an iron-containing catalyst. A basic flow diagram of the Haber–Bosch process is shown in Figure 17·23.

The Haber-Bosch process shifts the reaction equilibrium to favor the forward reaction. Since four moles of reactants form two moles of products, increased pressure favors the formation of ammonia. High temperatures and a catalyst are used to increase the rate of the reaction. Today ammonia plants use temperatures of 300°C to 600°C and pressures of 200 to 1000 atmospheres. The ammonia is separated from the reactants by dissolving it in water or by liquefaction. For their research efforts, both Haber and Bosch were independently awarded the Nobel Prize in Chemistry. Haber received the award in 1918, and Bosch in 1931.

Figure 17·23
In the Haber–Bosch process, nitrogen and hydrogen combine to form ammonia. The hydrogen is derived from the reaction of steam with methane. What is the source of the nitrogen?

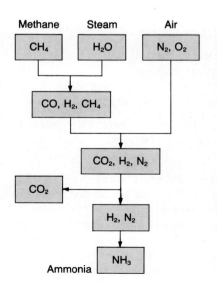

Key Terms

activated complex	17·1	intermediate	17·4
activation energy	17·1	law of disorder	17·5
change in Gibbs free		Le Châtelier's	
energy (ΔG)	17·9	principle	17·12
chemical		nonspontaneous	
equilibrium	17·10	reaction	17·7
elementary reaction	17·4	rate law	17·3
endergonic	17·8	reaction	
entropy	17·5	mechanism	17·4
equilibrium		reaction rate	17·1
constant	17·11	reversible reaction	17·10
equilibrium		specific rate	
position	17·10	constant	17·3
exergonic	17·8	spontaneous	
first-order reaction	17·3	reaction	17·7
free energy	17·8	standard entropy	
heterogeneous		(S^0)	17·6
reactions	17·2	transition state	17·1
inhibitor	17·2		

Chapter Summary

The speed, or rate, at which a chemical reaction goes is determined by an activation energy barrier. The activation energy is the minimum energy that reactants must have to go to products. (In the reverse direction, it is the minimum energy that products must have to go to reactants.) The higher the activation energy barrier, the slower the reaction. Chemists help reactants overcome the activation barrier in a number of ways. Their most effective method is to either increase the temperature at which the reaction is done or use a catalyst.

The natural tendency for all things to go to lower energy and greater disorder (entropy) determines whether a reaction will go. Reactions that actually occur as written are called spontaneous reactions. Equations for other reactions may be written, but the reactions are nonspontaneous. All spontaneous reactions release energy that becomes available to do work. This energy is called free energy.

In principle all reactions are reversible. That is, reactants go to products in the forward direction and products go to reactants in the reverse direction. In practice, however, some reactions are irreversible (go completely to products). Others are reversible, and products go to reactants only to a certain extent. The point at which the rate of conversion of reactants to products and vice versa is equal is the position of equilibrium. The equilibrium constant of a reversible reaction, K_{eq}, is useful for determining the position of equilibrium. It is essentially a measure of the ratio of products to reactants at equilibrium. The direction of change in the position of equilibrium may be predicted by applying Le Châtelier's principle.

Practice Questions and Problems

17. Explain the collision theory of reactions. *17·1*

18. Does every collision between reacting particles lead to products? What other factor is involved? *17·1*

19. What is meant by the rate of reaction? *17·2*

20. How does each of the following factors affect the rate of a chemical reaction? *17·2*
  a. temperature   **c.** particle size
  b. concentration   **d.** inhibitor

21. How is the rate of a spontaneous reaction influenced by a catalyst? How do catalysts make this possible? *17·2*

22. Which of these statements is true? *17·2*
  a. All chemical reactions can be speeded up by increasing the temperature.
  b. Once a chemical reaction gets started, the colliding particles no longer have to "climb over" the activation energy barrier.
  c. Enzymes are biological catalysts.

23. When the gas to a stove is turned on, the gas does not burn unless lit by a flame. Once lit, however, the gas burns until turned off. Explain these observations in terms of the effect of temperature on reaction rate.   *17·2*

24. What is meant by the following terms?   *17·3*
 a. specific rate constant
 b. first-order reaction
 c. rate law

25. The reaction $NO + O_3 \rightarrow NO_2 + O_2$ is first-order in NO and O_3, and second-order overall. Write the complete rate law for this reaction.   *17·3*

26. The reactant in a first-order reaction has half disappeared in 50 minutes. How many minutes are required for this particular reaction to be 75% complete?   *17·3*

27. Define each of the following terms as applied to chemical reactions.   *17·4*
 a. elementary reaction
 b. intermediate
 c. reaction mechanism

28. Sketch a potential energy diagram for the reaction whose mechanism is shown below.

$$2NO \longrightarrow N_2O_2 \text{ (fast)}$$
$$N_2O_2 + O_2 \longrightarrow 2NO_2 \text{ (slow)}$$

Write the balanced equation for the overall reaction.
                                                        17·4

29. What is the meaning of the term *entropy?* *17·5*

30. The products in a spontaneous process are more ordered than the reactants. Is this entropy change favorable or unfavorable?   *17·5*

31. Which system has the lower entropy?   *17·5*
 a. a completed jigsaw puzzle *or* the separate pieces?
 b. 50 mL of water *or* 50 mL of ice?
 c. 10 g of sodium chloride crystals *or* a solution containing 10 g sodium chloride?
 d. a house *or* a pile of bricks and lumber?

32. Predict the direction of the entropy change in each of the following reactions.   *17·5*
 a. $CaCO_3(s) \longrightarrow CaO(s) + CO_2(g)$
 b. $NH_3(g) + HCl(g) \longrightarrow NH_4Cl(s)$
 c. $2NaHCO_3(s) \longrightarrow$
         $\qquad Na_2CO_3(s) + H_2O(g) + CO_2(g)$
 d. $CaO(s) + CO_2(g) \longrightarrow CaCO_3(s)$

33. The standard entropies are given below for some substances at 25°C.

   $KBrO_3(s) \qquad S^0 = 149.2$ J/K-mol

   $KBr(s) \qquad S^0 = 96.4$ J/K-mol

   $O_2(g) \qquad S^0 = 205.0$ J/K-mol

Calculate ΔS^0 for this reaction.   *17·6*

$$KBrO_3(s) \longrightarrow KBr(s) + \tfrac{3}{2}O_2(g)$$

34. Based upon the data of Table 17·1, calculate ΔS^0 for conversion of the monoclinic form of sulfur to the rhombic form.   *17·6*

35. What is a spontaneous reaction? Give an example (not given in the text) of a spontaneous reaction.
                                                        17·7

36. What two factors together determine whether or not a reaction is spontaneous?   *17·7*

37. Is it true that all spontaneous processes are exothermic? Explain your answer.   *17·7*

38. What is free energy? How can knowing the free energy of a reaction help you predict whether or not the reaction will be spontaneous?   *17·8*

39. What is meant by the terms *endergonic* and *exergonic?*   *17·8*

40. The reaction of hydrogen and oxygen gas proceeds with a large decrease of free energy. The reaction is very slow at room temperature but occurs with explosive rapidity in the presence of a flame or ignition wire. Explain.   *17·8*

41. For the decomposition of $CaCO_3(s)$ to $CaO(s)$ and $CO_2(g)$ at 298 K the ΔH^0 is 178.5 kJ/mol and the ΔS^0 is 161.6 J/K-mol. Is the reaction spontaneous or nonspontaneous at this temperature? *17·9*

42. From the data of Table 17·4, calculate the standard free energy change for each of the following reactions, and say whether the reaction is spontaneous or nonspontaneous at standard conditions.
                                                        17·9
 a. $Cl_2(g) + H_2O(g) \longrightarrow 2HCl(g) + \tfrac{1}{2}O_2(g)$
 b. $4NH_3(g) + 7O_2(g) \longrightarrow 4NO_2(g) + 6H_2O(g)$
 c. $2CO_2(g) \longrightarrow 2CO(g) + O_2(g)$
 d. $2H_2S(g) \longrightarrow 2H_2(g) + 2S(s, rhombic)$

43. In your own words define a reversible reaction.
                                                        17·10

44. A reversible reaction has reached a state of dynamic chemical equilibrium. What does this information tell you?   *17·10*

45. How do the rates of the forward and reverse reactions compare at a state of dynamic chemical equilibrium? *17·10*

46. Write the expression for the equilibrium constant for each of the following. *17·11*
 a. $2HBr(g) \rightleftharpoons H_2(g) + Br_2(g)$
 b. $2SO_3(g) \rightleftharpoons 2SO_2(g) + O_2(g)$
 c. $CO_2(g) + H_2(g) \rightleftharpoons CO(g) + H_2O(g)$
 d. $4NH_3(g) + 5O_2(g) \rightleftharpoons 6H_2O(g) + 4NO(g)$

47. Write the expression for the equilibrium constant for each of the following. *17·11*
 a. $4H_2(g) + CS_2(g) \rightleftharpoons CH_4(g) + 2H_2S(g)$
 b. $PCl_5(g) \rightleftharpoons PCl_3(g) + Cl_2(g)$
 c. $2NO(g) + O_2(g) \rightleftharpoons 2NO_2(g)$
 d. $CO(g) + H_2O(g) \rightleftharpoons H_2(g) + CO_2(g)$

48. At a high temperature the following system reaches equilibrium.

$$N_2(g) + O_2(g) \rightleftharpoons 2NO(g)$$

An analysis of the equilibrium mixture in a 1-L flask gives the following results: nitrogen 0.50 mol; oxygen 0.50 mol; nitrogen monoxide 0.020 mol. Calculate K_{eq} for the reaction. *17·11*

49. At 750°C this reaction reaches an equilibrium in a 1-L container.

$$H_2(g) + CO_2(g) \longrightarrow H_2O(g) + CO(g)$$

An analysis of the equilibrium mixture gives the following results: hydrogen 0.053 mol; carbon dioxide 0.053 mol; water 0.047 mol; carbon monoxide 0.047 mol. Calculate K_{eq} for the reaction. *17·11*

50. Comment on the favorability of product formation in each of these reactions. *17·11*

 a. $H_2(g) + F_2(g) \rightleftharpoons 2HF(g)$ $K_{eq} = 1 \times 10^{13}$
 b. $SO_2(g) + NO_2(g) \rightleftharpoons NO(g) + SO_3(g)$
 $$K_{eq} = 1 \times 10^2$$
 c. $2H_2O(g) \rightleftharpoons 2H_2(g) + O_2(g)$
 $$K_{eq} = 6 \times 10^{-28}$$

51. What is Le Châtelier's principle? Use it to explain why carbonated drinks go flat when their containers are left open. *17·12*

52. Can a pressure change be used to shift the position of equilibrium in every reversible reaction? Explain your answer. *17·12*

53. Carbon disulfide can be made by the reaction of carbon dioxide and sulfur trioxide.

$$2SO_3(g) + CO_2(g) + heat \rightleftharpoons CS_2(g) + 4O_2(g)$$

Assuming that the reaction is at equilibrium, what effect do the following changes have on the equilibrium position? *17·12*
 a. addition of CO_2
 b. addition of heat
 c. decrease in the pressure
 d. removal of O_2
 e. addition of a catalyst

54. The industrial production of ammonia is described by this reversible reaction.

$$N_2(g) + 3H_2(g) \rightleftharpoons 2NH_3(g) + 92 \text{ kJ}$$

What effect do the following changes have on the equilibrium position? *17·12*
 a. addition of heat
 b. increase in pressure
 c. addition of catalyst
 d. removal of heat
 e. removal of NH_3

Mastery Questions and Problems

55. Consider the decomposition of N_2O_5 in carbon tetrachloride (CCl_4) at 45°C.

$$2N_2O_5 \longrightarrow 4NO_2 + O_2$$

The reaction is first order in N_2O_5, with the specific rate constant 6.08×10^{-4} per second. Calculate the reaction rate under these conditions.
 a. $[N_2O_5] = 0.200$ mole/liter
 b. $[N_2O_5] = 0.319$ mole/liter

56. Which statement or combination of statements is sufficient to determine whether a reaction will be spontaneous or not?
 a. The reaction is exothermic.
 b. Entropy is increased in the reaction.
 c. Free energy is released in the reaction.

57. For the reaction: $A + B \longrightarrow C$, the activation energy of the forward reaction is 5 kJ and the total energy change is -20 kJ. What is the activation energy of the reverse reaction?

58. Explain each of the following.
 a. A campfire is "fanned" to help get it going.
 b. An explosion at a grain elevator is blamed on dust.
 c. A pinch of powdered manganese dioxide causes hydrogen peroxide to explode even though the manganese dioxide is not changed.

59. Write the rate law for each of the elementary rate processes of Problem 28.

60. A large box is divided into two compartments with a door between them. Equal quantities of two different monoatomic gases are placed in each compartment, as shown in the overhead view in Figure A. The door between the compartments is opened and the gas particles immediately start to mix (Figure B). Why would it be highly unlikely for the situation in Figure B to progress to the situation shown in Figure C?

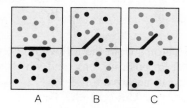

A   B   C

61. Would you expect the entropy to increase in each of the following reactions?
a. $C(s) + O_2(g) \longrightarrow CO_2(g)$
b. $Al_2O_3(s) \longrightarrow 2Al(s) + \frac{3}{2}O_2(g)$
c. $2N(g) \longrightarrow N_2(g)$
d. $N_2(g) \longrightarrow 2N(g)$

62. What would be the effect on the position of equilibrium by decreasing the volume in each of the following reactions?
a. $4HCl(g) + O_2(g) \rightleftharpoons 2Cl_2(g) + 2H_2O(g)$
b. $CO_2(s) \rightleftharpoons CO_2(g)$
c. $CaCO_3(s) \rightleftharpoons CaO(s) + CO_2(g)$

63. A mixture at equilibrium at 827°C contains 0.552 mol CO_2, 0.552 mol H_2, 0.448 mol CO, and 0.448 mol H_2O.

$$CO_2(g) + H_2(g) \rightleftharpoons CO + H_2O(g)$$

What is the value of K_{eq}?

64. Write the equilibrium constant expression for each of the following reactions.
a. $I_2(g) + Cl_2(g) \rightleftharpoons 2ICl(g)$
b. $2NO_2(g) \rightleftharpoons 2NO(g) + O_2(g)$
c. $2SO_2(g) + O_2(g) \rightleftharpoons 2SO_3(g)$
d. $Cl_2(g) + PCl_3(g) \rightleftharpoons PCl_5(g)$

65. The freezing of liquid water at 0°C can be written as follows.

$$H_2O(l,\ d = 1.00\ \text{g/cm}^3) \rightleftharpoons$$
$$H_2O(s,\ d = 0.92\ \text{g/cm}^3)$$

Explain why the application of pressure causes ice to melt.

66. Sketch an energy profile curve for this gas-phase reaction.

$$F + H_2 \longrightarrow HF + H$$

The reaction has an activation energy of 22 kJ and the total energy change is -103 kJ.

67. By means of a sketch, show that the activation energy of an endothermic reaction must be greater than or equal to the total energy change in the reaction. Does a similar relationship exist for exothermic reactions?

68. The decomposition of hydrogen peroxide is catalyzed by iodide ions. The mechanism is thought to be as follows.

$$H_2O_2 + I^- \longrightarrow H_2O + IO^- \text{ (slow)}$$
$$IO^- + H_2O_2 \longrightarrow H_2O + O_2 + I^- \text{ (fast)}$$

a. What is the reactive intermediate?
b. Does I^- qualify as a catalyst?
c. What is the minimum number of activated complexes needed to describe the reaction?
d. Which of the two reactions has the smallest specific rate constant?
e. Write the balanced equation for the entire reaction.

69. The entropy decreases in an endothermic reaction. Is the reaction spontaneous?

70. The reaction between diamond (carbon) and oxygen is spontaneous. What can you say about the speed of this reaction?

$$C(s,\ \text{diamond}) + O_2(g) \longrightarrow CO_2(g)$$

71. Predict what will happen if a catalyst is added to a slow reversible reaction. What happens to the equilibrium position?

72. Using the standard entropy data of Table 17·1, calculate the standard entropy change ΔS^0 for this reaction.

$$2NO(g) + 3H_2O(g) \longrightarrow 2NH_3(g) + \tfrac{5}{2}O_2(g)$$

73. Suppose equilibrium exists for the following reaction at 425 K.

$$Fe_3O_4(s) + 4H_2(g) \rightleftharpoons 3Fe(s) + 4H_2O(g)$$

How would the equilibrium concentration of H_2O be affected by these actions?
a. adding more H_2 to the mixture
b. adding more Fe(s)
c. removing H_2
d. adding a catalyst

Critical Thinking Questions

74. Choose the term that best completes the second relationship.
  a. reaction:intermediate   life span: _____
      (1) birth   (3) adolescence
      (2) products   (4) 75 years
  b. catalyst:inhibitor   sunshine: _____
      (1) clouds   (3) rain
      (2) photosynthesis (4) energy
  c. temperature:heat   entropy: _____
      (1) free energy   (3) reaction rate
      (2) disorder   (4) equilibrium

75. An increase in temperature raises the *energy* of the collisions between reactant molecules. An increase in the concentration of reactants increases the *number* of collisions. What is the effect of a catalyst on the collisions between reactant molecules?

76. Make a list of five things you did today that resulted in an increase in entropy.

Review Questions and Problems

77. Name each of the following ions and identify it as an anion or a cation.
  a. F^- **e.** Na^+
  b. Cu^{2+} **f.** I^-
  c. P^{3-} **g.** O^{2-}
  d. H^+ **h.** Mg^{2+}

78. Explain how crystalline and amorphous substances differ.

79. What is wrong with saying that solid potassium chloride is composed of KCl molecules?

80. Use ionic charges as a guide to write the formula for each of the following compounds.
  a. calcium chloride   **e.** barium carbonate
  b. aluminum sulfide   **f.** ammonium nitrate
  c. magnesium nitrate   **g.** potassium sulfate
  d. aluminum carbonate   **h.** aluminum phosphate

81. Name the following compounds and give the charge on the anion for each.
  a. $NaClO_4$ **d.** $MgCO_3$
  b. $KMnO_4$ **e.** Na_2SO_4
  c. $Ca_3(PO_4)_2$ **f.** $K_2Cr_2O_7$

82. List some properties that are typical of ionic compounds.

83. Write electron configurations and draw electron dot structures for the following elements.
  a. Ge   **d.** Ar
  b. Ca   **e.** Br
  c. O   **f.** P

84. Write the formulas for each of these ionic compounds using electron dot structures.
  a. LiBr   **c.** MgF_2
  b. $AlCl_3$ **d.** Na_2S

85. Write the electron configurations and draw electron dot structures for the following:
  a. Ca^{2+} **c.** Br^-
  b. Li^+ **d.** S^{2-}

86. Which atoms from the following list would you expect to form positive ions, and which would you expect to form negative ions?
  a. Cl   **e.** Cu   **h.** Fe
  b. Ca   **f.** Sn   **i.** N
  c. P   **g.** K   **j.** Ni
  d. Se

Challenging Questions and Problems

87. Hemoglobin is a protein molecule in the red blood cell that carries oxygen from the lungs to cells throughout the body. Carbon monoxide poisoning results from formation of a complex of carbon monoxide (CO) with the hemoglobin. This complex is 200 times stronger than the oxygen–hemoglobin complex. This makes it extremely difficult for the oxygen to compete with carbon monoxide in being carried to the cells. When too much hemoglobin is tied up by the carbon monoxide, death occurs by asphyxiation. This is the reaction.

Hemoglobin—O_2 + CO $\rightleftharpoons$

Hemoglobin—CO + O_2

What would be a logical treatment for a person suffering from carbon monoxide poisoning? Using your knowlege of the gas laws, suggest a way to increase the effectiveness of this treatment.

88. Is eating sugar candy really bad for your teeth? Tooth decay is the result of the dissolving of tooth enamel, $Ca_5(PO_4)_3OH$. In the mouth the following equilibrium is established.

$Ca_5(PO_4)_3OH(s) \rightleftharpoons$

$$5Ca^{2+}(aq) + 3PO_4^{3-}(aq) + OH^-(aq)$$

When sugar ferments on the teeth, H^+ is produced. What effect does this increased H^+ have on tooth enamel?

89. The following data were collected for the decomposition of compound AB into its elements. The reaction is first order in AB.

a. Make a graph of concentration versus time.

b. Using the method shown in Figure 17·8 determine the rate of this reaction at $t = 100$ seconds and $t = 250$ seconds.

[AB] (mol/L)	Time (s)
0.300	0
0.246	50
0.201	100
0.165	150
0.135	200
0.111	250
0.090	300
0.075	350

Research Projects

1. Investigate the effects of different catalysts on the decomposition of hydrogen peroxide.

2. Report on a catalyst used in industrial reactions. How does the catalyst operate on a molecular level?

3. What are the problems involved in the scale-up of a laboratory reaction to an industrial process?

4. What contributions did Josiah Gibbs make to chemical reaction thermodynamics? Why was his work relatively unrecognized for almost 20 years?

5. Discuss some common spontaneous reactions. What makes them spontaneous?

6. What effect do catalysts and temperature have on nonspontaneous reactions? Cite some examples.

7. How does a catalytic converter reduce pollution from automobile emissions?

8. What contributions did Henri Le Châtelier make to the understanding of equilibrium and reaction rates? How did his discoveries help industrial chemists?

9. What contributions did Gottlieb Kirchhoff make to industrial chemistry?

10. What chemical reactions are involved in smog formation? Compare the local, regional, and global effects of photochemical smog.

11. What are the major sources of air pollution in your community? How is it being controlled? What are the tangible and intangible costs of air pollution?

12. Review the effects of clean air standards since 1970.

13. Determine the effect of increasing water temperature on the rate of dissolving sugar.

Readings and References

Davenport, Derek A. "When Push Comes to Shove: Disturbing the Equilibrium." *Chem Matters* (American Chemical Society) (February 1985), p. 14.

Fuller, John G. *The Poison That Fell from the Sky.* New York: Random House, 1977.

O'Connor, Rod, and Charles Mickey. *Solving Problems in Chemistry: With Emphasis on Stoichiometry and Equilibrium.* New York: Harper & Row, 1974.

Siezen, Roland, "Pumping Oxygen." *ChemMatters* (February 1984), pp. 6–9.

Stoker, H. Stephen, and Spencer L. Seager. *Environmental Chemistry: Air and Water Pollution,* 2nd ed. Glenview, IL: Scott, Foresman, 1976.

Verbit, Lawrence P. "Chemistry's Speedy Servants." *Science Year* (1982), pp. 194–205.

Yohe, Brad. "Toothpaste." *ChemMatters* (February 1986), pp. 12–13.

18 Acids and Bases

Chapter Preview

18·1 Properties of Acids and Bases

18·A Etching in Art and Industry

18·2 Hydrogen Ions from Water

18·3 The pH Concept

18·4 Calculating pH Values

18·B Arrhenius Led the Way

18·5 Arrhenius Acids and Bases

18·6 Brønsted–Lowry Acids and Bases

18·C Measuring pH

18·7 Lewis Acids and Bases

18·8 The Strengths of Acids and Bases

18·9 Calculating Dissociation Constants

Acids and bases play a central role in much of the chemistry that affects our daily lives. They are widely used in manufacturing processes. They are also extremely important in the correct functioning of our bodies. This chapter covers the qualitative and quantitative aspects of these two classes of compounds. You will see how they ionize or dissociate in water. You will also see how the concept of pH is used to measure the strength of acidic and basic solutions.

18·1 Properties of Acids and Bases

Acids and bases are readily identified by their characteristic properties.

Compounds that are classified as acids or bases have many distinctive properties. Acidic compounds give foods a tart or sour taste. For example, vinegar contains acetic acid. Lemons contain citric acid. Aqueous solutions of acids will conduct electricity and thus are electrolytes. Some are strong electrolytes whereas others are weak electrolytes. Acids cause certain chemical dyes, called indicators, to change color. Many metals, such as zinc and magnesium, react with aqueous solutions of acids to produce hydrogen gas. Acids react with compounds containing hydroxide ions to form water and a salt.

Bases are compounds that react with acids to form water and a salt. Milk of magnesia (magnesium hydroxide) is a base. It is used to treat the problem of excess stomach acid. Aqueous solutions of bases have a bitter taste and feel slippery. Like acids, bases can be strong or weak electrolytes. Like acids, bases will change the color of an acid–base indicator.

Figure 18·1
The tangy taste of a fresh orange comes from the citric acid found in all citrus fruits.

431

Several theories explain the observed properties of acids and bases. Before discussing these theories, we must examine water and the ways in which the hydrogen-ion concentration is measured in solutions.

ChemDirections

Concentrated acids and bases are transported across the nation in specially designed tank trucks. Read **ChemDirections** *Trucking Chemicals,* page 659, and begin to recognize these materials in transit on your highways.

Acids and bases are among the most common and industrially important chemicals.

=== Science, Technology, and Society ===

18·A   Etching in Art and Industry

For centuries printmakers have used acids to make engravings. Initially, a metal plate is coated with a thin layer of wax or other material that is acid resistant. The artist draws in the wax so that areas of the metal plate are exposed. When the drawing is complete the waxed surface is dipped in an acid. Wherever the wax was removed from the plate the exposed metal is eaten away. After an appropriate amount of time the plate is washed to stop the etching process, and the wax is removed.

To make a print the plate is inked with a roller that forces the ink into the etched areas. A piece of paper is laid on the inked surface. Pressure transfers the ink from the etched metal surface to the paper.

Printmakers commonly use copper, zinc, or iron plates and either nitric acid (HNO_3) or hydrochloric acid (HCl). The various combinations of metals and acids produce different effects. For example, nitric acid "bites" an etched line more readily on the sides than on the bottom. This makes the line wider, and the final print takes on a more diffuse appearance. Hydrochloric acid bites deeper into the metal plate. It is preferred for etching details.

Glass is etched using hydrofluoric acid (HF). The resulting frosted and transparent areas can be used to create artistic designs. The process is also used to produce the fine markings on laboratory glassware and to frost the insides of light bulbs.

Figure 18·2
Artists use acids to etch patterns in glass and metals.

432   Chapter 18   Acids and Bases

Figure 18·3
All of these items contain acids or bases.

18·2 Hydrogen Ions from Water

Water molecules are highly polar. Even at room temperature they are in continuous motion. Occasionally, the collisions between water molecules are energetic enough that a hydrogen ion is transferred from one water molecule to another. *A water molecule that loses a hydrogen ion becomes a negatively charged* **hydroxide ion** (OH⁻). *A water molecule that gains a hydrogen ion becomes a positively charged* **hydronium ion** (H_3O^+).

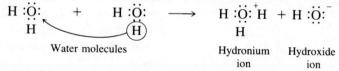

$$H \overset{..}{\underset{..}{O}}: \quad + \quad H \overset{..}{\underset{..}{O}}: \quad \longrightarrow \quad H \overset{..}{\underset{..}{O}}:^+ H \; + \; H \overset{..}{\underset{..}{O}}:^-$$

Water molecules · · · Hydronium ion · · Hydroxide ion

The reaction in which two water molecules react to give ions is the **self-ionization** *of water*. This reaction can be written as a simple dissociation.

$$H_2O \rightleftharpoons H^+ + OH^-$$
Hydrogen ion · · Hydroxide ion

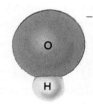

Figure 18·4
Which of these ions is a hydronium ion? Which is a hydroxide ion?

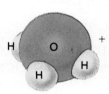

Water self-ionizes.

Brackets around a chemical formula [H₂O] are used to indicate concentrations of solutions in moles per liter (molarity).

In water or aqueous solution, hydrogen ions (H^+) are always joined to water molecules as hydronium ions (H_3O^+). The hydronium ions are themselves solvated to form species such as $H_9O_4^+$. Hydrogen ions in aqueous solution have several names. Some chemists call them protons. Others prefer to call them hydrogen ions, or hydronium ions, or solvated protons. In this book we use either H^+ or H_3O^+ to represent hydrogen ions in aqueous solution.

The self-ionization of water occurs to a very small extent. In pure water at 25°C the concentration of hydrogen ions, $[H^+]$, and the concentration of hydroxide ions, $[OH^-]$, are each only 1.0×10^{-7} mol/L. Note that the concentrations of H^+ and OH^- must be equal in pure water. These ions are produced in a 1:1 ratio. Chemists describe pure water as neutral. In fact, *any aqueous solution in which the $[H^+]$ and the $[OH^-]$ are 1.0×10^{-7} mol/L is described as a* **neutral solution.**

In any aqueous solution the $[H^+]$ and the $[OH^-]$ are interdependent. If the $[H^+]$ increases then the $[OH^-]$ decreases. If the $[H^+]$ decreases then

18·2 Hydrogen Ions from Water **433**

the [OH⁻] increases. Le Châtelier's principle applies here. If additional ions (either hydrogen ions or hydroxide ions) are added to the solution, the equilibrium will shift to decrease the other type of ion. More water molecules are formed in the process.

$$H^+ + OH^- \rightleftharpoons H_2O$$

For aqueous solutions the product of the hydrogen ion concentration, $[H^+]$, and the hydroxide ion concentration, $[OH^-]$, is equal to 1.0×10^{-14} (mol/L)².

$$[H^+] \times [OH^-] = 1.0 \times 10^{-14} \text{ (mol/L)}^2$$

The product of the concentrations of the hydrogen ions and hydroxide ions in water is K_w, the **ion-product constant for water.**

$$K_w = [H^+] \times [OH^-] = 1.0 \times 10^{-14} \text{ (mol/L)}^2$$

$K_w = [H^+] \times [OH^-]$ is true for every aqueous solution regardless of the presence of any other ions.

Not all solutions are neutral. When some substances dissolve in water they release hydrogen ions. For example, when hydrogen chloride dissolves in water it forms hydrochloric acid.

$$HCl(g) \xrightarrow{\text{water}} H^+(aq) + Cl^-(aq)$$

In such a solution the hydrogen ion concentration, $[H^+]$, is greater than the hydroxide ion concentration, $[OH^-]$. (The hydroxide ions are present from the self-ionization of water.) *In* **acidic solutions** *the $[H^+]$ is greater than the $[OH^-]$.* Therefore the $[H^+]$ of an acidic solution is greater than 1.0×10^{-7} mol/L.

When sodium hydroxide is dissolved in water it forms hydroxide ions.

$$NaOH(s) \xrightarrow{\text{water}} Na^+(aq) + OH^-(aq)$$

In such a solution the hydrogen ion concentration, $[H^+]$, is less than the hydroxide ion concentration, $[OH^-]$. (The hydrogen ions are present from the self-ionization of water.) *In* **basic solutions** *the $[H^+]$ is less than the $[OH^-]$.* Therefore the $[H^+]$ of a basic solution is less than 1.0×10^{-7} mol/L. *Basic solutions are also known as* **alkaline solutions.**

Figure 18·5
Which of these reagents would increase the hydrogen ion concentration when added to an aqueous solution? Which would increase the hydroxide ion concentration?

HCl   NaOH

Example 1

If $[H^+] = 1.0 \times 10^{-5}$ mol/L, is the solution acidic, basic, or neutral? What is the $[OH^-]$ of this solution?

Solution

The $[H^+]$ is 1.0×10^{-5} mol/L. Because this is greater than 1.0×10^{-7} mol/L, the solution is acidic. We know that $K_w = [H^+] \times [OH^-]$. Therefore the $[OH^-] = K_w/[H^+]$. Substituting the numerical values $K_w = 1.0 \times 10^{-14}$ (mol/L)² and $[H^+] = 1.0 \times 10^{-5}$ mol/L, we can compute $[OH^-]$.

$$[OH^-] = \frac{1.0 \times 10^{-14} \text{ (mol/L)}^2}{1.0 \times 10^{-5} \text{ (mol/L}} = 1.0 \times 10^{-9} \text{ mol/L}$$

Problem

1. If the hydroxide-ion concentration of an aqueous solution is 1.0×10^{-3} mol/L, what is the $[H^+]$ in the solution? Is the solution acidic, basic, or neutral?

Before reading this section you should review logarithms by doing Section B·5 of the Math Review in Appendix B.

The pH scale is used to express hydrogen-ion concentrations.

18·3 The pH Concept

The expression of hydrogen-ion concentration in moles per liter is cumbersome. A more widely used system is the pH scale. It was proposed in 1909 by the Danish scientist Søren Sørensen (1868–1939). *The* **pH** *of a solution is the negative logarithm of the hydrogen-ion concentration.*

$$pH = -\log[H^+]$$

In a neutral solution the $[H^+] = 1 \times 10^{-7}$ mol/L. The pH of a neutral solution is 7.

$$
\begin{aligned}
pH &= -\log(1 \times 10^{-7}) \\
&= -(\log 1 + \log 10^{-7}) \\
&= -(0.0 + (-7)) \\
&= 7.0
\end{aligned}
$$

For pH calculations, concentrations should be expressed in exponential form. For example, a hydrogen-ion concentration of $0.01M$ is rewritten as $1 \times 10^{-2}M$. The pH of this solution is 2.0. Neutral solutions have a pH of 7.0. A pH of 0 is strongly acidic. A pH of 14 is strongly basic.

In a similar way, the pOH of a solution equals the negative logarithm of the hydroxide ion concentration.

$$pOH = -\log[OH^-]$$

A neutral solution has a pOH of 7.0. A solution with a pOH of less than 7.0 is basic. A solution with a pOH of greater than 7.0 is acidic.

Figure 18·6
The pH scale shows the relationship between pH and the hydrogen-ion concentration. Notice that acids have lower pHs than bases.

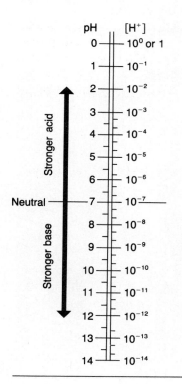

Example 2

The hydrogen-ion concentration of a solution is 1×10^{-10} mol/L. What is the pH of the solution?

Solution

We know that the $[H^+] = 1 \times 10^{-10}$ mol/L.

$$
\begin{aligned}
pH &= -\log[H^+] \\
&= -\log(1 \times 10^{-10}) \\
&= -(0.0 + (-10)) \\
&= 10.0
\end{aligned}
$$

A solution in which the $[H^+]$ is 1×10^{-10} mol/L has a pH of 10.0.

Problem

2. Determine the pH of the following solutions.
   a. $[H^+] = 1 \times 10^{-6}$ mol/L
   b. $[H^+] = 0.0001M$
   c. $[OH^-] = 1 \times 10^{-2}$ mol/L
   d. $[OH^-] = 1 \times 10^{-11}$ mol/L

Example 3

The pH of a solution is 6.0. What is the hydrogen-ion concentration?

Solution

To do this problem we work backward.

$$pH = -\log[H^+]$$
$$6.0 = -\log[H^+]$$
$$-6.0 = \log[H^+]$$

The number whose log is -6.0 is 1×10^{-6}. Therefore the $[H^+] = 1 \times 10^{-6}$ mol/L.

Problems

3. What are the hydrogen-ion concentrations for solutions with the following pH values?   a. 4.0   b. 11.0   c. 8.0

4. What are the hydroxide-ion concentrations for solutions with the following pH values?   a. 6.0   b. 9.0   c. 12.0

Chemical Connections

Gardeners often test the pH of their soil. Some plants, such as azaleas and watermelons, prefer slightly acidic soil. Other plants prefer basic or neutral soils.

Figure 18·7
You can determine the approximate pH of a substance by testing it with pH test paper and comparing the color with the standards. What is the pH of the sample tested with the pH test paper at the right? Is it an acid or base?

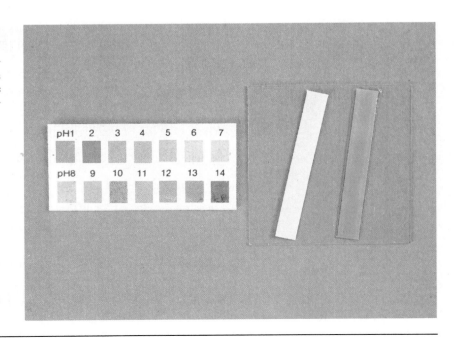

Table 18·1 Relationship between the $[H^+]$, the $[OH^-]$, and pH

	$[H^+]$ (mol/L)	$[OH^-]$ (mol/L)	pH	Aqueous system
Increasing acidity ↑	1×10^0	1×10^{-14}	0.0	$1M$ HCl (0.0)
	1×10^{-1}	1×10^{-13}	1.0	$0.1M$ HCl (1.0)
	1×10^{-2}	1×10^{-12}	2.0	Gastric juice (1.6–1.8)
	1×10^{-3}	1×10^{-11}	3.0	Lemon juice (2.3), vinegar (2.4–3.4)
	1×10^{-4}	1×10^{-10}	4.0	Soda water (3.8), tomato juice (4.2)
	1×10^{-5}	1×10^{-9}	5.0	Black coffee (5.0)
	1×10^{-6}	1×10^{-8}	6.0	Milk (6.3–6.6), urine (5.5–7.0)
Neutral	1×10^{-7}	1×10^{-7}	7.0	Pure water (7.0), saliva (6.2–7.4)
	1×10^{-8}	1×10^{-6}	8.0	Blood (7.35–7.45), bile (7.8–8.6)
Increasing basicity ↓	1×10^{-9}	1×10^{-5}	9.0	Sodium bicarbonate (8.4), sea water (8.4)
	1×10^{-10}	1×10^{-4}	10.0	Milk of magnesia (10.5)
	1×10^{-11}	1×10^{-3}	11.0	Household ammonia (11.5)
	1×10^{-12}	1×10^{-2}	12.0	Washing soda (12.0)
	1×10^{-13}	1×10^{-1}	13.0	$0.1M$ NaOH (13.0)
	1×10^{-14}	1×10^0	14.0	$1M$ NaOH (14.0)

The pH of pure water or a neutral aqueous solution is 7.0. A solution in which the $[H^+]$ is greater than 1×10^{-7} mol/L has a pH less than 7.0 and is acidic. A solution that has a pH greater than 7.0 is basic and has a $[H^+]$ of less than 1×10^{-7} mol/L.

- *Neutral solution:* pH equals 7.0; the $[H^+]$ equals 1×10^{-7} mol/L.

- *Acidic solution:* pH is less than 7.0; the $[H^+]$ is greater than 1×10^{-7} mol/L.

In an acidic solution $[H^+] > [OH^-]$. In a basic solution $[H^+] < [OH^-]$.

- *Basic solution:* pH is greater than 7.0; the $[H^+]$ is less than 1×10^{-7} mol/L.

Table 18·1 gives the pH values of a number of common aqueous solutions. It also summarizes the relationships between the $[H^+]$, the $[OH^-]$, and pH. You may notice that pH can sometimes be read easily from the value of the $[H^+]$. If the $[H^+]$ is written in standard exponential form and has a characteristic of exactly 1, the pH of the solution equals the exponent with the sign changed from minus to plus. For example, a solution with a $[H^+] = 1 \times 10^{-2}$ has a pH of 2.

Problem

5. Give the pH of each of the following solutions.
   a. $[H^+] = 1 \times 10^{-4}$
   b. $[H^+] = 0.001$
   c. $[H^+] = 1 \times 10^{-9}$
   d. $[H^+] = 100 \times 10^{-12}$

18·4 Calculating pH Values

Most pH values are not whole numbers. For example, Table 18·1 shows that milk of magnesia has a pH of 10.5. Using the definition of pH, this means that the $[H^+]$ must equal $1 \times 10^{-10.5}$. Therefore the $[H^+]$ must be less than 1×10^{-10} (pH 10.0) but greater than 1×10^{-11} (pH 11.0).

When the $[H^+]$ is written in standard exponential form but its coefficient is *not* 1, then a table of common logarithms is needed to calculate pH. A four-place log table is in Appendix B·5. The following example shows how such a calculation is made.

The $[H^+]$ and $[OH^-]$ can be calculated from a measurement of pH.

Example 4

What is the pH of a solution if the $[OH^-] = 4.0 \times 10^{-11}$ mol/L?

Solution

To calculate pH, first calculate the $[H^+]$. Use the definition of K_w.

$$K_w = [OH^-] \times [H^+]$$

$$[H^+] = \frac{K_w}{[OH^-]} = \frac{1.0 \times 10^{-14}}{4.0 \times 10^{-11}} = 0.25 \times 10^{-3}$$

$$= 2.5 \times 10^{-4} \text{ mol/L}$$

Then solve for the pH using the definition of pH.

$$pH = -\log[H^+]$$

$$pH = -\log (2.5 \times 10^{-4})$$

$$= -(\log 2.5 + \log 10^{-4})$$

log (a × b) = log a + log b

The log table is in Appendix B·5.
Some calculators also give logs.

Use the log table to find the value of log 2.5. By inspection, log $10^{-4} = -4$.

$$pH = -(0.40) - (-4)$$

$$= -0.40 + 4$$

$$= 3.60$$

Problem

6. Calculate the pH for each solution.
   a. $[H^+] = 5.0 \times 10^{-6}$ c. $[OH^-] = 2.0 \times 10^{-5}$
   b. $[H^+] = 8.3 \times 10^{-10}$ d. $[OH^-] = 4.5 \times 10^{-11}$

The hydrogen-ion concentration of a solution can be calculated if the pH is known. If the pH of a solution is 3.0, for example, then the $[H^+] = 1 \times 10^{-3}$ mol/L. When the pH is not a whole number, log tables are needed to calculate the hydrogen-ion concentration. For example, if the pH is 3.7 the hydrogen-ion concentration must be greater than 1×10^{-4} mol/L (pH 4.0) and less than 1×10^{-3} mol/L (pH 3.0). To get an accurate value use log tables.

Example 5

What is the $[H^+]$ of a solution if the pH $= 3.7$?

Solution

First rearrange the equation: $pH = -\log[H^+]$.

$$\log[H^+] = -pH = -3.7$$

You cannot use a log table directly to find a number whose log is negative. To get around this problem, add and subtract the whole number that is closest to and larger than the negative log. In this case, the negative log is 3.7, and the whole number is 4.

$$\log[H^+] = (-3.7 + 4) - 4$$
$$= 0.3 - 4$$
$$[H^+] = 10^{(0.3-4)}$$
$$= 10^{0.3} \times 10^{-4}$$

From the log table, the number whose log is 0.3 is 2. Thus the antilog of 0.3 is 2. The number whose log is -4 is 10^{-4}. Therefore the $[H^+] = 2 \times 10^{-4}$ mol/L.

Problem

7. Calculate the $[H^+]$ for each solution.
   a. pH = 5.0   c. pH = 12.20
   b. pH = 5.80   d. pH = 2.64

Figure 18·8
Svante Arrhenius received a Nobel Prize for his revolutionary work in chemistry.

Science, Technology, and Society

18·B Arrhenius Led the Way

Svante Arrhenius (1859–1927) studied the electrical conductivity of solutions. At the time the structure of the atom was completely unknown. The nature of the particles responsible for electricity was also unknown. Certain solutions were known to be capable of conducting electricity. The solutes of those solutions were called electrolytes. Solutes that would not conduct electricity while in solution were called nonelectrolytes.

Some chemists noticed that solutes caused a depression in the freezing point of solvents (Section 16·7). It was thought that this property was related to the number of particles in the solution. The electrolyte sodium chloride (NaCl) caused twice the expected freezing point depression. Barium chloride (BaCl$_2$) produced three times the expected freezing point depression.

From these observations Arrhenius proposed that electrolytic compounds broke into particles when they dissolved. Depending upon the composition of the solute some solute particles could break into two or more parts. This is called electrolytic dissociation. It was consistent with the idea that the freezing point depression depended on the number of particles. Arrhenius also reasoned that the constituent parts must be electrically charged. This accounted for two critical observations that had not been explained at the time. First, the elements from the compounds were not apparent in the solution. Second, the solution conducted electricity. Arrhenius later extended his concepts of electrolytic dissociation to explain certain behaviors of acids and bases.

The idea of electrically charged solute particles was not readily accepted. (This is often the case for new concepts.) Arrhenius was given the lowest possible grade for his Ph.D. defense of this theory in 1884. In the following decade, however, numerous discoveries undermined the idea that the atom was structureless. When Thomson discovered the negatively charged electron in the 1890s, Arrhenius' ideas gained acceptance. In 1903 Arrhenius' work was formally recognized when he received the Nobel Prize in Chemistry.

Arrhenius' theory is sometimes referred to as the theory of ionization.

18·5 Arrhenius Acids and Bases

In aqueous solutions acids release hydrogen ions, and bases release hydroxide ions.

In 1887 the young Swedish chemist Svante Arrhenius proposed a revolutionary way of thinking about acids and bases. He said that *acids are compounds containing hydrogen that ionize to yield hydrogen ions (H^+) in aqueous solution.* Similarly, he said that *bases are compounds that ionize to yield hydroxide ions (OH^-) in aqueous solution.*

Table 18·2 lists some important acids. Nitric acid (HNO_3) and *any acid that contains one ionizable hydrogen is called a* **monoprotic acid.** Sulfuric acid (H_2SO_4) and *any acid that contains two ionizable protons is a* **diprotic acid.** Phosphoric acid (H_3PO_4) and *any acid that contains three ionizable protons is a* **triprotic acid.** Do not assume that all compounds containing hydrogen are acids or that all the hydrogens in an acid are released as hydrogen ions. Only the hydrogens in very polar bonds are ionizable. These are bonds in which hydrogen is joined to a very electronegative element. Hydrogen ions are released when   molecules

Safety

Nitric acid is very reactive and should be stored separately from other acids.

The formula for acetic acid can also be written as CH_3CO_2H or CH_3COOH.

Table 18·2 Some Common Acids	
Name	Formula
Hydrochloric acid	HCl
Nitric acid	HNO_3
Sulfuric acid	H_2SO_4
Phosphoric acid	H_3PO_4
Acetic acid	$HC_2H_3O_2$
Carbonic acid	H_2CO_3

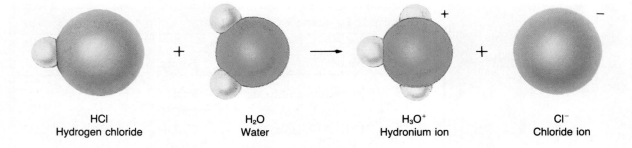

HCl	H₂O	H₃O⁺	Cl⁻

$$HCl \qquad H_2O \qquad H_3O^+ \qquad Cl^-$$

HCl — Hydrogen chloride $+$ H₂O — Water $\longrightarrow$ H₃O⁺ — Hydronium ion $+$ Cl⁻ — Chloride ion

Figure 18·9
Hydrochloric acid is actually hydrogen chloride dissolved in water. The ionized hydrogen forms hydronium ions and makes this compound an acid.

that contain such bonds dissolve in water. This is because the hydrogen ions are stabilized by solvation. The hydrogen chloride molecule, for example, is a polar covalent molecule. It ionizes to form hydrochloric acid in aqueous solution.

$$\overset{\delta+}{H}-\overset{\delta-}{Cl}(g) \xrightarrow{\text{water}} H^+(aq) + Cl^-(aq)$$

Hydrogen chloride $\qquad$ Hydrogen ion $\qquad$ Chloride ion
(hydrochloric acid)

By contrast, the four hydrogens in methane (CH_4) are in weakly polar C—H bonds. Methane has no ionizable hydrogens and is not an acid. Acetic acid ($HC_2H_3O_2$) is different. Although each molecule contains four hydrogens, acetic acid is a monoprotic acid. The structural formula shows why.

$$H-\underset{\underset{H}{|}}{\overset{\overset{H}{|}}{C}}-\overset{\overset{O}{\|}}{C}-O-H$$

Acetic acid
($HC_2H_3O_2$)

The three hydrogens attached to carbon are in weakly polar bonds. They do not ionize. Only the hydrogen attached to oxygen is ionizable. As you see more written formulas, you will be able to recognize ionizable hydrogens.

Problem

8. Identify the following acids as monoprotic, diprotic, or triprotic.
a. H_2CO_3 **b.** H_3PO_4 **c.** HCl

Table 18·3 lists some common bases. The base with which you are perhaps most familiar is sodium hydroxide (NaOH). This is ordinary lye. Sodium metal reacts with water to form sodium hydroxide.

$$2Na + 2H_2O \longrightarrow 2NaOH + H_2$$

Sodium metal $\qquad$ Water $\qquad$ Sodium hydroxide $\qquad$ Hydrogen

Figure 18·10
Hydrochloric acid, commonly known as muriatic acid, is used to clean stone buildings and swimming pools.

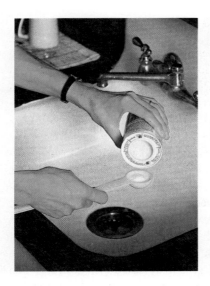

Figure 18·11
Sodium hydroxide is used to clean clogged drains because it is very corrosive.

Safety

Never taste bases or any other chemical in the laboratory.

Table 18·3 Some Common Bases		
Name	Formula	Solubility in water
Potassium hydroxide	KOH	High
Sodium hydroxide	NaOH	High
Calcium hydroxide	Ca(OH)$_2$	Very low
Magnesium hydroxide	Mg(OH)$_2$	Very low

Potassium reacts vigorously with water to produce potassium hydroxide (KOH). Both sodium hydroxide and potassium hydroxide are ionic solids. They dissociate completely into the metal ions and hydroxide ions when dissolved in water.

$$\underset{\substack{\text{Sodium}\\\text{Hydroxide}}}{\text{NaOH}} \xrightarrow{\text{water}} \underset{\substack{\text{Sodium}\\\text{ion}}}{\text{Na}^+(aq)} + \underset{\substack{\text{Hydroxide}\\\text{ion}}}{\text{OH}^-(aq)}$$

Sodium and potassium are Group 1A elements. They are named the alkali metals because they react with water to produce alkaline solutions. Metal oxides also react with water to produce alkaline solutions.

Both sodium hydroxide and potassium hydroxide are very soluble in water. Concentrated solutions of these compounds are readily prepared. Such solutions would have a bitter taste and feel slippery. Note, however, that they are extremely corrosive to the skin. They can cause painful, deep, slow-healing wounds if not immediately washed off. Calcium hydroxide, Ca(OH)$_2$, and magnesium hydroxide, Mg(OH)$_2$, are both hydroxides of Group 2A metals. They are not very soluble in water. Consequently, their solutions are always very dilute, even when saturated. The concentration of hydroxide ions in such solutions is correspondingly low.

A saturated solution of calcium hydroxide contains 0.165 g of Ca(OH)$_2$ per 100 g of water. Limewater is an aqueous suspension of solid calcium hydroxide. Magnesium hydroxide is much less soluble than calcium hydroxide. A saturated solution contains only 0.0009 g of Mg(OH)$_2$ per 100 g of water. Suspensions of solid magnesium hydroxide in water contain low concentrations of hydroxide ion. They are taken internally as "milk of magnesia," an antacid or mild laxative.

Problems

9. Write a balanced equation for each reaction.
  a. Potasssium metal reacts with water.
  b. Calcium metal reacts with water.

10. Write the reaction for the dissociation of each compound in water.
  a. potassium hydroxide   **b.** magnesium hydroxide

18·6 Brønsted–Lowry Acids and Bases

One problem with the Arrhenius theory is that it is not comprehensive enough. It does not explain the behavior of all compounds that behave as bases. For example, aqueous solutions of ammonia (NH_3) and sodium carbonate (Na_2CO_3) are basic. Neither of these compounds is a hydroxide. In 1923 Johannes Brønsted (Danish; 1879–1947) and Thomas Lowry (English; 1874–1936) independently proposed a new theory. *The Brønsted–Lowry theory defines an acid as a* **hydrogen-ion donor.** Similarly, *a Brønsted–Lowry base is a* **hydrogen-ion acceptor.** All the acids and bases included in the Arrhenius theory are also acids and bases according to the Brønsted–Lowry theory. Some compounds that were not included in the Arrhenius theory are classified as bases in the Brønsted–Lowry theory.

The behavior of ammonia as a base can be understood by the Brønsted–Lowry theory. Ammonia gas is very soluble in water. When ammonia dissolves it acts as a base because it accepts a hydrogen ion from water.

$$NH_3 \quad + \quad H_2O \quad \rightleftharpoons \quad NH_4^+ \quad + \quad OH^-$$

Ammonia (hydrogen-ion acceptor, Brønsted–Lowry base)   Water (hydrogen-ion donor, Brønsted–Lowry acid)   Ammonium ion   Hydroxide ion (makes the solution basic)

> A Brønsted–Lowry acid is a hydrogen-ion donor. A Brønsted–Lowry base is a hydrogen-ion acceptor.

In this reaction ammonia is the hydrogen-ion acceptor. Therefore it is a Brønsted–Lowry base. Water, the hydrogen-ion donor, is a Brønsted–Lowry acid. Hydrogen ions are transferred from water to ammonia. This causes the hydroxide-ion concentration to be greater than it is in pure water. As a result, solutions of ammonia are basic.

Heating an aqueous solution of ammonia drives off ammonia gas. This action causes the equilibrium in the ammonia dissolution equation to shift to the left. The ammonium ion, NH_4^+, reacts with OH^- to form NH_3 and H_2O. When the reaction goes from right to left, NH_4^+ gives up a hydrogen ion; it acts as a Brønsted–Lowry acid. The hydroxide ion accepts a H^+; it acts as a Brønsted–Lowry base. Overall then, this equilibrium has two acids and two bases.

$$NH_3 + H_2O \quad \rightleftharpoons \quad NH_4^+ \quad + \quad OH^-$$

Base   Acid   Conjugate acid   Conjugate base

Safety

Ammonia is a very irritating gas. Use care when smelling chemicals. Wave your hand over the open container to waft the aroma to your nose.

Figure 18·12
Ammonia dissolves in water to form ammonium ions and hydroxide ions. Why is ammonia not classified as an Arrhenius base?

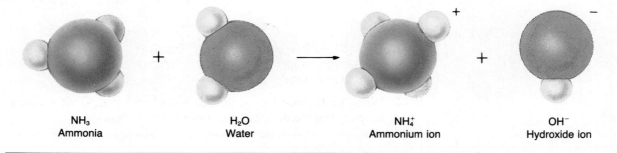

NH_3
Ammonia

H_2O
Water

NH_4^+
Ammonium ion

OH^-
Hydroxide ion

When ammonia dissolves, NH_4^+ is the conjugate acid of the base NH_3. *A* **conjugate acid** *is the particle formed when a base gains a hydrogen ion*. Similarly, OH^- is the conjugate base of the acid water. *A* **conjugate base** *is the particle that remains when an acid has donated a hydrogen ion*. Conjugate acids and bases are always paired with an acid and a base. *A* **conjugate acid-base pair** *is two substances that are related by the loss or gain of a single hydrogen ion*. The ammonia molecule and ammonium ion are a conjugate acid–base pair. The water molecule and hydroxide ion are also a conjugate acid–base pair.

The conjugate acid of a strong base is a weak acid. The conjugate base of a strong acid is a weak base.

$$NH_3 + H_2O \rightleftharpoons NH_4^+ + OH^-$$

Base   Acid   Conjugate Conjugate
                                acid   base

The Brønsted–Lowry theory is also applicable to acids. Consider the ionization of hydrogen chloride in water.

A strong Brønsted–Lowry acid gives up its hydrogen ion easily.

$$HCl + H_2O \rightleftharpoons H_3O^+ + Cl^-$$

Acid   Base   Conjugate Conjugate
                                acid   base

In this reaction hydrogen chloride is the hydrogen-ion donor. It is therefore a Brønsted–Lowry acid. Water is the hydrogen-ion acceptor. It is a Brønsted–Lowry base. The chloride ion is the conjugate base of the acid HCl. The hydronium ion is the conjugate acid of the base water.

You may have noticed that water has a split personality. Sometimes water receives a hydrogen ion. At other times it donates a hydrogen ion. *A substance that can act as both an acid and a base is termed* **amphoteric**. Water is amphoteric. In the reaction with HCl, water accepts a proton and is therefore a base. In the reaction with NH_3, water donates a proton and is an acid.

Problem

11. Write equations for the ionization of HNO_3 and Na_2CO_3 in water.
    a. Identify the hydrogen-ion donor and hydrogen-ion acceptor in each case.   **b.** Label the conjugate acid–base pairs.

Figure 18·13
When sulfuric acid dissolves in water it forms hydronium ions and hydrogen sulfate ions. Which ion is the conjugate acid and which is the conjugate base?

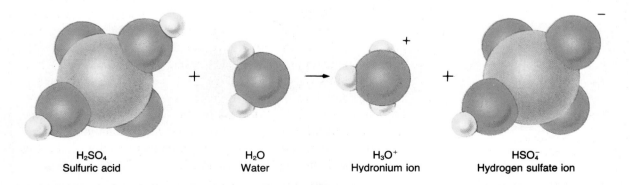

H_2SO_4
Sulfuric acid
    H_2O
Water
    H_3O^+
Hydronium ion
    HSO_4^-
Hydrogen sulfate ion

Figure 18·14
Acid–base indicators respond to pH changes over a specific range. Methyl orange (top row) is shown at pH 2, 4, and 6 (left to right). Bromothymol blue (bottom row) is shown at pH 5, 7, and 9.

Figure 18·15
Acid rain has a pH of less than 5.6. It damages stonework, soil, and vegetation and can kill many freshwater organisms.

18·C Measuring pH

Chemists must be able to measure the pH of the solutions they use. For preliminary pH measurements and for small volume samples, indicators are often used. For precise and continuous measurements, a pH meter is preferred.

An *indicator* (In) is a weak acid or base that undergoes dissociation in a known pH range. In this range the acid (or base) is a different color from its conjugate base (or acid). The following generalized equation represents this process.

$$\underset{\text{Acid}}{\text{HIn}} \; \underset{\text{H}^+}{\overset{\text{OH}^-}{\rightleftharpoons}} \; \underset{\substack{\text{Conjugate}\\\text{base}}}{\text{In}^-} \; + \; \text{H}_2\text{O}$$

Each indicator is only useful over a pH range of approximately two units. Many different indicators are needed to span the entire pH spectrum. Figure 18·16 shows some commonly used indicators.

Indicators are useful tools. They have certain characteristics, however, that limit their use. The listed pH values of indicators are usually given for 25°C. At other temperatures, an indicator may change color at a different pH. If the solution being tested has a color of its own, the color of the indicator may be distorted. Dissolved salts in a solution may also change the indicator's dissociation. Often, these problems can be overcome by using indicator strips. An indicator strip is a piece of paper or plastic that is impregnated with the indicator. This strip is dipped into the unknown solution and compared with a color chart. Some pH paper contains multiple indicators so that it can indicate pHs from 1 to 14.

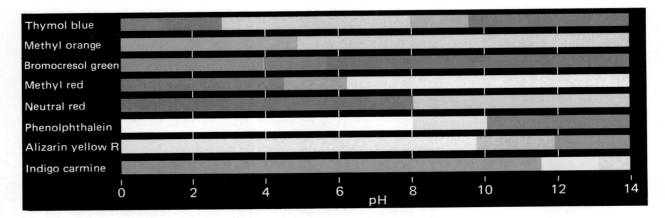

Thymol blue
Methyl orange
Bromocresol green
Methyl red
Neutral red
Phenolphthalein
Alizarin yellow R
Indigo carmine

pH scale: 0 2 4 6 8 10 12 14

Figure 18·16
Each indicator changes color at a different pH. What indicator would you choose to show that a reaction solution has changed from pH 3 to pH 6?

Most chemistry laboratories have a pH meter. It is used to make rapid, accurate measurements of pH. If a pH meter is connected to a strip chart recorder, it can be used to make a continuous recording of pH changes.

The structure of the pH meter is shown in Figure 18·17. It consists of two electrodes connected to a millivoltmeter. The reference electrode has a constant voltage. The voltage of the glass electrode changes with the $[H_3O^+]$ in the solution in which it is dipped. The pH meter makes an electrical measurement of pH by comparing the voltages of the two electrodes. The millivoltmeter gives a readout directly as pH.

Before a pH meter is used, it must be calibrated. The electrodes are dipped into a solution of known pH. The readout of the millivoltmeter is adjusted to that pH. The electrodes are then rinsed with distilled water and dipped into the solution of unknown pH. The electrodes may be left in the solution so that continuous readings of pH can be made.

Figure 18·17
A pH meter is used to measure hydrogen-ion concentrations. The instrument consists of two electrodes: a glass electrode and a reference electrode. A meter measures the voltage difference between the two electrodes and gives a pH reading. The meter may be connected to a strip-chart recorder.

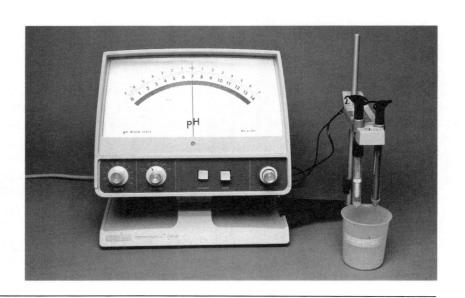

Safety

The concentration of a "concentrated" acid is too high to be measured on the pH scale. Concentrated HCl is $12M$, concentrated HNO_3 is $16M$, and concentrated H_2SO_4 is $18M$. Use these materials only with great care.

A pH meter is a valuable instrument. In many situations it is easier to use than liquid indicators or indicator strips. Values of pH obtained with a pH meter are typically accurate to within a 0.01 pH unit of the true pH. The color and cloudiness of the unknown solution do not affect the accuracy of the pH value obtained. Hospitals use pH meters to find small but meaningful changes of pH in blood and other body fluids (Figure 19·9). Sewage and industrial effluents are easily monitored with a pH meter.

18·7   Lewis Acids and Bases

A third theory of acids and bases was proposed by Gilbert Lewis (1875–1946). He focused on the donation or acceptance of a pair of electrons during a reaction. This concept is more general than either the Arrhenius theory or the Brønsted–Lowry theory. A **Lewis acid** *is a substance that can accept a pair of electrons to form a covalent bond. A* **Lewis base** *is a substance that can donate a pair of electrons to form a covalent bond.* Brønsted–Lowry acids and bases are also Lewis acids and bases. A hydrogen ion (Brønsted–Lowry acid) can accept a pair of electrons in forming a bond. A hydrogen ion, therefore, is also a Lewis acid. A substance that accepts a hydrogen ion (Brønsted–Lowry base) must have a pair of electrons available (Lewis base).

A Lewis acid is an electron-pair acceptor. A Lewis base is an electron-pair donor.

Consider the reaction of H^+ and OH^-.

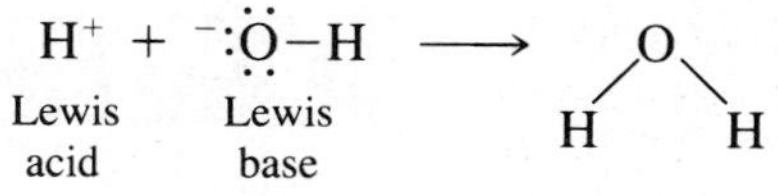

$$H^+ + {}^-:\overset{..}{\underset{..}{O}}-H \longrightarrow \underset{H \qquad H}{O}$$

Lewis acid   Lewis base

A hydroxide ion is a Lewis base. It is also a Brønsted–Lowry base. A hydrogen ion is both a Lewis acid and a Brønsted–Lowry acid. The Lewis theory also includes some compounds that are not classified as Brønsted-Lowry acids or bases.

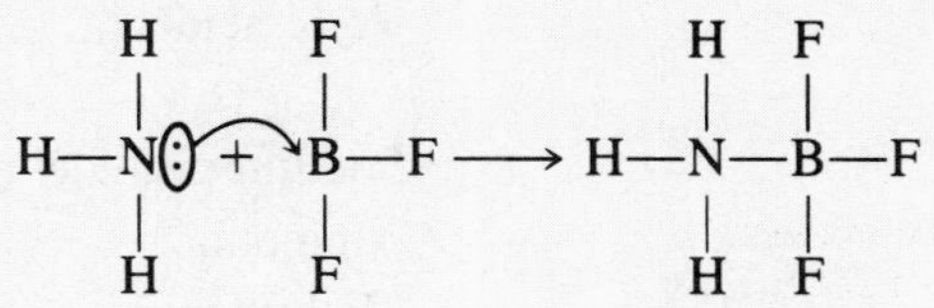

Note that this acid-base reaction does not involve the transfer of a hydrogen ion. A Lewis acid does not need to contain any hydrogen ions.

Example 6

Identify the Lewis acid and the Lewis base in this reaction.

$$\underset{H}{\overset{H}{H-N:}} + \underset{F}{\overset{F}{B-F}} \longrightarrow \underset{H \quad F}{\overset{H \quad F}{H-N-B-F}}$$

Solution

Ammonia is donating a pair of electrons. Ammonia is a Lewis base. Boron trifluoride is accepting a pair of electrons. It is a Lewis acid.

Problem

12. Identify the Lewis acid and Lewis base in each reaction.

a. $H^+ + H-\ddot{O}-H \longrightarrow H_3O^+$ **b.** $AlCl_3 + Cl^- \longrightarrow AlCl_4^-$

18·8 The Strengths of Acids and Bases

Chemists classify acids as strong or weak depending on the degree to which the acids ionize in water. For practical purposes, **strong acids** *are completely ionized in aqueous solution*. Hydrochloric acid and sulfuric acid are strong acids.

$$HCl \longrightarrow H^+ + Cl^- \text{ (100\% dissociated)}$$

Weak acids *ionize only slightly in aqueous solution*. The ionization of acetic acid, a typical weak acid, is not complete.

$$\underset{\substack{\text{Acetic} \\ \text{acid}}}{HC_2H_3O_2} + \underset{\text{Water}}{H_2O} \rightleftharpoons \underset{\substack{\text{Hydronium} \\ \text{ion}}}{H_3O^+} + \underset{\substack{\text{Acetate} \\ \text{ion}}}{C_2H_3O_2^-}$$

Over 99% of acetic acid molecules exist in solution as undissociated covalent molecules ($HC_2H_3O_2$). Less than 1% are ionized at any instant. Table 18·4 gives the relative strength of some common acids and bases.

<table>
<tr><th colspan="3">Table 18·4   Relative Strengths of Common Acids and Bases</th></tr>
<tr><th>Substance</th><th>Formula</th><th>Relative Strength</th></tr>
<tr><td>Hydrochloric acid</td><td>HCl</td><td rowspan="3">Strong acids</td></tr>
<tr><td>Nitric acid</td><td>HNO_3</td></tr>
<tr><td>Sulfuric acid</td><td>H_2SO_4</td></tr>
<tr><td>Phosphoric acid</td><td>H_3PO_4</td><td rowspan="6">Increasing strength ↑</td></tr>
<tr><td>Acetic acid</td><td>$HC_2H_3O_2$</td></tr>
<tr><td>Carbonic acid</td><td>H_2CO_3</td></tr>
<tr><td>Hydrosulfuric acid</td><td>H_2S</td></tr>
<tr><td>Hypochlorous acid</td><td>HClO</td></tr>
<tr><td>Boric acid</td><td>H_3BO_3</td></tr>
<tr><td colspan="2"></td><td>Neutral solution</td></tr>
<tr><td>Sodium cyanide</td><td>NaCN</td><td rowspan="4">Increasing strength ↓</td></tr>
<tr><td>Ammonia</td><td>NH_3</td></tr>
<tr><td>Methylamine</td><td>CH_3NH_2</td></tr>
<tr><td>Sodium silicate</td><td>Na_2SiO_3</td></tr>
<tr><td>Calcium hydroxide</td><td>$Ca(OH)_2$</td><td rowspan="3">Strong bases</td></tr>
<tr><td>Sodium hydroxide</td><td>NaOH</td></tr>
<tr><td>Potassium hydroxide</td><td>KOH</td></tr>
</table>

A strong acid or a strong base is completely dissociated in aqueous solution.

For strong acids or bases the equilibrium favors the products. For weak acids or bases the equilibrium favors the reactants.

Chemical Connections

Acidulants are weak acids used as food additives. They are used to preserve foods, add flavor, control pH, prevent oxidation, and change the food's melting characteristics.

Figure 18·18
Which of these pH test papers has been dipped into 1.0M hydrochloric acid and which has been dipped into 1.0M acetic acid?

 Issues in Chemistry

What problems does acid rain cause where you live? The Clean Air Act of 1990 requires emissions of acid rain precursors (the oxides of sulfur and nitrogen) to be significantly reduced. What changes, if any, will this lead to in your lifestyle?

The equilibrium-constant expression for the ionization of acetic acid can be written from the preceding equation.

$$K_{eq} = \frac{[H_3O^+] \times [C_2H_3O_2^-]}{[HC_2H_3O_2] \times [H_2O]}$$

For dilute solutions the concentration of water is a constant. It can be combined with K_{eq} to give an acid dissociation constant. *An **acid dissociation constant** (K_a) is the ratio of the concentration of the dissociated form of an acid to the undissociated form.* The dissociated form includes both the H^+ and the anion.

$$K_{eq} \times [H_2O] = K_a = \frac{[H^+] \times [C_2H_3O_2^-]}{[HC_2H_3O_2]}$$

The acid dissociation constant, K_a, for acetic acid at 25°C is 1.8×10^{-5}.

The acid dissociation constant, K_a, reflects that fraction of an acid that is in the ionized form. For this reason dissociation constants are sometimes called ionization constants. If the value of the ionization constant, K_a, is small then the dissociation of the acid in the solution is small. Weak acids have small K_a values. A larger value of K_a means that the dissociation, or ionization, of the acid is more complete. Strong acids have large K_a values. For example, nitrous acid has a K_a of 4.4×10^{-4} whereas acetic acid has a K_a of 1.8×10^{-5}. This means that nitrous acid is more ionized, or dissociated, in solution than acetic acid. Nitrous acid is a stronger acid than acetic acid.

Diprotic and triprotic acids lose their hydrogens one at a time. Each ionization reaction has a separate ionization constant, K_a. Thus phosphoric acid, H_3PO_4, has three ionization constants to go with its three ionizable hydrogens. Table 18·5 shows the ionization reactions and ionization constants for some common acids. The acids are ranked by the value of the first ionization constant for each acid.

Problems

13. Which acid in Table 18·4 would you expect to have the lowest ionization constant?

14. Based on the first ionization constant, which acid in Table 18·5 is the strongest? Which is weakest?

Table 18·5   Dissociation Constants of Weak Acids

Acid	Ionization	$K_a(25°C)$
Oxalic Acid	$H_2C_2O_4(aq) \rightleftharpoons H^+(aq) + HC_2O_4^-(aq)$	5.6×10^{-2}
	$HC_2O_4^-(aq) \rightleftharpoons H^+(aq) + C_2O_4^{2-}(aq)$	5.1×10^{-5}
Phosphoric Acid	$H_3PO_4(aq) \rightleftharpoons H^+(aq) + H_2PO_4^-(aq)$	7.5×10^{-3}
	$H_2PO_4^-(aq) \rightleftharpoons H^+(aq) + HPO_4^{2-}(aq)$	6.2×10^{-8}
	$HPO_4^{2-}(aq) \rightleftharpoons H^+(aq) + PO_4^{3-}(aq)$	4.8×10^{-13}
Formic Acid	$HCOOH(aq) \rightleftharpoons H^+(aq) + HCOO^-(aq)$	1.8×10^{-4}
Benzoic Acid	$C_6H_5COOH(aq) \rightleftharpoons H^+(aq) + C_6H_5COO^-(aq)$	6.3×10^{-5}
Acetic Acid	$CH_3COOH(aq) \rightleftharpoons H^+(aq) + CH_3COO^-(aq)$	1.8×10^{-5}
Carbonic Acid	$H_2CO_3(aq) \rightleftharpoons H^+(aq) + HCO_3^-(aq)$	4.3×10^{-7}
	$HCO_3^-(aq) \rightleftharpoons H^+(aq) + CO_3^{2-}(aq)$	4.8×10^{-11}

Just as there are strong acids and weak acids, there are also strong bases and weak bases. All the bases in Table 18·3 are classified as strong bases. **Strong bases** *dissociate completely into metal ions and hydroxide ions in aqueous solution.* (Some bases, such as calcium hydroxide and magnesium hydroxide, are not very soluble in water. The small amounts of those bases that do dissolve dissociate completely.) Unlike strong bases, **weak bases** *do not dissociate completely in aqueous solution.* The relative strengths of some bases are listed in Table 18·4.

Aqueous ammonia is an example of a weak base.

$$NH_3 + H_2O \rightleftharpoons NH_4^+ + OH^-$$

Ammonia   Water   Ammonium ion   Hydroxide ion

The equilibrium greatly favors NH_3 and H_2O. Approximately 99% of the ammonia is un-ionized, and only 1% is in the form of NH_4^+ and OH^-. Very little NH_4^+ and OH^- are available. Not even this small amount of NH_4OH can be isolated, however, because all the NH_3 and H_2O must first be removed. If the NH_3 and H_2O were removed, all the NH_4^+ and OH^- ions would revert to NH_3 and H_2O. Bottles of aqueous ammonia used in laboratories are usually labeled ammonium hydroxide, NH_4OH. Nevertheless, no chemist has ever isolated this compound.

The **base dissociation constant,** K_b, *is the ratio of the dissociated form of a base to the undissociated form.* It indicates the degree of dissociation. The base dissociation constant for ammonia is 1.8×10^{-5}.

$$K_b = \frac{[NH_4^+] \times [OH^-]}{[NH_3]} = 1.8 \times 10^{-5}$$

Weak acids and weak bases have relatively small K_a and K_b values, respectively.

Problem

15. From Table 18·4, name a weak base and a strong base.

Figure 18·19
Which of these pH test papers has been dipped into 1.0M sodium hydroxide? Which has been dipped into 1.0M sodium bicarbonate?

The words *concentrated* and *dilute* tell how much of an acid or base is dissolved in solution. These terms refer to the number of moles of the acid or base in a given volume. The words *strong* and *weak* refer to the extent of ionization of an acid or base. They indicate how many of the molecules dissociate into ions. Hydrochloric acid, HCl, is a strong acid. Gastric juice in the stomach is a dilute solution of hydrochloric acid. A relatively small number of HCl molecules are present in a given volume of gastric juice, but they are all dissociated into ions. A sample of hydrochloric acid added to a large volume of water becomes more dilute, but it is still a strong acid. Vinegar is a dilute solution of a weak acid, acetic acid. Pure acetic acid (glacial acetic acid) is still a weak acid. Aqueous ammonia is an example of a weak base. Solutions of ammonia can be concentrated or dilute depending on the amount of ammonia dissolved in a given volume of water. Any solution of ammonia, whether concentrated or dilute, will be a weak base. This is because the amount of dissociation will be small. Table 18·6 lists the concentrations of acids and bases commonly found in the laboratory.

Concentrated and dilute do not mean the same as strong and weak. For example, it is possible to have a concentrated weak acid or a dilute strong base.

Safety

Glacial acetic acid is 17M. It causes severe chemical burns. A concentrated weak acid can be very corrosive.

Table 18·6	Concentrations of Some Common Laboratory Acids and Bases	
	Concentration	
Acid or base	Moles/liter (molarity)	Grams/liter
Concentrated hydrochloric acid	12	438
Dilute hydrochloric acid	6	219
Concentrated sulfuric acid	18	1764
Dilute sulfuric acid	6	588
Concentrated phosphoric acid	15	1470
Concentrated nitric acid	16	1008
Dilute nitric acid	6	378
Glacial acetic acid	17	1020
Dilute acetic acid	6	360
Dilute sodium hydroxide	6	240
Concentrated aqueous ammonia	15	255
Dilute aqueous ammonia	6	102

18·9 Calculating Dissociation Constants

The acid dissociation constant, K_a, can be calculated from pH measurements.

The acid dissociation constant, K_a, of a weak acid can be calculated from experimental data. To do this, the equilibrium concentrations of all the substances present at equilibrium must be measured. For a weak acid these concentrations can be determined experimentally if we know two conditions. First, the initial molar concentration of the acid must be known. Second, the pH (or hydrogen-ion concentration) of the solution at equilibrium must be measured.

Example 7

A $0.100M$ solution of acetic acid is only partially ionized. Using a measure of pH, the $[H^+]$ is calculated as $1.34 \times 10^{-3}M$. What is the acid dissociation constant of acetic acid?

Solution

Write the equation for the ionization of acetic acid.

$$HC_2H_3O_2(aq) \rightleftharpoons H^+ + C_2H_3O_2^-$$

Each molecule of $HC_2H_3O_2$ that ionizes gives a H^+ and an $C_2H_3O_2^-$ ion. Therefore at equilibrium the $[H^+] = [C_2H_3O_2^-] = 1.34 \times 10^{-3}M$. The equilibrium concentration of $HC_2H_3O_2$ is the initial concentration changed by the ionization of the acid, $(0.100 - 0.00134)M = 0.09866M$. This data is summarized in the following table.

Concentrations	$[HC_2H_3O_2]$	$[H^+]$	$[C_2H_3O_2^-]$
Initial	0.100	0	0
Change	-1.34×10^{-3}	$+1.34 \times 10^{-3}$	$+1.34 \times 10^{-3}$
Equilibrium	0.0987	1.34×10^{-3}	1.34×10^{-3}

(All concentrations are in mol/L.)

The equilibrium values are now substituted into the expression for K_a.

$$K_a = \frac{[H^+] \times [C_2H_3O_2^-]}{[HC_2H_3O_2]}$$

$$= \frac{(1.34 \times 10^{-3}) \times (1.34 \times 10^{-3})}{0.0987}$$

$$= 1.82 \times 10^{-5}$$

Problem

16. A $0.200M$ solution of a weak acid has a $[H^+]$ of $9.86 \times 10^{-4}M$. **a.** What is the pH of this solution? **b.** What is the value of K_a for this acid?

18 Acids and Bases

Chapter Review

Key Terms

acid dissociation constant (K_a)	18·8	hydroxide ion (OH⁻)	18·2
acidic solution	18·2	ion-product constant for water (K_w)	18·2
alkaline solution	18·2	Lewis acid	18·7
amphoteric	18·6	Lewis base	18·7
base dissociation constant (K_b)	18·8	monoprotic acid	18·5
basic solution	18·2	neutral solution	18·2
conjugate acid	18·6	pH	18·3
conjugate base	18·6	self-ionization	18·2
conjugate acid–base pair	18·6	strong acid	18·8
diprotic acid	18·5	strong base	18·8
hydrogen-ion acceptor	18·6	triprotic acid	18·5
hydrogen-ion donor	18·6	weak acid	18·8
hydronium ion (H_3O^+)	18·2	weak base	18·8

Chapter Summary

Water molecules dissociate into hydrogen ions (H^+) and hydroxide ions (OH^-). The concentrations of these ions in pure water at 25°C are both equal to 1×10^{-7} mol/L.

The pH scale, which has a range from 0 to 14, is used to denote the hydrogen-ion concentration of a solution. On this scale 0 is strongly acidic, 14 is strongly basic, and 7 is neutral. Water at 25°C has a pH of 7. The pH of a solution is measured with acid–base indicators.

Compounds can be classified as acids or bases according to three different theories. An Arrhenius acid gives hydrogen ions in aqueous solution. An Arrhenius base gives hydroxide ions in aqueous solution. A Brønsted–Lowry acid is a proton donor. A Brønsted–Lowry base is a proton acceptor. In the Lewis theory an acid is an electron-pair acceptor. A Lewis base is an electron-pair donor.

The strength of an acid or base is determined by the degree of ionization of the substance in solution. The acid dissociation constant, K_a, is a quantitative measure of acid strength. A strong acid has a much larger K_a than a weak acid. The K_a of an acid is determined from measured pH values.

Hydrochloric acid and sulfuric acid are completely ionized in solution and are strong acids. Acetic acid, which is only about 1% ionized, is a weak acid. Sodium hydroxide and calcium hydroxide are strong bases. Ammonia is only slightly ionized in aqueous solution and is a weak base.

Practice Questions and Problems

17. List at least three characteristic properties of acids and three properties of bases. *18·1*

18. Write an equation showing the ionization of water. *18·2*

19. What are the concentrations of H^+ and OH^- in pure water at 25°C? *18·2*

20. What is true about the relative concentrations of hydrogen ions and hydroxide ions in each of these solutions? *18·2*
 a. basic **b.** acidic **c.** neutral

21. How is the pH of a solution calculated? *18·3*

22. Why is the pH of pure water at 25°C equal to 7.0? *18·3*

23. Calculate the pH for the following solutions and indicate whether the solution is acidic or basic.
 a. $[H^+] = 1 \times 10^{-2}$ mol/L *18·3*
 b. $[OH^-] = 1 \times 10^{-2}$ mol/L
 c. $[OH^-] = 1 \times 10^{-8}$ mol/L
 d. $[H^+] = 1 \times 10^{-6}$ mol/L

24. What are the hydroxide-ion concentrations for solutions that have the following pH values?
 a. 4.0 **b.** 8.0 **c.** 12.0 *18·3*

25. Classify each of these as an Arrhenius acid or an Arrhenius base.　　*18·5*
　a. Ca(OH)$_2$　　　　　**c.** KOH
　b. HNO$_3$　　　　　**d.** C$_2$H$_5$COOH

26. Write balanced equations for the reaction of each of these metals with water.　*18·5*
　a. lithium　**b.** barium

27. Identify each of the reactants in the following equations as a hydrogen-ion donor (acid) or a hydrogen-ion acceptor (base).　　*18·6*
　a. HNO$_3$ + H$_2$O $\rightleftharpoons$ H$_3$O$^+$ + NO$_3^-$
　b. HC$_2$H$_3$O$_2$ + H$_2$O $\rightleftharpoons$ H$_3$O$^+$ + C$_2$H$_3$O$_2^-$
　c. H$_2$O + NH$_3$ $\rightleftharpoons$ NH$_4^+$ + OH$^-$
　d. H$_2$O + C$_2$H$_3$O$_2^-$ $\rightleftharpoons$ HC$_2$H$_3$O$_2$ + OH$^-$

28. Label the conjugate acid–base pairs in each equation in Problem 27.　　*18·6*

29. What makes a substance amphoteric?　*18·6*

30. Write the equations for the ionization of the following acids and bases in water.　　*18·8*
　a. nitric acid　　　**c.** ammonia
　b. acetic acid　　　**d.** magnesium hydroxide

31. Define strong and weak acids and bases.　*18·8*

32. Identify each of the following compounds as a strong or weak acid or base.　　*18·8*
　a. NaOH　　**b.** HCl　　**c.** NH$_3$　　**d.** H$_2$SO$_4$

33. Would a strong acid have a large or a small K_a? Explain.　　*18·9*

34. Why are Mg(OH)$_2$ and Ca(OH)$_2$ called strong bases even though their saturated solutions are only mildly basic?　　*18·9*

35. Write the expression for K_a for each of these acids. Assume that only one hydrogen is ionized.　**a.** HI　**b.** H$_2$CO$_3$　　*18·9*

Mastery Questions and Problems

36. Calculate the pH or [H$^+$] for each solution.
　a. [H$^+$] = 2.4 × 10^{-6}　　**c.** pH = 13.2
　b. [H$^+$] = 9.1 × 10^{-9}　　**d.** pH = 6.7

37. Is it possible to have a concentrated weak acid? Explain.

38. Write equations that show that the hydrogen phosphate ion, HPO$_4^{2-}$, is amphoteric.

39. The pH of a 0.50M HNO$_2$ solution is 1.83. What is the K_a of this acid?

40. Write the formula and name of the conjugate base of each Brønsted–Lowry acid.
　a. HCO$_3^-$　**b.** HI　**c.** NH$_4^+$　**d.** H$_2$SO$_3$

41. Using Figure 18·16 and the visual evidence in the photo below, estimate the pH of **a.** the vinegar and **b.** the ammonia cleaner.

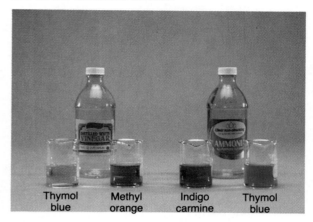

| Thymol blue | Methyl orange | Indigo carmine | Thymol blue |

42. Write the formula and name of the conjugate acid of each Brønsted–Lowry base.
　a. ClO$_2^-$　**b.** H$_2$PO$_4^-$　**c.** H$_2$O　**d.** NH$_3$

43. It is determined that 1.40% of a 0.080M solution of a weak acid is ionized. Calculate the K_a for this acid.

44. Calculate the [OH$^-$] or the pH of each solution.
　a. pH = 4.6　　**c.** [OH$^-$] = 1.8 × 10^{-2}
　b. pH = 9.3　　**d.** [OH$^-$] = 7.3 × 10^{-9}

45. Write the three equations for the stepwise dissociation of phosphoric acid.

Critical Thinking Questions

46. Choose the term that best completes the second relationship.
　a. sky:blue　　acid: _____
　　(1) metallic　　　　(3) sour
　　(2) basic　　　　　(4) bitter
　b. plants:oxygen　　base: _____
　　(1) electrons　　　　(3) hydronium ions
　　(2) hydroxide ions　　(4) protons
　c. thermometer:degrees　　indicator: _____
　　(1) pH　　　　　(3) H$^+$
　　(2) acid　　　　(4) base

47. Why might the hydrogen ions produced by an acid be corrosive to such materials as skin and metal?

48. Arrhenius, Brønsted-Lowry, and Lewis all developed theories to explain acids and bases.
  a. Which theory is easiest for you to understand?
  b. Which theory seems the best?
  c. How can three different theories all be accepted by chemists?

Temperature (°C)	K_w	pH
0	1.137×10^{-15}	_____
10	2.917×10^{-15}	_____
20	6.807×10^{-15}	_____
30	1.469×10^{-14}	_____
40	2.917×10^{-14}	_____
50	5.470×10^{-14}	_____

Review Questions and Problems

49. How would you prepare 400.0 mL of a $0.680M$ KOH solution?

50. How many liters of $8.0M$ HCl are needed to prepare 1.50 L of $2.5M$ HCl?

51. How would each change affect the position of equilibrium of this reaction?

$$2H_2(g) + O_2(g) \rightleftharpoons 2H_2O(g) + \text{heat}$$

  a. increasing the pressure
  b. adding a catalyst
  c. increasing the concentration of $H_2(g)$
  d. cooling the reaction mixture
  e. removing water vapor from the container

52. Write an equilibrium constant expression for each equation.
  a. $2CO_2(g) \rightleftharpoons 2CO(g) + O_2(g)$
  b. $N_2(g) + 3H_2(g) \rightleftharpoons 2NH_3(g)$

Challenging Questions and Problems

53. Calculate the pH of a $0.010M$ solution of sodium cyanide, NaCN. The K_b of CN^- is 2.1×10^{-5}.

54. Show that for any conjugate acid–base pair $K_a K_b = K_w$.

55. The K_w of water varies with temperature.
  a. Find the pH of water for each temperature in the table. Use this data to prepare a plot of pH versus temperature.
  b. What is the pH of water at 5°C?
  c. At what temperature is the pH of water approximately 6.85?

Research Projects

1. What are the steps involved in etching a silicon chip for computers? Why are scientists investigating the use of reactive ion plasmas to etch the chips?

2. The colored solutions that result from boiling the skins of a number of vegetables in water make good acid/base indicators. Determine the changes in color at different pH for a number of these natural indicators.

3. Use pH test paper to determine the pH of common substances such as orange juice, milk, shampoo, cleaning solutions, tap water, and rain water.

4. How is sodium hydroxide manufactured? What are the major uses for this compound?

5. How do meteorolgical processes contribute to acid rain? How have local industrial pollution control measures affected the dispersion of acid rain?

Readings and References

Gay, Kathlyn. *Acid Rain*. New York: Watts, 1983.

"If You Breathe, Don't Smoke." *Science 84* (January/February 1984), pp. 83–84.

Ostmann, Robert, Jr. *Acid Rain: A Plague upon the Waters*. Minneapolis: Dillon, 1982.

Sitwell, Nigel. "Our Trees Are Dying." *Science Digest* (September 1984), pp. 39–48.

Visich, Marian, Jr. "Acid from the Sky." *Science Year* (1984), pp. 40–53.

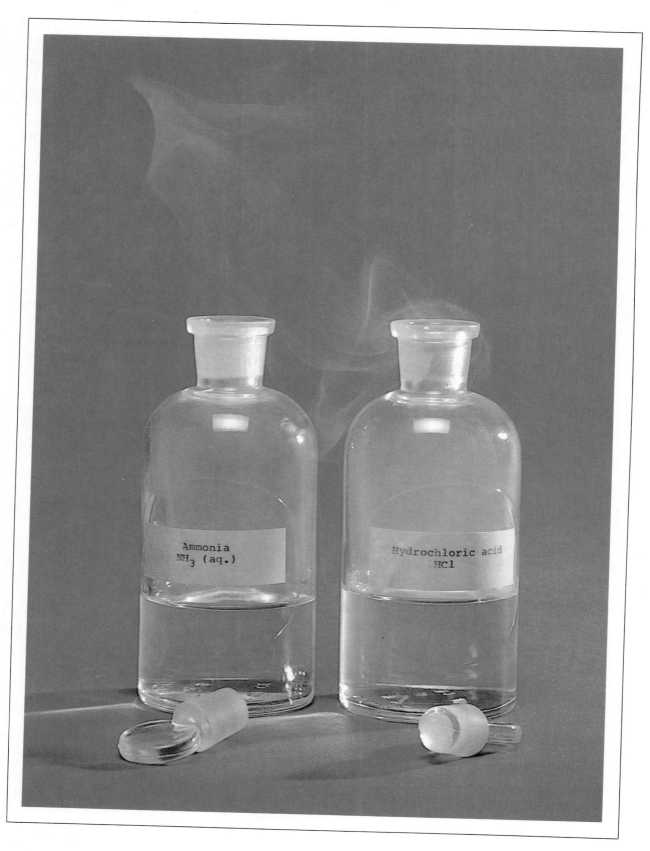

19 Neutralization and Salts

Chapter Preview

19·1 Neutralization Reactions
19·A Soil Chemistry for Farmers
19·2 Titration
19·3 Equivalents
19·4 Normality

19·5 Salt Hydrolysis
19·6 Buffers
19·7 The Solubility Product Constant
19·B Swimming Pool Chemistry
19·8 The Common Ion Effect

The properties of acids, bases, and salts help explain many diverse phenomena. For example, the usefulness of antacids depends on the process of acid–base neutralization. Farmers use a similar process to control the pH of their soil. The formation of caves and of stalactites in caves is caused by changes in the solubilities of salts. In a similar way certain conditions lead to the formation of kidney stones from salts within the body. Neutralization reactions and the solubilities of salts are the topics of this chapter.

19·1 Neutralization Reactions

Acids and bases react to produce salts and water.

If a solution containing hydronium ions (an acid) is mixed with a solution that has an equal amount of hydroxide ions (a base), a neutral solution results. The final solution has properties that are not characteristic of either an acidic or a basic solution. Consider these two examples.

$$HCl + NaOH \longrightarrow NaCl + H_2O$$
$$H_2SO_4 + 2KOH \longrightarrow K_2SO_4 + 2H_2O$$

In each reaction, a strong acid reacts with a strong base. If solutions of these substances are mixed in the mole ratios specified by the balanced equation, neutral solutions will result. Similar reactions of weak acids and/or weak bases do not usually produce neutral solutions. In general, however, *reactions in which an acid and a base react in an aqueous solution to produce a salt and water are called* **neutralization reactions.** They are all double-replacement reactions (Section 7·6).

Neutralization reactions are one way that pure samples of salts can be prepared. For example, potassium chloride could be prepared by mixing together equal molar quantities of hydrochloric acid and potassium

Figure 19·1
A salt is the result of an acid–base neutralization. The white cloud of ammonium chloride seen here forms from the reaction of hydrogen chloride and ammonia gases escaping from aqueous solutions of those gases.

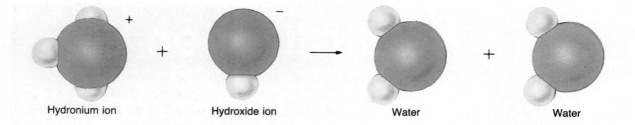

| Hydronium ion | Hydroxide ion | Water | Water |

Figure 19·2
In a neutralization reaction, hydrogen ions (as hydronium ions) combine with hydroxide ions to form neutral water.

hydroxide. An aqueous solution of potassium chloride would result. The water could be removed by evaporation. Table 19·1 lists some common salts and their applications.

Problem

1. Identify the products and write balanced equations for the following neutralization reactions.
   a. $HNO_3 + KOH \longrightarrow$
   b. $HCl + Mg(OH)_2 \longrightarrow$
   c. $H_2SO_4 + NaOH \longrightarrow$

Table 19·1   Some Salts and Their Applications

Name	Formula	Applications
Ammonium sulfate	$(NH_4)_2SO_4$	Fertilizer
Barium sulfate	$BaSO_4$	Gastrointestinal studies; white pigment
Calcium chloride	$CaCl_2$	De-icing roadways and sidewalks
Calcium sulfate dihydrate (gypsum)	$CaSO_4 \cdot 2H_2O$	Plasterboard
Copper sulfate pentahydrate (blue vitriol)	$CuSO_4 \cdot 5H_2O$	Dyeing; fungicide
Calcium sulfate sesquihydrate	$CaSO_4 \cdot \frac{1}{2}H_2O$	Plaster casts
Magnesium sulfate heptahydrate (Epsom salts)	$MgSO_4 \cdot 7H_2O$	Purgative
Potassium chloride	KCl	Sodium-free salt substitute
Potassium permanganate	$KMnO_4$	Disinfectant and fungicide
Silver nitrate	$AgNO_3$	Cauterizing agent
Silver bromide	$AgBr$	Photographic emulsions
Sodium bicarbonate (baking soda)	$NaHCO_3$	Antacid
Sodium carbonate decahydrate (washing soda)	$Na_2CO_3 \cdot 10H_2O$	Glass manufacture; water softener
Sodium chloride (table salt)	$NaCl$	Body electrolyte; chlorine manufacture
Sodium sulfate decahydrate (Glauber's salt)	$Na_2SO_4 \cdot 10H_2O$	Purgative
Sodium thiosulfate (hypo)	$Na_2S_2O_3$	Fixing agent in photographic process

Figure 19·3
Farmers and gardeners benefit from having their soil tested for pH and important nutrients. The appropriate fertilizers can then be chosen.

Issues in Chemistry

Would you be willing to pay more for "organically" grown vegetables, than those grown with "chemicals?" Is it even possible to grow all food organically?

ChemDirections

The ions present in common table salt are essential to human health, but they can also be dangerous in large quantities. These and other essential elements are described in **ChemDirections** *Mineral Nutrients*, page 652.

19·A   Soil Chemistry for Farmers

Soil is the complex mixture of minerals and organic matter that covers large portions of the earth's crust. It is formed from rocks and minerals weathered by wind and precipitation. Chemical and biological reactions play important secondary roles in the formation of soil. Soil is often considered lifeless and inert. Actually it undergoes constant change and is teeming with life. One square meter of soil only a few centimeters deep may contain well over one billion organisms!

Soil scientists classify soil into 10 to 16 categories depending on the history and chemistry of the soil. The soil's pH, the amount and type of clay it contains, and its nitrogen content are factors that determine its type. These are also the factors that are most important in agricultural chemistry.

Soils that contain a good deal of organic matter are generally acidic. As plants decay, acids are formed. If rainfall is frequent in the area, these acids are washed down through the soil. As they move through the soil, the hydrogen ions in the acid replace metal ions in the clay particles of the soil. Many clays are primarily silicates of aluminum. In other clays, some of the aluminum ions are replaced by sodium, magnesium, or other metals. These cations in the clay can be exchanged for other cations such as H^+, NH_4^+, and K^+. These cations are then held by the clay particles. The process of acidic water washing metal ions out of clay by replacing them with hydrogen ions is called *leaching*.

The pH of soil is a very important consideration in farming. It affects the solubilities of minerals and nutrients. It also affects the ease with which these substances reach the roots of growing plants. Most crops have adapted to grow within a specific pH range, between 5 and 7. Soils with a pH of less than 5 are suitable only for acid-tolerant plants such as those found in moist forested areas. Only a few plants are capable of surviving in soils with a pH greater than 7. Such alkaline soils are generally found in very arid regions.

If a soil is too acidic, its pH can be raised. Lime (calcium oxide or calcium hydroxide), or limestone (calcium carbonate) can be added to neutralize the soil. If the soil is too alkaline, its pH can be lowered by adding aluminum sulfate. This salt hydrolyzes to give the weak base aluminum hydroxide and hydrogen ions.

$$Al_2(SO_4)_3 + 6\ H_2O \longrightarrow 2Al(OH)_3 + 6\ H^+ + 3\ SO_4^{2-}$$

Fertilizers that contain ammonium ions increase soil acidity. So does increasing the amount of organic material in the soil by adding manure. Plowing under a plant crop also increases soil acidity.

For healthy growth, plants need a wide variety of nutrients. Nitrogen in the form of nitrate ions, phosphorus in the form of phosphate ions, and potassium ions are called macronutrients. They are needed in relatively large amounts. These materials are the three principal components of chemical fertilizers. The soil and air supply other necessary macronutrients. These are carbon, hydrogen, oxygen, sulfur, magnesium, calcium, and iron. Plants also need a number of other elements in trace amounts. Manganese, boron, copper, zinc, molybdenum and chlorine are necessary micronutrients. They are supplied by special fertilizers if they are not present in the soil in sufficient amounts. Good farming practices insure that all these nutrients are supplied to growing plants. Routine chemical tests can alert farmers if the level of any of these substances is low.

19·2 Titration

A titration is a convenient method of determining the concentration of an acid or a base.

It should be clear from the previous section that acids and bases sometimes, but not always, react in a 1:1 mole ratio.

$$HCl + NaOH \longrightarrow NaCl + H_2O$$
$$\text{1 mol} \quad \text{1 mol} \qquad \text{1 mol} \quad \text{1 mol}$$

When sulfuric acid reacts with sodium hydroxide, however, the ratio is 1:2. Two moles of base are required to neutralize one mole of H_2SO_4.

$$H_2SO_4 + 2NaOH \longrightarrow Na_2SO_4 + 2H_2O$$
$$\text{1 mol} \quad \text{2 mol} \qquad \text{1 mol} \quad \text{2 mol}$$

Similarly, calcium hydroxide and hydrochloric acid react in a $1:2$ ratio.

$$Ca(OH)_2 + 2HCl \longrightarrow CaCl_2 + 2H_2O$$
$$\text{1 mol} \quad \text{2 mol} \qquad \text{1 mol} \quad \text{2 mol}$$

Example 1

How many moles of sulfuric acid are required to neutralize 0.50 mol of sodium hydroxide?

Solution

Write a balanced equation for the reaction to find the acid–base mole ratio.

$$H_2SO_4 + 2NaOH \longrightarrow Na_2SO_4 + 2H_2O$$
$$\text{1 mol} \quad \text{2 mol} \qquad \text{1 mol} \quad \text{2 mol}$$

The ratio of H_2SO_4 to NaOH is 1:2. The necessary number of moles of H_2SO_4 is calculated using this ratio.

$$0.50 \; \cancel{\text{mol NaOH}} \times \frac{1 \text{ mol } H_2SO_4}{2 \; \cancel{\text{mol NaOH}}} = 0.25 \text{ mol } H_2SO_4$$

Problem

2. How many moles of sodium hydroxide are required to neutralize 0.20 mol of nitric acid?

Figure 19·4
Laboratory titrations may be performed with two burets (one filled with acid, the other filled with base) a flask, and an indicator such as phenolphthalein.

The amount of acid (or base) in a solution is determined by carrying out a neutralization reaction. An appropriate acid–base indicator is used to show when neutralization is completed. The steps in the process are as follows. **1.** A measured amount of an acid of unknown concentration is added to a flask. **2.** An appropriate indicator (such as phenolphthalein) is added to the solution. **3.** Measured amounts of a base of known concentration are mixed into the acid. *The solution of known concentration is called the* **standard solution.** The addition is carried out using a buret. This process is continued until the indicator shows that neutralization has occurred. *The point at which neutralization is achieved is the* **end point** *of the titration.* The process just described is a **titration,** *the addition of a known amount of solution to determine the volume (or concentration) of another solution.* The unknown solution can be a base instead of an acid; the process is the same.

Example 2

A 25 mL solution of H_2SO_4 is neutralized by 18 mL of $1.0M$ NaOH using phenolphthalein as an indicator. What is the concentration of the H_2SO_4 solution?

Solution

Write an equation for the neutralization.

$$H_2SO_4 + 2NaOH \longrightarrow Na_2SO_4 + 2H_2O$$

The H_2SO_4 has two ionizable hydrogens. Thus it will take twice as many moles of NaOH for neutralization to occur. First calculate the moles of NaOH needed for neutralization ($mL \rightarrow L$).

$$0.018 \text{ L NaOH} \times \frac{1.0 \text{ mol NaOH}}{1 \text{ L NaOH}} = 0.018 \text{ mol NaOH}$$

Next use the equation to find the moles of H_2SO_4 neutralized.

$$0.018 \text{ mol NaOH} \times \frac{1 \text{ mol } H_2SO_4}{2 \text{ mol NaOH}} = 0.0090 \text{ mol } H_2SO_4$$

Then calculate the concentration of the acid.

$$\text{Molarity} = \frac{\text{moles}}{\text{liters}} = \frac{0.0090 \text{ mol}}{0.025 \text{ L}} = 0.36M$$

The H_2SO_4 is $0.36M$.

Problems

3. How many milliliters of $0.45M$ hydrochloric acid must be added to 25.0 mL of $1.00M$ potassium hydroxide to make a neutral solution?

4. What is the molarity of phosphoric acid if 15.0 mL of the solution is neutralized by 38.5 mL of $0.15M$ NaOH?

Figure 19·5
The color of the indicator begins to change as the endpoint of the titration is approached.

Neutralization reactions require equivalent amounts of acid and base.

The equivalent mass is less than or equal to the formula mass.

19·3 Equivalents

In any neutralization reaction, one mole of hydrogen ions reacts with one mole of hydroxide ions. This does not mean that one mole of any acid will neutralize one mole of any base. For some acids, one mole of the acid can give one mole of hydrogen ions. HCl and HNO_3 are such acids. For other acids, one mole of the acid can give two or more moles of hydrogen ions. For example, one mole of H_2SO_4 gives two moles of hydrogen ions. One mole of H_3PO_4 gives three moles of hydrogen ions. The same is true of bases such as $Ca(OH)_2$ and $Al(OH)_3$. They give two and three moles of hydroxide ions respectively for one mole of base.

Chemists use a unit for the amount of an acid (or base) that will give one mole of hydrogen (or hydroxide) ions. This unit is called an equivalent of the acid (or base). *One* **equivalent** *is the amount of an acid (or base) that will give one mole of hydrogen (or hydroxide) ions.*

One mole of HCl is one equivalent of HCl. One mole of H_2SO_4 is two equivalents of H_2SO_4. One mole of NaOH is one equivalent of NaOH. One mole of $Ca(OH)_2$ is 2 equivalents of $Ca(OH)_2$. One mole of HCl will neutralize one mole of NaOH. This is because one mole of each compound is also one equivalent.

$$HCl + NaOH \longrightarrow H_2O + NaCl$$

One mole of HCl will not neutralize one mole of $Ca(OH)_2$.

$$2HCl + Ca(OH)_2 \longrightarrow 2H_2O + CaCl_2$$

In this example, two moles of acid are required to neutralize one mole of base. Two moles of HCl contains two equivalents of acid. One mole of $Ca(OH)_2$ contains two equivalents of base. In any neutralization reaction, one equivalent of acid will neutralize one equivalent of base.

The mass of one equivalent of a substance is called its **gram equivalent mass.** One mole of HCl is one equivalent of HCl. Its gram equivalent mass is equal to its gram molecular mass, 36.5 grams. One mole of H_2SO_4 is two equivalents of H_2SO_4. Its gram equivalent mass is only half of its gram molecular mass, or 49.0 grams.

Example 3

What is the mass of 1 equiv of calcium hydroxide?

Solution

The gram formula mass of $Ca(OH)_2$ is 74 g. One formula unit of $Ca(OH)_2$ has two hydroxide ions.

$$Ca(OH)_2 \longrightarrow Ca^{2+} + 2OH^-$$

$$\text{Gram equivalent mass of } Ca(OH)_2 = \frac{74 \text{ g/mol}}{2 \text{ equiv/mol}} = 37 \text{ g/equiv}$$

Calcium hydroxide is 2 equiv per mol and 1 equiv of $Ca(OH)_2$ is 37 g.

Example 4

How many equivalents is 4.8 g of sulfuric acid?

Solution

The gram formula mass of H_2SO_4 is 98 g. Therefore, 4.8 g of H_2SO_4 is less than 1 mol.

$$4.8 \text{ g } H_2SO_4 \times \frac{1 \text{ mol } H_2SO_4}{98 \text{ g } H_2SO_4} = 0.050 \text{ mol } H_2SO_4$$

The acid H_2SO_4 is 2 equiv per mol. Thus 0.050 mol H_2SO_4 is 0.10 equiv of H_2SO_4, and 4.8 g of H_2SO_4 is 0.10 equiv of H_2SO_4.

Problems

5. Determine the gram equivalent mass and the equivalents per mole for each compound. **a.** KOH **b.** HCl **c.** H_2SO_4

6. How many equivalents is each of the following?
   a. 3.7 g $Ca(OH)_2$ **b.** 189 g H_2SO_4 **c.** 9.8 g H_3PO_4

19·4 Normality

The concentrations of acids and bases can be stated in molarity. Chemists are usually more interested, however, in how many equivalents of acid or base a solution contains. Thus the concentrations of acids and bases are usually expressed as normalities. A solution containing 1.0 equiv of an acid or base per liter has a normality of 1.0. That is, the solution is 1.0 normal (1.0N). *The* **normality** *of a solution is the concentration expressed as the number of equivalents of solute in 1 L of solution.*

Concentrations of acids and bases are often expressed in normalities.

$$\text{Normality } (N) = \text{equiv/L}$$

The numerical values of normality and molarity are equal for acids and bases that give 1 equiv of H^+ or OH^- per mole. For example, a solution containing 1 mol of NaOH per liter is $1M$ and also $1N$. A solution containing 1 mol of H_2SO_4 per liter is $1M$, but it is $2N$. This is because H_2SO_4 contains 2 equiv per mole.

Normality = molarity × number of equivalents.

Problems

7. What is the normality of the following solutions?
   a. $2M$ HCl   **c.** $0.3M$ H_3PO_4
   b. $0.1M$ $HC_2H_3O_2$ **d.** $0.25M$ H_2SO_4

8. What is the normality of the following solutions?
   a. 20.0 g NaOH in 1.0 L of solution
   b. 4.9 g H_2SO_4 in 500 mL of solution

The number of equivalents of an acid or base in a known volume of a solution of known normality can be calculated.

$$\text{Number of equivalents} = \begin{matrix} \text{volume (liters)} \\ \text{of solution} \end{matrix} \times \begin{matrix} \text{normality} \\ \text{of solution} \end{matrix}$$

$$\text{Equiv} = V(\text{L}) \times N$$

Example 5

How many equivalents are in 2.5 L of $0.60N$ H_2SO_4?

Solution

Number of equivalents $= V(\text{L}) \times N$

$$= 2.5 \ \cancel{L} \times \frac{0.60 \ \text{equiv}}{\cancel{L}} = 1.5 \ \text{equiv}$$

Problem

9. How many equivalents are in the following?
  a. 0.55 L of $1.8N$ NaOH
  b. 1.6 L of $0.50N$ H_3PO_4
  c. 250 mL of $0.28N$ H_2SO_4

Solutions of known normality can be made less concentrated by diluting them with water. The changes in concentration can be calculated using this relationship.

$$N_1 \times V_1 = N_2 \times V_2$$

Here N_1 and V_1 are the initial solution's normality and volume. N_2 and V_2 are the final solution's normality and volume.

Example 6

You need to make 250 mL of $0.10N$ sodium hydroxide from a stock solution that is $2.0N$ sodium hydroxide. How many milliliters of the stock solution must you dilute to 250 mL to get the required solution?

Solution

Use $N_1 \times V_1 = N_2 \times V_2$. You know that $N_1 = 2.0N$, $N_2 = 0.10N$, and $V_2 = 250$ mL. To find V_1, rearrange the equation and then insert the values of N_1, N_2, and V_2.

$$V_1 = \frac{N_2 \times V_2}{N_1} = \frac{0.10 \ \cancel{N} \times 250 \ \text{mL}}{2.0 \ \cancel{N}} = 12.5 \ \text{mL} = 13 \ \text{mL}$$

(to two significant figures)

Dilute 13 mL of $2.0N$ NaOH to 250 mL to make $0.10N$ NaOH.

Figure 19·6
A solution with a given normality can be diluted to make a solution of a lower normality. Calculate the volume of concentrated solution needed to make the dilute solution. Then use accurate measuring devices such as a pipet and volumetric flask to make the dilution.

Problem

10. How would you prepare 500 mL of $0.20N$ sulfuric acid from a stock solution of $4.0N$ sulfuric acid?

Acid–base titrations are neutralization reactions.

Titration calculations are usually done in terms of normality instead of molarity. This is because normality takes into account the number of ionizable hydrogens in an acid whereas molarity does not. *In a titration, the point of neutralization is called the* **equivalence point.** At the equivalence point, the number of equivalents of acid and base are equal. It is thus possible to calculate the number of equivalents of acid or base in an unknown sample. Let N_A and N_B be the normalities of the acid and base solutions. Let V_A and V_B be the volumes of the acid and base solutions required to give a neutral solution.

$$\text{Equivalents of acid} = N_A \times V_A$$
$$\text{Equivalents of base} = N_B \times V_B$$

At the equivalence point of a titration the equivalents of acid equals the equivalents of base.

The number of equivalents of acid and base are equal at the equivalence point.

$$N_A \times V_A = N_B \times V_B$$

The volumes of V_A and V_B may be expressed in liters or milliliters, provided the same unit is used for both. For practical reasons, milliliters are usually used for reporting solution volumes in analyses.

Example 7

If 35.0 mL of $0.20N$ hydrochloric acid is required to neutralize 25.0 mL of an unknown base, what is the normality of the base?

Solution

$$N_B = \frac{V_A \times N_A}{V_B} = \frac{35.0\,\text{mL} \times 0.20N}{25.0\,\text{mL}} = 0.28N$$

Example 8

How many milliliters of $0.500N$ sulfuric acid are required to neutralize 50.0 mL of $0.200N$ potassium hydroxide?

Solution

$$V_A = \frac{V_B \times N_B}{N_A}$$

$$= \frac{50.0 \text{ mL} \times 0.200 \cancel{N}}{0.500 \cancel{N}} = 20.0 \text{ mL}$$

Problems

11. How many milliliters of $0.20N$ sodium hydroxide must be added to 75 mL of $0.050N$ hydrochloric acid to make a neutral solution?

12. What is the normality of a solution of a base if 25 mL is neutralized by 75 mL of $0.40N$ acid?

19·5 Salt Hydrolysis

Solutions of a salt are not neutral if the salt promotes hydrolysis.

Salts of weak acids and salts of weak bases promote hydrolysis.

A salt is made by neutralizing an acid with a base. Many solutions of salts are neutral, but some are acidic and others are basic. Solutions of sodium chloride and potassium sulfate are neutral. A solution of ammonium chloride is acidic. A solution of sodium acetate is basic. This happens because some salts promote hydrolysis. *In **salt hydrolysis,** the cations or anions of the dissociated salt accept hydrogen ions from water or donate hydrogen ions to water.* Depending on the direction of the hydrogen-ion transfer, solutions containing hydrolyzing salts may be either acidic or basic. Hydrolyzing salts are usually derived from a strong acid and a weak base or from a weak acid and a strong base.

Sodium acetate ($NaC_2H_3O_2$) is the salt of a weak acid (acetic acid, $HC_2H_3O_2$) and a strong base (sodium hydroxide, NaOH). In solution the salt is completely ionized.

$$NaC_2H_3O_2 \longrightarrow C_2H_3O_2^- + Na^+$$

Sodium acetate　　　　Acetate ion　　Sodium ion

The acetate ion is a Bronsted–Lowry base. It establishes an equilibrium with water, forming un-ionized acetic acid and hydroxide ions.

$$C_2H_3O_2^- + H_2O \rightleftharpoons HC_2H_3O_2 + OH^-$$

(hydrogen-ion acceptor, Brønsted–Lowry base)　　(hydrogen-ion donor, Brønsted–Lowry acid)　　　　　　(makes the solution basic)

Sodium acetate, $NaC_2H_3O_2$

Figure 19·7
The salt of a strong base and a weak acid will form a basic solution when added to water.

This process is called hydrolysis because it splits a hydrogen ion off a water molecule. The solution contains a hydroxide-ion concentration greater than the hydrogen-ion concentration. Thus the solution is basic.

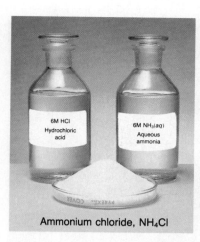
Ammonium chloride, NH₄Cl

Figure 19·8
Will the solution be basic or acidic when the salt of a strong acid and weak base is dissolved in water?

Buffers are solutions that resist changes in pH.

Both the HCO_3^-/H_2CO_3 and the $HPO_4^{2-}/H_2PO_4^-$ buffers help maintain the pH of blood.

Ammonium chloride (NH_4Cl) is the salt of a strong acid (hydrochloric acid, HCl) and a weak base (ammonia, NH_3). It is completely ionized in solution.

$$NH_4Cl \longrightarrow NH_4^+ + Cl^-$$

The ammonium ion (NH_4^+) is a strong enough acid to donate a hydrogen ion to a water molecule, although the equilibrium is strongly to the left.

$$NH_4^+ + H_2O \rightleftharpoons NH_3 + H_3O^+$$

(hydrogen-ion donor, Brønsted–Lowry acid) (hydrogen-ion acceptor, Brønsted–Lowry base) (makes the solution acidic)

This process is also called hydrolysis. It results in the formation of un-ionized ammonia and hydronium ions. The $[H_3O^+]$ is greater than the $[OH^-]$. Thus a solution of ammonium chloride is acidic.

19·6 Buffers

Buffers *are solutions in which the pH remains relatively constant when small amounts of acid or base are added.* A buffer is a solution of a weak acid and one of its salts. It could also be a solution of a weak base and one of its salts. The acetate buffer system is a typical example. A solution containing 0.20 mol/L of both acetic acid and sodium acetate has a pH of 4.76. When moderate amounts of either acid or base are added to this buffer system, the pH changes little. For example, the addition of 10 mL of $0.10M$ sodium hydroxide to 1 L of this buffer increases the pH by only 0.01 pH units (from 4.76 to 4.77). In contrast, the addition of 10 mL of $0.10M$ sodium hydroxide to 1 L of pure water increases the pH by 4.0 pH units (from 7.0 to 11.0).

The buffer solution is better able to resist drastic changes in pH than pure water. Acetic acid ($HC_2H_3O_2$) and its anion ($C_2H_3O_2^-$) act as reservoirs of neutralizing power. They react with any hydrogen ions or hydroxide ions that are added to the solution. For example, the sodium acetate ($NaC_2H_3O_2$) in the buffer solution is completely ionized.

$$NaC_2H_3O_2 \longrightarrow C_2H_3O_2^- + Na^+$$

Sodium acetate   Acetate ion   Sodium ion

When an acid is added to the solution, the acetate ions ($C_2H_3O_2^-$) act as a "hydrogen-ion sponge." This creates acetic acid, which does not dissociate extensively in water. Thus the pH does not change appreciably.

$$C_2H_3O_2^- + H^+ \longrightarrow HC_2H_3O_2$$

Acetate ion   Hydrogen ion   Acetic acid

A base is a source of hydroxide ions. When a base is added to the solution, the acetic acid and the hydroxide ions react to produce water.

$$HCH_3O_2 + OH^- \longrightarrow C_2H_3O_2^- + H_2O$$

Acetic acid   Hydroxide ion   Acetate ion   Water

The acetate ion is not a strong enough base to accept hydrogen ions from water extensively. Again the pH changes very little.

Figure 19·9
This device measures the pH of human blood. The carbonic acid–bicarbonate buffer system is responsible for maintaining blood pH within a very narrow range.

The manner in which a buffer resists changes in pH is an excellent illustration of Le Châtelier's principle.

An acetate buffer can not control the pH when too much acid is added. Then no more acetate ions are present to accept hydrogen ions. The buffer also becomes ineffective when too much base is added. Then no more acetic acid molecules are present to donate hydrogen ions. *The **buffer capacity** is the amount of acid or base that may be added to a buffer solution before a significant change in pH occurs.* When too much acid or base is added, the buffer capacity of a solution is exceeded. Two buffer systems are crucial in controlling pH in the human body. One is the carbonic acid–hydrogen carbonate buffer. The other is the monohydrogen phosphate–dihydrogen phosphate buffer. Table 19·2 lists several buffer systems.

Example 9

Show how the carbonic acid–hydrogen carbonate buffer can mop up hydrogen ions and hydroxide ions.

Solution

The carbonic acid–hydrogen carbonate buffer is a solution of carbonic acid (H_2CO_3) and hydrogen carbonate ions (HCO_3^-). When a base is added to this buffer, it reacts with H_2CO_3 to produce neutral water. The pH changes very little.

$$H_2CO_3 \ + \ OH^- \ \longrightarrow \ HCO_3^- \ + \ H_2O$$

Carbonic acid — Hydroxide ion — Hydrogen carbonate ion — Water

When an acid is added to the buffer, it reacts with HCO_3^- to produce undissociated carbonic acid. Again the pH changes very little.

$$HCO_3^- \ + \ H^+ \ \longrightarrow \ H_2CO_3$$

Hydrogen carbonate ion — Hydrogen ion — Carbonic acid

Table 19·2   Important Buffer Systems

Buffer name	Buffer species	Buffer pH (components 0.1M)
Acetic acid–acetate ion	$HC_2H_3O_2/C_2H_3O_2^-$	4.76
Dihydrogen phosphate ion– hydrogen phosphate ion	$H_2PO_4^-/HPO_4^{2-}$	7.20
Carbonic acid–hydrogen carbonate (solution saturated with CO_2)	H_2CO_3/HCO_3^-	6.46
Ammonium ion–ammonia	NH_4^+/NH_3	9.25

Problems

13. Write reactions to show what happens when the following occur.
   a. Acid is added to a solution of HPO_4^{2-}.
   b. Base is added to a solution of $H_2PO_4^-$.

14. Write equations that show what happens when acid is added to the following buffers.   **a.** acetic acid–acetate buffer   **b.** ammonium ion–ammonia buffer

19·7   The Solubility Product Constant

Salts differ in their solubilities. In general, compounds of the alkali metals are soluble in water. For example, over 35 g of sodium chloride will dissolve in 100 mL of water. Many classes of ionic compounds, however, are insoluble. For example, many compounds that contain phosphate, sulfide, sulfite, or carbonate ions are insoluble. Exceptions are compounds in which these ions are combined with ammonium ions or alkali metal ions. Table 19·3 summarizes the solubilities of many ionic compounds in water.

Most "insoluble" salts will dissolve to some extent in water. These salts are said to be slightly or sparingly soluble in water. For example, when the "insoluble" salt silver chloride is mixed with water, a very small amount of it dissolves.

$$AgCl(s) \rightleftharpoons Ag^+(aq) + Cl^-(aq)$$

An equilibrium expression can be written for this process.

$$K_{eq} = \frac{[Ag^+] \times [Cl^-]}{[AgCl]}$$

As long as some undissolved (solid) AgCl is present, the concentration of the AgCl is a constant. Thus, the concentration of AgCl can be combined with the equilibrium constant.

$$K_{eq} \times [AgCl] = [Ag^+] \times [Cl^-] = K_{sp}$$

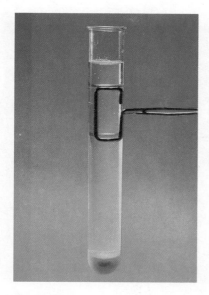

Figure 19·10
Silver chloride is slightly soluble in water.

Table 19·3   Solubilities of Ionic Compounds in Water

Compounds	Solubility	Exceptions
Salts of Group 1A metals and ammonia	Soluble	Some lithium compounds
Acetates, nitrates, chlorates, and perchlorates	Soluble	Few exceptions
Sulfates	Soluble	Compounds of Pb, Ag, Hg, Ba, Sr, and Ca
Chlorides, bromides, and iodides	Soluble	Compounds of Ag and some compounds of Hg and Pb
Sulfides and hydroxides	Most are insoluble	Alkali metal sulfides and hydroxides are soluble. Compounds of Ba, Sr, and Ca are slightly soluble.
Carbonates, phosphates, and sulfites	Insoluble	Compounds of the alkali metals and of ammonium ions

The concentration of ions of slightly soluble salts can be calculated from the solubility product, K_{sp}.

This new constant, called *the* **solubility product constant,** K_{sp}, *is equal to the product of the concentration terms each raised to the power of the coefficient of the substance in the dissociation equation.* The value of K_{sp} for silver chloride at 25°C is 1.8×10^{-10}.

$$K_{sp} = [Ag^+] \times [Cl^-] = 1.8 \times 10^{-10}$$

Example 10

What is the concentration of silver and chloride ions in a saturated silver chloride solution at 25°C? $K_{sp} = 1.8 \times 10^{-10}$.

Solution

The equation shows that for each Ag^+ ion formed, one Cl^- ion is formed.

$$AgCl \rightleftharpoons Ag^+ + Cl^-$$

Therefore at equilibrium $[Ag^+] = [Cl^-]$.
Write the expression for K_{sp}.

$$K_{sp} = [Ag^+] \times [Cl^-] = 1.8 \times 10^{-10}$$

Substitute for the $[Cl^-]$ then solve for the $[Ag^+]$.

$$[Ag^+] \times [Ag^+] = 1.8 \times 10^{-10}$$
$$[Ag^+]^2 = 1.8 \times 10^{-10}$$
$$[Ag^+] = 1.3 \times 10^{-5}M$$

When doing this kind of calculation be sure to express the exponent so that it will be evenly divisible when you take the square root.

The equilibrium concentration of Cl^- is also $1.3 \times 10^{-5}M$.

The smaller the value of K_{sp}, the lower the solubility of a salt.

Figure 19·11
Barium sulfate is used for X-rays of the digestive tract. This relatively insoluble salt is used in a suspension. X-rays do not pass through the salt thereby producing light areas on the film.

Table 19·4   Solubility Product Constants, K_{sp}, at 25°C

Salt	K_{sp}	Salt	K_{sp}
Halides		**Sulfides**	
AgCl	1.8×10^{-10}	NiS	4×10^{-20}
AgBr	5.0×10^{-13}	CuS	8×10^{-37}
AgI	8.3×10^{-17}	Ag_2S	8×10^{-51}
$PbCl_2$	1.7×10^{-5}	ZnS	3×10^{-23}
$PbBr_2$	2.1×10^{-6}	FeS	8×10^{-19}
PbI_2	7.9×10^{-9}	CdS	1×10^{-27}
PbF_2	3.6×10^{-8}	SnS	1.3×10^{-26}
CaF_2	3.9×10^{-11}	PbS	3×10^{-28}
Carbonates		**Hydroxides**	
$CaCO_3$	4.5×10^{-9}	$Al(OH)_3$	3×10^{-34}
$SrCO_3$	9.3×10^{-10}	$Zn(OH)_2$	3.0×10^{-16}
$ZnCO_3$	1.0×10^{-10}	$Ca(OH)_2$	6.5×10^{-6}
Ag_2CO_3	8.1×10^{-12}	$Mg(OH)_2$	7.1×10^{-12}
$BaCO_3$	5.0×10^{-9}	$Fe(OH)_2$	7.9×10^{-16}
Sulfates		**Chromates**	
$PbSO_4$	6.3×10^{-7}	$PbCrO_4$	1.8×10^{-14}
$BaSO_4$	1.1×10^{-10}	Ag_2CrO_4	1.2×10^{-12}
$CaSO_4$	2.4×10^{-5}		

Example 11

Calcium fluoride has a K_{sp} of 3.9×10^{-11} at 25°C. What is the fluoride ion concentration at equilibrium?

Solution

Write the equation: $CaF_2 \rightleftharpoons Ca^{2+} + 2F^-$.

Write the K_{sp} expression: $K_{sp} = [Ca^{2+}] \times [F^-]^2 = 3.9 \times 10^{-11}$.

When the $[Ca^{2+}]$ is x, the $[F^-]$ is 2x. (There are twice as many F^- ions as Ca^{2+} ions formed when CaF_2 dissociates.) Now substitute into the K_{sp} expression and solve for x.

$$K_{sp} = (x)(2x)^2 = 3.9 \times 10^{-11}$$

$$4x^3 = 3.9 \times 10^{-11}$$

$$x^3 = 9.8 \times 10^{-12}$$

$$x = 2.1 \times 10^{-4}$$

The $[Ca^{2+}] = 2.1 \times 10^{-4}M$. The $[F^-]$ is twice the $[Ca^{2+}]$. Thus $[F^-] = 4.2 \times 10^{-4}M$.

Problems

15. Lead(II) sulfide, PbS, has a K_{sp} of 3×10^{-28}. What is the concentration of lead(II) ion in a saturated solution of PbS?

16. The K_{sp} of silver sulfide, Ag_2S, is 8×10^{-51}. **a.** What is the silver-ion concentration of a saturated solution of silver sulfide? **b.** What is the sulfide-ion concentration of the same solution?

Figure 19·12
Chemicals are used to keep the pH of a swimming pool near neutral.

—— Science, Technology, and Society ——

19·B Swimming Pool Chemistry

The major problem in swimming pool maintenance is preventing the growth of algae and bacteria. Algae foul the water and can clog filters. Bacteria can cause illness. Chlorine compounds are usually used to disinfect the water in pools. The "liquid chlorine" sold for use in home pools is a solution of sodium hypochlorite, NaOCl. "Dry chlorine" is solid calcium hypochlorite, $Ca(OCl)_2$. When these ionic compounds are added to water, they undergo hydrolysis to form the weak acid, hypochlorous acid, HOCl.

$$OCl^- + H_2O \longrightarrow HOCl + OH^-$$

Chlorine gas, Cl_2, also forms hypochlorous acid when added to water. This method is sometimes used in large public pools.

$$Cl_2 + H_2O \rightleftharpoons Cl^- + H^+ + HOCl$$

For proper pool maintenance enough hypochlorous acid must be present in the water. It prevents the growth of bacteria and algae. The amount of undissociated hypochlorous acid in the water depends on the pH. If the pH is too high, the hydrolysis reaction shown in the first reaction will be shifted toward the reactants. Then the concentration of HOCl is reduced. This is according to LeChâtelier's principle. If the pH is too low, however, the acid content of the water can cause eye and skin irritation. Too much acid can also etch the plaster and corrode the metal piping and filters in the pool.

If the pool pH is too high, it can be lowered by adding an acid. This will neutralize the excess hydroxide ions. Muriatic acid (a solution of HCl) or solid sodium hydrogen sulfate can be used.

$$HCl + OH^- \longrightarrow H_2O + Cl^-$$
$$NaHSO_4 + OH^- \longrightarrow Na^+ + SO_4^{2-} + H_2O$$

If the pool pH is too low, it can be raised by neutralizing some of the acid. Sodium carbonate (also called soda ash) or sodium hydrogen carbonate (baking soda) can be used.

$$Na_2CO_3 + 2H^+ \longrightarrow 2Na^+ + H_2O + CO_2$$
$$NaHCO_3 + H^+ \longrightarrow Na^+ + H_2O + CO_2$$

19·8 The Common Ion Effect

In a saturated solution of silver chloride, an equilibrium is established between the solid silver chloride and its ions.

$$AgCl(s) \rightleftharpoons Ag^+(aq) + Cl^-(aq) \qquad K_{sp} = 1.8 \times 10^{-10}$$

What would happen if some silver nitrate was added to this solution? Immediately after the addition, the product of the $[Ag^+]$ and the $[Cl^-]$ would be greater than K_{sp}. Le Châtelier's principle (Section 17·12) applies here. The "stress" of the additional Ag^+ could be relieved if the reaction shifted to the left. Silver ions would combine with chloride ions to form additional solid AgCl. In fact, AgCl would precipitate until the product of the $[Ag^+]$ and the $[Cl^-]$ once again equalled 1.8×10^{-10}.

In this example, the silver ion is called a common ion. *A common ion is an ion that is common to both salts.* Adding silver nitrate to a saturated solution of AgCl causes the solubility of AgCl to be decreased. The solubility of AgCl is less in the presence of a solution of $AgNO_3$ than it is in pure water. *The lowering of the solubility of a substance by the addition of a common ion is called the* **common ion effect.**

In this example, the addition of sodium chloride would also give the common ion effect. The additional chloride ion (common ion) would cause the reaction to shift to the left. More AgCl would be formed, and the solubility of AgCl would again decrease.

▮ The solubility of an ionic solid is decreased when a common ion is present in the solution.

The common ion effect is not a unique phenomenon. It is merely another example of Le Châtelier's principle.

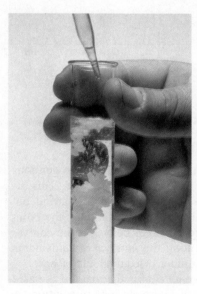

Figure 19·13
When silver nitrate is added to a saturated solution of silver chloride, additional silver chloride forms due to the common ion effect.

Example 12

The K_{sp} of silver iodide is 8.3×10^{-17}. What is the iodide ion concentration of a 1.00 L saturated solution of AgI to which 0.020 mol of $AgNO_3$ is added?

Solution

Write the equation for the equilibrium.

$$AgI(s) \rightleftharpoons Ag^+ + I^-$$

Write the expression for the solubility product.

$$K_{sp} = [Ag^+] \times [I^-] = 8.3 \times 10^{-17}$$

If the equilibrium concentration of iodide ion from the dissociation is x, then the equilibrium concentration of silver ion is x + 0.020. Because of the small value of K_{sp}, x will be small compared to 0.020. Therefore, the $[Ag^+]$ at equilibrium equals x + 0.020 ≈ 0.020. Substitute these values into the K_{sp} expression.

$$K_{sp} = [Ag^+] \times [I^-] = 8.3 \times 10^{-17}$$
$$(0.020)(x) = 8.3 \times 10^{-17}$$
$$x = 4.2 \times 10^{-15}$$

The equilibrium concentration of iodide ion is $4.2 \times 10^{-15}M$.

Problem

17. The K_{sp} of $SrSO_4$ is 3.2×10^{-7}. What is the equilibrium concentration of sulfate ion in a 1.0-L solution of strontium sulfate to which 0.10 mol of $Sr(C_2H_3O_2)_2$ has been added?

The K_{sp} is used to calculate whether precipitation will occur when solutions of ions are mixed.

The solubility product, K_{sp}, can be used to predict whether a precipitate will form when solutions are mixed together. If the ion product concentration of two ions in the mixture is greater than the K_{sp} of the compound, a precipitate will form. If the ion product equals K_{sp}, the solution is saturated, and no precipitate will form. If the ion product is less than K_{sp}, no precipitate will form, and the solution will be unsaturated.

Problem

18. A student prepares a solution by combining 0.025 mol $CaCl_2$ and 0.015 mol $Pb(NO_3)_2$ and adding water to make 1.0 L of solution. Will a precipitate of $PbCl_2$ form in this solution?

19 Neutralization and Salts
Chapter Review

Key Terms

buffer	19·6	neutralization	
buffer capacity	19·6	reaction	19·1
common ion	19·8	normality (N)	19·4
common ion effect	19·8	salt hydrolysis	19·5
endpoint	19·2	solubility product	
equivalence point	19·4	constant (K_{sp})	19·7
equivalent (equiv)	19·3	standard solution	19·2
gram equivalent mass		titration	19·2
of an acid	19·3		
gram equivalent mass			
of a base	19·3		

Chapter Summary

In the reaction of an acid with a base, hydrogen ions and hydroxide ions react to produce water. This reaction, called neutralization, is usually carried out by titration. The end point in a titration is the point at which the indicator changes color. At the equivalence point of a titration the number of equivalents of acid equals the number of equivalents of base.

An equivalent of an acid is the mass of the acid that provides 1 mol of hydrogen ions in solution. A solution that contains one equivalent of an acid or a base in a single liter of solution is a one normal ($1N$) solution.

A salt forms when an acid is neutralized by a base. Salts consist of an anion from the acid and a cation from the base. Salts of strong acid–strong base reactions produce neutral solutions with water. Salts formed from weak acids or weak bases hydrolyze water. They produce solutions that are acidic or basic.

Solutions that resist changes in pH are called buffer solutions. The pH of body fluids is kept within its normal range with buffers.

The solubility product constant, K_{sp}, is the equilibrium constant for the equilibrium between an ionic solid and its ions in solution. The solubility of a salt is decreased by the addition of a common ion.

Practice Questions and Problems

19. Write complete balanced equations for these acid–base reactions. Give the names of the salts produced. *19·1*
a. $H_2SO_4 + 2KOH \longrightarrow$
b. $HCl + Ca(OH_2) \longrightarrow$
c. $2H_3PO_4 + 3Ca(OH)_2 \longrightarrow$
d. $2HNO_3 + Mg(OH)_2 \longrightarrow$

20. How many moles of hydrochloric acid are required to neutralize these bases? *19·2*
a. 0.2 mol NaOH
b. 2 mol KOH
c. 0.1 mol $Ca(OH)_2$

21. What is true at the end point of a titration? *19·2*

22. What is the molarity of sodium hydroxide if 20.0 mL of the solution is neutralized by the following $1.00M$ solutions? *19·2*
a. 28.0 mL of HCl
b. 17.4 mL of H_3PO_4

23. What is the equivalent mass of an acid? *19·3*

24. What is the equivalent mass of a base? *19·3*

25. Determine the gram equivalent mass of each compound. *19·3*
a. H_3PO_4 **c.** $Mg(OH)_2$
b. NaOH   **d.** $HC_2H_3O_2$

26. Determine the number equivalents in each of the following. *19·3*
a. 20 g NaOH   **c.** 9.8 g H_3PO_4
b. 7.4 g $Ca(OH)_2$ **d.** 19.6 g H_2SO_4

27. How is the normality of a solution usually calculated? *19·4*

28. What is the normality of each of the following solutions? *19·4*
a. $1M$ NaOH   **c.** $0.2M$ KOH
b. $2M$ HNO_3 **d.** $0.1M$ H_2SO_4

29. Determine the normality of each solution. *19·4*
a. 250 mL of solution containing 10 g of NaOH
b. 750 mL of solution containing 4.9 g of H_2SO_4
c. 270 mL of solution containing 0.74 g of HCl
d. 2.80 L of solution containing 18.6 g of HNO_3
e. 7.3 g HCl in 250 mL of solution
f. 18.4 g HNO_3 in 1250 mL of solution

30. A student titrated several solutions of unknown concentration with various standard solutions to the point of neutralization. The volume of each unknown solution and the volume and normality of the standard solution used are given below. Calculate the normality for each unknown. *19·4*
a. 25.0 mL H_2SO_4 required 15.0 mL of $0.100N$ NaOH
b. 10.0 mL NaOH required 20.0 mL of $0.200N$ HCl
c. 17.5 mL NaOH required 25.0 mL of $0.120N$ HNO_3
d. 50.0 mL $HC_2H_3O_2$ required 39.6 mL of $0.0950N$ KOH
e. 29.2 mL $Ca(OH)_2$ required 50.0 mL of $0.152N$ HNO_3

31. What kinds of salts hydrolyze water? *19·5*

32. Explain why solutions of salts that hydrolyze water are not pH 7. *19·5*

33. Predict whether an aqueous solution of each salt will be acidic, basic, or neutral. *19·5*
a. $NaHCO_3$ **c.** KCl
b. NH_4NO_3 **d.** Na_2CO_3

34. Would a solution of HCl and NaCl be a good buffer? Explain. *19·6*

35. What does the solubility product constant, K_{sp} signify? *19·7*

36. Write the solubility product expression for each salt. *19·7*
a. NiS   **c.** $PbCl_2$
b. $BaCO_3$ **d.** Ag_2CrO_4

37. How does the addition of a common ion affect the solubility of another substance? *19·8*

Mastery Questions and Problems

38. What must be true about the concentration of two ions if precipitation occurs when solutions of the two ions are mixed?

39. Find the normality of each solution.
 a. 86.3 g $Mg(OH)_2$ in 2.5 L of solution.
 b. 5.6 g HBr in 450 mL of solution.
 c. 49.4 g H_2SO_3 in 1.5 L of solution.

40. The data below were collected in a titration of 50.00 mL of acetic acid, $HC_2H_3O_2$, of unknown concentration with $0.100N$ NaOH. Plot these data (pH on the y-axis) to obtain a titration curve.

Volume of NaOH (mL)	pH
0	
10.00	4.15
25.00	4.76
40.00	5.36
49.00	6.45
49.99	8.55
50.00	8.73
50.01	8.89
51.00	11.00
60.00	11.96
75.00	12.30
100.00	12.52

 a. What is the pH at the endpoint of this titration?
 b. Using Figure 18·16 on page 446, identify one or more indicators that could be used to determine the endpoint in this titration.

41. What is the concentration of hydroxide ions in a saturated solution of each salt?
 a. $Zn(OH)_2$
 b. $Ca(OH)_2$
 c. $Al(OH)_3$

42. Give the reactions for the addition of base to these buffers.
 a. ammonium ion–ammonia buffer
 b. dihydrogen phosphate ion–monohydrogen phosphate ion buffer

43. Write an equation to show that an aqueous solution of sodium acetate will be basic.

44. Arrange these solutions in order of decreasing acidity.
 a. $0.1N$ NaOH   **c.** $0.1M$ ammonium chloride
 b. $0.1N$ HCl   **d.** $0.1M$ sodium acetate

45. What is the equilibrium concentration of barium ion in a 1.0-L saturated solution of barium carbonate to which 0.25 mol of K_2CO_3 has been added?

46. Would precipitation occur when 500 mL of a $0.02M$ solution of $AgNO_3$ is mixed with 500 mL of a $0.001M$ solution of NaCl?

Critical Thinking Questions

47. Choose the term that best completes the second relationship.
 a. animal:dog   indicator: _____
   (1) pH   (3) titration
   (2) phenolphthalein (4) endpoint
 b. mole:atoms   equivalent: _____
   (1) molecules   (3) bases
   (2) acids   (4) ions
 c. molarity:moles normality: _____
   (1) normals   (3) equivalents
   (2) concentration   (4) ions

48. It is important for the pH of the blood to be maintained in the range of 7.35 to 7.45. The bicarbonate ion/carbonic acid buffer system is the most important of three buffer systems in the blood that help maintain this pH range. This system is represented by the following equation.

$$H_2O + CO_2 \underset{\text{lungs}}{\rightleftharpoons} H_2CO_3 \underset{\text{blood}}{\rightleftharpoons} H^+ + HCO_3^-$$

Given this equation, can you explain how abnormal breathing patterns can lead to acid–base imbalances in the blood that are called respiratory acidosis (abnormally low pH) and respiratory alkalosis (abnormally high pH)?

Too-rapid and too-deep breathing, a condition called hyperventilation, leads to respiratory alkalosis. Why? Conversely, hypoventilation, the result of too-shallow breathing, can cause respiratory acidosis. Explain.

49. The following equilibria are involved in the solubility of carbon dioxide in water.

$$CO_2(g) \rightleftharpoons CO_2(aq)$$
$$CO_2(aq) + H_2O \rightleftharpoons H_2CO_3(aq)$$
$$H_2CO_3(aq) \rightleftharpoons H^+(aq) + HCO_3^-(aq)$$
$$HCO_3^-(aq) \rightleftharpoons H^+(aq) + CO_3^{2-}(aq)$$

If seawater is slightly alkaline, would you expect the concentration of dissolved CO_2 to be higher or lower than in pure water? Explain.

Review Questions and Problems

50. Beakers A and B below both contain the same solution. Bromocresol green has been added to beaker A, and phenolphthalein has been added to beaker B.
  a. What is the approximate pH of the solution?
  b. How could you determine the pH more accurately?

51. Calculate the pH of solutions with the following hydrogen-ion concentrations.
  a. $4.6 \times 10^{-6}M$ **c.** $3.0 \times 10^{-1}M$
  b. $5.0 \times 10^{-12}M$ **d.** $9.8 \times 10^{-10}M$

52. How many grams of potassium chloride are in 45.0 mL of a 5.0% (by mass) solution?

53. Draw electron dot formulas for each compound.
  a. SiF_4 **b.** SCl_2 **c.** C_3H_8 **d.** H_2O_2

54. Write the formula for the conjugate base of each acid.
  a. H_2SO_4 **b.** HCN   **c.** H_2O **d.** NH_4^+

55. Which of the pair has the highest entropy?
  a. NaCl(s) or NaCl(aq)
  b. $CO_2(s)$ or $CO_2(g)$
  c. H_2O at 60°C or H_2O at 25°C

56. What is the molarity of the salt in the solution that results from the mixing of 200 mL of 1.00M NaOH and 200 mL of 1.00M HCl?

Challenging Questions and Problems

57. What is the normality of an H_2SO_4 solution if 80.0 mL of the solution reacts completely with 0.424 g of Na_2CO_3?

$$H_2SO_4 + Na_2CO_3 \longrightarrow H_2O + CO_2 + Na_2SO_4$$

58. An impure sample of $AgNO_3$ weighing 0.340 g was dissolved in water. After the addition of 10.0 mL of 0.200N HCl, 0.213 g of AgCl was recovered. Calculate the percentage of $AgNO_3$ in the sample.

59. Use the cyanate buffer HOCN/OCN$^-$ to illustrate how a buffer system works. Show by means of equations how the pH of a solution can be kept almost constant when small amounts of acid or base are added.

Research Projects

1. How do antacids work? Devise an experiment to compare and test their effectiveness.

2. Describe some common electrodes used on pH meters. How do they work?

3. Why are salts a problem in soil? What steps are being taken to reduce this problem?

4. Are different dyes needed for different types of fabric? Describe some examples.

5. Describe the buffering action of the bicarbonate ion in the bloodstream.

Readings and References

Bolton, Ruth, Elizabeth Lamphere, and Mario Menesini. *Action Chemistry.* New York: Holt, Rinehart and Winston, 1979.

Dombrink, Kathleen and David O. Tanis. "Swimming Pool Chemistry." *ChemMatters* (April 1983), pp. 4–5.

Tanis, David. "Underground Sculpture." *ChemMatters:* American Chemical Society (February 1984), pp. 10–11.

20 Oxidation–Reduction Reactions

Chapter Preview

20·1 Oxygen in Redox Reactions

20·2 Electron Transfer in Redox Reactions

20·A Corrosion

20·3 Assigning Oxidation Numbers

20·4 Oxidation Number Changes

20·5 Balancing Redox Equations

20·6 Identifying Redox Reactions

20·B Rocket Fuels and Household Bleach

20·7 Using Half-Reactions to Balance Redox Equations

20·C Photographic Chemistry

The *chemical changes that occur when electrons are transferred between reactants are known as* **oxidation–reduction reactions.** Oxidation reactions are the principal source of energy on earth. The combustion of gasoline in an automobile engine and the burning of wood in a fireplace are oxidation reactions. So is the "burning" of food by our bodies. All oxidation reactions are accompanied by reduction reactions. *Oxidation–reduction reactions are also called* **redox reactions.**

20·1 Oxygen in Redox Reactions

Elements and compounds combine with oxygen in oxidation reactions.

Oxidation *originally meant the combination of an element with oxygen to give oxides.* When iron slowly turns to rust, it oxidizes to iron(III) oxide (Fe_2O_3). When carbon burns in air, it oxidizes to carbon dioxide.

$$4Fe + 3O_2 \longrightarrow 2Fe_2O_3$$

$$C + O_2 \longrightarrow CO_2$$

Compounds can also be oxidized. Methane gas (CH_4) burns in oxygen. It oxidizes to form oxides of carbon and hydrogen.

$$CH_4 + 2O_2 \longrightarrow CO_2 + 2H_2O$$

Over the years, **reduction** *has meant the loss of oxygen from a compound.* The reduction of iron ore to metallic iron causes the removal of oxygen from iron(III) oxide. It is done by heating the ore with charcoal.

$$\underset{\text{Iron(III) oxide}}{2Fe_2O_3} + \underset{\text{Carbon}}{3C} \longrightarrow \underset{\text{Iron}}{4Fe} + \underset{\text{Carbon dioxide}}{3CO_2}$$

Figure 20·1
The corrosion of iron is an oxidation–reduction reaction that causes the formation of rust.

479

Figure 20·2
The iron in this piece of steel wool burns brightly in oxygen. The iron is oxidized to iron(III) oxide.

As iron oxide loses oxygen, it is reduced to metallic iron. The term *reduction* refers to the fact that when a metal oxide is reduced to the metal, there is a considerable decrease in volume. In other words, the amount of solid material has been reduced.

Oxidation and reduction occur simultaneously. As iron oxide is reduced to iron by losing oxygen, carbon is oxidized to carbon dioxide by gaining oxygen. No oxidation occurs without reduction and no reduction occurs without oxidation. This is true of all redox reactions.

20·2 Electron Transfer in Redox Reactions

In an oxidation–reduction reaction, electrons are transferred from one reactant to another.

Today, chemists have extended the concepts of oxidation and reduction to include all transfers or shifts of electrons. The advantage of the new definition is that it has much wider application than the old. Hence *oxidation is complete or partial loss of electrons or gain of oxygen. Reduction is complete or partial gain of electrons or loss of oxygen.*

ChemDirections

By far the most common oxidation products on earth are carbon dioxide and water, both formed during the combustion of coal, oil, and natural gas. Many scientists believe that the accumulation of carbon dioxide in our atmosphere is having a significant effect on the world's climate. **ChemDirections** *Greenhouse Effect*, page 664, explains how this gas effects our climate.

Oxidation	**Reduction**
Loss of electrons or gain of oxygen	Gain of electrons or loss of oxygen

Some examples will show what this means. In the reactions between a metal and nonmetal, electrons are transferred from atoms of the metal to atoms of the nonmetal. An ionic compound such as magnesium sulfide is produced.

$$\cdot Mg \cdot \; + \; \cdot \ddot{S} : \; \longrightarrow \; Mg^{2+} \; + \; : \ddot{S} :^{2-}$$

Magnesium atom   Sulfur atom   Magnesium ion   Sulfide ion

The net result of this reaction is a transfer of two electrons from a magnesium atom to a sulfur atom. The magnesium atom loses two electrons and is oxidized to a magnesium ion. Simultaneously the sulfur atom gains two electrons and is reduced to a sulfide ion.

Oxidation: $\cdot Mg \cdot \longrightarrow Mg^{2+} + 2e^-$ (loss of electrons)

Reduction: $\cdot \ddot{S} : \; + 2e^- \longrightarrow : \ddot{S} :^{2-}$ (gain of electrons)

It may be helpful to remember that "LEO the lion goes GER," where LEO = Losing Electrons in Oxidation and GER = Gaining Electrons in Reduction.

The substance in a redox reaction that donates electrons is a **reducing agent.** By *losing electrons,* magnesium reduces sulfur. Magnesium is a

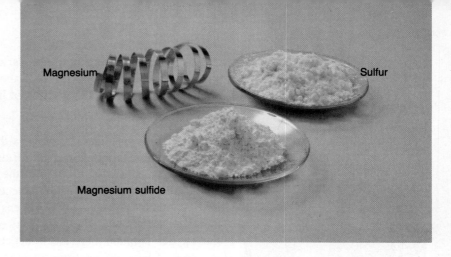

Magnesium

Sulfur

Magnesium sulfide

Figure 20·3
When magnesium and sulfur are heated together they undergo an oxidation–reduction reaction and form magnesium sulfide. Magnesium becomes more stable by losing electrons (oxidation). Sulfur becomes more stable by gaining electrons (reduction).

Safety

Oxidizing agents should be stored separately from other chemicals.

Reducing agents are themselves oxidized. Oxidizing agents are themselves reduced.

 Issues in Chemistry
Can you imagine pulling into a service station and asking for a fill up of hydrogen gas? The oxidation of hydrogen may indeed someday provide the energy that powers your car. Are there other alternative fuels for cars?

reducing agent. *The substance in a redox reaction that accepts electrons is an* **oxidizing agent.** By *accepting electrons*, sulfur oxidizes magnesium. Sulfur is an oxidizing agent.

$$\underset{\substack{\text{Reducing} \\ \text{agent}}}{\text{Mg}} + \underset{\substack{\text{Oxidizing} \\ \text{agent}}}{\text{S}} \longrightarrow \text{MgS}$$

(oxidized over Mg...S; reduced over S...MgS)

Example 1

What is oxidized and what is reduced in this single replacement reaction?

$$2AgNO_3(aq) + Cu(s) \longrightarrow Cu(NO_3)_2(aq) + 2Ag(s)$$

Solution

Begin by rewriting the equation and showing the ions.

$$2Ag^+ + 2NO_3^- + Cu \longrightarrow Cu^{2+} + 2NO_3^- + 2Ag$$

What has happened in this reaction? Two electrons have been transferred from a copper atom (Cu) to two silver ions (Ag^+).

Oxidation: $Cu \longrightarrow Cu^{2+} + 2e^-$ (loss of electrons)

Reduction: $2Ag^+ + 2e^- \longrightarrow 2Ag$ (gain of electrons)

The Ag^+ gains electrons and is reduced; Cu loses electrons and is oxidized. Hence Ag^+ is the oxidizing agent and Cu is the reducing agent.

Problem

1. Determine what is oxidized and what is reduced in each reaction. Identify the oxidizing agent and reducing agent in each case.
   a. $2Na + S \longrightarrow Na_2S$
   b. $2K + Cl_2 \longrightarrow 2KCl$
   c. $4Al + 3O_2 \longrightarrow 2Al_2O_3$

Figure 20·4
An oxyhydrogen torch can be used to weld metals like aluminum and iron. When hydrogen burns in oxygen, the redox reaction generates temperatures of about 2600°C.

Complete transfers of electrons are easy to see in ionic reactions such as those just examined. What about reactions that produce covalent compounds? Consider the reaction of hydrogen and oxygen.

$$2H_2 + O_2 \longrightarrow 2H_2O$$

The old definition of oxidation says that hydrogen is oxidized to water when it combines with oxygen. Electron transfer can also explain this process. Consider what happens to the bonding electrons in the reactants and in the products. The bonding electrons in the hydrogen molecule are shared equally between the hydrogens. In water, however, the bonding electrons are not shared equally between hydrogen and oxygen. (Recall that oxygen is much more electronegative than hydrogen.) The net result is a shift of bonding electrons away from hydrogen.

H—H electrons shared equally   H—O H → shift of bonding electrons away from hydrogen

Hydrogen is oxidized because it experiences a partial loss of electrons.

What about oxygen, the other reactant? The bonding electrons are shared equally between oxygens in the oxygen molecule, but they shift closer to oxygen in water.

O=O electrons shared equally   H—O H → shift of bonding electrons toward oxygen

Oxygen is therefore reduced because there is a partial gain of electrons.

There is always an oxidizing agent and a reducing agent in every redox reaction. In the reaction of hydrogen and oxygen to produce water, hydrogen is the reducing agent and oxygen is the oxidizing agent.

To summarize, oxidation–reduction reactions can be described on the basis of the addition or removal of oxygen. They can also be described on the basis of electron transfer. With compounds of carbon, the addition of oxygen or the removal of hydrogen is always oxidation. Conversely, the removal of oxygen or the addition of hydrogen is always reduction. Table 20·1 lists the various processes that constitute oxidation and reduction. The last entry in the table mentions oxidation numbers, another way that oxidation and reduction can be described.

Problem

2. In each of the following reactions decide which reactant is oxidized and which reactant is reduced. Which reactant is the reducing agent and which is the oxidizing agent? (To determine this, use the electronegativity values in Table 12·3.)

   a. $H_2 + Cl_2 \longrightarrow 2HCl$ d. $N_2 + 2O_2 \longrightarrow 2NO_2$
   b. $N_2 + 3H_2 \longrightarrow 2NH_3$ e. $2Li + F_2 \longrightarrow 2LiF$
   c. $S + Cl_2 \longrightarrow SCl_2$ f. $H_2 + S \longrightarrow H_2S$

Table 20·1	Processes Leading to Oxidation and Reduction	
Oxidation		**Reduction**
Complete loss of electrons (ionic reactions)		Complete gain of electrons (ionic reactions)
Shift of electrons away from an atom in a covalent bond		Shift of electrons toward an atom in a covalent bond
Gain of oxygen		Loss of oxygen
Loss of hydrogen by a covalent compound		Gain of hydrogen by a covalent compound
An increase in oxidation number		A decrease in oxidation number

Science, Technology, and Society

20·A Corrosion

Billions of dollars are spent worldwide each year to prevent and clean up after a special family of redox reactions, the corrosion of metals. Iron and steel, common construction metals, can be oxidized to metallic ions by water and oxygen in the environment. Oxygen is reduced by the reaction to the oxide or hydroxide ion. The corrosion of iron is described by these equations.

$$2Fe + O_2 + 2H_2O \longrightarrow 2Fe(OH)_2$$

$$4Fe(OH)_2 + O_2 + 2H_2O \longrightarrow 4Fe(OH)_3$$

Corrosion occurs more easily in the presence of salts or acids. These substances combine with the water present to make conducting solutions. This makes electron transfer easier.

Not all metals corrode easily. Gold and platinum are called the noble metals because they are very resistant to losing their electrons by corrosion. Other metals lose electrons easily but are protected by an oxide coating on their surface. Aluminum, for example, corrodes quickly in air to form a coating of very tightly packed aluminum oxide particles. This coating protects the aluminum object from any further corrosion. When iron corrodes, the coating of iron oxide is not tightly packed. Water and air can penetrate it and attack more of the iron metal below the coating. The corrosion continues until the iron object is only a pile of rust.

The corrosion of a shovel or a knife is a common but relatively small problem. By contrast, the corrosion of the support pillar of a bridge or the hull of an oil tanker is much more serious and costly to repair! Methods for controlling and preventing corrosion fall into two categories: surface protection and electrochemical protection.

Surface protection involves coating the metal surface with oil, paint, plastic, or another metal. These coatings exclude air and water,

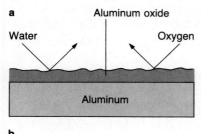

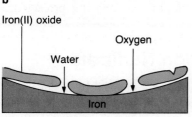

Figure 20·5
Oxidation causes the corrosion of some metals and not of others. How does aluminum oxide differ from iron(III) oxide?

a

b

Figure 20·6
a Painting a metal surface protects it from the corrosive effects of chemicals in the environment.
b Chromium metal also serves as a protective coating on automobile parts, while imparting an attractive, mirror-like finish. Like aluminum, chromium forms a corrosion-resistant oxide film on its surface.

and so prevent the corrosion reaction. If the coating is scratched or worn away, however, the bare metal will begin to corrode.

Electrochemical protection involves "sacrificing" one metal to save another. To protect an iron object, for example, a piece of magnesium or zinc is placed in electrical contact with it. Magnesium and zinc are more easily oxidized than iron, but they resist corrosion by oxygen and water because of oxide coatings. When oxygen and water attack an iron object, iron atoms lose electrons. Immediately, the magnesium or zinc transfers electrons to the iron ions, reducing them to neutral iron metal once again. This transfer is not prevented by the oxide coating, because there is electrical contact between the metals. Plates of magnesium or zinc are bolted to bridge pillars and ship hulls. Underground pipelines and storage tanks are protected by a block of magnesium buried nearby. It is connected to the pipe or tank by a conducting cable. Obviously, it is much easier to replace a block or plate of magnesium than to replace a bridge or pipeline. Galvanized iron is iron coated with zinc. It is used for roofing, plumbing, and large and small items made from sheet metal. The sacrificial metal is slowly consumed, but the iron object is protected.

20·3 Assigning Oxidation Numbers

Each atom in a substance can be assigned an oxidation number.

Oxidation numbers are a bookkeeping concept devised by chemists. *An **oxidation number** is a positive or negative number assigned to an atom according to a set of arbitrary rules.* As the next section will show, complex redox equations can be balanced by using oxidation-number changes. As a general rule, an oxidation number is the charge that an atom would have if the electrons in each bond were assigned to the atom of the more electronegative element. In binary ionic compounds such as NaCl and $CaCl_2$ the oxidation numbers of the atoms are equal to their ionic charges. The compound sodium chloride is composed of sodium ions, Na^{1+}, and chloride ions, Cl^{1-}. Thus, the oxidation number of sodium is $+1$, and that of chlorine is -1. Notice that when writing oxidation numbers, the sign is put before the number. Sodium has an ionic charge of $1+$ and an oxidation number of $+1$. In calcium fluoride, CaF_2, the oxidation number of calcium is $+2$, and that of fluorine, -1.

Because water is a molecular compound, no ionic charges are associated with its atoms. As was shown in the last section, however, oxygen is considered to be reduced in the formation of water. Oxygen is the more electronegative element. In water, the shared electrons in the bond are shifted closer to oxygen and away from hydrogen. Imagine that the two electrons contributed by the hydrogen atoms were completely transferred to the oxygen. The charges that would result from this transfer are the oxidation numbers of the elements. The oxidation number of oxygen is -2. The oxidation number of each hydrogen is $+1$.

Rules for Assigning Oxidation Numbers

1. The oxidation number of a monatomic ion is equal in magnitude and sign to its ionic charge. For example, the oxidation number of the bromide ion, Br^{1-}, is -1; that of the Fe^{3+} ion is $+3$.

2. The oxidation number of hydrogen in a compound is always $+1$ except in metal hydrides, for example, NaH, where it is -1.

3. The oxidation number of oxygen in a compound is always -2 except in peroxides, for example, H_2O_2, where it is -1.

4. The oxidation number of an uncombined element is zero. For example, the oxidation number of the potassium atoms in potassium metal, K, and of the nitrogen atoms in nitrogen gas, N_2, is zero.

5. For any neutral compound, the sum of the oxidation numbers of the atoms in the compound must equal zero.

6. For a polyatomic ion, the sum of the oxidation numbers must equal the ionic charge of the ion.

These last two rules can be used to determine the oxidation number of elements not covered in the first four rules.

Atoms in an uncombined element have an oxidation number of zero.

The sum of the oxidation numbers in a compound must be zero.

Figure 20·7
Oxidation–reduction reactions cause corrosion. The copper on this roof reacted with carbon dioxide and water vapor in the air to form a pale green film of hydrated copper carbonate.

Example 2

What is the oxidation number of each element in the following?
a. SO_2 **b.** CO_3^{2-} **c.** K_2SO_4

Solution

a. The oxidation number of oxygen is -2. There are two oxygens, and the sum of the oxidation numbers must be zero. Thus the oxidation number of sulfur is $+4$.

$$\overset{+4\,-2}{SO_2}$$

b. The oxidation number of oxygen is -2.

$$\overset{?\,-2}{CO_3^{2-}}$$

The sum of the oxidation numbers of the carbon and oxygen atoms must equal the ionic charge, $2-$. The oxidation number of carbon must be $+4$.

$$\overset{+4\,-2}{CO_3^{2-}}$$

c. The oxidation number of the potassium ion is $+1$, the same as its ionic charge. The oxidation number of oxygen is -2.

$$\overset{+1\ \ ?\,-2}{K_2SO_4}$$

For the sum of the oxidation numbers in the compound to be zero, the oxidation number of sulfur must be $+6$.

$$\overset{+1\ +6\,-2}{K_2\,S\,O_4}$$

Problem

3. Find the oxidation number of each element in these formulas.
   a. P_4O_{10} **b.** NH_4^+ **c.** $Na_2Cr_2O_7$ **d.** $Ca(OH)_2$.

20·4 Oxidation Number Changes

Oxidation numbers allow chemists to keep track of electrons during chemical reactions.

An *increase* in the oxidation number of an atom signifies *oxidation*. A *decrease* in the oxidation number of an atom signifies *reduction*. Look again at the equation in Example 1 to identify what is being oxidized and reduced on the basis of oxidation-number changes. Oxidation numbers have been added to this equation.

$$\overset{+1\,+5\,-2}{2AgNO_3}(aq) + \overset{0}{Cu}(s) \longrightarrow \overset{+2\ +5\,-2}{Cu(NO_3)_2}(aq) + \overset{0}{2Ag}(s)$$

The change in oxidation number is used to identify the substance being oxidized and the substance being reduced.

In this reaction, the oxidation number of silver decreases from $+1$ to 0. This is reduction. Silver is reduced from Ag^{+1} to Ag^0. On the other hand, copper is oxidized in this reaction. Its oxidation number increases from 0 to $+2$. Copper is oxidized from Cu^0 to Cu^{2+}. These results agree with those obtained by analyzing the reaction by using electron transfer.

Figure 20·8
When a copper wire is placed in a colorless silver nitrate solution **(a)**, crystals of silver coat the wire **(b)**. The solution slowly turns blue, as a result of copper(II) nitrate formation. What change occurs in the oxidation number of the silver? How does the oxidation number of the copper change?

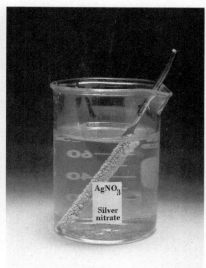

a   b

Safety

Many substances that are harmless in their reduced form may be toxic or corrosive in their oxidized form, or vice versa. Do not assume that all forms of a chemical are harmless just because one form is harmless.

Example 3

Use the change in oxidation number to identify which elements are oxidized and reduced in each of these reactions.

a. $Cl_2(g) + 2HBr(aq) \longrightarrow 2HCl(aq) + Br_2(l)$
b. $C(s) + O_2(g) \longrightarrow CO_2(g)$
c. $Zn(s) + 2MnO_2(s) + 2NH_4Cl(aq) \longrightarrow$
$$ZnCl_2(aq) + Mn_2O_3(s) + 2NH_3(g) + H_2O(l)$$

Solution

Assign oxidation numbers to each element. A decrease in oxidation number indicates reduction. An increase in oxidation number indicates oxidation.

a. $\overset{0}{Cl_2}(g) + 2\overset{+1-1}{HBr}(aq) \longrightarrow 2\overset{+1-1}{HCl}(aq) + \overset{0}{Br_2}(l)$

The element chlorine is reduced $(0 \rightarrow -1)$. The element bromine is oxidized $(-1 \rightarrow 0)$.

b. $\overset{0}{C}(s) + \overset{0}{O_2}(g) \longrightarrow \overset{+4-2}{CO_2}(g)$

The element carbon is oxidized $(0 \rightarrow +4)$. The element oxygen is reduced $(0 \rightarrow -2)$.

c. $\overset{0}{Zn}(s) + 2\overset{+4\ -2}{MnO_2}(s) + 2\overset{-3+1\ -1}{NH_4Cl}(aq) \longrightarrow$
$$\overset{+2\ -1}{ZnCl_2}(aq) + \overset{+3\ -2}{Mn_2O_3}(s) + 2\overset{-3+1}{NH_3}(g) + \overset{+1\ -2}{H_2O}(l)$$

The element zinc is oxidized $(0 \rightarrow +2)$. The element manganese is reduced $(+4 \rightarrow +3)$.

A reducing agent is an electron donor. An oxidizing agent is an electron acceptor.

Example 4

Identify the oxidizing agent and reducing agent in each of the equations in Example 3.

Solution

Recall from Section 20·2 that the oxidizing agent is reduced in the reaction and the reducing agent is oxidized.

a. Since chlorine is reduced, Cl_2 is the oxidizing agent. Since bromine (in HBr) is oxidized, Br^{1-} is the reducing agent.
b. Carbon is the reducing agent. Oxygen is the oxidizing agent.
c. Zinc is the reducing agent. Manganese (in MnO_2) is the oxidizing agent.

Problem

4. In each of the following reactions identify the element oxidized, the element reduced, the oxidizing agent, and the reducing agent.
a. $2Na(s) + Cl_2(g) \longrightarrow 2NaCl(s)$
b. $2HNO_3(aq) + 6HI(g) \longrightarrow 2NO(g) + 3I_2(s) + 4H_2O(l)$
c. $3H_2S(aq) + 2HNO_3(aq) \longrightarrow 3S(s) + 2NO(g) + 4H_2O(l)$
d. $2PbSO_4(s) + 2H_2O(l) \longrightarrow Pb(s) + PbO_2(s) + 2H_2SO_4(aq)$

20·5 Balancing Redox Equations

Simple redox equations can be balanced by inspection or by trial and error. Many other oxidation–reduction reactions are too complex, however, to balance in this way. Fortunately, two systematic methods are available. These are the oxidation–number change method and the half-reaction method. These methods are based on the fact that *the total number of electrons gained in reduction must equal the total number of electrons lost in oxidation.* The oxidation-number change method will be described in this section. The half-reaction method will be described in Section 20·7.

The change in the oxidation numbers of atoms in a chemical reaction can be used to balance redox equations.

With the **oxidation-number change method,** *a redox equation is balanced by comparing the increases and decreases in oxidation numbers.* To use this method, start with a skeleton equation for a redox reaction. We will use the reduction of iron ore as an example.

$$Fe_2O_3 + CO \longrightarrow Fe + CO_2$$

Step 1: Assign oxidation numbers to all the atoms in the equation. Write the number *above* the appropriate atoms.

$$\overset{+3 \ -2}{Fe_2O_3} + \overset{+2-2}{CO} \longrightarrow \overset{0}{Fe} + \overset{+4-2}{CO_2}$$

Note that the oxidation number is stated as the charge *per atom.* In Fe_2O_3 the total positive charge is $+6$, but the oxidation number of Fe is $+3$.

Figure 20·9
Iron ore, Fe_2O_3, (on the left) is converted to metallic iron (on the right) by an oxidation–reduction process involving carbon monoxide, CO. What is oxidized and what is reduced?

In every oxidation–reduction reaction the total increase in oxidation number must equal the total decrease in oxidation number.

Step 2: Identify which atoms are oxidized and which are reduced. In this reaction, iron decreases in oxidation number from +3 to 0, a change of −3. Therefore iron is reduced. Carbon increases in oxidation number from +2 to +4, a change of +2. Thus carbon is oxidized. Oxygen does not change in oxidation number (because neither elemental oxygen nor peroxides are involved).

Step 3: Use a line to connect the atoms that undergo oxidation and those that undergo reduction. Write the oxidation-number change at the mid-point of each line.

$$\overset{+3\ -2}{Fe_2O_3} + \overset{+2-2}{CO} \longrightarrow \overset{0}{Fe} + \overset{+4\ -2}{CO_2}$$

(with $+2$ connecting C atoms above, and -3 connecting Fe atoms below)

Step 4: Make the total increase in oxidation number equal to the total decrease in oxidation number by using appropriate coefficients. In this example, the oxidation number increase should be multiplied by 3 and the oxidation number decrease should be multiplied by 2. This gives an increase of +6 and a decrease of −6. This can be done in the equation by placing the coefficient 2 in front of Fe and the coefficient 3 in front of both CO and CO_2. The formula Fe_2O_3 does not need a coefficient because the formula indicates 2Fe.

$$Fe_2O_3 + 3CO \longrightarrow 2Fe + 3CO_2$$

(with $3 \times (+2) = +6$ above, $2 \times (-3) = -6$ below)

Step 5: Finally, check to be sure that the equation is balanced for both atoms and charge. If necessary, the remainder of the equation is balanced by inspection.

$$Fe_2O_3 + 3CO \longrightarrow 2Fe + 3CO_2$$

Example 5

Balance this redox equation by using the oxidation-number change method.

$$K_2Cr_2O_7(aq) + H_2O(l) + S(s) \longrightarrow$$
$$KOH(aq) + Cr_2O_3(aq) + SO_2(g)$$

Solution

Step 1: Assign oxidation numbers.

$$\overset{+1\ +6\ -2}{K_2Cr_2O_7}(aq) + \overset{+1\ -2}{H_2O}(l) + \overset{0}{S}(s) \longrightarrow$$
$$\overset{+1-2+1}{KOH}(aq) + \overset{+3\ -2}{Cr_2O_3}(aq) + \overset{+4-2}{SO_2}(g)$$

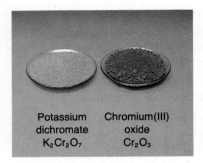

Potassium dichromate $K_2Cr_2O_7$ Chromium(III) oxide Cr_2O_3

Figure 20·10
Potassium dichromate is an orange crystalline substance. It reacts with water and sulfur to form chromium(III) oxide, the green compound shown here. What are the other products?

Steps 2 and 3: Connect the atoms that change in oxidation number. Indicate the sign and magnitude of the change.

$$\overset{+6}{K_2Cr_2O_7}(aq) + H_2O(l) + \overset{0}{S}(s) \longrightarrow$$
$$KOH(aq) + \overset{+3}{Cr_2O_3}(aq) + \overset{+4}{S}O_2(g)$$

(−3 connecting Cr atoms; +4 connecting S atoms)

Step 4: Balance the increase and decrease in oxidation numbers. Four chromium atoms must be reduced [$4 \times (-3) = -12$ decrease] for each three sulfur atoms that are oxidized [$3 \times (+4) = +12$ increase]. Put the coefficient 3 in front of S and SO_2 and the coefficient 2 in front of $K_2Cr_2O_7$ and Cr_2O_3.

$$(4)(-3) = -12$$
$$2\overset{+6}{K_2Cr_2O_7}(aq) + H_2O(l) + 3\overset{0}{S}(s) \longrightarrow$$
$$KOH(aq) + 2\overset{+3}{Cr_2O_3}(aq) + 3\overset{+4}{S}O_2(g)$$
$$(3)(+4) = +12$$

Step 5: Now finish balancing by inspection. The coefficient 4 in front of KOH balances potassium. The coefficient 2 in front of H_2O balances the hydrogen and oxygen.

$$2K_2Cr_2O_7(aq) + 2H_2O(l) + 3S(s) \longrightarrow$$
$$4KOH(aq) + 2Cr_2O_3(aq) + 3SO_2(g)$$

Problem

5. Balance each of these redox equations.
   a. $Al(s) + Cl_2(g) \longrightarrow AlCl_3(s)$
   b. $KClO_3(s) \longrightarrow KCl(aq) + O_2(g)$
   c. $PH_3(g) + I_2(s) + H_2O(l) \longrightarrow H_3PO_2(aq) + HI(aq)$
   d. $Cl_2(g) + KOH(aq) \longrightarrow KClO_3(aq) + KCl(aq) + H_2O(l)$
   e. $HNO_3(aq) + H_2S(g) \longrightarrow S(s) + NO(g) + H_2O(l)$
   f. $KIO_4(aq) + KI(aq) + HCl(aq) \longrightarrow KCl(aq) + I_2(s) + H_2O(l)$

20·6   Identifying Redox Reactions

In general, all chemical reactions can be assigned to one of two classes. In oxidation–reduction reactions electrons are transferred from one reacting species to another. In all other reactions electrons are not transferred. A majority of the reactions presented prior to this chapter do not involve electron transfer and are not redox reactions. Double-replacement reactions, and acid–base reactions are not redox reactions. Many single-replacement reactions, combination reactions, decomposition reactions, and combustion reactions are redox reactions.

An equation in which the oxidation numbers of at least two elements change is a redox reaction.

How can you determine if a reaction is a redox reaction? Use oxidation numbers. If the oxidation number of an element in a reacting species changes, then that element has undergone either oxidation or reduction. The change is part of a redox reaction.

Example 6

Use the change in oxidation number to identify which of these reactions are redox reactions. If a reaction is a redox reaction, name the element reduced, the element oxidized, the reducing agent, and the oxidizing agent.

a. $N_2O_4(g) \longrightarrow 2NO_2(g)$
b. $Cl_2(g) + 2NaBr(aq) \longrightarrow 2NaCl(aq) + Br_2(g)$
c. $PbCl_2(s) + K_2SO_4(aq) \longrightarrow 2KCl(aq) + PbSO_4(s)$
d. $2NaOH(aq) + H_2SO_4(aq) \longrightarrow Na_2SO_4(aq) + 2H_2O(l)$
e. $2K(s) + 2H_2O(l) \longrightarrow 2KOH(aq) + H_2(g)$

Solution

Assign oxidation numbers to each element. A decrease in oxidation number indicates reduction. An increase indicates oxidation.

a. $\overset{+4\ -2}{N_2O_4}(g) \longrightarrow 2\overset{+4-2}{NO_2}(g)$

This is a decomposition reaction. Neither oxygen nor nitrogen changes in oxidation number. Therefore it is not a redox reaction.

b. $\overset{0}{Cl_2}(g) + 2\overset{+1\ -1}{NaBr}(aq) \longrightarrow 2\overset{+1\ -1}{NaCl}(aq) + \overset{0}{Br_2}(l)$

This is a single-displacement reaction. The element chlorine is reduced $(0 \rightarrow -1)$. The element bromine is oxidized $(-1 \rightarrow 0)$. This is a redox reaction. Chlorine is the oxidizing agent; bromine is the reducing agent.

Figure 20·11
Potassium reacts violently with water. Is this a redox reaction?

c. $\overset{+2\ -1}{PbCl_2}(s) + \overset{+1\ +6\ -2}{K_2SO_4}(aq) \longrightarrow 2\overset{+1-1}{KCl}(aq) + \overset{+2\ +6-2}{PbSO_4}(s)$

This is a double-replacement reaction. None of the elements changes in oxidation number. This is not a redox reaction.

d. $2\overset{+1\ -2+1}{NaOH}(aq) + \overset{+1\ +6-2}{H_2SO_4}(aq) \longrightarrow \overset{+1\ +6\ -2}{Na_2SO_4}(aq) + 2\overset{+1-2}{H_2O}(l)$

This is an acid–base reaction. None of the elements changes in oxidation number. This is not a redox reaction.

e. $2\overset{0}{K}(s) + 2\overset{+1\ -2}{H_2O}(l) \longrightarrow 2\overset{+1-2+1}{KOH}(aq) + \overset{0}{H_2}(g)$

The element potassium is oxidized $(0 \rightarrow +1)$. The element hydrogen is reduced $(+1 \rightarrow 0)$. This is a redox reaction. Potassium is the reducing agent; hydrogen is the oxidizing agent. (Note that not all the hydrogens are reduced. Some of them do not change in oxidation number.)

Problem

6. Identify which of these are oxidation–reduction reactions. If a reaction is a redox reaction, name the element oxidized, the element reduced, the oxidizing agent, and the reducing agent.
   a. $BaCl_2(aq) + 2KIO_3(aq) \longrightarrow Ba(IO_3)_2(s) + 2KCl(aq)$
   b. $H_2CO_3(aq) \longrightarrow H_2O(l) + CO_2(g)$
   c. $Mg(s) + Br_2(l) \longrightarrow MgBr_2(s)$
   d. $NH_4NO_2(s) \longrightarrow N_2(g) + 2H_2O(l)$
   e. $2KClO_3(s) \longrightarrow 2KCl(s) + 3O_2(g)$
   f. $CaCO_3(s) + 2HCl(aq) \longrightarrow CaCl_2(aq) + H_2O(l) + CO_2(g)$
   g. $CuO(s) + H_2(g) \longrightarrow Cu(s) + H_2O(l)$
   h. $CaCO_3(s) \longrightarrow CaO(s) + CO_2(g)$
   i. $6Sb(s) + 10HNO_3(aq) \longrightarrow 3Sb_2O_5(s) + 10NO(g) + 5H_2O(l)$

Figure 20·12
Oxidation–reduction reactions provide the energy to launch space vehicles.

—— Science, Technology, and Society ——

20·B Rocket Fuels and Household Bleach

What could laundry bleach and rocket fuels have in common? Both do their work through oxidation–reduction reactions. Bleaches work by producing very reactive substances that are strong oxidizing agents. The most common household bleach is a solution of sodium hypochlorite. It decomposes to produce atomic oxygen.

$$NaClO \longrightarrow NaCl + O$$

Hydrogen peroxide is used commercially for bleaching. It also produces atomic oxygen.

$$H_2O_2 \longrightarrow H_2O + O$$

Atomic oxygen is a very strong oxidizing agent. It reacts with the organic molecules that give color to cloth and paper. The oxidized forms of these organic molecules are colorless.

The power of a rocket is also the result of oxidation–reduction reactions. Some fuels are mixtures of two liquids: a fuel and an oxidizer. Liquid hydrogen, for example, can be oxidized by liquid oxygen. They combine to form water.

$$2H_2(l) + O_2(l) \longrightarrow 2H_2O(g)$$

Hydrogen is oxidized, and oxygen is reduced in this reaction. The thrust of a rocket increases as the molecular mass of the exhaust gas decreases. Water vapor has a low molecular mass. The fuel gases also occupy a small volume. These facts make this cryogenic hydrogen/oxygen mixture the most commonly used fuel in space launch vehicles.

Some rocket fuels are self-oxidizing. The oxidation–reduction reactions of hydrogen peroxide and hydrazine are decompositions rather than combustion reactions. Hydrazine decomposes to molecular nitrogen and hydrogen.

$$N_2H_4 \longrightarrow N_2 + 2H_2$$

20·7 Using Half-Reactions to Balance Redox Equations

The half-reaction method involves balancing and adding oxidation and reduction half-reactions to obtain a balanced equation.

Electron-transfer reactions can be separated into two half-reactions.

Another method can be used to balance redox equations. *The half-reaction method is used to balance redox equations by balancing the oxidation and reduction half-reactions.* The procedure is different but the outcome is the same as with the oxidation-number change method. The first step in the half-reaction method is to divide the overall reaction into two parts. *A half-reaction is an equation showing either the reduction or the oxidation of a species in an oxidation–reduction reaction.* Then each half-reaction is balanced. The number of electrons gained by the reduction half-reaction must be equal to the number of electrons lost by the oxidation half-reaction. Finally, the two half-reactions are added to give a balanced equation. The half-reaction method is particularly useful for balancing equations for ionic reactions.

The oxidation of sulfur with nitric acid in aqueous solution can be our example.

$$S + HNO_3 \longrightarrow SO_2 + NO + H_2O$$

Step 1: Write the equation in ionic form. (Hint: Only HNO_3 is ionized. The products are covalent compounds.)

$$S + H^+ + NO_3^- \longrightarrow SO_2 + NO + H_2O$$

Step 2: Write separate half-reactions for the oxidation and reduction processes. (Hint: Sulfur is oxidized because its oxidation number increases from 0 to 4. Nitrogen is reduced because its oxidation number decreases from +5 to +2.)

$$\text{Oxidation: } S \longrightarrow SO_2 \qquad \text{Reduction: } NO_3^- \longrightarrow NO$$

Step 3: Balance the atoms in the half-reactions.

  a. *Balance the oxidation half-reaction.* Since this reaction takes place in acid solution, H_2O and H^+ can be used to balance oxygen and hydrogen as needed. Sulfur is already balanced. Two molecules of H_2O can be added to balance the oxygen.

$$2H_2O + S \longrightarrow SO_2$$

Oxygen is now balanced, but four hydrogens must be added to the right to balance those on the left.

$$2H_2O + S \longrightarrow SO_2 + 4H^+$$

This half-reaction is now balanced in terms of atoms.

  b. *Balance the reduction half-reaction.* Nitrogen is already balanced. Two molecules of H_2O can be added to balance the oxygen.

$$NO_3^- \longrightarrow NO + 2H_2O$$

Oxygen is balanced, but four hydrogens ($4H^+$) must be added to the left to balance hydrogen.

$$4H^+ + NO_3^- \longrightarrow NO + 2H_2O$$

This half-reaction is now balanced in terms of atoms.

Table 20·2 Oxidation Numbers of Sulfur in Different Compounds

Compound	Oxidation Number
H_2SO_4	+6
SO_3	+6
H_2SO_3	+4
SO_2	+4
$Na_2S_2O_3$	+2
SCl_2	+2
S_2Cl_2	+1
S	0
H_2S	−2

Electrons are as much a part of a half-reaction as the atoms themselves. They must be written into the equation.

Step 4: *Add sufficient electrons to one side of each half-reaction to balance the charges.* Four electrons are needed on the right side in the oxidation half-reaction. Three electrons are needed on the left side in the reduction half-reaction.

$$\text{Oxidation: } 2H_2O + S \longrightarrow SO_2 + 4H^+ + 4e^-$$

$$\text{Reduction: } 4H^+ + NO_3^- + 3e^- \longrightarrow NO + 2H_2O$$

Each half-reaction is now balanced with respect to both atoms and charge.

In every oxidation–reduction reaction the total number of electrons gained is equal to the total number of electrons lost.

Step 5: *Multiply each half-reaction by an appropriate number to make the electron changes equal.* In any redox reaction, the electrons lost in oxidation must be equal to the electrons gained in reduction. If the oxidation half-reaction is multiplied by 3 and the reduction half-reaction by 4, the number of electrons lost in oxidation and the number of electrons gained in reduction both equal 12.

$$\text{Oxidation: } 6H_2O + 3S \longrightarrow 3SO_2 + 12H^+ + 12e^-$$

$$\text{Reduction: } 16H^+ + 4NO_3^- + 12e^- \longrightarrow 4NO + 8H_2O$$

Step 6: *Add the half-reactions and subtract terms that appear on both sides of the equation.*

$$16H^+ + \overset{4}{\cancel{6H_2O}} + 4NO_3^- + 3S + \cancel{12e^-} \longrightarrow$$
$$3SO_2 + 4NO + \cancel{12H^+} + \cancel{12e^-} + \overset{2}{\cancel{8}}H_2O$$

The final equation should be checked to be sure that atoms are conserved, charge is conserved, and electrons have cancelled.

Now the balanced equation can be written.

$$4H^+ + 4NO_3^- + 3S \longrightarrow 3SO_2 + 4NO + 2H_2O$$

or

$$4HNO_3 + 3S \longrightarrow 3SO_2 + 4NO + 2H_2O$$

Example 7

Balance the reaction that occurs between permanganate ions, MnO_4^-, and chloride ions, Cl^-, in acid solution. The products are manganese (II) ions, Mn^{2+}, and chlorine gas, Cl_2.

Solution

Step 1: Write the unbalanced reaction in ionic form.

$$MnO_4^- + Cl^- \longrightarrow Mn^{2+} + Cl_2$$

Step 2: Write the half-reactions for oxidation and reduction. (Use oxidation numbers to determine the oxidation process and the reduction process.)

$$\text{Oxidation: } Cl^- \longrightarrow Cl_2$$

$$\text{Reduction: } MnO_4^- \longrightarrow Mn^{2+}$$

Figure 20·13
a A redox reaction occurs when hydrochloric acid, HCl, is added to potassium permanganate, KMnO₄.
b The products of this reaction are manganese(II) chloride, chlorine gas, water, and potassium chloride. Why should this reaction be done only in a fume hood?

a   b

Step 3: Balance the atoms. Because the solution is acidic, use H_2O and H^+ to balance the hydrogen and oxygen if necessary.

Oxidation: $2Cl^- \longrightarrow Cl_2$ (atoms balanced)

Reduction: $MnO_4^- + 8H^+ \longrightarrow Mn^{2+} + 4H_2O$
(atoms balanced)

Step 4: Balance the charges. Add electrons as needed.

Oxidation: $2Cl^- \longrightarrow Cl_2 + 2e^-$ (charges balanced)

Reduction: $MnO_4^- + 8H^+ + 5e^- \longrightarrow Mn^{2+} + 4H_2O$
(charges balanced)

Step 5: Make the electron charges equal. Multiply the oxidation half-reaction by 5 and the reduction half-reaction by 2. Ten electrons are lost in oxidation, and 10 electrons are gained in reduction.

Oxidation: $10Cl^- \longrightarrow 5Cl_2 + 10e^-$

Reduction: $2MnO_4^- + 16H^+ + 10e^- \longrightarrow 2Mn^{2+} + 8H_2O$

Step 6: Add the half-reactions and subtract terms that appear on both sides of the equation.

$$2MnO_4^- + 16H^+ + 10Cl^- + \cancel{10e^-} \longrightarrow$$
$$2Mn^{2+} + 5Cl_2 + 8H_2O + \cancel{10e^-}$$

The balanced ionic equation is the net equation.

$$2MnO_4^- + 16H^+ + 10Cl^- \longrightarrow 2Mn^{2+} + 5Cl_2 + 8H_2O$$

In this example, the permanganate ions could come from potassium permanganate, KMnO₄. The chloride ions were from HCl. As an additional step the spectator ions should be added to the equation.

Step 7: Add the spectator ions and balance. **Spectator ions** *are those which do not change oxidation number or composition during a reaction.* The K^+ and Cl^- ions can now be added.

$$2K^+ + 2MnO_4^- + 16H^+ + 16Cl^- \longrightarrow$$
$$2Mn^{2+} + 4Cl^- + 5Cl_2 + 8H_2O + 2K^+ + 2Cl^-$$

or

$$2KMnO_4 + 16HCl \longrightarrow 2MnCl_2 + 5Cl_2 + 8H_2O + 2KCl$$

Figure 20·14
a A redox reaction occurs when potassium chlorate is added to potassium iodide in acid solution. The chlorate ion is the oxidizing agent and the iodide ion is the reducing agent.
b The resulting solution is yellow due to dissolved iodine, a product of the reaction. See Problem **7.a.**

a   b

As a quick review, here is a summary of the steps for balancing redox equations by the ion–electron method.

Step 1: Write the equation in ionic form.

Step 2: Write separate half-reactions for the oxidation and reduction processes.

Step 3: Balance the atoms in the half-reactions. Use H_2O and H^+ to balance oxygen and hydrogen in an acid solution. Use H_2O and OH^- for reactions in a basic solution.

Step 4: Add electrons to one side of each half-reaction to balance the charges.

Step 5: Multiply each half-reaction by an appropriate number to make the electron changes equal.

Step 6: Add the half-reactions and subtract terms that appear on both sides of the equation.

Step 7: The final equation should be checked to be sure that atoms are conserved, charge is conserved, and all electrons have cancelled. If spectator ions are known, add them now and the equation will be balanced.

Problem

7. Balance each of these redox equations by the ion-electron method.
   a. $ClO_3^-(aq) + I^-(aq) \longrightarrow Cl^-(aq) + I_2(aq)$ [acid solution]
   b. $C_2O_4^{2-}(aq) + MnO_4^-(aq) \longrightarrow$
   $\qquad\qquad\qquad\qquad\qquad Mn^{2+}(aq) + CO_2(aq)$ [acid solution]
   c. $Br_2(l) + SO_2(g) \longrightarrow Br^-(aq) + SO_4^{2-}(aq)$ [acid solution]
   d. $MnO_2(s) + H^+(aq) + NO_2^-(aq) \longrightarrow$
   $\qquad\qquad\qquad\qquad NO_3^-(aq) + Mn^{2+}(aq) + H_2O(l)$
   e. $MnO_4^-(aq) + NO_2^-(aq) \longrightarrow$
   $\qquad\qquad\qquad MnO_2(s) + NO_3^-(aq)$ [basic solution]
   f. $Cl_2(g) \longrightarrow ClO_3^-(aq) + Cl^-(aq)$ [basic solution]

Chemical Connections

Color film consists of several emulsion layers, each sensitive to a different color of light. The final color print consists of a mixture of the images that appear in each of these layers. In instant color film, all the chemicals needed for both the negative and positive prints are packed together.

Figure 20·15
On a developed negative, the light and developer have caused silver bromide to be reduced and deposited as dark metallic silver. Areas of the film that received less light appear clear. The unchanged silver bromide crystals were dissolved by the fixer.

Figure 20·16
A positive print reverses the areas of light and dark that are recorded on a film negative.

20·C Photographic Chemistry

Making photographic prints and negatives depends upon oxidation-reduction reactions. Black-and-white photographic film consists of a clear plastic sheet coated with a thin layer of gelatin emulsion. The emulsion contains very fine crystals (grains) of silver bromide. (Color film is much more complex.) Silver bromide is sensitive to visible light. When the film is exposed to light, the crystals are activated. The activation of the crystals on the film is not visible to the naked eye.

To make the changes in the silver bromide visible, the film must be developed. This process must be done in complete darkness. The exposed film is placed in a solution containing a reducing agent. The agent, called the developer, is usually an organic chemical such as hydroquinone, $C_6H_4(OH)_2$. It reduces the silver ions in the activated silver bromide to metallic silver. It does not reduce the ions in the crystals that were not activated by the exposure. The metallic silver produced is black in color.

$$AgBr(s) \text{ (activated)} + C_6H_4(OH)_2(aq) \longrightarrow 2Ag(s) + C_6H_4O_2 + 2HBr(aq)$$

Where the film has been exposed to the most light, the greatest number of silver ions are reduced. These regions of the developed film become the darkest. For this reason, the lightest areas of the photographed image will be the darkest on the film.

The silver bromide that was not activated slowly reacts with the hydroquinone. It must now be removed to prevent the entire film from becoming black. The film is placed in a solution of "fixer" to remove the excess silver ions. Sodium thiosulfate, commonly called "hypo," is used for this purpose.

$$AgBr + 2Na_2S_2O_3 \longrightarrow NaBr + Na_3Ag(S_2O_3)_2$$

After a bath in fixer, the film is dark where it was exposed to light, and clear where it was not exposed. So, an image is created that reverses the light and dark areas of the original scene. For this reason, the film-image is called a negative.

A positive print can be make from a negative by shining light through it onto a sheet of photographic paper. Like the film, this paper is coated with a light-sensitive emulsion of silver bromide. Light passes through the clear areas of the negative but is stopped where the negative is dark. Wherever light strikes the paper, silver bromide is activated. When the print is developed, those silver ions are reduced to silver and regions appear dark. What was dark in the original scene and clear in the negative is again dark in the photographic print. Through this series of redox reactions, the final print resembles the scene that was photographed.

Light-sensitive chemical reactions are the subject of a special branch of chemistry called photochemistry.

20 Oxidation-Reduction Reactions
Chapter Review

Key Terms

half-reaction	20·7	oxidation–reduction	
half-reaction		reaction	20·0
method	20·7	oxidizing agent	20·2
oxidation	20·1	redox reaction	20·0
oxidation number	20·3	reducing agent	20·2
oxidation-number		reduction	20·1
change method	20·5	spectator ion	20·7

Chapter Summary

Oxidation–reduction, or redox, reactions are an important category of chemical reactions. The original meaning of oxidation was the chemical combination of a substance with oxygen. Reduction was originally the loss of oxygen. These definitions have long since been expanded. Oxidation is now considered to be any shift of electrons away from an atom. Reduction now includes any shift of electrons toward an atom. An oxidation reaction is always accompanied by a reduction reaction. The substance that does the oxidizing is called an oxidizing agent. It is reduced. The substance that does the reducing is called a reducing agent. It is oxidized.

An oxidation number can be assigned to an element in a substance according to a set of rules. The oxidation number of an element in an uncombined state is zero. The oxidation number of a monatomic ion is the same in magnitude and sign as its ionic charge. The sum of the oxidation numbers of the elements in a neutral compound is zero. In a polyatomic ion, however, the sum is equal to the charge of the ion. Oxidation numbers help us keep track of electrons in redox reactions. An oxidation number increase is oxidation. A decrease is reduction.

One method for balancing redox equations involves determining the change in oxidation number of the substances that are oxidized and reduced. Coefficients are then used to make the increase in oxidation-number equal to the decrease.

The half-reaction method is another way to write a balanced equation for a redox reaction. In this method the net ionic equation is first divided into two half-reactions. One is for the oxidation and the other is for the reduction. Each half-reaction is balanced independently for mass. H^+, OH^-, or H_2O are added as needed. The net charge on both sides is balanced by adding electrons. The half-reactions are then multiplied by factors to make the number of electrons the same in each. Finally the half-reactions are added.

Practice Questions and Problems

8. Balance each equation. Then classify each reaction as oxidation or reduction based on the loss or gain of oxygen. *20·1*
 a. $C_2H_4 + O_2 \longrightarrow CO_2 + H_2O$
 b. $KClO_3 \longrightarrow KCl + O_2$
 c. $CuO + H_2 \longrightarrow Cu + H_2O$
 d. $H_2 + O_2 \longrightarrow H_2O$

9. Define oxidation and reduction in these terms.
 a. Gain or loss of electrons. *20·2*
 b. Gain or loss of hydrogen by a covalent bond.
 c. Gain or loss of oxygen.
 d. The shift of electrons in a covalent bond.

10. Identify these reactions as either oxidation or reduction. *20·2*
 a. $Li \longrightarrow Li^+ + e^-$ **c.** $Zn^{2+} + 2e^- \longrightarrow Zn$
 b. $2I^- \longrightarrow I_2 + 2e^-$ **d.** $Br_2 + 2e^- \longrightarrow 2Br^-$

11. Which of the following would most likely be oxidizing agents and which would most likely be reducing agents? **a.** Mn^{4+} **b.** Cl_2 **c.** K
 d. O^{2-} **e.** K^+ *20·2*

12. Use electron transfer or electron shift to identify what is oxidized and what is reduced in each of these reactions. Use electronegativity values for molecular compounds. *20·2*
 a. $2Na + Br_2 \longrightarrow 2NaBr$
 b. $N_2 + 3H_2 \longrightarrow 2NH_3$

c. $S + O_2 \longrightarrow SO_2$
d. $Mg + Cu(NO_3)_2 \longrightarrow Mg(NO_3)_2 + Cu$

13. Identify the oxidizing agent and the reducing agent in each reactant in Problem 12.   20·2

14. How are oxidation numbers determined and used?   20·3

15. Determine the oxidation number of each element in these substances.   20·3
a. S_2O_3 **b.** O_2 **c.** $Al_2(SO_4)_3$ **d.** Na_2O_2

16. Use the oxidation number to identify which elements are oxidized and which are reduced in each of these reactions.   20·4
a. $2H_2(g) + O_2(g) \longrightarrow 2H_2O(l)$
b. $2KNO_3(s) \longrightarrow 2KNO_2(s) + O_2(g)$
c. $NH_4NO_2(s) \longrightarrow N_2(g) + 2H_2O(g)$
d. $PbO_2(aq) + 4HI(aq) \longrightarrow$
                        $I_2(aq) + PbI_2(aq) + 2H_2O(l)$

17. Identify the oxidizing agent and reducing agent in each equation in Problem 22.   20·4

18. Balance each of these redox equations using the oxidation-number change method.   20·5
a. $KClO_3 \longrightarrow KCl + O_2$
b. $HNO_2 + HI \longrightarrow NO + I_2 + H_2O$
c. $As_2O_3 + Cl_2 + H_2O \longrightarrow H_3AsO_4 + HCl$
d. $Bi_2S_3 + HNO_3 \longrightarrow$
                 $Bi(NO_3)_3 + NO + S + H_2O$
e. $MnO_2 + H_2SO_4 + H_2C_2O_4 \longrightarrow$
                    $MnSO_4 + CO_2 + H_2O$
f. $SbCl_5 + KI \longrightarrow SbCl_3 + KCl + I_2$

19. Identify the oxidizing agent and the reducing agent in each reaction in Problem 18.   20·5

20. Identify which of these equations represent redox reactions.   20·6
a. $Li(s) + H_2O(l) \longrightarrow LiOH(aq) + H_2(g)$
b. $K_2Cr_2O_7(aq) + HCl(aq) \longrightarrow$
     $KCl(aq) + CrCl_3(aq) + H_2O(l) + Cl_2(g)$
c. $Al(s) + HCl(aq) \longrightarrow AlCl_3(aq) + H_2(g)$
d. $P_4(s) + S_8(s) \longrightarrow P_2S_5(s)$
e. $MnO(s) + PbO_2(s) \longrightarrow$
           $MnO_4^-(aq) + Pb^{2+}(aq)$ [acidic]
f. $Cl_2(g) + H_2O(l) \longrightarrow HCl(aq) + HClO(aq)$
g. $I_2O_5(s) + CO(g) \longrightarrow I_2(s) + CO_2(g)$
h. $H_2O(l) + SO_3(g) \longrightarrow H_2SO_4(aq)$
i. $Bi(OH)_3 + K_2SnO_2(aq) \longrightarrow$
         $Bi(s) + H_2O(l) + K_2SnO_3(aq)$

21. For each redox equation in Problem 20, identify the oxidizing agent and reducing agent.   20·6

22. Write balanced ionic equations for the following reactions which occur in acid solution. Use the half-reaction method.   20·7
a. $Sn^{2+}(aq) + Cr_2O_7^{2-}(aq) \longrightarrow$
                $Sn^{4+}(aq) + Cr^{3+}(aq)$
b. $CuS(s) + NO_3^-(aq) \longrightarrow$
       $Cu(NO_3)_2(aq) + NO_2(g) + SO_2(g)$
c. $I^-(aq) + NO_3^-(aq) \longrightarrow I_2(s) + NO(g)$

23. The following reactions take place in basic solution. Use the half-reaction method to write a balanced ionic equation for each.   20·7
a. $MnO_4^-(aq) + I^-(aq) \longrightarrow MnO_2(s) + I_2(aq)$
b. $NiO_2(s) + S_2O_3^{2-}(aq) \longrightarrow$
         $Ni(OH)_2(s) + SO_3^{2-}(aq)$
c. $Zn(s) + NO_3^-(aq) \longrightarrow$
         $NH_3(aq) + Zn(OH)_4^{2-}(aq)$

Mastery Questions and Problems

24. Balance the equations in Problem 20 by the most appropriate method.

25. Determine the oxidation number of phosphorus in each substance or ion.
a. PCl_5 **d.** P_4O_6
b. PO_4^{3-} **e.** $H_2PO_4^-$
c. P_4O_{10} **f.** PO_3^{3-}

26. Sodium chlorite is a powerful bleaching agent that is used in the paper and textile industries. It is prepared by this reaction.

$4NaOH(aq) + Ca(OH)_2(aq) + C(s) + 4ClO_2(g)$
       $\longrightarrow 4NaClO_2(aq) + CaCO_3(s) + 3H_2O(l)$

a. Identify the element oxidized in this reaction.
b. What is the oxidizing agent?

27. Identify the element oxidized, the element reduced, the oxidizing agent, and the reducing agent in each of these redox reactions.
a. $MnO_2 + HCl \longrightarrow MnCl_2 + Cl_2 + H_2O$
b. $Cu + HNO_3 \longrightarrow Cu(NO_3)_2 + NO_2 + H_2O$
c. $P + HNO_3 + H_2O \longrightarrow NO + H_3PO_4$
d. $Bi(OH)_3 + Na_2SnO_2 \longrightarrow$
           $Bi + Na_2SnO_3 + H_2O$
e. $NaCrO_2 + NaClO + NaOH \longrightarrow$
           $Na_2CrO_4 + NaCl + H_2O$
f. $V_2O_5 + KI + HCl \longrightarrow$
           $V_2O_4 + KCl + I_2 + H_2O$

28. Balance each of the redox equations shown in Problem 27 by using the oxidation-number change method.

29. The following oxidation-reduction reactions take place in basic solution. Use the half-reaction method to write a balanced ionic equation for each reaction.

a. $MnO_4^-(aq) + ClO_2^-(aq) \longrightarrow$
$$MnO_2(s) + ClO_4^-(aq)$$

b. $Cr^{3+}(aq) + ClO^-(aq) \longrightarrow$
$$CrO_4^{2-}(aq) + Cl^-(aq)$$

c. $Mn^{3+}(aq) + I^-(aq) \longrightarrow Mn^{2+}(aq) + IO_3^-(aq)$

Critical Thinking Questions

30. Choose the term that best completes the second relationship.

a. oxidation:reduction   fill up:_____
(1) empty   (3) dry out
(2) overflow   (4) splash

b. plants:oxygen   reducing agent:_____
(1) reduction   (3) electrons
(2) rust   (4) oxygen

c. umbrella:water   chromium:_____
(1) corrosion   (3) salts
(2) oxygen   (4) paint

31. Many decomposition, single-replacement, combination, and combustion reactions are also redox reactions. Why is a double replacement reaction never a redox reaction?

32. Many redox equations can be balanced using the techniques you learned in Chapter 7. What advantages over these techniques do the oxidation-number change method and the half-reaction method offer when balancing redox equations?

Review Questions and Problems

33. Calculate the pH of solutions with the following hydrogen-ion or hydroxide-ion concentrations.

a. $[H^+] = 0.00001M$
b. $[OH^-] = 1 \times 10^{-4}M$
c. $[OH^-] = 1 \times 10^{-1}M$
d. $[H^+] = 3 \times 10^{-7}M$

34. Classify each of the solutions in the previous problem as acidic, basic, or neutral.

35. Identify the conjugate acid–base pairs in each of these equations.

a. $NH_4^+ + H_2O \rightleftharpoons NH_3 + H_3O^+$
b. $H_2SO_3 + NH_2^- \rightleftharpoons HSO_3^- + NH_3$
c. $HNO_3 + I^- \rightleftharpoons HI + NO_3^-$
d. $H_2O + ClO_4^- \rightleftharpoons HClO_4 + OH^-$

36. How many milliliters of $4.00M$ KOH are needed to neutralize 45.0 mL of $2.5M$ H_2SO_4?

37. What is the hydrogen-ion concentration of solutions with the following pH? **a.** 2.0 **b.** 11.0 **c.** 8.8

38. What is the normality of the solution prepared by dissolving 46.4 g of H_3PO_4 in enough water to make 1.25 L of solution?

39. The K_{sp} of lead(II) bromide at 25°C is 2.1×10^{-6}. What is the solubility of $PbBr_2$ in mol/L at this temperature?

40. How would you make 440 mL of $1.5M$ hydrochloric acid from a stock solution of $6.0M$ HCl?

41. Bottles containing $0.1M$ solutions of Na_2SO_4, $BaCl_2$, and NaCl have had their labels accidentally switched around. To discover which bottle contains the NaCl, you set up the following test. You place a clear saturated solution of $BaSO_4$ ($K_{sp} = 1.1 \times 10^{-10}$) into each of three test tubes. Then you add a few drops of each mislabeled solution to a different test tube. The results are shown below. To which tube was NaCl added? Explain your reasoning.

A   B   C

42. The complete combustion of hydrocarbons involves the oxidation of both carbon and hydrogen atoms. The carbon combines with oxygen to form carbon dioxide, and the hydrogen combines with oxygen to form water. The following table lists the moles of O_2 used and the moles of CO_2 and moles of H_2O produced when a series of hydrocarbons called alkanes are burned.

a. Complete the table.

b. Based on the data write a balanced generalized equation for the complete combustion of any alkane. Use the following form, in which the coefficients are written in terms of x and y:

$$C_xH_y + \_O_2 \rightarrow \_CO_2 + \_H_2O$$

Alkane burned	O_2 used (mol)	CO_2 produced (mol)	H_2O produced (mol)
CH_4	2	1	2
C_2H_6	3.5	2	3
C_3H_8	5	3	4
C_4H_{10}	___	___	___
C_5H_{12}	___	___	___
C_6H_{14}	___	___	___

Challenging Questions and Problems

43. How many grams of copper are needed to completely reduce the silver in 85.0 mL of $0.150M$ $AgNO_3(aq)$?

44. How many milliliters of $0.280M$ $K_2Cr_2O_7(aq)$ are needed to reduce 1.40 g of sulfur? First balance the equation.

$$K_2Cr_2O_7(aq) + H_2O(l) + S(s) \longrightarrow$$
$$SO_2(g) + KOH(aq) + Cr_2O_3(aq)$$

45. Carbon monoxide can be removed from the air by passing it over solid diiodine pentoxide.

$$CO(g) + I_2O_5(s) \longrightarrow I_2(s) + CO_2(g)$$

a. Balance the equation.

b. Identify the element being oxidized and the element being reduced.

c. How many grams of carbon monoxide can be removed from the air by 0.55 g of I_2O_5?

Research Projects

1. What did George Ernst Stahl contribute to corrosion science? Why did his phlogiston theory fail to account for mass changes in rusting?

2. What techniques are used to prevent corrosion in industrial pipes carrying gas, oil, or water?

3. Investigate the effect different dissolved solids have on the rate of rusting of iron nails in water.

4. When is metal corrosion caused by a liquid a problem? What can be done to prevent it?

5. What causes corrosion fatigue? When is it a problem?

6. What is hydrogen embrittlement?

7. Describe the oxidation–reduction reactions involved in rotting fruit.

8. Trace the development of rocket fuels. Why was the change from solid to liquid fuels significant?

9. What substances are used as flame retardants for cloth? How do they prevent cloth from burning?

10. Trace the history of photography beginning with the work of Joseph Niepce and Louis Daguerre.

11. Can the rusting of iron in a salt water solution be inhibited by the presence of other metals in the solution? Does it make any difference if the iron and other metal are touching? Devise an experiment to answer these questions.

Readings and References

Conkling, John A. "Chemistry of Fireworks." *Chemical and Engineering News* (June 29, 1981), pp. 24–32.

Hedgecoe, John. *The Photographer's Handbook*, 2nd ed. New York: Knopf, 1982.

Horenstein, Henry. *Black and White Photography; a Basic Manual*, 2nd ed. Boston: Little, Brown, 1983.

Light and Film. Alexandria, VA: Time-Life, 1981.

Low, Betty-Bright. "Pyrotechnic Paeans Have Been Flying High for 600 Years." *Smithsonian* (July 1980), pp. 84–92.

Robson, David P. "Sunken Treasure." *ChemMatters* (April 1987), pp. 4–9.

21 Electrochemistry

Chapter Preview

Electrochemistry has many applications in the home as well as in industry. Flashlight and automobile batteries are familiar examples. Others are the manufacture of sodium and aluminum metals and the silver plating of tableware. Biological systems also use electrochemistry to carry out nerve-impulse conduction. This chapter starts with a discussion of the relationship between redox reactions and electrochemistry.

21·1 Electrochemical Processes

The conversion of chemical energy into electrical energy and the conversion of electrical energy into chemical energy are **electrochemical processes.** All electrochemical processes involve redox reactions. The conversion of chemical energy into electrical energy will be discussed first.

When a strip of zinc metal is dipped into an aqueous solution of copper sulfate, the zinc becomes copper-plated (Figure 21·2). The net ionic equation involves only zinc and copper.

$$Zn(s) + Cu^{2+}(aq) \longrightarrow Zn^{2+}(aq) + Cu(s)$$

Electrons are transferred from zinc atoms to copper ions. This is a *redox reaction* and it occurs *spontaneously*. As the reaction proceeds, zinc atoms lose electrons as they are oxidized to zinc ions. The zinc metal slowly dissolves. At the same time, copper ions in solution gain electrons. They are reduced to copper atoms and deposit as metallic copper. As the copper ions are gradually replaced by zinc ions, the blue color of the solution fades. Balanced half-reactions for this redox reaction can be written as follows.

■ Redox reactions allow chemical energy to be converted into electrical energy and electrical energy into chemical energy.

In electrochemical reactions metal atoms are oxidized by ions of other metals.

Figure 21·1
Electroplating utilizes electrochemical processes to protect and beautify metal surfaces.

a

b

Figure 21·2
a A spontaneous redox reaction occurs when a zinc strip is immersed in a solution of copper(II) sulfate. **b** If the zinc remains in solution for a long time, the blue color of the solution fades and the zinc strip is badly corroded. What substance is oxidized? What is reduced?

Oxidation: $Zn(s) \longrightarrow Zn^{2+}(aq) + 2e^-$

Reduction: $Cu^{2+}(aq) + 2e^- \longrightarrow Cu(s)$

If you look at the activity series of metals (Table 21·1), you will see that zinc is above copper. This table is very useful in electrochemistry. For any two metals in the table, the metal that is the higher of the two is the most readily oxidized. As Figure 21·2 shows, when zinc is dipped in a copper sulfate solution the zinc becomes copper-plated. By contrast, when a copper rod is dipped into a solution of zinc sulfate, the copper does not become zinc-plated. This is because copper metal is not oxidized by zinc ions. Zinc is above copper in the activity series. Zinc plating on copper is a nonspontaneous process.

An electric current is a flow of electrons. When a zinc rod is dipped into a copper sulfate solution, electrons are transferred from zinc metal to copper ions. If a redox reaction is to be used as a source of electrical energy, however, the two half-reactions must be physically separated. That is, the electrons released by zinc must pass through an external circuit to reach the copper ions. Electrical energy is produced in an electrochemical cell. Alternatively, an electric current can also be used to produce a chemical change. *An* **electrochemical cell** *is any device that converts chemical energy into electrical energy or electrical energy into chemical energy.* Redox reactions occur in electrochemical cells.

Table 21·1	Activity Series of Metals, with Half-Reactions for Oxidation Processes		
	Element	Oxidation half-reactions	
Most active and most easily oxidized	Lithium	$Li(s) \longrightarrow Li^+(aq) + e^-$	
	Potassium	$K(s) \longrightarrow K^+(aq) + e^-$	
	Barium	$Ba(s) \longrightarrow Ba^{2+}(aq) + 2e^-$	
	Calcium	$Ca(s) \longrightarrow Ca^{2+}(aq) + 2e^-$	
	Sodium	$Na(s) \longrightarrow Na^+(aq) + e^-$	
	Magnesium	$Mg(s) \longrightarrow Mg^{2+}(aq) + 2e^-$	
	Aluminum	$Al(s) \longrightarrow Al^{3+}(aq) + 3e^-$	
	Zinc	$Zn(s) \longrightarrow Zn^{2+}(aq) + 2e^-$	
	Iron	$Fe(s) \longrightarrow Fe^{2+}(aq) + 2e^-$	
	Nickel	$Ni(s) \longrightarrow Ni^{2+}(aq) + 2e^-$	
	Tin	$Sn(s) \longrightarrow Sn^{2+}(aq) + 2e^-$	
	Lead	$Pb(s) \longrightarrow Pb^{2+}(aq) + 2e^-$	
	Hydrogen*	$H_2(g) \longrightarrow 2H^+(aq) + 2e^-$	
	Copper	$Cu(s) \longrightarrow Cu^{2+}(aq) + 2e^-$	
Least active and least easily oxidized	Mercury	$Hg(s) \longrightarrow Hg^{2+}(aq) + 2e^-$	
	Silver	$Ag(s) \longrightarrow Ag^+(aq) + e^-$	
	Gold	$Au(s) \longrightarrow Au^{3+}(aq) + 3e^-$	

*Hydrogen is included for reference purposes.

21·A Michael Faraday

Unquenchable enthusiasm for reading and the habit of taking careful notes launched the career of one of the greatest chemists of all time. Michael Faraday (1791–1867) was the son of a poor English blacksmith. He had little formal education and was apprenticed to a bookbinder when he was 14. Faraday first learned about the phenomenon of electricity from an article in an encyclopedia that was brought to his employer for rebinding. His interest in science was born, and he pursued it against all odds. He joined a group of young men in London who attended lectures on science. He persuaded one of this group to help him improve his writing to make up for his lack of schooling. He read everything he could find on electricity and science. Furthermore, he wrote complete notes on every book he read and every lecture he heard.

Eventually Faraday won a job at the prestigious Royal Institution by sending in the notes he had taken at one of the director's lectures. The job was menial, but Faraday advanced quickly for he soon became known for his careful laboratory work. He went on to become the principal lecturer at the Institution. In this role he did much to popularize science among the English upper class.

Faraday began working in analytical chemistry. He discovered benzene in 1825 and was the first person to produce compounds of carbon and chlorine in the laboratory. He is most famous for his work with electricity. Faraday gave the world electricity when he discovered that an electric current was induced in a coil of wire rotating in a magnetic field. He went on to prove that electricity generated in this way is identical to electricity produced by an electrochemical

Figure 21·3
Faraday began his work at the Royal Institution under Sir Humphrey Davy. In 1807 and 1808 Davy was the first to isolate several alkali metals. This was accomplished with a "trough battery." Troughs of baked wood held partitions of copper and zinc plates, soldered in pairs, with ammonium chloride solution filling the compartments between the plates.

ChemDirections

Solar cells are photovoltaic cells. They convert radiant energy of the sun into electrical energy as described in **Chem-Directions** *Solar Cells*, page 655.

Voltaic cells produce electrical energy from redox reactions.

The voltaic cell is also called a galvanic cell after Luigi Galvani (1737–1798), an Italian anatomist who discovered that the contraction of muscles is caused by electrochemical processes.

Figure 21·4
A voltaic cell can be constructed without a salt bridge. It will not function long, however, because one electrode will become plated.

cell. He performed many experiments measuring the changes that take place in such cells. His work established these laws of electrochemistry.

1. The amount of material decomposed or synthesized in an electrochemical cell is proportional to the amount of electricity that passes through the cell.

2. The amounts of substances deposited and dissolved in such a cell are proportional to their molecular masses.

Other scientists realized that these results showed that electricity must be made of particles. The amount of charge carried by one mole of electrons is called a faraday in honor of this great chemist.

21·2 Voltaic Cells

The first electrochemical cell was invented by the Italian physicist Alessandro Volta (1745–1827). He designed a cell that can be used to generate a direct electric current (DC current). **Voltaic cells** *are electrochemical cells that are used to convert chemical energy into electrical energy*. The energy is produced by *spontaneous* redox reactions.

A **half-cell** *is one part of a voltaic cell in which either oxidation or reduction occurs*. A half-cell consists of a metal rod or strip immersed in a solution of its ions. In a typical voltaic cell, one half-cell is a zinc rod immersed in a solution of zinc sulfate. The other half-cell is a copper rod immersed in a solution of copper sulfate. The two half-cells are separated by a porous partition. A **salt bridge,** *which is a tube containing a conducting solution,* may also be used. These dividers allow the passage of ions from one compartment to the other but prevent the solutions from mixing completely. A wire carries the electrons in the external circuit from zinc to copper. A voltmeter or light bulb can be connected in the circuit. The driving force of such a voltaic cell is the spontaneous redox reaction between zinc metal and copper(II) ions in solution.

The zinc and copper rods in a voltaic cell are the electrodes. *An* **electrode** *is a conductor in a circuit that carries electrons to or from a substance other than a metal*. The type of electrode reaction determines whether an electrode is labeled as a cathode or an anode. Note that the name of the electrode is *not* determined by its charge. *The electrode at which oxidation occurs is the* **anode.** Electrons are produced here. Thus the anode is labeled the *negative* electrode. *The electrode at which reduction occurs is the* **cathode.** Electrons are consumed at the cathode. As a result, it is labeled the *positive* electrode. Neither electrode is really charged. All parts of the voltaic cell remain balanced in terms of charge at all times. The moving electrons balance any charge that might build up as oxidation and reduction occur.

The electrochemical process that occurs in a zinc–copper voltaic cell can best be described in a number of steps. These steps all occur at the same time.

Figure 21·5
In all electrochemical cells, oxidation occurs at the anode and reduction occurs at the cathode. In this voltaic cell which electrode is positive and which is negative?
The anode is negative and the cathode is positive.

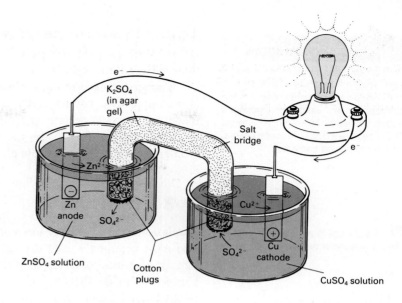

Safety

Even a low voltage source can have a dangerously high current if the resistance is very low. Treat all sources of electricity with care.

1. Electrons are produced at the zinc rod according to the *oxidation* half-reaction.

$$Zn(s) \longrightarrow Zn^{2+}(aq) + 2e^-$$

Because it is oxidized, the zinc rod is the anode, or negative electrode.

2. The electrons leave the zinc anode and pass through the external circuit to the copper rod. (If a bulb is in the circuit, it will light. If a voltmeter is present, it will indicate a voltage.)

3. Electrons enter the copper rod and pass to copper ions in solution. There the following *reduction* half-reaction occurs.

$$Cu^{2+}(aq) + 2e^- \longrightarrow Cu(s)$$

4. At the copper rod, copper ions are reduced. The copper rod is the cathode, or positive electrode. To complete the circuit, ions (both positive and negative) move through the aqueous solutions via the salt bridge. The two half-reactions can be summed to get the overall cell reaction. The electrons must cancel.

$$Zn(s) + Cu^{2+}(aq) \longrightarrow Zn^{2+}(aq) + Cu(s)$$

Chemists use a shorthand method to represent an electrochemical cell. The zinc–copper voltaic cell can be written as follows.

$$Zn(s) \mid ZnSO_4(aq) \parallel CuSO_4(aq) \mid Cu(s)$$

Oxidation and reduction reactions always accompany each other. An oxidation half-cell cannot function apart from a reduction half-cell.

The half-cell which undergoes oxidation (the anode) is written first at the left. The double vertical lines represent the salt bridge or porous partition that separates the anode compartment from the cathode compartment. The single vertical lines indicate boundaries of phases that are in contact. For example, the zinc rod, $Zn(s)$, and the zinc sulfate solution, $ZnSO_4(aq)$, are separate phases in physical contact. When the solutions in the voltaic cells are both $1.0M$, the cell generates an electrical

The *dry* cell is a voltaic cell that is filled with a *moist* paste.

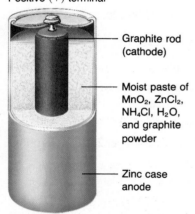

Positive (+) terminal

Graphite rod (cathode)

Moist paste of MnO_2, $ZnCl_2$, NH_4Cl, H_2O, and graphite powder

Zinc case anode

Negative (−) terminal (bottom of case)

Figure 21·6
A flashlight battery is a single electrochemical cell. It produces about 1.5 V. What is oxidized in this cell and what is reduced?

A 12-volt lead storage battery is composed of six voltaic cells connected in series.

potential of 1.10 volts (1.10 V). If different metals are used for the electrodes in the cell, the voltage will change. Different solution concentrations will also affect the voltage.

The zinc–copper voltaic cell is of great historical importance, but it is no longer used commercially. Nevertheless, this cell is a convenient model to use when describing the production of electrical energy from a chemical change. Today, the more practical and compact "dry" cell is usually chosen when a portable electrical energy source is required.

21·3 Dry Cells

A voltaic cell in which the electrolyte is a paste is known as a **dry cell.** The most commonly used dry cell is usually referred to as a flashlight battery. A zinc container is filled with a thick moist paste of manganese(IV) oxide (MnO_2), zinc chloride ($ZnCl_2$), ammonium chloride (NH_4Cl), and water. A graphite rod is embedded in the paste (Figure 21·6). The zinc container is the anode and the graphite rod is the cathode. The thick paste, surrounded by a paper liner, prevents the contents of the cell from freely mixing so that a salt bridge is not needed. The half-reactions for this cell are as follows.

Oxidation: $Zn(s) \longrightarrow Zn^{2+}(aq) + 2e^-$ (anode reaction)

Reduction: $2MnO_2(s) + 2NH_4^+(aq) + 2e^- \longrightarrow$

$$Mn_2O_3(s) + 2NH_3(aq) + H_2O(l) \text{ (cathode reaction)}$$

In this cell the graphite rod serves only as a conductor and does not undergo reduction, even though it is the cathode. The manganese in MnO_2 is the species that is reduced. The electrical potential of this cell usually begins at 1.5 V but falls steadily during use to about 0.8 V. Dry cells of this construction are not rechargeable. A voltaic cell that is rechargeable is the lead storage battery.

21·4 Lead Storage Batteries

A **battery** *is a group of cells that are connected together.* Lead storage batteries are used in automobiles. A 12-V storage battery consists of six voltaic cells connected together (Figure 21·7). Each cell, which produces about 2 V, contains two lead electrodes or grids. One of the grids, the anode, is packed with spongy lead. The other grid, the cathode, is packed with lead(IV) oxide, PbO_2. The grids are immersed in strong sulfuric acid and are separated by a perforated plate. The half-reactions are as follows.

Oxidation: $Pb + SO_4^{2-} \longrightarrow PbSO_4 + 2e^-$

Reduction: $PbO_2 + 4H^+ + SO_4^{2-} + 2e^- \longrightarrow PbSO_4 + 2H_2O$

When a lead storage battery discharges, it produces the electric power to start the car. The overall spontaneous redox reaction is the sum of the oxidation and reduction half-reactions.

Figure 21·7
A 12 V lead storage battery consists of six 2-V cells in series. The cells do not need to be in separate compartments, but this improves performance. While the battery is producing a current, lead at the anode and lead(IV) oxide at the cathode are both converted to lead sulfate. This decreases the sulfuric acid concentration in the battery. Recharging the battery reverses these reactions. This occurs whenever the engine in a car is running.

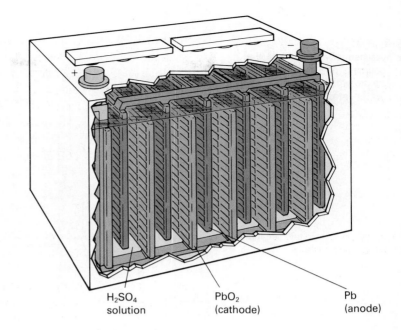

H$_2$SO$_4$ solution   PbO$_2$ (cathode)   Pb (anode)

Safety

The sulfuric acid in a lead storage battery is about 5M and is very corrosive. Handle with care.

$$Pb(s) \ + \ PbO_2(s) \ + \ 2H_2SO_4(aq) \longrightarrow \ 2PbSO_4(s) \ + \ 2H_2O(l)$$

This reaction shows that during discharge, lead sulfate slowly builds up on the plates, and the concentration of sulfuric acid decreases. The condition of a lead storage battery can be checked by measuring the specific gravity of the electrolyte, sulfuric acid. If it is much below 1.25, the battery should be recharged.

When a storage battery is recharged, the reverse reaction occurs.

$$2PbSO_4(s) \ + \ 2H_2O(l) \longrightarrow \ Pb(s) \ + \ PbO_2(s) \ + \ 2H_2SO_4(aq)$$

This reaction is *nonspontaneous*. To make it go, a direct current must be passed through the cell in the reverse direction. In theory, a lead storage battery can be recharged indefinitely, but in practice its life span is limited. This is because small amounts of lead sulfate continually fall from the electrodes and collect on the bottom of the cell. Eventually the electrodes lose so much lead sulfate that the recharging process is ineffective or the cell is shorted out. To help overcome this problem, fuel cells with renewable electrodes have been developed in recent years.

21·5 Fuel Cells

A voltaic cell in which a fuel is oxidized is known as a fuel cell.

Fuel Cells *are voltaic cells in which a fuel substance undergoes oxidation and from which electrical energy is obtained continuously.* Fuel cells do not have to be recharged. They can be designed to emit no air pollutants and to operate more quietly and more cheaply than a conventional electrical generator.

Perhaps the simplest fuel cell to visualize is the hydrogen–oxygen fuel cell. In this cell three compartments are separated from one another by two electrodes made of porous carbon (Figure 21·8). Oxygen (the oxidizer) is fed into the cathode compartment. Hydrogen (the fuel) is fed

Figure 21·8
The hydrogen–oxygen fuel cell is a clean source of power. What "waste" products are produced?
Pure water (steam).

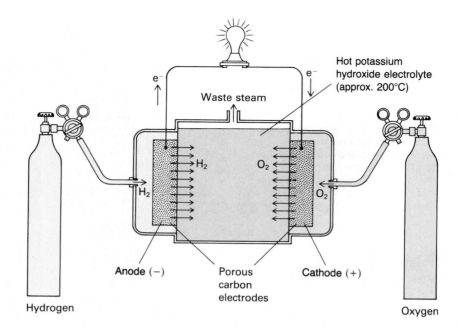

into the anode compartment. The gases diffuse slowly through the electrodes. The electrolyte in the central compartment is a hot concentrated solution of potassium hydroxide. Electrons from the oxidation reaction at the anode pass through an external circuit to enter the reduction reaction at the cathode.

Oxidation: $2H_2(g) + 4OH^-(aq) \longrightarrow 4H_2O(l) + 4e^-$ (anode)

Reduction: $O_2(g) + 2H_2O(l) + 4e^- \longrightarrow 4OH^-(aq)$ (cathode)

The equation for the overall reaction is the oxidation of hydrogen to form water.
$$2H_2(g) + O_2(g) \longrightarrow 2H_2O(l)$$

Other fuels, such as methane, CH_4, and ammonia, NH_3, can be used in place of hydrogen. Other oxidizers, such as chlorine, Cl_2, and ozone, O_3, can be used in place of oxygen. Fuel cells are currently being built as auxiliary power sources for submarines and other military vehicles. At present, however, they are too expensive for general use. Hydrogen–oxygen fuel cells, weighing approximately 100 kg each, were used in the Apollo spacecraft missions. Instead of producing toxic wastes like an internal combustion engine, these cells produce drinkable water.

21·6 Half-Cells

The **electrical potential** *of a voltaic cell is the ability of the cell to produce an electric current.* Electrical potential is usually measured in volts (V). The potential of an isolated half-cell cannot be measured. For example, we cannot measure the electrical potential of a zinc half-cell or of a copper half-cell separately. When these two half-cells are connected to form a voltaic cell, however, the difference in potential can be measured. The electrical potential of a $1.0M$ zinc–copper cell is $+1.10$ V.

The electrical potential of a cell results from a competition between the two half-cells for electrons. The half-cell with the greater tendency to acquire electrons will do so and will undergo reduction. The other half-cell loses electrons and is oxidized. The half-cell in which reduction occurs has a greater reduction potential than the half-cell in which oxidation occurs. *The **reduction potential** of a half-cell is a measure of the tendency of a given half-reaction to occur as a reduction. The difference between the reduction potentials of the two half-cells is called the* **cell potential.**

$$\text{Cell potential} = \left(\begin{array}{c} \text{reduction potential} \\ \text{of half-cell in which} \\ \text{reduction occurs} \end{array} \right) - \left(\begin{array}{c} \text{reduction potential} \\ \text{of half-cell in which} \\ \text{oxidation occurs} \end{array} \right)$$

or $\quad E^0_{cell} = E^0_{red} - E^0_{oxid}$

*The **standard cell potential*** (E^0_{cell}) *is the measured cell potential when the ion concentrations in the half-cells are 1.00M, gases are at a pressure of 1 atm, and the temperature is 25°C.* The symbols E^0_{red} and E^0_{oxid} are the standard reduction potentials for the reduction and oxidation half-cells respectively. Because half-cell potentials cannot be measured, scientists have chosen an arbitrary reference electrode. *The **standard hydrogen electrode** is used with other electrodes so that the reduction potentials of those cells can be measured.* The standard reduction potential of the hydrogen electrode has been assigned a value of 0.00 V.

The standard hydrogen electrode (Figure 21·9) consists of a platinum electrode immersed in a solution with a hydrogen ion concentration of 1.00M. The solution is at 25°C. The platinum electrode is a small square of platinum foil coated with finely divided platinum (known as *platinum black*). Hydrogen gas, at a pressure of 1 atm, is bubbled around the platinum electrode. The half-cell reaction that occurs at the platinum black surface is as follows.

$$2H^+(aq, 1M) + 2e^- \rightleftharpoons H_2(g, 1\text{ atm}) \quad E^0_{H^+} = 0.00 \text{ V.}$$

The symbol $E^0_{H^+}$ represents the standard reduction potential of H^+. The double arrows mean that the reaction is reversible. Whether this half-cell reaction occurs as a reduction or as an oxidation is determined by the reduction potential of the half-cell to which it is connected.

21·7　Standard Reduction Potentials

A voltaic cell can be constructed by connecting a standard hydrogen half-cell to a standard zinc half-cell (Figure 21·10). The first step in determining the overall reaction for this cell is to identify in which half-cell reduction takes place. Reduction takes place at the cathode and oxidation takes place at the anode in all electrochemical cells. A voltmeter gives a reading of +0.74 V when the zinc electrode is connected to the negative terminal and the hydrogen electrode is connected to the positive terminal. Thus, the zinc electrode must be the anode. Zinc is oxidized. The hydrogen electrode must be the cathode. Hydrogen ions are reduced. The half-reactions and cell reaction can now be written.

A half-cell with a greater tendency to acquire electrons (than another half-cell) has the greater reduction potential.

Standard reduction potentials are measured at 25°C when the concentrations are 1M and the pressure is 1 atm.

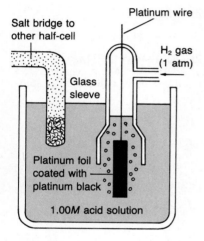

Figure 21·9
The standard hydrogen electrode is arbitrarily assigned a standard reduction potential of 0.00 V at 25°C.

Safety

When running an electrochemical cell, always consider whether any of the products is a flammable or toxic gas.

Figure 21·10
This voltaic cell consists of zinc and hydrogen half-cells. What is the measured cell potential?

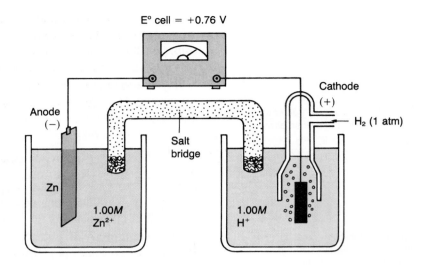

Oxidation: $Zn(s) \longrightarrow Zn^{2+}(aq) + 2e^-$

Reduction: $2H^+(aq) + 2e^- \longrightarrow H_2(g)$

Cell reaction: $Zn(s) + 2H^+(aq) \longrightarrow Zn^{2+}(aq) + H_2(g)$

The use of a standard hydrogen electrode allows us to calculate the standard reduction potential for the zinc half-cell.

$$E^0_{cell} = E^0_{red} - E^0_{oxid}$$

$$E^0_{cell} = E^0_{H^+} - E^0_{Zn^{2+}}$$

The cell potential was measured: $E^0_{cell} = +0.76$ V. The reduction potential of the hydrogen half-cell is a defined standard: $E^0_{H^+} = 0.00$ V.

$$+0.76 \text{ V} = 0.00 \text{ V} - E^0_{Zn^{2+}}$$

$$E^0_{Zn^{2+}} = -0.76 \text{ V}$$

The standard reduction potential for the zinc half-cell is -0.76 V. The value is negative because the tendency for zinc ions to be reduced in this cell is less than that of hydrogen ions (H^+). Consequently, zinc metal is oxidized.

Many different half-cells can be paired with the hydrogen half-cell in a similar manner. The standard reduction potential for each half-cell can be obtained. With a standard copper half-cell, for example, the measured standard cell potential is $+0.34$ V. Copper is the cathode (the positive electrode) and Cu^{2+} is reduced to Cu when the cell operates. The hydrogen half-cell is the anode (the negative electrode) and H_2 is oxidized to H^+. The standard reduction potential for copper is calculated as follows.

$$E^0_{cell} = E^0_{red} - E^0_{oxid}$$

$$= E^0_{Cu^{2+}} - E^0_{H^+}$$

$$+0.34 \text{ V} = E^0_{Cu^{2+}} - 0.00$$

$$E^0_{Cu^{2+}} = +0.34 \text{ V}$$

The reduction potential of a half-cell can be calculated even though a half-cell cannot operate independently.

These calculations show that the standard reduction potential of copper is +0.34 V. The value is positive because the tendency for copper ions to be reduced in this cell is greater than that of hydrogen ions.

Table 21·2 lists some standard reduction potentials at 25°C. These are arranged in increasing order. The half-reactions at the top of the table have the least tendency to occur as reductions. The half-reactions at the bottom of the table have the greatest tendency to occur as reductions.

21·8 Calculating Cell Potentials

In order to function, a cell must be constructed of two half-cells. The half-cell reaction having the *more positive* reduction potential occurs as a reduction in the cell. With this in mind it is possible to calculate cell potentials and write cell reactions for cells before they have been assembled. Standard reduction potentials for the various half-cells are used for this purpose.

Example 1

Calculate the standard cell potential of a cell composed of the half-cells $Zn \mid Zn^{2+}$ and $Cu \mid Cu^{2+}$. Write the half-cell reactions for the anode and cathode processes and the cell reaction.

Solution

Table 21·2 gives the following standard reaction potentials.

$$Zn^{2+}(aq) + 2e^- \longrightarrow Zn(s) \qquad E^0_{Zn^{2+}} = -0.76 \text{ V}$$

$$Cu^{2+}(aq) + 2e^- \longrightarrow Cu(s) \qquad E^0_{Cu^{2+}} = +0.34 \text{ V}$$

Because Cu^{2+} has the larger reduction potential, it occurs as a reduction in the cell. The zinc is oxidized.

The standard cell potential is the reduction potential of the reduction half-cell minus the reduction potential of the oxidation half-cell.

$$E^0_{cell} = E^0_{red} - E^0_{oxid}$$

$$= E^0_{Cu^{2+}} - E^0_{Zn^{2+}}$$

$$= +0.34 \text{ V} - (-0.76 \text{ V})$$

$$= +1.10 \text{ V}$$

In this cell Zn is oxidized at the anode and Cu^{2+} is reduced at the cathode. The standard cell potential is 1.10 V. The half-reactions are as follows.

Oxidation: $Zn(s) \longrightarrow Zn^{2+}(aq) + 2e^-$ (anode)

Reduction: $Cu^{2+}(aq) + 2e^- \longrightarrow Cu(s)$ (cathode)

The cell reaction is the sum of the half-reactions.

$$Cu^{2+}(aq) + Zn(s) \longrightarrow Cu(s) + Zn^{2+}(aq)$$

■ When two half-cells are connected, the one with the larger reduction potential acquires electrons from the half-cell with the smaller reduction potential.

Table 21·2   Reduction Potentials at 25°C with 1M Concentrations of Aqueous Species

Electrode	Half-Reaction	E^0 (V)
Li^+/Li	$Li^+ + e^- \longrightarrow Li$	-3.05
K^+/K	$K^+ + e^- \longrightarrow K$	-2.93
Ba^{2+}/Ba	$Ba^{2+} + 2e^- \longrightarrow Ba$	-2.90
Ca^{2+}/Ca	$Ca^{2+} + 2e^- \longrightarrow Ca$	-2.87
Na^+/Na	$Na^+ + e^- \longrightarrow Na$	-2.71
Mg^{2+}/Mg	$Mg^{2+} + 2e^- \longrightarrow Mg$	-2.37
Al^{3+}/Al	$Al^{3+} + 3e^- \longrightarrow Al$	-1.66
H_2O/H_2	$2H_2O + 2e^- \longrightarrow H_2 + 2OH^-$	-0.83
Zn^{2+}/Zn	$Zn^{2+} + 2e^- \longrightarrow Zn$	-0.76
Cr^{3+}/Cr	$Cr^{3+} + 3e^- \longrightarrow Cr$	-0.74
Fe^{2+}/Fe	$Fe^{2+} + 2e^- \longrightarrow Fe$	-0.44
H_2O/H_2 (pH 7)	$2H_2O + 2e^- \longrightarrow H_2 + 2OH^-$	-0.42
Cd^{2+}/Cd	$Cd^{2+} + 2e^- \longrightarrow Cd$	-0.40
$PbSO_4/Pb$	$PbSO_4 + 2e^- \longrightarrow Pb + SO_4^{2-}$	-0.36
Co^{2+}/Co	$Co^{2+} + 2e^- \longrightarrow Co$	-0.28
Ni^{2+}/Ni	$Ni^{2+} + 2e^- \longrightarrow Ni$	-0.25
Sn^{2+}/Sn	$Sn^{2+} + 2e^- \longrightarrow Sn$	-0.14
Pb^{2+}/Pb	$Pb^{2+} + 2e^- \longrightarrow Pb$	-0.13
Fe^{3+}/Fe	$Fe^{3+} + 3e^- \longrightarrow Fe$	-0.036
H^+/H_2	$2H^+ + 2e^- \longrightarrow H_2$	0.000
$AgCl/Ag$	$AgCl + e^- \longrightarrow Ag + Cl^-$	$+0.22$
Hg_2Cl_2/Hg	$Hg_2Cl_2 + 2e^- \longrightarrow 2Hg + 2Cl^-$	$+0.27$
Cu^{2+}/Cu	$Cu^{2+} + 2e^- \longrightarrow Cu$	$+0.34$
O_2/OH^-	$O_2 + 2H_2O + 4e^- \longrightarrow 4OH^-$	$+0.40$
Cu^+/Cu	$Cu^+ + e^- \longrightarrow Cu$	$+0.52$
I_2/I^-	$I_2 + 2e^- \longrightarrow 2I^-$	$+0.54$
Fe^{3+}/Fe^{2+}	$Fe^{3+} + e^- \longrightarrow Fe^{2+}$	$+0.77$
Hg_2^{2+}/Hg	$Hg_2^{2+} + 2e^- \longrightarrow 2Hg$	$+0.79$
Ag^+/Ag	$Ag^+ + e^- \longrightarrow Ag$	$+0.80$
O_2/H_2O (pH 7)	$O_2 + 4H^+ + 4e^- \longrightarrow 2H_2O$	$+0.82$
Hg^{2+}/Hg	$Hg^{2+} + 2e^- \longrightarrow Hg$	$+0.85$
Br_2/Br^-	$Br_2 + 2e^- \longrightarrow 2Br^-$	$+1.07$
O_2/H_2O	$O_2 + 4H^+ + 4e^- \longrightarrow 2H_2O$	$+1.23$
MnO_2/Mn^{2+}	$MnO_2 + 4H^+ + 2e^- \longrightarrow Mn^{2+} + 2H_2O$	$+1.28$
$Cr_2O_7^{2-}/Cr^{3+}$	$Cr_2O_7^{2-} + 14H^+ + 6e^- \longrightarrow 2Cr^{3+} + 7H_2O$	$+1.33$
Cl_2/Cl^-	$Cl_2 + 2e^- \longrightarrow 2Cl^-$	$+1.36$
PbO_2/Pb^{2+}	$PbO_2 + 4H^+ + 2e^- \longrightarrow Pb^{2+} + 2H_2O$	$+1.46$
MnO_4^-/Mn^{2+}	$MnO_4^- + 8H^+ + 5e^- \longrightarrow Mn^{2+} + 4H_2O$	$+1.51$
$PbO_2/PbSO_4$	$PbO_2 + 4H^+ + SO_4^{2-} + 2e^- \longrightarrow PbSO_4 + 2H_2O$	$+1.69$
F_2/F^-	$F_2 + 2e^- \longrightarrow 2F^-$	$+2.87$

Example 2

What is the cell reaction and the standard cell potential for a voltaic cell composed of the following half-cells?

$$Fe^{3+}(aq) + e^- \longrightarrow Fe^{2+}(aq) \quad E^0_{Fe^{3+}} = +0.77 \text{ V}$$

$$Ni^{2+}(aq) + 2e^- \longrightarrow Ni(s) \quad E^0_{Ni^{2+}} = -0.25 \text{ V}$$

Which half-cell is the cathode?

Solution

The half-cell with the more positive reduction potential occurs as a reduction. In this cell Fe^{3+} is reduced and Ni is oxidized. Because reduction takes place at the Fe^{3+} half-cell, this half-cell is the cathode. The half-cell reactions are as follows.

$$\text{Oxidation: } Ni(s) \longrightarrow Ni^{2+}(aq) + 2e^- \text{ (anode)}$$

$$\text{Reduction: } Fe^{3+}(aq) + e^- \longrightarrow Fe^{2+}(aq) \text{ (cathode)}$$

Before the half-reactions are added, care must be taken that the electrons cancel.

$$Ni(s) \longrightarrow Ni^{2+}(aq) + 2e^-$$
$$\underline{2[Fe^{3+}(aq) + e^- \longrightarrow Fe^{2+}(aq)]}$$
$$Ni(s) + 2Fe^{3+}(aq) \longrightarrow Ni^{2+}(aq) + 2Fe^{2+}(aq)$$

The standard cell potential can now be calculated.

$$E^0_{cell} = E^0_{red} - E^0_{oxid}$$
$$= E^0_{Fe^{3+}} - E^0_{Ni^{2+}}$$
$$= +0.77 \text{ V} - (-0.25 \text{ V})$$
$$= +1.02 \text{ V}$$

Note that the E^0 of a half-cell is not multiplied by any number even if one of the equations was multiplied by a coefficient to make the electrons cancel.

Problems

1. A voltaic cell is constructed using electrodes with the following half-reactions.

$$Cu^{2+}(aq) + 2e^- \longrightarrow Cu(s) \quad E^0_{Cu^{2+}} = +0.34 \text{ V}$$

$$Al^{3+}(aq) + 3e^- \longrightarrow Al(s) \quad E^0_{Al^{3+}} = -1.66 \text{ V}$$

What is the overall cell reaction and the standard cell potential?

2. From these half-reactions determine the cell reaction and the standard cell potential.

$$Ag^+(aq) + e^- \longrightarrow Ag(s) \quad E^0_{Ag^+} = +0.80 \text{ V}$$

$$Cu^{2+}(aq) + 2e^- \longrightarrow Cu(s) \quad E^0_{Cu^{2+}} = +0.34 \text{ V}$$

Issues in Chemistry

The sound of a whirling drill may put you on edge when visiting your dentist. Do you look forward to the day when cavities can be fixed without having the decayed tooth drilled away?

The use of standard reduction potentials makes it possible to predict whether or not a redox reaction will take place. The reactants do not need to be paired in a voltaic cell. The half-reaction with the more positive reduction potential always undergoes reduction. Therefore, the other half-reaction has to undergo oxidation.

If the cell potential for the redox reaction, as written, is positive, then the reaction is spontaneous. If the cell potential is negative, then the reaction is nonspontaneous. It will be spontaneous in the reverse direction and the cell potential will be equally positive.

Example 3

Is the following redox reaction spontaneous as written?

$$Ni(s) + Fe^{2+}(aq) \longrightarrow Ni^{2+}(aq) + Fe(s)$$

Solution

The half-reactions are as follows.

$$\text{Oxidation: } Ni(s) \longrightarrow Ni^{2+}(aq) + 2e^-$$
$$\text{Reduction: } Fe^{2+}(aq) + 2e^- \longrightarrow Fe(s)$$

The standard reduction potentials are found in Table 21·2.

$$Ni^{2+}(aq) + 2e^- \longrightarrow Ni(s) \quad E^0_{Ni^{2+}} = -0.25 \text{ V}$$
$$Fe^{2+}(aq) + 2e^- \longrightarrow Fe(s) \quad E^0_{Fe^{2+}} = -0.44 \text{ V}$$

The standard cell potential is the difference of the potentials of the nickel half-cell and the iron half-cell.

$$E^0_{cell} = E^0_{red} - E^0_{oxid}$$
$$= E^0_{Fe^{2+}} - E^0_{Ni^{2+}}$$
$$= -0.44 \text{ V} - (-0.25 \text{ V})$$
$$= -0.19 \text{ V}$$

Since the standard cell potential is a negative number, the redox equation is nonspontaneous as written. Energy would have to be applied in order to make this reaction occur.

Problems

3. Calculate the cell potential for this spontaneous redox reaction.

$$Co^{2+}(aq) + Fe(s) \longrightarrow Fe^{2+}(aq) + Co(s)$$

4. Determine which of these redox reactions will occur spontaneously and calculate the standard cell potential in each case.
a. $Cu(s) + 2H^+(aq) \longrightarrow Cu^{2+}(aq) + H_2(g)$
b. $2Ag(s) + Fe^{2+}(aq) \longrightarrow 2Ag^+(aq) + Fe(s)$
c. $3Zn^{2+}(aq) + 2Cr(s) \longrightarrow 3Zn(s) + 2Cr^{3+}(aq)$

Figure 21·11
In cells like these, magnesium metal and chlorine gas are produced by the electrolysis of magnesium chloride.

21·9 Electrolytic Cells

Electrolysis is the use of electrical energy to make a nonspontaneous chemical change occur.

The section on voltaic cells described how a spontaneous chemical reaction can be used to generate a flow of electrons (an electric current). This section will show how an electric current can be used to make a nonspontaneous redox reaction go. *The process in which electrical energy is used to bring about a chemical change is called* **electrolysis.** The apparatus in which electrolysis is carried out is an electrolytic cell.

Electrolytic cells *are electrochemical cells used to cause a chemical change through the application of electrical energy.* An electrolytic cell is any cell that uses electrical energy (DC current) to make a nonspontaneous redox reaction go to products.

In both voltaic and electrolytic cells, electrons flow from the anode to the cathode in the external circuit. In both types of cells, the electrode at which the reduction reaction occurs is the cathode. The electrode at which oxidation occurs is the anode. In an electrolytic cell, however, the flow of electrons is being "pushed" by an outside source such as a battery. The cathode is called the negative electrode of the electrolytic cell. This is because it is connected to the negative electrode of the battery, the anode. (Remember that in a voltaic cell, the anode is the *negative* electrode and the cathode is the *positive* electrode.) The anode in the electrolytic cell is called the positive electrode. This is because it is connected to the positive electrode of the battery (the battery's cathode). It is important to remember these conventions.

Pure water cannot be electrolyzed.

Electrolysis of Water. When a current is applied via two electrodes in pure water nothing happens. There is no current flow and no electrolysis. When an electrolyte is added, such as H_2SO_4 or KNO_3 in low concentration, the solution conducts, and electrolysis takes place (Figure 21·12). The products of the electrolysis of water are hydrogen and oxygen. Water is reduced to hydrogen at the cathode.

$$2H_2O(l) + 2e^- \longrightarrow H_2(g) + 2OH^-(aq)$$

Water is oxidized at the anode.

$$2H_2O(l) \longrightarrow O_2(g) + 4H^+(aq) + 4e^-$$

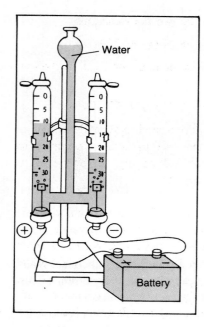

Figure 21·12
When an electric current is passed through water, it decomposes into oxygen and hydrogen. Which tube contains hydrogen gas and which tube contains oxygen gas? Hint: Note from the equation how many moles of hydrogen are produced for each mole of oxygen.

The region around the cathode turns basic due to an increase in OH^- ions. The region around the anode turns acidic due to an increase in H^+ ions. The overall cell reaction is obtained by adding the half-reactions.

$$2[2H_2O(l) + 2e^- \longrightarrow H_2(g) + 2OH^-(aq)]$$
$$\underline{2H_2O(l) \longrightarrow O_2(g) + 4H^+(aq) + 4e^-}$$
$$6H_2O(l) \longrightarrow 2H_2(g) + O_2(g) + 4H^+(aq) + 4OH^-(aq)$$

Some of the ions produced re-form into reactants.

$$4H^+(aq) + 4OH^-(aq) \longrightarrow 4H_2O(l)$$

These are not included in the net reaction.

$$2H_2O(l) \xrightarrow{\text{electrolysis}} 2H_2(g) + O_2(g)$$

In some cases, the electrolyte will be more easily oxidized or reduced than water. Then the products of electrolysis will be substances other than hydrogen and oxygen. An example is the electrolysis of brine.

Electrolysis of Brine. Chlorine gas, hydrogen gas, and sodium hydroxide are three important industrial chemicals. They are produced simultaneously by electrolysis of a concentrated aqueous sodium chloride solution (brine). The electrolytic cell for this process is shown in Figure 21·13. During electrolysis, chloride ions are oxidized to produce chlorine gas at the anode. Water is reduced to produce hydrogen gas at the cathode. Sodium ions are not reduced to sodium metal in this process because water molecules are more easily reduced than sodium ions. The reduction of water produces hydroxide ions as well as hydrogen gas. Thus the electrolyte in solution becomes sodium hydroxide.

Oxidation: $2Cl^-(aq) \longrightarrow Cl_2(g) + 2e^-$
(anode)

Reduction: $2H_2O(l) + 2e^- \longrightarrow H_2(g) + 2OH^-(aq)$
(cathode)

Figure 21·13
The electrolysis of a concentrated solution of sodium chloride produces which two gases?

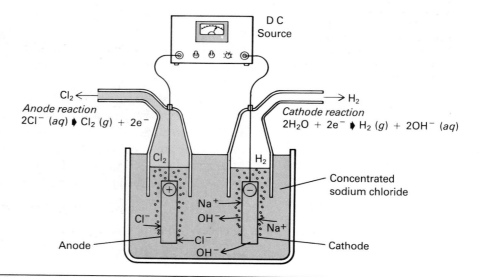

The overall ionic equation is the sum of the two half-reactions.

$$2Cl^-(aq) + 2H_2O(l) \longrightarrow Cl_2(g) + H_2(g) + 2OH^-(aq)$$

The spectator ion, Na^+, can be included in the equation to show the formation of sodium hydroxide.

$$2NaCl(aq) + 2H_2O(l) \longrightarrow Cl_2(g) + H_2(g) + 2NaOH(aq)$$

Chloride ions are eliminated from solution and hydroxide ions are formed. The electrolyte gradually changes from sodium chloride to sodium hydroxide. When the sodium hydroxide solution is about 10%, it is removed from the cell and processed further.

Electrolysis of brine is a very important industrial process. It is used to produce chlorine gas, hydrogen gas, and sodium hydroxide.

Electrolysis of Molten Sodium Chloride. Sodium and chlorine are both of commercial importance. Sodium is used in sodium vapor lamps and as the coolant in some nuclear power reactors. Chlorine, a toxic greenish–yellow gas, is used to sterilize drinking water and is important in the manufacture of polyvinyl chloride and various pesticides. These two elements are produced when molten sodium chloride is electrolyzed. Chlorine gas comes off at the anode and molten sodium collects at the cathode. The electrolytic cell in which this process is carried out commercially is called the Down's cell (Figure 21·14). Its design allows fresh sodium chloride to be added as required. The design also keeps the products apart so that they will not react to re-form sodium chloride. The half-reactions and overall cell reaction are as follows.

Oxidation: $2Cl^-(l) \longrightarrow Cl_2(g) + 2e^-$
(anode)

Reduction: $2Na^+(l) + 2e^- \longrightarrow 2Na(l)$
(cathode)

Cell reaction: $2NaCl(l) \longrightarrow 2Na(l) + Cl_2(g)$

Figure 21·14
The Down's cell is constructed to prevent sodium and chlorine from recombining after they are separated from molten sodium chloride by electrolysis. A Down's cell operates at about 600°C to maintain the salt in a molten state.

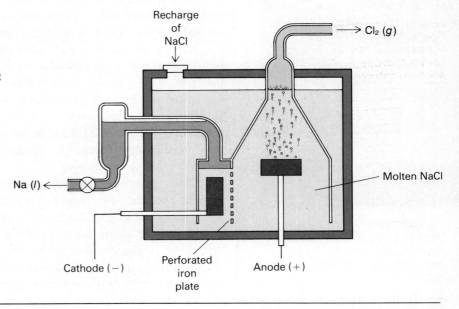

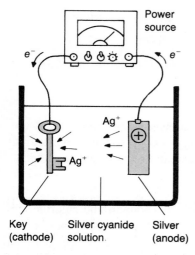

Key (cathode) | Silver cyanide solution | Silver (anode)

Figure 21·15
Electroplating is the coating of an object with a thin layer of metal. It is done in an electrolytic cell. The electrolyte contains ions of the metal that will be plated on the object that is made the cathode in the cell.

Figure 21·16
Aluminum is anodized to produce construction materials such as aircraft body parts, window frames, and rain gutters.

21·B Electroplating and Related Processes

Electroplating is the deposition of a thin layer of a metal on an object in an electrolytic cell. It has many important applications. An object may be electroplated to protect the surface of the base metal from corrosion or to make it more attractive. The layer of the deposited metal is very thin, usually from 5×10^{-5} cm to 1×10^{-3} cm thick. Metals commonly used for electroplating include gold, silver, copper, nickel, and chromium. An object that is to be silver-plated, for example, is made the cathode in a cell (Figure 21·15), and the anode is metallic silver. A silver cyanide solution is the electrolyte. When a DC current is applied, silver ions move to the object to be plated. They pick up electrons and are deposited as silver atoms.

$$\text{Cathode reaction: } Ag^+(aq) + e^- \longrightarrow Ag(s)$$

The net result is that silver is transferred from the silver electrode to the object being plated. Controlling this reaction is a fine art.

Many factors contribute to the quality of the metal coating that is formed. In the plating solution, the concentration of the cations to be reduced is carefully controlled. The solution must contain compounds to control the acidity, and to increase the conductivity. Some solutions contain compounds that form complex ions with the cation to be reduced. Other compounds make the metal coating brighter or smoother. In many plating cells, the anode must be shaped like the object at the cathode in order to plate all parts of the cathode evenly!

Several techniques involve the depositing of metal at the cathode of a cell. In *electroforming,* an object is reproduced by making a metal "mold" of it at the cathode. A phonograph record, for example, can be coated with metal so that it will conduct electricity. It is then electroplated to get a thick coating. This coating can be stripped off and used as a mold to produce copies of the record.

Impure metals can be purified in electrolytic cells. In *electrowinning,* the cations of molten salts or aqueous solutions are reduced at the cathode to give very pure metals. Electrowinning from molten salt is the only method for obtaining pure sodium and other very reactive metals. In *electrorefining,* a piece of impure metal is made the anode of the cell. It is oxidized to the cation, and then reduced to the pure metal at the cathode. This technique is used for obtaining ultrapure silver, lead, and copper.

Other electrolytic processes are centered on the anode rather than the cathode. In *electropolishing,* an object at the anode is selectively dissolved to give it a high polish. In *electromachining* a piece of metal at the anode dissolves to give an exact copy of the object that is the cathode. In *anodizing,* the metal object at the anode is oxidized to create an oxide deposit on its surface.

Key Terms

anode	21·2	electrolysis	21·9
battery	21·4	electrolytic cell	21·9
cathode	21·2	fuel cell	21·5
cell potential	21·6	half-cell	21·2
dry cell	21·3	reduction potential	21·6
electrical potential	21·6	salt bridge	21·2
electrochemical cell	21·1	standard cell potential	21·6
electrochemical process	21·1	standard hydrogen electrode	21·6
electrode	21·2	voltaic cell	21·2

Chapter Summary

Spontaneous redox reactions can be used to generate electrical energy in voltaic cells. Conversely, electrical energy can be used to make nonspontaneous reactions go in electrolytic cells. In either type of electrochemical cell, the electrode at which oxidation occurs is the anode, and the electrode at which reduction occurs is the cathode.

In a voltaic cell the half-reactions take place in half-cells. The half-cells are separated by a porous plate or salt bridge. This barrier prevents the contents of the two half-cells from mixing but permits the passage of ions between the half-cells. Reduction occurs at the cathode (the positively-charged electrode) and oxidation occurs at the anode (the negatively-charged electrode). Electrons transfer in the external circuit from the anode to the cathode. The half-cell with the higher reduction potential undergoes reduction, the other half-cell undergoes oxidation. The cell potential is the difference between the reduction potentials of the half-cells.

The reduction potential of an isolated half-cell cannot be measured. Values can be assigned, however, by comparing them to a reference electrode. The reference used is the standard hydrogen electrode which is given a reduction potential of 0.00 volts. Half-cells more easily reduced than the reference electrode have positive reduction potentials. Half-cells less easily reduced have negative reduction potentials.

Using our knowledge of electrochemistry we can design cells and batteries to provide electrical energy for specific purposes. The dry cell, the battery commonly used in flashlights, is not rechargeable. The lead storage battery, as used in automobiles, is rechargeable. Fuel cells do not have to be recharged. The fuels that they consume are fed into the cell continuously.

In electrolysis, a flow of electrons causes reduction at the cathode (the negatively-charged electrode) and oxidation at the anode (the positively-charged electrode). In an electrolytic cell electrical energy is used to bring about desirable redox reactions. Electrolytic cells are used in electroplating, in the refining of metals, and in the production of such important chemicals as sodium hydroxide, aluminum and sodium metals, and chlorine gas.

Practice Questions and Problems

5. Predict the result when a strip of copper is dipped into a solution of iron(II) sulfate. *21·1*

6. For each of the pairs of metals listed below decide which metal is the more readily oxidized. *21·1*
a. Hg and Cu   **c.** Ni and Mg   **e.** Pb and Zn
b. Ca and Al   **d.** Sn and Ag   **f.** Cu and Al

7. What is meant by the term "half-reaction"? Write the half-reactions that occur when a strip of Al is dipped into a solution of copper sulfate. *21·1*

8. Explain the function of the salt bridge in a voltaic cell. *21·2*

9. What material is **a.** the anode and **b.** the cathode made of in the flashlight battery? *21·3*

10. Describe the composition of the anode, cathode, and electrolyte in a fully charged lead storage battery. *21·4*

11. Explain why the specific gravity of the electrolyte in a lead storage battery decreases during the discharge process. *21·4*

12. List the advantages of a fuel cell over a lead storage battery. *21·5*

13. What is the purpose of using the standard hydrogen electrode as a reference electrode? *21·6*

14. What is the difference between a *cell potential* and a *standard cell potential*? *21·6*

15. What is the electric potential of a cell? *21·6*

16. The standard reduction potential for the cadmium half-cell is -0.40 volts. What does this mean? *21·7*

17. Explain how you would determine the standard reduction potential for the aluminum half-cell. *21·7*

18. Use the information in Table 21·2 to calculate standard cell potentials for these voltaic cells. *21·8*
 a. Ni | Ni²⁺ ‖ Cl₂ | Cl⁻   **b.** Sn | Sn²⁺ ‖ Ag⁺ | Ag

19. What processes occur at the anode and cathode in an electrolytic cell? *21·9*

20. Why is a *direct current* and not an *alternating current* used in the electroplating of metals? *21·9*

21. Describe briefly how you would electroplate a teaspoon with silver. *21·9*

Mastery Questions and Problems

22. Distinguish between voltaic and electrolytic cells.

23. Why is it not possible to measure the potential of an isolated half-cell?

24. Describe the composition of the anode, cathode, and electrolyte of a fully discharged lead storage battery.

25. Predict what will happen, if anything, when an iron nail is dipped into a solution of copper sulfate. Write the oxidation and reduction half-reactions for this process and the balanced equation for the overall reaction.

26. Calculate E^0_{cell} and write the overall cell reaction for these cells:
 a. Sn | Sn²⁺ ‖ Pb²⁺ | Pb   **b.** H₂ | H⁺ ‖ Br₂ | Br⁻

27. What property of lead(II) sulfate and lead dioxide makes it unnecessary to have salt bridges in a lead storage battery?

28. Complete this table of data for the electrolysis of water.

H₂O used	H₂ formed	O₂ formed
a. 2.0 mol	_____ mol	_____ mol
b. _____ mol	_____ g	16.0 g
c. _____ mL	10.0 g	_____ g
d. 44.4 g	_____ g	_____ g
e. _____ g	8.8 L (STP)	_____ L (STP)
f. 66.0 mL	_____ g	_____ L (STP)

Critical Thinking Questions

29. Choose the term that best completes the second relationship.
 a. battery:cells   herd:_____
  (1) birds   (3) trees
  (2) sheep   (4) fuel cells
 b. river:water   electrode:_____
  (1) electrons   (3) salt bridge
  (2) anode   (4) ions
 c. anode:electrons   fire hydrant:_____
  (1) fire   (3) dogs
  (2) firefighter   (4) water

30. Gold is not included in Table 21·2 on page 514. In what part of the table does gold belong?

31. Lead storage batteries can be recharged. Why are dry cells not rechargeable?

32. For any voltaic cell, chemists consider the electrode that produces electrons to be negative. They call it the anode. Most dictionaries, however, define the anode as the positively charged electrode. How can you account for this disagreement?

Review Questions and Problems

33. A sample of oxygen gas has a volume of 500 mL at 30°C. What is the new volume of the gas if the temperature is raised to 60°C while the pressure is kept constant?

34. Balance these equations:
- **a.** $H_2S + HNO_3 \longrightarrow S + NO + H_2O$
- **b.** $AgNO_3 + Pb \longrightarrow Pb(NO_3)_2 + Ag$
- **c.** $Cl_2 + NaOH \longrightarrow NaCl + NaClO_3 + H_2O$
- **d.** $CuO + NH_3 \longrightarrow N_2 + Cu + H_2O$
- **e.** $Mg(OH)_2 + HBr \longrightarrow MgBr_2 + H_2O$
- **f.** $Al_2O_3 + H_2SO_4 \longrightarrow Al_2(SO_4)_3 + H_2O$

35. Concentrated nitric acid is $16M$. How would you prepare 500 mL of $1.0M$ HNO_3?

36. What is the oxidation number of the italicized element in each of these formulas?
- **a.** $K_2Cr_2O_7$
- **b.** KIO_3
- **c.** MnO_4^-
- **d.** $FeCl_3$

Challenging Questions and Problems

37. Write the overall cell reactions and calculate E^0_{cell} for voltaic cells composed of the following sets of half-reactions.
- **a.** $AgCl + e^- \rightleftharpoons Ag + Cl^-$
  $Ni^{2+} + 2e^- \rightleftharpoons Ni$
- **b.** $Al^{3+} + 3e^- \rightleftharpoons Al$
  $Cl_2(g) + 2e^- \rightleftharpoons 2Cl^-$

38. Impure copper is purified in an electrolytic cell. Design an electrolytic cell that will allow you to carry out this process. Give the oxidation and reduction half-reactions and a balanced equation for the overall reaction.

39. This spontaneous redox reaction occurs in the voltaic cell illustrated below.
$$Ni^{2+}(aq) + Fe(s) \longrightarrow Ni(s) + Fe^{2+}(aq)$$
- **a.** Identify the anode and the cathode.
- **b.** Assign charges to the electrodes.
- **c.** Write the half-reactions.
- **e.** Calculate the standard cell potential when the half-cells are at standard conditions.

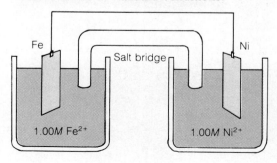

Research Projects

1. What contributions did Luigi Galvani make to the discovery of electric current? How is his name used to describe electrochemical processes?

2. How did Humphry Davy use electricity to identify elements?

3. Construct a simple voltage cell.

4. What contributions did William Nicholson make to the field of electrochemistry? How did his work relate to Alessandro Volta's work?

5. Measure the rate of movement of various ions in an agar salt bridge.

6. Why do some batteries last longer than others? Test several common batteries to see which lasts the longest under various conditions.

7. What did John Daniell do to improve the quality of batteries? What are some recent improvements in batteries?

8. Compare the voltage produced by a pure metal to the voltage produced by an alloy of the metal.

9. How do electrostatic precipitators in industrial smokestacks work? Draw a diagram of this device.

10. How are fuel cells used? What are the prospects for widespread use? Describe the commerical production of sodium and chlorine.

11. Using strips of zinc, copper, and other metals try making a fruit or vegetable battery. Insert the two metals with wires attached into the fruit or vegetable. Hold the other ends of the wire close together on your tongue to complete the circuit and detect an electrical current. Experiment with different combinations of metals in different fruits and vegetables.

12. How can hot brines be used commercially?

Readings and References

McDougall, Angus. *Fuel Cells.* New York: Halsted/Wiley, 1977.

Wilt, Rachel. "Leaf Jewelry." *ChemMatters* (December 1987), pp. 14–15.

22 The Chemistry of Metals

Chapter Preview

22·1   The Alkali Metals
22·2   The Alkaline Earth Metals
22·A   The Solvay Process
22·3   Aluminum for the Masses
22·4   Tin and Lead
22·5   The Transition Metals

22·6   Iron, Cobalt, and Nickel
22·7   Copper, Silver, and Gold
22·B   Toxic Metals
22·8   Metal Ores
22·9   The Blast Furnace
22·10 Steelmaking

Approximately 75 of the more than one hundred known elements are metals. These elements exhibit a wide range of properties. They have common physical and chemical characteristics, however, that distinguish them from nonmetals. This chapter describes some of the distinctive properties of metals and their practical applications.

22·1   The Alkali Metals

The alkali metals and their compounds show a striking similarity. For this reason, sodium will be discussed as a typical member of the group. Table 22·1 lists some physical properties of alkali metals. Along with their Group 2A neighbors, alkali metal atoms are among the largest atoms known. Even though their atoms are closely packed, these metals have low densities because of their large atomic sizes. Atomic size also

The physical properties of the alkali metals follow definite trends. As atomic number increases, their melting points, boiling points, hardnesses, specific heats, and ionization energies all decrease, while their densities increase.

Figure 22·1
Molten metals are poured into molds in which they will solidify into ingots.

Table 22·1	Some Physical Properties of the Alkali Metals			
Element	Melting point (°C)	Boiling point (°C)	Density (g/cm³)	Atomic radius (nm)
Lithium	179	1336	0.53	0.123
Sodium	98	883	0.97	0.157
Potassium	64	758	0.86	0.203
Rubidium	39	700	1.53	0.216
Cesium	28	670	1.90	0.235

525

Safety

Alkali metals should not be used except under the direct supervision of the teacher.

Because the alkali metals are so reactive they occur in nature as salts instead of as the free metals.

helps explain the low ionization energies of these elements. Because they lose a single valence electron so easily, alkali metals are powerful reducing agents.

Although sodium is the best known alkali metal, few people are familiar with its appearance. It is so active chemically that it never occurs in nature as the free metal. As a group, the 1A elements are the most reactive metals known. Within the group, the larger atoms, cesium and rubidium, are the most reactive.

A lump of sodium has the consistency of stiff modeling clay and is easily cut (Figure 12·14). When freshly cut, it has the typical silvery luster of a metal, but it tarnishes quickly in air. Like most metals, sodium is a good conductor of heat and electricity.

Rich deposits of alkali metal salts are found all over the world. The Bonneville Salt Flats in Utah and deposits at Searles Lake, California contain huge quantities of sodium chloride and other alkali salts. The evaporation of ancient seas left large deposits which are now underground near the Gulf Coast shores of Texas and Louisiana.

Alkali metal salts are very soluble in water. Rainwater leaches them out of the soil, and rivers carry them to the sea. Seawater is about 3% (by mass) alkali salts.

Sodium is the only alkali metal produced on a large scale. In most industrial processes it serves as well as the more expensive members of the group. To produce the free metal, sodium ions must be reduced. This is usually done by electrolysis. Production of metallic sodium by the electrolysis of molten sodium chloride is described in Section 21·9. Chlorine gas is a valuable by-product of this process.

Sodium is used in sodium vapor lamps and in the production of many chemicals. Liquid sodium is an excellent conductor of heat. It is often used in nuclear reactors to remove heat from the reactor core.

Figure 22·2
Large amounts of sodium chloride are produced by the evaporation of sea water.

Sodium reacts vigorously with most nonmetals. When sodium burns in a stream of dry oxygen, sodium peroxide, Na_2O_2, forms.

$$2Na(s) + O_2(g) \longrightarrow Na_2O_2(s)$$

In water, metal peroxides are powerful oxidizing agents used in bleaching and disinfecting. The reaction between sodium peroxide and water produces sodium hydroxide and hydrogen peroxide.

$$Na_2O_2(s) + 2H_2O(l) \longrightarrow 2NaOH(aq) + H_2O_2(aq)$$

The freshly formed hydrogen peroxide decomposes rapidly in the strongly basic solution to yield oxygen and water.

Sodium metal reacts violently with cold water, forming sodium hydroxide (lye) and releasing hydrogen gas.

$$2Na(s) + 2H_2O(l) \longrightarrow 2NaOH(aq) + H_2(g)$$

The reactions of the alkali metals with water are so rapid and exothermic that in many instances the hydrogen burns as it is produced (Figure 7·12).

Sodium hydroxide is one of the most widely produced strong bases. It is used in the production of soap, petroleum, synthetic textiles, and paper pulp. The electrolysis of brine (a concentrated aqueous solution of sodium chloride) is the major source of sodium hydroxide for the chemical industry. Chlorine and hydrogen gases are also produced.

$$2NaCl(aq) + 2H_2O(l) \xrightarrow{\text{electrical energy}} 2NaOH(aq) + H_2(g) + Cl_2(g)$$

Problem

1. Write a balanced equation for the reaction of potassium with water.

22·2 The Alkaline Earth Metals

Calcium and magnesium are the most abundant metals of Group 2A.

The alkaline earth metals are also highly reactive and are not found free in nature. They are less reactive, however, than the alkali metals. Barium is the most reactive element in the group. Group 2A metals are somewhat harder than those in Group 1A. Some physical properties of the alkaline earth metals are listed in Table 22·2.

Alkaline earth salts are less soluble than the corresponding alkali salts. Nevertheless, the sea is a rich source of magnesium and calcium ions. Calcium ions in seawater are used by shellfish in building their calcium carbonate shells. Coral animals carry out a similar process in forming reefs.

Some of the alkaline earth carbonates and sulfates are insoluble enough to resist weathering and the leaching action of rainwater. Limestone is the most common mineral form of calcium carbonate (Figure 22·3). It is often made up of compressed layers of seashells. The White Cliffs of Dover on the southern coast of England are another form of calcium carbonate. Shells of microscopic marine animals have been pressed

Figure 22·3
Calcium carbonate occurs naturally in several forms. The rectangular slab is marble. Continuing clockwise, calcite, limestone, and a cross-section of a stalactite are shown.

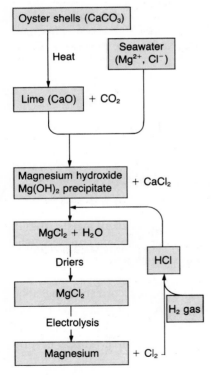

Figure 22·4
Sea water is the main source of magnesium compounds. Most commercial magnesium is prepared electrolytically from magnesium chloride. Each ton of sea water yields about 3 kg of magnesium.

into a natural chalk. Marble is limestone that has been transformed to metamorphic rock by the action of heat and pressure deep within the earth. Another important calcium mineral is the hydrate gypsum, $CaSO_4 \cdot 2H_2O$, from which plaster of Paris is made.

Igneous rocks such as granite and basalt are composed of insoluble silicate salts of almost all the Group 1A and 2A metals. The slow weathering of these rocks is responsible for the presence of alkali and alkaline earth metal ions in fertile soil.

Calcium and magnesium are the most important alkaline earth metals. Calcium is produced by the electrolysis of molten calcium chloride in a reaction similar to that used to produce sodium. By contrast, magnesium is prepared from seawater (Figure 22·4). Lime is added to precipitate the magnesium as magnesium hydroxide. Then hydrochloric acid is added to convert the hydroxide to magnesium chloride. Magnesium metal is recovered from molten magnesium chloride by electrolysis.

Magnesium is the only Group 2A metal widely used in the free state. It is an important structural material and is the chief component in a number of high tensile-strength, low-density alloys. These properties make these alloys valuable in aircraft and spacecraft construction. Magnesium ribbon or powder is used in photoflash bulbs.

With the exception of beryllium and magnesium, the alkaline earth metals tarnish quickly in air. They may continue to corrode until completely converted to the oxide, hydroxide, or carbonate. Beryllium and magnesium also oxidize readily. They form a tough oxide film, however, that prevents further reactions at room temperature.

Calcium, strontium, and barium react with cold water but more slowly than Group 1A metals do (Figure 22·5). Beryllium and magnesium react only with hot water or steam.

$$Ca(s) + 2H_2O(l) \longrightarrow Ca(OH)_2(aq) + H_2(g)$$

Table 22·2	Some Physical Properties of the Alkaline Earth Metals			
Element	Melting point (°C)	Boiling point (°C)	Density (g/cm³)	Atomic radius (nm)
Beryllium	1280	1500	1.86	0.089
Magnesium	651	1107	1.75	0.136
Calcium	851	1487	1.55	0.174
Strontium	800	1366	2.60	0.191
Barium	850	1537	3.59	0.198

Note that there are no clear-cut trends in the melting points, boiling points, and densities of the alkaline earth metals. They have higher melting points and are harder, however, than the alkali metals.

Alkaline earth oxides are used more widely than the pure metals. They are white solids with very high melting points. They react with water to form hydroxides and with carbon dioxide to produce carbonates. The reaction with water, called slaking, is exothermic.

$$CaO(s) + H_2O(l) \longrightarrow Ca(OH)_2(aq) + heat$$
$$\text{Lime} \qquad\qquad\qquad \text{Slaked lime}$$

$$CaO(s) + CO_2(s) \longrightarrow CaCO_3(s)$$

Calcium oxide, commonly called **lime,** or quicklime, is an important industrial chemical. It is made by the high-temperature decomposition of limestone in a lime kiln.

$$CaCO_3(s) \xrightarrow{\;900°C\;} CaO(s) + CO_2(g)$$
$$\text{Limestone} \qquad\qquad \text{Lime}$$

Water is added to quicklime to get **slaked lime,** *calcium hydroxide*. It is used to make plaster and mortar and to neutralize acid soils. Mortar is a mixture of slaked lime, sand, and water. The setting process involves drying and crystallization. These processes are followed by slow conversion of the slaked lime to calcium carbonate by the action of atmospheric carbon dioxide.

Magnesium oxide is manufactured from magnesium carbonate in a process that is similar to the manufacture of lime. Magnesium oxide is used for making firebrick and for insulating steam pipes.

Group 2A hydroxides are strong bases similar to those of Group 1A, although less soluble in water. Calcium hydroxide is used in large quantities in the manufacture of building mortar and bleaching powder, and in water softening. A suspension of magnesium hydroxide, $Mg(OH)_2$, in water is known as milk of magnesia. It is used as a laxative and antacid.

Safety

Calcium oxide should be kept in airtight and waterproof containers when not in use. Large amounts of heat are released when it reacts with water.

Figure 22·5
Alkaline earth metals are less reactive than the alkali metals. Here, calcium reacts with water. What are the products of the reaction?

Problem

2. Write a balanced equation for the reaction between magnesium oxide and carbon dioxide.

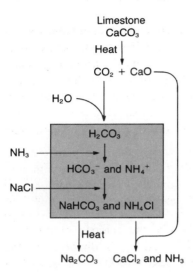

Limestone
CaCO₃

Heat

CO₂ + CaO

H₂O

H₂CO₃

NH₃

HCO₃⁻ and NH₄⁺

NaCl

NaHCO₃ and NH₄Cl

Heat

Na₂CO₃   CaCl₂ and NH₃

Figure 22·6
Large amounts of sodium carbonate
are manufactured by the Solvay
process. This material is used in the
manufacture of glass, soap, and
paper, in cleaning compounds, and
for water softening.

Chemical Connections

Oxides of metals such as aluminum,
iron, and cerium are used as abrasives,
which are materials used to reduce,
clean, or polish the surface of other substances. Abrasives are important in
many industrial processes, such as pol-
ishing glass lenses.

22·A   The Solvay Process

Sodium carbonate, Na_2CO_3, is an important industrial chemical used
in the manufacture of glass, soap, and paper. Commonly called soda
ash, it is also an ingredient in cleaning compounds and water softeners. Sodium carbonate is obtained commercially from natural salt
lakes. The demand for it outstrips this supply. Thus more than five
million metric tons are produced in North America each year by the
Solvay process.

The Belgian chemist Ernest Solvay (1838–1922) invented the
ammonia–soda process around 1860. It uses sodium chloride, calcium carbonate, and ammonia as raw materials (Figure 22·6). Limestone is heated to produce calcium oxide and carbon dioxide, as described in Section 22·2. The carbon dioxide is then bubbled into a
cold aqueous solution of sodium chloride and ammonia. In the solution, carbonic acid is formed in equilibrium with the dissolved carbon dioxide. The ammonia reacts with carbonic acid. Ammonium
and hydrogen carbonate ions are produced.

$$CO_2(aq) + H_2O(l) \rightleftharpoons H_2CO_3(aq)$$

$$NH_3(aq) + H_2CO_3(aq) \longrightarrow NH_4^+(aq) + HCO_3^-(aq)$$

At the low reaction temperature, sodium hydrogen carbonate precipitates from the solution and is filtered off. When heated, it yields
sodium carbonate. Carbon dioxide from the reaction is recycled.

$$NH_4^+(aq) + HCO_3^-(aq) + Na^+(aq) + Cl^-(aq) \longrightarrow$$
$$NaHCO_3(s) + NH_4Cl(aq)$$

$$2NaHCO_3(s) \xrightarrow{heat} Na_2CO_3(s) + CO_2(g) + H_2O(g)$$

Ammonium chloride is also recovered from the reaction liquor. It is
combined with lime to give a second major product, calcium chloride. The ammonia that forms is recycled.

$$2NH_4Cl(s) + CaO(s) \longrightarrow CaCl_2(s) + 2NH_3(g) + H_2O(g)$$

22·3   Aluminum for the Masses

Aluminum is present in many rocks and minerals and is the most abundant metal in the earth's crust. As a pure metal it has a low density, good
electrical conductivity, and corrosion resistance. These properties make it
valuable as a structural metal, as well as in the manufacture of many essential products. Freshly formed aluminum is highly reactive. It will react with oxygen, water, acids, and bases. Fortunately, when aluminum is
exposed to air, an adherent film of oxide quickly forms, protecting it from
corrosion from oxygen and water. Until the end of the last century, however, aluminum sold for the same price as silver. At that time only the rich
could afford aluminum items.

Figure 22·7
Our chief source of aluminum is bauxite, a reddish-brown ore. When this ore is heated with coke (the black substance), aluminum oxide (the white powder) is formed. Aluminum metal is obtained from aluminum oxide by electrolysis.

Aluminum is refined electrolytically from bauxite by the Hall–Heroult process.

Aluminum was expensive because chemists were unable to produce it by the electrolysis of molten aluminum halides. These compounds have relatively low boiling points. They distill out of the electrolyzing baths at the high operating temperatures necessary. The melting point of another potential raw material, aluminum oxide, was over 2000°C. No practical way of melting it was available.

The first commercial process for the production of aluminum was invented almost simultaneously by Paul Heroult, a Frenchman, and Charles Hall, an American. Hall's chemistry professor at Oberlin College challenged class members to develop a cheap way to produce aluminum. After graduation, Hall set up a laboratory in a woodshed to tackle this problem. Within a year, he discovered that cryolite, Na_3AlF_6, would dissolve aluminum oxide. A low-melting solution formed, from which aluminum could be obtained by electrolysis.

This process is essentially the method used for aluminum production today. Modern production is a continuous process, however, using long lines of electrolytic cells connected in series. The structure of an individual cell is shown in Figure 22·8. In a cell, cryolite is melted in a large iron tank lined with carbon. The carbon is the cathode of the cell. Refined bauxite, an aluminum ore rich in Al_2O_3, is dissolved in the melted cryolite at a temperature slightly below 1000°C. Large carbon rods serve as the anode. When a current passes through the cell, aluminum ions are reduced at the cathode. Molten aluminum sinks to the bottom of the tank where it is periodically drawn off. Oxygen is liberated at the anode. It reacts with the carbon, producing carbon dioxide. Thus the anodes must be replaced at regular intervals.

Cathode reaction: $Al^{3+} + 3e^- \longrightarrow Al$

Anode reaction: $2O^{2-} \longrightarrow O_2 + 4e^-$

Net cell reaction: $4Al^{3+} + 6O^{2-} \longrightarrow 4Al + 3O_2$

To make 1 kg of aluminum metal requires 2 kg of aluminum oxide and 0.5 kg of carbon anode. Eight to ten kilowatt hours of electricity are needed per kilogram of product. Aluminum plants are often located near hydroelectric power plants where the cost of electricity is relatively low.

Figure 22·8
In the Hall–Heroult process, aluminum is produced by passing an electric current through an electrolyte of fused cryolite and aluminum oxide. The molten aluminum is withdrawn from the bottom of the tank.

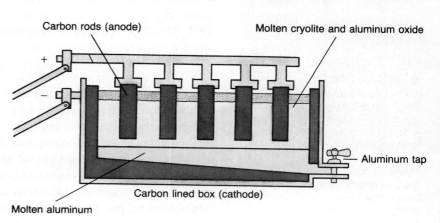

Figure 22·9
Fusible alloys that have melting points lower than the boiling point of water are used in automatic sprinkling systems for buildings. These alloys are usually made of lead, tin, cadmium, and bismuth.

Tin and lead are components of many alloys.

The transition metals have typical metallic properties.

Figure 22·10
Many jet engine parts are made from hard, corrosion-resistant titanium alloys.

22·4   Tin and Lead

Tin and lead are the metals of Group 4A. Tin is usually found in nature as its oxide, SnO_2. The mineral is reduced to the free metal by heating with coke.

$$SnO_2(s) + C(s) \longrightarrow Sn(s) + CO_2(g)$$

Lead is found chiefly as the sulfide ore galena, PbS. The ore is roasted in air to produce the oxide, which is then reduced to metallic lead by heating with carbon.

$$2PbS(s) + 3O_2(g) \longrightarrow 2PbO(s) + 2SO_2(g)$$

$$2PbO(s) + C(s) \longrightarrow 2Pb(s) + CO_2(g)$$

Because tin and lead are not very reactive, they have many uses as the free metal. Both are widely utilized in alloys. **Bronze** *is an alloy of tin, copper and zinc.* Solder is an alloy of lead and tin.

Tin is applied as a thin coating to iron or steel to make tin cans. The tin protects the iron against corrosion, particularly that caused by acidic foods. A tin compound, stannous fluoride (SnF_2), is an important ingredient in some toothpastes.

Since Roman times lead has been employed in plumbing. It is also used to make lead storage batteries and in the manufacture of tetraethyl lead, $Pb(C_2H_5)_4$. In small quantities this compound improves the octane-rating of gasoline. The current switch to lead-free gasolines began when scientists discovered the toxic effects of lead from exhaust gases.

22·5   The Transition Metals

The transition metals exhibit typical metallic properties. In general, they are ductile, malleable, and good conductors of heat and electricity. With the exception of copper and gold, they have a silvery luster. Some important uses of these metals are shown in Table 22·3.

Titanium is an important structural metal. It is strong and corrosion resistant. Moreover, its density is about half that of iron. Titanium alloys are used in high performance aircraft engines and missiles.

The most important compound of titanium is its white dioxide, TiO_2, commonly used as a pigment in white paint. Titanium tetrachloride, $TiCl_4$, is a clear, colorless liquid. Because of its rapid reaction with damp air, it is employed by the military to make dense smoke screens.

$$TiCl_4(l) + 2H_2O(l) \longrightarrow TiO_2(s) + 4HCl(g)$$

Chromium is a hard, brittle metal with a white luster. Its extreme resistance to corrosion makes it an ideal coating for iron and steel objects. Chromium is an important ingredient in stainless steel. Typical stainless steel contains about 18% chromium and 10% nickel, with trace amounts of carbon, manganese, and phosphorus.

Like most transition metals, chromium has more than one oxidation state. The most common oxidation states of chromium are $+2$, $+3$, and $+6$. The metal has an oxidation number of $+6$ in two polyatomic ions:

Table 22·3 Common Transition Metals and Their Uses

Metal	Uses
Cadmium (Cd)	Batteries; control rods for nuclear reactors
Chromium (Cr)	Plating; making stainless steel
Cobalt (Co)	Alloys; treatment of cancer (Co-60 only)
Copper (Cu)	Electrical wiring; plumbing; coinage
Gold (Au)	Jewelry; ornaments; standard of wealth
Iron (Fe)	Steel; magnets
Manganese (Mn)	Steel; nonferrous alloys
Mercury (Hg)	Lamps; switches; thermometers; barometers
Nickel (Ni)	Hardens steel; plating; catalyst
Platinum (Pt)	Catalyst; electronics; lab-ware; jewelry
Silver (Ag)	Mirrors; jewelry; photography; coins
Tantalum (Ta)	Surgery; corrosion resistant equipment
Titanium (Ti)	Combustion chambers for rockets and jets
Tungsten (W)	Filaments for light bulbs; alloys
Vanadium (V)	Shock resistant steel alloys; catalyst
Zinc (Zn)	Galvanized iron; brass; dry cells

chromate, CrO_4^{2-}, and dichromate, $Cr_2O_7^{2-}$. All compounds in which chromium is in the $+6$ oxidation state are powerful oxidizing agents, particularly in acid solution.

Zinc is one of industry's most useful metals. It is obtained from the sulfide ore zinc blende (sphalerite). Large amounts of zinc are used to produce galvanized iron. This is iron coated with a thin layer of zinc to protect it from corrosion. Zinc is also a component of several alloys. Brass is an alloy of copper and zinc; bronze is an alloy of tin, copper, and zinc. In ordinary dry cell batteries the zinc lining acts as the anode.

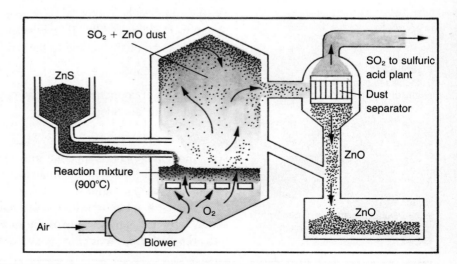

Figure 22·11
When hot zinc sulfide is treated with compressed air, it reacts to form zinc oxide dust and sulfur dioxide. Later carbon monoxide is used to reduce the zinc oxide to zinc metal.

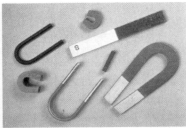

Figure 22·12
Strong permanent magnets are made from alloys of iron with nickel, cobalt, and small amounts of other metals.

Iron, cobalt, and nickel are the magnetic metals.

Copper, silver, and gold are less reactive than most other metals.

Figure 22·13
Turquoise is one of the many beautiful copper-containing ores.

22·6　Iron, Cobalt, and Nickel

Iron, cobalt, and nickel share a number of common properties. Their densities are similar, as are their melting and boiling points. Their most striking physical property, however, is magnetism. Iron is the most strongly magnetic, and nickel is the least magnetic of the three metals.

Making up about 5% of the earth's crust, iron is the second most abundant metal. (Aluminum is the most abundant.) Common iron minerals are Fe_2O_3 (the red oxide, hematite) and Fe_3O_4 (magnetite).

The world's largest nickel and cobalt deposits are in Ontario, Canada. The two metals occur as sulfide ores, together with copper, silver, iron, platinum, paladium, and iridium. Most of the nickel and cobalt produced today is used in the manufacture of steels. Nickel makes steel more ductile and corrosion resistant.

Unlike iron, nickel and cobalt are resistant to atmospheric corrosion. Nickel is often used to electroplate iron and steel objects before they are plated with chromium. **Monel metal,** *a strong, corrosion-resistant alloy of nickel and copper,* is used for propeller shafts of seagoing vessels. Stellite, an alloy of cobalt, chromium, and tungsten, retains its hardness when hot. It is useful for high-speed cutting tools and drill bits. Both cobalt and nickel are used as catalysts in industrial processes. For example, finely divided nickel catalyzes the hydrogenation of vegetable oils.

22·7　Copper, Silver, and Gold

Copper, silver, and gold have low chemical activities and often occur naturally in the free state. They were the first metals collected and worked by humans. In the Middle East, copper and gold artifacts dating back to about 7000 B.C. have been found.

Copper was probably the first metal to be reduced from its ore. Some of the most important copper ores are chalcopyrite, $CuFeS_2$, chalcocite, Cu_2S, and cuprite, Cu_2O. Low grade sulfide ores are processed into copper of high purity by reduction and electrolytic refining. The crushed ore, which may be less than 1% copper, is concentrated by flotation (Figure 22·19). In the smelter furnaces, oxygen-enriched air converts some of the sulfide to the oxide.

$$2Cu_2S(s) + 3O_2(g) \longrightarrow 2Cu_2O(s) + 2SO_2(g)$$

The mixture of copper(I) sulfide and copper(I) oxide reacts, forming copper and sulfur dioxide.

$$2Cu_2O(s) + Cu_2S(s) \longrightarrow 6Cu(s) + SO_2(g)$$

The molten copper is poured into slabs called "blister copper" with 98.5% to 99.5% purity. The final step is electrolytic refining, which produces 99.99% pure copper.

Copper is the most widely-used metal for electrical wiring. Only silver is a better conductor of electricity. Copper is also widely used in plumbing for pipes that carry hot or cold water.

a

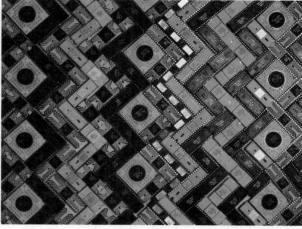

b

Figure 22·14
Copper, silver, and gold are excellent conductors. **a** Copper tubing is also used for plumbing. **b** Gold is often used for the contacts on computer chips.

The tarnish on silverware is a thin coating of black Ag_2S.

Silver occurs as the free metal and as the sulfide ore argentite, Ag_2S. Today, however, most commercial silver is produced as a by-product in the processing of copper, lead, and zinc ores. The muddy layers that collect on the bottom of electrolysis tanks in copper refining are the best source of silver. The mud is treated with sulfuric acid, forming silver sulfate. Then copper rods are dipped into the solution and silver deposits form on the copper.

$$Ag_2SO_4(aq) + Cu(s) \longrightarrow CuSO_4(aq) + 2Ag(s)$$

Because of its high luster, silver reflects light extremely well and has long been used to coat the backs of mirrors. The light-sensitive silver halides are utilized in photographic processes.

Gold occurs chiefly as small particles of the free metal in veins of quartz. About 5 g of gold is produced from a metric ton of gold-bearing rock. Gold is also recovered as a by-product of copper refining. It is the most malleable and ductile of all metals and can be pounded into sheets so thin that they transmit light. Gold leaf, as it is called, is used in decorative lettering and other forms of ornamentation. Corrosion resistance coupled with high electrical and thermal conductivity makes gold valuable in the high-tech industries. It is used to plate the contacts in microcircuits and to cover the external surfaces of satellite components.

Figure 22·15
The last step in producing copper is electrolysis. Blister copper is placed at the anode. Pure copper accumulates at the cathode. About six metric tons of rock is mined to produce ten kilograms of copper metal.

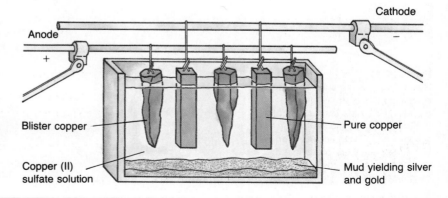

Figure 22·16
Mercury pollution has forced the closing of many waterways to fishing.

 Issues in Chemistry

Do you think the toxicity of a metal varies with the type of exposure (inhalation, ingestion, skin contact)? Is the toxicity of an element the same in all compounds?

22·B Toxic Metals

Metals are among the industrial waste products that contaminate the environment. Some of these elements are essential for human nutrition in trace amounts. Nevertheless, higher concentrations are toxic (Table 22·4). Others belong to a group of heavy metals that have no known biological function in humans. For example, low concentrations of lead, cadmium, and mercury produce toxic effects. Within the body, heavy-metal ions eventually cause irreversible damage to the central nervous system (Table 22·5).

Metals vary greatly in their toxic properties. Copper, for example, has a relatively low toxicity. A deficiency of copper in the diet causes anemia and nerve cell degeneration. Only 2 mg ingested daily prevents these effects. If the copper intake is increased to 50 mg, however, vomiting and diarrhea occur.

By contrast, mercury is highly toxic and is a serious environmental pollutant. Scientists have discovered that bacteria convert insoluble forms of mercury, once thought harmless, into highly toxic dimethyl mercury, $(CH_3)_2Hg$. This volatile liquid can enter the body through the skin or via the lungs.

For decades mercury and mercury salts have been used in industrial processes. As a result, large quantities of these substances have been discarded into the environment. Much of this mercury has collected in lake and river sediments. Here, dimethyl mercury accumulates in the tissues of aquatic plants and small animals. Like other heavy metals, it is eliminated from living tissues very slowly. It moves up the food chain, becoming more concentrated with each step (Figure 22·17). This sequence makes some fish unfit for human consumption and creates a long-term environmental hazard.

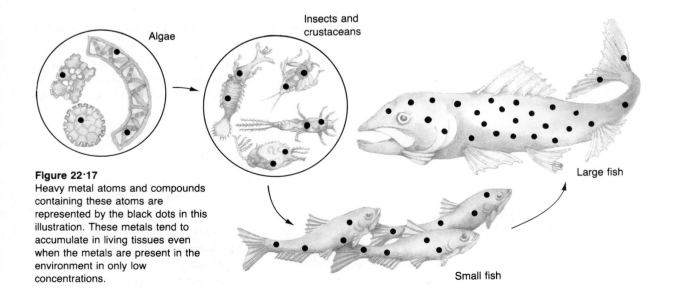

Figure 22·17
Heavy metal atoms and compounds containing these atoms are represented by the black dots in this illustration. These metals tend to accumulate in living tissues even when the metals are present in the environment in only low concentrations.

Table 22·4	Some Essential Elements that Produce Toxic Effects		
Element	Required for	Toxic effect	Recommended limit in drinking water (mg/L)
Arsenic	Reproduction	Mental disturbance, kidney failure	0.05
Copper	Component of several enzymes, formation of hemoglobin	Liver damage	1.0
Selenium	Normal liver function	Mental disturbance, liver damage	0.01

Figure 22·18
Most metals are refined from ores. Those shown here (clockwise from the upper left) are: cinnabar (mercury), sphalerite (zinc), wulfenite (lead, molybdenum), malachite (copper), and rutile (titanium) in the center.

Most metals are obtained from ore deposits.

Table 22·5	Toxic Metals	
Element	Toxic effect	Recommended limit in drinking water (mg/L)
Cadmium	Kidney damage, loss of red blood cells	0.01
Lead	Mental retardation in children, convulsions, kidney failure	0.05
Mercury	Birth defects, nerve damage, paralysis, brain damage	0.002

22·8 Metal Ores

Although some metals are extracted from seawater, most come from mineral deposits in the earth. *Minerals that are used for the commercial production of metals are called* **ores.** Some common metal ores are listed in Table 22·6.

Metallurgy *involves the various procedures used to separate metals from their ores.* Many of these techniques were developed over the centuries by trial and error. Separating metals from their ores requires three steps.

1. concentrating the ore
2. chemically reducing the ore to the metal
3. refining and purifying the metal

Freshly mined ore contains various amounts of worthless sand and rock called **gangue.** If the quantity of gangue is high, the ore is crushed and ground to separate the mineral and gangue. The ore is then enriched by physical processes such as washing, froth flotation, or magnetic separation. These processes remove the gangue. Washing the crushed ore with a turbulent stream of water removes lighter gangue from a mineral.

a

b

Figure 22·19
a An entire mountain has been dug away to get copper out of the world's largest open pit copper mine in Bingham Canyon, Utah.
b Finely crushed copper ore is agitated with detergent and oil in the copper flotation process. Oil-coated ore particles stick to air bubbles rising to the surface where they can be skimmed off.

Washing helps enrich some iron ores. *The* **froth flotation** *method is commonly used to enrich the sulfide ores of copper and lead* (Figure 22·19). Finely crushed ore is agitated with a detergent and oil. The oil sticks to the particles of sulfide, but not to the gangue. When air is blown into the mixture, oil-coated ore particles stick to the bubbles, and rise to the surface where they can be skimmed off. The gangue settles to the bottom and is removed at intervals.

Many ores need a second pretreatment. **Roasting** *is used to purify ores by heating the ore with air.* Enriched sulfide ores are usually roasted to drive off volatile impurities. Roasting converts the sulfides to oxides.

$$2PbS(s) + 3O_2(g) \longrightarrow 2PbO(s) + 2SO_2(g)$$

$$2ZnS(s) + 3O_2(g) \longrightarrow 2ZnO(s) + 2SO_2(g)$$

The sulfur dioxide formed is a serious air pollutant. It is removed from exhaust gases by reacting it with calcium oxide to form calcium sulfite.

$$CaO(s) + SO_2(g) \longrightarrow CaSO_3(s)$$

Smelting *involves melting an ore to remove impurities.* It is another type of refining process. During smelting, metal ions in an ore are reduced at a high temperature to the molten metal. A **flux** *can be added to a furnace to combine with undesired material in an ore.* The flux (usually lime) combines with the gangue (usually sand) at high temperatures. A molten material called slag is formed. **Slag** *is the fused waste material from a smelting process.*

$$\underset{\text{Flux}}{CaO(s)} + \underset{\text{Gangue}}{SiO_2(s)} \longrightarrow \underset{\text{Slag}}{CaSiO_3(s)}$$

The less dense slag floats on the surface and is removed. Slag helps prevent contact between the molten metal and oxygen in the air.

Table 22·6	Metal Ores	
Metal	Ore	Process
Chromium	chromite, $FeCr_2O_4$	reduction, electrolysis
Cobalt	cobalite, $CoAsS$	roasting, reduction
Copper	chalcopyrite, $CuFeS_2$ chalcocite, Cu_2S cuprite, Cu_2O	flotation, reduction, electrolysis
Lead	galena, PbS	flotation, roasting, reduction
Nickel	pentlandite, NiS	flotation, roasting, reduction
Tin	cassiterite, SnO_2	flotation, reduction
Titanium	rutile, TiO_2	convert to $TiCl_4$, reduction
Zinc	sphalerite, ZnS	roasting, reduction

Figure 22·20
The refining of iron is an important industrial process.

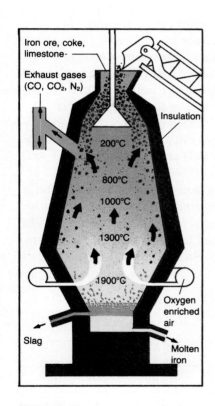

Iron ore, coke, limestone·
Exhaust gases (CO, CO₂, N₂)
Insulation
200°C
800°C
1000°C
1300°C
1900°C
Oxygen enriched air
Slag
Molten iron

Figure 22·21
A blast furnace is about 30 m high and 10 m in diameter. Iron ore, limestone, and coke (a form of carbon) are added at the top. During the process, carbon, the reducing agent, is oxidized to CO_2. The limestone converts sand in the ore to silicate slag.

22·9 The Blast Furnace

The metals in most ores have positive oxidation numbers. Thus production of the free metal from an ore requires reduction. Oxides of many metals can be reduced by carbon. *A **blast furnace** is a towerlike furnace in which carbon is used to reduce iron ores to metallic iron.* Once fired up, these huge vessels may operate continuously for several years.

*The **charge** is a load of material added to a furnace.* A mixture of iron ore, limestone, and coke enters the top of the blast furnace through a hopper. A typical ore contains iron oxide (Fe_2O_3) and gangue (sand). When heated, the limestone produces a flux (CaO) which reacts with the sand to form a slag (Section 22·8). Coke is almost pure carbon that is produced when coal is heated and the volatile matter is driven off. In the blast furnace, coke reduces the iron ore to the free metal.

To get the reaction going, a blast of hot air is blown up through the furnace. Rapid combustion of coke with the hot air occurs first.

$$C(s) + O_2(g) \longrightarrow CO_2(g) + \text{heat}$$

This reaction is very exothermic. It raises the temperature to about 2000°C. As the carbon dioxide rises up through the furnace, it is reduced by the hot coke to carbon monoxide.

$$CO_2(g) + C(s) \longrightarrow 2CO(g)$$

This reaction is endothermic and causes the temperature in the upper part of the furnace to drop to about 1300°C. As the carbon monoxide rises through the charge, it reacts with the iron oxide to form metallic iron. A series of redox reactions occurs in which the iron compounds are reduced and carbon monoxide is oxidized.

$$3Fe_2O_3(s) + CO(g) \longrightarrow 2Fe_3O_4(s) + CO_2(g)$$
$$Fe_3O_4(s) + CO(g) \longrightarrow 3FeO(s) + CO_2(g)$$
$$FeO(s) + CO(g) \longrightarrow Fe(s) + CO_2(g)$$

Iron ores are reduced to metallic iron in a blast furnace.

The reduction of iron ore to iron takes place in the middle and upper portions of the furnace. It is often summarized as a single equation.

$$Fe_2O_3(s) + 3CO(g) \longrightarrow 2Fe(s) + 3CO_2(g)$$

The molten iron trickles down through the charge and collects in a well at the bottom of the furnace. About every six hours, the furnace must be tapped to drain off the iron. The molten slag also trickles down through the charge and collects on the surface of the iron. It is drawn off periodically when the furnace is tapped.

Pig iron is crude iron as it comes from a blast furnace. The molten metal is cast in sand molds which resemble a pig feeding her litter.

Some blast furnaces produce up to 2500 tons of pig iron a day. *This* **pig iron** *still contains impurities and usually undergoes further treatment to produce steel.* To produce a ton of pig iron requires two tons of iron ore, one ton of coke, 0.3 ton of limestone, and four tons of air. Besides the iron, 0.6 ton of slag and six tons of flue gas are produced. The slag is a valuable by-product, used in the manufacture of Portland cement.

22·10 Steelmaking

Pig iron has little practical use. The impurities make it brittle and reduce its tensile strength. These impurities include 3%–5% carbon and smaller amounts of phosphorus, manganese, silica, and sulfur. *When pig iron is melted with scrap iron, the product is called* **cast iron.** Cast iron cannot be rolled, forged, or welded, but articles can be made from it by casting them in a mold. Cast iron has limited value as a structural material.

Most pig iron is converted to steel. This requires decreasing the carbon content (to 0.02%–2.0%) and the removal of other impurities. At the same time, specific amounts of other metals are added to produce a steel with the desired qualities (Table 22·7).

Steel has a much lower carbon content than cast iron.

ChemDirections

Stainless steels were among the first materials to be successfully implanted in the human body. To learn more about implant materials turn to **ChemDirections** *Biomedical Implants*, page 667.

One of the first methods of making steel was invented in England in 1856 by Sir Henry Bessemer. *In the* **Bessemer process,** *a blast of hot air is bubbled through molten pig iron in a huge egg-shaped converter.* The hot air oxidizes the excess carbon in each batch of about 25 tons of iron. A vigorous reaction between the carbon and air takes place. Because of this, careful adjustment of the percentages of carbon and other alloying elements is not possible.

Table 22·7	The Composition of Some Alloy Steels*		
Name	Composition	Characteristics	Uses
Nickel steel	Ni 3.5%	Tough, hard	Crankshafts
Chrome steel	Cr 1.0%	Tough, strong	Springs
Stainless steel	Cr 12% − 30%	Hard, resists corrosion	Surgical implements
Manganese steel	Mn 12% − 14%	Holds temper	Safes
Tungsten steel	W 5%	Maintains hardness when hot	High speed cutting tools

*The carbon content is typically 0.1 to 1.0 percent.

Figure 22·22
In an open hearth furnace two burners alternate to produce bursts of hot air and burning gases. The chambers on each side of the hearth store the heat.

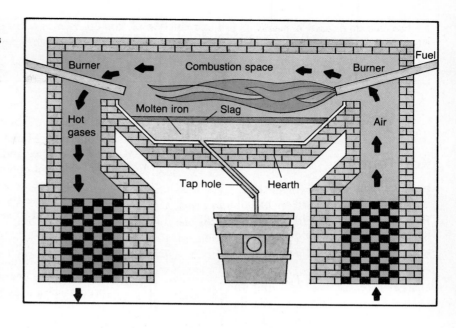

Figure 22·23
In a basic oxygen furnace, scrap steel is mixed with molten iron from a blast furnace. Oxygen is introduced through the lance to oxidize the impurities in the charge.

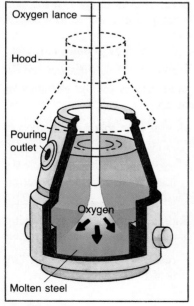

The **open-hearth method** *uses a shallow lined vessel to produce high quality steel in a five to eight hour process.* The hearth, or floor of the open hearth furnace, is a shallow vessel. It is approximately 13 m long, 7 m wide, and 1 m deep. The hearth is lined with a basic or an acidic lining, depending on the type of pig iron being purified. In North America, iron ores usually have acid impurities (sulfur or phosphorus). Therefore basic linings of calcium oxide and magnesium oxide are used. An arched roof of firebrick covers the hearth. The floor of the hearth is exposed to the blast of hot air and the burning gases used to heat the steel. Charges of 100 tons or more include pig iron, rusty scrap iron or smaller amounts of iron ore, and materials for special alloys. As the charge melts, chemical reactions convert the molten mass into steel. Carbon dioxide gas bubbles out of the melt, and the oxides of sulfur and phosphorus combine with the basic oxides of the lining to form slag. Since the process is slow, each batch of steel can be analyzed as needed to ensure quality control. This is the chief advantage of this method over the Bessemer process. Alloying metals such as manganese, nickel, chromium, molybdenum, tungsten, and vanadium may be added as needed before the charge is poured. When the steel is ready, the furnace is tapped and the steel is ready for fabrication.

The **basic oxygen process** *uses calcium oxide and pressurized oxygen to produce high quality steel.* Most steel today is made by this process which usually takes about one hour per batch (Figure 22·23). A batch of steel is about 300 metric tons. Molten pig iron, scrap iron, and powdered limestone are placed in a converter lined with a basic oxide such as calcium oxide. Pure oxygen, blown through a narrow tube called a lance at about 10 atmospheres pressure, oxidizes the impurities in the charge. No external heat source is required. The rapid, highly exothermic reactions cause the contents to remain mixed and molten.

22 The Chemistry of Metals
Chapter Review

Key Terms

basic oxygen process	22·10	metallurgy	22·8
Bessemer process	22·10	Monel metal	22·6
blast furnace	22·9	open-hearth method	22·10
bronze	22·4	ore	22·8
cast iron	22·10	pig iron	22·9
charge	22·9	roasting	22·8
froth flotation	22·8	slag	22·8
flux	22·8	slaked lime	22·2
gangue	22·8	smelting	22·8
lime	22·2		

Chapter Summary

The alkali metals have large atomic radii. This helps explain their low densities, low ionization energies, and high reactivities. Because they ionize so readily, the alkali metals do not occur naturally as free metals. Important sodium compounds are sodium hydroxide and sodium carbonate.

The alkaline earth metals are also not found free in nature. They are less reactive, however, than the alkali metals. Calcium and magnesium are the two most important alkaline earth metals. Alkaline earth oxides such as calcium oxide (lime) and calcium hydroxide (slaked lime) have many industrial uses.

Aluminum, in Group 3A, is the most abundant metal in the earth's crust. It is not found as a free metal. Its value in industry derives from its corrosion-resistance and good electrical conductivity.

Tin and lead are the metals of Group 4A. They have numerous uses as both the free metal and in alloys such as bronze and solder. Tin is used in plating "tin cans," and lead is used in plumbing.

Many transition metals, such as titanium, chromium, and zinc, are used to make alloys. Iron, cobalt, and nickel are also prized for their magnetic properties. Monel metal is a corrosion-resistant alloy of copper and nickel. Copper, silver, and gold are excellent conductors.

Ores are minerals used for the commercial production of metals. Metallurgy involves various procedures used to separate metals from their ores. In the purifying of metals the worthless gangue must be removed from the ore.

A blast furnace is used to reduce iron ore contained in the charge to metallic iron. This smelting process produces pig iron, which is further refined into steel.

Practice Questions and Problems

3. Why do the alkali metals have low densities and low ionization energies? 22·1

4. When sodium is exposed to oxygen, it forms a peroxide. When this compound is added to water, it gives a solution of sodium hydroxide. The same solution can be obtained by adding sodium to water. Write balanced equations for these reactions. 22·1

5. What are the group names for the elements in Group 1A and Group 2A? 22·2

6. What is lime? How is lime made from calcium carbonate, $CaCO_3$? 22·2

7. Which alkaline earth metals react with cold water and which do not? Write a typical equation that shows such a reaction. 22·2

8. Why do the alkaline earth metals as a group have higher densities than the alkali metals? 22·2

9. How does the reactivity of the Group 2A metals compare to that of the Group 1A metals? 22·2

10. Give two reactions by which $Ca(OH)_2$ can be made. 22·2

11. Write the equation that represents the setting of mortar. 22·2

12. Name an aluminum ore rich in Al_2O_3. *22·3*

13. Describe the physical properties of aluminum that make it a commercially valuable metal. *22·3*

14. Aluminum is produced by electrolysis. Write the electrode reactions for the process. *22·3*

15. What are some uses for tin and lead? *22·4*

16. Write equations for the preparation of lead from galena, its sulfide ore. *22·4*

17. What are the common oxidation states of chromium? *22·5*

18. What is the most important compound of titanium? Why is titanium commercially useful? *22·5*

19. What are hematite and magnetite? *22·6*

20. Write formulas for the oxides of iron. *22·6*

21. Calculate the mass percent of metal in each of the following mineral compounds.
  a. Fe_3O_4 **b.** Fe_2O_3 **c.** CoAsS *22·6*

22. Nickel is an important metal. Why? What is *monel* metal? *22·6*

23. What are the properties of cobalt and how is it used? *22·6*

24. Copper, silver, and gold are among the few metals that have been used for thousands of years. Explain. *22·7*

25. What is an ore? *22·7*

26. Name an ore that contains each element.
  a. silver **b.** copper *22·7*

27. How does froth flotation separate an ore mineral from its impurities? *22·8*

28. Reduction and not oxidation is required to convert the metal in a compound to the free metal. Why? *22·8*

29. List the steps involved in extracting a metal from its ore. Briefly describe the significance of each step. *22·8*

30. Write an equation that summarizes the reactions that take place in the upper regions of a blast furnace. *22·9*

31. How is slag formed in a blast furnace? *22·9*

32. What is steel? How does steel differ from pig iron and from cast iron? *22·10*

33. Describe the open hearth method for making steel. *22·10*

34. Name three different types of steel. Comment on their composition, characteristics, and use. *22·10*

Mastery Questions and Problems

35. Calculate the mass percent of aluminum in cryolite, Na_3AlF_6.

36. A sample of magnetite contains Fe_3O_4 76.0%, sand (SiO_2) 11.0%, and other impurities 13.0%. Calculate the percent by mass of iron in the ore.

37. The data table below shows the quantities of iron ore (Fe_2O_3) and coke (C) that react in a blast furnace to produce pig iron.
  a. Complete the table.
  b. Plot moles of iron ore (y-axis) versus moles of carbon (x-axis).
  c. From the slope of the line, determine the mole ratio by which iron reacts with carbon in a blast furnace.

Mass of Fe_2O_3 (kg)	Amount of Fe_2O_3 (mol)	Mass of C (kg)	Amount of C (mol)
8.65×10^4	___	1.95×10^4	___
1.26×10^5	___	2.84×10^4	___
2.01×10^5	___	4.54×10^4	___
6.56×10^5	___	1.48×10^5	___
9.61×10^5	___	2.17×10^5	___

38. Complete and balance the following equations.
  a. $Na + O_2 \longrightarrow$
  b. $Na + H_2O \longrightarrow$
  c. $Cs + O_2 \longrightarrow$

39. At which electrode is the free metal produced in the electrolysis of a metal compound? Explain your answer.

40. Describe the *basic oxygen* process. Why has this process largely replaced the open hearth method for steel production?

41. A commercially important source of titanium is ilmenite, $FeTiO_3$. Calculate the mass percent of titanium in ilmenite.

42. Describe and illustrate the electrolytic refining of copper.

43. Calculate the percentage of cobalt in cobaltite containing 72.6% CoAsS by mass. How many kilograms of cobalt would be contained in 1000 kg of the ore?

44. Describe the processes used to obtain the following elements from their natural sources.
a. Al   **b.** Cu   **c.** Mg

Critical Thinking Questions

45. Choose the term that best completes the second relationship.
a. automobile:exhaust   smelting: _____
   (1) flux   (3) metals
   (2) froth   (4) slag
b. sister:brother   beryllium: _____
   (1) cesium   (3) barium
   (2) lead   (4) sodium
c. wheat:flour   chalcopyrite: _____
   (1) cuprite   (3) cobalt
   (2) copper   (4) electrolysis

46. What advantage does the recycling of aluminum have over the production of aluminum from bauxite ore?

47. The period of human history called the Bronze Age began about 3000 B.C., and was followed by the Iron Age. Why was bronze widely used for tools and weapons before iron?

Review Questions and Problems

48. Write the chemical formula for each of the following ionic compounds.
a. potassium sulfide   **d.** sodium iodide
b. barium bromide   **e.** calcium oxide
c. aluminum fluoride

49. Write complete electron configurations for the following.
a. Fe   **e.** Cu
b. V   **f.** Ni^{2+}
c. Ag   **g.** Zn
d. Fe^{3+} **h.** Ag^+

50. How many outer-shell d electrons are in each of these transition metal ions?
a. Mo^{3+} **c.** Ir^{3+}
b. Cu^{2+} **d.** Fe^{3+}

51. A sample of spring water contains 46.0 mg of magnesium ions per liter.
a. What is the molarity of Mg^{2+} in the sample?
b. What volume of this water contains 1.00 mol Mg^{2+}?

52. What are the ions produced when the following substances are dissolved in water?
a. LiBr   **c.** KI
b. $Ca(OH)_2$ **d.** $CdBr_2$

53. Classify each of the following compounds as a nonelectrolyte, a weak electrolyte, or a strong electrolyte.
a. CH_3COCH_3 (acetone)   **d.** KNO_3
b. C_2H_5OH (ethyl alcohol) **e.** Na_2SO_3
c. CH_3COOH (acetic acid) **f.** $CaCl_2$

54. Balance the following equation.
$$IO_3^-(aq) + Cr(OH)_4^-(aq) \longrightarrow$$
$$I^-(aq) + CrO_4^{2-}(aq) \text{ (basic)}$$

55. Determine the oxidation number of nitrogen in each of the following.
  a. N_2 **b.** NO_3^- **c.** NO_2 **d.** N_2O **e.** NO_2^-

56. Identify the oxidation and reduction half-reactions in the following equation.
$$2Ti^{2+}(aq) + Co^{2+}(aq) \longrightarrow 2Ti^{3+}(aq) + Co(s)$$

57. Identify the oxidizing agent and the reducing agent in the following reaction.
$$I_2O_5(s) + 5CO(g) \longrightarrow 5CO_2(g) + I_2(s)$$

58. Give the chemical formula for each of these compounds.
a. copper(I) bromide   **c.** cadmium sulfide
b. zinc nitride   **d.** silver iodide

Challenging Questions and Problems

59. Calculate the mass in kilograms of sulfur dioxide produced in the roasting of one thousand kilograms of chalcocite ore containing Cu_2S 7.2%, Ag_2S 0.6%, and no other sulfur compounds.

60. A 0.50 g sample of metallic sodium is converted to NaCl by reaction with chlorine gas. What volume of Cl_2 at 20°C and 740 mm Hg will be required to react completely with the sodium?

61. Azurite, a dark blue copper mineral shown below, has the following composition: Cu 55.3%, C 6.97%, O 37.1%, H 0.585%. Determine the simplest formula of this mineral.

62. Write the equation for the reaction that occurs when $Mg(OH)_2$ dissolves in hydrochloric acid. Use this equation to answer the following questions.
  a. How many moles of H^+ are required to react with 3.20 g $Mg(OH)_2$?
  b. Calculate the volume of 3.00 M HCl required to react with 3.20 g $Mg(OH)_2$?

63. Barium peroxide decomposes above 700°C according to this equation.

$$2BaO_2(s) \longrightarrow 2BaO(s) + O_2(g)$$

The oxygen liberated when 20.0 g of barium peroxide is heated, is collected in a 1-L flask at 25°C. Calculate the pressure in the flask.

64. How many grams of Co_3O_4 must react with an excess of aluminum to produce 500 g of cobalt metal, assuming 75% yield? The equation for the reaction is as follows.

$$3Co_3O_4 + 8Al \xrightarrow{\text{heat}} 9Co + 4Al_2O_3$$

Research Projects

1. How do growing plants use potassium? Why is it that other alkali metals cannot be substituted?

2. How did Charles Hall and Paul Heroult improve on the work of Henri Sainte Claire Deville?

3. How is aluminum recycled? What proportion of the aluminum used today is recycled aluminum?

4. How can rare metals be produced in space? How does this compare with their production on earth?

5. How has the search for gold been significant in history?

6. How are silver mirrors made?

7. Make a diagram of how mercury gets into the food chain.

8. How were iron tools developed in the Iron Age?

9. Make a timeline of the historical developments in metallurgy.

10. Collect samples of various small metallic objects. Use density determinations to decide if the objects are pure metals or alloys.

11. Devise a method to remove the iron from a sample of iron-fortified cereal.

12. Why were the nineteenth century advances in metallurgy important?

13. How does water pollution occur from mining mineral deposits? What can be done to prevent it?

14. Report on a land reclamation project in a strip-mined area.

Readings and References

Coombs, Charles. *Gold and Other Precious Metals.* New York: Morrow, 1981.

Hartmann, William R. "Mines in the Sky Promise Riches, a Greener Earth." *Smithsonian* (September 1982), pp. 70–77.

Kerrod, Robin. *Metals.* Morristown, NJ: Silver Burdett, 1982.

Noguchi, Thomas. "The Wrong Knife." *ChemMatters* (December 1985), p. 14.

Orna, Mary Virginia. "Paintmaking Adventures." *ChemMatters:* (December 1984), pp. 12–13.

Page, Jake. "Moly Be-Damned." *Science 84* (July/August 1984), pp. 48–51.

23 The Chemistry of Nonmetals

Chapter Preview

23·1 Hydrogen and Its Compounds

23·A Reaction Alternatives: Preparing Hydrogen

23·2 Oxygen

23·3 Compounds of Oxygen

23·B The Frasch Process

23·4 Sulfur and Its Compounds

23·C Sulfuric Acid

23·5 Nitrogen

23·6 Compounds of Nitrogen

23·7 Preparation of the Halogens

23·8 Properties of the Halogens

The nonmetallic elements are found in the upper right corner of the periodic table. Unlike metals, nonmetals are poor conductors of heat and electricity. Bonding between nonmetal atoms is primarily covalent. Pairs of electrons are shared and the compounds formed consist of molecules. The electronegativities of nonmetals are high. Compounds formed between metals and nonmetals tend to be ionic, with metals as cations and nonmetals as anions.

23·1 Hydrogen and Its Compounds

Hydrogen is a common component of many compounds.

Hydrogen and helium are the most abundant elements in the universe. Free, or elemental, hydrogen is very rare on Earth, but compounds of hydrogen are common. They account for about one percent of the earth's crust. Water is the most abundant hydrogen compound. Combined with carbon and oxygen, hydrogen is present in all sugars, starches, fats, and proteins. These complex compounds are abundant in living tissues. Coal, natural gas, and petroleum products such as gasoline, kerosene, and lubricating oils also contain hydrogen.

Alchemists undoubtedly produced hydrogen long ago when they reacted certain metals with acids and obtained a colorless, odorless gas. Sir Henry Cavendish is generally given credit, however, for the discovery of hydrogen in 1766. At that time, he thought hydrogen came from the metal rather than from the acid. Cavendish called the gas "inflammable air." He recognized that it burned in air to produce water. Antoine Lavoisier showed that hydrogen is a component of water. He named it *hydrogen* from the Greek words *hydro,* water, and *genes,* forming.

Figure 23·1
The halogens have many similar chemical properties. From left to right are chlorine, bromine, and iodine.

Figure 23·2
Hydrogen is seldom found free in nature but is present in a large number of compounds. All these items contain compounds of hydrogen.

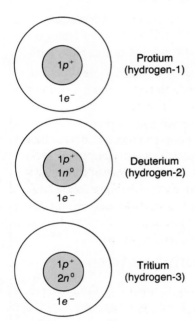

Figure 23·3
Most naturally occurring hydrogen atoms are hydrogen-1 or protium. About one in every 6,000 hydrogen atoms in sea water is deuterium. Tritium is rarely found in nature but can be produced artificially by certain nuclear reactions. How do these three isotopes differ?

Naturally occurring hydrogen is composed of three isotopes (Figure 23·3). Protium (hydrogen-1) and deuterium (hydrogen-2) account for 99.98% and 0.02% of a sample, respectively. The most abundant isotope is protium. It is commonly called hydrogen. Tritium, an unstable form of hydrogen, is present in extremely small amounts.

Hydrogen burns readily in oxygen with a hot, colorless flame. A mixture of hydrogen and air in a test tube produces a small explosion when brought near an open flame. This result confirms the presence of hydrogen gas in a laboratory experiment.

Hydrogen combines directly with a number of metallic and non-metallic elements. With chlorine, it forms hydrogen chloride (HCl); with calcium, calcium hydride (CaH_2); with nitrogen (at 1000 atm and 500°C); ammonia (NH_3). Because it loses its single electron easily, hydrogen is a good reducing agent. When passed over heated calcium oxide, for example, it reduces the metal oxide to the free metal.

$$CaO(s) + H_2(g) \xrightarrow{\Delta} Ca(s) + H_2O(g)$$

The major use of hydrogen is in the manufacture of ammonia. It is also used to reduce certain metal oxides to the free metal. Metals that can be produced from their oxide by hydrogen reduction include copper, silver, tungsten, and iron.

Large volumes of hydrogen are used in the conversion of vegetable oils such as peanut and coconut oil into solid fats. This process, called **hydrogenation,** *involves treating an oil with hydrogen at high temperature and pressure in the presence of a catalyst.* Finely divided nickel or platinum is commonly used. Solid shortenings and margarine are produced in this way.

Liquid hydrogen has become an important rocket fuel because of its high energy and light weight. Hydrogen gas was once used to fill lighter-than-air balloons and dirigibles. Helium, a nonflammable gas, is now used instead. The oxyhydrogen torch is used to cut and weld metals. It utilizes the intense heat produced when hydrogen burns in oxygen.

Figure 23·4
Mixtures of hydrogen and oxygen explode violently when ignited by a spark. The Hindenburg was only one of 73 hydrogen-filled dirigibles that burned. The nonflammable gas helium is now used to inflate balloons and blimps.

Safety

Do not ignite any quantity of hydrogen larger than the volume of a small test tube. If oxygen has mixed with the hydrogen, the explosion could shatter the container, particularly if it is scratched or weakened from use.

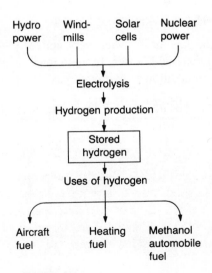

Figure 23·5
Hydrogen could become a more widely used source of energy if cheaper methods of producing it are devised. It could be produced during off-peak periods of electricity generation.

23·A Reaction Alternatives: Preparing Hydrogen

Supplies of hydrogen compounds are almost limitless. The gas can be produced from water, natural gas, coal, crude oil, and many other substances. The cost of production usually determines the method to be used commercially.

The electrolysis of water can produce very pure hydrogen. This method is only practical, however, in small scale operations. The raw material is cheap and plentiful, but the cost of electricity makes the process expensive. The apparatus is shown in Figure 21·12. A small amount of acid or base is added to the water to make it an electrical conductor. When the current flows, bubbles of hydrogen gas collect at the cathode and oxygen collects at the anode.

$$2H_2O(l) \xrightarrow{\text{electrical energy}} 2H_2(g) + O_2(g)$$

The volume of hydrogen produced is slightly more than twice that of oxygen. A little less oxygen than expected is produced because oxygen is more soluble in water than hydrogen.

Hydrogen can be prepared commercially from water by the Bosch process. In this method, steam is passed over red hot iron filings. The iron combines with the oxygen in steam and hydrogen gas is liberated.

$$3Fe(s) + 4H_2O(g) \longrightarrow Fe_3O_4(s) + 4H_2(g)$$

Hydrogen has great potential as a non-polluting fuel. Thus a great deal of research is being directed toward finding a cheaper method of producing hydrogen from water.

More than half of the hydrogen produced in North America is produced by the reaction between steam and natural gas. This method is known as steam reforming. It is the least expensive method to produce large volumes of this gas. The raw material is methane, the chief component of natural gas. It is reacted with steam over a finely divided nickel catalyst at 700°C to 1000°C.

$$CH_4(g) + H_2O(g) \xrightarrow{\text{catalyst}} CO(g) + 3H_2(g)$$

The carbon monoxide is removed by cooling and compressing the mixtures of gases. Hydrogen remains a gas after the carbon monoxide liquefies out of the mixture.

The first commercial process developed for hydrogen production involved passing steam over white-hot coke. This method also produces a mixture of carbon monoxide and hydrogen known as water gas.

$$C(s) + H_2O(g) \xrightarrow{\Delta} CO(g) + H_2(g)$$

The carbon monoxide is removed by oxidizing it to carbon dioxide, using steam and a catalyst. Additional hydrogen is formed.

$$CO(g) + H_2O(g) \xrightarrow{\text{catalyst}} CO_2(g) + H_2(g)$$

The mixture passes through a cold-water scrubber, which dissolves the carbon dioxide in water under moderate pressure. Production of hydrogen from coke declined as natural gas became readily available. Today, the process is more expensive than steam reforming and is practical only in small scale operations.

Small quantities of hydrogen are prepared in the laboratory by the reaction between certain metals and an acid. The apparatus for this method is shown in Figure 23·6. Dilute hydrochloric acid is poured down the thistle tube onto pieces of zinc in the flask. Hydrogen gas bubbles off as long as the metal and acid are in contact. The gas is usually collected over water.

$$Zn(s) + 2HCl(aq) \longrightarrow H_2(g) + ZnCl_2(aq)$$

Hydrogen is also produced in the laboratory when very active metals such as sodium and potassium react with water. In these reactions, the metal hydroxide is formed in addition to hydrogen gas.

$$2Na(s) + 2H_2O(l) \longrightarrow 2NaOH(aq) + H_2(g)$$

Safety

A hydrogen generator flask should be wrapped in a towel to contain the glass in case of an explosion. Never ignite hydrogen coming from a generator.

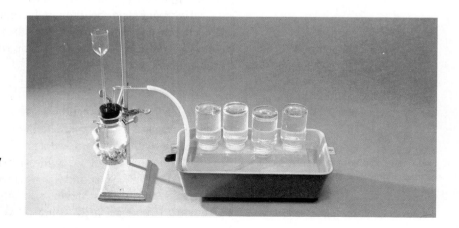

Figure 23·6
Hydrogen is prepared in the laboratory by adding dilute hydrochloric acid ⁀ ...ubbles of hydrogen displace ...collecting bottle. What is ...uct of the reaction?

549

23   The Chemistry of Nonmetals

Figure 23·7
Oxygen is often administered to emergency medical patients who have difficulty breathing.

23·2 Oxygen

Of the elements that make up the earth's crust, oxygen is by far the most abundant. Its total mass is nearly equal to all of the other elements combined. Free oxygen exists as diatomic covalent molecules, O_2. The gas is a major component of the atmosphere, accounting for 21% of the air, by volume. Most of this oxygen is the product of photosynthesis. In this process, green plants manufacture carbohydrates by combining carbon dioxide and water with the aid of sunlight. Glucose, a simple sugar, is one product; oxygen is the other.

$$6CO_2(g) \ + \ 6H_2O(l) \ \xrightarrow{\text{light energy}} \ C_6H_{12}O_6(aq) \ + \ 6O_2(g)$$

Most of the oxygen on earth, however, is in compounds that are the constituents of rocks and soils. Quartz and sand (SiO_2), limestone and marble ($CaCO_3$), and gypsum ($CaSO_4$) are common oxygen-containing minerals. Clay has a complex composition of aluminum, silicon, and oxygen. Water and the organic molecules that make up living tissues also contain oxygen.

Karl Wilhelm Scheele, a Swedish pharmacist, and Joseph Priestley, an English clergyman, carried out early experiments with oxygen. In 1771, Scheele prepared oxygen by heating a mixture of concentrated sulfuric acid and manganese dioxide.

$$2MnO_2(s) \ + \ 2H_2SO_4(l) \ \longrightarrow \ 2MnSO_4(aq) \ + \ 2H_2O(l) \ + \ O_2(g)$$

Priestley decomposed mercury(II) oxide by heating it, in 1774 (Figure 2·4). When the French chemist Antoine Lavoisier learned of their work, he carried out further experiments (Section 2·A). Lavoisier gave oxygen its name, which means "acid forming."

Small quantities of oxygen can be generated in the laboratory by heating unstable oxygen-containing compounds. These include mercury(II) oxide and potassium chlorate (Figure 23·8).

$$2KClO_3(s) \ \xrightarrow{\Delta} \ 2KCl(s) \ + \ 3O_2(g)$$

Figure 23·8
When potassium chlorate is heated, it decomposes to yield potassium chloride and oxygen. Manganese dioxide is used as a catalyst.

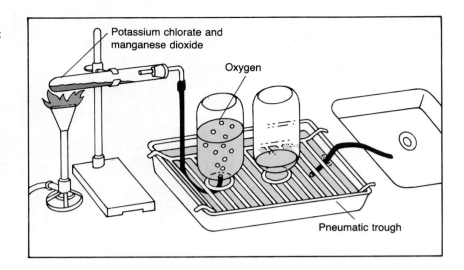

Potassium chlorate and manganese dioxide

Oxygen

Pneumatic trough

Such methods are unsuitable for large scale oxygen production because the starting materials are expensive. The primary source of commercial oxygen is air. Relatively pure oxygen is separated from the other gases present by liquefying air and then fractionally distilling the liquid. Nitrogen (b.p. $-196°C$) comes off first, followed by the noble gas argon (b.p. $-186°C$). Oxygen distills off at $-183°C$.

When cooled, oxygen condenses to a clear blue liquid that freezes at $-218°C$. In liquid form, oxygen is attracted to a magnet. This property, called paramagnetism, is caused by the presence of unpaired electrons in the oxygen molecule.

LOX is the trade name for liquid oxygen.

$$:\overset{..}{\underset{..}{O}}:\overset{..}{\underset{..}{O}}:$$

The major commercial use of oxygen is in the basic oxygen process for the manufacture of steel. Oxygen is also used in oxyhydrogen and oxyacetylene torches for welding metals and glassworking. Fuels burn in oxygen at a higher temperature than they do in air. The flame temperature in an oxyacetylene torch is about 2500°C. The fuel and oxygen are brought together in a specially designed torch (Figure 20·4). Acetylene is a clean fuel because it burns to produce carbon dioxide and water.

 Issues in Chemistry
A kerosene space heater uses up the oxygen from the air in the room where it is burning. Apart from oxygen depletion, are there other concerns that must be addressed when using a kerosene heater?

$$2C_2H_2(g) \ + \ 5O_2(g) \ \longrightarrow \ 4CO_2(g) \ + \ 2H_2O(g) \ + \ heat$$

Oxygen tanks are carried in airplanes for use at high altitudes. Medical emergency teams administer oxygen to victims of smoke inhalation, electrical shock, or drowning. For certain medical conditions such as pneumonia, emphysema, or gas poisoning, a patient may need to breathe air enriched with oxygen for long periods of time.

Oxygen exists in a second form, ozone, which has the molecular formula O_3. The two forms are referred to as allotropes of oxygen. **Allotropes** *of an element are two or more different molecular forms in the same physical state.* Ozone is formed in the earth's upper atmosphere by the action of ultraviolet light on oxygen. This decreases the amount of ultraviolet light which penetrates the atmosphere. Without this process,

Oxygen exists in two allotropic forms.

Figure 23·9
Large quantities of ozone are produced for industrial use. This form of oxygen is used as a bleaching agent and as a sterilizing agent.

ChemDirections

Compounds used on earth have begun to damage the ozone layer. As described in **ChemDirections** *Ozone Layer,* page 665, ozone, an allotrope of oxygen, is important in protecting the earth from the harmful ultraviolet radiation of the sun.

Ozone is an extremely powerful oxidant.

living things would be damaged by the additional ultraviolet light that would reach the earth's surface. The process of ozone formation occurs in two stages.

$$O_2 \xrightarrow{\text{ultraviolet light}} 2O$$

$$O_2 + O \longrightarrow O_3$$

Ozone is also produced near high-voltage generators and during electrical storms.

Ozone is a pale blue gas with a characteristic odor. When cooled to $-112°C$, it condenses into a deep blue, explosive liquid. Ozone is a strong oxidizing agent. It is used commercially to bleach flour, oil, and delicate fabrics. It is also used to sterilize water, and to destroy odors.

Because ozone is unstable, it must be generated where it will be used. It is prepared by passing oxygen through an electrical discharge. The structure of an industrial ozonator is shown in Figure 23·10. Air is passed through glass tubes partially covered with metal foil. The tubes are placed one within the other, and the pieces of foil are connected to a high-voltage source. When the current is on, electrical discharges pass between the layers of foil. Some of the oxygen moving through the generator is converted to ozone.

Figure 23·10
Ozone is produced by passing air through concentric tubes that are covered with aluminum foil. High voltage discharges between the tubes convert some of the oxygen to ozone.

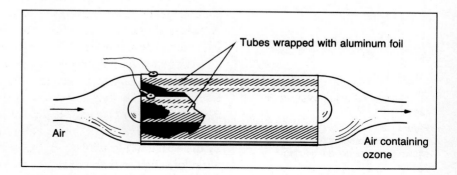

23·3 Compounds of Oxygen

When substances combine chemically with oxygen, the process is called oxidation and the substance is oxidized. Substances oxidize at different rates. A lighted candle oxidizes rapidly, but iron oxidizes (rusts) slowly. The product of an oxidation reaction is called an oxide. In general, oxides of metals are solids while oxides of nonmetals may be solid, liquid, or gas. The oxides of metals and nonmetals differ chemically from each other in one important respect. Many metallic oxides react with water to form bases. Many nonmetallic oxides, however, react with water to form acids.

Oxides of all elements, except the noble gases and a few inactive metals, can be prepared simply by heating the element with oxygen. Here are some examples.

$$2Hg(l) + O_2(g) \longrightarrow 2HgO(s)$$
$$3Fe(s) + 2O_2(g) \longrightarrow Fe_3O_4(s)$$
$$S(s) + O_2(g) \longrightarrow SO_2(g)$$

A few oxides such as mercury(II) oxide are easily decomposed by heating. Others like calcium oxide are stable even at a temperature of 3000°C.

While most metals yield oxides in reaction with oxygen, a few also yield peroxides. Peroxides are compounds that contain the peroxide ion, O_2^{2-}. For example, barium reacts with oxygen to form barium peroxide.

$$Ba(s) + O_2(g) \longrightarrow BaO_2(s)$$

Hydrogen peroxide, H_2O_2, is one of the most important peroxides. It is a colorless, viscous, highly unstable liquid. Dilute aqueous solutions of hydrogen peroxide, usually 3% by volume, are used in homes as a bleach and antiseptic. The Lewis structure of hydrogen peroxide is shown here.

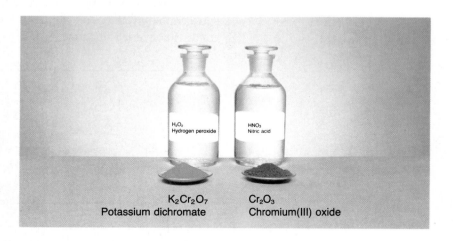

Figure 23·11
Oxygen-rich compounds like those shown here are good oxidizing agents.

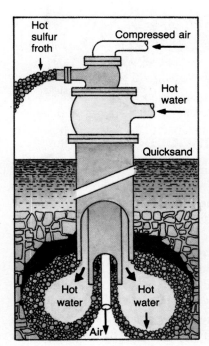

Figure 23·12
Sulfur deposits deep within the earth are mined by the Frasch process. The sulfur, melted by hot water and whipped into a froth by compressed air, rises to the surface.

A metric ton is 1000 kg.

23·B The Frasch Process

Some of the world's richest deposits of elemental sulfur are on the Gulf Coast in Louisiana and Texas. These deposits are not easily mined because they are buried under several hundred feet of quicksand. A German–American engineer, Herman Frasch (1851–1914), devised a clever method for getting this sulfur out of the ground. The method takes advantage of sulfur's low melting point (119°C). Eighty percent of the world's sulfur output is produced by the Frasch process.

Wells are drilled into the sulfur bed. Then an arrangement of three concentric tubes (with 2.5 cm, 7.5 cm, and 15 cm diameters) is installed (Figure 23·12). Superheated water (180°C) under pressure is pumped down the outside tube to melt the sulfur. Compressed air is pumped down the center tube. A frothy mixture of air, water, and molten sulfur rises up the third tube. The molten sulfur, 99.5% pure, is pumped into large storage vats where it cools and solidifies into huge blocks. These blocks may be as much as 130 m long, 70 m wide, and 30 m high. For shipping, the sulfur is loosened by dynamite blasting and loaded into freight cars.

Large amounts of sulfur are also obtained from hydrogen sulfide gas, a product of petroleum refining. Some of the hydrogen sulfide is burned in air to produce sulfur dioxide. The sulfur dioxide is then treated with more hydrogen sulfide. This can be summarized in a single reaction.

$$2H_2S(g) + SO_2(g) \longrightarrow 2H_2O(g) + 3S(s)$$

The annual production of sulfur in North American is more than ten million metric tons.

Figure 23·13
After being mined, sulfur is dried and stored in huge blocks until it is shipped.

Crystalline sulfur

Molten sulfur

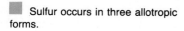

Sulfur occurs in three allotropic forms.

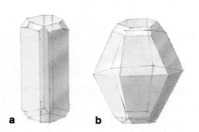

23·4 Sulfur and Its Compounds

Sulfur has been known since ancient times to be a pale yellow, tasteless, odorless, brittle solid. It is insoluble in water, but dissolves in carbon disulfide. Today, sulfur is an extremely important raw material in the chemical industry. Large amounts of sulfur are used in the vulcanization of rubber and the manufacture of paper from wood pulp. It is also used in the preparation of fungicides, fertilizers, matches, fireworks, insecticides, explosives, drugs, paints, plastics, and dyes.

Sulfur occurs in several allotropic forms. When molten sulfur cools below 119°C, it forms needle-like crystals of *monoclinic* sulfur. In this allotrope, eight-membered rings of covalently bonded sulfur atoms (S_8) are arranged in monoclinic crystals (Figure 23·15). Below 95.5°C, the sulfur changes to *rhombic* sulfur, another allotrope. Rhombic sulfur is also composed of S_8 units, but it has orthorhombic crystals. Rhombic sulfur is stable at ordinary temperatures.

As molten sulfur is heated, it becomes a deep red-brown color and its viscosity increases. At about 160°C, it is so thick it cannot be poured. The viscosity decreases, however, as the temperature nears the boiling point (445°C). Apparently, as sulfur melts, the S_8 rings begin to break, producing short chains. With continued heating, they join to produce longer chains. The longer chains become entangled so that the viscosity increases, reaching a maximum at 160°C. Above this temperature, the long chains begin to break up and the viscosity decreases.

A third form of sulfur is the dark brown rubber-like allotrope known as amorphous or "plastic sulfur." This can be produced by pouring molten sulfur into cold water when the sulfur is near its boiling point. Amorphous solids do not have a distinct crystal form. Within hours, the amorphous sulfur loses its elasticity as it converts to rhombic crystals.

Sulfur can be purified by boiling it and condensing the vapor. Tiny rhombic crystals form in flower-like patterns on the walls of the condensation chamber. This sulfur powder is called "flowers of sulfur."

Sulfur is a reactive element. When heated, it burns readily in air or oxygen, producing an irritating, toxic gas, sulfur dioxide. In the laboratory, sulfur dioxide is prepared by the action of an acid on a sulfite.

$$Na_2SO_3(aq) + 2HCl(aq) \longrightarrow 2NaCl(aq) + H_2O(l) + SO_2(g)$$

Sulfur dioxide will combine with more oxygen in the presence of a catalyst to form sulfur trioxide.

$$2SO_2(g) + O_2(g) \xrightarrow{\text{catalyst}} 2SO_3(g)$$

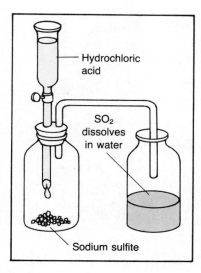

Figure 23·17
Sulfur dioxide can be prepared by the action of hydrochloric acid on sodium sulfite. Sulfurous acid, H_2SO_3, is formed when the SO_2 dissolves in and reacts with water.

Sulfuric acid is the most important manufactured chemical.

Safety

Never add water to concentrated sulfuric acid. The heat generated may cause the water to boil, spattering the acid.

This reaction of sulfur dioxide with oxygen is the basis for the manufacture of sulfuric acid. Sulfur dioxide is also used as a bleaching agent and a fumigant.

When heated with metals (except gold and platinum), sulfur forms compounds called sulfides.

$$Fe(s) + S(s) \longrightarrow \underset{\text{iron(II) sulfide}}{FeS(s)}$$

$$2Cu(s) + S(s) \longrightarrow \underset{\text{copper(I) sulfide}}{Cu_2S(s)}$$

With active metals like magnesium and zinc, the reaction is highly exothermic.

Hydrogen combines with sulfur to form hydrogen sulfide, a poisonous gas that smells like rotten eggs. In the laboratory, hydrogen sulfide is prepared by the action of an acid on a sulfide.

$$FeS(s) + 2HCl(aq) \longrightarrow FeCl_2(aq) + H_2S(g)$$

Science, Technology, and Society

23·C Sulfuric Acid

The principal use of sulfur is in the manufacture of sulfuric acid. In North America alone, more than 8×10^7 metric tons of sulfuric acid are produced annually. The two main methods of making sulfuric acid are the lead chamber process and the contact process. Both processes convert sulfur dioxide to sulfur trioxide. The latter forms sulfuric acid when dissolved in water. Most of the sulfuric acid produced in this country is made by the contact process, described in Section 17·A. The product from the lead chamber process is less pure, but satisfactory for some commercial uses.

Pure, anhydrous sulfuric acid is a dense, colorless, oily liquid. The concentrated acid used in laboratories is 98% H_2SO_4 and 2% H_2O. Sulfuric acid has the chemical properties of a typical acid. The dilute acid reacts with metals, oxides, hydroxides, or carbonates to form salts called sulfates. The reaction with metals also releases hydrogen gas. Sulfuric acid displaces other acids from their salts. For example, hydrogen chloride can be produced from sulfuric acid and sodium chloride. When dissolved in water it forms hydrochloric acid.

$$H_2SO_4(l) + 2NaCl(s) \longrightarrow Na_2SO_4(s) + 2HCl(g)$$

Concentrated sulfuric acid dissolves readily in water, liberating a large amount of heat. This strong affinity for water makes sulfuric acid a powerful dehydrating agent. Thus it must be handled with great care. For example, when the concentrated acid is added to sugar, a vigorous reaction takes place. A large volume of charred material is produced as the acid removes water, leaving only carbon.

$$C_{12}H_{22}O_{11}(s) + 11H_2SO_4(l) \longrightarrow 12C(s) + 11H_2SO_4 \cdot H_2O(aq)$$

Figure 23·18
Sulfuric acid was used to manufacture the paper of this book and many other objects you use daily.

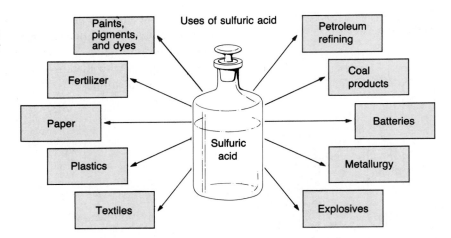

Uses of sulfuric acid

Paints, pigments, and dyes · Petroleum refining · Fertilizer · Coal products · Paper · Batteries · Plastics · Sulfuric acid · Metallurgy · Textiles · Explosives

Hot concentrated sulfuric acid is an oxidizing agent. It reacts with certain metals to generate sulfur dioxide.

$$2H_2SO_4(l) + Cu(s) \longrightarrow CuSO_4(aq) + SO_2(g) + 2H_2O(l)$$

Practically every industry uses some sulfuric acid. Almost half the world's production goes into the manufacture of fertilizers such as ammonium sulfate and superphosphate. Large amounts of sulfuric acid are used in the metal industry for pickling iron and steel. *Pickling* involves putting a metal in a chemical bath to remove oxides on the metal surface. Sulfuric acid is also essential in petroleum refining. Other uses are shown in Figure 23·18.

23·5   Nitrogen

Air is about 80% elemental nitrogen by volume. Nitrogen is isolated commercially by two processes. In one method, nitrogen is fractionally distilled from liquefied air. Since liquid nitrogen boils at a lower temperature than liquid oxygen, nitrogen distills off first. In the second method, air is passed over red-hot coke. The carbon combines with oxygen to form carbon dioxide, while the nitrogen remains unchanged.

Free nitrogen is a colorless, odorless, tasteless gas, composed of diatomic molecules, N_2. It is slightly soluble in water. Its melting point is $-210°C$ and its boiling point is $-196°C$.

Molecular nitrogen is extremely stable. With an electronegativity of 3.0, we would expect nitrogen to be a reactive element. Molecular nitrogen resists reaction, however, because of the strong attraction nitrogen atoms have for each other. In a molecule of elemental nitrogen, the two atoms share three pairs of electrons, forming a triple covalent bond. The bond energy is very large: 226 kcal/mol.

Despite the stability of atmospheric nitrogen, numerous nitrogen compounds are found in nature. Large deposits of sodium nitrate (Chile saltpeter) and potassium nitrate (saltpeter) exist, and soils may contain ammonium compounds. In addition, nitrogen is an essential component

Figure 23·19
Liquid nitrogen boils away quickly when poured from an insulated flask into a beaker at room temperature. The smoke-like effect in the beaker is caused by the condensation of water vapor from the air.

The nitrogen molecule, N_2, is extremely stable.

of many organic compounds found in both plants and animals. These organic compounds include amino acids, proteins, and nucleic acids.

In the laboratory, nitrogen can be prepared by heating certain unstable nitrogen compounds. When an aqueous solution containing ammonium ions and nitrite ions is carefully heated, it decomposes, yielding nitrogen gas and water.

$$NH_4^+(aq) + NO_2^-(aq) \xrightarrow{\Delta} N_2(g) + 2H_2O(l)$$

Solid ammonium nitrite is unstable and is potentially explosive.

23·6 Compounds of Nitrogen

In its compounds, nitrogen exhibits more oxidation states than any other element. It can have all possible oxidation states between +5 and −3.

The most important use of atmospheric nitrogen is in the manufacture of nitrogen compounds. Two important industrial processes use nitrogen: the Haber–Bosch process for making ammonia (Section 17·C) and the Ostwald process for preparing nitric acid.

Ammonia is a colorless gas with a strong odor. The gas is extremely soluble in water. At room temperature, about 700 volumes of ammonia dissolve in one volume of water! The resulting solution is called aqueous ammonia. Ammonia is easily liquefied by cooling. Liquid ammonia boils at −33°C and it solidifies at −78°C. Liquid ammonia exhibits many water-like properties because ammonia molecules are polar. Hydrogen bonds can form between the hydrogen atoms of one ammonia molecule and the nitrogen atoms of a nearby molecule. Evidence of this hydrogen bonding is ammonia's high boiling and melting points and its high heat of vaporization for such a small molecule.

Ammonia and nitric acid are the two most important nitrogen compounds.

Liquid ammonia, aqueous ammonia, and several ammonium salts are used as fertilizers. Liquid ammonia is also called anhydrous ammonia. It is sometimes applied under pressure directly to the soil, or it may be bubbled into irrigation water. Ammonium sulfate is the most important solid fertilizer in the world today.

Figure 23·20
Pure, liquid ammonia is called anhydrous ammonia and is used extensively as a fertilizer (below left). Ammonia used for agricultural purposes is kept under pressure in a tank (below right).

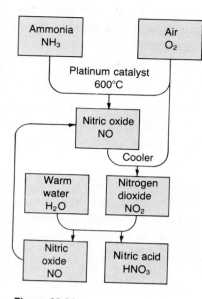

Figure 23·21
The Ostwald process produces nitric acid by a two-stage oxidation of ammonia to nitrogen dioxide. In the third step, the nitrogen dioxide is passed through warm water, forming nitric acid.

Liquid ammonia is used as a refrigerant, particularly in the frozen food industry. Aqueous ammonia is weakly basic and is a component of many cleaning products. Household ammonia is approximately a two molar aqueous solution of ammonia with detergent added. About one-quarter of all the ammonia produced is made into nitric acid. This acid is used in etching processes and the production of fertilizers and dyes. It is also an important raw material in the manufacture of explosives. Nitro-cellulose (gun cotton), nitroglycerine, and trinitrotoluene (TNT) are all made from nitric acid.

Nitric acid is made in the laboratory by the action of hot concentrated sulfuric acid on sodium nitrate.

$$NaNO_3(s) + H_2SO_4(l) \longrightarrow NaHSO_4(s) + HNO_3(g)$$

Large amounts of nitric acid are made commercially by the Ostwald method. The process was invented by a German chemist, Wilhelm Ostwald (1854–1932). It involves three steps (Figure 23·21).

1. Ammonia is oxidized by air in the presence of a catalyst to form nitric oxide (nitrogen monoxide). The mixture of air and ammonia is heated to about 600°C before contact with the catalyst, a cylinder of platinum gauze.

$$4NH_3(g) + 5O_2(g) \xrightarrow{600°C/Pt} 4NO(g) + 6H_2O(g)$$

2. More air is introduced into the system. It oxidizes the nitric oxide to nitrogen dioxide.

$$2NO(g) + O_2(g) \longrightarrow 2NO_2(g)$$

3. After cooling, the nitrogen dioxide is passed through water, forming a solution of nitric acid.

$$3NO_2(g) + H_2O(l) \longrightarrow 2HNO_3(aq) + NO(g)$$

The nitric oxide, a by-product of this third reaction, is recycled.

23·7 Preparation of the Halogens

The halogens form a large number of widely distributed compounds. Most of these compounds are soluble in water. Thus halide ions are abundant in seawater and in salt beds formed by the evaporation of salt water (Table 23·1).

Fluorine is the most electronegative element. Although it had long been known in its combined form, the free element was not isolated until 1886. French chemist Henri Moissan made it by electrolyzing an ice-cold solution of potassium fluoride in hydrogen fluoride. Fluorine is still made this way today. The electrolytic cell is made of a Monel metal and steel with carbon electrodes.

Chlorine gas is made commercially by the electrolysis of a sodium chloride solution (see Section 21·9). In a separate process, the hydrogen also produced is burned in the chlorine to make hydrogen chloride gas.

Except for fluorine, the halogens are prepared from the salts of sea water or salt beds.

Table 23·1
Concentration of Halide Ions in Seawater

Ion	g/L
F^-	1.3×10^{-3}
Cl^-	1.9×10^{1}
Br^-	6.5×10^{-2}
I^-	5.0×10^{-5}

This is an important source of hydrochloric acid. In the laboratory, chlorine is made by heating concentrated hydrochloric acid and manganese dioxide (Figure 23·22).

$$4HCl(aq) + MnO_2(s) \longrightarrow MnCl_2(aq) + 2H_2O(l) + Cl_2(g)$$

Bromine is obtained commercially from seawater and salt-well brines. Sodium chloride in the water is allowed to crystallize out, leaving a solution that contains the more soluble bromides. Chlorine gas, more electronegative than bromine, is then used to displace bromide ions from the solution.

$$2NaBr(aq) + Cl_2(g) \longrightarrow 2NaCl(aq) + Br_2(l)$$

In the laboratory, bromine is prepared by adding hot concentrated sulfuric acid to a mixture of potassium bromide and manganese dioxide (Figure 23·22). Hydrogen bromide is produced, then oxidized to bromine.

$$KBr(s) + H_2SO_4(l) \longrightarrow KHSO_4(s) + HBr(aq)$$

$$4HBr(aq) + MnO_2(s) \longrightarrow MnBr_2(aq) + 2H_2O(l) + Br_2(g)$$

At one time, iodine was extracted from the ashes of certain seaweeds, which concentrate iodine from seawater. Now it is produced commercially from sodium iodate, which occurs as an impurity in deposits of sodium nitrite.

$$2NaIO_3(aq) + 5NaHSO_3(aq) \longrightarrow$$
$$2Na_2SO_4(aq) + 3NaHSO_4(aq) + H_2O(l) + I_2(s)$$

The laboratory preparation of iodine is similar to that of bromine. A mixture of potassium iodide and manganese dioxide is treated with hot concentrated sulfuric acid.

$$2NaI(s) + MnO_2(s) + 2H_2SO_4(l) \longrightarrow$$
$$Na_2SO_4(s) + MnSO_4(s) + 2H_2O(l) + I_2(g)$$

The iodine vapor condenses on a cold surface, forming black crystals.

Figure 23·22
The preparation of both chlorine and bromine should be carried out in a hood. The chlorine is collected by upward displacement of air. What does this tell you about the density of chlorine compared to that of air?

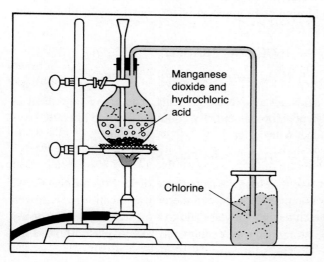

Manganese dioxide and hydrochloric acid

Chlorine

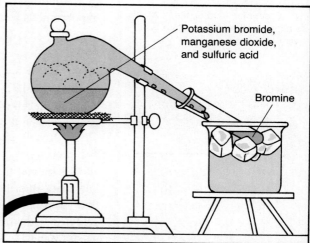

Potassium bromide, manganese dioxide, and sulfuric acid

Bromine

Figure 23·23
A heated piece of steel wool reacts vigorously in chlorine gas. The "smoke" is actually fine particles of the iron(III) chloride which form. Would the reaction be more or less vigorous in bromine? in fluorine?

The halogens are very reactive. Their reactivities decrease as the atomic mass increases.

23·8 Properties of the Halogens

Although the halogens are similar in many ways, their appearances vary considerably (Figure 23·1). Bromine and iodine are not gases at room temperature, but they vaporize easily. When heated, iodine crystals sublime: they change directly from a solid to a gas, forming a dense cloud of purple vapor. Bromine is slightly soluble in cold water and is often available in the laboratory as bromine water. Iodine is only slightly soluble in water, but dissolves readily in alcohol, forming a dark brown solution.

The halogens are among the most reactive nonmetals. They are always found combined with other elements in nature. Fluorine is the most electronegative and chemically reactive of all the nonmetals. It is also the strongest elemental oxidizing agent known. Fluorine forms compounds with all elements except helium, neon, and argon. The reactivity of the other halogens decreases as their size and mass increase. Thus iodine is the least reactive of the common halogens.

Halogens react readily with most metals to form ionic salts. For example, copper is oxidized by chlorine to form copper(II) chloride.

$$Cu(s) + Cl_2(g) \longrightarrow Cu^{2+} + 2Cl^- \qquad (CuCl_2(s))$$

With nonmetals the halogens bond covalently to form molecules.

Halogens react with hydrogen to form the corresponding hydrogen halides. All of these compounds are colorless gases.

$$H_2(g) + Br_2(g) \longrightarrow 2HBr(g)$$

The reaction with fluorine and chlorine is explosive; bromine and iodine react more slowly. With the exception of hydrogen fluoride, the hydrogen halides are highly ionized in water, forming strong acids. Hydrofluoric acid is a weak acid because it is weakly ionized in water. The large electronegativity difference between hydrogen and fluorine ($4.0 - 2.1 = 1.9$) explains why. Hydrogen fluoride molecules are very polar and strongly hydrogen-bonded to one another. The other hydrogen halides exhibit this effect to a much lesser degree.

Solutions of chlorine or bromine in water contain hypochlorous and hypobromous acids, respectively.

$$Cl_2(g) + H_2O(l) \rightleftharpoons HClO(aq) + HCl(aq)$$

$$Br_2(l) + H_2O(l) \rightleftharpoons HBrO(aq) + HBr(aq)$$

In solution, these acids act as oxidizing agents. The active ingredient in chlorine bleach is hypochlorous acid. By contrast, fluorine liberates oxygen gas when it reacts with water.

$$2F_2(g) + 2H_2O(l) \longrightarrow O_2(g) + 4HF(aq)$$

Elemental fluorine is seldom used because of its extreme reactivity, but many fluorine compounds have commercial applications. Hydrogen fluoride is used to etch designs in glass and to frost light bulbs. For many years, it had to be stored in wax containers. Today, plastic bottles are used instead.

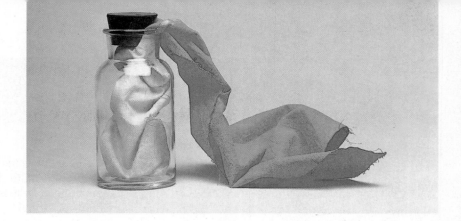

Figure 23·24
What happens to cloth when it is placed in a bottle of chlorine gas?

Teflon (polytetrafluoroethylene) is a heat-resistant plastic used in electrical insulators and for coating non-stick cookware. It is a polymer made up of fluorine and carbon. The volatile liquids used as coolants in most household refrigerators are fluorine compounds like dichlorodifluoromethane, CCl_2F_2. Unlike most fluorine compounds, they are relatively non-toxic.

The enrichment of uranium is another important use of fluorine. In the process for separating the isotopes of uranium, the metal is converted to uranium hexafluoride (UF_6), a gas. The diffusion of this gas allows the separation of U-235, the fissionable isotope, from the nonfissionable isotope, U-238.

Small quantities of sodium fluoride are added to drinking water to reduce tooth decay. The concentration is very low, about $5 \times 10^{-5}M$. Fluorides are also added to toothpaste to make teeth more decay resistant.

Large amounts of chlorine are used in the purification of city water supplies, swimming pools, and sewage. Because chlorine in solution is a powerful oxidizing agent, it kills disease-causing bacteria. Hypochlorous acid and its salts, the hypochlorites, are strong disinfectants and bleaches. Chlorine reacts with lime to produce a white compound known as *bleaching powder,* or chlorinated lime. It is a mixture of calcium chloride and calcium hypochlorite, $Ca(OCl)_2$, usually assigned the formula $Ca(ClO)Cl$. In the presence of dilute acid, bleaching powder releases chlorine gas. Thus it is frequently used for bleaching cloth.

An important use of chlorine is in the synthesis of vinyl chloride, $CH_2{=}CHCl$, which reacts to form polyvinyl chloride (PVC). PVC, a plastic, is used for floor coverings, phonograph records, and other "vinyl" products. With hot, concentrated alkalis, chlorine forms chlorates, which are used in making matches and as a weed-killer.

$$6KOH(aq) + 3Cl_2(g) \longrightarrow 5KCl(aq) + KClO_3(aq) + 3H_2O(l)$$

Chlorine gas was used in World War I, the first poisonous gas to be used in modern warfare.

Iodine is essential for the correct functioning of the thyroid gland. Small amounts of potassium iodide are added to table salt to insure adequate iodine in the diet. Silver iodide, like silver bromide, is used in making photographic film. Tincture of iodine, a solution of iodine in alcohol, is a well-known antiseptic.

Figure 23·25
Teflon is the trade name for a heat resistant plastic composed of fluorine and carbon. It is used as a non-stick coating.

23·8 Properties of the Halogens **563**

23 The Chemistry of Nonmetals

Chapter Review

Key Terms

allotrope	*23·2*
hydrogenation	*23·1*

Chapter Summary

The nonmetallic elements are found in the upper right corner of the periodic table. They have very high electronegativities. Bonding between nonmetal atoms is primarily covalent. By contrast, compounds formed between metals and nonmetals are usually ionic.

Free hydrogen is rare on earth. Hydrogen containing compounds, however, such as water, are abundant. The three isotopes of hydrogen are protium, deuterium, and tritium. Hydrogen is used as a reducing agent, in ammonia production, and in the hydrogenation of vegetable oils.

Oxygen is the most abundant element on the earth's crust. It exists freely in the atmosphere as O_2. Most of the earth's oxygen, however, is found in numerous compounds. Ozone, O_3, is an allotrope of oxygen formed in the upper atmosphere. Oxygen's primary industrial use is in the production of steel.

Elemental sulfur occurs in several allotropic forms, including monoclinic and rhombic sulfur. Sulfur readily forms compounds with most metals and nonmetals. Some uses of sulfur are the manufacture of sulfuric acid, rubber, and paper.

Nitrogen is found in the free state as N_2. It comprises 80% of the air by volume. Two important nitrogen compounds are ammonia and nitric acid. Ammonia is used to make cleaning products and fertilizers. Nitric acid is used to make fertilizers and explosives.

The halogens do not occur in the free state. They exist in nature as halide ions found in seawater and salt beds. Fluorine, the most electronegative element, reacts with nearly all of the other elements.

Its uses include the enrichment of uranium and the inhibition of tooth decay. Chlorine, a powerful oxidizing agent, is used in disinfectants and the manufacture of "vinyl" products. It is also used to make hydrochloric acid. Iodine and bromine are two other halogens with numerous uses.

Practice Questions and Problems

1. What is the largest industrial use of hydrogen? *23·1*

2. What are the names, atomic symbols, and mass numbers of the isotopes of hydrogen? *23·1*

3. What is a catalyst? Name one of the catalysts used in hydrogenation. *23·1*

4. In what three major forms is oxygen present on earth? *23·2*

5. What is the volume precentage of oxygen in the earth's atmosphere? What are the other components of air? *23·2*

6. Explain the term paramagnetism. Use liquid oxygen as your example. *23·2*

7. What is the largest industrial use of oxygen? *23·2*

8. Give three physical properties of each substance. **a.** oxygen **b.** ozone *23·2*

9. Where is ozone produced naturally? *23·2*

10. Write the name of each compound. *23·3*
a. SiO_2 **d.** Fe_2O_3
b. H_2O_2 **e.** BaO_2
c. SO_3 **f.** CO_2

11. Write the formula of each compound. *23·3*
a. calcium oxide **d.** aluminum oxide
b. mercury(II) oxide **e.** sulfur dioxide
c. carbon monoxide **f.** sodium peroxide

12. List some uses of hydrogen peroxide. *23·3*

13. Complete and balance these equations *23·3*
 a. $Mg(s) + O_2(g) \longrightarrow$
 b. $H_2(g) + O_2(g) \longrightarrow$
 c. $S(s) + O_2(g) \longrightarrow$

14. Name four physical properties of sulfur. *23·4*

15. Describe the common allotrope of sulfur that is stable at room temperature. *23·4*

16. What are some uses of sulfur? *23·4*

17. Complete and balance these equations. *23·4*
 a. $Mg + S \longrightarrow$ **c.** $H_2 + S \longrightarrow$
 b. $Al + S \longrightarrow$ **d.** $Na + S \longrightarrow$

18. How is elemental nitrogen prepared for industrial use? *23·5*

19. In what two major forms is nitrogen present on Earth? *23·5*

20. List major uses of ammonia in industry. *23·6*

21. Give four physical properties of ammonia. *23·6*

22. Nitric acid is produced commercially from ammonia by the Ostwald process. Write equations for the individual steps in this process. *23·6*

23. Give the names and formulas of two substances that are used as fertilizers. *23·6*

24. What are three uses of nitric acid? *23·6*

25. Name two nitrogen-containing explosives. *23·6*

26. How is chlorine gas normally prepared for commercial use? *23·7*

27. Describe how iodine is obtained commercially. *23·7*

28. By what process is fluorine made? *23·7*

29. Give the names and molecular formulas of the halogens. *23·8*

30. The halogens do not exist in nature in the free state. Explain. *23·8*

31. How do chlorine and bromine react with water? Write equations. *23·8*

32. Write the chemical equation to show how fluorine reacts with water. *23·8*

33. How would you make *bleaching powder*? Give equations. *23·8*

34. Why is chlorine added to drinking water and swimming pools? *23·8*

35. Name the two allotropic forms of the element oxygen. Describe one method of preparation for each.

36. Explain why superheated water and compressed air are used in sulfur recovery by the Frasch process. What conditions do not allow sulfur to be mined, like coal, for example?

37. Give the equation for the laboratory preparation of chlorine from manganese dioxide and hydrochloric acid. What mass of HCl is required for the production of 1 mol of Cl_2?

38. What is the oxidation state (number) of nitrogen in each form?
 a. NO_2 **f.** NH_3
 b. N_2O **g.** NO
 c. NO_3^- **h.** N_2O_3
 d. N_2 **i.** NH_4^+
 e. HNO_3 **j.** NO_2^-

39. Large volumes of hydrogen are used in the hydrogenation of vegetable oils. Describe the process and comment on its commercial value.

40. Give three important physical properties of each of the halogens.

41. Calculate the mass percent of oxygen in these compounds.
 a. H_2O **c.** SiO_2
 b. $CaCO_3$ **d.** $CaSO_4$

42. Distinguish between oxygen and ozone.

43. Write formulas for these substances.
 a. nitrous oxide   **d.** hypochlorous acid
 b. ozone   **e.** sodium periodate
 c. bleaching powder   **f.** vinyl chloride

44. Describe how iodine can be prepared in the laboratory.

45. Write a balanced equation for the reaction of hydrogen with these elements.
 a. nitrogen **b.** chlorine **c.** calcium

46. List all of the halogens in order of increasing electronegativity.

47. List some of the major uses of sulfuric acid in industry.

48. Explain why ammonia exhibits many water-like properties.

49. Draw Lewis structures for the oxide ion, O^{2-}, and the peroxide ion, O_2^{2-}.

50. List the nonmetals that occur as diatomic molecules.
a. gases **b.** liquids **c.** solids

51. Name a nonmetallic oxide that reacts with water to form an acid and a metallic oxide that reacts with water to form a base. Write a balanced equation for each reaction.

52. The data table below lists the melting and boiling points of the halogens.
a. Describe the trends you observe.
b. Explain these trends. Are they likely to occur with other groups in the periodic table?

Element	Melting Point (°C)	Boiling Point (°C)
F	−219	−188
Cl	−107	−34
Br	−7	58
I	113	184

Critical Thinking Questions

53. Choose the term that best completes the second relationship.
a. oxygen gas:ozone   monoclinic sulfur: _____

(1) molten sulfur   (3) sulfur dioxide
(2) rhombic sulfur   (4) flowers of sulfur
b. nitrogen:ammonia   sulfur: _____
(1) sulfuric acid   (3) nitric acid
(2) sulfur trioxide   (4) allotrope
c. protium:deuterium   nitrogen-14: _____
(1) isotope   (3) carbon-14
(2) tritium   (4) nitrogen-15

54. Hydrogen and helium are by far the most abundant elements in the universe. Why are the free, or elemental, forms of hydrogen and helium very rare on earth?

55. When chlorine gas bleaches fabric, is it being used up or is it acting as a catalyst? How could you find out?

56. Compare the methods used to produce hydrogen. What are the disadvantages of each for producing hydrogen as a fuel?

Review Questions and Problems

57. Explain these statements.
a. Group 1 metals form 1+ ions.
b. The atomic radius increases going down a group in the periodic table.

58. Give uses for these compounds.
a. $CaSO_4 \cdot 2H_2O$
b. $BaSO_4$
c. $CaCO_3$
d. $MgSO_4 \cdot 7H_2O$

59. State the outer electron configuration.
a. Pb   **c.** I
b. Bi   **d.** Ba

60. Give chemical equations for the reactions involved in the analytical test for Ag^+.

61. Calcium carbonate is used as an antacid to neutralize HCl in the stomach. Write the equation for the neutralization reaction.

62. Use Lewis dot structures to predict the products of the following reactions.
a. beryllium and oxygen gas
b. calcium and sulfur
c. aluminum and chlorine gas

63. Determine the oxidation state of carbon in each of the following compounds.
a. CH_3OH **d.** HCO_2H
b. H_2CO **e.** CO_2
c. CH_4 **f.** CO

64. Using electron configurations, predict the product of the reaction between each of the following pairs of elements.
a. magnesium and bromine
b. calcium and bromine
c. cesium and selenium

65. Why is carbon such a useful industrial reducing agent? How is *coke* made?

66. Write a balanced equation for each reaction.
a. magnesium is exposed to steam
b. $CaSO_4 \cdot 2H_2O$ is heated
c. calcium carbonate is heated

Challenging Questions and Problems

67. How many kilograms of Br⁻ are there in a cubic mile of seawater?

68. How many grams of sulfur dioxide are produced when 200 g of sulfur is burned in oxygen? What volume would the sulfur dioxide occupy at STP?

69. What is the principal commercial method for producing hydrogen gas? Write a balanced chemical equation for the process. Identify each species.
  a. the reducing agent
  b. the oxidizing agent
  c. the substance reduced
  d. the substance oxidized
Use the changes in oxidation state to illustrate your answer.

70. Hydrogen peroxide decomposes according to the equation.

$$2H_2O_2 \longrightarrow 2H_2O + O_2$$

A solution is 3.00% H_2O_2 by volume and has a density of 1.44 g/cm₃. How many liters of oxygen gas, measured at STP, is produced when 1 L of 3.00% H_2O_2 is decomposed?

71. Coal contains sulfur. When coal is burned the sulfur is oxidized to sulfur dioxide, SO_2. In the graph below, the SO_2 emissions of coal-burning electric plants in a recent year are shown to vary according to generating capacity and plant age. In this year all coal-burning electric plants emitted approximately 13 million tons of SO_2.
  a. What are the three types of plants with the largest SO_2 emissions?
  b. Approximately what percent of the year's emissions were due to the three types of plants identified in part a?

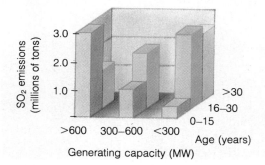

72. Give the reactions for the production of sulfuric acid from elemental sulfur by the contact process. How many kilograms of H_2SO_4 will be obtained from 2.00×10^4 kg of sulfur if its purity is 99.6%?

Research Projects

1. How did Henry Cavendish discover hydrogen?

2. How is deuterium used in scientific research?

3. Report on the discoveries of Joseph Priestly and the results of his collaborations with Benjamin Franklin and Antoine Lavoisier.

4. Compare the reactivities of various allotropic forms of sulfur.

5. Make a diagram of the lead chamber process for sulfuric acid production.

6. What is "nitrogen fixation"? Describe one natural and one industrial process of nitrogen fixation.

7. Why was Fritz Haber's method of nitrogen fixation important? How did his work with gases change the course of history? Why was his Nobel prize denounced by some other scientists?

8. Devise a method to detect halide ions in seawater.

9. Investigate the effect of varying temperature on the ability of bleach to remove stains.

10. How is Freon used to keep food cold in modern refrigerators?

11. What are some of the most commonly used propellants in aerosols? How do they compare to chlorofluorocarbons?

Readings and References

Asimov, Isaac. *Building Blocks of the Universe.* New York: Abelard-Schuman, 1974.

Renmore, C. D. *Silicon Chips and You: The Magical Mineral in Your Telephone, Calculator, Toys, Automobile, Hospital, Air Conditioning, Factory, Furnace, Sewing Machine, and Countless Other Future Inventions.* New York: Beaufort, 1980.

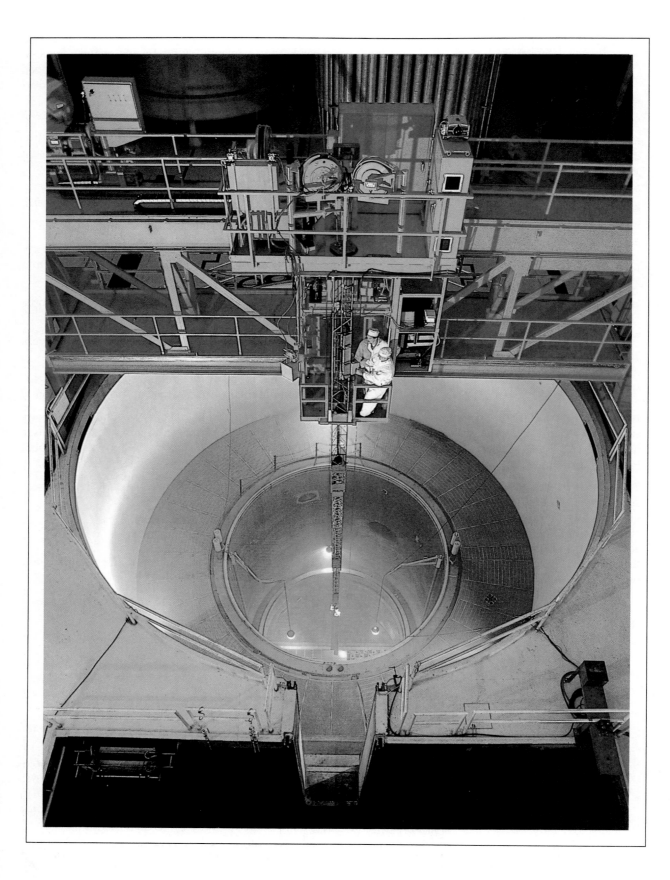

24 Nuclear Chemistry

Chapter Preview

24·1 Radioactivity
24·2 Types of Radiation
24·3 Nuclear Stability
24·4 Half-Life
24·A Carbon-14 Dating
24·5 Transmutation Reactions

24·B Particle Accelerators
24·6 Nuclear Fission
24·7 Nuclear Fusion
24·8 Detecting Radiation
24·C Radioisotopes in Research and Medicine

Figure 24·1
Nuclear reactions release tremendous amounts of energy. Nuclear reactors are designed to control these reactions so that the energy can be used safely.

In the preceding chapters the focus has been on chemical reactions. In such reactions atoms gain stability by attaining stable electron configurations. This chapter deals with nuclear reactions. In nuclear reactions certain isotopes called radioisotopes gain stability by making changes within their nuclei. These changes are accompanied by the emission of large amounts of radiation energy. Unlike a chemical reaction, a nuclear reaction is not affected by changes in temperature or pressure, and it cannot be turned off.

Atoms with unstable nuclei are radioactive.

24·1 Radioactivity

In 1896 the French chemist Antoine Henri Becquerel (1852–1908) accidentally discovered that uranium ores emit invisible rays. Becquerel had left some ore on top of photographic plates that were wrapped with light-tight paper. The plates became fogged. At that time two of his graduate students were Marie Curie (1867–1934) and Pierre Curie (1859–1906). The Curies were able to show that the fogging was caused by rays emitted by the uranium in the ore. Marie named *the process by which uranium gives off rays* **radioactivity.** *The penetrating rays emitted by a radioactive source are one type of* **radiation.** Pierre assisted his wife in the isolation of several radioactive elements. Together with Becquerel in 1903, they won the Nobel Prize in physics for this work.

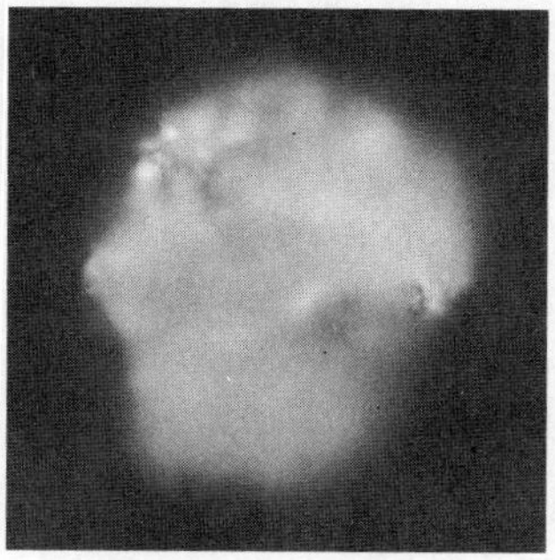

Figure 24·2
Radiation caused the exposure of this film. Becquerel's accidental discovery of this phenomenon was the first time anyone observed the effects of radioactivity.

Becquerel's discovery of radioactivity was the deathblow to Dalton's theory of an indivisible atom. A radioactive atom undergoes drastic changes as it emits radiation. *Certain isotopes, called* **radioisotopes,** *are radioactive because they have unstable nuclei*. The stability of the nucleus depends on its ratio of protons to neutrons. Too many or too few neutrons lead to an unstable nucleus. *An unstable nucleus loses energy by*

569

Figure 24·3
An electric field has different effects on the three types of radiation. Alpha particles and beta particles are deflected in opposite directions. Gamma rays are undeflected.

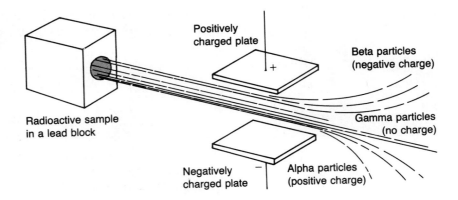

emitting radiation during the process of **radioactive decay.** Eventually unstable nuclei achieve a more stable state when they are transformed into atoms of a different element. Radioactive decay is spontaneous and does not require any input of energy.

24·2 Types of Radiation

Radioisotopes decay by emitting alpha radiation, beta radiation, and/or gamma radiation.

Several types of radiation can be emitted during radioactive decay. They include alpha radiation (Greek α), beta radiation (Greek β), and gamma radiation (Greek γ). Table 24·1 summarizes the properties of different types of radiation. The different types of radiation from a radioactive source can be separated by an electric or magnetic field (Figure 24·3).

Safety

Substances that emit alpha particles are very dangerous if ingested. Never eat, drink, chew gum, etc. when handling any radioisotopes.

Alpha radiation *consists of helium nuclei that have been emitted from a radioactive source.* **Alpha particles** *contain two protons and two neutrons and have a double positive charge.* In writing nuclear chemical reactions, an alpha particle is written as $^4_2\mathrm{He}$ or α. The charge is omitted.

Table 24·1 Characteristics of Some Ionizing Radiations			
Property	Alpha radiation	Beta radiation	Gamma radiation
Composition	Alpha particle (helium nucleus)	Beta particle (electron)	High-energy electromagnetic radiation
Symbol	α, $^4_2\mathrm{He}$	β, $^{\;\,0}_{-1}e$	γ
Charge	2+	1−	0
Mass (amu)	4	1/1837	0
Common source	Radium-226	Carbon-14	Cobalt-60
Approximate energy	5MeV*	0.05 to 1 MeV	1 MeV
Penetrating power	Low (0.05 mm body tissue)	Moderate (4 mm body tissue)	Very high (penetrates body easily)
Shielding	Paper, clothing	Metal foil	Lead, concrete (incompletely shields)

*(1 MeV = 1.60×10^{-13} J)

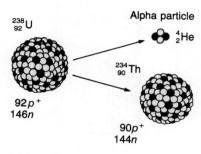

Figure 24·4
What particle is emitted when uranium-238 decays to thorium-234?

The radioisotope uranium-238 releases alpha radiation. The more stable (but still radioactive) isotope thorium-234 is produced.

$$\underset{\text{Uranium-238}}{^{238}_{92}\text{U}} \xrightarrow{\substack{\text{radioactive}\\\text{decay}}} \underset{\text{Thorium-234}}{^{234}_{90}\text{Th}} + \underset{\substack{\text{Alpha}\\\text{particle}}}{^{4}_{2}\text{He}} \qquad (\alpha \text{ emission})$$

When an atom loses an alpha particle, the atomic number of the product atom is lower by two and its mass number is lower by four. Because of their large mass and charge, alpha particles do not travel very far and are not very penetrating. They are easily stopped by a sheet of paper or by the dead cells on the surface of the skin. They are dangerous when ingested, however, because they easily penetrate soft tissue.

Problem

1. The disintegration of the radioisotope radium-226 produces an isotope of the element radon and alpha radiation. The atomic number of radium (Ra) is 88; the atomic number of radon (Rn) is 86. Write a balanced equation for this transmutation.

Beta radiation *consists of fast-moving electrons formed by the decomposition of a neutron of an atom.* The neutron breaks into a proton and an electron. *These fast-moving electrons are called* **beta particles.**

$$\underset{\text{Neutron}}{^{1}_{0}n} \longrightarrow \underset{\text{Proton}}{^{1}_{1}\text{H}} + \underset{\substack{\text{Electron}\\\text{(beta particle)}}}{^{0}_{-1}e}$$

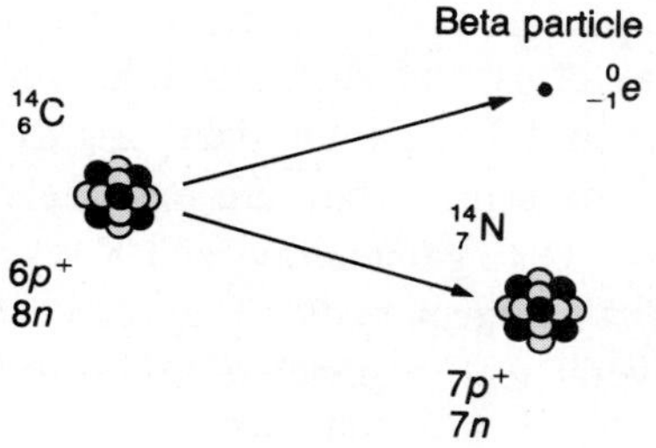

The proton stays in the nucleus and the electron is ejected from the atom. Carbon-14 emits a beta particle as it undergoes radioactive decay to form nitrogen-14.

$$\underset{\substack{\text{Carbon-14}\\\text{(radioactive)}}}{^{14}_{6}\text{C}} \longrightarrow \underset{\substack{\text{Nitrogen-14}\\\text{(stable)}}}{^{14}_{7}\text{N}} + \underset{\substack{\text{Beta}\\\text{particle}}}{^{0}_{-1}e} \qquad (\beta \text{ emission})$$

The mass number of the nitrogen-14 that is produced is the same as that of carbon-14. Meanwhile the atomic number has increased by 1. The nucleus now contains an additional proton and one less neutron.

Beta particles are much smaller than alpha particles, and they have half as much charge. Consequently, they are much more penetrating. Beta particles are stopped by aluminum foil or thin pieces of wood.

Figure 24·5
What particle is emitted when carbon-14 decays to nitrogen-14?

Problem

2. A radioisotope of the element lead (Pb) decays to an isotope of the element bismuth (Bi) by emission of a beta particle. Complete the equation for the decay process.

$$^{210}\text{Pb} \longrightarrow {}_{83}\text{Bi} + {}_{-1}^{0}e$$

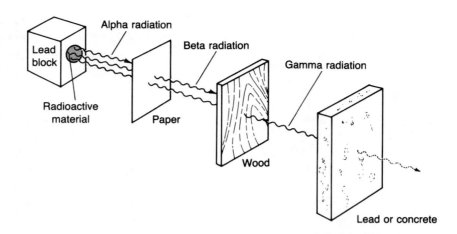

Figure 24·6
Alpha particles have the least penetrating power because they have a large mass and charge. Gamma rays have no mass or charge and are very penetrating.

Alpha radiation
Beta radiation
Gamma radiation
Lead block
Radioactive material
Paper
Wood
Lead or concrete

ChemDirections

Food stays fresh longer and potatoes don't sprout when exposed to gamma radiation. To learn more about this alternate method of food preservation read **ChemDirections** *Irradiated Foods*, page 657.

Safety

It is easier to prevent spills than to decontaminate a spill area. Keep radioisotope containers in secondary containers to prevent spills. Work in trays rather than directly on the lab bench.

The neutron-to-proton ratio determines the stability of a nucleus.

Gamma radiation *is electromagnetic radiation.* Visible light (the light we see) is also electromagnetic radiation but of much lower energy. Gamma rays are often emitted by the nuclei of disintegrating radioactive atoms along with alpha or beta radiation.

$$^{226}_{88}\text{Ra} \longrightarrow {}^{222}_{86}\text{Rn} + {}^{4}_{2}\text{He} + \gamma$$

Radium-226   Radon-222   Alpha particle   Gamma ray

$$^{234}_{90}\text{Th} \longrightarrow {}^{234}_{91}\text{Pa} + {}^{0}_{-1}e + \gamma$$

Thorium-234   Protactinium-234   Beta particle   Gamma ray

Gamma rays have no mass and no electrical charge. Thus the emission of only gamma radiation does not alter the atomic number or mass number of an atom.

X-radiation is produced by the decay of excited electrons in atoms. (It is not emitted during radioactive decay.) Except for their origins, gamma rays and X-rays are essentially the same. They are extremely penetrating and potentially very dangerous. Both gamma rays and X-rays pass easily through paper, wood, and the human body. They can be stopped (though not completely) by several feet of concrete or several inches of lead (Figure 24·6).

Problem

3. List three types of radiation in order of increasing penetrating power.

24·3 Nuclear Stability

About 1500 different nuclei are known. Of these only 264 are stable. Stable nuclei do not decay or change with time. The stability of a nucleus depends on its neutron-to-proton ratio. For elements of low atomic number, below about 20, this ratio is 1. This means that the nuclei have equal numbers of neutrons and protons. For example, the isotopes $^{12}_{6}\text{C}$, $^{14}_{7}\text{N}$,

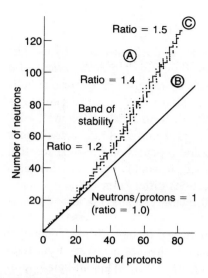

Figure 24·7
On a neutron versus proton plot, stable nuclei form a pattern known as a band of stability. For isotopes of small atomic number, the stable ratio is 1.0; for the heavier isotopes it increases to about 1.5. In region A unstable nuclei undergo beta decay. In region B they convert a proton to a neutron. In region C they undergo alpha particle emission. (The solid line is a reference line for a neutron-to-proton ratio equal to 1.0.)

and $^{16}_{8}O$ are stable. Above atomic number 20 the stable nuclei have more neutrons than protons. The neutron-to-proton ratio reaches about 1.5 with heavy elements. The lead isotope $^{206}_{82}Pb$, for example, with 124 neutrons and 82 protons, is stable. Its ratio is $\frac{206}{82} \approx 1.5$.

A plot of the number of neutrons versus the number of protons for the known stable nuclei is given in Figure 24·7. *The stable nuclei on a neutron-vs-proton plot are located in a* **band of stability.** The nuclei that fall outside the band of stability undergo spontaneous radioactive decay. The type of decay that occurs depends upon the position of the nucleus with respect to the band of stability.

Nuclei in *region A,* to the left of the band of stability, have too many neutrons. They turn a neutron into a proton by emitting a beta particle or electron from the nucleus.

$$^{1}_{0}n \longrightarrow {}^{1}_{1}H + {}^{0}_{-1}e$$

This process is known as *beta decay* or *beta emission*. It produces a simultaneous increase in the number of protons and decrease in the number of neutrons.

$$^{66}_{29}Cu \longrightarrow {}^{66}_{30}Zn + {}^{0}_{-1}e$$

$$^{14}_{6}C \longrightarrow {}^{14}_{7}N + {}^{0}_{-1}e$$

Nuclei in *region B,* to the right of the band of stability, have too many protons. They increase their stability by converting a proton to a neutron. An electron is captured in this process.

$$^{59}_{28}Ni + {}^{0}_{-1}e \longrightarrow {}^{59}_{27}Co$$

$$^{37}_{18}Ar + {}^{0}_{-1}e \longrightarrow {}^{37}_{17}Cl$$

A **positron** *is a particle with the mass of an electron but a positive charge,* $_{+1}^{0}e$. A positron may be emitted as a proton changes to a neutron.

$$^{8}_{5}B \longrightarrow {}^{8}_{4}Be + {}^{0}_{+1}e$$

$$^{15}_{8}O \longrightarrow {}^{15}_{7}N + {}^{0}_{+1}e$$

When a proton is converted to a neutron the atomic number decreases by one, the number of neutrons increases by one, and the mass number remains the same.

Region C is above the upper end of the band of stability. The nuclei in this region are especially heavy. They have both too many neutrons and too many protons. They emit alpha particles. Alpha emission results in an increase in the neutron-to-proton ratio. This is a favorable change and increases nuclear stability.

$$^{204}_{82}Pb \longrightarrow {}^{200}_{80}Hg + {}^{4}_{2}He$$

In alpha emission, the mass number decreases by four and the atomic number decreases by two. *All* nuclei with atomic number greater than 82 are radioactive. A majority of these undergo alpha emission.

$$^{226}_{88}Ra \longrightarrow {}^{222}_{86}Rn + {}^{4}_{2}He$$

$$^{232}_{90}Th \longrightarrow {}^{228}_{88}Ra + {}^{4}_{2}He$$

24·4 Half-Life

The rate of decay of a radioisotope is measured in terms of half-life.

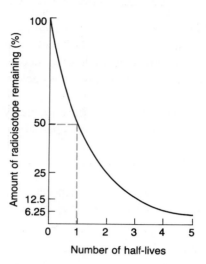

Figure 24·8
This decay curve for a radioactive element shows that after each half-life, one-half of the remaining original radioactive atoms have decayed into atoms of a new element.

Every radioisotope has a characteristic rate of decay measured by its half-life. *A **half-life** ($t_{1/2}$) is the time required for one-half of the atoms of a radioisotope to emit radiation and to decay to products.* After one half-life, one-half of the original radioactive atoms have decayed into atoms of a new element. The other one-half remain unchanged (Figure 24·8). After a second half-life only one-quarter of the original radioactive atoms remain. If the masses of the reactants and products of a nuclear reaction were determined with a sensitive balance, it would appear that the mass was conserved. However, nuclear reactions differ from ordinary chemical reactions. An infinitesimally small quantity of mass is lost. The lost mass is converted to radiation energy.

The stability of a radioisotope is indicated by its half-life. The longer the half-life, the more stable the isotope. Half-lives may vary from fractions of a second to millions of years (Table 24·2). Many artificially produced radioisotopes are highly unstable and have very short half-lives. This feature is a great advantage in nuclear medicine. This is because rapidly decaying isotopes are not long-term biological radiation hazards.

One isotope, uranium-238, decays through a complex series of radioactive intermediates to the stable isotope lead-206 (Figure 24·9). The age of certain rocks is determined by measuring the ratio of uranium-238 to lead-206. The half-life of uranium-238 is 4.5×10^9 years. Thus it is possible to use this method to date rocks as old as our solar system (4.6×10^9 years).

$$^{238}_{92}U \xrightarrow[4 \times 10^9 \text{ yr}]{\alpha} {}^{234}_{90}Th \xrightarrow[25 \text{ days}]{\beta, \gamma} {}^{234}_{91}Pa \xrightarrow[7 \text{ hr}]{\beta, \gamma} {}^{234}_{92}U$$

$^{234}_{92}U \xrightarrow{2.7 \times 10^5 \text{ yr}} \alpha$

$$^{218}_{84}Po \xleftarrow[4 \text{ days}]{\alpha} {}^{222}_{86}Rn \xleftarrow[2.3 \times 10^3 \text{ yr}]{\alpha, \gamma} {}^{226}_{88}Ra \xleftarrow[8 \times 10^4 \text{ yr}]{\alpha, \gamma} {}^{230}_{90}Th$$

$^{218}_{84}Po \xrightarrow{3 \text{ min}} \alpha$

$$^{214}_{82}Pb \xrightarrow[27 \text{ min}]{\beta, \gamma} {}^{214}_{83}Bi \xrightarrow[20 \text{ min}]{\beta, \gamma} {}^{214}_{84}Po \xrightarrow[1.6 \times 10^{-4} \text{ s}]{\alpha} {}^{210}_{82}Pb$$

$^{210}_{82}Pb \xrightarrow{22 \text{ yr}} \beta, \gamma$

$$\text{Stable isotope } {}^{206}_{82}Pb \xleftarrow[138 \text{ days}]{\alpha} {}^{210}_{84}Po \xleftarrow[5 \text{ days}]{\beta} {}^{210}_{83}Bi$$

Figure 24·9
Uranium-238 decays through a complex series of radioactive intermediates, including radon gas (Rn). What is the stable end product of this series?

Half-life values are determined experimentally.

Safety

All radioisotopes must be isolated from living systems until their radioactivity has been reduced to safe levels. For rapidly decaying isotopes this time will be relatively short.

 Issues in Chemistry

The long half-lives of many of the radioactive isotopes found in the wastes of nuclear power plants complicate the disposal of these materials. Is the disposal of radioactive wastes an issue in your community? Could it become an issue?

Table 24·2   Half-lives and Radiation of Some Naturally Occurring Radioisotopes

Isotope	Half-life	Radiation emitted
Carbon-14	5.73×10^3 yr	β
Potassium-40	1.28×10^9 yr	β, γ
Radon-222	3.8 days	α
Radium-226	1.6×10^3 yr	α, γ
Thorium-230	8.0×10^4 yr	α, γ
Thorium-234	25 days	β, γ
Uranium-235	7.1×10^8 yr	α, γ
Uranium-238	4.5×10^9 yr	α

Example 1

Nitrogen-13 emits beta radiation and decays to carbon-13 with $t_{1/2}$ = 10 min. Assume a starting mass of 2 g of nitrogen-13.
a. How long is four half-lives?
b. How many grams of that iostope will remain after four half-lives?

Solution

a. One half-life is 10 min. Four half-lives is 4×10 min. = 40 min.

b. After four half-lives, 0.0625 g remains out of each initial gram.

$$2g \times \frac{0.0625 \text{ g}}{1g} = 0.125 \text{ g}$$

Problem

4. Manganese-56 is a beta emitter with a half-life of 2.6 hr. What is the mass of manganese-56 in a 1-mg sample of the isotope after 10.4 hr?

───── Science, Technology, and Society ─────

24·A   Carbon-14 Dating

Archaeologists use the half-lives of naturally occurring radioisotopes to establish the ages of fossils and other ancient artifacts. Carbon-14, with a half-life of 5730 years, has been used most extensively. When cosmic rays in the upper atmosphere strike atoms, the atoms disintegrate into electrons, protons, neutrons, and other particles. Sometimes a free neutron collides with a nitrogen atom and causes it to lose a proton. The radioactive product is carbon-14.

$$^{14}_{7}\text{N} \quad + \quad ^{1}_{0}n \quad \longrightarrow \quad ^{14}_{6}\text{C} \quad + \quad ^{1}_{1}\text{H}$$

Nitrogen-14   Neutron   Carbon-14   Proton

Carbon-14 combines with oxygen, just as carbon-12 does, to form carbon dioxide.

$$^{14}\text{C} + \text{O}_2 \longrightarrow {}^{14}\text{CO}_2$$
Carbon-14 ⠀ Oxygen ⠀⠀ Carbon dioxide (radioactive)

$$^{12}\text{C} + \text{O}_2 \longrightarrow {}^{12}\text{CO}_2$$
Carbon ⠀ Oxygen ⠀⠀ Carbon dioxide (nonradioactive)

Plants use carbon dioxide to make carbon-containing compounds. Animals eat plants containing these compounds. Thus, all living organisms contain carbon-12 and carbon-14. The carbon-14 slowly decays in the organism's tissues. It is continuously replaced, however, by more carbon-14. Therefore, while the organism is alive, its ratio of carbon-12 to carbon-14 is constant. After the organism dies, however, it does not incorporate any new carbon. The carbon-12 to carbon-14 ratio changes as the carbon-14 decays to nitrogen-14.

$$^{14}_{6}\text{C} \longrightarrow {}^{14}_{7}\text{N} + {}^{0}_{-1}e$$

In the 1940s, the American chemist Willard Libby developed a method using carbon-14 to determine the age of an artifact. He measured the ratio of carbon-14 to carbon-12 in wood and textile samples from parchments and mummies. Then, he compared this ratio with the carbon-14 to carbon-12 ratio in recently-dead samples of similar materials. Using the half-life of carbon-14, he was able to calculate the age of the artifacts. For example, an object that had once been living might now have a carbon-14 concentration of one-half its original value. This object must therefore be about 5730 years old (one half-life). This technique can be used to determine the ages of objects between 200 and 50 000 years old. The accuracy of this method is good. For example, the Dead Sea Scrolls were determined to be 1940 ± 70 years old. This compares well with the historical record. Libby was awarded the 1960 Nobel Prize in chemistry for his work.

One of two techniques is used to measure the carbon-14 in an object. The most traditional method involves burning a small sample of the object to convert all the carbon to carbon dioxide. The carbon dioxide is trapped and frozen with liquid air. The carbon-14 present is determined by measuring the radioactivity of the carbon dioxide with a radiation counter. A newer technique involves ionizing the sample and passing it through a particle accelerator. The isotopes are separated by a magnetic field and counted directly by a detector.

Carbon-14 dating is a good example of how chemistry helps researchers in other areas. Before this technique was used, archaeologists had to use time-consuming and sometimes imprecise methods to date objects. With carbon-14 dating, they have been able to accurately determine the age of objects. This assists in dating significant events, such as the initial migration of people to this continent.

Figure 24·10
The age of ancient artifacts such as the Dead Sea scrolls can be determined by carbon-14 dating. The half-life of carbon-14 is 5730 years. Scientists can date objects up to 50 000 years old by using this system.

24·5 Transmutation Reactions

The conversion of an atom of one element to an atom of another element is called **transmutation.** Radioactive decay (Figure 24·4) is one way in which transmutations occur. A transmutation can also occur when high-energy particles bombard the nucleus of an atom. The high-energy particles may be protons, neutrons, or alpha particles. Some transmutations occur in nature. For example, the production of carbon-14 from nitrogen-14 occurs in the upper atmosphere. Many other transmutations are done in laboratories or in nuclear reactors. The earliest artificial transmutation was performed in 1919 by Ernest Rutherford. He bombarded nitrogen gas with alpha particles to produce an unstable isotope of fluorine.

$$^{14}_{7}\text{N} + ^{4}_{2}\text{He} \longrightarrow ^{18}_{9}\text{F}$$

Nitrogen-14   Alpha particle   Fluorine-18

The fluorine isotope quickly decomposes to a stable isotope of oxygen and a proton. This experiment eventually led to the discovery of the proton.

$$^{18}_{9}\text{F} \longrightarrow ^{17}_{8}\text{O} + ^{1}_{1}\text{H}$$

Fluorine-18   Oxygen-17   Proton

James Chadwick's discovery of the neutron in 1932 also involved a transmutation experiment. The neutrons were produced when beryllium-9 was bombarded with alpha particles.

$$^{9}_{4}\text{Be} + ^{4}_{2}\text{He} \longrightarrow ^{12}_{6}\text{C} + ^{1}_{0}n$$

Beryllium-9   Alpha Particle   Carbon-12   Neutron

The elements in the periodic table with atomic numbers above 92 are called the **transuranium elements.** None of them occur in nature and all of them are radioactive. These elements have been synthesized in nuclear reactors and nuclear accelerators, which accelerate the bombarding particles to very high speeds.

Chemical Connections

Some transuranic elements have practical applications. Fast neutrons from a californium isotope are more effective than X-rays in treating some forms of cancer. Isotopes of americium are used in smoke detectors and various electronic devices.

Figure 24·11
In 1919, Ernest Rutherford carried out the first artificial transmutation when he bombarded nitrogen gas with alpha particles. What particles were formed?

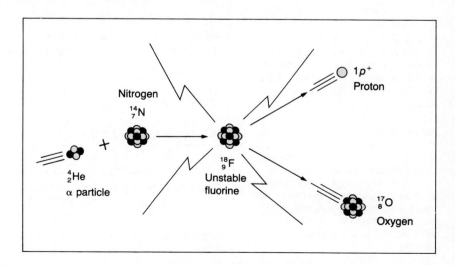

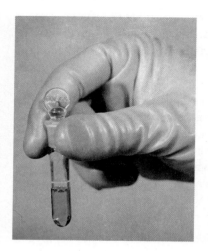

Figure 24·12
All of the elements with atomic numbers above 92 are synthetic. These elements are prepared by bombarding nuclei of uranium or more complex atoms with neutrons, alpha particles, or other nuclear "bullets". Plutonium, (shown here in acid solution) and neptunium were the first elements to be produced artificially. They are both transuranium elements.

When uranium-238 is bombarded with slow neutrons from a nuclear reactor, some uranium nuclei capture neutrons to produce uranium-239.

$$^{238}_{92}U + ^{1}_{0}n \longrightarrow ^{239}_{92}U$$

Uranium-239 is radioactive and emits a beta particle. The product isotope is the artificial, radioactive element neptunium with atomic number 93.

$$^{239}_{92}U \longrightarrow ^{239}_{93}Np + ^{0}_{-1}e$$

Neptunium is unstable and emits a beta particle to produce a second artificial element, plutonium. Plutonium has an atomic number of 94.

$$^{239}_{93}Np \longrightarrow ^{239}_{94}Pu + ^{0}_{-1}e$$

Plutonium and neptunium are both transuranium elements. They do not occur in nature. Each has an atomic number greater than 92. Neptunium and plutonium, the first artificial elements ever made, were synthesized in 1940. Since that time 15 more transuranium elements, atomic numbers 95 through 109, have been synthesized.

Science, Technology, and Society

24·B Particle Accelerators

Since the 1930s, physicists have used particle accelerators to probe the mysteries of the atom. In a particle accelerator, sub-atomic particles, usually electrons or protons, travel at close to the speed of light. A beam of these particles has enough energy to smash a nucleus into fragments (hence the name "atom smashers"). The relativity effect at such high energies results in the conversion of energy into matter. As a result, new particles are created. These particles leave a trail of bubbles in a detection device called a bubble chamber. Electronic detectors may also be used in conjunction with a computer that records the event on a printout.

Accelerators use electric fields and electromagnetic guides to accelerate charged particles to very high speeds. The simplest accelerators use a steady electric field. The field is created by mounting a metal sphere on a long pipe and charging it to a very high voltage. This causes a beam of particles to move faster and faster down the

Figure 24·13
A linear accelerator is made up of a long series of tubes connected to a source of high-frequency alternating current. As the charge on the tubes alternates, a particle is repelled by the tube it is leaving and attracted to the next tube. A high vacuum reduces collisions that would slow down the particles.

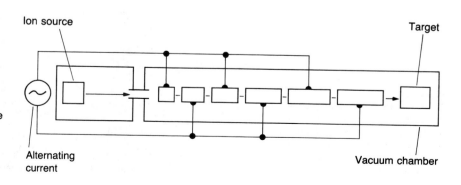

length of the pipe and to strike a target. This design is capable of energies of up to 10 million electron volts (1×10^7 eV). The linear accelerator, or linac, later improved on this design. It uses an alternating electric field to reach energies of up to 5×10^{10} eV. A linac boosts an electron's speed to nearly the speed of light. Large magnets keep the electrons from veering off course. The two mile long Stanford Linear Accelerator in California is currently the largest linac in North America.

The cyclotron has at its center a flat hollow cylinder, cut in half to form two D-shaped chambers. The particles to be accelerated enter the cyclotron midway between the two "D's" and move in a spiral outward. They are accelerated at each turn of the spiral by an electrical jolt, while an electromagnet guides them. The largest cyclotron has reached an energy of 7.2×10^8 eV. Oval synchrotrons accelerate particles in a pathway resembling a race track. A linac in one or more of the straight sections increases the particles' speed. Particles can also be sent in opposite directions to produce high-speed collisions. The synchrotron's unique design has made energies of 1×10^{12} eV possible.

Today's giant accelerators have opened the door to new research. They have revealed, for example, that protons and neutrons consist of even smaller particles called quarks (Section 4·C). Accelerator research has had practical benefits as well. Particle beams are being used by medical researchers to destroy tumors and create radioactive tracers. Accelerator technology is also being applied in the development of superspeed trains.

Figure 24·14
Scientists use huge particle accelerators to produce high energy particles for bombardment. The synchrotron at Fermi National Accelerator laboratory near Chicago has a circumference of 6.3 km.

Figure 24·15

In nuclear fission, uranium-235 breaks into two fragments. Energy and more neutrons are released. The released neutrons can cause other uranium-235 atoms to split.

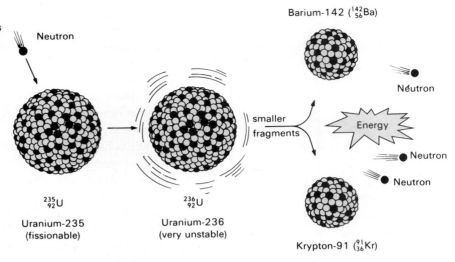

Neutron

Barium-142 ($^{142}_{56}$Ba)

Neutron

smaller fragments

Energy

Neutron

Neutron

$^{235}_{92}$U
Uranium-235
(fissionable)

$^{236}_{92}$U
Uranium-236
(very unstable)

Krypton-91 ($^{91}_{36}$Kr)

24·6 Nuclear Fission

The nucleus of an atom splits on nuclear fission.

When the nuclei of certain isotopes are bombarded with neutrons, they undergo **fission** *which is the splitting of a nucleus into smaller fragments*. Uranium-235 and plutonium-239 are fissionable materials. A fissionable atom such as uranium-235 breaks into two fragments of roughly the same size when it is struck by a slow-moving neutron. At the same time, more neutrons are released (Figure 24·15). These neutrons react with the nuclei of other uranium-235 atoms, continuing fission by a chain reaction. In a chain reaction, some of the neutrons produced react with other atoms, producing more neutrons that react with still more atoms.

Einstein stated the relationship between the mass lost and the energy released. $E = mc^2$

Here c is the speed of light. The loss of only 1 gram of mass will produce 9×10^{10} kilojoules of energy. Thus in a nuclear reaction matter and energy can be interconverted.

Fission unleashes enormous amounts of energy. The fission of 1 kg of uranium-235 releases an amount of energy equal to that generated in the explosion of 20000 tons of dynamite. If a nuclear chain reaction is uncontrolled, the energy release is nearly instantaneous. Atomic bombs are devices that start uncontrolled nuclear chain reactions.

Fission can be controlled so that energy is released more slowly. Nuclear reactors use controlled fission to produce useful energy (Figure 24·16). In the controlled fission reaction within the nuclear reactor, much of the energy generated appears as heat. The heat is removed from the reactor core by a suitable coolant fluid (usually liquid sodium or water). The heat is used to generate steam to drive a turbine. The spinning turbine generates electricity.

The control of fission in a nuclear reactor involves two steps.

Moderators and control rods are used to regulate the fission process in a nuclear reactor.

1. Neutron moderation *slows down the neutrons so that they can be captured by the reactor fuel to continue the chain reaction*. The reactor fuel is usually uranium-235. Most of the neutrons produced by uranium are moving so fast that they will pass right through the nucleus without being absorbed. Water and carbon are good moderators. They slow down the neutrons so that the chain reaction can be sustained.

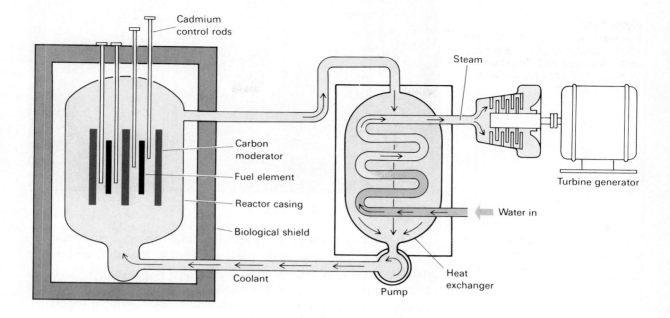

Figure 24·16
All nuclear reactors have five basic components: fuel, moderator, control rods, coolant, and shielding. Heat produced by the fission process is removed from the reactor and is used to turn water to steam. A steam driven turbine is used to generate electricity.

Chemical Connections

Uranium fuel rods are "spent," or used up, after only a few years in a reactor. Then they must be replaced. The spent fuel is still highly radioactive, however, and must be stored for up to thousands of years. A satisfactory program for disposal of nuclear wastes has not yet been developed.

Fusion unites the nuclei of small atoms.

Fusion reactions release more energy than fission reactions.

2. Neutron absorption *decreases the number of slow neutrons.* To prevent the chain reaction from going too fast, some of the slowed neutrons must be trapped before they hit fissionable atoms. Neutron absorption is carried out by the use of control rods made of a material such as cadmium which absorbs neutrons. When the control rods extend all the way into the reactor core, they absorb many neutrons and fission occurs slowly. As the rods are pulled out, they absorb fewer neutrons and the fission process speeds up. If the chain reaction were to go too fast, heat might be produced faster than it could be removed by the coolant. In that case, the reactor core would overheat. This could lead to mechanical failures and a release of radioactive materials into the atmosphere. Ultimately, a meltdown of the reactor core may occur. The fuel in a reactor is never concentrated enough, however, to produce a nuclear explosion.

Once a nuclear reactor is started, it remains highly radioactive for many generations. Shields are used to protect the reactor structure from radiation damage. Walls of high-density concrete are designed to protect the operating personnel.

24·7 Nuclear Fusion

The sun is an extraordinary energy source. The energy released from the sun is the result of a nuclear fusion, or thermonuclear, reaction. **Fusion** *occurs when two nuclei combine to produce a nucleus of heavier mass.* In solar fusion, hydrogen nuclei (protons) are fused to make helium nuclei. The reaction requires two beta particles.

$$4^1_1\text{H} + 2^{\ 0}_{-1}e \longrightarrow {}^4_2\text{He} + \text{energy}$$

Fusion takes place only at temperatures that exceed 40 000 000°C. It releases an even greater amount of heat!

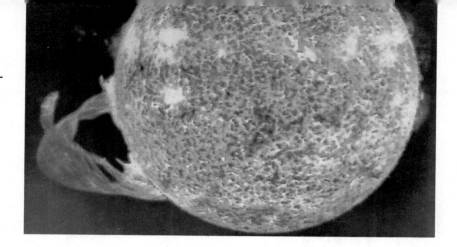

Figure 24·17
Thermonuclear fusion reactions occurring in the sun have provided the earth with energy for billions of years.

There is enough deuterium and tritium in the water in our oceans to fill virtually any future energy need we could imagine.

Controlled nuclear fusion is appealing as an energy source on earth. The potential fuels are inexpensive and readily available, and fusion products are usually not radioactive. The problems with fusion are in achieving the high temperatures necessary to start the reaction and in containing the reaction. One reaction scientists are studying is the combination of a deuterium nucleus and a tritium nucleus to form a helium nucleus.

$$^2_1H + {}^3_1H \longrightarrow {}^4_2He + {}^1_0n + \text{energy}$$

The high temperatures required to initiate fusion reactions have been achieved by using an atomic bomb. It is the triggering device for setting off the hydrogen bomb, which is an uncontrolled fusion device. It is clearly of no use, however, as a controlled generator of power.

To date, no large scale controlled fusion reactions have been achieved. At the high temperatures involved, matter exists as a *plasma*. Containing the plasma is a formidable task. No known structural material can withstand the high temperatures and corrosive effects of plasmas.

Figure 24·18
This experimental tokamak is a nuclear reactor in which magnets are used to confine a hydrogen plasma within a doughnut-shaped torus. Thirty-two D-shaped coils create a magnetic field that holds the hot corrosive plasma away from the walls of the vacuum vessel. The inner and outer field coils are used to adjust the shape of the plasma.

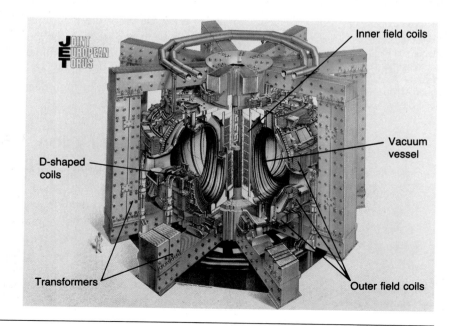

Figure 24·19
We cannot see, hear, feel, or smell radiation. Thus we must depend on detection instruments and signs to alert us of the presence of radiation and to monitor its levels.

■ Radiation can be detected with Geiger counters, scintillation counters, and film badges.

Chemical Connections

Certain isotopes that occur naturally in the world around us give off radiation. This background radiation varies from place to place but is always present to some extent.

Some success has been achieved with magnetic confinement using "magnetic bottles". In this approach, specially shaped magnets keep the reacting nuclei trapped in dense free-flowing plasma. Use of powerful lasers to generate the temperatures required for fusion has also been the subject of much recent research. Tremendous technical difficulties will have to be resolved, however, before fusion becomes a practical source of energy.

24·8 Detecting Radiation

The radiation emitted by radioisotopes (and X-rays) is ionizing radiation. **Ionizing radiation** *knocks electrons off some atoms of the bombarded substance to produce ions.* Ionizing radiation cannot be detected with any of our senses. Instead, various instruments and devices are used to do the job.

A **Geiger counter** *uses a gas-filled metal tube to detect radiation* (Figure 24·20). The tube is connected to a power supply and has a central wire electrode. When ionizing radiation penetrates the thin window at one end of the tube, the gas inside becomes ionized. When charged ions and free electrons are produced, the gas becomes an electrical conductor. Each time a Geiger tube is exposed to radiation, a current flows. The bursts of current drive electronic counters or cause audible clicks from a built-in speaker. Geiger counters are used primarily to detect beta radiation. Alpha particles cannot pass through the end window. Most gamma rays and X-rays pass through the gas, causing few ionizations.

A **scintillation counter** *uses a phosphor to detect radiation.* Ionizing radiation striking the specially coated surface, or phosphor, produces bright flashes of light (scintillations). (This device is similar to a television screen coated with zinc sulfide, ZnS, as the phosphor. The electrons from the TV gun striking the phosphor of the screen also produce scintillations. The pattern of scintillations is the TV picture.) The number of flashes and their respective energies are detected electronically. They are then converted into electronic pulses that are measured and recorded. Scintillation counters have been designed to detect all types of ionizing radiation.

Figure 24·20
The Geiger counter is widely used to detect beta radiation.

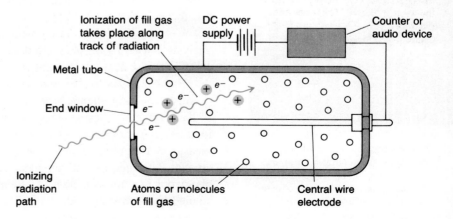

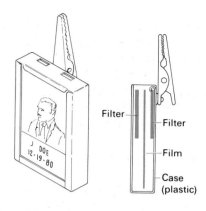

Figure 24·21
The film badge detector is worn by people who work with radiation. The film must be developed to be read. It darkens in proportion to the amount of radiation it receives.

Film badges are the most important radiation detectors for persons near radiation sources. *A **film badge** consists of several layers of photographic film covered with black lightproof paper encased in a plastic or metal holder* (Figure 24·21). It is worn all the time a person is at work. At frequent specific intervals, depending on the type of work involved, the film is removed and developed. The strength and type of radiation exposure are determined from any darkening of the film. Records are kept of the results. Film badges do not protect a person from radiation exposure. They merely serve as precautionary monitoring devices. The only protection against radiation is to keep a safe distance away and use adequate shielding.

Problem

5. A radioactive solution containing an alpha emitter accidentally gets on your hands. You wash them with soap and water and then check them with a Geiger counter for residual radioactive contamination. The Geiger counter does not register any radioactivity. Are your hands positively free from radioactive contamination? Explain.

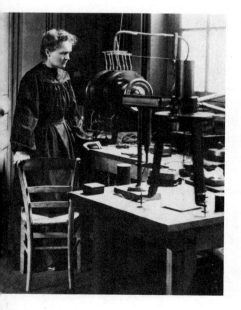

Figure 24·22
Both Marie Curie (shown here) and her daughter, Irene, died of leukemia. This was probably the result of overexposure to radiation.

Science, Technology, and Society

24·C Radioisotopes in Research and Medicine

Although radiation can be harmful and should be handled with care, it can be used safely and is important in many scientific procedures.

Neutron activation analysis is used to detect trace amounts of elements in samples. In this technique, a sample is bombarded with neutrons from a radioactive source. This causes the atoms in the sample to become radioactive. The half-life and type of radiation emitted by the radioisotopes are detected and processed by a computer. Since this information is characteristic for each element, scientists can tell what radioisotopes are produced and what elements were originally present in the sample. This is one of the most sensitive techniques for detecting trace amounts of elements. It is capable of measuring 10^{-9} g of an element in a sample. Neutron activation analysis is used by museums to detect art forgeries and by crime laboratories to analyze gunpowder residues.

Radioisotopes called tracers are used by chemists and biochemists to study chemical reactions and molecular structures. One of the reactants, labeled with a radioisotope, is added to the reaction mixture. After the reaction is over, the radiation of the product is measured to determine the uptake of the tracer. By comparing this amount with the amount originally added, scientists can learn much about the reaction mechanism. Reactions with many steps can be studied using this method.

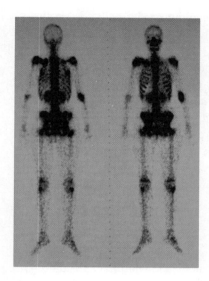

Figure 24·23
A radioisotope scan can be used to diagnose some suspected diseases without using exploratory surgery.

Agricultural researchers use radioisotopes to test the effects of herbicides, pesticides, and fertilizers. The tracer is introduced into the substance being tested to make it radioactive. Next, the plants are treated with the radioactive substance. Then, the radioactivity of the plants is measured to determine the location of the substance. Often the tracer is also monitored in animals that consume the plants, in water, and in soil. This information helps scientists determine the effects of using the substance.

Radioisotopes are used to diagnose some diseases. Iodine-131 is used to detect thyroid problems. The thyroid gland extracts iodide ions from the bloodstream and uses them to make thyroxine. To diagnose thyroid disease, the patient is given a small amount of iodine-131 in a liquid. After about two hours, the amount of iodide uptake is measured by scanning the patient's throat with a radiation detector. In a similar way, technetium-99^m is used to detect brain tumors, and phosphorus-32 is used to detect skin cancer.

Radiation is used to treat some cancers. This group of diseases is characterized by rapidly dividing abnormal cells. These cells are more sensitive to radiation than are normal cells. The cancerous area can be treated with radiation to kill the cancer cells. Some normal cells are also killed, however, and cancer cells at the center of the tumor may be radiation resistant. Therefore, the benefits of killing the cancer cells and the risks to the patient must be evaluated.

In teletherapy, a narrow beam of high intensity gamma radiation is directed at the cancerous tissue. Cobalt-60 and cesium-137 are commonly used as radiation sources. To minimize damage to healthy tissue, the patient is positioned so that only the cancerous region is within the radiation beam at all times. The unit rotates so that the radiation dose to the skin and surrounding normal tissue is distributed in a belt all the way around the patient.

Radioisotope seeds are radioisotope salts that are sealed in gold tubes and directly implanted in tumors. The radioisotopes emit beta

Figure 24·24
Radiation therapy is a commonly used method for treating cancer. This cobalt-60 unit emits a narrow, intense beam of radiation that destroys the ability of cells to reproduce. Cancer cells are more sensitive to radiation than normal cells because they divide more rapidly. Cells are most vulnerable when they are dividing.

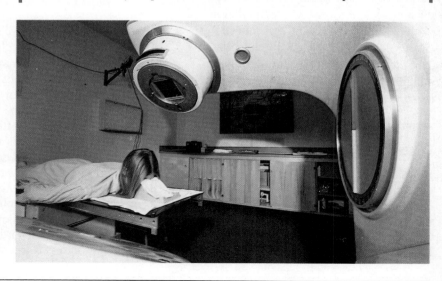

and gamma rays which kill the surrounding cancer cells. Since the radioisotope is in a sealed container, the excretions of patients undergoing this therapy are not radioactive.

Radioactive pharmaceuticals containing radioisotopes of gold, iodine, or phosphorus are sometimes given in radiation therapy. A larger dose of iodine-131 than that given to detect thyroid disease can be given to treat the disease. The radioactive iodine collects in the thyroid and emits beta and gamma rays to provide therapy. Since the radioisotope is not sealed, the patient's excretions are radioactive and must be specially handled.

24 Nuclear Chemistry
Chapter Review

Key Terms

alpha particle	24·2	neutron	
alpha radiation	24·2	moderation	24·6
band of stability	24·3	positron	24·3
beta particles	24·2	radiation	24·1
beta radiation	24·2	radioactive decay	24·1
film badge	24·8	radioactivity	24·1
fission	24·6	radioisotopes	24·1
fusion	24·7	scintillation	
gamma radiation	24·2	counter	24·8
Geiger counter	24·8	transmutation	24·5
half-life ($t_{1/2}$)	24·4	transuranium	
ionizing radiation	24·8	element	24·5
neutron absorption	24·6		

Chapter Summary

Isotopes with unstable nuclei are radioactive and are called radioisotopes. The nuclei of radioisotopes decay to stable nuclei plus radiation. The radiation may be alpha, beta, or gamma. Alpha radiation consists of alpha particles (positively charged helium nuclei) that are easily stopped by a sheet of paper. Beta radiation is composed of fast-moving beta particles, which are electrons. Beta radiation is more penetrating than alpha radiation; it is stopped by aluminum foil. Gamma radiation is electromagnetic radiation similar to visible light, but much more energetic. Gamma radiation has no mass or charge. It is extremely penetrating. Lead bricks and concrete reduce the intensity of gamma radiation but do not completely stop it. Specially built instruments produce X-radiation, which is essentially the same as gamma radiation.

Every radioisotope decays at a characteristic rate. A half-life is the time required for one-half of the nuclei in a radioisotope to decay. The product nuclei may or may not be radioactive. Half-lives vary from fractions of a second to millions of years.

The decay of radioisotopes is represented by nuclear equations. When a radioactive nucleus emits an alpha particle, its atomic number decreases by 2 and its mass number decreases by 4. When a beta particle is emitted, the atomic number increases by 1, and the mass number stays the same. When a gamma ray is emitted, the atomic number and the mass number stay the same.

Nuclear fission occurs when fissionable isotopes are bombarded with neutrons. The isotopes split into two fragments of about the same size, and in the process they release more neutrons. Some of these neutrons strike other nuclei, releasing more neutrons, which split more nuclei. This is called a chain reaction. Fission releases enormous amounts of energy. Controlled fission is the energy source in every nuclear power plant. In nuclear fusion small nuclei fuse to make heavier nuclei. The sun's energy is released when hydrogen nuclei fuse to make helium nuclei. Fusion releases even more energy than fission.

Radiation may be detected with a Geiger counter or a scintillation counter. The film badge is the most important personnel radiation detector. The radiation its wearer receives is detected by darkening of the encased film.

Practice Questions and Problems

6. Explain the difference between the terms *isotope* and *radioisotope*. *24·1*

7. What part of an atom undergoes change during radioactive decay? *24·1*

8. Write a symbol and state the charge for each.
a. an alpha particle
b. a beta particle
c. a gamma ray *24·2*

9. Alpha radiation is emitted during the disintegration of the following isotopes. Write balanced nuclear equations for their decay processes. Name the element produced in each case. *24·2*
a. uranium-238 ($^{238}_{92}U$)   **c.** uranium-235 ($^{235}_{92}U$)
b. thorium-230 ($^{230}_{90}Th$)   **d.** radon-222 ($^{222}_{86}Rn$)

10. The following radioisotopes are beta emitters. Write balanced nuclear equations for their decay processes. *24·2*
a. carbon-14 ($^{14}_{6}C$)   **c.** potassium-40 ($^{40}_{19}K$)
b. strontium-90 ($^{90}_{38}Sr$)   **d.** nitrogen-13 ($^{13}_{7}N$)

11. How are the atomic number and mass of a nucleus affected by the loss of the following? *24·2*
a. a beta particle
b. an alpha particle
c. a gamma ray

12. Tell how alpha, beta, and gamma radiation are distinguished on the basis of the following. *24·2*
a. mass   **b.** charge   **c.** penetrating power

13. The following radioactive nuclei decay by emitting alpha particles. Write the product of the decay process for each. *24·2*
a. $^{238}_{94}Pu$ **b.** $^{210}_{83}Bi$ **c.** $^{210}_{84}Po$ **d.** $^{230}_{90}Th$

14. Identify the most stable isotope in each of the following pairs. *24·3*
a. $^{14}_{6}C$, $^{13}_{6}C$ **c.** $^{16}_{8}O$, $^{18}_{8}O$
b. $^{3}_{1}H$, $^{1}_{1}H$ **d.** $^{14}_{7}N$, $^{15}_{7}N$

15. Name the elements represented by the following symbols and indicate which of them would have *no* stable isotopes. *24·3*
a. Pt   **e.** Xe
b. Th   **f.** Cf
c. Fr   **g.** V
d. Tc   **h.** Pd

16. Explain the term *half-life*. *24·4*

17. The mass of cobalt-60 in a sample is found to have decreased from 0.8 g to 0.2 g in a period of 10.5 years. From this information, calculate the half-life of cobalt-60. *24·4*

18. A patient is administered 20 mg of iodine-131. How much of this isotope will remain in the body after 40 days if the half-life for iodine-131 is 8 days? *24·4*

19. Explain the process of transmutation. Write at least three nuclear equations to illustrate your answer. *24·5*

20. What are the transuranium elements? Why are they unusual? *24·5*

21. What are the differences between natural and artificial radioactivity? *24·5*

22. Complete and balance the equations for the following nuclear reactions. *24·5*
a. $^{27}_{13}Al + ^{4}_{2}He \longrightarrow ^{30}_{14}Si +$
b. $^{214}_{83}Bi \longrightarrow ^{4}_{2}He +$
c. $^{27}_{14}Si \longrightarrow ^{0}_{-1}e +$
d. $^{66}_{29}Cu \longrightarrow ^{66}_{30}Zn +$

23. Describe the process of nuclear fission. *24·6*

24. Explain the term *nuclear chain reaction*. *24·6*

25. How is the chain reaction in a nuclear reactor controlled? *24·6*

26. Name one fissionable material that occurs in nature and one that is artificial. *24·6*

27. Fusion reactions produce enormous amounts of energy. Why don't we see power stations operating fusion devices? *24·7*

28. Why are X-rays and the radiations emitted by radioisotopes called *ionizing radiation*? *24·8*

29. What is the purpose of wearing a film badge when working with radiation sources? *24·8*

Mastery Questions and Problems

30. Write nuclear equations for these conversions.
a. $^{30}_{15}P$ to $^{30}_{14}Si$ b. $^{13}_{6}C$ to $^{14}_{6}C$

31. What is the difference between the nuclear reactions taking place in the sun and the nuclear reactions taking place in a nuclear reactor?

32. Complete these nuclear reactions.
a. $^{32}_{15}P \longrightarrow$ _____ $+ \ ^{0}_{-1}e$
b. _____ $\longrightarrow \ ^{14}_{7}N + \ ^{0}_{-1}e$
c. $^{238}_{92}U \longrightarrow \ ^{234}_{90}Th +$ _____
d. $^{141}_{56}Ba \longrightarrow$ _____ $+ \ ^{0}_{-1}e$

33. Write nuclear equations for the beta decay of the following isotopes.
a. $^{90}_{38}Sr$ b. $^{14}_{6}C$ c. $^{137}_{55}Cs$ d. $^{239}_{93}Np$

34. The graph below shows the radioactive decay curve for thorium-234. Use the graph to answer these questions.
a. What percent of the isotope remains after 60 days?
b. How many grams of a 250-g sample of thorium-234 would remain after 40 days?
c. How many days would pass while 44 g of thorium-234 decayed to 4.4 g of thorium-234?
d. What is the half-life of thorium-234?

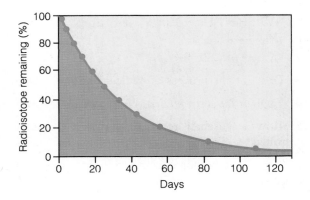

35. Write a nuclear reaction for each of these word equations.
a. Radon-222 emits an alpha particle to form polonium-218.
b. Radium-230 is produced when thorium-234 emits an alpha particle.
c. When polonium-210 emits an alpha particle the product is lead-206.

36. Describe the various contributions the following people made to the fields of nuclear and radiation chemistry.
a. Marie Curie   c. James Chadwick
b. Antoine Becquerel   d. Ernest Rutherford

Critical Thinking Questions

37. Choose the term that best completes the second relationship.
a. decomposition:combination   fission: _____

   (1) energy   (3) radioactivity
   (2) nuclei   (4) fusion
b. female:male   electron: _____
   (1) positron   (3) beta particle
   (2) neutron   (4) alpha particle
c. umbrella:rain   wood: _____
   (1) radioactive decay (3) beta radiation
   (2) gamma radiation (4) X-radiation

38. Which kinds of radiation can be affected by a particle accelerator?

39. Why does the large mass and size of an alpha particle limit its penetrating power?

40. Why are radioactive isotopes of C, N, and O especially dangerous to living creatures?

Review Questions and Problems

41. What is the Pauli exclusion principle? What is Hund's rule?

42. Balance the following equations.
a. $Ca(OH)_2 + HCl \longrightarrow CaCl_2 + H_2O$
b. $Fe_2O_3 + H_2 \longrightarrow Fe + H_2O$
c. $NaHCO_3 + H_2SO_4 \longrightarrow$
$$Na_2SO_4 + CO_2 + H_2O$$
d. $C_2H_6 + O_2 \longrightarrow CO_2 + H_2O$

43. You have a $0.30M$ solution of sodium sulfate. What volume in milliliters must be measured to give 0.0020 mol of sodium sulfate?

44. How many protons, neutrons, and electrons are in an atom of each isotope?
 a. iron-59   **b.** uranium-235   **c.** chromium-52

45. Identify the bonds between the following pairs of atoms as ionic or covalent.
 a. carbon and silicon   **c.** sulfur and nitrogen
 b. calcium and fluorine   **d.** bromine and cesium

46. How many cm^3 of hydrogen gas, at STP, will be produced when 10.00 g of magnesium metal reacts with an excess of sulfuric acid? How many moles is this?

Challenging Questions and Problems

47. The radioisotope cesium-137 has a half-life of 30 years. A sample decays at the rate of 544 counts/min (544 cpm) in 1985. In what year will the decay rate be 17 cpm?

48. Bismuth-211 is a radioisotope. It decays by alpha emission to yield another radioisotope which emits beta radiation as it decays to a stable isotope. Write equations for the nuclear reactions and name the decay products.

49. What isotope remains after three beta particles and five alpha particles are lost from the thorium-234 isotope. (Refer to the uranium-238 decay series to check your answer.)

Research Projects

1. Why is it said that the atomic age began with Enrico Fermi?

2. How did Marie Curie's discovery of radium affect science and industry?

3. How did the Curies obtain large quantities of radium?

4. Design and build a simple diffusion cloud chamber. Use your chamber to detect radiation.

5. How has nuclear science helped scientists produce elements?

6. What safeguards exist at nuclear power plants?

7. Make a diagram of what happens in a nuclear meltdown.

8. Report on the Three Mile Island or Chernobyl incidents. What steps should have been taken to prevent the incident? How likely are the prospects for a similar incident in the future?

9. What are some of the technical difficulties in using nuclear fusion as an energy source? Do you believe it should be developed? Why or why not?

10. What are the environmental hazards of fast breeder reactors?

11. What is radioactive fallout? How is it measured? What are its effects?

12. Measure background radiation at various sites in your community and under various weather conditions.

Readings and References

Atwood, Charles H. "Nuclear Diagnosis." *Chem-Matters* (December 1985), pp. 4–7.

Kunetka, James W. *City of Fire: Los Alamos and the Birth of the Atomic Age, 1943–1945*. Englewood Cliffs, NJ: Prentice-Hall, 1978.

Murray, Raymond L. *Understanding Radioactive Waste*. Columbus, OH: Battelle Press, 1982.

Olsen, Steve. "Nuclear Undertakers." *Science 84* (September 1984), pp. 50–59.

Patterson, Walter C. *Nuclear Power*, 2nd ed. New York: Penguin, 1983.

Reid, Robert. *Marie Curie*. New York: Saturday Review Press/Dutton, 1974.

Romer, Alfred. *The Restless Atom: The Awakening of Nuclear Physics*. New York: Dover, 1982.

Walker, Charles A., Leroy C. Gould, and Edward J. Woodhouse, eds. *Too Hot To Handle? Social and Policy Issues in the Management of Radioactive Wastes*. New Haven, CT: Yale University Press, 1983.

West, Susan. "Hot." *Science 84* (December 1984), pp. 28–37.

25 Hydrocarbon Compounds

Chapter Preview

25·1 Hydrocarbon Bonds
25·2 Continuous-Chain Alkanes
25·3 Branched-Chain Alkanes
25·4 Properties of Alkanes
25·5 Alkenes and Alkynes
25·A Goodyear's Discovery

25·6 Geometric Isomers
25·7 Stereoisomers
25·8 Cyclic and Aromatic Hydrocarbons
25·9 Natural Gas and Petroleum
25·B Coal

The element carbon comprises only two-tenths of 1% of the elements found in the earth's crust. Oxygen, silicon, aluminum, and iron are much more abundant, but carbon is the basis of life on earth.

Scientists of 150 years ago believed the ability to produce carbon compounds rested exclusively with living things. The creation of carbon compounds was thought to be directed by a mysterious "vital force." Vitalism was rudely shattered in 1828 by a German chemist named Friedrich Wöhler (1800–1882). Wöhler discovered one way to make urea in the absence of any living agent. Urea is a carbon-containing compound found in urine.

Since Wöhler's day the definition of organic chemistry has been extended to include the study of all carbon compounds, regardless of their origin. Organic chemists have discovered how carbon compounds can be synthesized, or built, from simpler materials by ordinary chemical reactions. *Organic compounds that contain only carbon and hydrogen are called* **hydrocarbons.** They provide a logical introduction to organic chemistry.

25·1 Hydrocarbon Bonds

The simplest organic molecules are the **alkanes** *which are hydrocarbons that contain only single covalent bonds.* Methane is the simplest alkane. It is a gas at standard temperature and pressure. Methane is the major component of natural gas. It is sometimes called "marsh gas" because it is formed by the action of bacteria on decaying vegetation in swamps and other marshy areas. The methane molecule, which contains four hydrogens and one carbon, is a good example of carbon–hydrogen bonding.

Figure 25·1
In a petroleum refinery, useful hydrocarbons such as gasoline, kerosene, and heating oil are distilled from crude oil.

591

Figure 25·2
In this painting by Ford Madox Brown, John Dalton is collecting "marsh gas" (methane), which he used in many of his experiments. Methane is one product of the decomposition of plant materials by decay bacteria.

The covalent bonding in alkanes involves the sharing of valence electrons between carbon and hydrogen or between two carbons.

A carbon atom has four valence electrons. Four hydrogen atoms, each with one valence electron, form four covalent carbon–hydrogen bonds. This combination is a molecule of methane.

$$\overset{\cdot}{\underset{\cdot}{C}}\cdot \quad + \quad 4H\cdot \quad \longrightarrow \quad H:\overset{\cdot\cdot}{\underset{\cdot\cdot}{C}}:H$$

Carbon atom   Hydrogen atoms   Methane molecule

This illustrates an important principle. *Because a carbon atom contains four valence electrons, it always forms four covalent bonds.* Remembering this principle will help you write complete and correct structures for organic molecules.

For simplicity, organic chemists often abbreviate bonding electron pairs as short lines. The line between the atomic symbols represents two bonding electrons.

$$H-\overset{\displaystyle H}{\underset{\displaystyle H}{C}}-H \qquad \text{Line represents shared electron pair}$$

Methane molecule

Structural formulas are convenient to write on a page. Keep in mind though that they are only two-dimensional representations of three-dimensional molecules. Molecular models represent the shapes of molecules more accurately. These shapes are predicted by VSEPR theory and hybrid orbital theory (Sections 14·8 and 14·9). For example, methane has a tetrahedral shape (Figure 25·3).

Carbon has the unique ability to make stable carbon–carbon bonds and to form chains. This is the major reason for the vast number of

Figure 25·3
This three-dimensional model shows the tetrahedral shape of the methane molecule.

organic molecules. Silicon also forms short chains, but they are unstable in an oxygen environment.

Ethane, C_2H_6, is the simplest alkane containing a carbon–carbon bond. Like methane, ethane is a gas at standard temperature and pressure. When ethane is formed from carbon and hydrogen, two carbon atoms share a pair of electrons. A carbon-carbon covalent bond is formed. The remaining six valence-shell electrons form bond pairs with the electrons from six hydrogen atoms.

$$2 \cdot \overset{\cdot}{\underset{\cdot}{C}} \cdot \; + \; 6H\cdot \; \longrightarrow \; \cdot \overset{\cdot}{C} - \overset{\cdot}{C} \cdot \; + \; 6H\cdot \; \longrightarrow$$

Ethane

25·2   Continuous-Chain Alkanes

Continuous-chain alkanes are hydrocarbons with the general formula C_nH_{2n+2}.

Continuous-chain alkanes *contain any number of carbon atoms in a straight chain.* They are similar to ethane. To draw a structural formula, just write the symbol for carbon as many times as necessary to get the proper chain length. Then fill in with hydrogens and lines representing covalent bonds. Remember that carbon has four covalent bonds. Table 25·1 shows the continuous-chain alkanes containing up to ten carbons. Note that the names of alkanes always end with *-ane*.

The continuous-chain alkanes are an example of a homologous series. *A group of compounds forms a* **homologous series** *if there is a constant increment of change in molecular structure from one compound in the series to the next.* The increment of change in the continuous-chain alkanes is the $-CH_2-$ group. For example, propane and butane are *homologs* of each other. Note that as the number of carbons in this series increases, so does the boiling point. This is also true of the melting point.

$-CH_2-$ is shorthand notation for a carbon with two hydrogens.

Table 25·1	Structural Formulas of the First Ten Continuous-Chain Alkanes		
Name	Molecular formula	Structural formula	Boiling point (°C)
Methane	CH_4	CH_4	−161.0
Ethane	C_2H_6	CH_3CH_3	−88.5
Propane	C_3H_8	$CH_3CH_2CH_3$	−42.0
Butane	C_4H_{10}	$CH_3CH_2CH_2CH_3$	0.5
Pentane	C_5H_{12}	$CH_3CH_2CH_2CH_2CH_3$	36.0
Hexane	C_6H_{14}	$CH_3CH_2CH_2CH_2CH_2CH_3$	68.7
Heptane	C_7H_{16}	$CH_3CH_2CH_2CH_2CH_2CH_2CH_3$	98.5
Octane	C_8H_{18}	$CH_3CH_2CH_2CH_2CH_2CH_2CH_2CH_3$	125.6
Nonane	C_9H_{20}	$CH_3CH_2CH_2CH_2CH_2CH_2CH_2CH_2CH_3$	150.7
Decane	$C_{10}H_{22}$	$CH_3CH_2CH_2CH_2CH_2CH_2CH_2CH_2CH_2CH_3$	174.1

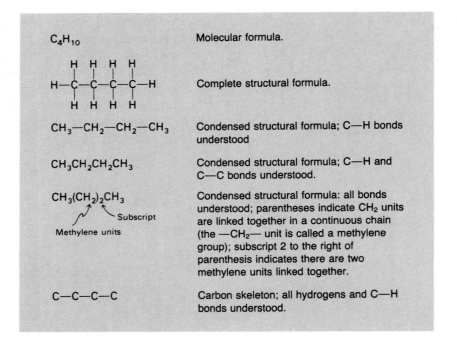

C_4H_{10}	Molecular formula.
Complete structural formula (H—C—C—C—C—H)	Complete structural formula.
CH_3—CH_2—CH_2—CH_3	Condensed structural formula; C—H bonds understood
$CH_3CH_2CH_2CH_3$	Condensed structural formula; C—H and C—C bonds understood.
$CH_3(CH_2)_2CH_3$ — Methylene units / Subscript	Condensed structural formula: all bonds understood; parentheses indicate CH_2 units are linked together in a continuous chain (the —CH_2— unit is called a methylene group); subscript 2 to the right of parenthesis indicates there are two methylene units linked together.
C—C—C—C	Carbon skeleton; all hydrogens and C—H bonds understood.

 Issues in Chemistry

Much of the world's oil is transported by ship in some part of its journey to a refinery. Is the risk of oil spills worth the benefit of using crude oil to make plastics, alcohols, gasoline, and other consumer products?

Complete structural formulas show all the atoms and bonds in a molecule. Sometimes, however, shorthand or condensed structural formulas work just as well. **Condensed structural formulas** *leave out some bonds and/or atoms from the structural formula*. The reader must understand that these bonds and atoms are there. Figure 25·4 shows several ways to draw condensed structural formulas for butane.

Example 1

Draw complete structural formulas for the continuous-chain alkanes that have three and four carbons.

Solution

$$
\begin{array}{ccc}
H & H & H \\
| & | & | \\
H-C-C-C-H \\
| & | & | \\
H & H & H
\end{array}
\qquad
\begin{array}{cccc}
H & H & H & H \\
| & | & | & | \\
H-C-C-C-C-H \\
| & | & | & | \\
H & H & H & H
\end{array}
$$

Problems

1. Draw complete structural formulas for the continuous-chain alkanes with five and six carbons.

2. Draw condensed structural formulas for pentane and hexane. Assume that the C—H and C—C bonds are understood.

Figure 25·5
Pressurized tanks of propane and butane are used as heating and cooking fuels for campers. They are also used in rural areas where gas pipelines are not available.

The names listed in Table 25·1 are recommended by the International Union of Pure and Applied Chemistry (IUPAC). You should memorize these names. They are the basis of *a precise, internationally accepted system of naming organic compounds called the* **IUPAC system.** Note, however, that organic chemists still rely on a mixture of systematic, semisystematic, and common names. This is in spite of the precision of the IUPAC system.

Problem

3. Name the following alkanes.

a. H—C—C—C—H (with H atoms)

c. H—C—C—C—C—C—C—H (with H atoms)

b. H—C—C—C—C—C—H (with H atoms)

ChemDirections

A large portion of the hazardous chemicals transported on the nation's highways are highly flammable hydrocarbons. How do you know what a tank truck is carrying? Read **Chem Directions** *Trucking Chemicals*, page 659, to find the answer.

Many organic compounds are best known by their common names. These names bear no relation to the IUPAC name or to the molecular structure of the compounds. Nevertheless, these names have been used for a long time. Many organic compounds isolated from nature have common names that reflect their origin. For example, penicillin is named from the mold *Penicillium notatum*. In most instances, common names are simpler than IUPAC names, which can become long and cumbersome. Scientists also use semisystematic names which mix IUPAC and common names. This is similar to the way that immigrants lapse into their native tongues when they have important ideas to express. Like any living language, the language of science is constantly changing. IUPAC, semisystematic, and common names are used in this book.

25·3 Branched-Chain Alkanes

One or more hydrogens on a continuous-chain alkane can be substituted by other hydrocarbon groups to form branched-chain hydrocarbons.

An atom or group of atoms, called a **substituent,** *can take the place of a hydrogen atom on a parent hydrocarbon molecule.* Common substituents are the halogens and groups of atoms including carbon, hydrogen, oxygen, nitrogen, sulfur, or phosphorus.

C←Substituent
C—C—C
Parent alkane (propane)

C C C ←Substituents
C—C—C—C—C—C
Parent alkane (hexane)

A hydrocarbon substituent is called an **alkyl group.** It can be one or several carbons long. Three common alkyl groups are the methyl group (CH_3—), the ethyl group (CH_3CH_2—), and the propyl group ($CH_3CH_2CH_2$—). As you can see, an alkyl group consists of an alkane with one terminal hydrogen removed. Alkyl groups are sometimes referred to as radicals. They are named by removing the *-ane* ending from the parent hydrocarbon name and adding *-yl*.

An alkane with one or more alkyl groups is a **branched-chain alkane.** The IUPAC rules for naming branched-chain alkanes are quite straightforward. The following compound can be used as an example.

$$\overset{7}{CH_3}-\overset{6}{CH_2}-\overset{5}{CH_2}-\overset{4}{CH}-\overset{3}{CH}-\overset{2}{CH}-\overset{1}{CH_3}$$
$$| \quad\quad | \quad\quad |$$
$$CH_2 \quad CH_3 \quad CH_3$$
$$|$$
$$CH_3$$

1. *Find the longest continuous chain of carbons in the molecule.* This chain is used as the parent structure. In the example, the longest continuous chain contains seven carbons. Therefore the parent hydrocarbon structure is heptane.

2. *Number the carbons in the main chain in sequence.* In doing this start at the end that will give the groups attached to the chain the *smallest numbers.* This has already been done in the preceding example. In this instance the numbers go from right to left, which places the substituent groups at carbon atoms 2, 3, and 4. If the chain were numbered from left to right, the groups would be at positions 4, 5, and 6. These are higher numbers and therefore violate the rule.

3. *Add numbers to the names of the substituent groups to identify their positions on the chain.* These numbers become prefixes to the name of the parent alkane. In this example the substituents and positions are 2-methyl, 3-methyl, and 4-ethyl.

4. *Use prefixes to indicate the appearance of a group more than once in the structure.* Common prefixes are *di* (twice), *tri* (three times), *tetra* (four), and *penta* (five). This example has two methyl substituents. Thus the word *dimethyl* will be part of the complete name.

5. *List the names of alkyl substituents in alphabetical order.* For purposes of alphabetizing, the prefixes *di, tri,* and so on are ignored. In this example, the 4-ethyl group is listed before the 2-methyl and 3-methyl groups (which are combined as 2,3-dimethyl in the name).

6. *Use proper punctuation.* This is very important in writing the names of organic compounds in the IUPAC system. Commas are used to separate numbers. Hyphens are used to separate numbers and words. *The name of the alkane is written as one word.* The demonstration compound, then, would be written as dimethylheptane, *not* dimethyl heptane.

According to the IUPAC rules, the name of this compound is 4-ethyl-2,3-dimethylheptane.

Figure 25·6
This is a ball and stick model of 4-ethyl-2,3-dimethylheptane. The parent structure forms a diagonal zig-zag chain from the lower right to the upper left.

Example 2

Name these compounds using the IUPAC system. (Note: The longest continuous chain is *not* written in a straight line in molecule **a.**)

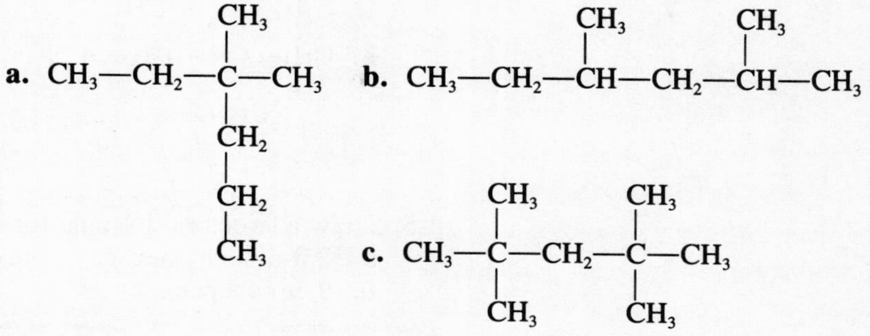

a. $CH_3-CH_2-\underset{\underset{\underset{\underset{CH_3}{|}}{CH_2}}{\overset{\overset{\overset{CH_3}{|}}{|}}{C}}-CH_3$

b. $CH_3-CH_2-\underset{}{\overset{\overset{CH_3}{|}}{CH}}-CH_2-\overset{\overset{CH_3}{|}}{CH}-CH_3$

c. $CH_3-\underset{\underset{CH_3}{|}}{\overset{\overset{CH_3}{|}}{C}}-CH_2-\underset{\underset{CH_3}{|}}{\overset{\overset{CH_3}{|}}{C}}-CH_3$

Solution

a. 3,3-dimethylhexane   b. 2,4-dimethylhexane
c. 2,2,4,4-tetramethylpentane

With an alkane name and knowledge of the IUPAC rules, it is easy to reconstruct the structural formula.

1. *Find the root word* (ending in *-ane*) *in the hydrocarbon name.* Then write the longest carbon chain to create the parent structure.
2. *Number the carbons on this parent carbon chain.*
3. *Identify the substituent groups.* Attach the substituents to the numbered parent chain at the proper positions.
4. *Add hydrogens as needed.*

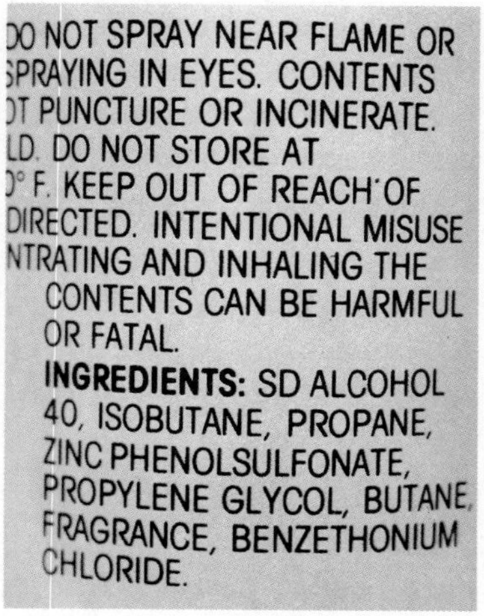

Example 3

Draw complete structural formulas for the following compounds.
a.   3,3-dimethylpentane   c.   2,2,4-trimethylpentane
b.   3-ethylhexane

Solution

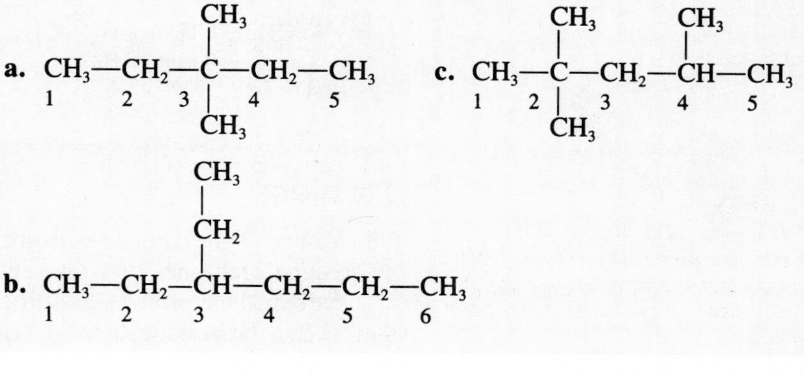

a. $\underset{1}{CH_3}-\underset{2}{CH_2}-\underset{3}{\underset{\underset{CH_3}{|}}{\overset{\overset{CH_3}{|}}{C}}}-\underset{4}{CH_2}-\underset{5}{CH_3}$

c. $\underset{1}{CH_3}-\underset{2}{\underset{\underset{CH_3}{|}}{\overset{\overset{CH_3}{|}}{C}}}-\underset{3}{CH_2}-\underset{4}{\overset{\overset{CH_3}{|}}{CH}}-\underset{5}{CH_3}$

b. $\underset{1}{CH_3}-\underset{2}{CH_2}-\underset{3}{\underset{}{\overset{\overset{CH_3}{|}}{\underset{\overset{CH_2}{|}}{CH}}}}-\underset{4}{CH_2}-\underset{5}{CH_2}-\underset{6}{CH_3}$

Figure 25·7
Alkanes such as isobutane, propane, and butane are commonly used as propellants in aerosol sprays.

Problems

4. Name the following compounds according to the IUPAC system.

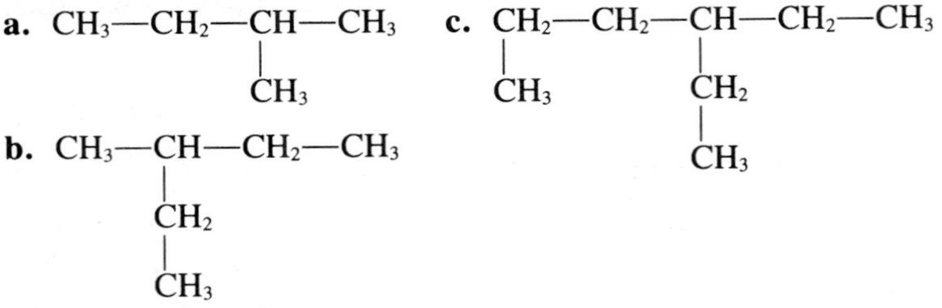

    a. CH_3—CH_2—CH—CH_3
    $\qquad\qquad\quad |$
    $\qquad\qquad\ CH_3$

    c. CH_2—CH_2—CH—CH_2—CH_3
    $\qquad |\qquad\qquad\quad |$
    $\qquad CH_3\qquad\quad CH_2$
    $\qquad\qquad\qquad\qquad |$
    $\qquad\qquad\qquad\ CH_3$

    b. CH_3—CH—CH_2—CH_3
    $\qquad\qquad |$
    $\qquad\quad CH_2$
    $\qquad\qquad |$
    $\qquad\quad CH_3$

5. Draw a structural formula for each of the following compounds.
    a. 2,3-dimethylhexane
    c. 3-ethyl-2,4-dimethyloctane
    b. 3-ethylheptane

25·4 Properties of Alkanes

The electron pair in a carbon–hydrogen or a carbon–carbon bond is about equally shared by the nuclei of the elements involved. Therefore hydrocarbon molecules such as the alkanes are nonpolar. The nonpolar attractions that hold hydrocarbon molecules together are the very weak van der Waals forces. Thus hydrocarbons of low formula weight tend to be gases or low-boiling liquids.

Nonpolar organic molecules such as the hydrocarbons are not attracted to water. For example, oil and water do not mix. A good rule of thumb is that "like dissolves like." That is, two nonpolar compounds will form a solution as will two polar compounds. By contrast, a nonpolar compound and a polar compound will not form a solution.

It is possible to draw the structures of two or more alkanes that have the same molecular formula but different molecular structures. *Compounds that have the same molecular formula but different molecular structures are called* **structural isomers.** For example, two different molecules have the formula C_4H_{10}: butane and 2-methylpropane. They are isomers of each other.

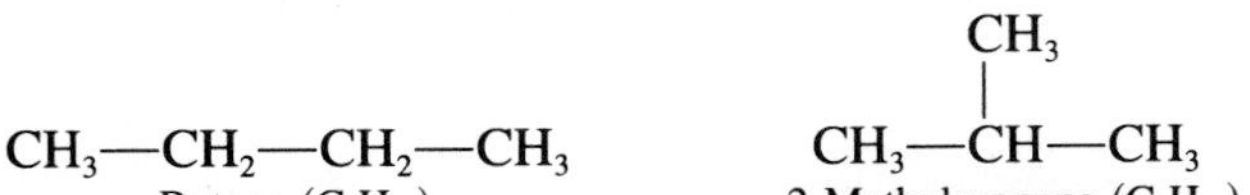

CH_3—CH_2—CH_2—CH_3
Butane (C_4H_{10})
(bp 0°C)

$\qquad\qquad\qquad CH_3$
$\qquad\qquad\qquad |$
CH_3—CH—CH_3
2-Methylpropane (C_4H_{10})
(bp −10.2°C)

Problem

6. Draw all the structural isomers with the molecular formula C_6H_{14}. Name each one. (For convenience, you may wish to draw only the carbon skeleton for each structure.)

■ The alkanes are gases or low-boiling, greasy liquids which are insoluble in water.

The physical properties of structural isomers are different.

Structural isomers differ in their physical properties such as boiling points and melting points. They also have different chemical reactivities. In general, the more highly branched the hydrocarbon structure, the lower its boiling point compared with its other structural isomers. For example, 2-methylpropane has a lower boiling point than butane.

25·5 Alkenes and Alkynes

The alkenes have the general formula C_nH_{2n}, and they contain carbon–carbon double bonds.

The carbon–carbon bonds in alkanes are examples of single bonds. Multiple bonds between carbons also exist. *Organic compounds containing carbon–carbon double bonds are called* **alkenes.** This is the carbon–carbon double bond found in alkenes.

$$\underset{/}{\overset{\backslash}{C}}=\underset{\backslash}{\overset{/}{C}}$$

Organic compounds that contain double and triple carbon–carbon bonds are called **unsaturated compounds.** This is because they contain fewer than the maximum number of hydrogens in their structure. *The alkanes (which contain the maximum number of hydrogens) are called* **saturated compounds.**

Figure 25·8
Plants produce ethene, which plays a vital role in their growth and development. Farmers treat tomato plants with a compound that breaks down into ethene inside the plants. This makes the tomatoes ripen at the same time so they can be harvested more efficiently.

To name an alkene by the IUPAC system, find the longest continuous chain in the molecule that contains the double bond. This chain is the parent alkene. It gets the root name of the alkane with the same number of carbons plus the ending *-ene.* The chain is numbered so that the carbon atoms of the double bond get the lowest possible numbers. Substituents on the chain are named and numbered the same way as for the alkanes. Ethene and propene are the simplest alkenes. They are often called by the common names ethylene and propylene. Here are some examples of the structures and IUPAC names of simple alkenes.

Ethene
(ethylene, the simplest alkene)

$$CH_3-\overset{\overset{\displaystyle H}{|}}{C}=\overset{\overset{\displaystyle H}{|}}{C}-H$$

Propene
(propylene)

$$H_2C=CH-CH_2-CH_3$$

1-Butene

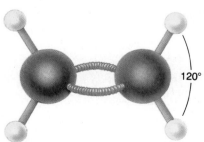

Figure 25·9
The six atoms of ethene lie in one plane. Rotation does not occur about the double bond.

$$CH_3-\overset{\overset{\displaystyle H}{|}}{C}=\overset{\overset{\displaystyle H}{|}}{C}-CH_3$$

2-Butene
(1-butene and 2-butene are structural isomers of C_4H_8)

$$CH_3-\overset{\overset{\displaystyle CH_3}{|}}{C}H-\overset{\overset{\displaystyle H}{|}}{C}=\overset{\overset{\displaystyle H}{|}}{C}-CH_3$$

4-Methyl-2-pentene

No rotation occurs about a carbon–carbon double bond. The four hydrogens that project from the double-bonded carbons in ethene lie in a plane and are 120° apart (Figure 25·9). This is the maximum separation of atoms that can be attained without breaking bonds.

Figure 25·10
This is a ball-and-stick model of ethyne, the simplest alkyne. Rotation about the triple bond does not occur.

Safety

Alkenes and alkynes are very reactive because of the presence of double and triple bonds.

The physical properties of alkanes and alkenes of similar molecular mass are similar.

Organic compounds containing carbon–carbon triple bonds are called **alkynes.** Like alkenes, alkynes are unsaturated compounds. This is the carbon–carbon triple bond found in alkynes.

$$-C\equiv C-$$

Alkynes are not plentiful in nature. The simplest alkyne is the gas ethyne, C_2H_2. The common name for ethyne is acetylene. It is the fuel burned in oxyacetylene torches used in welding. In open flame lamps that were used for mining, acetylene was produced by reacting calcium carbide with water.

$$CaC_2(s) + 2H_2O(l) \longrightarrow C_2H_2(g) + Ca(OH)_2(aq)$$

The single bonds that extend from the carbons involved in the carbon–carbon triple bond of ethyne are separated by the maximum angle of 180° (Figure 25·10). Thus ethyne is a linear molecule.

The major attractions between alkanes, alkenes, or alkynes are weak van der Waals forces. As a result, the introduction of a double or triple bond into a hydrocarbon does not have a dramatic effect on physical properties such as the boiling point (Table 25·2).

Table 25·2	Boiling Points of Homologous Alkanes, Alkenes, and Alkynes	
Name	Molecular structure	Boiling Point(°C)
C_2		
Ethane	CH_3-CH_3	−88.5
Ethene	$CH_2=CH_2$	−103.9
Ethyne	$CH\equiv CH$	−81.8
C_3		
Propane	$CH_3CH_2CH_3$	−42.0
Propene	$CH_3CH=CH_2$	−47.0
Propyne	$CH_3C\equiv CH$	−23.3
C_4		
Butane	$CH_3CH_2CH_2CH_3$	−0.5
1-Butene	$CH_3CH_2CH=CH_2$	−6.3
1-Butyne	$CH_3CH_2C\equiv CH$	8.6
C_5		
Pentane	$CH_3CH_2CH_2CH_2CH_3$	36.0
1-Pentene	$CH_3CH_2CH_2CH=CH_2$	30.0
1-Pentyne	$CH_3CH_2CH_2C\equiv CH$	40.0

Figure 25·11
The vulcanization of rubber was accidentally discovered by Charles Goodyear. It gives automobile tires and other rubber products strength, durability, and weather-resistance.

Figure 25·12
The isoprene unit of natural rubber is shown here in blue. When a mixture of sulfur and rubber is heated, sulfur atoms form bonds between the isoprene units, strengthening the rubber.

25·A Goodyear's Discovery

Natural rubber comes from rubber trees found in moist, tropical climates. Rubber's usefulness derives mostly from its excellent elasticity, or ability to stretch. It is also airtight, water resistant, and shock absorbing. Natural rubber, however, has its limitations. It becomes sticky in hot weather and brittle in cold weather. The solution to this problem was accidentally discovered in 1839 by the American inventor Charles Goodyear. While conducting an experiment, Goodyear spilled a mixture of rubber and sulfur on a hot stove. He was surprised to find that the rubber had been improved by the accident. It could now keep its shape and elasticity even in extreme heat and cold. This method of heating a sulfur–rubber mixture became known as vulcanization. It is named after Vulcan, the Roman god of fire.

A look at the chemistry of rubber reveals why vulcanization strengthens rubber. In natural rubber, thousands of isoprene (C_5H_8) molecules link to form giant polymers. In unstretched rubber, the polymers fold back on each other much like an accordion. Stretching the rubber straightens the chain. It returns to its folded position when released. In vulcanization, sulfur chemically combines with the rubber to form cross-links between the polymers. This binds the chain more firmly, giving the rubber greater elasticity and strength. If harder rubber is desired, more sulfur is added.

Many improvements have been made in rubber since Goodyear's time. Synthetic rubber is more widely used now than natural rubber. Synthetic rubber has hydrocarbons other than isoprene as its building blocks. Some compounds used are butadiene (C_4H_6), styrene ($C_6H_5CHCH_2$), and chloroprene (C_4H_5Cl). All of these substances are easier to polymerize than isoprene. Other forms of synthetic rubber are being made from silicon rather than carbon compounds.

Current research is focused on the production of synthetic rubbers that can withstand greater temperature extremes. This has become critical in the age of nuclear plants and space travel. Other researchers are looking for new ways to break down and recycle rubber. Mounds of abandoned tires are a monument to rubber's durability. They are also eyesores. At present, some used rubber is digested in a bath of zinc chloride. This removes all nonrubber components. The reclaimed rubber can then be used for new products.

Isoprene unit

$$\text{\textasciitilde CH}_2-\underset{\underset{CH_3}{|}}{C}=CH-CH_2-CH_2-\underset{\underset{CH_3}{|}}{C}=CH-CH_2\text{\textasciitilde}$$

$$\text{\textasciitilde CH}_2-\underset{\overset{|}{CH_3}}{C}=CH-CH_2-CH_2-\underset{\overset{|}{CH_3}}{C}=CH-CH_2\text{\textasciitilde}$$

Natural rubber

Sulfur and heat →

$$\text{\textasciitilde CH}-\underset{\underset{S}{|}}{\overset{\overset{CH_3}{|}}{C}}=CH-CH_2-CH-\underset{}{\overset{\overset{CH_3}{|}}{C}}=CH-CH_2\text{\textasciitilde}$$

$$\text{\textasciitilde CH}-\underset{\overset{|}{CH_3}}{C}=CH-CH_2-CH_2-\underset{\overset{|}{CH_3}}{C}=CH-CH\text{\textasciitilde}$$

Vulcanized rubber

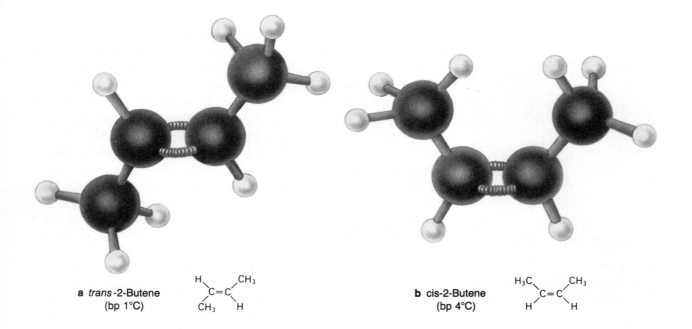

a *trans*-2-Butene
(bp 1°C)

b cis-2-Butene
(bp 4°C)

25·6 Geometric Isomers

The lack of rotation around carbon–carbon double bonds has an important structural implication. Look at the structure of 2-butene in Figure 25·13. Two arrangements are possible for the methyl groups with respect to the rigid double bond. *In the* **trans** *configuration the substituted groups are on opposite sides of the double bond. In the* **cis** *configuration the substituted groups are on the same side of the double bond. Trans*-2-butene and *cis*-2-butene are geometric isomers. **Geometric isomers** *differ only in the geometry of their substituted groups*. Like other structural isomers, isomeric 2-butenes are distinguishable by their different physical and chemical properties. The groups on the carbons of the double bond need not be the same. Geometric isomerism is possible whenever each carbon of the double bond has at least one substituent.

trans-2-Pentene   *cis*-2-Pentene   2-Methyl-1-butene
(no *cis*, *trans* isomers)

Problem

7. Draw structural formulas for the following alkenes. If a compound has geometric isomers, draw both the *cis* and *trans* forms.
   a. 1-pentene   **c.** 2-methyl-2-hexene
   b. 2-hexene   **d.** 2,3-dimethyl-2-butene

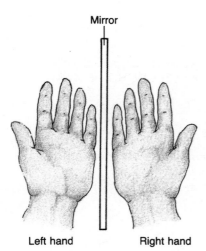

Mirror

Left hand   Right hand

Figure 25·14
The reflected image of a right hand in a mirror is a left hand. Note the position of the thumb relative to the other fingers.

▮ Many objects in nature, including molecules, exist in right-handed and left-handed forms.

25·7 Stereoisomers

Placing an object in front of a mirror can give two different results. If the object is symmetrical, like a ball, then its mirror image is superimposable. That is, the appearance of the ball and its reflection are indistinguishable. By contrast, a pair of hands are distinguishable even though they consist of identical parts. The right hand reflects as a left hand and the left hand reflects as a right hand (Figure 25·14). Your hands are examples of nonsuperimposable mirror images. They are mirror images which cannot be placed on top of each other to obtain a match. Many pairs of ordinary objects like ears, feet, shoes, and bird wings are similarly related.

Recall that four groups attached to a carbon by single covalent bonds form a tetrahedron with the carbon at the center. In Figure 25·15 the carbon has four *different* groups attached: —F, —H, —Cl, and —Br. *A carbon with four different groups attached is called an* **asymmetric carbon.** Compounds whose molecules contain an asymmetric carbon have handedness. For these compounds, two kinds of molecules exist that are related to one another in much the same way as a pair of hands. Figure 25·15 shows the mirror images of CHFClBr. Like hands, these mirror images are nonsuperimposable. Unless bonds are broken, these molecules cannot be superimposed. The four atoms attached to the carbon will not all match at once. **Stereoisomers** *are molecules of the same molecular structure that differ only in the arrangement of the atoms in space.* The mirror images of CHFClBr are stereoisomers.

Problem

8. Which of the following objects would have a nonsuperimposable mirror image? (Ignore designs or other markings.)
   a. clam shell   **b.** cup   **c.** ball   **d.** car   **e.** wood screw
   f. fingerprint   **g.** baseball bat

Mirror

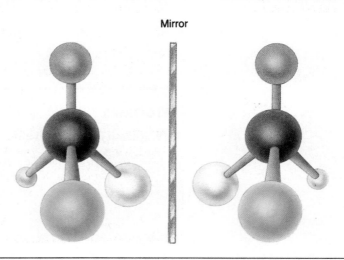

Figure 25·15
Stereoisomers are mirror image molecules that cannot be superimposed on each other.

Example 4

Which of the following compounds have an asymmetric carbon?
a. CH_3CHCH_3 **b.** $CH_3CHCH_2CH_3$ **c.** CH_3CHCHO **d.** CH_3CHOH
      | | | |
      OH   OH   OH   OH

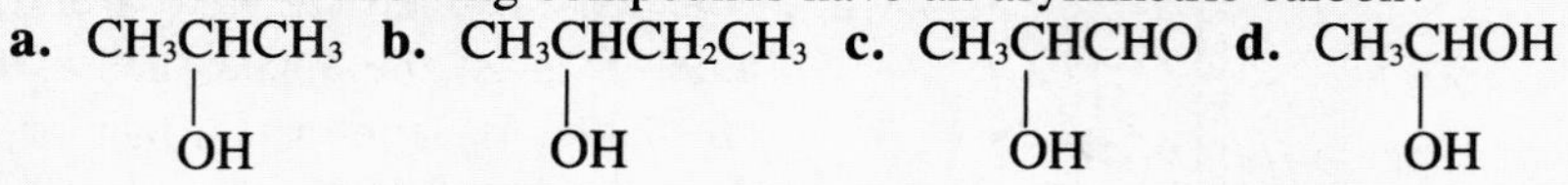

Solution

An asymmetric carbon has four different groups attached. It may help to draw the structures in a more complete form.

a. The central carbon has one H, one OH, and two CH_3 groups attached. It is not asymmetric.

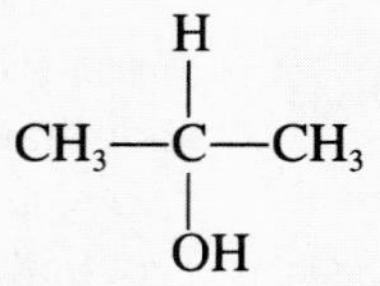

$$CH_3-\overset{\overset{\displaystyle H}{|}}{\underset{\underset{\displaystyle OH}{|}}{C}}-CH_3$$

b. The central carbon has one H, one OH, one CH_3, and one CH_2CH_3 group attached. Because these four groups are different, the central carbon is asymmetric. It can be marked with an asterisk. None of the other carbons in this molecule is asymmetric (check to be sure).

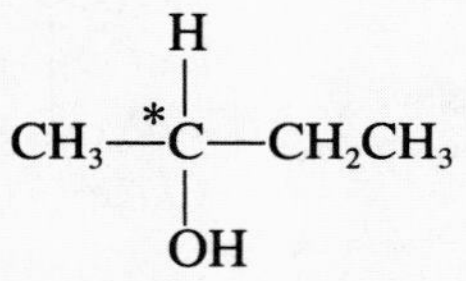

$$CH_3-\overset{\overset{\displaystyle H}{|}}{\underset{\underset{\displaystyle OH}{|}}{\overset{*}{C}}}-CH_2CH_3$$

c. This molecule also has an asymmetric carbon. It is marked with an asterisk.

$$CH_3-\overset{\overset{\displaystyle H}{|}}{\underset{\underset{\displaystyle OH}{|}}{\overset{*}{C}}}-CHO$$

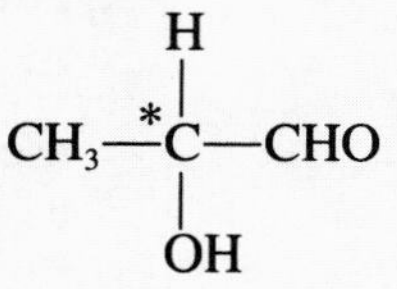

d. This molecule does not have an asymmetric carbon.

$$CH_3-\overset{\overset{\displaystyle H}{|}}{\underset{\underset{\displaystyle OH}{|}}{C}}-OH$$

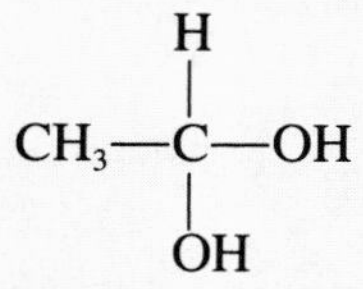

Problem

9. Identify the asymmetric carbon, if there is one, in each of the following structures.

a. CH_3CH_2CHO **b.** 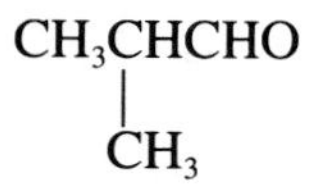 CH_3CHCHO **c.** 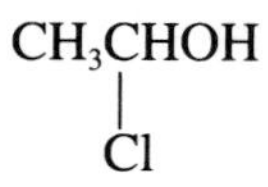 CH_3CHOH
                            | |
                           CH_3 Cl

d. 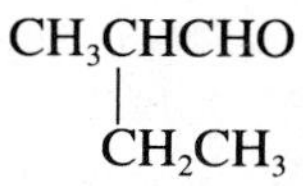 CH_3CHCHO
      |
      CH_2CH_3

25·8 Cyclic and Aromatic Hydrocarbons

Carbon forms rings as well as open-chain molecules.

In some hydrocarbon compounds, the two ends of a carbon chain are attached to form a ring. *Compounds that contain a hydrocarbon ring are* **cyclic hydrocarbons.** The structures of some cyclic hydrocarbons are shown in Figure 25·16. Rings containing from 3 to 20 carbons are found in nature. Five- and six-membered rings are the most abundant. *All hydrocarbon compounds which do not contain rings are known as* **aliphatic compounds.** They include compounds with both short and long carbon chains.

A special group of unsaturated cyclic hydrocarbons are known as **arenes.** These compounds contain single rings or groups of rings. The arenes were originally called aromatic compounds because many of them have pleasant odors. Benzene, C_6H_6, is the simplest arene. *Today the term* **aromatic compound** *is applied to any substance in which the bonding is like that of benzene.*

The benzene molecule has a six-membered carbon ring with a hydrogen attached to each carbon. This leaves one electron from each carbon free to participate in a double bond. Two different structures with double bonds can be written for benzene.

These structural formulas show only the extremes in electron sharing between any two adjacent carbons in benzene. One extreme is a normal single bond. The other extreme is a normal double bond. *Resonance occurs when two or more equally valid structures can be drawn for a molecule.* The benzene molecule exhibits resonance. Benzene and other molecules that exhibit resonance are more stable than similar molecules that do not exhibit resonance. Thus benzene is not as reactive as six-carbon alkenes.

Because of resonance, the bonding in benzene and related arenes is unique. The benzene molecule is perfectly flat. Bending or twisting would disrupt the electron sharing and the molecule's stability. Some chemists use a circle to show the presence of resonance in a benzene ring.

The inscribed circle is a good way to represent the nature of resonance bonding. However, it does not show the number of electrons involved. For this reason, the traditional structure (shown on the right) will be used.

Cyclopropane (bp −34.4°C)

Cyclobutane (bp −13°C)

Cyclopentane (bp 49.5°C)

Cyclohexane (bp 81.4°C)

Cycloheptane (bp 118°C)

Cyclooctane (bp 149°C)

Figure 25·16
Cycloalkanes are named after the parent alkane. Cycloalkenes are similar but are not shown here.

Compounds containing substituents attached to the benzene ring are named as derivatives of benzene. Sometimes the benzene ring is named as a substitutent on an alkane. In such instances the C_6H_5— group is called a phenyl group.

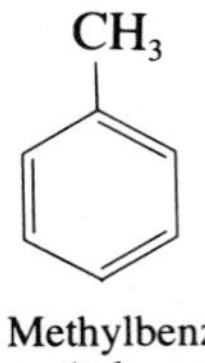

Methylbenzene (toluene)

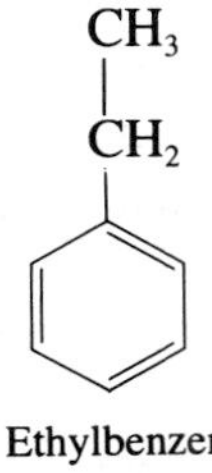

Ethylbenzene

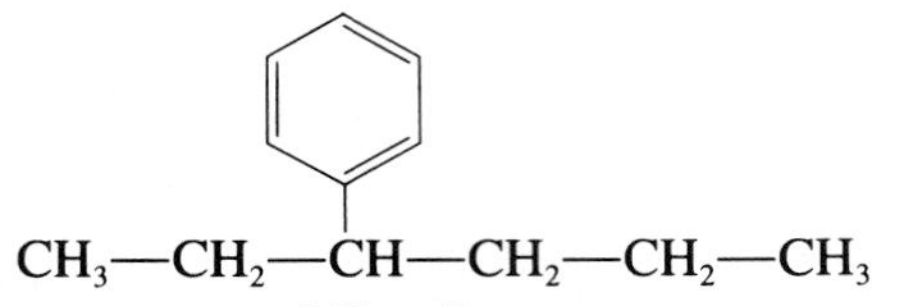

$CH_3-CH_2-CH-CH_2-CH_2-CH_3$
3-Phenylhexane

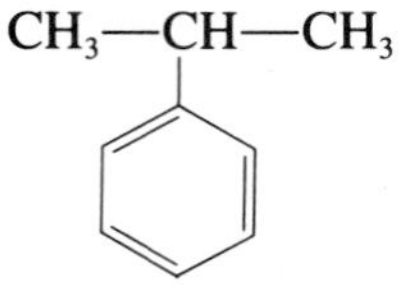

Problem

10. Name the following compounds.

a. 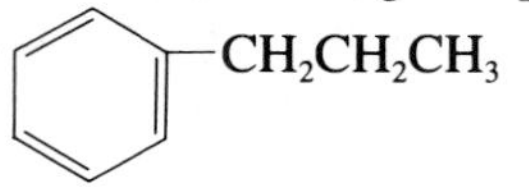 ⬡—$CH_2CH_2CH_3$

b. 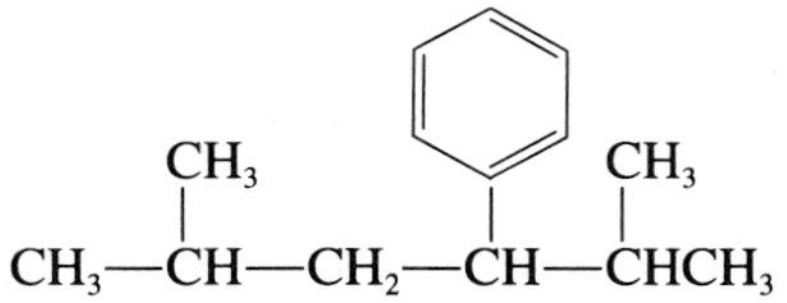 $CH_3-CH-CH_2-CH-CHCH_3$ (with CH_3 groups and phenyl substituent)

c. $CH_3-CH-CH_3$ (with phenyl substituent)

Some derivatives of benzene have two substituents. Such derivatives are called *disubstituted* benzenes. Three different structural isomers occur for the liquid aromatic compound dimethylbenzene, $C_6H_4(CH_3)_2$. Once again, the physical properties of structural isomers are different, as indicated by their boiling points.

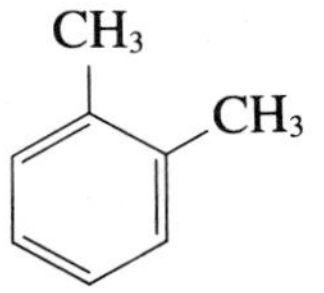

1,2-Dimethylbenzene
(*o*-dimethylbenzene, *o*-xylene)
(bp 144°C)

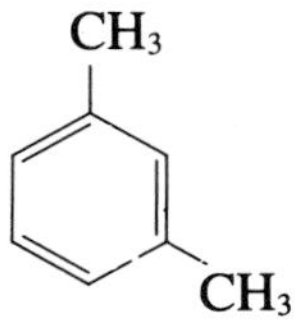

1,3-Dimethylbenzene
(*m*-dimethylbenzene, *m*-xylene)
(bp 139°C)

1,4-Dimethylbenzene
(*p*-dimethylbenzene, *p*-xylene)
(bp 138°C)

In the IUPAC naming system, the possible positions of two substituents in disubstituted benzene are designated as 1,2; 1,3; or 1,4. Common names for disubstituted benzenes use the terms *ortho, meta,* and *para* (abbreviated *o, m,* and *p*) in place of numbers. The dimethylbenzenes are also called xylenes.

25·9 Natural Gas and Petroleum

Petroleum, formed from the effects of heat and pressure on decaying vegetation over time, is an important source of energy and organic chemicals.

Much of the world's need for energy is supplied by burning fossil fuels. These fuels are organic in nature because they are derived from the decay of organisms in geologic history. Natural gas and petroleum are two important fossil fuels. They contain mostly aliphatic, or open-chain, hydrocarbons.

Petroleum, or crude oil, was first discovered seeping out of rocks. Petroleum and natural gas had their origin in marine life buried in the sediments of the oceans millions of years ago. Heat, pressure, and the action of bacteria changed this residue into petroleum and natural gas. The natural gas is often found overlying the oil deposits or in separate pockets in rock (Figure 25·18).

Natural gas is an important source of alkanes of low molecular mass. Typically, natural gas is composed of about 80% methane, 10% ethane, 4% propane, and 2% butane. The remaining 4% consists of nitrogen and higher molecular mass hydrocarbons. Natural gas also contains a small amount of the noble gas helium, and is one of its major sources. Methane is the major constituent of natural gas. It is especially prized for combustion because it burns with a hot, clean flame.

$$CH_4 + 2O_2 \longrightarrow CO_2 + 2H_2O + heat$$

Figure 25·17
Chemicals derived from petroleum and natural gas are called petrochemicals. They are the raw materials from which literally millions of essential products are made. Petrochemicals were used in the production of all of these items.

Chemical Connections

Since our supplies of coal, oil, and natural gas are limited, scientists are exploring alternative sources of energy. Among these is biomass conversion, a method for producing energy from plants, plant products, and organic wastes.

Figure 25·18
Natural gas and oil are typically found in dome-shaped geological formations. Sometimes the gas is under pressure and will force the oil up the well pipe, but pumping is usually required. What are the components of natural gas?

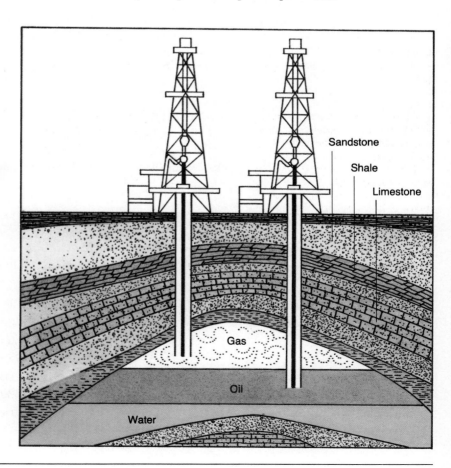

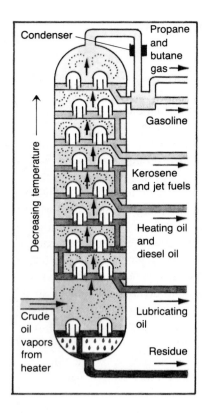

Figure 25·19
Petroleum is refined by fractional distillation. First the crude oil is heated so that it vaporizes and rises through the column. Compounds with the highest boiling points condense first near the bottom. Compounds with the lowest boiling points condense last near the top.

Propane and butane are also good heating fuels. They are separated from the other gases by liquefaction. They are sold in liquid form in pressurized tanks as liquid petroleum gas (LPG).

A sufficient supply of oxygen is necessary to oxidize a hydrocarbon fuel completely and to obtain the greatest amount of heat. Complete combustion of a hydrocarbon gives a blue flame. Incomplete combustion gives a yellow flame. This is due to the formation of small, glowing carbon particles that are deposited as soot when they cool. Carbon monoxide, a toxic gas, is also formed along with carbon dioxide and water during incomplete combustion.

$$6CH_4 + 9O_2 \longrightarrow 2C + 2CO + 2CO_2 + 12H_2O$$

The organic compounds found in petroleum are more complex than those in natural gas. Most of its hydrocarbons are continuous-chain and branched-chain alkanes. Petroleum also contains small amounts of aromatic compounds, and small amounts of sulfur-, oxygen-, and nitrogen-containing organic compounds.

Vast deposits of petroleum were discovered in the United States in 1859, and in the Middle East in 1908. It has since been found in other parts of the world as well. Crude oil must be refined before it is commercially useful. Petroleum refining consists of distilling crude oil to divide it into fractions according to boiling point (Figure 25·19). Each fraction contains several different hydrocarbons. The fractions, which have a variety of uses, are shown in Table 25·3.

The fractions containing compounds of higher molecular mass can be "cracked" to produce the more useful short chain components of gasoline and kerosene. **Cracking** *is a controlled process by which hydrocarbons are broken down or rearranged into smaller, more useful molecules.* Hydrocarbons are cracked with the aid of a catalyst or with heat. By this process, petroleum is the principal source of raw materials for the organic chemicals industry. For example, the low molecular mass alkanes are starting materials for the manufacture of paints and plastics.

Table 25·3 Fractions Obtained from Crude Oils			
Fraction	Composition of carbon chains	Boiling range (°C)	Percent of crude oil
Natural gas	C_1 to C_4	Below 20	
Petroleum ether (solvent)	C_5 to C_6	30 to 60	10%
Naphtha (solvent)	C_7 to C_8	60 to 90	
Gasoline	C_6 to C_{12}	75 to 200	40%
Kerosene	C_{12} to C_{15}	200 to 300	10%
Fuel oils, mineral oil	C_{15} to C_{18}	300 to 400	30%
Lubricating oil, petroleum jelly, greases, paraffin wax, asphalt	C_{16} to C_{24}	Over 400	10%

Figure 25·20
Peat and three forms of coal are shown here. They are arranged clockwise, starting from the top: peat, lignite, bituminous or soft coal, and anthracite or hard coal.

Figure 25·21
Coal is chemically similar to both natural gas and crude oil. All three are composed chiefly of hydrocarbons. When heat and pressure are applied in the right way, coal can be converted into synthetic gas and crude oil. This two kilogram chunk of coal will yield about one liter of synthetic crude oil or about one cubic meter of synthetic gas.

25·B Coal

Geologists believe that coal had its origin some 300 million years ago. At that time, huge tree ferns and mosses grew abundantly in swampy tropical regions. When the plants died, they formed thick layers of decaying vegetation. They were eventually covered by layer after layer of soil and rock, which caused a build-up of intense pressure. This pressure together with the heat from the interior of the earth slowly turned these plant remains into coal.

The first stage in the formation of coal is an intermediate material known as *peat*. Peat is a soft, brown, spongy, fibrous material rather like decayed and compressed garden refuse. It has a very high water content when first dug out of a "peat bog." After it has been allowed to dry, it produces a low cost but smoky fuel. If peat is left in the ground, it continues to change. After a long period of time it loses most of its fibrous texture and becomes *lignite,* or brown coal. Lignite is much harder than peat and has a higher carbon content (about 50%). The water content, however, is still high. Continued pressure and heat slowly change lignite to *bituminous,* or soft coal. Bituminous coal has a lower water content and higher carbon content (70–80%) than lignite. In some regions of the earth's crust, even greater pressures have been exerted. In those places, such as in eastern Pennsylvania, coal has been changed into *anthracite,* or hard coal. Anthracite has a carbon content that exceeds 80%, making it an excellent fuel source.

Coal is obtained from both underground and surface mines. It is usually found in seams from one to three meters thick. In North America, coal mines are usually less than 100 meters underground. Much of the coal is so close to the surface that it is strip-mined. By contrast, many coal mines in Europe and other parts of the world go down 1000 to 1500 meters.

Coal consists largely of condensed ring compounds of very high molecular mass. These compounds have a very high proportion of carbon compared with hydrogen. Due to the high proportion of these aromatic compounds, coal leaves more soot upon burning than do the more aliphatic fuels obtained from petroleum. The majority of the coal burned in North America contains about 7% sulfur which burns to form SO_2 and SO_3, which are major pollutants.

Coal may be distilled to obtain a variety of products: coal gas, coal tar, ammonia, and coke. Coal gas consists mainly of hydrogen, methane, and carbon monoxide. All of these are flammable. Coal tar can be distilled further into benzene, toluene, naphthalene, phenol, and pitch. Coke is used as a fuel in many industrial processes. Because it is almost pure carbon coke produces intense heat and little or no smoke when it burns. The ammonia from distilled coal is converted to ammonium sulfate for use as a fertilizer.

Key Terms

aliphatic compound	25·8	cracking	25·9
alkane	25·1	cyclic hydrocarbon	25·8
alkene	25·5	geometric isomer	25·6
alkyl group	25·3	hydrocarbon	25·0
alkyne	25·5	homologous series	25·2
arene	25·8	IUPAC system	25·2
aromatic compound	25·8	saturated compound	25·5
asymmetric carbon	25·7	stereoisomer	25·7
branched-chain alkane	25·3	structural isomer	25·4
		substituent	25·3
cis configuration	25·6	trans configuration	25·6
condensed structural formula	25·2	unsaturated compound	25·5
continuous-chain alkane	25·2		

Chapter Summary

The branch of chemistry that deals with carbon compounds is called organic chemistry. Carbon makes stable covalent bonds with other carbons to form chain and ring compounds. Hydrocarbons are compounds containing only carbon and hydrogen. Many hydrocarbons exhibit structural isomerism. Structural isomers have the same molecular formula but different molecular structures.

Alkanes contain only carbon–carbon single bonds. The groups attached to single bonds in continuous-chain alkanes rotate freely about the bonds at room temperature. Alkenes are unsaturated hydrocarbons. That is, they contain one or more carbon–carbon double bonds. Alkenes may exhibit geometric isomerism. Geometric isomers are *cis* or *trans* according to whether substituent groups are on the same side or on opposite sides of the double bond. Alkynes are also unsaturated compounds. They contain one or more carbon–carbon triple bonds. Rotation about the multiple bonds of alkenes and alkynes does not occur under ordinary conditions. Some organic molecules exhibit stereoisomerism. Stereoisomers are related in much the same way that the right hand is related to the left. Stereoisomerism can occur if four different groups are attached to carbon.

Aromatic hydrocarbons or arenes are related to the hydrocarbon benzene. Benzene is rather unusual among hydrocarbons. As a result of resonance, the interior bonds of the benzene ring are somewhere between ordinary single bonds and double bonds. Benzene is less reactive than ordinary alkenes because of this unusual bonding.

Aliphatic, or open-chain, hydrocarbons come mainly from petroleum. Aromatic hydrocarbons come mainly from coal. Many hydrocarbons are obtained directly from crude petroleum by distillation. The molecular structures of the hydrocarbons present in natural petroleum can be reorganized into other useful products by the cracking process.

Practice Questions and Problems

11. What is the number of covalent bonds formed by carbon? 25·1

12. Name the alkanes that have the following formulas. 25·2
  a. $CH_3CH_2CH_3$
  b. $CH_3(CH_2)_6CH_3$
  d. C_3H_8
  e. $CH_3CH_2CH_2CH_2CH_3$

  c.

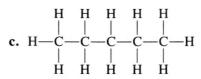

13. Draw structural formulas for these compounds.
  a. pentane 25·3
  b. 3-ethylpentane
  c. 4-ethyl-2,3,4,-trimethylnonane
  d. 3,5-diethyl-2,3-dimethyl-5-propyldecane

14. Write structures for the alkyl groups derived from methane, ethane, and propane. $25 \cdot 3$

15. Why are the following names incorrect? What are the correct names? $25 \cdot 3$
  a. 2-dimethylpentane   **c.** 3-methylbutane
  b. 1,3-dimethylpropane   **d.** 3,4-dimethylbutane

16. Give the IUPAC name for each compound. $25 \cdot 3$

  a. $CH_3—CH—CH_2$
            $|$ $|$
          CH_3 CH_3

  b. $CH_3—CH—CH—CH_3$
            $|$ $|$
          CH_3 CH_3

  c. $CH_3—CH—CH_2—CH_2$
            $|$ $|$
          CH_2 CH_3
            $|$
          CH_3

17. Draw one structural isomer of each of the following compounds. $25 \cdot 4$

  a.
           CH_3
            $|$
    $CH_3—C—CH_3$
            $|$
           CH_3

  b.
               CH_3
                $|$
    $CH_3—CH_2—C—CH_3$
                $|$
              CH_2
                $|$
              CH_3

  c.
        CH_3 CH_3
         $|$ $|$
  $CH_3—CH—CH_2—CH—CH$
                 CH_2 CH_2
                  $|$ $|$
                 CH_3 CH_3

18. Explain the difference between saturated and unsaturated hydrocarbons. $25 \cdot 5$

19. Give a systematic name for these alkenes. $25 \cdot 5$

  a. $CH_3CH{=}CH_2$

  b.
    CH_3 H
        $\diagdown$ $\diagup$
          $C{=}C$
        $\diagup$ $\diagdown$
     H CH_2CH_3

  c. $CH_3CHCH_2CH{=}CH_2$
           $|$
          CH_3

  d.
    CH_2 CH_2CH_3
       $\diagdown$ $\diagup$
         $C{=}C$
       $\diagup$ $\diagdown$
    CH_3 CH_2CH_3

  e. $CH_2{=}CHCH_2CH_2$
                 $|$
               CH_2CH_3

20. Draw a structural formula for each alkene with the molecular formula C_5H_{10}. Name each of these compounds. $25 \cdot 5$

21. Draw all the alkenes with the molecular formula C_4H_8. Name each compound. $25 \cdot 5$

22. Show how lack of rotation about a carbon–carbon double bond leads to geometric isomerism. Use the isomers of 2-pentene to illustrate your answer. $25 \cdot 6$

23. Draw a structural formula or carbon skeleton for each of the following alkenes. Include both *cis* and *trans* forms if the compound has geometric isomers. **a.** 2-pentene **b.** 2-methyl-2-pentene **c.** 3-methyl-2-pentene $25 \cdot 6$

24. For which of the following molecular formulas can mirror image molecules be drawn? Why? $25 \cdot 7$
  a. CH_2Cl_2 **c.** $CH_3CF_2CH_3$

  b.
        F
        $|$
   $HS—C—OH$
        $|$
        F

  d.
             CH_3
              $|$
   $CH_3CH_2—C—Br$
              $|$
             F

25. Draw a structure for each compound. $25 \cdot 8$
  a. *p*-diethylbenzene
  b. 2-methyl-3-phenylpentane
  c. *p*-xylene
  d. toluene

26. The seven organic chemicals produced in the largest amounts in the United States in 1987 are listed in the table on the next page. Answer the following questions based on the data given for the amount of each compound produced. $25 \cdot 8$
  a. How many billion kilograms of aromatic compounds on the list were produced?
  b. Of the total mass of all seven compounds produced, what percent by mass was made up of aliphatic compounds?

c. The molecules of many of the compounds on this list have double bonds. Propose an explanation for this fact.

Chemical	Amount produced (kg $\times 10^9$)
Ethylene	15.9
Propylene	8.4
Urea	6.8
Ethylene dichloride	6.3
Benzene	5.3
Ethyl benzene	4.3
Vinyl chloride	3.7

27. Explain why both of these structures represent 1,2-diethylbenzene. *25·8*

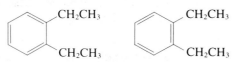

28. Does crude oil contain mostly aliphatic or aromatic hydrocarbons? *25·9*

29. Define *petroleum refining* and *cracking*. *25·9*

30. Write a balanced equation for the complete combustion of pentane. *25·9*

Mastery Questions and Problems

31. For each hydrocarbon pictured below, identify the type of bonding and name the compound.

32. Write structural formulas for each of the following compounds.
 a. propyne
 b. cyclohexane
 c. 2-phenylpropane
 d. 2,2,4-trimethylpentane
 e. 2,3-dimethylpentane
 f. 1,1-diphenylhexane

33. Name the next three higher homologs of ethane.

34. Draw electron dot structures for each of these compounds. **a.** ethene **b.** propane **c.** ethyne **d.** cyclobutane

35. Compare these three molecular structures. Which would you expect to be most stable? Why?

Critical Thinking Questions

36. Choose the term that best completes the second relationship.
 a. tree:branch   hydrocarbon chain: _____
   (1) alkyl group   (3) isomer
   (2) methane   (4) double bond
 b. cyclic:aliphatic   circle: _____
   (1) oval   (3) square
   (2) parallelogram   (4) line
 c. propane:three   hexene: _____
   (1) seven   (3) five
   (2) six   (4) four

37. Why do most cyclic hydrocarbons have higher boiling points than alkanes with the same number of carbons?

38. How could the composition of gasoline be altered to increase the ease of starting a car in cold weather?

Review Questions and Problems

39. What are the pH values for aqueous solutions containing each of the following hydroxide ion concentrations?
 a. $1.00 \times 10^{-4}M$ c. $0.01M$
 b. $3.92 \times 10^{-7}M$ d. $0.005M$

40. Give the oxidation number of each element in the following compounds.
 a. $CaCO_3$ c. $LiIO_3$
 b. Cl_2 d. Na_2SO_3

41. Draw electron dot structures and predict the shapes of the following molecules.
 a. PH_3 b. CO c. CS_2 d. CF_4

42. Write equilibrium constant expressions for the following reactions.

a. $Cl_2(g) + I_2(g) \rightleftharpoons 2ICl(g)$
b. $2HBr(g) \rightleftharpoons H_2(g) + Br_2(g)$
c. $2S_2Cl_2(g) + 2H_2O(g) \rightleftharpoons$
$$4HCl(g) + 3S(g) + SO_2(g)$$
d. $N_2(g) + 3H_2(g) \rightleftharpoons 2NH_3(g)$

43. Identify any incorrect formulas among the following compounds.

a. H_2O_2 **d.** CaS_2
b. $NaIO_4$ **e.** $CaHPO_4$
c. SrO **f.** $BaOH$

44. A colorless solution of unknown pH turns blue when tested with the acid–base indicator bromthymol blue. It remains colorless when tested with phenolphthalein.

a. What is the approximate pH of the solution?
b. How could you determine the pH more accurately?

Challenging Questions and Problems

45. Use the data of Table 25·1 to make a graph of boiling point versus number of carbons for the first ten continuous-chain alkanes. Does the graph describe a straight line? On the basis of this graph, what would you predict to be the boiling point of undecane, the continuous-chain alkane containing 11 carbons? Use a chemistry handbook to find the actual boiling point of undecane. Compare the boiling point found by experiment with your prediction.

46. The knocking properties of fuels for gasoline engines are based on standards of heptane, which causes severe engine knocking, and 2,2,4-trimethylpentane ("isooctane"), which is better than most other fuels. The *octane number* of gasoline is the percentage of 2,2,4-trimethylpentane in heptane that is required to match the knocking properties of the fuel. What are the octane numbers of gasolines available in your neighborhood? To what percentages of 2,2,4-trimethylpentane in heptane do they correspond? How might it be possible to achieve octane numbers greater than 100? On the basis of the behavior of 2,2,4-trimethylpentane and heptane, would you expect pentane or 2,2-dimethylpropane to give the best performance as a fuel in a gasoline engine?

Research Projects

1. How did Friedrich Wöhler refute the vitalism theory?

2. What was Friedrich Beilstein's major contribution to organic chemistry? How is his work used today?

3. Determine the density of a sample of natural gas. How does the value compare with the density of methane?

4. Devise a method to distinguish two geometric isomers. How would you distinguish stereoisomers?

5. How did Friedrich Kekulé discover the structure of benzene? What are some commercial uses for this compound?

6. What environmental problems are associated with petroleum drilling and refining?

7. Do you believe it is wise to remain dependent on fossil fuels? Support your position with statistics and projections.

8. Describe some uses of chemicals derived from coal. What contributions did John Kidd make in this area?

9. Make a diagram of a coal distillation process. How does this process compare to petroleum refining?

Readings and References

Kaplan, William A., and Melvyn Lebowitz. *The Student Scientist Explores Energy and Fuels*. New York: Rosen, 1981.

Kieffer, William F. *Chemistry: A Cultural Approach*. New York: Harper & Row, 1971.

Lyttle, Richard B. *Shale Oil and Tar Sands: The Promises and Pitfalls*. New York: Watts, 1982.

Stine, William R. *Chemistry for the Consumer*. Boston: Allyn and Bacon, 1978.

Tver, David F., and Richard W. Berry. *The Petroleum Dictionary*. New York: Van Nostrand Reinhold, 1980.

Wood, Clair G. "Natural Dyes." *ChemMatters* (December 1986), pp. 4–8.

26 Functional Groups and Organic Reactions

Chapter Preview

26·1 Functional Groups
26·2 Halocarbons
26·3 Substitution Reactions
26·4 Alcohols
26·5 Properties of Alcohols
26·A Percy Julian and Natural Product Chemistry
26·6 Addition Reactions
26·7 Ethers
26·8 Aldehydes and Ketones
26·9 Carboxylic Acids

26·10 Oxidation–Reduction Reactions
26·11 Esters
26·12 Polymerization
26·B Synthetic Fibers
26·13 Carbohydrates
26·14 Lipids
26·15 Amino Acids, Peptides, and Proteins
26·C Enzymes
26·16 Nucleic Acids

The previous chapter introduced hydrocarbon chains and rings. These are essential components of every organic compound. Yet in most chemical reactions involving organic molecules, the hydrocarbon skeletons of the molecules are chemically inert. Thus the chemistry of the alkanes is very limited. Most organic chemistry involves substituents attached to hydrocarbon chains. This chapter discusses some of these substituents and their chemical reactions.

26·1 Functional Groups

The substituents of organic molecules often contain oxygen, nitrogen, sulfur, or phosphorus. They are called functional groups because they are the chemically functional parts of the molecules. *A **functional group** is a specific arrangement of atoms in an organic compound that is capable of characteristic chemical reactions.* Most organic chemistry is functional group chemistry. Thus organic chemists classify organic compounds into categories according to their functional groups. The symbol R is used to represent any carbon chains or rings attached to the functional group. The double and triple bonds of alkenes and alkynes are chemically reactive and are considered functional groups. Table 26·1 lists the other functional groups covered in this book. You can refer to this table when you encounter a non-familiar group.

A functional group is the chemically reactive part of an organic molecule.

Figure 26·1
Nylon forms at the interface between the two organic liquids in this beaker: a dicarboxylic acid and a diamine.

Table 26·1 Organic Compounds Classified by Functional Group

Compound type	Compound structure	Functional group
Halocarbon	R—X (X = F, Cl, Br, I)	Halogen
Alcohol	R—OH	Hydroxyl
Ether	R—O—R	Ether
Aldehyde	$\underset{\displaystyle}{R-\overset{\displaystyle O}{\overset{\|}{C}}-H}$	Carbonyl
Ketone	$R-\overset{\displaystyle O}{\overset{\|}{C}}-R$	Carbonyl
Carboxylic acid	$R-\overset{\displaystyle O}{\overset{\|}{C}}-OH$	Carboxyl
Ester	$R-\overset{\displaystyle O}{\overset{\|}{C}}-O-R$	Ester
Amine	$R-NH_2$	Amino

Organic compounds are classified according to the kind of functional groups they contain, instead of the kind of chains or rings to which the functional group is attached.

Problem

1. Identify the functional group in each of the following structures.

   a. $CH_3—OH$ **d.** $CH_3—CH_2—CH_2—Br$

   b. $CH_3—CH_2—NH_2$ **e.** $CH_3—CH_2—O—CH_2—CH_2—CH_3$

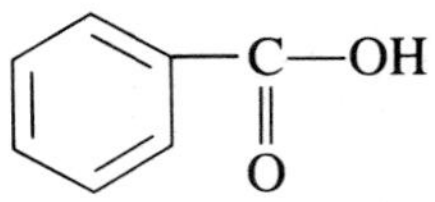

   c. $\text{(benzene ring)}-\overset{\displaystyle}{\underset{\displaystyle O}{\overset{\|}{C}}}-OH$ **f.** $C_6H_{13}-\underset{\displaystyle O}{\overset{\|}{C}}-C_3H_7$

26·2 Halocarbons

◼ Halocarbons result from the substitution of a halogen for hydrogen in a hydrocarbon.

Halocarbons *are a class of organic compounds containing covalently bonded fluorine, chlorine, bromine, or iodine.* Very few halocarbons are found in nature. Nevertheless they are readily prepared and used for many purposes. For example, they are widely used as anesthetics and insecticides.

The IUPAC rules for naming halocarbons are based on the name of the parent hydrocarbon. The halogen groups are simply added as substituents. Here are some examples of IUPAC names for simple halocarbons. (Common names are in parentheses.)

Safety

Many halogenated hydrocarbons are extremely toxic, carcinogenic, flammable, or all three.

CH₃—Cl

Chloromethane
(methyl chloride)

$$CH_3-\underset{\underset{Br}{|}}{\overset{\overset{CH_3}{|}}{C}}-CH_3$$

2-Bromo-2-methylpropane
(*tert*-butyl bromide)

Chloroethene
(vinyl chloride)

$$Cl-\underset{\underset{Cl}{|}}{\overset{\overset{Cl}{|}}{C}}-H$$

Trichloromethane
(chloroform)

Chlorobenzene
(phenyl chloride)

ChemDirections

Some organic fluorocompounds have a unique physical property. To discover what it is and why it is important read **ChemDirections** *Blood Substitutes*, page 666.

Common names of halocarbons consist of two parts. The first part names the hydrocarbon part of the molecule as an alkyl group, such as methyl or ethyl. The second part gives the halogen an *-ide* ending. Methyl chloride, CH₃Cl, is an example. Remember, however, that the bonding in a halocarbon is covalent, not ionic.

On the basis of their common names, *halocarbons in which a halogen is attached to a carbon of an aliphatic chain are called* **alkyl halides.** Table 26·2 gives names for alkyl groups besides those of methyl, ethyl, and propyl. *Halocarbons in which a halogen is attached to a carbon of an arene ring are called* **aryl halides.**

Table 26·2 Names of Some Common Alkyl Groups		
Name	Alkyl group	Remarks
Isopropyl	$CH_3-\underset{\underset{H}{|}}{\overset{\overset{CH_3}{|}}{C}}-$	The prefix *iso-* is reserved for carbon chains that are continuous except for the presence of a methyl group on the carbon second from the unsubstituted end of the longest chain.
Isobutyl	$CH_3-\underset{\underset{}{|}}{\overset{\overset{CH_3}{|}}{CH}}-CH_2-$ (primary carbon)	Note the use of the prefix *iso-*. The carbon joining this alkyl group to another group is bonded to one other carbon; it is a *primary carbon*.
Secondary butyl (*sec*-butyl)	$CH_3-CH_2-\underset{\underset{}{|}}{CH}-CH_3$ (secondary carbon)	The carbon joining this alkyl group to another group is bonded to two other carbons; it is a *secondary carbon*.
Tertiary butyl (*tert*-butyl)	$CH_3-\underset{\underset{CH_3}{|}}{\overset{\overset{CH_3}{|}}{C}}-$ (tertiary carbon)	The carbon joining this alkyl group to another group is bonded to three carbons; it is a *tertiary carbon*.
Vinyl		When used as an alkyl group in giving compounds common names, this group, derived from ethene, is called *vinyl*.
Phenyl		This group is derived from benzene.

Table 26·3 Some Halocarbons and Their Uses

Halocarbon	Use
CH_3-CH_2-Cl Chloroethane (ethyl chloride)	A local anesthetic. Its rapid evaporation on the skin (bp 13°C) cools nerve endings and cuts down transmission of pain.
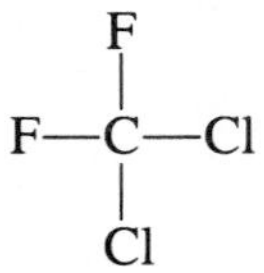 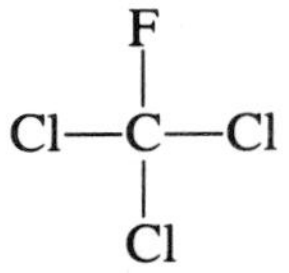 Dichlorodifluoromethane (Freon 12) and Trichlorofluoromethane (Freon 11)	Freon is the Dupont trade name for fluorinated compounds of this type. They are used as refrigerants. Freons are nontoxic, odorless, and nonflammable. Some are also used in specialized fire extinguishers.
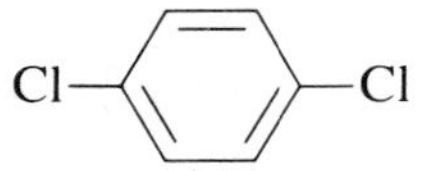 p-Dichlorobenzene	Used as a moth repellent.
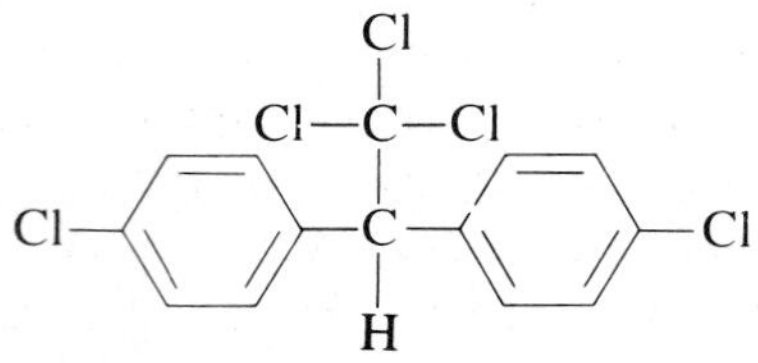 Dichlorodiphenyltrichloroethane (DDT)	A persistent pesticide. DDT was widely used as an insecticide from about 1950 to 1970. Its use is now limited because of its persistence in the environment. It is nonbiodegradable.

Table 26·3 gives the structures and uses of some halocarbons. The attractions between halocarbon molecules are primarily the result of the weak van der Waals interactions called dispersion forces. These attractions increase with the degree of halogen substitution. More highly halogenated organic compounds therefore have higher boiling points, as shown for the chloromethanes in Table 26·4.

Problems

2. Write IUPAC names for each of these halocarbons.

a. 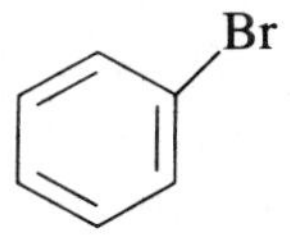 (benzene ring with Br)   **b.** CH_3CH_2Cl **c.** 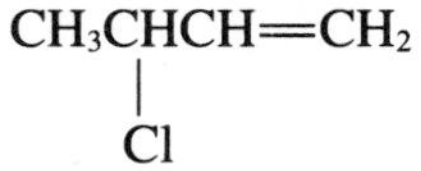 $CH_3CHCH=CH_2$ with Cl substituent

3. Give the structural formula for each of the following compounds.
a. isopropyl chloride   **b.** 1-iodo-2,2-dimethylpentane
c. p-bromotoluene

Table 26·4   Molecular Masses and Boiling Points of the Chloromethanes

Molecular structure	Name	Formula mass	Boiling point(°C)
CH_4^*	Methane	16	−161
CH_3Cl	Chloromethane (methyl chloride)	50.5	−24
CH_2Cl_2	Dichloromethane (methylene chloride)	85.0	40
$CHCl_3$	Trichloromethane (chloroform)	129.5	61
CCl_4	Tetrachloromethane (carbon tetrachloride)	154	74

*Included for purposes of comparison.

■ Organic substitution reactions involve the replacement of hydrogen or a functional group by another functional group.

26·3   Substitution Reactions

Organic reactions often proceed more slowly than the reactions of inorganic molecules and ions. This is because reactions of organic molecules commonly involve the breaking of relatively strong covalent bonds. Chemists are therefore constantly seeking new catalysts and improved procedures for conducting organic reactions. Many organic reactions are complex. They often produce a mixture of products. The desired product must then be separated by distillation, crystallization, or other means. A common type of organic reaction is **substitution,** *the replacement of an atom or group of atoms by another atom or group of atoms*.

A halogen can replace the hydrogen on an alkane to produce a halocarbon. The symbol X stands for a halogen in this generalized equation.

$$R\text{—}H \; + \; X_2 \; \longrightarrow \; R\text{—}X \; + \; HX$$

Alkane · · · · Halogen · · · · Halocarbon · · · · Hydrogen halide

Issues in Chemistry

Farmers and gardeners battle every year with yield-reducing infestations of weeds, insects, and fungi. Are there alternative weapons other than poisonous pesticides to fight these battles?

Figure 26·2
Some halocarbons are used as pesticides. Short lived pesticides break down within about a week. They are preferred over pesticides which break down slowly.

Sunlight or another source of ultraviolet radiation is usually a sufficient catalyst for this reaction. From the generalized equation, a specific one can be written.

$$CH_4 + Cl_2 \longrightarrow CH_3Cl + HCl$$

Methane · Chlorine · Chloromethane · Hydrogen chloride

Even under controlled conditions, this simple halogenation reaction produces a mixture of the mono-, di-, tri-, and tetrachloromethanes.

If benzene is treated with a halogen in the presence of a catalyst, substitution of a ring hydrogen occurs. Iron compounds are often used as catalysts for aromatic substitution reactions. A rusty nail dropped in the reaction flask works fine.

Benzene · Bromine $\xrightarrow{\text{catalyst}}$ Bromobenzene (phenyl bromide) · Hydrogen bromide

Halogens on carbon chains are readily displaced by hydroxide ions to produce an alcohol and a salt. The general reaction is as follows.

$$R\!-\!X + OH^- \xrightarrow[100°C]{H_2O} R\!-\!OH + X^-$$

Halocarbon · Hydroxide ion · Alcohol · Halide ion

Chemists usually use aqueous solutions of sodium or potassium hydroxide as the source of hydroxide ions. Fluoro groups are not easily displaced. Hence fluorocarbons are seldom, if ever, used to make alcohols. Here are two specific examples.

$$CH_3I + KOH \xrightarrow[100°C]{H_2O} CH_3OH + KI$$

Iodomethane (methyl iodide) · Potassium hydroxide · Methanol · Potassium iodide

$$CH_3CH_2Br + NaOH \xrightarrow[100°C]{H_2O} CH_3CH_2OH + NaBr$$

Bromoethane (ethyl bromide) · Sodium hydroxide · Ethanol · Sodium bromide

26·4 Alcohols

Alcohols *are organic compounds with an —OH group.*

$$\underset{\text{Alcohol molecule}}{R \overset{\overset{\displaystyle \cdot\cdot}{O}}{\diagdown} H}$$

The —OH functional group in alcohols is called a **hydroxyl group** *or* hydroxy function. It is not a hydroxide ion because the oxygen is covalently bonded to carbon. Chemists often arrange aliphatic alcohols into structural categories according to the number of R groups attached to the carbon with the hydroxyl group.

Alcohols are organic compounds in which a hydrogen in H_2O has been replaced by an R group.

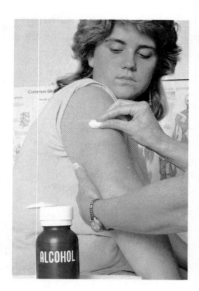

Figure 26·3
Isopropyl alcohol is an effective disinfectant.

Primary alcohol	$R-CH_2-OH$	Only one R group is attached to C—OH of a primary (abbreviated 1°) alcohol.
Secondary alcohol	$R-\underset{\underset{R}{\vert}}{C}H-OH$	Two R Groups are attached to C—OH of a secondary (2°) alcohol.
Tertiary alcohol	$R-\underset{\underset{R}{\vert}}{\overset{\overset{R}{\vert}}{C}}-OH$	Three R groups are attached to C—OH of a tertiary (3°) alcohol.

Both IUPAC and common names are used for alcohols. To name continuous-chain and substituted alcohols by the IUPAC system, drop the *-e* ending of the parent alkane name and add the ending *-ol*. The parent alkane is the longest continuous chain of carbons that includes the carbon attached to the hydroxyl group. In numbering the longest continuous chain, the position of the hydroxyl group is given the lowest possible number. Alcohols containing two, three, and four —OH substituents are named diols, triols, and tetrols.

Common names of aliphatic alcohols are written in the same way as those for the halocarbons. The alkyl group methyl, for example, is named and followed by the word alcohol, as in methyl alcohol. Compounds with more than one —OH substituent are called glycols. Here are some simple aliphatic alcohols along with their IUPAC and common names.

CH_3-OH
Methanol
(methyl alcohol)

CH_3-CH_2-OH
Ethanol
(ethyl alcohol)

$CH_3-CH_2-CH_2-OH$
1-Propanol
(propyl alcohol)

$CH_3-\underset{\underset{OH}{\vert}}{CH}-CH_3$
2-Propanol
(isopropyl alcohol)

$CH_3-\underset{\underset{CH_3}{\vert}}{CH}-CH_2-OH$
2-Methyl-1-propanol
(isobutyl alcohol)

$CH_3-CH_2-\underset{\underset{OH}{\vert}}{CH}-CH_3$
2-Butanol
(*sec*-butyl alcohol)

$CH_3-\underset{\underset{OH}{\vert}}{\overset{\overset{CH_3}{\vert}}{C}}-CH_3$
2-Methyl-2-propanol
(*tert*-butyl alcohol)

H_2C-CH_2
$\;\;|\;\;\;\;\;|$
$OH\;\;OH$
1,2-Ethandiol
(ethylene glycol)

$CH_3-CH-CH_2$
$\;\;\;\;\;\;|\;\;\;\;\;|$
$\;\;\;\;\;OH\;\;OH$
1,2-Propanediol
(propylene glycol)

$CH_2-CH-CH_2$
$\;\;|\;\;\;\;\;|\;\;\;\;\;|$
$OH\;\;OH\;\;OH$
1,2,3-Propantriol
(glycerol)

The common names of some diols have an *-ene* ending but the molecules contain no double bond.

Phenols are compounds in which a hydroxyl group is attached directly to an aromatic ring. Phenol is the parent compound. Cresol is the common name for the *o, m,* and *p* structural isomers of methylphenol.

Phenol	2-Methylphenol (*o*-cresol)	3-Methylphenol (*m*-cresol)	4-Methylphenol (*p*-cresol)

Problems

4. Give the following alcohols IUPAC names.

a. $CH_3CH_2CH_2CH_2OH$ **b.** $CH_3\overset{\overset{\displaystyle CH_3}{|}}{C}HOH$ **c.** $CH_3CH_2\overset{\overset{\displaystyle CH_3}{|}}{C}HCH_2OH$

5. Classify each of the alcohols in Problem 4 as primary, secondary, or tertiary.

26·5   Properties of Alcohols

The relatively high boiling points and high water solubility of alcohols are the result of intermolecular hydrogen bonding.

Like water, alcohols are capable of intermolecular hydrogen-bonding. Alcohols therefore boil at a higher temperature than alkanes and halocarbons containing comparable numbers of atoms.

Since alcohols are derivatives of water, you might expect them to be soluble in water. To a point that is correct. Alcohols of up to four carbons are soluble in water in all proportions. The solubility of alcohols with four or more carbons in the chain is usually much less. The reason is that alcohols consist of two parts: the carbon chain and the hydroxyl

Figure 26·4
The main ingredient in many antifreezes is 1,2-ethandiol (common name, ethylene glycol). It has a higher boiling point and lower freezing point than water.

group. These parts are in opposition to each other. The carbon chain is nonpolar and is not attracted to water. The polar hydroxyl group, however, strongly interacts with water through hydrogen-bonding.

Many aliphatic alcohols are used in laboratories, clinics, and industry. Isopropyl alcohol, a colorless, nearly odorless liquid (bp 82°C), is called rubbing alcohol. It is used for massages and as a base for perfumes, creams, lotions, and other cosmetics. Ethylene glycol (bp 197°C) is the principal ingredient of certain antifreezes. Its advantages over other high-boiling liquids are its solubility in water and a freezing point of -17.4°C. If water is added to ethylene glycol, the mixture freezes at an even lower temperature. A 50% (v/v) aqueous solution of ethylene glycol freezes at -36°C. Glycerol is a viscous, sweet-tasting, water-soluble liquid. It is used as a moistening agent in cosmetics, foods, and drugs. Glycerol is also an important component of fats and oils.

Ethyl alcohol, which has a boiling point of 78.5°C, is also called grain alcohol. It is an important industrial chemical. Some ethyl alcohol is still produced by yeast fermentation of sugar. **Fermentation** *is the production of ethanol from sugars by the action of yeast or bacteria*. The enzymes of the yeast or bacteria serve as catalysts for the transformation. The breakdown of the sugar glucose, $C_6H_{12}O_6$, is an important fermentation reaction.

$$C_6H_{12}O_6 \longrightarrow 2CH_3CH_2OH + 2CO_2$$

| Glucose | Ethanol (Ethyl alcohol) | Carbon Dioxide |

Ethyl alcohol is the intoxicating substance in alcoholic beverages. It damages the liver and causes behavior changes.

In order to protect its revenues, the government demands that ethyl alcohol for industrial use be denatured. **Denatured alcohol** *is ethanol with an added substance to make it toxic*. Methyl alcohol is often the denaturant. It is sometimes called wood alcohol because before 1925 it was prepared by the distillation of wood. Wood alcohol is extremely toxic. As little as 10 mL has been reported to cause permanent blindness and as little as 30 mL, death.

Figure 26·5
Ethanol is prepared commercially by the hydration of ethene or by the fermentation of sugars.

Figure 26·6
Julian's work with soybeans provided many useful substances, including hormones. His discovery of an inexpensive way to produce cortisone, a hormone used in treating rheumatoid arthritis, made this drug available to many people.

26·A Percy Julian and Natural Product Chemistry

Percy Julian was an important contributor to natural products chemistry. Natural products are organic substances derived from plants and animals. They are processed and used in the preparation of flavors, fragrances, resins, and pharmaceuticals. Julian, an industrial research chemist, was granted over 100 patents, many of them for substances he extracted from soybeans.

Perhaps Julian's most important contribution was the inexpensive method he developed for isolating sterols from soybeans. Sterols are unsaturated cyclic alcohols. Cortisone, a sterol derivative, is used by arthritis sufferers. As a result of Julian's research, cortisone became more widely available.

In 1935, Julian synthesized physostigmine, a drug used to treat glaucoma. A fire-fighting solution Julian developed from a soybean protein is credited with saving many lives during World War II. Julian also synthesized the hormones progesterone and testosterone. He did research on amino acids, indoles, and anti-fatigue drugs. In 1947, he was awarded the NAACP Spingarn Award, for outstanding achievement by a Black American.

26·6 Addition Reactions

Carbon–carbon single bonds are not easy to break. One of the bonds in an alkene double bond is somewhat weaker, however, than a carbon–carbon single bond. Thus, it is sometimes possible for a compound of general structure X–Y to add to a double bond. *In an* **addition reaction** *a substance is added at the double or triple bond of an alkene or alkyne.* Addition reactions are an important method of introducing new functional groups into organic molecules. In this general reaction, X and Y stand for the two parts of the compound which are added.

■ Addition reactions are the most important method of introducing a functional group into a hydrocarbon.

The breaking of a double bond opens *two* sites where addition occurs.

$$>\!C\!=\!C\!< \ + \ X\!-\!Y \longrightarrow -\overset{X}{\underset{|}{C}}-\overset{Y}{\underset{|}{C}}-$$

The addition of water to an alkene is a **hydration reaction.** Hydration reactions usually occur when the alkene and water are heated to about 100°C in the presence of a trace of strong acid. The acid serves as a catalyst for the reaction. Hydrochloric or sulfuric acid is generally used. The addition of water to ethene is a typical hydration reaction.

$$\underset{\text{Ethene}}{\overset{H}{\underset{H}{>}}C\!=\!C\overset{H}{\underset{H}{<}}} \ + \ \underset{\text{Water}}{H\!-\!OH} \ \xrightarrow{H^+} \ \underset{\text{Ethanol}}{H\!-\!\overset{H}{\underset{H}{C}}\!-\!\overset{OH}{\underset{H}{C}}\!-\!H}$$

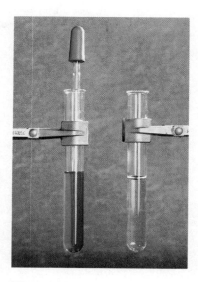

Figure 26·7
When a few drops of bromine solution (test tube on left) are added to an unsaturated organic compound, the bromine color disappears (right). This forms the basis of a test for unsaturated compounds.

When the reagent X–Y is a halogen molecule like chlorine or bromine, the product of the reaction is a disubstituted halocarbon.

$$CH_2{=}CH_2 + Br{-}Br \longrightarrow H{-}CHBr{-}CHBr{-}H$$

Ethene (colorless)   Bromine (brownish orange)   1,2-Dibromoethane (colorless)

The addition of bromine to carbon–carbon multiple bonds is often used as as a chemical test for unsaturation in an organic molecule. Bromine has a brownish-orange color, but most organic compounds of bromine are colorless. The test for unsaturation is done by adding a few drops of a 1% solution of bromine in carbon tetrachloride (CCl_4) to the suspected alkene. Loss of the orange color is a positive test for unsaturation.

Hydrogen halides such as HBr or HCl can also add to a double bond. The product contains only one substituent. It is called a monosubstituted halocarbon. The addition of hydrogen chloride to ethene is an example.

$$CH_2{=}CH_2 + H{-}Cl \longrightarrow H{-}CH_2{-}CHCl{-}H$$

Ethene (ethylene)   Hydrogen chloride   Chloroethane (ethyl chloride)

In a **hydrogenation reaction,** *hydrogen is added to a carbon–carbon double bond to give an alkane.* A hydrogenation reaction usually requires a catalyst. Finely divided platinum (Pt) or palladium (Pd) is often used. Such a process is used to hydrogenate unsaturated oils to make margarine.

$$CH_2{=}CH_2 + H{-}H \xrightarrow{Pt} H{-}CH_2{-}CH_2{-}H$$

Ethene   Hydrogen   Ethane

$$\text{Cyclohexene} + H{-}H \xrightarrow{Pt} \text{Cyclohexane}$$

Cyclohexene   Hydrogen   Cyclohexane

The hydrogenation of a double bond is a reduction reaction. Ethene is reduced to ethane, for example, and cyclohexene is reduced to cyclohexane.

Benzene resists hydrogenation. It also resists the addition of a halogen or a hydrogen halide. At high temperatures and high pressures of hydrogen, however, three molecules of hydrogen reduce one molecule of benzene to cyclohexane.

Benzene + 3H—H $\xrightarrow{Pt}$ Cyclohexane

Benzene   Hydrogen   Cyclohexane

Problems

6. Write the structure for the expected product from each reaction.

   a. $CH_2=CHCH_2CH_3 + Br_2 \longrightarrow$

   b. $CH_3CH=CHCH_3 + I_2 \longrightarrow$

   c. $CH_3CH=CHCH_3 + H_2 \xrightarrow{Pt}$

   d. (cyclohexene) $+ Cl_2 \longrightarrow$

7. Give the structure for the expected organic product from each of these reactions.

   a. CH_3\ /H C=C /H \CH_3 + HBr $\longrightarrow$

   b. (benzene) $+ Cl_2 \xrightarrow{catalyst}$

   c. (chlorobenzene) $+ 3H_2 \xrightarrow[\text{pressure}]{\text{catalyst}}$

26·7 Ethers

Ethers are organic compounds in which both hydrogens of H_2O have been replaced by R groups.

Ethers *are compounds in which oxygen is bonded to two carbon groups:* $R—O—R$. Ethers are easy to name. The alkyl groups attached to the ether linkage are named in alphabetical order and are followed by the word *ether*.

R /Ö\ R $CH_3CH_2—O—CH_3$
Ether molecule   Ethylmethyl ether

Ethylmethyl ether and methylphenyl ether are nonsymmetric. The R groups attached to the ether oxygen are different. When both R groups are the same, the ether is symmetric. Symmetric ethers are named by using the prefix *di-*. Sometimes, however, the prefix *di-* is dropped and a compound such as diethyl ether is simply called ethyl ether.

Figure 26·8
Halocarbons such as halothane, enflurane and isoflurane are now used as general anesthetics. What was the first reliable general anesthetic?

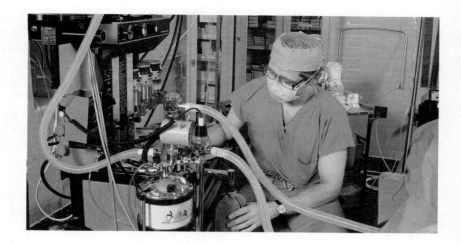

The structural formulas of ether molecules are usually written in linear fashion. However, because the central atom in an ether molecule is an oxygen atom, ether molecules are bent to form an angle that is nearly 105°.

$$CH_3CH_2—O—CH_2CH_3$$

Diethyl ether
(ethyl ether)

$$CH_3—O—$$

Methylphenyl ether
(anisole)

Diphenyl ether
(phenyl ether)

Diethyl ether, a volatile liquid (bp 35°C), was the first reliable general anesthetic. Originally reported in 1842 by Crawford W. Long, an American physician, diethyl ether was used by doctors for over a century. It has been replaced by other anesthetics because it is highly flammable and often causes nausea.

Ethers are usually lower-boiling than alcohols of comparable formula mass. They are higher-boiling than comparable hydrocarbons and halocarbons. Ethers are more soluble than hydrocarbons and halocarbons but less soluble than alcohols. The reason is that the oxygens in ethers are hydrogen acceptors, but ethers have no hydroxyl hydrogens to donate in hydrogen-bonding.

26·8   Aldehydes and Ketones

A carbonyl group *consists of a carbon atom and an oxygen atom joined by a double bond*. It is found in two groups of compounds called aldehydes and ketones.

Aldehydes *are organic compounds in which the carbon of the carbonyl group is always joined to at least one hydrogen*. The general formula for an aldehyde is RCHO.

Ketones *are organic compounds in which the carbon of the carbonyl group is joined to two other carbons*. The abbreviated form for a ketone is RCOR.

Because their carbonyl functional groups are identical, aldehydes and ketones have similar chemical and physical properties.

$$R—\overset{\overset{\displaystyle O}{\|}}{C}—H \}\ \longleftarrow \text{Carbonyl group}$$
Aldehyde

$$R—\overset{\overset{\displaystyle O}{\|}}{C}—R \}\ \longleftarrow \text{Carbonyl group}$$
Ketone

The IUPAC system may be used for naming aldehydes and ketones. For either class, the longest hydrocarbon chain that contains the carbonyl group must first be identified. The -e ending of the hydrocarbon is replaced by -al to designate an aldehyde. In the IUPAC system, the continuous-chain aldehydes are named methanal, ethanal, propanal, butanal, and so forth.

Ketones are named by changing the ending of the longest continuous carbon chain that contains the carbonyl group from -e to -one. Table 26·5 demonstrates this. If the carbonyl group could occur at several places on the chain, then its position is designated by the lowest possible number.

Problem

8. Give the IUPAC name for each of the following aldehydes and ketones.

a. CH_3CH_2CHO

b. $CH_3CH_2CH_2\overset{\displaystyle O}{\overset{\|}{C}}CH_2CH_3$

c. $CH_3CH_2\overset{\displaystyle CH_3}{\overset{|}{C}H}CH_2CHO$

d. $CH_3\overset{\displaystyle O}{\overset{\|}{C}}CH_2CH_3$

Aldehydes and ketones can form weak hydrogen bonds between the carbonyl oxygen and the hydrogens of water. The lower members of the series (formaldehyde, acetaldehyde, and acetone) are soluble in water in all proportions. As the length of the hydrocarbon chain increases, water solubility decreases. When the carbon chain exceeds five or six carbons, solubility of both aldehydes and ketones is very low. As might be expected, all aldehydes and ketones are soluble in nonpolar solvents.

Aldehydes and ketones cannot form intermolecular hydrogen bonds because they lack hydroxyl (—OH) groups. Consequently, they have boiling points lower than those of the corresponding alcohols. The aldehydes and ketones can attract one another, however, through polar-polar interactions of their carbonyl groups. Consequently, their boiling points are higher than those of the corresponding alkanes. These attractive forces account for the fact that nearly all aldehydes and ketones are either liquids or solids at room temperature. The exception is formaldehyde, which is an irritating pungent gas. Table 26·6 compares the boiling points of alkanes, aldehydes, and alcohols of similar formula mass.

A wide variety of aldehydes and ketones have been isolated from plants and animals. Many of them, particularly those with high formula masses, have fragrant or penetrating odors. They are usually known by their common names, which indicate their natural sources or perhaps a characteristic property. Aromatic aldehydes are often used as flavoring agents. Benzaldehyde is the simplest aromatic aldehyde. Also known as "oil of bitter almond," benzaldehyde is a constituent of the almond. It is a colorless liquid with a pleasant almond odor. Cinnamaldehyde imparts

Figure 26·9
The seeds from this orchid are the source of natural vanilla flavor for ice cream and many other foods. Vanillin is also produced synthetically.

Table 26·5   Some Common Aldehydes and Ketones

Condensed formula	Structural formula	IUPAC name	Common name
Aldehydes			
HCHO		Methanal	Formaldehyde
CH_3CHO		Ethanal	Acetaldehyde
C_6H_5CHO		Benzaldehyde	Benzaldehyde
$C_6H_5CH=CHCHO$		3-Phenyl-2-propenal	Cinnamaldehyde
$CH_3O(OH)C_6H_3CHO$		4-Hydroxy-3-methoxybenzaldehyde	Vanillin
Ketones			
CH_3COCH_3		Propanone	Acetone (dimethyl ketone)
$C_6H_5COC_6H_5$		Diphenylmethanone	Benzophenone (diphenyl ketone)

Table 26·6   Boiling Points of Some Compounds with One and Two Carbons

Compound	Formula	Formula mass	Boiling point (°C)	Comments
One carbon				
Methane	CH_4	16	-161	No hydrogen-bonding or polar-polar interactions
Formaldehyde	HCHO	26	-21	Polar-polar interactions
Methanol	CH_3OH	32	65	Hydrogen-bonding
Two carbons				
Ethane	C_2H_6	30	-89	No hydrogen-bonding or polar-polar interactions
Acetaldehyde	CH_3CHO	44	20	Polar-polar interactions
Ethanol	CH_3CH_2OH	46	78	Hydrogen-bonding

the characteristic odor of oil of cinnamon. Vanillin, which is responsible for the popular vanilla flavor, was at one time obtainable only from the podlike capsules of certain climbing orchids. Today much vanillin is synthetically produced.

The simplest aldehyde, formaldehyde, is very important industrially but inconvenient to handle in the gaseous state. Formaldehyde is usually available as a 40% aqueous solution, known as formalin. Formalin is used to preserve biological specimens. The formaldehyde in solution combines with protein in tissues to make them hard and insoluble in water. This process prevents the specimen from decaying. The greatest use of formaldehyde is in the manufacture of synthetic resins.

The most important industrial ketone is acetone. It is a colorless volatile liquid that boils at 56°C. It is used as a solvent for resins, plastics, and varnishes, and is often found in nail polish removers. Moreover, it is miscible with water in all proportions.

26·9 Carboxylic Acids

A **carboxyl group** *consists of a carbonyl group attached to a hydroxyl group.*

$$-\underset{\substack{\text{Carboxyl group}\\ \text{(also written } -CO_2H \text{ or } -COOH)}}{\overset{\overset{\displaystyle O}{\|}}{C}}-OH$$

Carbonyl group · Hydroxyl group

Carboxylic acids *are compounds with a carboxyl group*. The general formula of a carboxylic acid is RCOOH. Carboxylic acids are weak acids because they ionize slightly in solution to give a carboxylate ion and a proton.

$$R-\overset{\overset{\displaystyle O}{\|}}{C}-OH \rightleftharpoons R-\underset{\substack{\text{Carboxylate}\\ \text{ion}}}{\overset{\overset{\displaystyle O}{\|}}{C}}-O^- + \underset{\text{Proton}}{H^+}$$

In the IUPAC system for naming carboxylic acids, the *-e* ending of the parent alkane is replaced by the ending *-oic acid*. The parent alkane is the hydrocarbon with the longest continuous carbon chain containing the carboxyl group. Table 26·7 lists the names and formulas of some common aliphatic carboxylic acids.

Carboxylic acids are abundant and widely distributed in nature. Many have common names derived from a Greek or Latin word describing their natural sources. For example, the common name for ethanoic acid is acetic acid. Acetic acid is produced when wine turns sour and becomes vinegar. The pungent aroma of vinegar comes from its acetic acid. *Many continuous-chain carboxylic acids were first isolated from fats and are called* **fatty acids.** Propionic acid, the three-carbon acid, literally means "first fatty acid". Common names are used more often than IUPAC names for carboxylic acids.

The presence of the carboxyl group in an organic compound makes it behave as a weak acid.

Latin: *acetum* = vinegar

Greek: *protos* = first
       pion = fat

Figure 26·10
These candles are being hand-dipped to give them a final coating of colored wax. Standard commercial candles are a mixture of 60% paraffin, 35% stearic acid, and 5% beeswax.

Latin: *butyrum* = butter

Greek: *stear* = tallow

Latin: *formica* = ant

The low-formula-mass members of the aliphatic carboxylic acid series are colorless, volatile liquids. They have sharp, unpleasant odors. The smells of rancid butter and dirty feet are due in part to butyric acid. The higher members of the series are nonvolatile, low-melting, waxy solids. Stearic acid (C–18) is obtained from beef fat. It is used to make cheap wax candles. Stearic acid and other long-chain fatty acids have very little odor.

Like alcohols, carboxylic acids can form intermolecular hydrogen bonds. Because of this, carboxylic acids are higher boiling and higher melting than other compounds of similar formula mass. All aromatic carboxylic acids are crystalline solids at room temperature.

The carboxyl group in carboxylic acids is polar, and it readily forms hydrogen-bonds with water molecules. Formic, acetic, propionic, and butyric acids are completely miscible with water. The solubility of carboxylic acids of higher formula mass drops off sharply. Most carboxylic acids dissolve in organic solvents like ethanol or acetone.

Table 26·7 Saturated Aliphatic Carboxylic Acids

Formula	Carbon atoms	IUPAC name	Common name	Melting point (°C)
$HCOOH$	1	Methanoic acid	Formic acid	8
CH_3COOH	2	Ethanoic acid	Acetic acid	17
CH_3CH_2COOH	3	Propanoic acid	Propionic acid	−22
$CH_3(CH_2)_2COOH$	4	Butanoic acid	Butyric acid	−6
$CH_3(CH_2)_4COOH$	6	Hexanoic acid	Caproic acid	−3
$CH_3(CH_2)_6COOH$	8	Octanoic acid	Caprylic acid	16
$CH_3(CH_2)_8COOH$	10	Decanoic acid	Capric acid	31
$CH_3(CH_2)_{10}COOH$	12	Dodecanoic acid	Lauric acid	44
$CH_3(CH_2)_{12}COOH$	14	Tetradecanoic acid	Myristic acid	58
$CH_3(CH_2)_{14}COOH$	16	Hexadecanoic acid	Palmitic acid	63
$CH_3(CH_2)_{16}COOH$	18	Octadecanoic acid	Stearic acid	70

26·10 Oxidation–Reduction Reactions

Remember that oxidation is the gain of oxygen, loss of hydrogen, or loss of electrons. Reduction is just the reverse: loss of oxygen, gain of hydrogen, or gain of electrons. Oxidation and reduction reactions are coupled. One does not occur without the other.

In organic chemistry, the number of oxygens and hydrogens attached to carbon indicates the degree of oxidation of a compound. The fewer the number of hydrogens on a carbon-carbon bond, the more oxidized the bond. For example, ethane (an alkane) can be oxidized to ethene (an alkene) and then to ethyne (an alkyne). *The loss of hydrogen is a* **dehydrogenation reaction.** Strong heating and a catalyst are usually necessary to make dehydrogenation reactions occur. The loss of each molecule of hydrogen involves the loss of two electrons.

Many organic oxidation–reduction reactions involve a change in the degree of unsaturation of a carbon–carbon bond or the oxidation state of an oxygen-containing functional group.

Increasing oxidation: alkane, alkene, alkyne.

These reactions can be reversed. Alkynes can be reduced to alkenes, and alkenes can be reduced to alkanes by the addition of hydrogen to a double bond.

Oxidation in organic chemistry also involves the number and degree of oxidation of the oxygens attached to carbon. For example, methane, a saturated hydrocarbon, may be oxidized in steps to carbon dioxide. This occurs if it alternately gains oxygens and loses hydrogens. Methane is oxidized to methanol, then to formaldehyde, then to formic acid, and finally to carbon dioxide. The same sequence of oxidations occurs for other alkanes. Each series of compounds consists of an alkane, alcohol, aldehyde (or ketone), carboxylic acid, and carbon dioxide. The carbon dioxide is most oxidized or least reduced, and the alkane is least oxidized or most reduced.

Oxidation reactions are energy-releasing. *The more reduced a carbon compound, the more energy it can release upon its complete oxidation to carbon dioxide.* The energy-releasing properties of oxidation reactions are extremely important for energy production in tems. This is also why the combustion of hydrocarbons such as is a good source of energy.

Primary alcohols can be oxidized to aldehydes and secondary alcohols can be oxidized to ketones.

Increasing oxidation: alcohol, aldehyde or ketone, carboxylic acid.

Primary alcohol → Aldehyde   Secondary alcohol → Ketone

Tertiary alcohols cannot be oxidized because no hydrogen is present on the carbon bearing the hydroxyl group.

The primary alcohols methanol and ethanol can be oxidized to aldehydes by warming them at about 50°C with acidified potassium dichromate ($K_2Cr_2O_7$). In these reactions, methanol produces formaldehyde and ethanol produces acetaldehyde.

Methanol (methyl alcohol) (bp 65°C) → Methanal (formaldehyde) (bp −21°C)

Ethanol (ethyl alcohol) (bp 78°C) → Ethanal (acetaldehyde) (bp 21°C)

The preparation of an aldehyde by this method is often a problem because aldehydes are easily oxidized to carboxylic acids.

Aldehyde → Carboxylic acid

Further oxidation is not a problem, however, with aldehydes that have low boiling points, such as acetaldehyde. This is because the product can be distilled from the reaction mixture as it is formed.

Oxidation of the secondary alcohol 2-propanol by warming it with acidified potassium dichromate produces acetone.

2-Propanol (isopropyl alcohol) → Propanone (acetone)

Ketones are resistant to further oxidation. There is no need to remove them from the reaction mixture during the course of the reaction.

Problems

9. What products are expected when these compounds are oxidized?

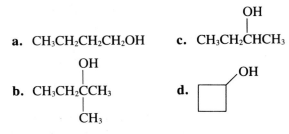

a. $CH_3CH_2CH_2CH_2OH$

b. $CH_3CH_2\overset{\underset{|}{OH}}{\underset{\underset{|}{CH_3}}{C}}CH_3$

c. $CH_3CH_2\overset{\underset{|}{OH}}{C}HCH_3$

d. (cyclobutane with OH)

10. Give the name and structure of the alcohol that must be oxidized to make the following compounds.

a. CH_3CH_2CHO

b. $CH_3CH_2\overset{\overset{O}{\|}}{C}CH_3$

c. $CH_3CH_2\overset{\underset{|}{CH_3}}{C}HCHO$

Chemists have taken advantage of the ease with which an aldehyde can be oxidized to develop several tests for their detection. **Benedict's** and **Fehling's tests** *are commonly used for aldehyde detection.* Benedict's and Fehling's reagents are deep blue alkaline solutions of copper sulfate of slightly differing composition. When an aldehyde is oxidized with Benedict's or Fehling's reagent, a bright red precipitate of cuprous oxide (Cu_2O) is obtained.

$$CH_3\overset{\overset{O}{\|}}{C}\text{—H} + 2Cu^{2+} + 5OH^- \longrightarrow CH_3\overset{\overset{O}{\|}}{C}\text{—O}^- + Cu_2O(s) + 3H_2O$$

Acetaldehyde / Cupric ion complex (blue solution) / Acetic acid (as acetate ion) / Cuprous oxide (red precipitate)

The aldehyde is oxidized to acetic acid. Cupric ions (Cu^{2+}) are reduced to cuprous ions (Cu^+).

Figure 26·11
When an aldehyde is mixed with Fehling's reagent (left) and heated, a brick-red precipitate forms (right). The blue copper(II) ions in Fehling's reagent are reduced to copper(I) ions in the red precipitate of copper(I) oxide, Cu_2O.

26·11 Esters

An ester is produced from the combination of a carboxylic acid and an alcohol.

Esters *are derivatives of carboxylic acids in which the —OH of the carboxyl group has been replaced by an —OR from an alcohol.* They contain a carbonyl group and an ether link to the carbonyl carbon.

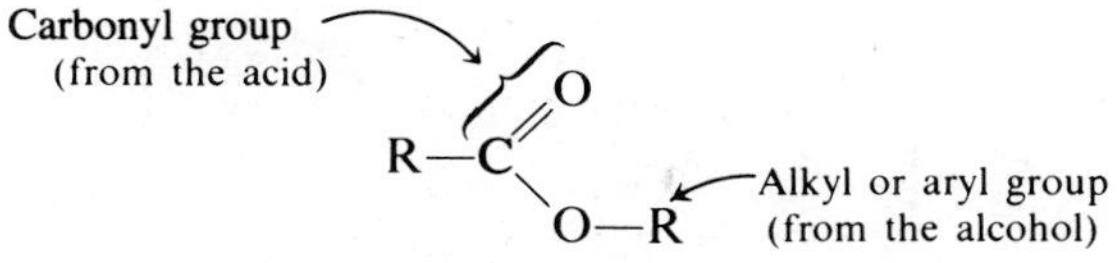

Carbonyl group (from the acid)

$$R-C \overset{O}{\underset{O-R}{\Big\langle}}$$

Alkyl or aryl group (from the alcohol)

The abbreviated formula for a carboxylate ester is RCOOR. The R group can be short chains or long chains, aliphatic (alkyl) or aromatic (aryl), saturated or unsaturated.

Simple esters are neutral substances. The molecules are polar but cannot form hydrogen bonds with one another. This is because they contain no hydrogen attached to oxygen or another electronegative atom. As a result, only weak attractions hold ester molecules to each other. They are much lower boiling than the strongly hydrogen-bonded carboxylic acids from which they are derived. The low-formula-mass esters are somewhat soluble in water. Esters containing more than four or five carbons have very limited solubility.

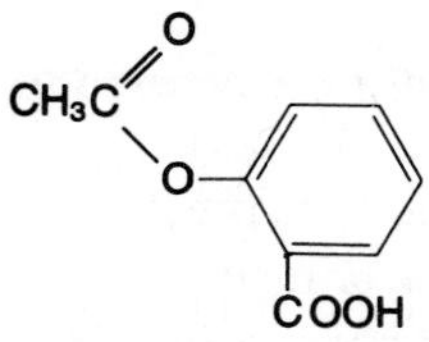

If an ester is heated with water for several hours, usually very little happens. In strong acid or base solutions, however, the ester breaks down. An ester is hydrolyzed by the addition of water to produce a carboxylic acid and an alcohol. The reaction is rapid in acidic solution because it is catalyzed by hydrogen ions.

$$CH_3-\overset{O}{\overset{\|}{C}}-OCH_2CH_3 + H-OH \underset{}{\overset{H^+}{\rightleftharpoons}} CH_3-\overset{O}{\overset{\|}{C}}-OH + HOCH_2CH_3$$

Ethyl acetate   Acetic acid   Ethanol

Hydroxide ions also catalyze this reaction. The usual agent for ester hydrolysis is an aqueous solution of sodium hydroxide or potassium hydroxide. Because many esters do not dissolve in water, a solvent like ethanol is added to make the solution homogeneous. The reaction mixture is usually heated. All the ester is converted to products. The carboxylic acid product is in solution as its sodium or potassium salt.

$$CH_3-\overset{O}{\overset{\|}{C}}-OCH_2CH_3 + H_2O \overset{NaOH}{\longrightarrow} CH_3-\overset{O}{\overset{\|}{C}}-O^-Na^+ + HOCH_2CH_3$$

Ethyl acetate   Sodium acetate   Ethanol

If the reaction mixture is acidified, the carboxylic acid is formed.

$$CH_3-\overset{O}{\overset{\|}{C}}-O^-Na^+ + HCl \longrightarrow CH_3-\overset{O}{\overset{\|}{C}}-OH + NaCl$$

Sodium acetate   Acetic acid

Figure 26·12
An ester of great commercial importance is acetyl salicylic acid, commonly called aspirin. The R groups in its structure are a methyl group and a benzoic acid group.

Figure 26·13
Esters give fruits their characteristic flavors and odors.

Problem

11. Write the structures of the expected products for these reactions.

a. $CH_3CH_2COOCH_2CH_3 \xrightarrow{\text{NaOH}}$

b. $CH_3COO-\langle\text{phenyl}\rangle \xrightarrow{\text{KOH}}$

c. $CH_3CH_2COOCH_2\underset{\underset{CH_3}{|}}{C}HCH_3 \xrightarrow{\text{HCl}}$

Figure 26·14
To create new flavors for food products a flavor chemist can select from a large number of natural and synthetic raw materials.

Giant molecules called polymers result from either addition polymerization or condensation polymerization.

Esters may be prepared from an acid and an alcohol. Esterification is the formation of an ester from a carboxylic acid and a primary or secondary alcohol. The reactants are heated with a trace of mineral acid as a catalyst. The reaction is reversible.

$$R-C\overset{O}{\underset{OH}{\big<}} + RO-H \xrightleftharpoons{H^+} R-C\overset{O}{\underset{OR}{\big<}} + H-OH$$

Carboxylic acid   Alcohol   Carboxylate ester   Water

For example, acetic acid and ethanol react to form ethyl acetate, an ester.

$$CH_3-C\overset{O}{\underset{OH}{\big<}} + CH_3CH_2O-H \xrightleftharpoons{H^+} CH_3-C\overset{O}{\underset{OCH_2CH_3}{\big<}} + H_2O$$

Acetic acid   Ethanol   Ethyl Acetate   Water

Chapter 17 showed how the equilibrium in a reversible reaction can be disturbed to improve the product yield. Esterification reactions can be pushed to completion in two different ways. An excess of one of the reactants (acid or alcohol) can be used. Alternatively, the water from the reaction mixture can be removed as it is produced.

26·12 Polymerization

Most of the reactions that have been examined so far have involved reactants and products of low molecular mass. Some of the most important organic compounds made by chemists, however, are giant molecules called polymers. A **polymer** is a large molecule formed by the covalent bonding of repeating smaller molecules. Most polymerization reactions require a catalyst.

Monomers are molecules that combine to form the repeating unit of a polymer. Some polymers contain only one type of monomer. Others contain two or more types of monomers. The two most common ways for monomers to be joined are addition polymerization and condensation polymerization.

Figure 26·15
Polyvinyl chloride (PVC) is one of the most versatile polymers. It is used to make all of the items shown here, and many more.

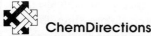

ChemDirections

Plastics recycling helps conserve our nonrenewable resources. To discover why, read **ChemDirections** *Plastics Recycling*, page 663.

The substances we call "plastics" consist of polymers.

$$x\,CH_3CH{=}CH_2 \longrightarrow \underset{\text{Polypropylene}}{{+}\underset{\displaystyle CH_3}{CH}{-}CH_2{+}_x}$$

Propene (propylene) → Polypropylene

$$x\,CH_2{=}CH \longrightarrow {+}CH_2{-}CH{+}_x$$
(with Cl substituents)

Chloroethene (vinyl chloride) → Polyvinyl chloride (PVC)

$$x\,CF_2{=}CF_2 \longrightarrow {+}CF_2{-}CF_2{+}_x$$

Tetrafluoroethene → Teflon

$$x\,CH_2{=}CH \longrightarrow {+}CH_2{-}CH{+}_x$$
(with phenyl substituents)

Vinyl benzene (styrene) → Polystyrene

Figure 26·16
The structures of some important addition polymers of alkenes are shown here.

Addition polymerization occurs when unsaturated monomers react to form a polymer. It is a specific type of addition reaction. Ethene undergoes addition polymerization. The molecules bond one to another to form the long-chain polymer polyethylene.

$$x \quad \underset{H}{\overset{H}{\diagdown}}C{=}C\underset{H}{\overset{H}{\diagup}} \longrightarrow H{+}CH_2{-}CH_2{+}_x H$$

x is number of ethylene units that combine to form long chain

Ethene (ethylene) → Polyethylene

x is number of repeating —CH₂—CH₂— units in polymer; parentheses identify the repeating unit

Polyethylene is an important industrial product because it is chemically resistant and easy to clean. It is used to make refrigerator dishes, plastic milk bottles, laboratory wash bottles, and many other familiar items found in homes and laboratories. By shortening or lengthening the carbon chains, chemists can control the physical properties of polyethylene. Polyethylene containing relatively short chains ($x = 100$) has the consistency of paraffin wax. Polyethylene with long chains ($x = 1000$) is harder and more rigid.

Polymers of substituted ethenes can also be prepared. Many of these polymers have useful properties (Figure 26·16).

Condensation polymers are formed by the head-to-tail joining of monomer units. This is usually accompanied by the loss of a small molecule, such as water. The formation of polyesters is an example of condensation. Polyesters are high-formula-mass polymers consisting of many repeating units of dicarboxylic acids and dihydroxy alcohols joined by ester bonds. The formation of a polyester is represented by a block diagram. Note that condensation polymerization always requires that there be *two* functional groups on each molecule.

$$x\,HO{-}\overset{\displaystyle O}{\overset{\|}{C}}{-}\Box{-}\overset{\displaystyle O}{\overset{\|}{C}}{-}OH + x\,HO{-}\bigcirc{-}OH \longrightarrow {+}\overset{\displaystyle O}{\overset{\|}{C}}{-}\Box{-}\overset{\displaystyle O}{\overset{\|}{C}}{-}O{-}\bigcirc{-}O{+}_x + 2x\,H_2O$$

Dicarboxylic acid Dihydroxy alcohol Representative polymer unit of a polyester

26·B Synthetic Fibers

In 1883 the Englishman Joseph Swan discovered that nitrocellulose could be made into silk-like filaments. He used these filaments in electric lamps. One year later, Louis Chardonnet, a Frenchman, produced similar material when he forced nitrocellulose through tiny holes. At the 1891 Paris Exposition, Chardonnet's synthetic material was called "rayon". The shiny fibers seemed to produce rays of light.

The widespread production of synthetic material began in the 1930s with Wallace Carothers, an American chemist. He made a polymer of diamines and dicarboxylic acids. The fibers he produced from this polymer had many of the properties of silk but were much stronger. This material, called nylon, was first used for toothbrush bristles. During World War II, nylon was reserved exclusively for parachutes. After the war, nylon became a popular material for women's hosiery.

Although there are many different synthetic fibers today, the production methods for most of these materials are similar. First the polymer is either melted or dissolved in a solvent. Then, the melted or dissolved polymer is forced through holes in a device called a spinneret. The fibers are solidified by cooling the mixture or evaporating the solvent. Finally, the fibers are stretched to make the polymer molecules lie parallel to each other. This increases the strength of the fibers.

Synthetic fibers can be divided into two groups depending on the starting materials. Cellulosic fibers use cellulose derived from wood pulp or cotton as starting material. Rayon and acetate are cellulosic fibers. Noncellulosic fibers are manufactured from petrochemicals. Nylon, polyester, and acrylic are noncellulosic polymers.

ChemDirections

Plastics and other polymeric materials are playing an increasing role in the field of biomedical implants. Discover some of the recent advances in the use of these materials in **ChemDirections** *Biomedical Implants*, page 667.

Figure 26·17
Synthetic fibers are now more widely used than natural fibers.

26·13 Carbohydrates

Carbohydrates are naturally occurring polyhydroxy aldehydes and ketones.

Carbohydrates abound in nature. Among the many forms are starch, table sugar, cotton, and wood. **Carbohydrates** *are monomers and polymers of aldehydes and ketones that have numerous hydroxy groups attached.* The name carbohydrate comes from the early observation that many of these compounds have the general formula $C_n(H_2O)_n$. As a result they appear to be "hydrates of carbon." An examination of the molecular structures will show that this is erroneous. Nevertheless, the name has stuck.

The simplest carbohydrate molecules are simple sugars called **monosaccharides.** These names are used interchangeably. Glucose and fructose are examples of simple sugars. They are structural isomers because they both have the molecular formula $C_6H_{12}O_6$. Glucose has an aldehyde functional group. Fructose has a ketone functional group. They undergo many of the same reactions as ordinary aldehydes and ketones.

Glucose and its cyclic form

Fructose and its cyclic form

In aqueous solution, sugars such as glucose and fructose exist in dynamic equilibrium in both the cyclic and straight chain forms. The cyclic form predominates. Glucose is abundant in plants and animals. Depending on the source, it has been called corn sugar, grape sugar, and blood sugar. Fructose occurs in a large number of fruits and in honey.

The cyclic forms of two simple sugars can be linked together with the loss of water. The linkage of glucose and fructose by means of a condensation reaction gives sucrose, common table sugar. *Sugars which are formed from two monosaccharides are known as* **disaccharides.** Sucrose is a disaccharide.

Figure 26·18
Carbohydrates are the most abundant and least expensive sources of energy in food. **a** These foods all contain natural sugars. **b** These foods are good sources of starch.

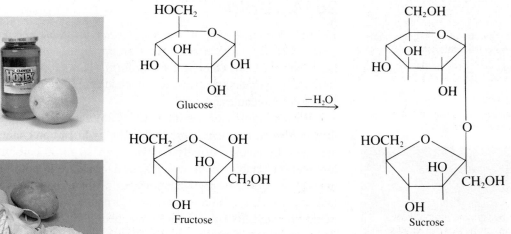

Glucose

Fructose

$-H_2O \rightarrow$

Sucrose

Sucrose is obtained mainly from the juice of sugar cane and sugar beets. The world's production from these sources exceeds 7×10^9 metric tons per year.

The formation of a disaccharide is sometimes the first step in a condensation polymerization. *The linkage of many monosaccharide monomers produces* **polysaccharides.** Starches, the major storage form of glucose in plants, are polymers consisting of glucose monomers. A typical linear starch molecule contains hundreds of the glucose monomers (Figure 26·19). Some starches are branched molecules, with each branch containing about a dozen glucose units. Glycogen is the animal form of starch. It is more highly branched than plant starches.

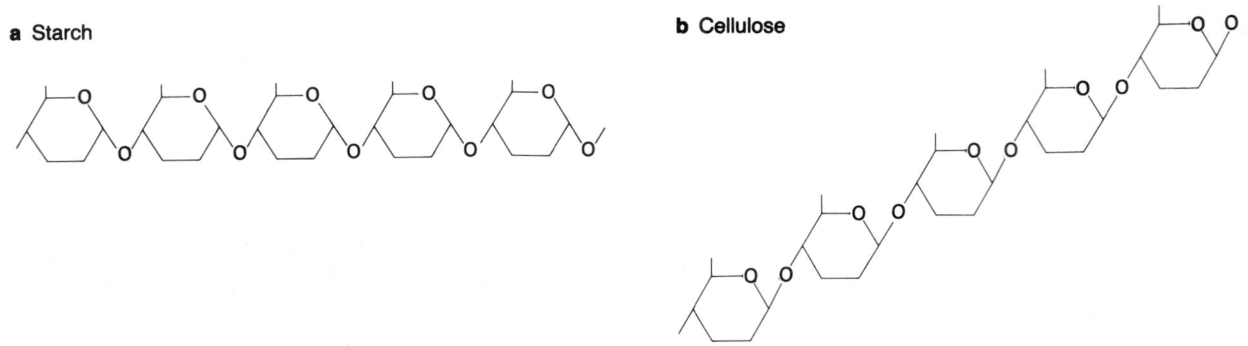

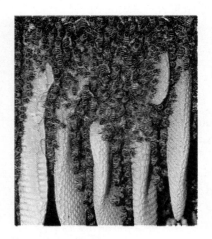

a Starch

b Cellulose

Figure 26·19
Starch and cellulose are similar polymers made up of hundreds of glucose monomers. They differ in the orientation of the bond between the glucose units. This causes starch to be readily digestible and cellulose to be nondigestible by most organisms.

Cellulose is probably the most abundant biological molecule. As shown in Figure 26·19, cellulose is also a polymer of glucose. The orientation of the bond linking the glucose monomers is different than it is in starch. Starch is edible and partially soluble in water. Cellulose can be digested by only a few microorganisms such as those that live in the digestive tract of cattle and termites. It is insoluble in water and forms rigid structures with other cellulose molecules. Cellulose is therefore an important structural polysaccharide. Plant cell walls, as in wood, are made of cellulose. Cotton is about 80% pure cellulose.

26·14 Lipids

Lipids are the major organic components of waxes, fats, and oils.

Lipids *are a large class of relatively water-insoluble compounds that includes fats, oils, and waxes.* As a group, lipids tend to dissolve in organic solvents like ether and chloroform. This property sets them apart from carbohydrates, proteins, and nucleic acids, the other great classes of biological molecules.

Waxes are part of the lipid family. **Waxes** *are esters of long-chain fatty acids and long-chain alcohols.* The hydrocarbon chains for both the acid and the alcohol usually contain from 10 to 30 carbons. Waxes are low-melting, stable solids which appear in nature in both plants and animals. A wax coat protects the surfaces of many plant leaves from water loss and attack by microorganisms. Carnauba wax is a major ingredient in car wax and floor polish. It comes from the leaves of a South American palm tree. Other waxes coat skin, hair, and feathers and help keep them pliable and waterproof. Beeswax is largely myricyl palmitate. It is the ester of myricyl alcohol and palmitic acid.

$$CH_3(CH_2)_{14}-\overset{\overset{\displaystyle O}{\|}}{C}-O-(CH_2)_{29}CH_3$$
Myricyl palmitate

Figure 26·20
Bees construct their honeycomb from beeswax, myricyl palmitate. Waxes are esters of a long-chain fatty acid and a long-chain alcohol.

Natural **triglycerides** *are triesters of long-chain fatty acids (C_{12} through C_{24}) and glycerol.* They are the major components of animal fats and oils. Triglycerides are simple lipids. They are important as the storage form of fat in the human body. The following equation shows the general reaction for the formation of triglycerides.

$$
\begin{array}{ccc}
\underset{\text{Glycerol}}{\begin{array}{c} CH_2OH \\ | \\ CHOH \\ | \\ CH_2OH \end{array}}
+
\underset{\substack{\text{3 Fatty acid} \\ \text{molecules}}}{\begin{array}{c} O \\ \parallel \\ HO-C-R \\ O \\ \parallel \\ HO-C-R \\ O \\ \parallel \\ HO-C-R \end{array}}
\longrightarrow
\underset{\substack{\text{Triglyceride} \\ \text{(triester of glycerol)}}}{\begin{array}{c} O \\ \parallel \\ CH_2-O-C-R \\ O \\ \parallel \\ CH-O-C-R \\ O \\ \parallel \\ CH_2-O-C-R \end{array}}
+
\underset{\text{Water}}{3H_2O}
\end{array}
$$

From the organic chemist's point of view, fats and oils are simply esters. Like other esters, they are easily hydrolyzed in the presence of acids and bases (Section 26·11). *The hydrolysis of oils or fats by boiling them with aqueous sodium hydroxide is called* **saponification.** This process is used to make soap. Soaps are the alkali metal (Na, K, or Li) salts of fatty acids.

$$
\underset{\substack{\text{Tristearin} \\ \text{(a triester)}}}{\begin{array}{c} O \\ \parallel \\ CH_2-O-C-(CH_2)_{16}CH_3 \\ O \\ \parallel \\ CH-O-C-(CH_2)_{16}CH_3 \\ O \\ \parallel \\ CH_2-O-C-(CH_2)_{16}CH_3 \end{array}}
+ 3NaOH \longrightarrow
\underset{\text{Glycerol}}{\begin{array}{c} CH_2-OH \\ | \\ CH-OH \\ | \\ CH_2-OH \end{array}}
+
\underset{\substack{\text{Sodium stearate} \\ \text{(soap)}}}{3CH_3(CH_2)_{16}\overset{\displaystyle O}{\overset{\parallel}{C}}-O^-Na^+}
$$

Soap is made by heating beef tallow or coconut oil in large kettles with an excess of sodium hydroxide. When sodium chloride is added to the saponified mixture, the sodium salts of the fatty acids separate as a thick curd of crude soap. Glycerol is an important by-product of the reaction. It is recovered by evaporating the water layer. The crude soap is then purified. Coloring agents and perfumes are added according to market demands.

Figure 26·21
Soap molecules dissociate in water. The anionic end of the soap molecule (the carboxylate ion, —COO⁻) becomes hydrogen bonded to water molecules. The long nonpolar hydrocarbon chain of the soap molecule dissolves in organic materials such as oils. The soap-oil mixture forms droplets that disperse in water and are readily washed away.

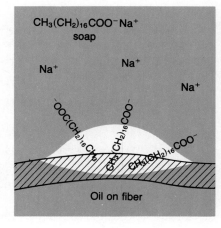
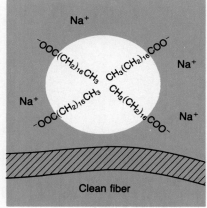

26·15 Amino Acids, Peptides, and Proteins

An **amino acid** *is a compound which contains amino (—NH₂) and carboxylic acid (—COOH) groups in the same molecule.* To most chemists and biochemists, however, the term is usually reserved for amino acids which are formed and used in living organisms. All 20 amino acids common in nature have a skeleton consisting of a carboxylic acid group and an amino group covalently bonded to a central atom. The remaining two groups on the central carbon are hydrogen and an R group which constitutes the amino acid side chain.

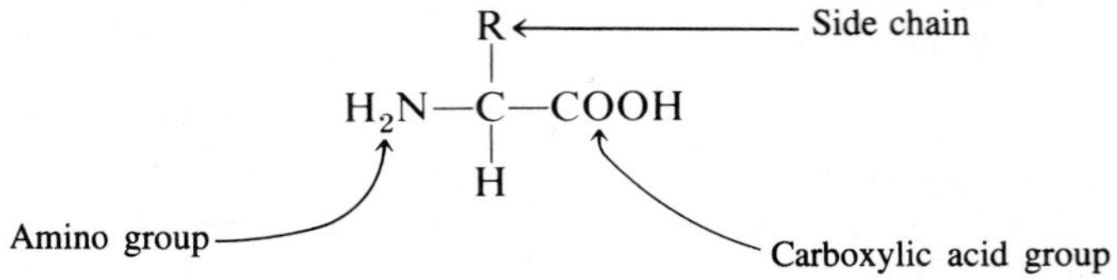

$$H_2N-\underset{\underset{H}{|}}{\overset{\overset{R \longleftarrow \text{Side chain}}{|}}{C}}-COOH$$

Amino group —

Carboxylic acid group —

The chemical natures of the side-chain groups account for differences in amino acid properties. Some side chains are aliphatic or aromatic hydrocarbons. Others are acidic or basic. Still others are neutral but polar. Figure 26·22 shows some of these amino acid structures. Table 26·8 gives all of the names with three-letter abbreviations.

The structures of the amino acids shown in Figure 26·22 are not actually correct. The acid–base properties of carboxylic acids and amino

Table 26·8 Abbreviations for Amino Acids	
Amino acid	Abbreviation
alanine	Ala
arginine	Arg
asparagine	Asn
aspartic acid	Asp
cysteine	Cys
glutamine	Gln
glutamic acid	Glu
glycine	Gly
histidine	His
isoleucine	Ile
leucine	Leu
lysine	Lys
methionine	Met
phenylalanine	Phe
proline	Pro
serine	Ser
threonine	Thr
tryptophan	Try
tyrosine	Tyr
valine	Val

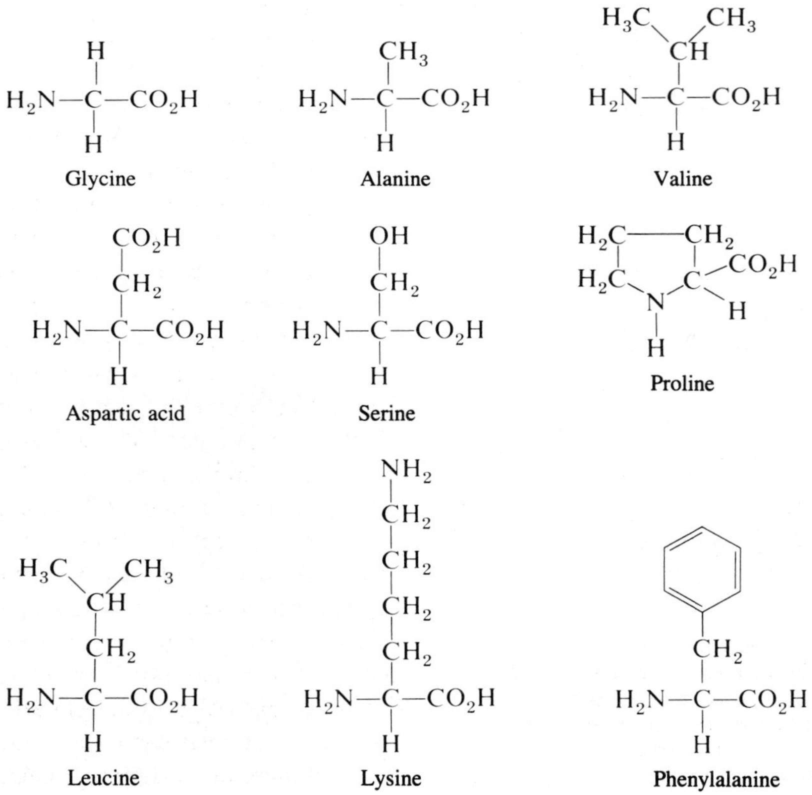

Glycine
Alanine
Valine
Aspartic acid
Serine
Proline
Leucine
Lysine
Phenylalanine

Figure 26·22
These are some of the 20 naturally-occurring amino acids.

642   Chapter 26 Functional Groups and Organic Reactions

groups alter these structures. The weakly acidic proton of the carboxylic acid group easily transfers to the amino group to form an internal salt. Thus amino acids are really ionic compounds. As previously discussed, ionic compounds generally have much higher melting points than molecular compounds. *The internal salts of amino acids are called* **zwitterions.** In the pure solid state and in aqueous solution near neutral pH, the amino acids exist almost completely as zwitterions.

$$R \diagdown \underset{\underset{H_2N}{}}{C} \diagup H \quad \quad R \diagdown \underset{\underset{H_3N^+}{}}{C} \diagup H$$
$$\text{Amino acid} \longrightarrow \text{Zwitterion}$$

A **peptide** *is any combination of amino acids in which the amino group of one acid is united with the carboxylic acid group of another.* The bonds between the amino acids always involve the central amino and central carboxylic acid groups. The side chains are not involved.

$$H_2N-\underset{\underset{H}{|}}{\overset{\overset{R}{|}}{C}}-\overset{\overset{O}{\|}}{C}-OH \; + \; H-\underset{\underset{H}{|}}{\overset{\overset{R}{|}}{N}}-\underset{\underset{H}{|}}{\overset{\overset{}{}}{C}}-\overset{\overset{O}{\|}}{C}-OH \longrightarrow H_2N-\underset{\underset{H}{|}}{\overset{\overset{R}{|}}{C}}-\overset{\overset{O}{\|}}{C}-\underset{\underset{H}{|}}{\overset{}{N}}-\underset{\underset{H}{|}}{\overset{\overset{R}{|}}{C}}-\overset{\overset{O}{\|}}{C}-OH \; + \; H_2O$$

Amino acid   Amino acid   Peptide

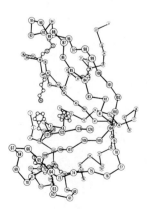

Figure 26·23
Each protein has a specific amino acid sequence. The order of the 124 amino acids in ribonuclease and the folding of the polypeptide chain are shown here.

More amino acids may be added in the same fashion to form long chains. This is another example of condensation polymerization. *The bond between the carbonyl group of one amino acid and the nitrogen of the next amino acid in the peptide chain is called a* **peptide bond.** It is also called a peptide link.

As more amino acids are added, a backbone common to all peptide molecules is formed. *Any peptide with more than ten amino acid residues is called a* **polypeptide.** In theory, the process of adding amino acids to a peptide chain may be continued indefinitely. *A peptide with more than about 100 amino acids is called a* **protein.** On the average, a molecule of 100 amino acids has a molecular mass of 10 000 amu.

The order in which the amino acids of a peptide or protein molecule are linked is the amino acid sequence of that molecule. Differences in the chemical and physiological properties of peptides and proteins result from differences in the amino acid sequence. The number of ways in which 20 amino acids can be linked in a protein molecule is very large. For example 20^{100} amino acid sequences are possible for a protein of 100 amino acids containing all 20 amino acids.

The long peptide chains of proteins are folded into relatively stable shapes. Sections of chain may coil into a regular spiral known as a helix. Peptide chains may also be arranged side by side to form a wavy sheet. Much irregular folding of the chains also occurs. Protein shape is maintained by hydrogen bonds between adjacent parts of the folded chains. Covalent bonds also form between sulfur atoms in cysteine amino acids that are folded near each other. In this way separate polypeptide chains may be joined into a single protein.

26·C Enzymes

Enzymes are biological catalysts. They increase the rates of chemical reactions in living things. In 1926, the American chemist James B. Sumner reported the first isolation and crystallization of an enzyme. The enzyme he isolated was urease. It is able to hydrolyze urea, a constituent of urine, into ammonia and carbon dioxide. (The strong smell of dirty diapers allowed to stand for a time is the result of the action of bacteria that contain this enzyme.)

$$\underset{\text{Urea}}{NH_2\text{—}\overset{\overset{\textstyle O}{\|}}{C}\text{—}NH_2} + \underset{\text{Water}}{H_2O} \xrightarrow{\text{urease}} \underset{\text{Ammonia}}{2NH_3} + \underset{\substack{\text{Carbon}\\\text{dioxide}}}{CO_2}$$

Sumner demonstrated that urease is a protein. The idea that a protein could be a catalyst was disputed initially. Since then, hundreds of enzymes have been isolated and structurally characterized as proteins.

Besides being able to promote reactions, enzymes have two other properties of true catalysts. First they are unchanged by the reaction they catalyze. Second, enzymes do not change the normal position of chemical equilibrium. The same amount of product is eventually formed whether or not an enzyme is present. Few reactions in cells ever reach equilibrium, however. The products are rapidly converted to another substance in a further enzyme-catalyzed reaction. According to Le Châtelier's principle, the removal of a reaction product pulls the reaction toward completion.

Most of the chemical changes that occur in the cell are catalyzed by enzymes. *Substrates* are the molecules on which an enzyme acts. As in other reactions, substrates are transformed into products by bond-making and bond-breaking processes. In an enzymatic reaction, the substrate interacts with side chains of the amino acids on the enzyme. These interactions cause the making and breaking of bonds.

Figure 26·24
An enzyme speeds up a chemical reaction. Each enzyme has a distinctively shaped active site. Only molecules with complementary shapes can attach to the enzyme. Some enzymes catalyze decomposition reactions. Others catalyze combination reactions.

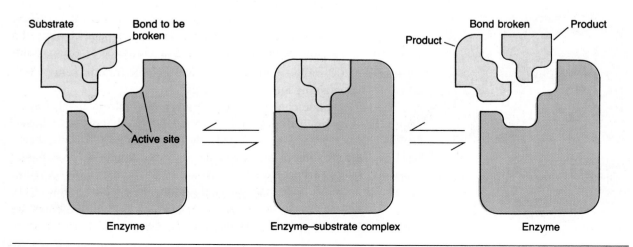

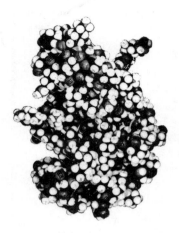

Figure 26·25
This is a space-filling model of the enzyme ribonuclease. The distinctive shape of the enzyme is important for its function as a biological catalyst.

To get some idea of the tremendous efficiency of enzymes, consider carbonic anhydrase. This enzyme catalyzes only one reversible reaction. It breaks down carbonic acid to water and carbon dioxide.

$$H_2CO_3 \overset{\text{carbonic}}{\underset{}{\overset{\text{anhydrase}}{\rightleftharpoons}}} CO_2 + H_2O$$

A single molecule of carbonic anhydrase can catalyze the breakdown of about 36 million molecules of carbonic acid in one minute!

A substrate molecule must contact an enzyme molecule before it can be transformed into product. Once the substrate has made contact, it must bind to the enzyme. The *active site* is the place on an enzyme where processes that convert substrates to products can take effect. The active site is usually a pocket or crevice formed by folds in the peptide chains of the protein. The active site of each enzyme has a distinctive shape. Only the substrate molecule for that enzyme will fit into it, much as only one key will fit into a certain lock. Thus each enzyme catalyzes only one chemical reaction with only one substrate. An *enzyme–substrate complex* is formed when an enzyme molecule and substrate molecule are joined.

Some enzymes can catalyze the transformation of biological substrates without assistance. Other enzymes need nonprotein coenzymes, also called cofactors, to assist the transformation. *Coenzymes are metal ions or small organic molecules that must be present for an enzyme-catalyzed reaction to occur.* The metal ions are usually magnesium, potassium, iron, or zinc.

26·16   Nucleic Acids

Nucleic acids *are polymers found primarily in cell nuclei.* They are indispensable components of every living thing. The kinds of nucleic acids found in cells are *d*eoxyribo*n*ucleic *a*cid (DNA) and *r*ibo*n*ucleic *a*cid (RNA). DNA stores the information needed to make proteins. It governs the reproduction and growth of cells and new organisms. RNA has a key role in the transmission of the information stored in DNA.

The monomers that make up DNA and RNA are called **nucleotides.** Nucleic acids are therefore polynucleotides. Each nucleotide consists of a phosphate group, a five-carbon sugar, and a nitrogen-containing compound called a nitrogen base.

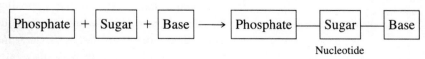

The sugar units in the nucleotides of DNA are the five-carbon monosaccharide deoxyribose. The base may be one of four compounds. Adenine and guanine are each composed of a double ring. Thymine and cytosine are each composed of a single ring. These bases are abbreviated as A, G, T, and C (Figure 26·27). Ribose, which has one more

Figure 26·26
James Watson and Francis Crick are shown here with their original model of the DNA molecule. The model was based on chemical analyses and X-ray diffraction studies.

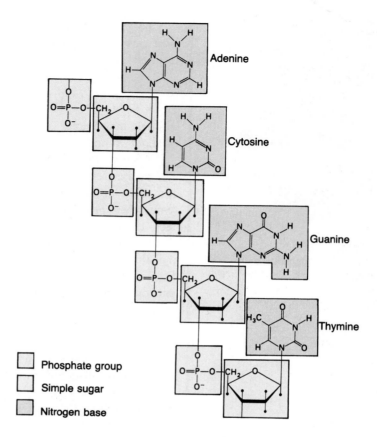

Figure 26·27
The nucleotide monomers of DNA are linked together through their phosphate groups.

Adenine

Cytosine

Guanine

Thymine

☐ Phosphate group
☐ Simple sugar
☐ Nitrogen base

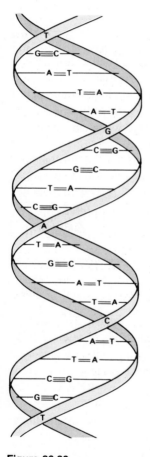

Figure 26·28
The DNA molecule is often referred to as a double helix. The nitrogen bases form hydrogen bonds to hold the two polynucleotide strands together.

▨ DNA and RNA molecules, which carry the blueprint of life, are condensation polymers of nucleotide monomers.

oxygen than deoxyribose, is the sugar in the nucleotide monomers of RNA. The base thymine is never found in RNA. Instead, it is replaced by a fifth base, uracil, abbreviated U.

Chemists studying nucleic acids have discovered that the amount of A in DNA is always equal to the amount T (A = T). Similarly, the amount of G is always equal to the amount of C (G = C). The significance of this fact was not apparent until 1953. At that time, James Watson and Francis Crick proposed that DNA molecules consist of two polynucleotide chains wrapped into a spiral shape (Figure 26·28). This is the famous double helix of DNA. In order for bases to fit neatly into the double helix, every double-ringed base must be matched with a single-ringed base on the opposing strand. The pairing of A with T and G with C provides the best possible fit. It also gives the maximum number of hydrogen bonds between the opposing bases. Thus, A—T and G—C makes the most stable arrangement in the double helix.

The molecular masses of DNA molecules reach into the millions and possibly billions. Even with only four bases, the number of possible sequences of nucleotides in a DNA chain is enormous. The order of the nitrogen bases A, T, G, and C in the DNA of an organism constitutes the genetic plan for that organism. Differences in the number and order of the bases in DNA ultimately are responsible for the diversity of living creatures.

26 Functional Groups and Organic Reactions
Chapter Review

Key Terms

addition reaction	26·6	halocarbon	26·2
alcohol	26·4	hydration reaction	26·6
aldehyde	26·8	hydrogenation	
alkyl halide	26·2	reaction	26·6
amino group	26·1	hydroxyl group	26·4
amino acid	26·16	ketone	26·8
aryl halide	26·2	lipid	26·14
Benedict's test	26·10	monomer	26·12
carbohydrate	26·13	monosaccharide	26·13
carbonyl group	26·8	nucleic acid	26·16
carboxyl group	26·9	nucleotide	26·16
carboxylic acid	26·9	peptide	26·15
dehydrogenation		peptide bond	26·15
reaction	26·10	polymer	26·12
denatured alcohol	26·5	polypeptide	26·15
disaccharide	26·13	polysaccharides	26·13
ester	26·11	protein	26·15
ether	26·7	saponification	26·14
fatty acid	26·9	substitution	26·3
Fehling's test	26·10	triglyceride	26·14
fermentation	26·5	wax	26·14
functional group	26·1	zwitterion	26·15

Common functional groups containing oxygen include ethers, aldehydes, ketones, carboxylic acids, and esters. The latter four types of functional groups contain the carbon–oxygen double bond or carbonyl group. Alcohols may be oxidized to aldehydes or ketones. Aldehydes may be oxidized to carboxylic acids. Fehling's and Benedict's solutions can be used to detect the presence of aldehydes. Esters result from the combination of carboxylic acids and alcohols. Hydrolysis of an ester produces these components. Condensation polymerization of monomer units containing hydroxy and carboxyl groups in the same molecule gives polyesters.

Carbohydrates, lipids, proteins, and nucleic acids are the great classes of biological molecules. The carbohydrates are often found in nature as simple molecules containing one or a few sugar units. Condensation polymers containing hundreds or thousands of sugar units, as in starch and glycogen, are also found. Lipids are the substances of fats and oils. The simplest lipids are triglycerides, esters of long chain fatty acids and glycerol. Soaps are the alkali metal salts of fatty acids. Peptides and proteins are condensation polymers of amino acids. Likewise, the nucleic acids are condensation polymers of monomer units called nucleotides. These giant molecules carry the information needed for reproduction and growth of living organisms.

Chapter Summary

There are few useful reactions of saturated hydrocarbons. The chemical reactions of most organic compounds involve functional groups. A common functional group is the carbon–carbon double bond of an alkene. It undergoes the addition of water to form an alcohol. It can add a hydrogen halide or halogen to form a halocarbon, and it can add hydrogen to form an alkane. Alkenes also undergo addition polymerization to form useful materials such as polyethylene. The benzene ring usually undergoes substitution rather than addition.

Practice Questions and Problems

12. Write a general structure for each of these types of compounds. *26·1*
a. halocarbon   c. ester
b. ketone   d. amide

13. Write a structural formula for each of the following compounds. *26·2*
a. 1,2,2-trichlorobutane
b. 1,3,5-tribromobenzene
c. 1,2-dichlorocyclohexane

14. Name the following halocarbons. *26·2*

   a. $CH_2{=}CHCH_2Cl$

   b. $CH_3CHCH_2CHCH_2Cl$ with CH_3 and Cl substituents

   c. 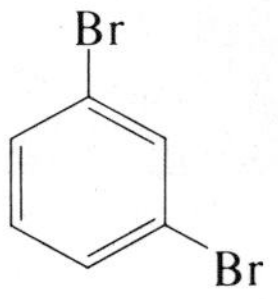 benzene ring with two Br substituents (meta)

15. Write structural formulas and give IUPAC names for all the isomers of these compounds. *26·2*
   a. $C_3H_6Cl_2$
   b. C_4H_9Br

16. What organic products are formed in the following reactions? *26·3*

   a. 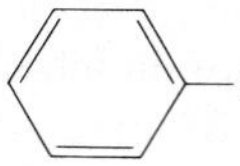 benzene ring—$ONa + CH_3Br \longrightarrow \underline{\quad} + NaBr$

   b. 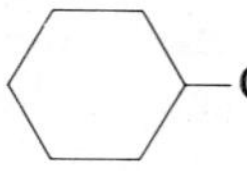 cyclohexane ring—$Cl + NaOH \longrightarrow \underline{\quad} + NaCl$

   c. $CH_3CHONa + CH_3CH_2Br \longrightarrow \underline{\quad} + NaBr$, with CH_3 substituent

   d. 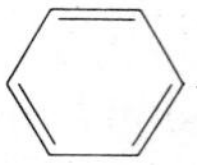 benzene ring $+ Br_2 \xrightarrow{\text{catalyst}} \underline{\quad} + HBr$

17. Write the name (IUPAC, common, or both) for each of the following alcohols. *26·4*

   a. $CH_3CH_2CHCH_3$ with OH substituent

   b. CH_3CHCH_2OH with CH_3 substituent

   c. $CH_3CHCH_2CH_2OH$ with CH_3 substituent

   d. $CH_3CH_2CCH_2CH_3$ with CH_3 and OH substituents

18. Identify each of the alcohols in Problem 17 as primary, secondary, or tertiary. *26·4*

19. Show how hydrogen bonds form between molecules of these pairs of compounds. *26·5*
   a. water–water
   b. water–methanol
   c. methanol–methanol

20. Write structures and names of the products obtained upon addition of each of the following reagents to ethene.   **a.** HBr   **b.** Cl_2 **c.** H_2O
   d. H_2 **e.** HCl   *26·6*

21. Name the following ethers. *26·7*

   a. $CH_3OCH_2CH_3$

   b. 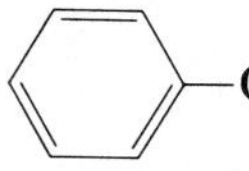 benzene ring—$O{-}CH_2CH_3$

   c. $CH_2{=}CHOCH{=}CH_2$

   d. $CH_3CHOCHCH_3$ with two CH_3 substituents

22. Explain why diethyl ether is more soluble in water than is dihexyl ether. Would you expect propane to be more soluble than diethyl ether in water? Why? *26·7*

23. Explain why 1-butanol has a higher boiling point than diethyl ether. Which compound would you expect to be more soluble in water? Why? *26·7*

24. What is a carbonyl group? Explain why a carbon–carbon double bond is nonpolar but a carbon–oxygen double bond is very polar. *26·8*

25. Name these aldehydes and ketones. *26·8*

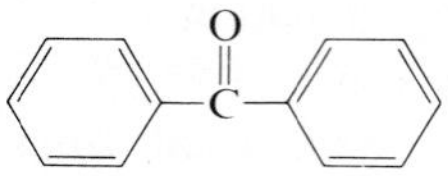

   a. $CH_3\overset{\text{O}}{\overset{\|}{C}}CH_3$

   b. CH_3CHCH_2CHO with CH_3 substituent

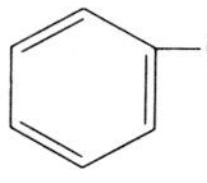

   c. benzene ring—CH_2CHO

   d. diphenyl ketone (benzene—$\overset{\text{O}}{\overset{\|}{C}}$—benzene)

   e. CH_3CHO

   f. $CH_3CH_2\overset{\text{O}}{\overset{\|}{C}}CH_2CH_2CH_3$

26. Propane $(CH_3CH_2CH_3)$ and acetaldehyde (CH_3CHO) have the same molecular mass, but propane boils at $-42°C$ and acetaldehyde boils at $20°C$. Account for this difference. *26·8*

27. What are the products of each of the following reactions? *26·9*
   a. $HCOOH + KOH \longrightarrow$
   b. $CH_3CH_2COOH + NaOH \longrightarrow$

28. Give common names for each of the following carboxylic acids. *26·9*
   a. $HCOOH$ **c.** CH_3CH_2COOH
   b. CH_3COOH **d.** $CH_3(CH_2)_{16}COOH$

29. Complete the reactions by writing the structure of the expected products. *26·10*
   a. $CH_3CH_2OH \xrightarrow{K_2Cr_2O_7}$

b. $CH_3CH_2CHO \xrightarrow{K_2Cr_2O_7}$

c. $CH_3CH_2\underset{\underset{CH_3}{|}}{C}HOH \xrightarrow{K_2Cr_2O_7}$

d. (benzene ring)$-CH_2CHO \xrightarrow{K_2Cr_2O_7}$

e. $CH_3CH_2\overset{\overset{O}{\parallel}}{C}CH_3 \xrightarrow{K_2Cr_2O_7}$

30. Write the name and structure for the alcohol that must be oxidized to make each of the following carbonyl compounds. *26·10*

a. HCHO

b. $CH_3\overset{\overset{O}{\parallel}}{C}HCH_3$

c. $CH_3\underset{\underset{CH_3}{|}}{C}HCHO$

d. (cyclohexanone structure with =O)

31. Complete the following reactions by writing the structures of the expected products and by naming each of the reactants and products. *26·11*

a. $CH_3COOCH_3 + H_2O \xrightarrow{HCl}$

b. $CH_3CH_2CH_2COOCH_2CH_2CH_3 + H_2O \xrightarrow{NaOH}$

c. $HCOOCH_2CH_3 + H_2O \xrightarrow{KOH}$

32. Write the structure and name of the ester that could be produced from each of the following reactions. *26·11*
  a. formic acid + methanol $\longrightarrow$
  b. butyric acid + ethanol $\longrightarrow$
  c. acetic acid + propanol $\longrightarrow$

33. What is the structure of the repeating units in a polymer in which the monomer is each compound?
  a. 1-butene **b.** 1,2-dichloroethene *26·12*

34. Give a source and use for each.
  a. starch
  b. cellulose
  c. glycogen *26·13*

35. A solution of glucose is treated with Benedict's solution. What should happen? Why? *26·13*

36. Name the two classes of organic compounds that are produced when waxes are hydrolyzed. *26·14*

37. Write the structure for glyceryl tristearate, a simple triglyceride. *26·14*

38. What is a soap? Give the name and formula of a typical soap molecule. *26·14*

39. The following compound is hydrolyzed by boiling with sodium hydroxide. What are the saponification products? *26·14*

$$CH_2-O-\overset{\overset{O}{\parallel}}{C}-(CH_2)_{14}CH_3$$
$$CH-O-\overset{\overset{O}{\parallel}}{C}-(CH_2)_{10}CH_3$$
$$CH_2-O-\overset{\overset{O}{\parallel}}{C}-(CH_2)_{16}CH_3$$

40. Write the structure and identify the R group for these amino acids. *26·15*
  a. alanine **b.** serine **c.** phenylalanine

41. Define the term *zwitterion*. Draw the amino acid glycine as a zwitterion. *26·15*

42. What is the name given to the amide bond connecting two amino acids in a peptide chain? *26·15*

43. Consider the tripeptide seryl-glycyl-phenylalanine. How many peptide bonds does this molecule have? *26·15*

44. What is meant by the amino acid sequence of a protein? *26·15*

45. Cells contain two types of nucleic acids. What are they called? *26·16*

46. What is the structural difference between the sugar unit in RNA and the sugar unit in DNA? *26·16*

47. What are the components of a nucleotide? *26·16*

48. Explain the difference between a nucleotide and a nucleic acid. *26·16*

49. What type of bonding holds a DNA double helix together? *26·16*

50. Which of the following base pairs are found in a DNA molecule? *26·16*
  a. A—A
  b. A—T
  c. C—G
  d. G—C
  e. G—A

Mastery Questions and Problems

51. Predict which of the following compounds has the highest boiling point. Formula masses are given in parentheses.
  a. CH_3CHO (44)
  b. CH_3CH_2OH (46)
  c. $CH_3CH_2CH_3$ (44)

52. Classify each of these compounds as an alcohol, a phenol, or an ether.

a. 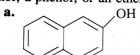   **d.**

b. 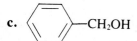   **e.** $CH_3CH_2\underset{\underset{CH_3}{|}}{C}HOH$

c.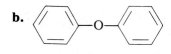

53. Write the structural formulas for the products from these reactions.

  a. $CH_3CH_2CH{=}CH_2 + Cl_2 \longrightarrow$

  b. $CH_3CH_2CH{=}CH_2 + Br_2 \longrightarrow$

  c. (cyclohexene) $+ HBr \longrightarrow$

54. For each compound pictured in a model below, identify the functional group and name the compound. The red atoms represent oxygen.

55. Write the structure and name of the expected products for each of the following reactions.

  a. $CH_3COOH + CH_3OH \xrightarrow{H^+}$

  b. $CH_3CH_2CH_2COOCH_2CH_3 + H_2O \xrightarrow{NaOH}$

  c. $CH_3CH_2OH \xrightarrow{K_2Cr_2O_7}$

  d. $CH_3CH_2CH_2COOH + NaOH \longrightarrow$

Critical Thinking Questions

56. Choose the term that best completes the second relationship.
  a. animals:carbon dioxide   yeast: _____
    (1) oxygen   (3) ethanol
    (2) methanol   (4) sugar
  b. polymer:monomer   protein: _____
    (1) monosaccharide   (3) nucleic acid
    (2) fatty acid   (4) amino acid
  c. cow:horse   aldehyde: _____
    (1) ketone   (3) ester
    (2) alcohol   (4) ether

57. Benzene is poisonous and a proven carcinogen. Yet many compounds containing benzene rings, such as benzaldehyde, are common in foods we eat. Why are some organic compounds with phenyl groups entirely safe to eat?

58. The outside layer of an animal cell, called the cell membrane, is composed mainly of a double layer of lipid molecules. What properties important to cells does such a membrane possess?

Review Questions and Problems

59. A solution is made by diluting 250 mL of $0.210M$ $Ca(NO_3)_2$ solution with water to a final volume of 450 mL. Calculate the molarity of $Ca(NO_3)_2$ in the diluted solution.

60. In a saturated solution containing undissolved solute, the solute is continually dissolving, but the solution concentration remains constant. Explain.

61. What is the maximum number of orbitals in the p sublevel of an atom?
  a. 1   **b.** 3   **c.** 5   **d.** 9

62. Using electron dot structures, illustrate the formation of F^- from a fluorine atom and a hydroxide ion from atoms of hydrogen and oxygen.

63. Calculate the mass, in grams, of one liter of SO_2 at standard temperature and pressure.

64. Name the following acids and bases.
a. HNO_3
b. H_2SO_3
c. HNO_2
d. $Ca(OH)_2$
e. $HClO_4$
f. H_3PO_4

65. Give the oxidation numbers of the following atoms.
a. Mn in $(MnO_4)^-$
b. Cr in Cr_2O_3
c. C in $(C_2O_4)^{2-}$
d. P in PH_3

Challenging Questions and Problems

66. Five continuous-chain organic compounds are listed in the table below. Although the molecular masses of these compounds are similar, their boiling points are not. Propose an explanation.

Structure of Compound	Molecular mass	Boiling point (°C)
$CH_3CH_2CH_2CH_2CH_3$	72	36
$CH_3CH_2-O-CH_2CH_3$	74	35
$CH_3CH_2CH_2Cl$	79	47
$CH_3CH_2CH_2CHO$	72	76
$CH_3CH_2CH_2CH_2OH$	74	118

67. Cholesterol is a compound that is in our diet and is also synthesized in the liver. Sometimes it is deposited on the inner walls of blood vessels, causing hardening of the arteries. Describe the structural features and functional groups of this important molecule.

68. Hydrocarbons from petroleum are an important source of raw material for the chemical industry. Using reactions covered in this chapter and any required inorganic chemicals, propose a scheme for the manufacture of ethylene glycol, a major component of antifreeze, from petrochemical ethene.

Research Projects

1. How were the substances isolated by Pierre Pelletier and Joseph Caventou important to the pharmaceutical industry?

2. Devise a method to isolate the ester components of common fruits.

3. The protein in egg white denatures when it is heated. After reading about protein denaturation, examine the effect of adding isopropyl alcohol (rubbing alcohol) and strong tea to different samples of egg white. How do the results compare to heating egg white? Explain.

4. Compare the saturations of various cooking oils.

5. Produce soaps from different fats and compare their properties.

6. Use paper chromatography to separate amino acids in a mixture. Identify each amino acid.

7. What are the structures for some common pesticides? How do these compounds operate? What problems are associated with their use?

Readings and References

Baltimore, David. "The Brain of a Cell." *Science 84* (November 1984), pp. 149–151.

Emsley, John, "Artificial Sweeteners." *ChemMatters* (February 1988), pp. 4–8.

Freidel, Robert. "The Plastics Man." *Science 84* (November 1984), pp. 49–51.

Twigg, John. *Looking at Plastics and Other Big Molecules*. North Pomfret, Vermont: B. T. Batsford, 1986.

MINERAL NUTRIENTS
Essential Elements of Life

What could you build using iron, chromium, cobalt, copper and zinc? How about a human being? Of course, there is much more to a human body than a collection of metals. But these metals, along with about 20 other elements, are essential for maintaining a healthy human body.

Oxygen, carbon, hydrogen and nitrogen are the most abundant elements in humans, as well as other organisms. Together, these four elements account for 99.4% of the atoms in the human body, and 96% of the body mass. About half of the mass of a human body is water; therefore, it is not surprising that oxygen and hydrogen are so abundant. Carbon is plentiful because it provides the framework for all organic molecules. Nitrogen is an important component of all proteins and nucleic acids.

The remaining 4% of human body mass consists of several minerals—each of which is present in tiny amounts. The mineral composition of all animals is very similar to the composition of sea water. Many scientists believe that this similarity exists because life on Earth originated in the sea.

Although the mineral nutrients are essential for human health, many of these minerals become toxic in larger doses. In fact, some of the trace elements, such as selenium and copper, were known for their toxic effects long before their nutritional value was recognized. Toxic consumption of trace elements can result from eating crops grown on soil containing high levels of these minerals. A less common source of toxic mineral consumption is a condition called pica. People with pica eat unusual things such as dirt, clay, ashes or laundry starch. This condition may be caused by a natural need for iron or other minerals.

Nutrition is a rapidly evolving area of scientific research. Much of what is known was unheard of just a few years ago. It is also very likely that notions that are widely believed today will be disproven in the near future. Because of the changeable nature of nutritional theories, the field is vulnerable to what the American Dietetic Association calls "nutritional quackery." Nutritional fraud has become a billion-dollar business. To avoid nutritional fraud, people should ignore fads, eat a balanced variety of foods, control the intake of fats, sugar, and salt, and eat plenty of fresh fruits and vegetables. Combined with an effective exercise program, sensible nutrition is a wise habit to develop.

Symbol	Essential Role	Toxicity Effects
Ca	Bones, teeth, muscle contractions, blood clotting, nerve impulse transmission	Poor absorption of other minerals
P	Bones, teeth, energy release, nucleic acids	Can lead to Ca deficiency
K	Cellular cation	Paralysis, abnormal heart rhythm
S	Component of some B vitamins and proteins	Unknown
Na	Extracellular cation, muscle and nerve functions	Swelling, high blood pressure
Cl	Extracellular anion	Disturbed acid-base balance
Mg	Bones, body temperature, nerve impulses	Poor nervous system function
I	Production of thyroid hormones	Unknown
Fe	Production of hemoglobin, red blood cell function	Fe buildup in heart, liver, and pancreas
Cr	Breakdown of glucose	Unknown
Co	Component of vitamin B_{12}	Goiter, excessive red blood cell production
Cu	Fe absorption, component of enzymes	Vomiting, diarrhea
F	Resistance to tooth decay	Spotted teeth, brittle bones
Mn	Enzyme production	Blurred speech, poor coordination
Mo	Component of enzymes	Gout, loss of Cu
Se	Antioxidant, liver function	Paralysis, blindness, hair loss
Zn	Nucleotide production	Nausea, abdominal pain, premature birth

DRUG TESTING
Truth or Consequences

The chemist watched the recording pen on the chart jump. "Cocaine," she murmured. The chart is the read-out of a gas chromatograph. The chemist is testing the urine of an athlete for drug abuse. The results will keep the athlete out of an upcoming event, and possibly off the team permanently.

Many employers feel it is beneficial to screen employees for drug abuse. They believe that drug abusers are more likely to miss work and cause accidents. In some professions, such as air traffic controllers, drug abusers can endanger the public. Drug testing is also a common practice in competitive sports, such as the Olympic Games. Although many opponents view drug testing as an invasion of privacy, the demand for an accurate, reliable drug test continues to grow.

A test to identify an abused substance in the body must be extremely accurate. A false-positive result could ruin a career. A false-negative result could endanger lives. The best method currently available to test for drug abuse is the **g**as **c**hromatography/**m**ass **s**pectrometer system, or GC/MS. Gas chromatography separates a chemical mixture into its individual components. The sample to be tested, such as urine or blood serum, is vaporized, and the resulting mixture of gas is forced through a separation column. The separation column is packed with a material that interacts physically with the gases. The time required for a particular gaseous component to pass through the column is called the retention time. Retention time can be used to identify compounds. By measuring the time needed for each component to exit the column, the composition of a mixture can be determined.

Alone, the GC method is not very reliable. Components of similar molecular structure may have similar retention times. To help identify the components, the gaseous mix can be swept from the GC into a mass spectrometer (MS) for further analysis. The MS splits gas molecules into ions of different masses and exposes them to a magnetic field. The degree of deflection of an ion in a magnetic field is related to its mass. The identity of a component can be determined by comparing patterns of ionic deflection to the MS "fingerprint" of a known substance.

Used together, the GC/MS method of drug testing is very

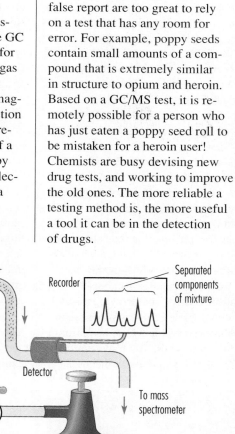

reliable–near 100%. But to some critics, almost 100% is not good enough. The consequences of a false report are too great to rely on a test that has any room for error. For example, poppy seeds contain small amounts of a compound that is extremely similar in structure to opium and heroin. Based on a GC/MS test, it is remotely possible for a person who has just eaten a poppy seed roll to be mistaken for a heroin user! Chemists are busy devising new drug tests, and working to improve the old ones. The more reliable a testing method is, the more useful a tool it can be in the detection of drugs.

The gas chromatograph (GC) is used to separate components of mixtures.

SEMICONDUCTORS
Big Jobs for Little Devices

The year is 1946. An eight-year-old named Agnes is sitting on the floor of the living room facing a large wooden box full of wires and glowing red tubes: the family radio. She is listening to her favorite show "Jack Armstrong: All American Boy." Dinner won't be served until her dad gets home from work at the University of Pennsylvania. There, a team of engineers have just finished building the **E**lectronic **N**umerical **I**ntegrator **A**nd **C**alculator (ENIAC). The ENIAC is a machine designed to perform high speed calculations for the government. The machine occupies 2000 square feet of floor space, almost half the size of a basketball court. Like Agnes's radio, the ENIAC is a mass of wires and vacuum tubes that processes electricity into complex circuits. Agnes has no idea that within 15 years she will be listening to rock-and-roll on a radio that fits in her pocket. Nor can she imagine that the calculations performed by the University's giant computer will someday be done on a device as small as a business card. The dizzying pace of technological progress would soon be made possible by the invention of the transistor. The transistor, a device that amplifies an electrical

signal, was the first practical use of semiconductor technology.

Semiconductors are elements that conduct electricity better than insulators, but less well than true conductors. Silicon (Si), germanium (Ge), and gallium (Ga) are semiconductors. The value of a semiconductor lies in the dramatic increase in conductivity that occurs when it is mixed with tiny amounts of other atoms. Adding other atoms, usually arsenic (As) or boron (B), to the semiconductor is called doping. By selectively doping various parts of a semiconductor, miniature electronic components can be constructed. Transistors are just one of the components that can be made from semiconductors. Components can be joined to create miniature integrated circuits. Circuits containing thousands of components can then be put on a semiconductor smaller than the fingernail on your pinky! The semiconductor is then sliced into individual "chips." A single integrated circuit chip can contain up to one million transistors.

Miniature electronic circuits have revolutionized the electronics and computer industries. Electronic equipment is now smaller and less

expensive than it could ever have been using vacuum tubes and conventional wires. Also, many items that are available today could not have been bought years ago for any price.

Agnes plans to retire in a few years. Today she keeps track of her investments and finances on a computer smaller than the radio she listened to as a child.

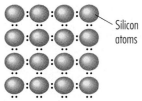

In a pure silicon crystal, each atom contains four electrons in the outer shell.

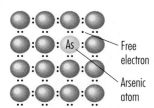

In a donor, or n-type semiconductor, some silicon atoms are replaced with arsenic, which contains five electrons in the outer shell. The extra electron remains unbound and mobile.

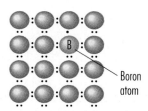

In an acceptor, or p-type semiconductor, some silicon is replaced with boron. Because boron has only three electrons in the outer shell, the acceptor has positive "holes." These holes are also mobile.

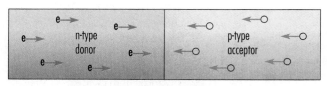

Semiconductor components are constructed from alternating layers of n-type and p-type semiconductor materials. When energized by an electric current, the free electrons move from donor to acceptor, while the positive holes migrate in the opposite direction.

SOLAR CELLS
Harnessing the Power of Apollo

Where would life on Earth be without energy from the sun? Without energy from the sun, plants could not make the food and oxygen that all living things depend on. People have always depended on the sun to provide warmth and light. Now, thanks to technology, energy from the sun can be used to power everything from calculators to cars. Harnessing the sun's energy is accomplished with solar, or photovoltaic, cells that convert solar energy into electricity.

Solar cells were first invented in the mid-1950s, but did not receive much attention. During the 1950s, fossil fuels, such as gasoline and oil, were cheap and abundant. Solar cells, on the other hand, were expensive, cumbersome and inefficient. Therefore, solar technology was largely ignored. During the energy crisis of the 1970s, interest in solar energy was renewed. By 1986, a gigantic solar

University City High School in California uses electricity produced by solar cells.

power station called the Solar Electric Generating System (SEGS) had been completed in the Mohave Desert of California. SEGS covers 400 acres and generates enough electricity to light almost 100,000 homes. In 1987, an American-built solar-powered car won an 1,867-mile race across Australia, at a record-breaking speed of 48 mph!

Using solar energy to generate

electricity has many compelling benefits. Solar energy is unlimited, pollution-free, silent, and odorless. It does not contribute to global warming, ozone destruction, or acid rain. Solar power does not have to be mined or imported, and it is most abundant in the summer when electricity demand is often at its peak. Unfortunately, solar energy also has a big problem, namely the inefficient conversion of sunlight into electricity. At an average of 25% efficiency, it takes a lot of solar cells to generate a little electricity. Recent advances in technology, however, have helped to overcome this problem. For example, researchers have found that the synthetic material gallium arsenide can sometimes be used instead of silicon. Because electrons flow through gallium arsenide more easily than through silicon, solar cells made from this new material are more efficient. With further improvements, the world can look towards solar power to help meet future energy needs.

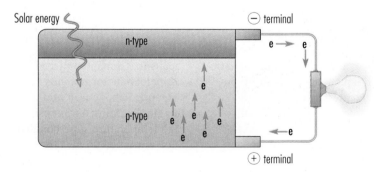

Solar cells are made by placing a thin layer of n-type on top of a layer of p-type semiconductor. When energized by the sun, free electrons flow through an external circuit, attracted by the positive holes of the p-type layer. As electrons vacate the n-type layer and accumulate in the p-type layer, the n-type layer develops a positive charge, and the p-type layer becomes more negative. The electrons are drawn upwards, attracted by the positive charge in the n-type layer. The circuit is now complete.

SUPERCONDUCTORS
A Theory in Search of a Substance

Trains that float in air? Miniature supercomputers? They don't exist yet, but floating trains and tiny high-speed computers may be some of the inventions made possible by superconductor technology. Superconductors are materials that conduct electricity without any resistance. When an electrical current flows through an ordinary conductor, such as a copper wire, it encounters resistance. The resistance slows the electrons, and causes the conductor to heat up. Toasters, space heaters, and electric blankets work by generating heat from electrical resistance. In electronic devices such as computers, getting rid of the heat is a big problem. The resistance also reduces

Train operating on magnetic tracks.

the speed at which calculations can be performed. Building electronic circuits out of superconductors would solve these problems.

An important feature of a superconductor is its critical temperature. Critical temperature, or T_c, is the temperature at which the superconducting material actually achieves zero resistance. Above T_c, the material loses its superconducting properties. Unfortunately, the T_c of all known superconductors is extremely low. The earliest superconductors all had a T_c near absolute zero (0 K). These materials had to be kept cold with liquid helium which boils at 4 K. The expense and inconvenience of the liquid helium limited the uses of these superconductors.

More recently, new superconductors have been developed with higher T_c's. For example, $Tl_2Ca_2Ba_2Cu_3O_{10}$ (thallium-calcium-barium-copper-oxide) has a T_c of 125 K. Instead of liquid helium, liquid nitrogen can be used to cool the new superconductors. Liquid nitrogen boils at 77 K, and is much less expensive and more readily available than liquid helium.

Although the newer superconductors represent a big advance in technology, their uses are still very limited. With a T_c of 125 K, they still are not practical for everyday use. The search is now on for a superconductor with a T_c near room temperature. A high temperature superconductor could dramatically increase the speed and efficiency of computers and other electronic devices. Unfortunately, the new higher

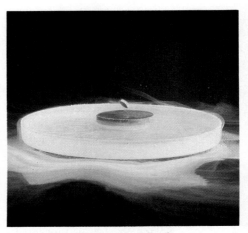

Levitation of a magnet above an icy pellet of superconductor.

T_c superconductors are too brittle to be useful. Brittleness prevents these new materials from being drawn into wires or shaped in circuit boards.

Besides conducting electricity without resistance, superconductors have another interesting property. They are diamagnetic. A diamagnet is a material that repels magnetic fields with a force exactly equal to the force of the field it is repelling. Diamagnetism results in the phenomenon called levitation. Levitation means that a magnet placed on a superconductor remains suspended above it in mid-air. Imagine building a train using high temperature superconductors. Such a train could be almost frictionless, suspended in air above magnetic tracks. Although high temperature superconductors have not been discovered yet, there will be many exciting inventions ready to make use of them when they arrive!

IRRADIATED FOODS
"Zapped" Today, Fresh Tomorrow

Irradiation! The word means many things to many people. Some people picture nuclear war. Others associate irradiation with cancer and mutations. Still others see irradiation as a way to help provide healthful, nutritious food to a hungry world. Can they all be right?

Irradiation dramatically increases the shelf life of food in two ways. First, the harmful effect of ionizing radiation destroys most food spoiling organisms. Mold, bacteria, insects, and even the deadly Trichinella worm that can infest pork products are controlled by irradiation. Secondly, irradiated vegetables such as potatoes and onions do not sprout during storage.

During irradiation, food is exposed to ionizing gamma radiation produced by cobalt-60 decay. The amount of the exposure, or dose, varies with the type of food and the desired effect. The dose can reach up to 1 million rads, or the equivalent of 20 million chest X-rays. The radiation penetrates the food, killing microbes and insects. The food itself does not retain the radiation and does not become radioactive.

The FDA requires this symbol to appear on irradiated retail foods.

Food-borne illnesses cost Americans billions of dollars each year in medical bills and lost wages. In developing countries, pests and spoilage can claim up to 30% of the harvest. Chemical preservatives and pesticides can reduce food losses, but these solutions may involve potential health risks as well. In an effort to reduce the cost of food spoilage, the United Nations, the World

Health Organization, and over 30 nations including the United States have approved the use of food irradiation. This procedure is currently being used in many nations, including Canada, Mexico, Japan, and South Africa.

Although irradiation was approved in the United States in the 1980s, the food industry has not made widespread use of the technique. Reluctance to use irradiated products is very strong among American consumers. Critics of food irradiation have many reasons for their disapproval. Irradiation alters the food somewhat, especially at high doses. Substances produced in the process, called radiolytic products, can give food a bad taste or smell. These products appear to be harmless, but they occur in such small amounts that they are difficult to study. Vitamins can also be destroyed, and the texture of protein can be affected. Because there is no reliable test to determine whether food has been irradiated, regulation is difficult. Opponents fear that illegal, repeated irradiation may be used as a substitute for proper food handling. Disposal of radioactive cobalt-60 could also create problems if it is not handled properly.

People initially mistrusted canning, freezing, and microwaving of food. Perhaps someday food irradiation will be as common as other processing techniques. Would seeing the irradiated food symbol on a package influence your shopping habits?

Food	Purpose
Herbs and spices	Retard microbe growth/kill insects
Potatoes	Retard sprouting and extend shelf-life
Wheat and flour	Kill insects
Fresh fruits	Control mold and delay maturation (overripening)
Pork	Control parasite that causes trichinosis
Poultry	Control illness-causing microorganisms

Some foods approved by the FDA for irradiation.

HAZARDOUS WASTES
Countdown to Cleanup

Nobody knows who owned the Paxton Avenue Lagoon south of Chicago. There are no records to show what was dumped there. But in 1985, an investigation of the lagoon showed the surrounding land to be heavily contaminated with toxic wastes. The contamination reached three meters below the surface of the soil. The toxic wastes included paints, solvents, cyanide, polychlorinated biphenyls, lead, mercury,

Sometimes the people responsible for the contamination undertake the cleanup. But in many cases, like the Paxton Avenue Lagoon, there is no one to hold responsible. In these situations, the government must step in and assume the burden of the cleanup. In 1980, Congress passed the **C**omprehensive **E**nvironmental **R**esponsibility, **C**ompensation and **L**iability **A**ct (CERCLA) to address the problem of hazardous waste cleanup. Also

states are cleaning up an additional 1000 sites.

Cleaning up a hazardous waste site is a complex project. One project may bring together experts from the fields of chemistry, biology, hydrogeology, engineering, medicine, toxicology, law, and politics. The experts study the site and decide upon the best cleanup approach to take. Traditional cleanup methods include incineration, chemical extraction, physical containment, and groundwater "pump and treat."

Bioremediation is a new method currently being developed to destroy toxic wastes. Bioremediation uses bacteria that naturally degrade hazardous materials. The cleanup site is sprayed with special fertilizers that encourage bacterial growth. The degradation process can be accelerated up to one million times the rate of normal degradation. Bioremediation has been shown to be effective in treating oil spills and city sewage. This method may also become a useful tool in cleaning toxic wastes. With a blend of old and new technology, plus plenty of time and money, contaminated land and water can be restored to their pristine state.

Paxton Avenue Lagoon before clean-up

and banned pesticides. Officials feared that the contamination would seep through the groundwater into nearby Lake Calumet, and then into Lake Michigan. The Illinois Environmental Protection Agency is currently working to clean up the lagoon. Eventually the cleaned land will be capped with clay and replanted—if funding remains available.

In the United States, there are thousands of hazardous waste sites in need of cleanup. Cleaning up waste sites is an extremely time-consuming and expensive process.

known as Superfund, this law authorizes the Environmental Protection Agency to clean up abandoned waste sites.

When a potential Superfund site is identified, it must first be investigated to see if it poses a threat to human health or to the environment. If so, the site is placed on a national priority cleanup list. By 1990, one decade after the passage of Superfund, approximately 1200 sites had been tabbed for the priority list. About 50 of the sites on the priority list had been cleaned up. Individual

TRUCKING CHEMICALS
An Order of Hazards to Go

A truck overturns on a highway, spilling a mysterious white powder onto the road. Within minutes, the road is closed. Dressed in protective gear, an emergency team arrives to begin the cleanup task. Rush hour traffic backs up for miles in both directions, while a sample of the powder is rushed to a laboratory for identification. Hours later, the powder is found to be laundry soap, and the road is reopened. The residents of the area were lucky this time. But without information about the contents of the truck, the delay was a necessary precaution. Next time, the chemical danger might be real.

Trucks carry approximately 900 million tons of hazardous materials along the highways of the United States each year. More than 99.99% of these hazardous cargos reach their destinations safely. Nevertheless, the few accidents that do result in chemical spills present a serious threat. In a recent survey of environmental policies, state legislators placed transportation of hazardous wastes at the top of their list of concerns.

In the United States, the Federal Department of Transportation encourages the use of the United Nations classification system. In this system, shipped materials are assigned an internationally recognized name and number. Placards displaying the identification number and any potential hazard are placed on the sides and ends of tank trucks. In the event of an accident or spill, an

emergency response crew can call the **Chem**ical **Tr**ansportation **E**mergency **C**enter (CHEMTREC). Only a phone call away, the people at CHEMTREC can use the number to identify the material and inform the crew of proper emergency procedures. The resources of CHEMTREC can provide information about health hazards, risk of fire or explosion, and cleanup procedures. With this information readily available, spills can be handled swiftly and efficiently, minimizing the risks to people and to the environment.

Of course, the identification of hazardous materials cannot prevent accidents. But the United Nations classification system can help to reduce the consequences of the few accidents involving chemical spills that do occur.

Material Name	U.N. Number	Hazard
Air, compressed	1002	Cylinder may explode in fire
Ammonia, anhydrous	1005	Poison; irritant
Liquefied petroleum gas	1075	Flammable; causes frostbite
Vinyl chloride	1086	Flammable; inhalation hazard
Acetone	1090	Flammable
Gasoline, petrol	1203	Flammable
Sodium, metal	1428	Flammable; water reactive
Hydrochloric acid	1789	Corrosive; inhalation hazard
Sodium hydroxide, lye	1823	Corrosive; burns skin/eyes
Sulfuric acid	1830	Corrosive; burns skin/eyes
Dry ice, solid CO_2	1845	Contact causes frostbite
Hydrogen, liquefied	1966	Flammable; causes frostbite
Propane, liquefied	1978	Flammable; causes frostbite
Sulfur, molten	2448	Combustible; emits poison gas

United Nations identification numbers for some common hazardous materials.

CERAMICS
From Bathtubs to Boom Boxes

Bathtubs, spacecraft, ancient vases, mosaic tiles, and stereo equipment! What do all these things have in common? The answer is ceramics! From the ancient vases of prehistoric people to the nose cone of the Space Shuttle, the history of human civilization can be traced through the art and science of ceramics.

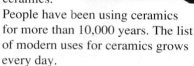

People have been using ceramics for more than 10,000 years. The list of modern uses for ceramics grows every day.

What are the unique properties of ceramics that make them so valuable and versatile? Ceramics have very high melting temperatures and are good thermal insulators. For example, the ceramic tiles on the Space Shuttle protect it from the intense heat of re-entering the earth's atmosphere.

Ceramics also are good electrical insulators, making them valuable in the electronics industry. Ceramics are very hard and able to withstand compression. These properties have made ceramic tile a popular material for the construction of floors and walkways. Ceramics are chemically unreactive and resistant to corrosion. Therefore, many ceramic vases and mosaics unearthed by archeologists are as beautiful today as they were thousands of years ago.

The properties of a ceramic result from its complex, crystalline structure. Although they may be different in appearance and the way they are used, ceramics share some common components. All ceramics are made of inorganic powders, such as sand or dry clay. At some point, the powders are subjected to high temperature treatments. Some ceramics, such as porcelain and brick, are molded and dried before heating. A new type of ceramic called glass ceramic is made from glass that is cooled slowly to produce a crystalline structure. These glass ceramics are very shatter-resistant.

Ceramic engineers can control the structure, and therefore the properties, of a ceramic by varying the powder components and heat treatment used to manufacture it. Most ceramics contain oxides, such as silicon dioxide (SiO_2). Oxides cannot react with oxygen, and therefore resist corrosion. Silicon nitride (Si_3N_4) is a ceramic that does not contain

Ceramics in the nose cone and tiles of the Space Shuttle protect it from the heat of re-entry

oxides, yet it is still resistant to corrosion. In the future, Si_3N_4 may be used to build automobile engines. Ceramic engines could run hotter and more efficiently than the metal engines in use today. Without a doubt, ceramic engineers will find many more innovative applications for ceramics in the world of the future.

GLASS
Viewing Stars Through a Grain of Sand

Imagine inventing a material with the molecular structure of a liquid, but the hardness of a solid. Imagine further that it can be made into almost any size or shape. This material could take on any color you chose, or be completely colorless. What if it were also a good thermal and electrical insulator, and waterproof too? Impossible, right? Not at all! People have been using this material for over 4000 years. It is called glass.

Glass is made from molten materials that have been cooled without crystallizing. Solids are typically made up of molecules arranged in a crystalline structure. In glass, the molecules remain disordered, like a liquid, even after cooling. Silica (SiO_2), the main component of sand, is one of the few substances that can

The telescope at the Mt. Hamilton Observatory in Northern California

be cooled without crystallizing. Most glass contains SiO_2 as its main ingredient.

Many other materials can be added to the silica base to change the properties of the glass. For example, the color of glass can be changed by adding small amounts of transition metal oxides. Adding iron oxide (Fe_2O_3) gives glass the green tint used in some beverage bottles and tinted windows. Blue glass contains CuO or CoO, orange glass contains Cr_2O_3, and red glass contains colloidal gold. Adding lead oxide (PbO) makes glass

sparkle and easier to polish. Leaded glass is popular for making chandeliers and cut glass vases. Eyeglass lenses sometimes contain silver chloride ($AgCl$). In bright light, silver chloride breaks down into chlorine atoms and silver atoms. The opaque silver atoms darken the glass and reduce the amount of light that can penetrate the lens. In the dark, the atoms reassociate and the glass becomes clear again.

Optical glass is the clearest kind of glass. It is used for microscopes, telescopes, and other instruments that require maximum light transmission. Glass can also be drawn into long, thin, flexible rods called optical fibers. Optical fibers can transmit light over long distances and even around corners. Like tiny periscopes, optical fibers can be used to view tissues deep within the human body. From examining a fetus in a mother's womb to charting the distant stars of the galaxy, glass has greatly expanded the limits of human vision.

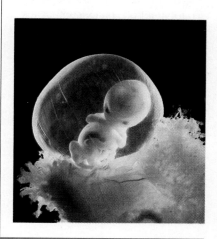

Composites
Greater Than the Sum of Their Parts

Have you noticed that cars have been "losing weight" lately? No, they have not been dieting or getting thinner. However, thanks to new uses for a family of materials called composites, cars have been getting lighter in weight. The reason for the "weight loss" is that composites are less dense than the metals traditionally used to manufacture cars.

Composites are made by imbedding some kind of fibers in a plastic base. The fibers may be made of glass, graphite, or a synthetic fabric such as polyester. The strength of the fibers, combined with

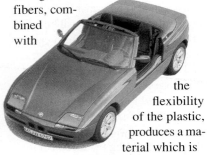

the flexibility of the plastic, produces a material which is remarkably strong, yet low in density. In addition, composites are noncorrosive and they absorb vibrations. To top it all off, many composites are inexpensive to manufacture. Some can even be made out of recycled plastic beverage containers. All these properties combine to make composites unique and valuable materials.

Think of all the advantages of manufacturing cars out of composites instead of steel or other metals. Cars would be lighter weight, so they would use less fuel. They would also be stronger, providing better protection for the passengers.

Because composites absorb vibrations, the cars would be quieter and smoother riding than steel-framed models. In addition, cars made of composites would not rust or corrode. Many sports cars are now made almost entirely of composites. Most car manufacturers have been replacing certain steel parts with composites in their newer models.

The unique properties of composites make them valuable for a wide variety of other applications, too. Because they are strong, but low in density, composites are now being used to build satellites,

military aircraft, and even personal airplanes. Their shock-absorbing properties make composites ideal for manufacturing sports equipment such as tennis racquets, baseball bats, and golf clubs. Thanks to composites, artificial limbs can be made that are less dense and stronger than ever before.

Surprisingly, composites are not really a new idea. Fiberglass, one of the first composites to be manufactured, has been around for many years. Invented during World War I as a substitute for asbestos, fiberglass is commonly used to build boat hulls that do not rot or splinter. Material scientists are finding new uses for composites every day. One of the newest kinds of composites is composite "lumber." Beams made of this material look like wood, can be cut and nailed like wood, but are stronger and safe from termites. Composite "lumber" can even be made from recycled plastics— saving money, trees, and the environment, too!

Composites give the stealth fighter an extremely long range.

PLASTICS RECYCLING
"Gold Mine" in a Garbage Dump

You walk across a carpet made from old plastic bottles. You put on a jacket made from old plastic bottles. You stroll to the park, where you sit on a bench made from...more old plastic bottles? What's going on? The answer is plastics recycling.

Before...

Plastics are among the most important materials to recycle. Made from polymers of organic molecules, some can be readily melted and reprocessed. Yet only about 1% of the plastics used in the U.S. are recycled. The rest end up in the garbage, making up about 18% by volume of the material in landfills.

Plastic recycling is important for three reasons. One reason is that plastics are made from crude oil, a non-renewable resource that is too precious to waste. Another reason to recycle plastics is that incineration of plastics can be dangerous. Some release toxic gases, such as hydrogen cyanide (HCN) or hydrogen chloride (HCl), when they are

burned. Finally, plastics do not degrade. In fact, resistance to the degrading effects of sunlight, moisture, and microorganisms is the property that makes plastic such a valuable packaging material. Unfortunately, their resistance to degradation means that plastics can remain unchanged in dumps and landfills for decades.

Plastics must be sorted before they can be recycled. There are many types of plastics, and not all are compatible with one another. Mixed plastics produce a low grade recycled product sold as "plastic lumber." "Plastic lumber" can be used as a wood substitute for decks and park benches. Careful sorting yields a higher quality recycled product useful for carpet yarn, toys, and fiberfill insulation in clothing. To assist sorting, the plastics industry has devised a number code to identify certain commonly used plastics. The code usually appears inside a triangle on the bottom of the item. Number 1 refers to polyethylene terephthalate, which is used for soft-drink bottles. Number 2 corresponds to high-density polyethylene, which is used for milk jugs and shampoo bottles.

Besides recycling, the other possible solution to the problem of plastic garbage is to produce plastics that degrade naturally. Researchers have developed plastics which contain up to 60% starch. Microbes consume the starch, leaving behind a plastic

polymer powder. Because the polymer itself remains, these plastics are not truly biodegradable. Plastics containing starch are commonly used to make supermarket sacks. Although the development of truly degradable plastics may offer a solution to the plastics problem, recycling remains a sound, sustainable method of garbage and resource management.

Now, will that be paper or plastic for you today?

...After

GREENHOUSE EFFECT
Too Much of a Good Thing

Have you ever noticed how warm it gets inside a car that has been parked in the sun for a while? Like a greenhouse, the interior of a car stays warm even on a cold day. In a similar process, the atmosphere of Earth keeps the planet warm. Because the atmosphere warms Earth in a way that is similar to the glass walls of a greenhouse, the warming process is called the greenhouse effect. Mars, frozen and barren, has no such effect. Venus swelters under too much of it. A delicate balance of gases in the atmosphere keeps the greenhouse effect on Earth just right. Too much, and all the water on Earth could boil away. Too little, and the water would freeze. Without liquid water, there can be no life as we know it. Could human activities upset this delicate balance? Many scientists say "yes."

Earth's atmosphere surrounds the planet like a blanket. Within the atmosphere are several trace gases known as greenhouse gases. Solar energy passes through these gases easily, reaching the earth's surface. Some of this energy is reflected off the surface as infrared heat waves. These heat waves do not pass through the greenhouse gases easily, but are instead absorbed and radiated back to Earth. By trapping the infrared heat waves, greenhouse gases keep the earth's surface about 33°C warmer than it would be otherwise.

Carbon dioxide (CO_2) is the most prevalent greenhouse gas.

Normally, interactions between the atmosphere, the oceans, and organisms keep the atmospheric CO_2 in check. Living plants remove CO_2 from the air through photosynthesis. When plants are burned, however, carbon is released into the air in the form of CO_2. By burning forests and fossil fuels, experts estimate that more than 8 billion metric tons of CO_2 are being pumped into the atmosphere every year! At the same time, an important CO_2 removal system is being destroyed as forests are cleared for agriculture.

Carbon dioxide is not the only culprit. Other gases, such as methane (CH_4), nitrous oxide (N_2O), and chlorofluorocarbons (CFCs), exist in much smaller quantities but produce a more powerful greenhouse effect than CO_2. The atmospheric content of these gases is also rising as a result of human activities.

According to current scientific models, a rise of only a few degrees in the earth's temperature could cause global problems. In

the worst possible scenario, weather patterns would change, ice caps would melt, and sea levels would rise. Farmlands would become deserts, and coastal cities would be under water. Solving the greenhouse problem will require the cooperation of nations throughout the world. Non-polluting energy sources must be developed. Destruction of the world's forests must end. Most important of all, people must realize that the atmosphere is intricately tied to the delicate, complex web of life on Earth.

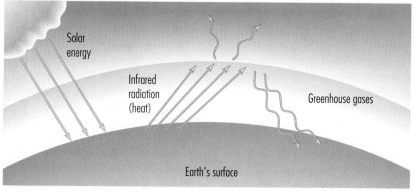

Greenhouse gases trap infrared radiation.

OZONE LAYER
Oxygen by Any Other Name

Afrightening scene appears on a movie screen as the curtain rises. Starving animals stumble about blindly. Sickly humans are covered with blisters and tumors. Sea creatures wash up on the shore as their food supply dwindles. Nightmarish science fiction, indeed, but could it really happen? Hopefully not, but this scene is a possibility some scientists foresee if the ozone layer continues to be destroyed at the present rate. Ozone (O_3) is a form of oxygen produced when a single oxygen atom joins a molecule of O_2. The ozone diffused in the stratosphere is known as the ozone layer. The ozone layer is important because it screens out 99% of the sun's ultraviolet (UV) radiation. UV radiation is extremely harmful to all living things. It can cause mutations, cancer, blindness, and crop damage.

During the 1970s, some scientists began to suspect that the ozone layer might be in serious danger, based on laboratory models. In 1985, their suspicions were confirmed when British researchers discovered that an ozone "hole" was forming over Antarctica every spring. The amount of ozone in the area of the hole had been reduced by as much as 50%. A similar thin spot has also been identified over the Arctic. The destruction of the ozone layer is being caused by a category of molecules called chlorofluorocarbons (CFCs). Once used primarily as propellants in aerosol spray cans, CFCs were banned from that use in the United States. They are still used as coolants in refrigerators and air conditioners, cleaners and solvents for computer chips, and in the manufacture of insulated foam containers such as coffee cups and "food-to-go" cartons. CFCs are highly stable and inert in the lower atmosphere, but they eventually drift up into the stratosphere. Once in the stratosphere, CFCs can be broken down by solar radiation, releasing chlorine atoms. These chlorine atoms drift through the stratosphere like atmospheric bandits, destroying ozone molecules in their path. A single chlorine atom can destroy as many as 100 000 molecules of ozone. Ice crystals in the clouds enhance the ozone-destroying reaction. Therefore, ozone depletion is most evident at the poles during the coldest time of the year.

The good news is that ozone depletion is a reversible process. If CFC production were reduced by 85%, ozone depletion could be reversed in the next century. At a historic conference held in Montreal, Canada in 1987, more than 40 nations signed an agreement to reduce global production of CFCs. The target was 50% reduction by 1998. In an 1990 revision, the goal was changed to 100% elimination by the year 2000. With the invention of safe substitutes for CFCs, and the cooperation of the global community, a world with an ozone-depleted stratosphere can remain the stuff of science fiction movies.

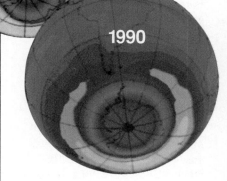

The ozone "hole" over Antarctica. Yellow areas have abundant ozone. Blue and purple areas indicate "holes."

BLOOD SUBSTITUTES
A Bloodless Coup

Centuries before the age of modern medicine, blood was thought to be a magical source of wisdom and immortality. Although everyone now knows that it is not magical, blood is vital to human life for many reasons. Blood transports nutrients, hormones and metabolic wastes throughout the body. Blood also helps regulate body temperature. Perhaps the most important function of blood, however, is to deliver oxygen (O_2) from the lungs to every cell in the body, and carry carbon dioxide (CO_2) from the cells to the lungs. Without the O_2 tranported by the blood, the brain and other organs would soon die.

Blood lost due to severe injuries, surgery, or battle wounds must be replaced immediately. Lost blood has traditionally been replaced by tranfusions. Blood transfusions have saved millions of lives, but can involve complications. Determining the recipient's blood type is time consuming in an emergency. Furthermore, blood transfusions can transmit diseases such as AIDS and hepatitis. Researchers are currently testing a blood substitute capable of transporting O_2 and CO_2, but free of the complications of real blood transfusions.

Artificial blood is made with chemicals called perfluorocarbons (PFCs). PFCs are organic compounds in which all the hydrogen has been replaced by fluorine. These compounds have the unique ability to dissolve and transport large quantities of O_2 and CO_2. Artificial blood

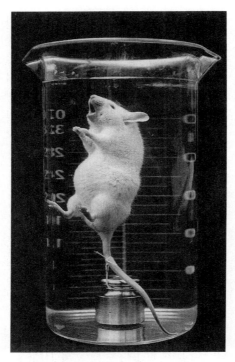

Mouse submerged in oxygen-saturated perfluorocarbon survives because it breathes oxygen from liquid that fills its lungs. The mouse was retrieved unharmed!

contains 20% PFCs by weight, and a blend of salts and emulsifiers that mimic the composition of real blood.

Perfluorocarbons can sometimes be better than real blood. Because they transport O_2 without cells, PFCs can deliver O_2 to areas where blood flow has been restricted. This property may be valuable in the treatment of tumors, sickle-cell anemia, and injuries in which blood vessels have been narrowed. PFC-based blood substitutes can also be stored frozen for years. Real blood must be refrigerated and used within weeks.

Doctors in Japan have been using PFC-based blood substitutes during surgery for more than a decade. Although artificial blood has already saved hundreds of lives in other countries, it is still being tested in the United States. One concern about artificial blood is that PFCs leave the tissues of the body very slowly. It is not yet known what the long-term effect of these chemicals in the body may be. Blood substitutes made from starch suspensions or hemoglobin extracts are also being tested. These alternatives, however, do not transport O_2 or CO_2 as well as the perfluorocarbons.

Because blood is so complex, scientists will probably never find a substitute to perform all its functions. Enzymes, hormones and clotting factors are important blood components that cannot yet be duplicated. Therefore, there will always be a need for donations of human blood.

Perfluorodecalin is the most abundant perfluorocarbon in artificial blood.

BIOMEDICAL IMPLANTS
Spare Parts for the Human Machine

Artificial body parts are nothing new. Amputees were fitted with artificial limbs long before Robert Louis Stevenson wrote about the legendary pirate Long John Silver in Treasure Island. Bioengineering research continues to improve upon the quality of artificial limbs. But recent research has focused on items that can be surgically installed within the body. These items, called biomedical implants, are made possible by the invention and application of new synthetic materials. Hip and knee joints, blood vessels, and skin are just a few of the biomedical implants in use today.

The development of materials for biomedical implants poses a complex chemical problem. Because implanted materials should last the lifetime of the recipient, they must not corrode, wear out, or fail under mechanical stress. Implants must also be biocompatible. A biocompatible material will not be rejected by the immune system of the recipient. Lack of biocompatibility has always been a major problem in living organ transplants. In addition, the implanted material must not promote blood clotting.

Some of the first surgically implanted materials were ferrous alloys, or stainless steels. Made of iron, chromium, and nickel, they were used to make plates and screws for joining broken bones. More recently, titanium- and cobalt- based alloys have been used instead. These new alloys are less dense, yet stronger than steel.

Polymer plastics have many uses in biomedical implants. For example, tubes woven from polyester fibers can be used to replace damaged blood vessels. Teflon®, a polymer often used to coat non-stick cookware, can also be used for implants. This particular polymer provides a smooth, gliding surface for replacement joints such as hips and knees. Nylon and polyurethane are two polymer plastics that can be molded and used for many types of implants, such as those used in "plastic" cosmetic surgery.

Recently, biomedical implant research has turned toward the development of bioactive materials. Bioactive substances are so much like the natural materials they replace that they become incorporated into the living tissue. For example, hydroxyapatite is a calcium phosphate material very much like human teeth and bones. When heated, hydroxyapatite powder becomes a porous ceramic. When artificial joints are made of this ceramic, new bone can permeate. The end result is an implant with the strength and resilience of natural bone.

Researchers are now developing bioactive artificial proteins. These proteins have great potential for use in artificial skin and blood vessels. The list of replaceable body parts is getting longer all the time. Are you ready for the age of the cyborg?

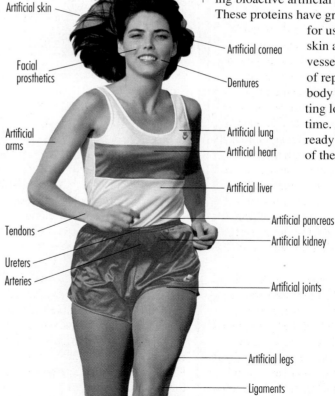

Artificial skin
Facial prosthetics
Artificial arms
Tendons
Ureters
Arteries
Artificial cornea
Dentures
Artificial lung
Artificial heart
Artificial liver
Artificial pancreas
Artificial kidney
Artificial joints
Artificial legs
Ligaments

Appendix A   Reference Tables

Table A·1   Some Properties of the Elements

Element	Symbol	Atomic number	Atomic mass	Melting point (°C)	Boiling point (°C)	Density (g/cm³) (gases at STP)	Major oxidation states
Actinium	Ac	89	(227.0482)	1050	3200	10.07	+3
Aluminum	Al	13	26.98154	660.37	2467	2.6989	+3
Americium	Am	95	243	994	2607	13.67	+3, +4, +5, +6
Antimony	Sb	51	121.75	630.74	1950	6.691	−3, +3, +5
Argon	Ar	18	39.948	−189.2	−185.7	0.0017837	
Arsenic	As	33	74.9216	817	613	5.73	−3, +3, +5
Astatine	At	85	(210)	302	337	—	
Barium	Ba	56	137.33	725	1640	3.5	+2
Berkelium	Bk	97	(247)	986	—	14.78	
Beryllium	Be	4	9.01218	1278	2970	1.848	+2
Bismuth	Bi	83	208.9804	271.3	1560	9.747	+3, +5
Boron	B	5	10.81	2079	3675	2.34	+3
Bromine	Br	35	79.904	−7.2	58.78	3.12	−1, +1, +5
Cadmium	Cd	48	112.41	320.9	765	8.65	+2
Calcium	Ca	20	40.08	839	1484	1.55	+2
Californium	Cf	98	(251)	900	—	14	
Carbon	C	6	12.011	3550	4827	2.267	−4, +2, +4
Cerium	Ce	58	140.12	799	3426	6.657	+3, +4
Cesium	Cs	55	132.9054	28.40	669.3	1.873	+1
Chlorine	Cl	17	35.453	−100.98	−34.6	0.003214	−1, +1, +5, +7
Chromium	Cr	24	51.996	1857	2672	7.18	+2, +3, +6
Cobalt	Co	27	58.9332	1495	2870	8.9	+2, +3
Copper	Cu	29	63.546	1083.4	2567	8.96	+1, +2
Curium	Cm	96	(247)	1340	—	13.51	+3
Dysprosium	Dy	66	162.50	1412	2562	8.550	+3
Einsteinium	Es	99	(252)	—	—	—	
Erbium	Er	68	167.26	159	2863	9.066	+3
Europium	Eu	63	151.96	822	1597	5.243	+2, +3
Fermium	Fm	100	(257)	—	—	—	
Fluorine	F	9	18.998403	−219.62	−188.54	0.001696	−1
Francium	Fr	87	(223)	27	677	—	+1
Gadolinium	Gd	64	157.25	1313	3266	7.9004	+3
Gallium	Ga	31	69.72	29.78	2403	5.904	+3
Germanium	Ge	32	72.59	937.4	2830	5.323	+2, +4
Gold	Au	79	196.9665	1064.43	3080	19.3	+1, +3
Hafnium	Hf	72	178.49	2227	4602	13.31	+4
Helium	He	2	4.00260	−272.2	−268.934	0.001785	
Holmium	Ho	67	164.9304	1474	2695	8.795	+3
Hydrogen	H	1	1.00794	−259.14	−252.87	0.00008988	+1
Indium	In	49	114.82	156.61	2080	7.31	+1, +3
Iodine	I	53	126.9045	113.5	184.35	4.93	−1, +1, +5, +7
Iridium	Ir	77	192.22	2410	4130	22.42	+3, +4
Iron	Fe	26	55.847	1535	2750	7.874	+2, +3
Krypton	Kr	36	83.80	−156.6	−152.30	0.003733	
Lanthanum	La	57	138.9055	921	3457	6.145	+3
Lawrencium	Lr	103	(260)	—	—	—	+3
Lead	Pb	82	207.2	327.502	1740	11.35	+2, +4
Lithium	Li	3	6.941	180.54	1342	0.534	+1
Lutetium	Lu	71	174.967	1663	3395	9.840	+3
Magnesium	Mg	12	24.305	648.8	1090	1.738	+2
Manganese	Mn	25	54.9380	1244	1962	7.32	+2, +3, +4, +7
Mendelevium	Md	101	257	—	—	—	+2, +3
Mercury	Hg	80	200.59	−38.842	356.58	13.546	+1, +2
Molybdenum	Mo	42	95.94	2617	4612	10.22	+6

Table A·1 Some Properties of the Elements (cont.)

Element	Symbol	Atomic number	Atomic mass	Melting point (°C)	Boiling point (°C)	Density (g/cm³) (gases at STP)	Major oxidation states
Neodymium	Nd	60	144.24	1021	3068	6.90	+3
Neon	Ne	10	20.179	−248.67	−246.048	.0008999	
Neptunium	Np	93	237.0482	640	3902	20.25	+3, +4, +5, +6
Nickel	Ni	28	58.69	1453	2732	8.902	+2, +3
Niobium	Nb	41	92.9064	2468	4742	8.57	+3, +5
Nitrogen	N	7	14.0067	−209.86	−195.8	.0012506	−3, +3, +5
Nobelium	No	102	(259)	—	—	—	+2, +3
Osmium	Os	76	190.2	3045	5027	22.57	+3, +4
Oxygen	O	8	15.9994	−218.4	−182.962	.001429	−2
Palladium	Pd	46	106.42	1554	2970	12.02	+2, +4
Phosphorus	P	15	30.97376	44.1	280	1.82	−3, +3, +5
Platinum	Pt	78	195.08	1772	3827	21.45	+2, +4
Plutonium	Pu	94	(244)	641	3232	19.84	+3, +4, +5, +6
Polonium	Po	84	(209)	254	962	9.32	+2, +4
Potassium	K	19	39.0982	63.25	760	.862	+1
Praseodymium	Pr	59	140.9077	931	3512	6.64	+3
Promethium	Pm	61	(145)	1168	2460	7.22	+3
Protactinium	Pa	91	231.0359	1560	4027	15.37	+4, +5
Radium	Ra	88	226.0254	700	1140	5.5	+2
Radon	Rn	86	(222)	−71	−61.8	.00973	
Rhenium	Re	75	186.207	3180	5627	21.02	+4, +6, +7
Rhodium	Rh	45	102.9055	1966	3727	12.41	+3
Rubidium	Rb	37	85.4678	38.89	686	1.532	+1
Ruthenium	Ru	44	101.07	2310	3900	12.41	+3
Samarium	Sm	62	150.36	1077	1791	7.520	+2, +3
Scandium	Sc	21	44.9559	1541	2831	2.989	+3
Selenium	Se	34	78.96	217	684.9	4.79	−2, +4, +6
Silicon	Si	14	28.0855	1410	2355	2.33	−4, +2, +4
Silver	Ag	47	107.8682	961.93	2212	10.50	+1
Sodium	Na	11	22.98977	97.81	882.9	0.971	+1
Strontium	Sr	38	87.62	769	1384	2.54	+2
Sulfur	S	16	32.06	112.8	444.7	2.07	−2, +4, +6
Tantalum	Ta	73	180.9479	2996	5425	16.654	+5
Technetium	Tc	43	(98)	2172	4877	11.50	+4, +6, +7
Tellurium	Te	52	127.60	449.5	989.8	6.24	−2, +4, +6
Terbium	Tb	65	158.9254	1356	3123	8.229	+3
Thallium	Tl	81	204.383	303.5	1457	11.85	+1, +3
Thorium	Th	90	232.0381	1750	4790	11.72	+4
Thulium	Tm	69	168.9342	1545	1947	9.321	+3
Tin	Sn	50	118.69	231.968	2270	7.31	+2, +4
Titanium	Ti	22	47.88	1660	3287	4.54	+2, +3, +4
Tungsten	W	74	183.85	3410	5660	19.3	+6
Uranium	U	92	238.0289	1132.3	3818	18.95	+3, +4, +5, +6
Vanadium	V	23	50.9415	1890	3380	6.11	+2, +3, +4, +5
Xenon	Xe	54	131.29	−111.9	−107.1	.005887	
Ytterbium	Yb	70	173.04	819	1194	6.965	+2, +3
Yttrium	Y	39	88.9059	1522	3338	4.469	+3
Zinc	Zn	30	65.38	419.58	907	7.133	+2
Zirconium	Zr	40	91.22	1852	4377	6.506	+4
Element 104	(Rf)	104					
Element 105	(Ha)	105					
Element 106		106					
Element 107		107					
Element 109		109					

Table A·2 Electron Configurations of the Elements

Elements		$1s$	$2s$	$2p$	$3s$	$3p$	$3d$	$4s$	$4p$	$4d$	$4f$	$5s$	$5p$	$5d$	$5f$	$6s$	$6p$	$6d$	$6f$	$7s$
												Sublevels								
1	hydrogen	1																		
2	helium	2																		
3	lithium	2	1																	
4	beryllium	2	2																	
5	boron	2	2	1																
6	carbon	2	2	2																
7	nitrogen	2	2	3																
8	oxygen	2	2	4																
9	fluorine	2	2	5																
10	neon	2	2	6																
11	sodium	2	2	6	1															
12	magnesium	2	2	6	2															
13	aluminum	2	2	6	2	1														
14	silicon	2	2	6	2	2														
15	phosphorus	2	2	6	2	3														
16	sulfur	2	2	6	2	4														
17	chlorine	2	2	6	2	5														
18	argon	2	2	6	2	6														
19	potassium	2	2	6	2	6		1												
20	calcium	2	2	6	2	6		2												
21	scandium	2	2	6	2	6	1	2												
22	titanium	2	2	6	2	6	2	2												
23	vanadium	2	2	6	2	6	3	2												
24	chromium	2	2	6	2	6	5	1												
25	manganese	2	2	6	2	6	5	2												
26	iron	2	2	6	2	6	6	2												
27	cobalt	2	2	6	2	6	7	2												
28	nickel	2	2	6	2	6	8	2												
29	copper	2	2	6	2	6	10	1												
30	zinc	2	2	6	2	6	10	2												
31	gallium	2	2	6	2	6	10	2	1											
32	germanium	2	2	6	2	6	10	2	2											
33	arsenic	2	2	6	2	6	10	2	3											
34	selenium	2	2	6	2	6	10	2	4											
35	bromine	2	2	6	2	6	10	2	5											
36	krypton	2	2	6	2	6	10	2	6											
37	rubidium	2	2	6	2	6	10	2	6			1								
38	strontium	2	2	6	2	6	10	2	6			2								
39	yttrium	2	2	6	2	6	10	2	6	1		2								
40	zirconium	2	2	6	2	6	10	2	6	2		2								
41	niobium	2	2	6	2	6	10	2	6	4		1								
42	molybdenum	2	2	6	2	6	10	2	6	5		1								
43	technetium	2	2	6	2	6	10	2	6	5		2								
44	ruthenium	2	2	6	2	6	10	2	6	7		1								
45	rhodium	2	2	6	2	6	10	2	6	8		1								
46	palladium	2	2	6	2	6	10	2	6	10										
47	silver	2	2	6	2	6	10	2	6	10		1								
48	cadmium	2	2	6	2	6	10	2	6	10		2								
49	indium	2	2	6	2	6	10	2	6	10		2	1							
50	tin	2	2	6	2	6	10	2	6	10		2	2							
51	antimony	2	2	6	2	6	10	2	6	10		2	3							
52	tellurium	2	2	6	2	6	10	2	6	10		2	4							
53	iodine	2	2	6	2	6	10	2	6	10		2	5							
54	xenon	2	2	6	2	6	10	2	6	10		2	6							

Table A·2 Electron Configurations of the Elements (cont.)

	Elements	$1s$	$2s$	$2p$	$3s$	$3p$	$3d$	$4s$	$4p$	$4d$	$4f$	$5s$	$5p$	$5d$	$5f$	$6s$	$6p$	$6d$	$6f$	$7s$
55	cesium	2	2	6	2	6	10	2	6	10		2	6			1				
56	barium	2	2	6	2	6	10	2	6	10		2	6			2				
57	lanthanum	2	2	6	2	6	10	2	6	10		2	6	1		2				
58	cerium	2	2	6	2	6	10	2	6	10	2	2	6			2				
59	praseodymium	2	2	6	2	6	10	2	6	10	3	2	6			2				
60	neodymium	2	2	6	2	6	10	2	6	10	4	2	6			2				
61	promethium	2	2	6	2	6	10	2	6	10	5	2	6			2				
62	samarium	2	2	6	2	6	10	2	6	10	6	2	6			2				
63	europium	2	2	6	2	6	10	2	6	10	7	2	6			2				
64	gadolinium	2	2	6	2	6	10	2	6	10	7	2	6	1		2				
65	terbium	2	2	6	2	6	10	2	6	10	9	2	6			2				
66	dysprosium	2	2	6	2	6	10	2	6	10	10	2	6			2				
67	holmium	2	2	6	2	6	10	2	6	10	11	2	6			2				
68	erbium	2	2	6	2	6	10	2	6	10	12	2	6			2				
69	thulium	2	2	6	2	6	10	2	6	10	13	2	6			2				
70	ytterbium	2	2	6	2	6	10	2	6	10	14	2	6			2				
71	lutetium	2	2	6	2	6	10	2	6	10	14	2	6	1		2				
72	hafnium	2	2	6	2	6	10	2	6	10	14	2	6	2		2				
73	tantalum	2	2	6	2	6	10	2	6	10	14	2	6	3		2				
74	tungsten	2	2	6	2	6	10	2	6	10	14	2	6	4		2				
75	rhenium	2	2	6	2	6	10	2	6	10	14	2	6	5		2				
76	osmium	2	2	6	2	6	10	2	6	10	14	2	6	6		2				
77	iridium	2	2	6	2	6	10	2	6	10	14	2	6	7		2				
78	platinum	2	2	6	2	6	10	2	6	10	14	2	6	9		1				
79	gold	2	2	6	2	6	10	2	6	10	14	2	6	10		1				
80	mercury	2	2	6	2	6	10	2	6	10	14	2	6	10		2				
81	thallium	2	2	6	2	6	10	2	6	10	14	2	6	10		2	1			
82	lead	2	2	6	2	6	10	2	6	10	14	2	6	10		2	2			
83	bismuth	2	2	6	2	6	10	2	6	10	14	2	6	10		2	3			
84	polonium	2	2	6	2	6	10	2	6	10	14	2	6	10		2	4			
85	astatine	2	2	6	2	6	10	2	6	10	14	2	6	10		2	5			
86	radon	2	2	6	2	6	10	2	6	10	14	2	6	10		2	6			
87	francium	2	2	6	2	6	10	2	6	10	14	2	6	10		2	6			1
88	radium	2	2	6	2	6	10	2	6	10	14	2	6	10		2	6			2
89	actinium	2	2	6	2	6	10	2	6	10	14	2	6	10		2	6	1		2
90	thorium	2	2	6	2	6	10	2	6	10	14	2	6	10		2	6	2		2
91	protactinium	2	2	6	2	6	10	2	6	10	14	2	6	10	2	2	6	1		2
92	uranium	2	2	6	2	6	10	2	6	10	14	2	6	10	3	2	6	1		2
93	neptunium	2	2	6	2	6	10	2	6	10	14	2	6	10	4	2	6	1		2
94	plutonium	2	2	6	2	6	10	2	6	10	14	2	6	10	6	2	6			2
95	americium	2	2	6	2	6	10	2	6	10	14	2	6	10	7	2	6			2
96	curium	2	2	6	2	6	10	2	6	10	14	2	6	10	7	2	6	1		2
97	berkelium	2	2	6	2	6	10	2	6	10	14	2	6	10	9	2	6			2
98	californium	2	2	6	2	6	10	2	6	10	14	2	6	10	10	2	6			2
99	einsteinium	2	2	6	2	6	10	2	6	10	14	2	6	10	11	2	6			2
100	fermium	2	2	6	2	6	10	2	6	10	14	2	6	10	12	2	6			2
101	mendelevium	2	2	6	2	6	10	2	6	10	14	2	6	10	13	2	6			2
102	nobelium	2	2	6	2	6	10	2	6	10	14	2	6	10	14	2	6			2
103	lawrencium	2	2	6	2	6	10	2	6	10	14	2	6	10	14	2	6	1		2
104	unnilquadium	2	2	6	2	6	10	2	6	10	14	2	6	10	14	2	6	2		2?
105	unnilpentium	2	2	6	2	6	10	2	6	10	14	2	6	10	14	2	6	3		2?
106	unnilhexium	2	2	6	2	6	10	2	6	10	14	2	6	10	14	2	6	4		2?
107	unnilseptium	2	2	6	2	6	10	2	6	10	14	2	6	10	14	2	6	5		2?
108	unniloctium	2	2	6	2	6	10	2	6	10	14	2	6	10	14	2	6	6		2?
109	unnilennium	2	2	6	2	6	10	2	6	10	14	2	6	10	14	2	6	7		2?

Table A·3 Symbols of Common Elements

Ag	silver	Cu	copper	O	oxygen		
Al	aluminum	F	fluorine	P	phosphorus		
As	arsenic	Fe	iron	Pb	lead		
Au	gold	H	hydrogen	Pt	platinum		
Ba	barium	Hg	mercury	S	sulfur		
Bi	bismuth	I	iodine	Sb	antimony		
Br	bromine	K	potassium	Sn	tin		
C	carbon	Mg	magnesium	Sr	strontium		
Ca	calcium	Mn	manganese	Ti	titanium		
Cl	chlorine	N	nitrogen	U	uranium		
Co	cobalt	Na	sodium	W	tungsten		
Cr	chromium	Ni	nickel	Zn	zinc		

Table A·4 Symbols of Common Polyatomic Ions

$C_2H_3O_2^-$	acetate	$Cr_2O_7^{2-}$	dichromate	NH_4^+	ammonium
ClO^-	hypochlorite	HCO_3^-	hydrogen carbonate (bicarbonate)	NO_3^-	nitrate
ClO_2^-	chlorite			NO_2^-	nitrite
ClO_3^-	chlorate	H_3O^+	hydronium	O_2^{2-}	peroxide
ClO_4^-	perchlorate	HPO_4^{2-}	hydrogen phosphate	OH^-	hydroxide
CN^-	cyanide	HSO_3^-	hydrogen sulfite	PO_4^{3-}	phosphate
CO_3^{2-}	carbonate	HSO_4^-	hydrogen sulfate	SO_3^{2-}	sulfite
CrO_4^{2-}	chromate	MnO_4^-	permanganate	SO_4^{2-}	sulfate

Table A·5 Other Symbols and Abbreviations

α	alpha rays	gmm	gram molecular mass	m	mass
β	beta rays	H	enthalpy	m	molality
γ	gamma rays	H_f	heat of formation	mL	milliliter (*volume*)
Δ	change in	h	hour	mm	millimeter (*length*)
δ^-, δ^+	partial ionic charge	h	Planck's constant	mol	mole (*amount*)
λ	wavelength	Hz	hertz (*frequency*)	mp	melting point
π	pi bond	J	joule (*energy*)	N	normality
σ	sigma bond	K	kelvin (*temperature*)	n^0	neutron
ν	frequency	K_a	acid dissociation constant	n	number of moles
amu	atomic mass unit	K_b	base dissociation constant	n	principal quantum number
(*aq*)	aqueous solution	K_b	molal boiling point elevation constant	P	pressure
atm	atmosphere (*pressure*)			p^+	proton
bp	boiling point	K_{eq}	equilibrium constant	Pa	pascal (*pressure*)
°C	degree Celsius (*temperature*)	K_f	molal freezing point depression constant	R	ideal gas constant
c	speed of light in a vacuum			S	entropy
cm	centimeter (*length*)	K_w	ion product constant for water	s	second
D	density			(*s*)	solid
E	energy	K_{sp}	solubility product constant	SI	International System of Units
e^-	electron	kcal	kilocalorie (*energy*)		
fp	freezing point	kg	kilogram (*mass*)	STP	standard temperature and pressure
G	Gibb's free energy	kPa	kilopascal (*pressure*)		
g	gram (*mass*)	L	liter (*volume*)	T	temperature
(*g*)	gas	(*l*)	liquid	$t_{1/2}$	half-life
gam	gram atomic mass	M	molarity	V	volume
gfm	gram formula mass	m	meter (*length*)	v	velocity, speed

Problem

1. Express these numbers in standard exponential form.
   - **a.** 20 000
   - **b.** 543.6
   - **c.** 0.000 005
   - **d.** 34.1
   - **e.** 0.000 55
   - **f.** 0.003 45
   - **g.** 40 230 000
   - **h.** 0.099
   - **i.** 628 000

When working problems, your calculated answer may not be in standard exponential form. Exponential numbers can be changed to standard form by using the rules just given.

$$504.2 \times 10^6 = 5.042 \times 10^8$$
2 places

$$0.15 \times 10^2 = 1.5 \times 10^1$$
1 place

$$42.2 \times 10^{-5} = 4.22 \times 10^{-4} \ (10^{-4} \text{ is larger than } 10^{-5})$$
1 place

$$0.0089 \times 10^{-2} = 8.9 \times 10^{-5}$$
3 places

Problem

2. Express these numbers in standard exponential form.
   - **a.** 34.5×10^5
   - **b.** 0.004×10^2
   - **c.** 180×10^{-1}
   - **d.** 0.72×10^{-2}
   - **e.** 765×10^7
   - **f.** 0.029×10^{-6}

Multiplication and Division

To multiply numbers written in exponential form, multiply the coefficients and add the exponents.

$$(3 \times 10^4) \times (2 \times 10^2) = (3 \times 2) \times 10^{4+2}$$
$$= 6 \times 10^6$$
$$(2.1 \times 10^3) \times (4.0 \times 10^{-7}) = (2.1 \times 4.0) \times 10^{3+(-7)}$$
$$= 8.4 \times 10^{-4}$$
$$(8.0 \times 10^{-2}) \times (7.0 \times 10^{-5}) = (8.0 \times 7.0) \times 10^{-2+(-5)}$$
$$= 56 \times 10^{-7}$$
$$= 5.6 \times 10^{-6} \text{ (standard form)}$$

Problem

3. Answer each problem in standard exponential form.
   - **a.** $(2 \times 10^9) \times (4 \times 10^3)$
   - **b.** $(6.2 \times 10^{-3}) \times (1.5 \times 10^1)$
   - **c.** $10^{-4} \times 10^8 \times 10^{-2}$
   - **d.** $(3.4 \times 10^{-3}) \times (2.5 \times 10^{-5})$
   - **e.** $(0.10 \times 10^5) \times (4.9 \times 10^{-2})$
   - **f.** $(88 \times 10^2) \times (0.15 \times 10^4)$

To divide numbers written in exponential form, divide the coefficients and subtract the exponent in the denominator from the exponent in the numerator.

$$\frac{3.0 \times 10^5}{6.0 \times 10^2} = \left(\frac{3.0}{6.0}\right) \times 10^{5-2} = 0.50 \times 10^3$$

$$= 5.0 \times 10^2 \text{ (standard form)}$$

$$\frac{8.4 \times 10^{-6}}{2.1 \times 10^2} = \left(\frac{8.4}{2.1}\right) \times 10^{-6-(+2)} = 4.0 \times 10^{-8}$$

$$\frac{4.56 \times 10^5}{2.93 \times 10^{-3}} = \left(\frac{4.56}{2.93}\right) \times 10^{5-(-3)} = 1.56 \times 10^8$$

Problem

4. Do each problem and express the answer in standard exponential form.

a. $\dfrac{8.8 \times 10^6}{2.2 \times 10^1}$ e. $\dfrac{0.40 \times 10^{-4}}{5.7 \times 10^1}$

b. $\dfrac{5.2 \times 10^2}{1.3 \times 10^{-7}}$ f. $\dfrac{6.8 \times 10^{12}}{0.22 \times 10^{-4}}$

c. $\dfrac{10^8 \times 10^{-3}}{10^{-4} \times 10^5}$ g. $\dfrac{13.6 \times 10^{12}}{8.00 \times 10^{15}}$

d. $\dfrac{1 \times 10^{-7}}{1 \times 10^{-9}}$ h. $\dfrac{753 \times 10^6}{0.300 \times 10^{-8}}$

Addition and Subtraction

Before numbers in exponential form are added or subtracted, the exponents must be the same. For example, what is the sum of 5.4×10^3 and 6.0×10^2? Rewrite the second number so that the exponent is a three, $6.0 \times 10^2 = 0.60 \times 10^3$. Now add the numbers.

$$(5.4 \times 10^3) + (0.60 \times 10^3) = (5.4 + 0.60) \times 10^3$$

$$= 6.0 \times 10^3$$

Similarly,

$$(3.42 \times 10^{-5}) - (2.5 \times 10^{-6}) = (3.42 \times 10^{-5}) - (0.25 \times 10^{-5})$$

$$= (3.42 - 0.25) \times 10^{-5}$$

$$= 3.17 \times 10^{-5}$$

Problem

5. Do each problem and express the answer in standard exponential form.

a. $5.2 \times 10^4 + 2.7 \times 10^4$ e. $23.4 \times 10^5 + 9.5 \times 10^4$

b. $9.4 \times 10^{-2} - 2.1 \times 10^{-2}$ f. $568 \times 10^3 + 2 \times 10^5$

c. $6.6 \times 10^{-8} - 4.0 \times 10^{-9}$ g. $3.75 \times 10^5 + 653 \times 10^3$

d. $6.7 \times 10^{-2} - 3.0 \times 10^{-3}$ h. $0.0073 \times 10^5 - 61000 \times 10^{-2}$

B·2 Algebraic Equations

Many relationships in chemistry can be expressed by simple algebraic equations. Often an equation is not given in the form that is most useful for a particular problem. In such a case you must first solve the equation for the unknown quantity. Solving an equation means rearranging it so that the unknown quantity is on one side of the equation and all the known quantities are on the other side. For instance, consider the following equation.

$$K = {}^{\circ}C + 273$$

It states the relationship between the Kelvin and Celsius temperature scales (Chapter 2). Can this equation be used to find the Celsius temperature equivalent to 400 K? Yes, it can, if the equation is first solved for the unknown quantity, $^{\circ}C$.

An equation is solved using the laws of equality. The **laws of equality** are summarized as follows. *If equals are added to, subtracted from, multiplied by, or divided by equals, the results are equal.* In other words, you can perform any of these mathematic operations on an equation and not destroy the equality, *as long as you do the same thing to both sides of the equation.* The laws of equality apply to any legitimate mathematic operation, including squaring, taking square roots, and taking the logarithm.

To solve the equation in the example for $^{\circ}C$, subtract 273 from both sides of the equation.

$$K = {}^{\circ}C + 273$$

$$K - 273 = {}^{\circ}C + \cancel{273} - \cancel{273}$$

$$^{\circ}C = K - 273$$

Once you have solved an equation, substitute in the known quantities and calculate the value of the unknown quantity. A temperature of 400 K is converted to $^{\circ}C$ as follows.

$$^{\circ}C = K - 273 = 400 - 273 = 127{}^{\circ}C$$

The relationship between the Fahrenheit and Celsius temperature scales is $^{\circ}F = 1.8 \times {}^{\circ}C + 32$. Use this relationship to find the Celsius temperature equivalent to 392°F.

To solve for $^{\circ}C$, both the 1.8 and the 32 must be moved to the other side of the equation. Since the 32 is added to the quantity (1.8 times $^{\circ}C$), first subtract 32 from both sides of the equation.

$$^{\circ}F = (1.8 \times {}^{\circ}C) + 32$$

$$^{\circ}F - 32 = (1.8 \times {}^{\circ}C) + \cancel{32} - \cancel{32}$$

$$^{\circ}F - 32 = (1.8 \times {}^{\circ}C)$$

Now divide each side of the equation by 1.8.

$$\frac{^{\circ}F - 32}{1.8} = \frac{\cancel{1.8} \times {}^{\circ}C}{\cancel{1.8}}$$

$$°C = \frac{°F - 32}{1.8}$$

Finally, substitute the given quantity.

$$°C = \frac{392 - 32}{1.8} = \frac{360}{1.8} = 200°C$$

An equation that shows how the volume of a gas changes with a change in temperature (Chapter 10) is as follows.

$$\frac{V_1}{T_1} = \frac{V_2}{T_2}$$

What is the value of T_2 when $V_1 = 5.0$ L, $V_2 = 15$ L, and $T_1 = 200$ K? Solve for T_2.

$$T_2 = \frac{V_2 \times T_1}{V_1}$$

Substitute the given quantities into the solved equation.

$$T_2 = \frac{15 \cancel{L} \times 200 \text{ K}}{5.0 \cancel{L}} = 600 \text{ K}$$

Problems

6. Solve for T: $P \times V = n \times R \times T$

7. Solve for V_1: $\dfrac{P_1 \times V_1}{T_1} = \dfrac{P_2 \times V_2}{T_2}$

8. Solve for v: $E = (m \times v^2)/2$

9. Solve for P_B: $P_T = P_A + P_B + P_C$.

10. Solve each of these equations for h. Calculate a value for h if $g = 12$, $k = 0.4$, and $m = 1.5$.

   a. $kh = \dfrac{g}{m}$ c. $gh - k = m$

   b. $\dfrac{g - m}{h} = k$ d. $\dfrac{mk}{g + h} = 2$

B·3 Percents

Percent means "parts of 100" or "parts per 100 parts." In slightly different terms, a percent is a part of a whole expressed in hundredths. The idea of a percent should be very familiar to you. Many times examination grades are expressed as a percent. What does a grade of 88% mean? Since the word percent (%) means "per 100 parts," 88% = 88/100, or 88 questions correct (the part) out of a possible 100 questions (the whole).

You are generally evaluated on exams that do not have 100 questions. Your grade can still be expressed as a percent. What is your grade if you get 24 questions correct on a 30-question exam? The part over the whole, expressed as a percent is calculated as follows.

$$\frac{\text{Part}}{\text{Whole}} = \frac{\text{number correct}}{\text{number possible}} = \frac{24 \text{ questions correct}}{30 \text{ questions possible}} = 0.80 \text{ or } \frac{80}{100} \text{ or } 80\%$$

Based on the definition of percent, 0.80, 80/100, and 80% all express the same idea: eighty hundredths or 80 "per 100 parts." Mathematically, 0.80 can be shown to be equal to 80/100 because multiplying by 100/100 does not change the value of a number.

$$\frac{0.80}{1} \times \frac{100}{100} = \frac{80}{100}$$

An alternate way to calculate a percent is to multiply the ratio of the part over the whole by 100%.

$$\text{Percent} = \frac{\text{part}}{\text{whole}} \times 100\%$$

Problem

11. Calculate a percent grade for each exam.
    a. Eleven questions correctly answered on a 15-question exam.
    b. Forty-one questions correctly answered on a 50-question exam.
    c. Seventeen questions correctly answered on a 30-question exam.

A percent represents a relationship between two quantities and, as such, it can be used as a conversion factor (Chapter 3). For example, a friend tells you that she got a grade of 95% on a 40-question examination. How many questions did she answer correctly? A grade of 95% means 95 questions correct for each 100 questions possible.

$$40 \text{ questions possible} \times \frac{95 \text{ questions correct}}{100 \text{ questions possible}} = 38 \text{ questions correct}$$

As a final example of using a percent, six students are absent from class. If 80% of the students are present, what is the class enrollment? Since 80% of the class is present, the six students absent must represent 20% of the class. This absence rate of 20% means that 20 students are absent for each 100 students enrolled. Starting with the given, the number of students enrolled is calculated.

$$6 \text{ students absent} \times \frac{100 \text{ students enrolled}}{20 \text{ students absent}} = \frac{600}{20} = 30 \text{ students enrolled}$$

Problems

12. The antiseptic hydrogen peroxide is often sold as a 3.0% (by mass) solution, the rest being water. How many grams of hydrogen peroxide are in 750 g of this solution?

13. A nighttime cold medicine is 22% alcohol (by volume). How many mL of alcohol are in a 250 mL bottle of this medicine?

B·4 Graphing

The relationship between two variables in an experiment is often determined by graphing the experimental data. A graph is a "picture" of the data. Once a graph is constructed, additional information can be derived about the variables.

In constructing a graph we must first label the axes. The independent variable is plotted on the *x-axis* (*abscissa*). This is the horizontal axis. The independent variable is generally controlled by the experimenter. When the independent variable is changed, a corresponding change in the dependent variable is measured. The dependent variable is plotted on the *y-axis* (*ordinate*). This is the vertical axis. The label on each axis should include the unit of the quantity being graphed.

Before data can be plotted on a graph, each axis must be scaled. The scale must take into consideration the smallest and largest values of each quantity. Each interval (square on the graph paper) on the scale must represent the same amount. To make it easy to find numbers along the scale, the interval chosen is usually a multiple of 1, 2, 5, or 10. Although each scale can start at zero, this is not always practical.

Data is plotted by putting a point at the intersection of corresponding values of each pair of measurement. This is illustrated in the next example.

Once the data has been plotted, the points are connected by a smooth curve. This is not the same as "connecting-the-dots," which is an incorrect approach to drawing a line. A smooth curve comes as close as possible to all the plotted points. It may in fact not touch any of them.

Depending on the relationship between two variables the curve may or may not be a straight line. Two common curves are shown in graphs A and B.

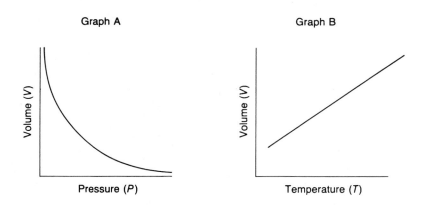

Graphs similar to these are found in Chapter 10 of the text. Graph A is typical of an inverse proportionality. As the independent variable (P) increases, the dependent variable (V) decreases. The product of the two variables at any point on the curve of an inverse proportionality is a constant. For graph A, $V \times P = $ constant.

The straight line in graph B is typical of a direct proportionality. As the independent variable (T) increases, there is a corresponding increase in the dependent variable (V). A straight line is represented by this general equation.

$$y = mx + b$$

Here y and x are the variables plotted on the vertical and horizontal axes respectively; m is the slope of the line; and b is the intercept on the y-axis.

The y intercept, b, is the value of y when x is zero. The slope, m, is the ratio of the change in y (Δy) for a corresponding change in x (Δx). This relationship is often symbolized in the following way.

$$m = \frac{\Delta y}{\Delta x}$$

As an example, consider this data about a bicyclist's trip. Assume that the bicyclist rode at a constant speed.

distance from home (km)	15	25	35	50	75
time (hours)	1	2	3	4.5	7

Let's graph this data using time as the independent variable, and then use the graph to answer the following questions.

a. How far from home was the bicyclist at the start of the trip?

b. How many hours did it take the bicyclist to get 40 km from home?

c. What was the bicyclist's average speed, in kilometers per hour, on the trip?

The plotted points are shown in Figure 1. Each point was plotted by finding the value of time on the x-axis, then moving up vertically to the value of the other variable (distance). A smooth curve, which in this case happens to be a straight line, has been drawn through the points.

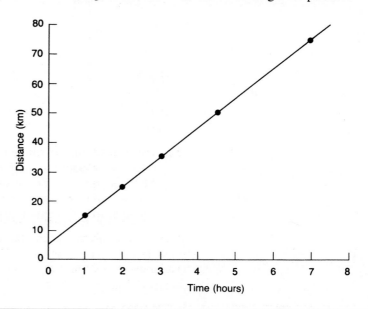

a. The graph in Figure 2 shows that the bicyclist started the trip 5 km from home. This is the value of the vertical axis (distance) when the time elapsed is zero (point a on the graph).

b. Find the given value, 40 km, on the vertical axis in Figure 2. Move to the right (horizontally) in the graph until you reach the line. Drop down vertically and read the value of time at this point (point b). It takes the bicyclist 3.5 hr to get 40 km from home. As another, similar example, how far is the bicyclist from home after riding 5 hr? Using the graph in Figure 2, start at 5 hr and go up to the line. Then move horizontally to the left. The distance, point c, is 55 km.

c. Speed is distance/time. The average speed of the bicyclist is the slope of the line. Calculate the slope using the data points from the previous part of this problem.

$$m = \frac{\Delta y}{\Delta x} = \frac{55 \text{ km} - 40 \text{ km}}{5 \text{ hr} - 3.5 \text{ hr}} = \frac{15 \text{ km}}{1.5 \text{ hr}} = 10 \text{ km/hr}$$

The equation for this line shows the relationship between time and distance traveled by the bicyclist.

$$\text{Distance} = (10 \text{ km/hr})(\text{time}) + 5 \text{ km}$$

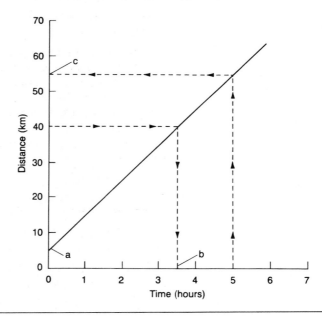

Problems

14. Use the data in the following table to draw a graph that shows the relationship between the Fahrenheit and Celsius temperature scales. Make degrees Fahrenheit the dependent variable. Use the graph to derive an equation between °F and °C. Then use the graph or the equation to complete the data table.

°F	50	212	356	−4	__	70	400
°C	10	100	180	−20	70	__	__

15. A bicyclist wants to ride 100 kilometers. The data below show the time required to ride 100 kilometers at different average speeds.

Time (hours)	4	5	8	10	15	20
Average speed (km/hr)	25	20	12.5	10	6.6	5

a. Graph the data, using average speed as the independent variable.

b. Is this a direct or inverse proportionality?

c. Use the graph to determine what average speed must be maintained to complete the ride in 12 hours.

d. If a bicyclist maintains an average speed of 18 km/hr, how many hours does it take to ride 100 km?

16. The pressure of a fixed volume of gas varies with the temperature. A student collected the following data.

Temperature (°C)	10	20	40	70	100
Pressure (mm Hg)	726	750	800	880	960

a. Graph the data, using pressure as the dependent variable.

b. Is this a direct or inverse proportionality?

c. At what temperature is the gas pressure 722 mm Hg?

d. What is the pressure of this gas at a temperature of 0°C?

e. How does the pressure change with a change in temperature? That is, what is the slope of the line?

f. Write an equation relating pressure and temperature for this gas.

B·5 Logarithms

A logarithm is the exponent to which a fixed number (base) must be raised in order to produce a given number. Common logarithms use 10 as the base. A logarithm to the base 10 of a number is the exponent to which 10 must be raised to obtain the number. If $x = 10^y$, then log $x = y$. Here x is the number and y is the logarithm of x to the base 10.

Common logarithms must not be confused with natural logarithms which use base e, where $e = 2.71828$. When base 10 is used, logarithm is abbreviated $\log_{10}$, or often simply log. When the base e is used, logarithm is abbreviated ln.

The logarithm of a number has two parts, the characteristic, or whole-number part, and the mantissa, or decimal part. In the example log $421.6 = 2.6249$, the characteristic is 2 and the mantissa is 0.6249.

For numbers written in standard exponential form, the characteristic corresponds to the exponent of 10.

$$\log 10^1 = 1, \log 10^2 = 2, \log 10^3 = 3, \log 10^{-2} = -2.$$

The mantissa is the decimal part of a logarithm and can be looked up in a "log table." The number of significant figures in a logarithm is the number of figures in the mantissa.

As an example, the logarithm of 176 is found as follows.

1. Write 176 in exponential notation as 1.76×10^2.

2. Locate the number 1.7 in the column labeled N; then under the column headed 6 read the mantissa as 0.2455.

3. Because 1.76×10^2 contains only three significant figures, the mantissa is rounded off to three digits. Thus the mantissa is 0.246.

4. The characteristic is the exponent, 2.

5. The logarithm of 176 is 2.246 or log 176 = 2.246.

Logarithms can also be found using the log key of a calculator. Simply enter the number and press the key labeled "log".

Problem

17. Write the logarithms of the following numbers.
    a. 3.45   **b.** 0.0087   **c.** 456   **d.** 3.11×10^{-5} **e.** 1.48×10^9

The reverse process of converting a logarithm into a number is referred to as obtaining the antilogarithm. The antilogarithm of the logarithm of a number is the number itself. Here x represents any number.

$$\text{antilog }(\log x) = x$$

Antilogarithms can be obtained from a table of common logarithms. They can also be obtained with a calculator by using the "antilog" key, the "10^x" key, or the "inverse" key in conjuction with the "log" key.

For example, what is the antilog of the logarithm 4.618? Look in the body of the log table for the mantissa, 0.618. This value is found in the row with an N of 4.1 and in the column labeled 5. Thus the antilog of 0.618 is 4.15. The characteristic is 4. This corresponds to an exponential term of 10^4. The number whose log is 4.618 is 4.15×10^4.

Problem

18. Use a log table or a calculator to find the numbers (antilogarithms) which have the following logarithms.
    a. 2.56   **b.** 6.11   **c.** −3.55   **d.** 1.03   **e.** 0.962

There are rules based on the laws of exponents that must be followed when using logarithms for calculations. The logarithm of the product of two numbers is the sum of the number's logs.

$$\log(a \times b) = \log a + \log b$$

The log of the ratio of two numbers is the difference between the log of the denominator subtracted from the log of the numerator.

$$\log(a/b) = \log a - \log b.$$

Table B-5 Common Logarithms

x	0	1	2	3	4	5	6	7	8	9
1.0	.0000	.0043	.0086	.0128	.0170	.0212	.0253	.0294	.0334	.0374
1.1	.0414	.0453	.0492	.0531	.0569	.0607	.0645	.0682	.0719	.0755
1.2	.0792	.0828	.0864	.0899	.0934	.0969	.1004	.1038	.1072	.1106
1.3	.1139	.1173	.1206	.1239	.1271	.1303	.1335	.1367	.1399	.1430
1.4	.1461	.1492	.1523	.1553	.1584	.1614	.1644	.1673	.1703	.1732
1.5	.1761	.1790	.1818	.1847	.1875	.1903	.1931	.1959	.1987	.2014
1.6	.2041	.2068	.2095	.2122	.2148	.2175	.2201	.2227	.2253	.2279
1.7	.2304	.2330	.2355	.2380	.2405	.2430	.2455	.2480	.2504	.2529
1.8	.2553	.2577	.2601	.2625	.2648	.2672	.2695	.2718	.2742	.2765
1.9	.2788	.2810	.2833	.2856	.2878	.2900	.2923	.2945	.2967	.2989
2.0	.3010	.3032	.3054	.3075	.3096	.3118	.3139	.3160	.3181	.3201
2.1	.3222	.3243	.3263	.3284	.3304	.3324	.3345	.3365	.3385	.3404
2.2	.3424	.3444	.3464	.3483	.3502	.3522	.3541	.3560	.3579	.3598
2.3	.3617	.3636	.3655	.3674	.3692	.3711	.3729	.3747	.3766	.3784
2.4	.3802	.3820	.3838	.3856	.3874	.3892	.3909	.3927	.3945	.3962
2.5	.3979	.3997	.4014	.4031	.4048	.4065	.4082	.4099	.4116	.4133
2.6	.4150	.4166	.4183	.4200	.4216	.4232	.4249	.4265	.4281	.4298
2.7	.4314	.4330	.4346	.4362	.4378	.4393	.4409	.4425	.4440	.4456
2.8	.4472	.4487	.4502	.4518	.4533	.4548	.4564	.4579	.4594	.4609
2.9	.4624	.4639	.4654	.4669	.4683	.4698	.4713	.4728	.4742	.4757
3.0	.4771	.4786	.4800	.4814	.4829	.4843	.4857	.4871	.4886	.4900
3.1	.4914	.4928	.4942	.4955	.4969	.4983	.4997	.5011	.5024	.5038
3.2	.5051	.5065	.5079	.5092	.5105	.5119	.5132	.5145	.5159	.5172
3.3	.5185	.5198	.5211	.5224	.5237	.5250	.5263	.5276	.5289	.5307
3.4	.5315	.5328	.5340	.5353	.5366	.5378	.5391	.5403	.5416	.5428
3.5	.5441	.5453	.5465	.5478	.5490	.5502	.5514	.5527	.5539	.5551
3.6	.5563	.5575	.5587	.5599	.5611	.5623	.5635	.5647	.5658	.5670
3.7	.5682	.5694	.5705	.5717	.5729	.5740	.5752	.5763	.5775	.5786
3.8	.5798	.5809	.5821	.5832	.5843	.5855	.5866	.5877	.5888	.5899
3.9	.5911	.5922	.5933	.5944	.5955	.5966	.5977	.5988	.5999	.6010
4.0	.6021	.6031	.6042	.6053	.6064	.6075	.6085	.6096	.6107	.6117
4.1	.6128	.6138	.6149	.6160	.6170	.6180	.6191	.6201	.6212	.6222
4.2	.6232	.6243	.6253	.6263	.6274	.6284	.6294	.6304	.6314	.6325
4.3	.6335	.6345	.6355	.6365	.6375	.6385	.6395	.6405	.6415	.6425
4.4	.6435	.6444	.6454	.6464	.6474	.6484	.6493	.6503	.6513	.6522
4.5	.6532	.6542	.6551	.6561	.6571	.6580	.6590	.6599	.6609	.6618
4.6	.6628	.6637	.6646	.6656	.6665	.6675	.6684	.6693	.6702	.6712
4.7	.6721	.6730	.6739	.6749	.6758	.6767	.6776	.6785	.6794	.6803
4.8	.6812	.6821	.6830	.6839	.6848	.6857	.6866	.6875	.6884	.6893
4.9	.6902	.6911	.6920	.6928	.6937	.6946	.6955	.6964	.6972	.6981
5.0	.6990	.6998	.7007	.7016	.7024	.7033	.7042	.7050	.7059	.7067
5.1	.7076	.7084	.7093	.7101	.7110	.7118	.7126	.7135	.7143	.7152
5.2	.7160	.7168	.7177	.7185	.7193	.7202	.7210	.7218	.7226	.7235
5.3	.7243	.7251	.7259	.7267	.7275	.7284	.7292	.7300	.7308	.7316
5.4	.7324	.7332	.7340	.7348	.7356	.7364	.7372	.7380	.7388	.7396

Table B-5 Common Logarithms (cont.)

x	0	1	2	3	4	5	6	7	8	9
5.5	.7404	.7412	.7419	.7427	.7435	.7443	.7451	.7459	.7466	.7474
5.6	.7482	.7490	.7497	.7505	.7513	.7520	.7528	.7536	.7543	.7551
5.7	.7559	.7566	.7574	.7582	.7589	.7597	.7604	.7612	.7619	.7627
5.8	.7634	.7642	.7649	.7657	.7664	.7672	.7679	.7686	.7694	.7701
5.9	.7709	.7716	.7723	.7731	.7738	.7745	.7752	.7760	.7767	.7774
6.0	.7782	.7789	.7796	.7803	.7810	.7818	.7825	.7832	.7839	.7846
6.1	.7853	.7860	.7868	.7875	.7882	.7889	.7896	.7903	.7910	.7917
6.2	.7924	.7931	.7938	.7945	.7952	.7959	.7966	.7973	.7980	.7987
6.3	.7993	.8000	.8007	.8014	.8021	.8028	.8035	.8041	.8048	.8055
6.4	.8062	.8069	.8075	.8082	.8089	.8096	.8102	.8109	.8116	.8122
6.5	.8129	.8136	.8142	.8149	.8156	.8162	.8169	.8176	.8182	.8189
6.6	.8195	.8202	.8209	.8215	.8222	.8228	.8235	.8241	.8248	.8254
6.7	.8261	.8267	.8274	.8280	.8287	.8293	.8299	.8306	.8312	.8319
6.8	.8325	.8331	.8338	.8344	.8351	.8357	.8363	.8370	.8376	.8382
6.9	.8388	.8395	.8401	.8407	.8414	.8420	.8426	.8432	.8439	.8445
7.0	.8451	.8457	.8463	.8470	.8476	.8482	.8488	.8494	.8500	.8506
7.1	.8513	.8519	.8525	.8531	.8537	.8543	.8549	.8555	.8561	.8567
7.2	.8573	.8579	.8585	.8591	.8597	.8603	.8609	.8615	.8621	.8627
7.3	.8633	.8639	.8645	.8651	.8657	.8663	.8669	.8675	.8681	.8686
7.4	.8692	.8698	.8704	.8710	.8716	.8722	.8727	.8733	.8739	.8745
7.5	.8751	.8756	.8762	.8768	.8774	.8779	.8785	.8791	.8797	.8802
7.6	.8808	.8814	.8820	.8825	.8831	.8837	.8842	.8848	.8854	.8859
7.7	.8865	.8871	.8876	.8882	.8887	.8893	.8899	.8904	.8910	.8915
7.8	.8921	.8927	.8932	.8938	.8943	.8949	.8954	.8960	.8965	.8971
7.9	.8976	.8982	.8987	.8993	.8998	.9004	.9009	.9015	.9020	.9025
8.0	.9031	.9036	.9042	.9047	.9053	.9058	.9063	.9069	.9074	.9079
8.1	.9085	.9090	.9096	.9101	.9106	.9112	.9117	.9122	.9128	.9133
8.2	.9138	.9143	.9149	.9154	.9159	.9165	.9170	.9175	.9180	.9186
8.3	.9191	.9196	.9201	.9206	.9212	.9217	.9222	.9227	.9232	.9238
8.4	.9243	.9248	.9253	.9258	.9263	.9269	.9274	.9279	.9284	.9289
8.5	.9294	.9299	.9304	.9309	.9315	.9320	.9325	.9330	.9335	.9340
8.6	.9345	.9350	.9555	.9360	.9365	.9370	.9375	.9380	.9385	.9390
8.7	.9395	.9400	.9405	.9410	.9415	.9420	.9425	.9430	.9435	.9440
8.8	.9445	.9450	.9455	.9460	.9465	.9469	.9474	.9479	.9484	.9489
8.9	.9494	.9499	.9504	.9509	.9513	.9518	.9523	.9528	.9533	.9538
9.0	.9542	.9547	.9552	.9557	.9562	.9566	.9571	.9576	.9581	.9586
9.1	.9590	.9595	.9600	.9605	.9609	.9614	.9619	.9624	.9628	.9633
9.2	.9638	.9643	.9647	.9652	.9657	.9661	.9666	.9671	.9675	.9680
9.3	.9685	.9689	.9694	.9699	.9703	.9708	.9713	.9717	.9722	.9727
9.4	.9731	.9736	.9741	.9745	.9750	.9754	.9759	.9763	.9768	.9773
9.5	.9777	.9782	.9786	.9791	.9795	.9800	.9805	.9809	.9914	.9818
9.6	.9823	.9827	.9832	.9836	.9841	.9845	.9850	.9854	.9859	.9863
9.7	.9868	.9872	.9877	.9881	.9886	.9890	.9894	.9899	.9903	.9908
9.8	.9912	.9917	.9921	.9926	.9930	.9934	.9939	.9943	.9948	.9952
9.9	.9956	.9961	.9965	.9969	.9974	.9978	.9983	.9987	.9991	.9996

Appendix C   Solutions to In-Text Problems

Chapter 1

1. **a.** hypothesis (Further testing with improved experiments would prove that this is a faulty hypothesis (Section 1·12).) **b.** law **c.** experiment **d.** theory (Chapter 4)
2. **a.** solid **b.** liquid **c.** gas **d.** solid **e.** liquid
3. **a.** heterogeneous **b.** heterogeneous **c.** homogeneous **d.** homogeneous **e.** homogeneous
4. It is a mixture. The original liquid was a solution of a solid dissolved in a liquid. The components were separated by physical means.
5. **a.** tin **b.** copper **c.** nitrogen **d.** cadmium **e.** phosphorus **f.** chlorine
6. **a.** element **b.** compound **c.** mixture **d.** mixture **e.** element
7. Radiant energy (sun) to chemical energy (plants) to mechanical energy (person) to mechanical energy (bike) to kinetic energy (moving wheel) to mechanical energy (generator) to electrical energy (generator) to light energy (light bulb).
8. **a.** chemical **b.** physical **c.** chemical **d.** chemical **e.** physical **f.** chemical **g.** physical **h.** physical

Chapter 2

1. **a.** 4 **b.** 4 **c.** 2 **d.** 5 **e.** 1 **f.** 2 **g.** 3 **h.** 5
2. **a.** 8.71×10^1 m, 9×10^1 m **b.** 4.36×10^8 m, 4×10^8 m **c.** 1.55×10^{-2} m, 2×10^{-2} m **d.** 9.01×10^3 m, 9×10^3 m **e.** 1.78×10^{-3} m, 2×10^{-3} **f.** 6.30×10^2 m, 6×10^2 m
3. **a.** 79.15 m = 79.2 m **b.** 7.329 m = 7.33 m **c.** 11.53 m **d.** 17.31 m = 17.3 m
4. **a.** $10.126 = 1.0 \times 10^1$ m² **b.** $5.22 \times 10^{-5} = 5.2 \times 10^{-5}$ m² **c.** 674.56 = 675 m **d.** 0.20784 = 0.21 m
5. **a.** 2mm **b.** 16.4 cm **c.** 25.3 cm **d.** 1.9 cm
6. Density $= \dfrac{\text{mass}}{\text{volume}} = \dfrac{612 \text{ g}}{245 \text{ cm}^3} = 2.50 \text{ g/cm}^3$

   Aluminum has a density of 2.70 g/cm³. The metal is not aluminum.
7. Density $= \dfrac{\text{mass}}{\text{volume}} = \dfrac{15.8 \text{ g}}{19.7 \text{ cm}^3} = 0.802 \text{ g/cm}^3$

   The ball will sink because its density is greater than the density of gasoline (0.70 g/cm³).
8. The mass of an object does not change with a change in the force of gravity. Therefore the density of a person is the same regardless of location.
9. **a.** 2.70 **b.** 13.5 **c.** 0.917
10. K = 170 + 273 = 443 K
11. °C = 87 − 273 = − 186°C
12. The increase in temperature is 55°C. To increase the temperature of 1 gram of water by this amount would require 55 cal of heat energy. The energy required for 32.0 grams is: 32.0 g × 55 cal/g = 1760 cal.
13. Specific heat $= \dfrac{435 \text{ J}}{3.4 \text{ g} \times 64°C} = \dfrac{2.0 \text{ J}}{\text{g} \times °C}$

Chapter 3

1. **a.** Mass = volume × density

   $= 0.87 \text{ cm}^3 \times \dfrac{19.3 \text{ g}}{1 \text{ cm}^3} = 17 \text{ g}$

   b. $17 \text{ g} \times \dfrac{\$12}{1 \text{ g}} = = \$204$
2. Heat(J) = mass $\times \Delta T \times$ specific heat

   $J = 10.0 \text{ g} \times 15.0°C \times \dfrac{4.18 \text{ J}}{\text{g} \times °C} = 627 \text{ J}$
3. $\Delta T = \dfrac{\text{heat(cal)}}{\text{mass} \times \text{specific heat}} = \dfrac{8.5 \times 10^2 \text{ cal}}{140 \text{ g} \times \dfrac{0.12 \text{ cal}}{\text{g} \times °C}} = 51°C$

   $T_{final} = T_{inital} + \Delta T = 21°C + 51°C = 72°C$
4. Find the mass of the added mercury using the density.

   $\text{Mass} = \text{volume} \times \text{density} = 15.0 \text{ mL} \times \dfrac{13.5 \text{ g}}{1.0 \text{ mL}}$
   $= 203 \text{ g Hg}$

   Add the mass of the mercury to the mass of the beaker.
   203 g + 56.7 g = 259.7 = 2.60×10^2 g
5. The density of silicon is 2.33 g/cm³.

   $\text{Volume} = \dfrac{\text{mass}}{\text{density}} = \dfrac{62.9 \text{ g}}{2.33 \text{ g/cm}^3} = 62.9 \text{ g} \times \dfrac{1 \text{ cm}^3}{2.33 \text{ g}}$
   $= 27.0 \text{ cm}^3$
6. **a.** $\dfrac{1 \text{ m}}{10^6 \text{ } \mu m}$ and $\dfrac{10^6 \text{ } \mu m}{1 \text{ m}}$ **b.** $\dfrac{1 \text{ g}}{10^2 \text{ cg}}$ and $\dfrac{10^2 \text{ cg}}{1 \text{ g}}$

   c. $\dfrac{1 \text{ g H}_2\text{O}}{1 \text{ mL H}_2\text{O}}$ and $\dfrac{1 \text{ mL H}_2\text{O}}{1 \text{ g H}_2\text{O}}$ **d.** $\dfrac{10^3 \text{ J}}{1 \text{ kJ}}$ and $\dfrac{1 \text{ kJ}}{10^3 \text{ J}}$
7. $\dfrac{1 \text{ g}}{100 \text{ cg}}, \dfrac{100 \text{ cg}}{1 \text{ g}}, \dfrac{1 \text{ g}}{10^6 \text{ } \mu g}, \dfrac{10^6 \text{ } \mu g}{1 \text{ g}}, \dfrac{100 \text{ cg}}{10^6 \text{ } \mu g},$ and $\dfrac{10^6 \text{ } \mu g}{100 \text{ cg}}$
8. $85.4°F \text{ temp. drop} \times 1 \dfrac{°C \text{ temp. drop}}{1.8°F \text{ temp. drop}} = 47.4°C$
9. There is 26.0 g of silver in 40.0 g of the amalgam.

   $15.0 \text{ g amalgam} \times \dfrac{26.0 \text{ g silver}}{40.0 \text{ g amalgam}} = 9.75 \text{ g silver}$
10. **a.** $0.73 \text{ km} \times \dfrac{10^3 \text{ m}}{1 \text{ km}} = 0.73 \times 10^3 \text{ m} = 7.3 \times 10^2 \text{ m}$

    b. $1.75 \text{ g} \times \dfrac{10^3 \text{ mg}}{1 \text{ g}} = 1.75 \times 10^3 \text{ mg}$

    c. $4.7 \text{ nm} \times \dfrac{1 \text{ m}}{10^9 \text{ nm}} = 4.7 \times 10^{-9} \text{ m}$

    d. $0.072 \text{ m} \times \dfrac{10^2 \text{ cm}}{1 \text{ m}} = 7.2 \text{ cm}$

    e. $53 \text{ cm}^3 \times \dfrac{1 \text{ dm}^3}{10^3 \text{ cm}^3} = 5.3 \times 10^{-2} \text{ dm}^3$

    f. $8.34 \text{ J} \times \dfrac{1 \text{ kJ}}{10^3 \text{ J}} = 8.34 \times 10^{-3} \text{ kJ}$

    g. $6.7 \text{ ms} \times \dfrac{1 \text{ s}}{10^3 \text{ ms}} = 6.7 \times 10^{-3} \text{ s}$

    h. $45.3 \text{ kcal} \times \dfrac{10^3 \text{ cal}}{1 \text{ kcal}} = 4.53 \times 10^4 \text{ cal}$

11. a. $23.6 \text{ g B} \times \dfrac{1 \text{ cm}^3 \text{ B}}{2.34 \text{ g B}} = 10.1 \text{ cm}^3 \text{ B}$

b. $1.4 \text{ L Ar} \times \dfrac{1.78 \text{ g Ar}}{1 \text{ L Ar}} = 2.5 \text{ g Ar}$

c. $8.96 \text{ g Hg} \times \dfrac{1 \text{ cm}^3 \text{ Hg}}{13.5 \text{ g Hg}} = 0.664 \text{ cm}^3 \text{ Hg}$

12. a. $0.57 \text{ } \mu\text{g} \times \dfrac{1 \text{ g}}{10^6 \text{ } \mu\text{g}} \times \dfrac{10^2 \text{ cg}}{1 \text{ g}} = 0.57 \times 10^{-4}$
$= 5.7 \times 10^{-5} \text{ cg}$

b. $27.6 \text{ } \mu\text{L} \times \dfrac{1 \text{ L}}{10^6 \text{ } \mu\text{L}} \times \dfrac{10^3 \text{ mL}}{1 \text{ L}} = 27.6 \times 10^{-3}$
$= 2.76 \times 10^{-2} \text{ mL}$

c. $583 \text{ nm} \times \dfrac{1 \text{ m}}{10^9 \text{ nm}} \times \dfrac{10^3 \text{ mm}}{1 \text{ m}} = 583 \times 10^{-6}$
$= 5.83 \times 10^{-4} \text{ mm}$

d. $0.0023 \text{ kcal} \times \dfrac{10^3 \text{ cal}}{1 \text{ kcal}} \times \dfrac{10^1 \text{ dcal}}{1 \text{ cal}} = 0.0023 \times 10^4$
$= 2.3 \times 10^1 \text{ dcal}$

e. $678 \text{ cJ} \times \dfrac{1 \text{ J}}{10^2 \text{ cJ}} \times \dfrac{1 \text{ kJ}}{10^3 \text{ J}} = 678 \times 10^{-5}$
$= 6.78 \times 10^{-3} \text{ kJ}$

f. $5.75 \times 10^5 \text{ cg} \times \dfrac{1 \text{ g}}{10^2 \text{ cg}} \times \dfrac{10^9 \text{ ng}}{1 \text{ g}} = 5.75 \times 10^{12} \text{ ng}$

13. a. $\text{Mass} = \dfrac{\text{cal}}{(\Delta T)(\text{specific heat})}$
$7.5 \text{ kcal} = 7.5 \times 10^3 \text{ cal}$
$\text{Mass} = \dfrac{7.5 \times 10^3 \text{ cal}}{25°\text{C} \times 1.00 \text{ cal}/(\text{g} \times °\text{C})} = 3.0 \times 10^2 \text{ g}$

b. $3.0 \times 10^2 \text{ g} \times \dfrac{1 \text{ kg}}{10^3 \text{ g}} = 3.0 \times 10^{-1} \text{ kg}$

14. a. $\dfrac{4.65 \text{ kg}}{1 \text{ L}} \times \dfrac{10^3 \text{ g}}{1 \text{ kg}} \times \dfrac{1 \text{ L}}{10^3 \text{ cm}^3} = 4.65 \text{ g/cm}^3$

b. $\dfrac{0.74 \text{ kcal}}{1 \text{ min}} \times \dfrac{10^3 \text{ cal}}{1 \text{ kcal}} \times \dfrac{1 \text{ min}}{60 \text{ s}} = 12 \text{ cal/s}$

c. $10 \text{ mm} = 1 \text{ cm}; \ 10^2 \text{ mm}^2 = 1 \text{ cm}^2$
$\dfrac{1.42 \text{ g}}{1 \text{ cm}^2} \times \dfrac{10^3 \text{ mg}}{1 \text{ g}} \times \dfrac{1 \text{ cm}^2}{10^2 \text{ mm}^2} = 1.42 \times 10^1 \text{ mg/mm}^2$

Chapter 4

1. Six protons and six electrons.
2. Sodium.
3.

Atomic number	Mass number	Protons	Neutrons	Electrons	Symbol
7	14	7	7	7	N
9	19	9	10	9	F
19	39	19	20	19	K
27	59	27	32	27	Co

4. Oxygen-16: $^{16}_{8}\text{O}$; oxygen-18: $^{18}_{8}\text{O}$.
5. a. 6 **b.** 8 **c.** 138
6. Calculate the weighted average by multiplying the percent abundance of each isotope by that isotope's mass.

$^{63}\text{Cu: } 63 \text{ amu} \times \dfrac{69.1}{100} = 43.5 \text{ amu}$

$^{65}\text{Cu: } 65 \text{ amu} \times \dfrac{30.9}{100} = \dfrac{20.1 \text{ amu}}{63.6 \text{ amu}}$

Chapter 5

1. a. metal **b.** nonmetal **c.** metal **d.** nonmetal **e.** metal
2. Potassium, boron, and iodine.
3. a. Ca^{2+}, calcium ion **b.** one electron gained, fluoride ion **c.** Al, 3 electrons lost, aluminum ion **d.** Se^{2-}, selenide ion
4. Find the lead-to-oxygen mass ratio for each compound.

$\dfrac{2.98 \text{ g Pb}}{0.461 \text{ g O}} = \dfrac{6.46 \text{ g Pb}}{1 \text{ g O}} \qquad \dfrac{9.89 \text{ g Pb}}{0.763 \text{ g O}} = \dfrac{12.96 \text{ g Pb}}{1 \text{ g O}}$

Calculate the mass ratio of lead to 1 g of oxygen.

$\dfrac{12.96 \text{ g Pb (compound 2)}}{6.46 \text{ g Pb (compound 1)}} = \dfrac{2}{1}$

5. a. I^- **b.** Hg^{2+} **c.** P^{3-} **d.** Ba^{2+} **e.** Sn^{4+} **f.** Ag^+
6. a. copper(II) or cupric ion **b.** oxide ion **c.** lithium ion **d.** lead(II) or plumbous ion **e.** fluoride ion **f.** hydrogen ion
7. a. Li_2S **b.** SnO_2 **c.** HCl **d.** Mg_3N_2
8. a. aluminum fluoride **b.** iron(II) oxide or ferrous oxide **c.** copper(I) sulfide or cuprous sulfide **d.** calcium selenide
9. a. BaSO_4 **b.** $\text{Al(HCO}_3)_3$ **c.** NaClO **d.** $\text{Pb(CrO}_4)_2$
10. a. chromium(II) nitrate or chromous nitrate **b.** magnesium phosphate **c.** copper(I) hydrogen phosphate or cuprous hydrogen phosphate **d.** lithium chromate
11. a. Al(OH)_3 **b.** potassium silicate **c.** lithium fluoride **d.** KClO_2 **e.** tin(II) phosphate or stannous phosphate **f.** $\text{Zn(HSO}_4)_2$
12. a. carbon tetrabromide **b.** dichlorine heptoxide **c.** dinitrogen pentoxide **d.** boron trichloride **e.** chromium(III) chloride
13. a. CS_2 **b.** N_2H_4 **c.** CCl_4 **d.** P_2O_3
14. a. hydrofluoric acid **b.** acetic acid **c.** sulfuric acid **d.** nitrous acid
15. a. H_2CrO_4 **b.** HI **c.** HClO_2 **d.** HClO_4
16. a. calcium carbonate **b.** potassium permanganate **c.** lead(II) chromate **d.** calcium hydrogen phosphate **e.** tin(II) dichromate **f.** magnesium phosphide **g.** ammonium sulfate **h.** iodine chloride **i.** nitrous acid **j.** strontium bromide
17. a. Sn(OH)_2 **b.** BaF_2 **c.** I_4O_9 **d.** $\text{Fe}_2(\text{C}_2\text{O}_4)_3$ **e.** H_2S **f.** $\text{Al(HCO}_3)_3$ **g.** Na_3PO_4 **h.** KClO_4 **i.** S_2O_3 **j.** $\text{Mg(NO}_3)_2$

Chapter 6

1. $1.20 \times 10^{25} \text{ atoms P} \times \dfrac{1.00 \text{ mol P}}{6.02 \times 10^{23} \text{ atoms P}}$
$= 0.199 \times 10^2 \text{ mol P} = 1.99 \times 10^1 \text{ mol P}$

2. $0.750 \text{ mol Zn} \times \dfrac{6.02 \times 10^{23} \text{ atoms Zn}}{1 \text{ mol Zn}}$
$= 4.52 \times 10^{23} \text{ atoms Zn}$

3. $0.400 \text{ mol N}_2\text{O}_5 \times \dfrac{6.02 \times 10^{23} \text{ molecules N}_2\text{O}_5}{1 \text{ mol N}_2\text{O}_5}$
$= 2.41 \times 10^{23} \text{ molecules N}_2\text{O}_5$

4. $1.20 \times 10^{24} \text{ molecules CO}_2 \times$
$\dfrac{1 \text{ mol CO}_2}{6.02 \times 10^{23} \text{ molecules CO}_2} = 1.99 \text{ mol CO}_2$

5. $0.036 \text{ mol (NH}_4)_3\text{PO}_4 \times \dfrac{6.02 \times 10^{23} \text{ formula units}}{1 \text{ mol (NH}_4)_3\text{PO}_4}$
$\times \dfrac{3 \text{ NH}_4^+ \text{ ions}}{1 \text{ formula unit}} = 6.50 \times 10^{22} \text{ NH}_4^+ \text{ ions}$

6. Add the number of carbon atoms in each compound.

$$3.00 \text{ mol } C_2H_2 \times \frac{6.02 \times 10^{23} \text{ molecules } C_2H_2}{1 \text{ mol } C_2H_2} \times$$

$$\frac{2 \text{ atoms C}}{1 \text{ molecule } C_2H_2} = 36.1 \times 10^{23} \text{ atoms C}$$

$$= 3.61 \times 10^{24} \text{ atoms C}$$

$$0.700 \text{ mol } CO \times \frac{6.02 \times 10^{23} \text{ molecules } CO}{1 \text{ mol } CO} \times$$

$$\frac{1 \text{ atom C}}{1 \text{ molecule } CO} = 4.21 \times 10^{23} \text{ atoms C}$$

Total number carbon atoms $= 36.1 \times 10^{23} + 4.21 \times 10^{23}$

$$= 40.3 \times 10^{23}$$

$$= 4.03 \times 10^{24} \text{ atoms C}$$

7. Round off the gram atomic mass to tenths.
a. 23.0 g
b. 74.9 g
c. 238.0 g

8. a. $\quad 1 \text{ mol } C \times \frac{12.0 \text{ g C}}{1 \text{ mol } C} = 12.0 \text{ g C}$

$$4 \text{ mol } Cl \times \frac{35.5 \text{ g Cl}}{1 \text{ mol } Cl} = 142.0 \text{ g Cl}$$

gfm $CCl_4 = 154.0$ g

b. $\quad 1 \text{ mol } P \times \frac{31.0 \text{ g P}}{1 \text{ mol } P} = 31.0 \text{ g P}$

$$3 \text{ mol } Cl \times \frac{35.5 \text{ g Cl}}{1 \text{ mol } Cl} = 106.5 \text{ g Cl}$$

gfm $PCl_3 = 137.5$ g

c. $\quad 8 \text{ mol } C \times \frac{12.0 \text{ g C}}{1 \text{ mol } C} = 96.0 \text{ g C}$

$$18 \text{ mol } H \times \frac{1.0 \text{ g H}}{1 \text{ mol } H} = 18.0 \text{ g H}$$

gfm $C_8H_{18} = 114.0$ g

d. $2 \text{ mol } N \times \frac{14.0 \text{ g N}}{1 \text{ mol } N} = 28.0 \text{ g N}$

$$5 \text{ mol } O \times \frac{16.0 \text{ g O}}{1 \text{ mol } O} = 80.0 \text{ g O}$$

gfm $N_2O_5 = 108.0$ g

9. a. $1 \text{ mol } Sr \times \frac{87.6 \text{ g Sr}}{1 \text{ mol } Sr} = 87.6 \text{ g Sr}$

$$2 \text{ mol } Cl \times \frac{35.5 \text{ g Cl}}{1 \text{ mol } Cl} = 71.0 \text{ g Cl}$$

gfm $SrCl_2 = 158.6$ g

b. $2 \text{ mol } Na \times \frac{23.0 \text{ g Na}}{1 \text{ mol } Na} = 46.0 \text{ g Na}$

$$1 \text{ mol } C \times \frac{12.0 \text{ g C}}{1 \text{ mol } C} = 12.0 \text{ g C}$$

$$3 \text{ mol } O \times \frac{16.0 \text{ g O}}{1 \text{ mol } O} = 48.0 \text{ g O}$$

gfm $Na_2CO_3 = 106.0$ g

c. $2 \text{ mol } Al \times \frac{27.0 \text{ g Al}}{1 \text{ mol } Al} = 54.0 \text{ g Al}$

$$3 \text{ mol } S \times \frac{32.1 \text{ g S}}{1 \text{ mol } S} = 96.3 \text{ g S}$$

$$12 \text{ mol } O \times \frac{16.0 \text{ g O}}{1 \text{ mol } O} = 192.0 \text{ g O}$$

gfm $Al_2(SO_4)_3 = 342.3$ g

d. $1 \text{ mol } Ca \times \frac{40.1 \text{ g Ca}}{1 \text{ mol } Ca} = 40.1 \text{ g Ca}$

$$2 \text{ mol } C \times \frac{12.0 \text{ g C}}{1 \text{ mol } C} = 24.0 \text{ g C}$$

$$2 \text{ mol } N \times \frac{14.0 \text{ g N}}{1 \text{ mol } N} = 28.0 \text{ g N}$$

gfm $Ca(CN)_2 = 92.1$ g

10. a. $10.0 \text{ mol } Cr \times \frac{52.0 \text{ g Cr}}{1 \text{ mol } Cr} = 520 = 5.20 \times 10^2 \text{ g Cr}$

b. $3.32 \text{ mol } K \times \frac{39.1 \text{ g K}}{1 \text{ mol } K} = 129.8 = 1.30 \times 10^2 \text{ g K}$

c. $2.20 \times 10^{-3} \text{ mol } Sn \times \frac{118.7 \text{ g Sn}}{1 \text{ mol } Sn} = 0.261 \text{ g Sn}$

d. $0.720 \text{ mol } Be \times \frac{9.01 \text{ g Be}}{1 \text{ mol } Be} = 6.49 \text{ g Be}$

e. $2.40 \text{ mol } N_2 \times \frac{28.0 \text{ g N}_2}{1 \text{ mol } N_2} = 67.2 \text{ g N}_2$

f. $0.160 \text{ mol } H_2O_2 \times \frac{34.0 \text{ g H}_2O_2}{1 \text{ mol } H_2O_2} = 5.44 \text{ g H}_2O_2$

g. $5.08 \text{ mol } Ca(NO_3)_2 \times \frac{164.1 \text{ g Ca(NO}_3)_2}{1 \text{ mol } Ca(NO_3)_2} =$

$$8.34 \times 10^2 \text{ g Ca(NO}_3)_2$$

h. $15.0 \text{ mol } H_2SO_4 \times \frac{98.1 \text{ g H}_2SO_4}{1 \text{ mol } H_2SO_4} =$

$$1.47 \times 10^3 \text{ g H}_2SO_4$$

i. $4.52 \times 10^{-3} \text{ mol } C_{20}H_{42} \times \frac{282 \text{ g C}_{20}H_{42}}{1 \text{ mol } C_{20}H_{42}} =$

$$1.27 \text{ g C}_{20}H_{42}$$

j. $0.0112 \text{ mol } K_2CO_3 \times \frac{138.2 \text{ g K}_2CO_3}{1 \text{ mol } K_2CO_3} = 1.55 \text{ g K}_2CO_3$

11. a. $72.0 \text{ g Ar} \times \frac{1 \text{ mol Ar}}{39.9 \text{ g Ar}} = 1.80 \text{ mol Ar}$

b. $3.70 \times 10^{-1} \text{ g B} \times \frac{1 \text{ mol B}}{10.8 \text{ g B}} = 3.43 \times 10^{-2} \text{ mol B}$

c. $187 \text{ g Al} \times \frac{1 \text{ mol Al}}{27.0 \text{ g Al}} = 6.93 \text{ mol Al}$

d. $333 \text{ g SnF}_2 \times \frac{1 \text{ mol SnF}_2}{156.7 \text{ g SnF}_2} = 2.13 \text{ mol SnF}_2$

e. $7.21 \times 10^{-2} \text{ g He} \times \frac{1 \text{ mol He}}{4 \text{ g He}} = 1.80 \times 10^{-2} \text{ mol He}$

f. $27.4 \text{ g TiO}_2 \times \frac{1 \text{ mol TiO}_2}{79.9 \text{ g TiO}_2} = 0.343 \text{ mol TiO}_2$

g. $5.00 \text{ g H}_2 \times \frac{1 \text{ mol H}_2}{2.0 \text{ g H}_2} = 2.5 \text{ mol H}_2$

h. $0.000 \, 264 \text{ g Li}_2HPO_4 \times \frac{1 \text{ mol Li}_2HPO_4}{109.8 \text{ g Li}_2HPO_4} =$

$$2.40 \times 10^{-6} \text{ mol Li}_2HPO_4$$

i. $11.0 \text{ g CH}_4 \times \frac{1 \text{ mol CH}_4}{16.0 \text{ g CH}_4} = 0.688 \text{ mol CH}_4$

j. $847 \text{ g (NH}_4)_2CO_3 \times \frac{1 \text{ mol (NH}_4)_2CO_3}{96.0 \text{ g (NH}_4)_2CO_3} =$

$$8.82 \text{ mol (NH}_4)_2CO_3$$

12. **a.** $5.40 \text{ mol } O_2 \times \dfrac{22.4 \text{ L } O_2}{1 \text{ mol } O_2} = 121 \text{ L } O_2$

b. $3.20 \times 10^{-2} \text{ mol } CO_2 \times \dfrac{22.4 \text{ L } CO_2}{1 \text{ mol } CO_2} = 0.717 \text{ L } CO_2$

c. $0.960 \text{ mol } CH_4 \times \dfrac{22.4 \text{ L } CH_4}{1 \text{ mol } CH_4} = 21.5 \text{ L } CH_4$

13. **a.** $89.6 \text{ L } SO_2 \times \dfrac{1 \text{ mol } SO_2}{22.4 \text{ L } SO_2} = 4.00 \text{ mol } SO_2$

b. $1.00 \times 10^3 \text{ L } C_2H_6 \times \dfrac{1 \text{ mol } C_2H_6}{22.4 \text{ L } C_2H_6} = 44.6 \text{ mol } C_2H_6$

c. $5.42 \times 10^{-1} \text{ mL } Ne \times \dfrac{1 \text{ L } Ne}{10^3 \text{ mL } Ne} \times \dfrac{1 \text{ mol } Ne}{22.4 \text{ L } Ne} = 2.42 \times 10^{-5} \text{ mol } Ne$

14. At STP the gfm of any gas occupies a volume of 22.4 L.

a. gas A: $22.4 \text{ L } \times \dfrac{1.25 \text{ g}}{1 \text{ L}} = 28.0 \text{ g (nitrogen)}$

b. gas B: $22.4 \text{ L } \times \dfrac{2.86 \text{ g}}{1 \text{ L}} = 64.1 \text{ g (sulfur dioxide)}$

c. gas C: $22.4 \text{ L } \times \dfrac{0.714 \text{ g}}{1 \text{ L}} = 16.0 \text{ g (methane)}$

15. $1 \text{ atom } Ni \times \dfrac{1.00 \text{ mol } Ni}{6.02 \times 10^{23} \text{ atoms } Ni} \times \dfrac{58.7 \text{ g } Ni}{1.00 \text{ mol } Ni} = 9.75 \times 10^{-23} \text{ g } Ni$

16. $6.00 \text{ L } CO_2 \times \dfrac{1.00 \text{ mol } CO_2}{22.4 \text{ L } CO_2} \times \dfrac{6.02 \times 10^{23} \text{ molec. } CO_2}{1 \text{ mol } CO_2} = 1.61 \times 10^{23}$ molecules of CO_2 or of any gas at STP.

17. **a.** Total mass $= 29.0 \text{ g } Ag + 4.30 \text{ g } S = 33.3 \text{ g}$

$\%Ag = \dfrac{29.0 \text{ g}}{33.3 \text{ g}} \times 100\% = 87.1\% \text{ Ag}$

$\%S = \dfrac{4.30 \text{ g}}{33.3 \text{ g}} \times 100\% = 12.9\% \text{ S}$

b. $9.03 \text{ g } Mg + 3.48 \text{ g } N = 12.51 \text{ g total}$

$\%Mg = \dfrac{9.03 \text{ g}}{12.51 \text{ g}} \times 100\% = 72.2\% \text{ Mg}$

$\%N = \dfrac{3.48 \text{ g}}{12.51 \text{ g}} \times 100\% = 27.8\% \text{ N}$

c. $222.6 \text{ g } Na + 77.4 \text{ g } O = 300.0 \text{ g total}$

$\%Na = \dfrac{222.6 \text{ g}}{300.0 \text{ g}} \times 100\% = 74.2\% \text{ Na}$

$\%O = \dfrac{77.4 \text{ g}}{300.0 \text{ g}} \times 100\% = 25.8\% \text{ O}$

18. **a.** gfm $C_3H_8 = 44.0 \text{ g}$

$\%C = \dfrac{36.0 \text{ g}}{44.0 \text{ g}} \times 100\% = 81.8\% \text{ C}$

$\%H = \dfrac{8.0 \text{ g}}{44.0 \text{ g}} \times 100\% = 18.2\% \text{ H}$

b. gfm $NaHSO_4 = 120.1 \text{ g}$

$\%Na = \dfrac{23.0 \text{ g}}{120.1 \text{ g}} \times 100\% = 19.2\% \text{ Na}$

$\%H = \dfrac{1.0 \text{ g}}{120.1 \text{ g}} \times 100\% = 0.83\% \text{ H}$

$\%S = \dfrac{32.1 \text{ g}}{120.1 \text{ g}} \times 100\% = 26.7\% \text{ S}$

$\%O = \dfrac{64.0 \text{ g}}{120.1 \text{ g}} \times 100\% = 53.3\% \text{ O}$

c. gfm $Ca(C_2H_3O_2)_2 = 158.1 \text{ g}$

$\%Ca = \dfrac{40.1 \text{ g}}{158.1 \text{ g}} \times 100\% = 25.4\% \text{ Ca}$

$\%C = \dfrac{48.0 \text{ g}}{158.1 \text{ g}} \times 100\% = 30.4\% \text{ C}$

$\%H = \dfrac{6.0 \text{ g}}{158.1 \text{ g}} \times 100\% = 3.8\% \text{ H}$

$\%O = \dfrac{64.0 \text{ g}}{158.1 \text{ g}} \times 100\% = 40.5\% \text{ O}$

d. gfm $HCN = 27.0 \text{ g}$

$\%H = \dfrac{1.0 \text{ g}}{27.0 \text{ g}} \times 100\% = 3.7\% \text{ H}$

$\%C = \dfrac{12.0 \text{ g}}{27.0 \text{ g}} \times 100\% = 44.4\% \text{ C}$

$\%N = \dfrac{14.0 \text{ g}}{27.0 \text{ g}} \times 100\% = 51.9\% \text{ N}$

19. **a.** Percent of H in C_3H_8 is 18.2%, or 18.2 g H/100 g C_3H_8.

$350 \text{ g } C_3H_8 \times \dfrac{18.2 \text{ g } H}{100 \text{ g } C_3H_8} = 63.7 \text{ g } H$

b. $20.2 \text{ g } NaHSO_4 \times \dfrac{0.83 \text{ g } H}{100 \text{ g } NaHSO_4} = 0.17 \text{ g } H$

c. $124 \text{ g } Ca(C_2H_3O_2)_2 \times \dfrac{3.8 \text{ g } H}{100 \text{ g } Ca(C_2H_3O_2)_2} = 4.71 \text{ g } H$

d. $378 \text{ g } HCN \times \dfrac{3.7 \text{ g } H}{100 \text{ g } HCN} = 14 \text{ g } H$

20. **a.** Calculate the lowest whole-number ratio of moles.

$79.9 \text{ g } C \times \dfrac{1.00 \text{ mol } C}{12.0 \text{ g } C} = 6.66 \text{ mol } C$

$\dfrac{6.66 \text{ mol } C}{6.66} = 1 \text{ mol } C$

$20.1 \text{ g } H \times \dfrac{1.00 \text{ mol } H}{1.00 \text{ g } H} = 20.1 \text{ mol } H$

$\dfrac{20.1 \text{ mol } H}{6.66} = 3 \text{ mol } H$

Empirical formula $= CH_3$

b. $67.6 \text{ g } Hg \times \dfrac{1.00 \text{ mol } Hg}{200.6 \text{ g } Hg} = 0.337 \text{ mol } Hg$

$\dfrac{0.337 \text{ mol } Hg}{0.336} = 1 \text{ mol } Hg$

$10.8 \text{ g } S \times \dfrac{1.00 \text{ mol } S}{32.1 \text{ g } S} = 0.336 \text{ mol } S$

$\dfrac{0.336 \text{ mol } S}{0.336} = 1 \text{ mol } S$

$21.6 \text{ g } O \times \dfrac{1.00 \text{ mol } O}{16.0 \text{ g } O} = 1.35 \text{ mol } O$

$\dfrac{1.35 \text{ mol } O}{0.336} = 4 \text{ mol } O$ Empirical formula $= HgSO_4$

c. $94.1 \text{ g } O \times \dfrac{1.00 \text{ mol } O}{16.0 \text{ g } O} = 5.88 \text{ mol } O$

$\dfrac{5.88 \text{ mol } O}{5.88} = 1 \text{ mol } O$

$5.9 \text{ g } H \times \dfrac{1.00 \text{ mol } H}{1.0 \text{ g } H} = 5.9 \text{ mol } H$

$\dfrac{5.9 \text{ mol } H}{5.88} = 1 \text{ mol } H$ Empirical formula $= HO$

d. $17.6 \text{ g Na} \times \dfrac{1.00 \text{ mol Na}}{23.0 \text{ g Na}} = 0.765 \text{ mol Na}$

$\dfrac{0.765 \text{ mol Na}}{0.763} = 1 \text{ mol Na}$

$39.7 \text{ g Cr} \times \dfrac{1.00 \text{ mol Cr}}{52.0 \text{ g Cr}} = 0.763 \text{ mol Cr}$

$\dfrac{0.763 \text{ mol Cr}}{0.763} = 1 \text{ mol Cr}$

$42.7 \text{ g O} \times \dfrac{1.00 \text{ mol O}}{16.0 \text{ g O}} = 2.67 \text{ mol O}$

$\dfrac{2.67 \text{ mol O}}{0.763} = 3.50 \text{ mol O}$

The result is $Na_1Cr_1O_{3.5}$. To get the lowest whole-number ratio, multiply through by 2. The empirical formula is $Na_2Cr_2O_7$.

e. $27.59 \text{ g C} \times \dfrac{1.00 \text{ mol C}}{12.0 \text{ g C}} = 2.30 \text{ mol C}$

$\dfrac{2.30 \text{ mol C}}{1.15} = 2 \text{ mol C}$

$1.15 \text{ g H} \times \dfrac{1.00 \text{ mol H}}{1.00 \text{ g H}} = 1.15 \text{ mol H}$

$\dfrac{1.15 \text{ mol H}}{1.15} = 1 \text{ mol H}$

$16.09 \text{ g N} \times \dfrac{1.00 \text{ mol N}}{14.0 \text{ g N}} = 1.15 \text{ mol N}$

$\dfrac{1.15 \text{ mol N}}{1.15} = 1 \text{ mol N}$

$55.17 \text{ g O} \times \dfrac{1.00 \text{ mol O}}{16.0 \text{ g O}} = 3.45 \text{ mol O}$

$\dfrac{3.45 \text{ mol O}}{1.15} = 3 \text{ mol O}$

Empirical formula = C_2HNO_3

21. Calculate the relative number of moles from the given masses.

a. $29.0 \text{ g Ag} \times \dfrac{1.00 \text{ mol Ag}}{107.9 \text{ g Ag}} = 0.269 \text{ mol Ag}$

$\dfrac{0.269 \text{ mol Ag}}{0.134} = 2 \text{ mol Ag}$

$4.30 \text{ g S} \times \dfrac{1.00 \text{ mol S}}{32.1 \text{ g S}} = 0.134 \text{ mol S}$

$\dfrac{0.134 \text{ mol S}}{0.134} = 1 \text{ mol S}$

Empirical formula = Ag_2S

b. $9.03 \text{ g Mg} \times \dfrac{1.00 \text{ mol Mg}}{24.3 \text{ g Mg}} = 0.372 \text{ mol Mg}$

$\dfrac{0.372 \text{ mol Mg}}{0.249} = 1.5 \text{ mol Mg}$

$3.48 \text{ g N} \times \dfrac{1.00 \text{ mol N}}{14.0 \text{ g N}} = 0.249 \text{ mol N}$

$\dfrac{0.249 \text{ mol N}}{0.249} = 1 \text{ mol N}$

The result is $Mg_{1.5}N_1$. Multiply through by 2.
Empirical formula = Mg_3N_2

c. $222.6 \text{ g Na} \times \dfrac{1.00 \text{ mol Na}}{23.0 \text{ g Na}} = 9.68 \text{ mol Na}$

$\dfrac{9.68 \text{ mol Na}}{4.84} = 2 \text{ mol Na}$

$77.4 \text{ g O} \times \dfrac{1.00 \text{ mol O}}{16.0 \text{ g O}} = 4.84 \text{ mol O}$

$\dfrac{4.84 \text{ mol O}}{4.84} = 1 \text{ mol O}$

Empirical formula = Na_2O

22. Calculate the empirical formula.

$58.8 \text{ g C} \times \dfrac{1.00 \text{ mol C}}{12.0 \text{ g C}} = 4.90 \text{ mol C}$

$\dfrac{4.90 \text{ mol C}}{1.96} = 2.50 \text{ mol C}$

$9.8 \text{ g H} \times \dfrac{1.00 \text{ mol H}}{1.00 \text{ g H}} = 9.8 \text{ mol H}$

$\dfrac{9.80 \text{ mol H}}{1.96} = 5.00 \text{ mol H}$

$31.4 \text{ g O} \times \dfrac{1.00 \text{ mol O}}{16.0 \text{ g O}} = 1.96 \text{ mol O}$

$\dfrac{1.96 \text{ mol O}}{1.96} = 1 \text{ mol O}$

The result is $C_{2.5}H_5O_1$. Multiply through by 2. The empirical formula is $C_5H_{10}O_2$. This is also the molecular formula since the gfm of $C_5H_{10}O_2$ is 102 g, which is the gmm of the compound.

23. Calculate the empirical formula. The mass of hydrogen is: 7.36 g − 6.93 g = 0.43 g H

$6.93 \text{ g O} \times \dfrac{1.00 \text{ mol O}}{16.0 \text{ g O}} = 0.433 \text{ mol O}$

$\dfrac{0.433 \text{ mol O}}{0.43} = 1 \text{ mol O}$

$0.43 \text{ g H} \times \dfrac{1.0 \text{ mol H}}{1.0 \text{ g H}} = 0.43 \text{ mol H}$

$\dfrac{0.43 \text{ mol H}}{0.43} = 1 \text{ mol H}$

The empirical formula is HO. The gfm of HO is 17 g; the gmm of the compound is 34.0. It takes 34 g/17 g, or 2 empirical formula units to make the molecular formula. The molecular formula is H_2O_2.

Chapter 7

1. a. $Al(s) + O_2(g) \longrightarrow Al_2O_3(s)$
b. $HgS(s) + O_2(g) \longrightarrow Hg(l) + SO_2(g)$
c. $KClO_3(s) \xrightarrow{MnO_2} KCl(s) + O_2(g)$

2. a. Aqueous solutions of barium chloride and sulfuric acid when mixed produce a precipitate of barium sulfate and aqueous hydrochloric acid. **b.** Gaseous ammonia and oxygen react in the presence of platinum to produce mononitrogen monoxide gas and water vapor. **c.** The gas dinitrogen trioxide reacts with water to produce an aqueous solution of nitrous acid.

3. a. $2Al + N_2 \longrightarrow 2AlN$
b. $2NaCl + H_2SO_4 \longrightarrow Na_2SO_4 + 2HCl$
c. $2Al + 3CuSO_4 \longrightarrow Al_2(SO_4)_3 + 3Cu$
d. $4P + 5O_2 \longrightarrow 2P_2O_5$
e. $2Fe(OH)_3 \longrightarrow Fe_2O_3 + 3H_2O$

4. **a.** $2Na + 2H_2O \longrightarrow 2NaOH + H_2$
   b. $H_2 + S \longrightarrow H_2S$
   c. $2FeCl_3 + 3Ca(OH)_2 \longrightarrow 2Fe(OH)_3 + 3CaCl_2$
   d. $2C + O_2 \longrightarrow 2CO$
   e. $2KNO_3 \longrightarrow 2KNO_2 + O_2$
5. **a.** $Ca + S \longrightarrow CaS$ **d.** $N_2O_5 + H_2O \longrightarrow 2HNO_3$
   b. $2Fe + O_2 \longrightarrow 2FeO$ **e.** $Na_2O + H_2O \longrightarrow 2NaOH$
   c. $4P + 5O_2 \longrightarrow P_4O_{10}$ **f.** $2Mg + O_2 \longrightarrow 2MgO$
6. **a.** $NiCO_3 \longrightarrow NiO + CO_2$
   b. $2Ag_2O \longrightarrow 4Ag + O_2$
   c. $NH_4NO_3 \longrightarrow N_2O + 2H_2O$
7. **a.** $2Al + 3H_2SO_4 \longrightarrow Al_2(SO_4)_3 + 3H_2$
   b. $Cl_2 + 2KI \longrightarrow 2KCl + I_2$
   c. $Cu + FeSO_4 \longrightarrow$ no reaction
   d. $2Li + 2H_2O \longrightarrow 2LiOH + H_2$
8. **a.** $3H_2SO_4 + 2Al(OH)_3 \longrightarrow Al_2(SO_4)_3 + 6H_2O$
   b. $3KOH + H_3PO_4 \longrightarrow K_3PO_4 + 3H_2O$
   c. $SrBr_2 + (NH_4)_2CO_3 \longrightarrow SrCO_3 + 2NH_4Br$
9. **a.** $C_6H_{12}O_6 + 6O_2 \longrightarrow 6CO_2 + 6H_2O$
   b. $2C_8H_{18} + 25O_2 \longrightarrow 16CO_2 + 18H_2O$
10. **a.** $Mg + H_2SO_4 \longrightarrow MgSO_4 + H_2$ (single-replacement)
   b. $3Fe + 2O_2 \longrightarrow Fe_3O_4$ (combination)
   c. $2Pb(NO_3)_2 \longrightarrow 2PbO + 4NO_2 + O_2$ (decomposition)
   d. $2C_2H_6 + 7O_2 \longrightarrow 4CO_2 + 6H_2O$ (combustion)
   e. $Pb(NO_3)_2 + 2NaI \longrightarrow PbI_2 + 2NaNO_3$
     (double-replacement)
   f. $(NH_4)_2SO_4 + 2NaOH \longrightarrow 2NH_3 + 2H_2O + Na_2SO_4$
     (double-replacement, then decomposition of NH_4OH)

Chapter 8
1. **a.** $2CO(g) + O_2(g) \longrightarrow 2CO_2(g)$
   2 molecules CO + 1 molecule $O_2 \longrightarrow$ 2 molecules CO_2
   2 mol CO + 1 mol $O_2 \longrightarrow$ 2 mol CO_2
   $(2 \times 28 \text{ g}) + 32 \text{ g} = 2 \times 44 \text{ g}$
   88 g = 88 g (law of conservation of mass)
   44.8 L CO + 22.4 L $O_2 \longrightarrow$ 44.8 L CO_2
   b. $2Na(s) + 2H_2O(l) \longrightarrow 2NaOH(aq) + H_2(g)$
   2 atoms Na + 2 molecules $H_2O \longrightarrow$
     2 formula units NaOH + 1 molecule H_2
   2 mol Na + 2 mol $H_2O \longrightarrow$ 2 mol NaOH + 1 mol H_2
   $(2 \times 23 \text{ g}) + (2 \times 18 \text{ g}) = (2 \times 40 \text{ g}) + 2 \text{ g}$
   82 g = 82 g (law of conservation of mass)
   c. $2C_2H_2(g) + 5O_2(g) \longrightarrow 4CO_2(g) + 2H_2O(g)$
   2 molecules C_2H_2 + 5 molecules $O_2 \longrightarrow$
     4 molecules CO_2 + 2 molecules H_2O
   2 mol C_2H_2 + 5 mol $O_2 \longrightarrow$ 4 mol CO_2 + 2 mol H_2O
   $(2 \times 26 \text{ g}) + (5 \times 32 \text{ g}) = (4 \times 44 \text{ g}) + (2 \times 18 \text{ g})$
   212 g = 212 g (law of conservation of mass)
   44.8 L + 112 L $\longrightarrow$ 89.6 L + 44.8 L

2. **a.** $\dfrac{4 \text{ mol Al}}{3 \text{ mol } O_2} \quad \dfrac{3 \text{ mol } O_2}{4 \text{ mol Al}} \quad \dfrac{4 \text{ mol Al}}{2 \text{ mol } Al_2O_3} \quad \dfrac{2 \text{ mol } Al_2O_3}{4 \text{ mol Al}}$

$\dfrac{3 \text{ mol } O_2}{2 \text{ mol } Al_2O_3} \quad \dfrac{2 \text{ mol } Al_2O_3}{3 \text{ mol } O_2}$

   b. $2.3 \text{ mol } Al_2O_3 \times \dfrac{4 \text{ mol Al}}{2 \text{ mol } Al_2O_3} = 4.6 \text{ mol Al}$

c. $0.84 \text{ mol Al} \times \dfrac{3 \text{ mol } O_2}{4 \text{ mol Al}} = 0.63 \text{ mol } O_2$

d. $17.2 \text{ mol } O_2 \times \dfrac{2 \text{ mol } Al_2O_3}{3 \text{ mol } O_2} = 11.5 \text{ mol } Al_2O_3$

3. **a.** $13.0 \text{ g } C_2H_2 \times \dfrac{1 \text{ mol } C_2H_2}{26.0 \text{ g } C_2H_2} \times \dfrac{5 \text{ mol } O_2}{2 \text{ mol } C_2H_2} \times$

$\dfrac{32.0 \text{ g } O_2}{1 \text{ mol } O_2} = 40.0 \text{ g } O_2$

b. $13.0 \text{ g } C_2H_2 \times \dfrac{1 \text{ mol } C_2H_2}{26.0 \text{ g } C_2H_2} \times \dfrac{4 \text{ mol } CO_2}{2 \text{ mol } C_2H_2} \times$

$\dfrac{44.0 \text{ g } CO_2}{1 \text{ mol } CO_2} = 44.0 \text{ g } CO_2$

$13.0 \text{ g } C_2H_2 \times \dfrac{1 \text{ mol } C_2H_2}{26.0 \text{ g } C_2H_2} \times \dfrac{2 \text{ mol } H_2O}{2 \text{ mol } C_2H_2} \times$

$\dfrac{18.0 \text{ g } H_2O}{1 \text{ mol } H_2O} = 9.0 \text{ g } H_2O$

c. Mass reactants = mass products
   13.0 g + 40.0 g = 44.0 g + 9.0 g
     53.0 g = 53.0 g

4. **a.** $5.00 \text{ g } CaC_2 \times \dfrac{1 \text{ mol } CaC_2}{64.1 \text{ g } CaC_2} \times \dfrac{1 \text{ mol } C_2H_2}{1 \text{ mol } CaC_2} \times$

$\dfrac{26.0 \text{ g } C_2H_2}{1 \text{ mol } C_2H_2} = 2.03 \text{ g } C_2H_2$

b. $98.0 \text{ g } H_2O \times \dfrac{1 \text{ mol } H_2O}{18.0 \text{ g } H_2O} \times \dfrac{1 \text{ mol } CaC_2}{2 \text{ mol } H_2O} =$

$2.72 \text{ mol } CaC_2$

c. $5.34 \text{ mol } C_2H_2 \times \dfrac{1 \text{ mol } Ca(OH)_2}{1 \text{ mol } C_2H_2} \times \dfrac{74.1 \text{ g } Ca(OH)_2}{1 \text{ mol } Ca(OH)_2} =$

$396 \text{ g } Ca(OH)_2$

5. **a.** $\text{vol } G \text{ (L)} \times \dfrac{1 \text{ mol } G}{22.4 \text{ L } G} \times \dfrac{b \text{ mol } W}{a \text{ mol } G} \times \dfrac{\text{gfm } W}{1 \text{ mol } W}$

$\longrightarrow \text{ mass of } W$

b. $\begin{array}{c}\text{representative} \\ \text{particles } G\end{array} \times \dfrac{1 \text{ mol } G}{6.02 \times 10^{23}} \times \begin{array}{c}\text{representative} \\ \text{particles } G\end{array}$

$\dfrac{b \text{ mol } W}{a \text{ mol } G} \longrightarrow \text{ mol } W$

c. $\text{mass of } G \times \dfrac{1 \text{ mol } G}{\text{gfm } G} \times \dfrac{b \text{ mol } W}{a \text{ mol } G} \times$

$\dfrac{6.02 \times 10^{23} \text{ particles } W}{1 \text{ mol } W} \longrightarrow \text{ particles } W$

6. **a.** $7.42 \times 10^{24} \text{ molecules HF} \times \dfrac{1 \text{ mol HF}}{6.02 \times 10^{23} \text{ molecules HF}} \times$

$\dfrac{1 \text{ mol } SnF_2}{2 \text{ mol HF}} \times \dfrac{156.7 \text{ g } SnF_2}{1 \text{ mol } SnF_2} = 9.66 \times 10^2 \text{ g } SnF_2$

b. $23.4 \text{ g } Sn \times \dfrac{1 \text{ mol } Sn}{118.7 \text{ g } Sn} \times \dfrac{1 \text{ mol } H_2}{1 \text{ mol } Sn} \times \dfrac{22.4 \text{ L } H_2}{1 \text{ mol } H_2} =$

$4.42 \text{ L } H_2$

c. $14.2 \text{ L } H_2 \times \dfrac{2 \text{ L HF}}{1 \text{ L } H_2} = 28.4 \text{ L HF}$

d. $80.0 \text{ L HF} \times \dfrac{1 \text{ mol HF}}{22.4 \text{ L HF}} \times \dfrac{1 \text{ mol } H_2}{2 \text{ mol HF}} \times$

$\dfrac{6.02 \times 10^{23} \text{ molecules } H_2}{1 \text{ mol } H_2} = 1.08 \times 10^{24} \text{ molecules } H_2$

7. 1. Complete combustion:

a. $2.70 \text{ mol } C_2H_4 \times \dfrac{3 \text{ mol } O_2}{1 \text{ mol } C_2H_4} = 8.10 \text{ mol } O_2$ required;
O_2 is limiting

b. $6.30 \text{ mol } O_2 \times \dfrac{2 \text{ mol } H_2O}{3 \text{ mol } O_2} = 4.20 \text{ mol } H_2O$

c. Calculate the moles of C_2H_4 that react with 6.30 mol O_2.

$6.30 \text{ mol } O_2 \times \dfrac{1 \text{ mol } C_2H_4}{3 \text{ mol } O_2} = 2.10 \text{ mol } C_2H_4$ used;

excess $C_2H_4 = 2.70 - 2.10 = 0.60 \text{ mol } C_2H_4$

2. Incomplete combustion:

a. $2.70 \text{ mol } C_2H_4 \times \dfrac{2 \text{ mol } O_2}{1 \text{ mol } C_2H_4} = 5.40 \text{ mol } O_2$ required;
O_2 is excess, C_2H_2 is limiting

b. $2.70 \text{ mol } C_2H_4 \times \dfrac{2 \text{ mol } H_2O}{1 \text{ mol } C_2H_4} = 5.40 \text{ mol } H_2O$

c. excess $O_2 = 6.30 - 5.40 = 0.90 \text{ mol } O_2$

8. **a.** Calculate the moles of each reactant.

$4.00 \text{ g HCl} \times \dfrac{1 \text{ mol HCl}}{36.5 \text{ g HCl}} = 0.110 \text{ mol HCl}$

$3.00 \text{ g Mg} \times \dfrac{1 \text{ mol Mg}}{24.3 \text{ g Mg}} = 0.123 \text{ mol Mg}$

HCl is limiting since the equation shows that twice as many moles of HCl are needed per mole of Mg.

$0.110 \text{ mol HCl} \times \dfrac{1 \text{ mol } H_2}{2 \text{ mol HCl}} \times \dfrac{2.0 \text{ g } H_2}{1 \text{ mol } H_2} = 0.110 \text{ g } H_2$

b. $0.110 \text{ g } H_2 \times \dfrac{1 \text{ mol } H_2}{2.0 \text{ g } H_2} \times \dfrac{22.4 \text{ L } H_2}{1 \text{ mol } H_2} = 1.23 \text{ L } H_2$

9. $1.87 \text{ g Al} \times \dfrac{1 \text{ mol Al}}{27.0 \text{ g Al}} \times \dfrac{3 \text{ mol Cu}}{2 \text{ mol Al}} \times \dfrac{63.5 \text{ g Cu}}{1 \text{ mol Cu}} =$

$6.60 \text{ g Cu (theoretical yield)}$

Percent yield $= \dfrac{3.74 \text{ g}}{6.60 \text{ g}} \times 100 = 56.7\%$

10. **a.** $583 \text{ g SO}_3 \times \dfrac{1 \text{ mol SO}_3}{80.1 \text{ g SO}_3} \times \dfrac{129.6 \text{ kJ}}{1 \text{ mol SO}_3} = 943 \text{ kJ}$

b. $943 \text{ kJ} \times \dfrac{1.00 \text{ kcal}}{4.18 \text{ kJ}} = 226 \text{ kcal}$

11. $22.2 \text{ kJ} \times \dfrac{2 \text{ mol CO}_2}{43.9 \text{ kJ}} \times \dfrac{6.02 \times 10^{23} \text{ molecules CO}_2}{1 \text{ mol CO}_2}$
$= 6.09 \times 10^{23} \text{ molecules CO}_2$

12. $\dfrac{44.0 \text{ kJ}}{1 \text{ mol } H_2O} \times \dfrac{1 \text{ mol } H_2O}{18.0 \text{ g } H_2O} \times \dfrac{1.00 \text{ kcal}}{4.18 \text{ kJ}} \times \dfrac{10^3 \text{ cal}}{1 \text{ kcal}} =$

$585 \dfrac{\text{cal}}{\text{g } H_2O}$

13. $\Delta H = H_f^\circ(\text{products}) - H_f^\circ(\text{reactants})$

a. $\Delta H = -393.5 \text{ kJ} + (2 \text{ mol})(-285.8 \text{ kJ/mol}) -$
(-74.86 kJ)
$= -393.5 \text{ kJ} - 571.6 \text{ kJ} + 74.86 \text{ kJ}$
$= -890.2 \text{ kJ}$

b. $\Delta H = (2 \text{ mol})(-393.5 \text{ kJ/mol}) -$
$(2 \text{ mol})(-110.5 \text{ kJ/mol})$
$= -787 \text{ kJ} - (-221 \text{ kJ})$
$= -566 \text{ kJ}$

Chapter 9

1. If the temperature does not change, the average kinetic energy is unaffected.
2. At the same temperature, the average kinetic energies of the particles in any two samples of matter are the same.
3. An increase in the temperature causes an increase in the average kinetic energy of the gas particles.
4. If the absolute temperature is tripled, the average kinetic energy is tripled.
5. $190 \text{ mm Hg} \times \dfrac{1 \text{ atm}}{760 \text{ mm Hg}} = 0.25 \text{ atm}$
6. $253 \text{ mm Hg} \times \dfrac{1 \text{ atm}}{760 \text{ mm Hg}} = 0.333 \text{ atm}$
7. One mole of any gas at STP occupies a volume of 22.4 L. One mole of any substance contains 6.02×10^{23} particles.

$11.2 \text{ L } H_2 \times \dfrac{1 \text{ mol } H_2}{22.4 \text{ L } H_2} = \dfrac{11.2}{22.4} = 0.500 \text{ mol } H_2$

$0.500 \text{ mol } H_2 \times \dfrac{6.02 \times 10^{23} \text{ molecules } H_2}{1 \text{ mol } H_2} =$
$3.01 \times 10^{23} \text{ molecules } H_2$

8. $0.25 \text{ mol} \times \dfrac{22.4 \text{ L}}{1 \text{ mol}} = 5.6 \text{ L}$
9. The balanced equation shows that 2 mol of hydrogen reacts with 1 mol of oxygen.
Moles $H_2 \longrightarrow$ moles $O_2 \longrightarrow$ liters O_2

$1.0 \text{ mol } H_2 \times \dfrac{1 \text{ mol } O_2}{2 \text{ mol } H_2} \times \dfrac{22.4 \text{ L } O_2}{1 \text{ mol } O_2} = 11.2 \text{ L } O_2$

10. Change the given mass to moles, then to volume in liters.
Mass $\longrightarrow$ moles $\longrightarrow$ volume

$8.8 \text{ g CO}_2 \times \dfrac{1 \text{ mol CO}_2}{44.0 \text{ g CO}_2} \times \dfrac{22.4 \text{ L CO}_2}{1 \text{ mol CO}_2} = 4.5 \text{ L CO}_2$

Chapter 10

1. If the amount of gas is to be doubled while the pressure is kept the same, the volume of the container must be doubled.
2. A balloon shrinks when the temperature of the gas inside decreases. The pressure inside the balloon decreases because the average kinetic energy of the gas particles inside the balloon is lower at the lower temperature. The reduced kinetic energy results in fewer and less forceful collisions with the inside walls of the balloon.
3. The total pressure of a mixture of gases is equal to the sum of the partial pressure of each gas in the mixture.
$P_T = P_{O_2} + P_{N_2} + P_{He}$
$= 150 \text{ mm Hg} + 350 \text{ mm Hg} + 200 \text{ mm Hg}$
$= 700 \text{ mm Hg}$
Since 760 mm Hg equals 1.0 atm, we have:
$P_T = 700 \text{ mm Hg} \times \dfrac{1 \text{ atm}}{760 \text{ mm Hg}} = 0.92 \text{ atm}$
4. $P_1 \times V_1 = P_2 \times V_2$ (Boyle's law)
$V_2 = \dfrac{P_1 \times V_1}{P_2} = \dfrac{760 \text{ mm Hg} \times 2.50 \text{ L}}{304 \text{ mm Hg}} = 6.25 \text{ L}$
check: Because the pressure decreases, the volume should increase.

5. $\frac{V_1}{T_1} = \frac{V_2}{T_2}$ (Charles' law)

$V_2 = \frac{V_1 \times T_2}{T_1} = \frac{6.8 \text{ L} \times 300 \text{K}}{600 \text{ K}} = 3.4 \text{ L}$

(27°C + 273; 327°C + 273)

check: Because the temperature decreases, the volume should decrease.

6. $\frac{P_1}{T_1} = \frac{P_2}{T_2}$ (Gay-Lussac's law)

$P_2 = \frac{P_1 \times T_2}{T_1} = \frac{50.0 \text{ mm Hg} \times 200 \text{ K}}{540 \text{ K}} = 18.5 \text{ mm Hg}$

check: Because the temperature decreases, the pressure should decrease.

7. $\frac{P_1 \times V_1}{T_1} = \frac{P_2 \times V_2}{T_2}$ To use Charles' law, hold the pressure constant.

$\frac{V_1}{T_1} = \frac{V_2 \times P_2}{T_2 \times P_1}$

P_1 and P_2 cancel. Thus $\frac{V_1}{T_1} = \frac{V_2}{T_2}$ (Charles' law).

8. $\frac{P_1 \times V_1}{T_1} = \frac{P_2 \times V_2}{T_2}$

To use Gay-Lussac's law, hold the volume constant.

$\frac{P_1}{T_1} = \frac{P_2 \times V_2}{T_2 \times V_1}$

V_1 and V_2 cancel, and $\frac{P_1}{T_1} = \frac{P_2}{T_2}$ (Gay-Lussac's law).

9. $\frac{P_1 \times V_1}{T_1} = \frac{P_2 \times V_2}{T_2}$ (Combined gas law)

$V_2 = \frac{P_1 \times V_1 \times T_2}{P_2 \times T_1} = \frac{1.5 \text{ atm} \times 1.0 \text{ L} \times 373 \text{ K}}{6.0 \text{ atm} \times 298 \text{ K}}$

(100°C + 273; 25°C + 273)

$= 0.31 \text{ L}$

check: The pressure increase (volume decrease) has a greater effect than the temperature increase (volume increase).

10. $P \times V = n \times R \times T$

$n = \frac{P \times V}{R \times T} = \frac{18 \text{ atm} \times 680 \text{ L}}{0.082 \frac{\text{L} \times \text{atm}}{\text{K} \times \text{mol}} \ 600 \text{ K}} = 2.5 \times 10^2 \text{ mol}$

11. $P \times V = n \times R \times T$

$n = \frac{P \times V}{R \times T} = \frac{1 \text{ atm} \times 2.2 \text{ L}}{0.082 \frac{\text{L} \times \text{atm}}{\text{K} \times \text{mol}} \ 310 \text{ K}} = 0.087 \text{ mol air}$

(37°C + 273)

$0.087 \text{ mol air} \times \frac{29 \text{ g air}}{1 \text{ mol air}} = 2.5 \text{ g air}$

12. The molar volume of an *ideal* gas is 22.41 L at STP. The molar volume of a *real* gas is less than 22.41 L. Intermolecular forces hold the gas particles together and reduce the volume. The decrease in volume depends upon the strength of the intermolecular attractive forces. Nonpolar O_2 and N_2 molecules behave like ideal gases. Their attractive forces are insignificant and their molar volumes are 22.41 L. The bonds in CH_4 are weakly polar but the molecule is nonpolar. Very weak intermolecular attractive forces reduce the molar volume to 22.38 L. The bonds in

CO_2 are moderately polar. The attractive forces between CO_2 molecules decrease the molar volume to 22.26 L. Ammonia molecules are polar and also form intermolecular hydrogen bonds. The strong attractive forces between the molecules reduce the molar volume to 22.06 L.

13. $\frac{\text{Rate H}_2}{\text{Rate Cl}_2} = \frac{\sqrt{\text{formula mass}_{Cl_2}}}{\sqrt{\text{formula mass}_{H_2}}} = \frac{\sqrt{71}}{\sqrt{2}} = \frac{8.4}{1.4} = 6.0$

Hydrogen diffuses six times faster than chlorine.

Chapter 11

1. **a.** 3 **b.** 1 **c.** 7 **d.** 3 **e.** 5 **f.** 9
2. $3d$, $4s$, $3p$, $3s$, $2p$
3. **a.** Boron has an atomic number of 5, an electron configuration of $1s^2 2s^2 2p^1$, and one unpaired electron.
   b. Fluorine has an atomic number of 9, an electron configuration of $1s^2 2s^2 2p^5$, and one unpaired electron.
4. **a.** Barium has an atomic number of 56. Its electron configuration is $1s^2 2s^2 2p^6 3s^2 3p^6 3d^{10} 4s^2 4p^6 4d^{10} \ 5s^2 5p^6 6s^2$. It has two electrons in its highest occupied energy level.
   b. Sodium has an atomic number of 11. Its electron configuration is $1s^2 2s^2 2p^6 3s^1$. It has one electron in its highest occupied energy level.
   c. Aluminum has an atomic number of 13. Its electron configuration is $1s^2 2s^2 2p^6 3s^2 3p^1$. It has three electrons in its highest occupied energy level.
   d. Oxygen has an atomic number of 8. Its electron configuration is $1s^2 2s^2 2p^4$. It has six electrons in its highest occupied energy level.
5. $\nu = \frac{c}{\lambda} = \frac{3.00 \times 10^{10} \text{ cm s}^{-1}}{5.00 \times 10^{-6} \text{ cm}}$
   $= 6.00 \times 10^{15} \text{ s}^{-1}$ (ultraviolet)
6. $\nu = \frac{c}{\lambda} = \frac{3.00 \times 10^{10} \text{ cm s}^{-1}}{6.56 \times 10^{-5} \text{ cm}}$
   $= 4.57 \times 10^{14} \text{ s}^{-1}$ (visible, orange-red)
7. Classical physics views energy changes as continuous. In the quantum concept energy changes occur in discrete units called quanta.
8. $E = h \times \nu$
   $= 6.62 \times 10^{-34} \text{ J s} \times 3.20 \times 10^{11} \text{ s}^{-1}$
   $= 2.12 \times 10^{-22} \text{ J}$
9. The electron of the hydrogen atom is raised (excited) to the next highest energy level.

Chapter 12

1. **a.** Boron has 5 electrons. Reading the periodic table from left to right we have the first period, $1s^2$, and the second period, $2s^2 2p^1$. There is one electron in the $2p$ sublevel because boron is the third element in the p block.
   b. Magnesium has 12 electrons. The first two periods are $1s^2 2s^2 2p^6$. Next we have $3s^2$, which completes the electron configuration. **c.** Vanadium has 26 electrons. We go through the first three periods, $1s^2 2s^2 2p^6 3s^2 3p^6$, then $4s^2$ and finally $3d^3$. Vanadium is the third element in the d block, and the complete electron configuration is $1s^2 2s^2 2p^6 3s^2 3p^6 3d^3 4s^2$. **d.** Strontium, with 38 electrons, is in the fifth period. To get to strontium we pass through $1s^2 2s^2 2p^6 3s^2 3p^6 4s^2 3d^{10} 4p^6$, and finally $5s^2$. The electron configuration for strontium is $1s^2 2s^2 2p^6 3s^2 3p^6 \ 3d^{10} 4s^2 4p^6 5s^2$.

2. **a.** The inert gas in period three is argon: $1s^2 2s^2 2p^6 3s^2 3p^6$.
   b. The element in group 4A, period 4 is germanium with 32 electrons, $1s^2 2s^2 2p^6 3s^2 3p^6 3d^{10} 4s^2 4p^2$. **c.** Group 2A, period 6 is barium with 56 electrons, $1s^2 2s^2 2p^6 3s^2 3p^6 3d^{10} 4s^2 4p^6 4d^{10} 5s^2 5p^6 6s^2$.

3. Fluorine has nine protons and nine electrons, $1s^2 2s^2 2p^5$, and oxygen has eight protons and eight electrons, $1s^2 2s^2 2p^4$. In each element the outer electrons are shielded from the nucleus by two $1s$ electrons. The effect of the increased nuclear charge on fluorine pulls the outer electrons tighter than in oxygen, and the atomic radius decreases. Chlorine, with 17 protons and electrons, $1s^2 2s^2 2p^6 3s^2 3p^5$, has its outer electrons shielded from the nucleus by a layer of electrons in the second energy level. The third energy level of electrons experiences less nuclear attraction and is consequently larger. Therefore chlorine has a larger atomic radius than fluorine.

4. Electron affinities of nonmetals such as the halogens are large negative numbers. This is because these elements acquire noble gas configurations by gaining one or more electrons.

5. **a.** sodium **b.** phosphorus

6. **a.** Ge, germanium **b.** Mg, magnesium **c.** Nb, niobium **d.** P, phosphorus **e.** Ti, titanium **f.** F, fluorine

Chapter 13

1. **a.** 1 **b.** 4 **c.** 2 **d.** 6

2. Cu^+: $1s^2 2s^2 2p^6 3s^2 3p^6 3d^{10}$
   Au^+: $1s^2 2s^2 2p^6 3s^2 3p^6 3d^{10} 4s^2 4p^6 4d^{10} 4f^{14} 5s^2 5p^6 5d^{10}$
   Cd^{2+}: $1s^2 2s^2 2p^6 3s^2 3p^6 3d^{10} 4s^2 4p^6 4d^{10}$
   Hg^{2+}: $1s^2 2s^2 2p^6 3s^2 3p^6 3d^{10} 4s^2 4p^6 4d^{10} 4f^{14} 5s^2 5p^6 5d^{10}$

3. **a.** lose 2 **b.** gain 1 **c.** lose 3 **d.** gain 2

4. **a.** S^{2-} **b.** Na^+ **c.** F^- **d.** Ba^{2+}

5. **a.** K^+ :I:$^-$, KI **b.** Ca^{2+} :S:$^{2-}$, CaS

   c. :O:$^{2-}$
   Al^{3+}
   :O:$^{2-}$
   Al^{3+}
   :O:$^{2-}$, Al_2O_3

   d. Na^+
   Na^+ :P:$^{3-}$, Na_3P
   Na^+

6. **a.** potassium iodide **b.** calcium sulfide **c.** aluminum oxide **d.** sodium phosphide

7. **a.** K^+I^- **b.** $Ca^{2+}S^{2-}$ **c.** $Al^{3+}O^{2-}$ **d.** Na^+P^{3-}
   KI   CaS Al_2O_3 Na_3P

Chapter 14

1. **a.** :Cl:Cl: **b.** :Br:Br: **c.** :I:I:

2. **a.** H:S:H **b.** H:P:H **c.** :Cl:F:
                 H

3. $\left[\begin{array}{c} :O: \\ :O:S:O: \\ :O: \end{array} \right]^{2-}$ $\left[\begin{array}{c} :O: \\ :O:C::O: \end{array} \right]^{2-}$

4. A sigma bond is formed by the overlap of two s orbitals, the overlap of an s orbital with a p orbital, or end-to-end overlap of two p orbitals.

           +   ⟶

1 s orbital   1s orbital   sigma bond

5. The boron in BF_3 shares only six valence electrons with the fluorines. An additional F^- can form a coordinate covalent bond with the BF_3. The geometry of the ion BF_4^- would be tetrahedral according to VSEPR theory, since this keeps the fluorines the maximum distance apart.

6.
Elements (electronegativities)	Electronegativity difference	Type of bond
a. H(2.1), Br(2.8)	0.7	Covalent (moderately polar)
b. K(0.8), Cl(3.0)	2.2	Ionic
c. C(2.5), O(3.5)	1.0	Covalent (very polar)
d. Cl(3.0), F(4.0)	1.0	Covalent (very polar)
e. Li(1.0), O(3.5)	2.5	Ionic
f. Br(2.8), Br(2.8)	0.0	Covalent (nonpolar)

7. The most polar bond will have the greatest electronegativity difference between the interacting atoms. In order of increasing polarity:

Elements	Electronegativity difference
e. F(4.0), F(4.0)	0.0
c. H(2.1), S(2.5)	0.4
d. H(2.1), C(2.5)	0.4
b. H(2.1), Br(2.8)	0.7
a. H(2.1), Cl(3.0)	0.9

8. From Table 14·4, the bond dissociation energy of a C—H bond is 393 kilojoules/mole.

$$0.1 \text{ mole } CH_4 \times \frac{4 \text{ mole C—H}}{1 \text{ mole } CH_4} \times \frac{393 \text{ kJ}}{1 \text{ mole C—H}} = 157 \text{ kJ}$$

9.
   H   H   H   H
   | | | |
H—N:---H—N: :O:---H—N:
   | | | |
   H   H   H   H

Chapter 15

1. Hydrogen-bonding between ammonia molecules occurs because the nitrogen on one ammonia molecule can attract a hydrogen from a neighboring ammonia molecule.

       N
     / | \
   H   H   H

Methane is a nonpolar molecule. It can have no hydrogen-bonding. Substances whose molecules form intermolecular hydrogen bonds tend to have higher boiling points.

2. Water is a polar solvent. Solutes that are ionic or polar generally dissolve in water. **a.** HCl is polar and will dissolve. **b.** NaI is ionic and will dissolve. **c.** NH_3 is polar and will dissolve. **d.** $MgSO_4$ is ionic and will dissolve. **e.** CH_4 is nonpolar and will not dissolve. **f.** $CaCO_3$ is ionic but will not dissolve. Some ionic compounds are insoluble in water because of the strong attractive forces within the solid crystal. **g.** Gasoline is nonpolar and will not dissolve.

3. First calculate the gram formula mass of the hydrate and of the attached water. Then calculate the percent of water.
   Cu: $1 \times 63.5 = 63.5$
   S: $1 \times 32.1 = 32.1$
   O: $4 \times 16.0 = 64.0$
   H_2O: $5 \times 18.0 = \underline{90.0}$
               249.6 g/mol

$$\text{Percent } H_2O = \frac{\text{mass of } H_2O}{\text{mass of hydrate}} = \frac{90.0 \text{ g}}{249.6 \text{ g}} = 36.1\%$$

Chapter 16

1. The solubility of NaCl is 36.2 g NaCl per 100 mL of solution at 25°C.

$$750 \text{ mL} \times \frac{36.2 \text{ g NaCl}}{100 \text{ mL solution}} = 271.5 = 272 \text{ g NaCl}$$

2. Henry's law relates the solubility of a gas to the pressure of the gas above the liquid.

$$\frac{S_1}{P_1} = \frac{S_2}{P_2}$$

$$P_2 = \frac{S_2 P_1}{S_1}$$

(This is like solving Charles' law for T_2.)

$$P_2 = \frac{(9.5 \text{ g/L})(1.0 \text{ atm})}{(3.6 \text{ g/L})} = 2.6 \text{ atm}$$

This answer makes sense. To increase the solubility of a gas, the pressure of the gas above the liquid should be increased.

3. Molarity $(M) = \dfrac{\text{number of moles of solute}}{\text{number of liters of solution}}$

Change the given volume to the unit liters:

$$250 \text{ mL} \times \frac{1 \text{L}}{1000 \text{ mL}} = 0.25 \text{ L}$$

$$\text{Molarity} = \frac{0.70 \text{ mol}}{0.25 \text{ L}} = 2.8M$$

4. $\text{Moles glucose} = \dfrac{36.0 \text{ g}}{180 \text{ g/mol}} = 0.20 \text{ mol}$

$$\text{Molarity} = \frac{0.20 \text{ mol}}{2.0 \text{ L}} = 0.10M$$

5. $2.0M$ CaCl$_2$ means 2.0 mol CaCl$_2$ per 1.0 L.

$$250 \text{ mL} \times \frac{1.0 \text{ L}}{1000 \text{ mL}} \times \frac{2.0 \text{ mol CaCl}_2}{1.0 \text{ L}} = 0.50 \text{ mol CaCl}_2$$

To find the mass of CaCl$_2$, you need its gram formula mass (gfm).

Ca: $1 \times 40.1 = 40.1$ g
Cl: $2 \times 35.5 = \underline{71.0 \text{ g}}$
111.1 g/mol

$$\text{Mass of CaCl}_2 = 0.50 \text{ mol CaCl}_2 \times \frac{111.1 \text{ g CaCl}_2}{1 \text{ mol CaCl}_2}$$
$$= 55.5 = 56 \text{ g CaCl}_2$$

6. For dilution problems: $M_1 \times V_1 = M_2 \times V_2$
The volume of stock solution needed is

$$V_1 = \frac{M_2 \times V_2}{M_1} = \frac{0.2 \text{ } M \times 250 \text{ mL}}{1.0 \text{ } M} = 50 \text{ mL}$$

Use a pipet to transfer 50 mL of the $1.0M$ solution to a 250-mL volumetric flask. Then add distilled water up to the mark.

7. Percent by volume $= \dfrac{\text{volume of solute}}{\text{volume of solution}} \times 100\%$

$$= \frac{10 \text{ mL}}{200 \text{ mL}} \times 100\% = 5.0\% \text{ (v/v)}$$

8. 1.6% MgSO$_4$ (m/v) means that there is 1.6 g MgSO$_4$ in 100 of mL solution.

$$250 \text{ mL} \times \frac{1.6 \text{ g MgSO}_4}{100 \text{ mL}} = 4.0 \text{ g MgSO}_4$$

9. Percent (m/v) $= \dfrac{\text{mass of solute (g)}}{\text{volume of solution (mL)}} \times 100\%$

$$= \frac{2.7 \text{ g}}{75 \text{ mL}} \times 100\% = 3.6\% \text{ CuSO}_4 \text{ (m/v)}$$

10. Molality $= \dfrac{\text{moles of solute}}{1000 \text{ g of solvent}}$

Use a conversion factor to convert 10.0 g NaCl in 600 g H$_2$O to grams of NaCl in 1000 g H$_2$O.

$$1000 \text{ g H}_2\text{O} \times \frac{10.0 \text{ g NaCl}}{600 \text{ g H}_2\text{O}} = 16.7 \text{ g NaCl}$$

Convert 16.7 g NaCl to moles.
gfm NaCl = 23.0 + 35.5 = 58.5 g

$$16.7 \text{ g NaCl} \times \frac{1 \text{ mol NaCl}}{58.5 \text{ g NaCl}} = 0.285 \text{ mol NaCl}$$

When 10.0 g NaCl is dissolved in 600 g H$_2$O, the resulting solution is 0.285 molal NaCl.

11. Convert 300 g C$_6$H$_5$OH to moles.
gfm C$_2$H$_5$OH = 24.0 + 5.0 + 16.0 + 1.0 = 46.0 g

$$300 \text{ g C}_2\text{H}_5\text{OH} \times \frac{1 \text{ mol C}_2\text{H}_5\text{OH}}{46.0 \text{ g C}_2\text{H}_5\text{OH}} = 6.52 \text{ mol C}_2\text{H}_5\text{OH}$$

Convert 500 g H$_2$O to moles.
gfm H$_2$O = 2.0 + 16.0 = 18.0 g

$$500 \text{ g H}_2\text{O} \times \frac{1 \text{ mol H}_2\text{O}}{18.0 \text{ g H}_2\text{O}} = 27.8 \text{ mol H}_2\text{O}$$

The mole fraction of ethanol, X_E, in the solution is the number of moles of ethanol divided by the total number of moles in solution.

$$X_E = \frac{n_E}{n_E + n_{H_2O}} = \frac{6.52 \text{ mol}}{6.52 \text{ mol} + 27.8 \text{ mol}} = 0.190$$

The mole fraction of water, X_{H_2O}, is the number of moles of water divided by the total number of moles in solution.

$$X_{H_2O} = \frac{n_{H_2O}}{n_{H_2O} + n_E} = \frac{27.8 \text{ mol}}{27.8 \text{ mol} + 6.52 \text{ mol}} = 0.810$$

Total = 1.000

12. The solution contains 0.150 moles of NaCl in 1000 g H$_2$O. Use a conversion factor to convert 1000 g H$_2$O to moles of H$_2$O.
gfm H$_2$O = 2.0 + 16.0 = 18.0 g

$$1000 \text{ g H}_2\text{O} \times \frac{1 \text{ mol H}_2\text{O}}{18.0 \text{ g H}_2\text{O}} = 55.6 \text{ mol H}_2\text{O}$$

The mol fraction of sodium chloride, X_{NaCl}, is the number of moles of NaCl divided by the total number of moles in solution.

$$X_{NaCl} = \frac{n_{NaCl}}{n_{NaCl} + n_{H_2O}} = \frac{0.150}{0.150 + 55.6} = 0.00269$$

The mol fraction of water, X_{H_2O}, is the number of moles of H$_2$O divided by the total number of moles in solution.

$$X_{H_2O} = \frac{n_{H_2O}}{n_{H_2O} + n_{NaCl}} = \frac{55.6}{55.6 + 0.150} = 0.99731$$

Total = 1.00000

13. $X_{CH_3OH} = 0.40 = \dfrac{n_{CH_3OH}}{n_{CH_3OH} + n_{H_2O}}$

$$X_{H_2O} = 0.60 = \frac{n_{H_2O}}{n_{H_2O} + n_{CH_3OH}}$$

$$X_{CH_3OH} = \frac{0.6}{0.4} X_{H_2O} = 1.5 \times X_{H_2O}$$

Therefore: $n_{H_2O} = 1.5 \, n_{CH_3OH}$
Substitute in first two equations to check:
$n_{H_2O} = 1.5$; $n_{CH_3OH} = 1.0$

$$X_{CH_3OH} = \frac{1}{1 + 1.5} = \frac{1}{2.5} = 0.4$$

$$X_{H_2O} = \frac{1.5}{1.5 + 1} = \frac{1.5}{2.5} = 0.6$$

To make a solution 0.40 mole fraction in CH_3OH we need a ratio of 1.5 mol H_2O to 1.0 mol CH_3OH.

gfm H_2O = 2.0 + 16.0 = 18.0 g

gfm CH_3OH = 12.0 + 3.0 + 16.0 + 1.0 = 32.0 g

Add 1.5 × 18.0 g = 27.0 g H_2O to 32.0 g CH_3OH to give an aqueous solution of methanol in which the mole fraction of methanol is 0.40.

14. Determine the molality of the solution.

$$\frac{mol\ CaCl_2}{1000\ g\ H_2O} = \frac{1.25\ mol\ CaCl_2}{1400\ g\ H_2O} \times \frac{1000\ g\ H_2O}{1\ kg\ H_2O}$$
$$= 0.893\ m$$

One mol of $CaCl_2$ in water gives 3 mol of particles.

$CaCl_2(s) \rightarrow Ca^{2+}(aq) + 2\ Cl^-(aq)$

The molality of *total* particles is:

$3 \times 0.893\ m = 2.68\ m$

The boiling elevation is:

$\Delta T_b = K_b m = 0.512\ °C/m \times 2.68\ m$
$= 1.37\ °C$

The boiling point of this solution is:

$100°C + 1.37°C = 101.37°C$

15. a. Determine the molality of the solution.

$$\frac{mol\ Na_2SO_4}{1000\ g\ H_2O} = \frac{1.40\ mol\ Na_2SO_4}{1750\ g\ H_2O} \times \frac{1000\ g\ H_2O}{1\ kg\ H_2O}$$
$$= 0.800\ m$$

One mol of Na_2SO_4 in water gives 3 mol of particles.

$Na_2SO_4(s) \rightarrow 2\ Na^+(aq) + SO_4^{2-}(aq)$

The molality of *total* particles is:

$3 \times 0.800\ m = 2.40\ m$

The freezing point depression is:

$\Delta T_f = K_f m = 1.86\ °C/m \times 2.40\ m$
$= 4.46\ °C$

The freezing point of this solution is:

$0\ °C - 4.46\ °C = -4.46\ °C$

b. The molality of the solution is:

$$\frac{mol\ MgSO_4}{1000\ g\ H_2O} = \frac{0.60\ mol\ MgSO_4}{1300\ g\ H_2O} \times \frac{1000\ g\ H_2O}{1\ kg\ H_2O}$$
$$= 0.46\ m$$

One mole of $MgSO_4$ gives 2 mol of particles.

$MgSO_4(s) \rightarrow Mg^{2+}(aq) + SO_4^{2-}(aq)$

The molality of total particles is:

$2 \times 0.46\ m = 0.92\ m$

The freezing point depression is:

$\Delta T_f = K_f m = 1.86\ °C/m \times 0.92\ m$
$= 1.71\ °C$

The freezing point of this solution is:

$0\ °C - 1.71\ °C = -1.71\ °C$

16. The boiling point elevation ($4 \times 0.52°C = 2.08°C$) and freezing point depression ($4 \times -1.86 = -7.44°C$) both show that the compound gave 4 mol of particles in solution. The formula must be $FeCl_3$.

17. The molal freezing point content for H_2O is:

$$K_f = 1.86\ °C/m = 1.86\ \frac{°C\ kg\ of\ H_2O}{mol\ of\ solute}$$

The observed freezing point depression is:

$\Delta T_f = 0.390\ °C$

First calculate the molarity of the solution.

$\Delta T_f = K_f \times m$

$$m = \frac{\Delta T_f}{K_f} = \frac{0.390\ °C}{1.86\ °C\ \frac{kg\ of\ H_2O}{mol\ of\ solute}} = 0.210\ \frac{mol\ of\ solute}{kg\ of\ H_2O}$$

Now calculate moles of solute in solution.

$$0.210\ \frac{mol\ of\ solute}{kg\ of\ H_2O} \times 475\ g\ H_2O \times \frac{1\ kg\ of\ H_2O}{1000\ g\ H_2O}$$
$$= 0.0998\ mol\ solute$$

Finally, use number of moles of solute and its mass to determine the molecular mass of the solute.

$$Molecular\ mass\ of\ solute = \frac{mass\ of\ solute}{moles\ of\ solute}$$
$$= \frac{3.90\ g}{0.0998\ mol}$$
$$= 39.1\ g/mol$$

Chapter 17

1. Reactions conducted by bacteria in food lead to spoilage. These reactions are faster at room temperature than in the cold. Bacteria also multiply faster in a warm environment.

2. At least four factors contribute to reaction rates: temperature (heating enables more reactant molecules to cross the activation energy barrier); concentration (more reactants in a given volume increase the chance for productive collisions); particle size (the smaller the particles, the more surface of reactant is exposed to reaction); and catalysis (catalysts provide a path of lower activation energy).

3. For a reaction first-order in species A:

Rate = k[A]

The rate is in units of concentration per unit of time (e.g. moles per liter per second), [A] is in moles per liter.

Thus: Units of k = rate/[A]
$\quad$ = moles per liter/time × liters per mole
$\quad$ = 1/time¹

4. Since the rate of a first-order reaction is directly proportional to the concentration of one reactant, the rate will be 0.25 mol/L × s when one-half the starting material remains, and 0.125 mol/L × s when one-fourth of the starting material remains.

5. For the hypothetical second-order reaction:

A + B → C

Rate = k[A][B]

Units of k = rate/[A][B]

$$= \frac{mol}{A \times s} \times \frac{A}{mol} \times \frac{L}{mol} = L/(mol \times s)$$

6. When NO_2^- is doubled with NH_4^+ held constant, the rate doubles. The reaction is first-order in NO_2^-. When NH_4^+ is doubled with NO_2^- held constant, the rate also doubles. The reaction is also first-order with respect to NH_4^+, and the overall reaction is second-order.

7. a. playing cards in use $\qquad$ **c.** a cup of water at 50°C
$\quad$ **b.** a dissolved sugar cube $\qquad$ **d.** 1 g of powdered salt

8. $\Delta S^0 = S^0(products) - S^0(reactants)$

Values of $S°$ for products and reactants are from Table 17·1. Units are J/K.

a. $(39.75 + 213.6) - (88.7) = 165$
b. $(2 \times 69.94) - ((2 \times 130.6) + 205.0) = -326.3$
c. $(2 \times 186.7) - (130.6 + 23) = 9.8$

9. Calculate ΔG^0 to determine if it is negative (spontaneous reaction) or positive (nonspontaneous reaction). Use the ΔG_f of Table 17·4. All units are kJ/mol.

$\Delta G^0 = \Delta G_f(products) - \Delta G_f(reactants)$

a. $(2 \times -384.03) - (2 \times 0.00 + 0.00) = -768.06$
b. $(2 \times -1576.4) - (4 \times 0.00 + 3 \times 0.00) = -3152.8$

The reactions are spontaneous.

10. Recall that: $\Delta G^0 = \Delta H^0 - T\Delta S^0$
$$\Delta S^0 = (2 \times 192.5) - (3 \times 130.6 + 191.5)$$
$$= -198.3 \text{ J/K}$$
$$= -0.198 \text{ kJ/K}$$
$$\Delta H^0 = \Delta G^0 + T\Delta S^0$$
At 298 K:
$$\Delta H^0 = 2(-16.64) + (298 \times -0.198)$$
$$= -92.2 \text{ kJ/2 mol NH}_3 \text{ formed}$$
$$\Delta H_f^0 = -92.2 \text{ kJ} \div 2$$
$$= -46.1 \text{ kJ/mol NH}_3$$

11. $K_{eq} = \dfrac{[\text{NH}_3]^2}{[\text{H}_2]^3 \times [\text{N}_2]}$

12. $K_{eq} = \dfrac{[0.10]}{[0.15]^3 \times [0.25]}$
$= 12$ (to two significant figures)

13. $K_{eq} = \dfrac{[\text{H}_2] \times [\text{I}_2]}{[\text{HI}]^2}$

14. From problem 13:
$$K_{eq} = \dfrac{[\text{H}_2] \times [\text{I}_2]}{[\text{HI}]^2}$$
If $[\text{H}_2] = 0.50$ mol, $[\text{I}_2]$ must $= 0.50$ mol according to the balanced equation.
$$[\text{HI}]^2 = \dfrac{[0.50] \times [0.50]}{0.020}$$
$$= 12.5$$
$$[\text{HI}] = \sqrt{12.5} = 3.5 \text{ (to two significant figures)}$$

15. For K_{eq} greater than 1.0, the products are usually favored.
    a. products b. reactants
    c. products d. reactants

16. a. Removing heat shifts equilibrium position toward reactants. (Heat is needed to make the reaction go.)
    b. Increasing pressure pushes equilibrium toward reactants.
    c. Removing H_2 draws equilibrium toward products.

Chapter 18

1. $[\text{H}^+] = \dfrac{K_w}{[\text{OH}^-]} = \dfrac{1 \times 10^{-14}}{1.0 \times 10^{-3}} = 1 \times 10^{-11} \text{ mol/L}$
Since $[\text{OH}^-]$ is greater than $[\text{H}^+]$, the solution is basic.

2. By definition, pH $= -\log[\text{H}^+]$,
   a. pH $= -\log(1 \times 10^{-6}) = -(0.0 - 6) = 6.0$
   b. pH $= -\log(1 \times 10^{-4}) = (0.0 - 4) = 4.0$
   If the $[\text{OH}^-]$ concentration is given, first find the $[\text{H}^+]$ using K_w; then find the pH.
   c. $[\text{H}^+] = \dfrac{K_w}{[\text{OH}^-]} = \dfrac{1 \times 10^{-14}}{1 \times 10^{-2}} = 1 \times 10^{-12} \text{ mol/L}$
   pH $= -\log(1 \times 10^{-12}) = -(0.0 - 12) = 12.0$
   d. $[\text{H}^+] = \dfrac{K_w}{[\text{OH}^-]} = \dfrac{1 \times 10^{-14}}{1 \times 10^{-11}} = 1 \times 10^{-3} \text{ mol/L}$
   pH $= -\log(1 \times 10^{-3}) = -(0.0 - 3) = 3.0$

3. The pH of a solution with a hydrogen-ion concentration of $1 \times 10^{-x}M$ is x. a. pH $= 4.0$, $[\text{H}^+] = 1 \times 10^{-4}M$
   b. pH $= 11.0$, $[\text{H}^+] = 1 \times 10^{-11}M$ c. pH $= 8.0$, $[\text{H}^+] = 1 \times 10^{-8}M$

4. First find the $[\text{H}^+]$ as in the previous example. Then use K_w to find $[\text{OH}^-]$.
   a. pH $= 6.0$, $[\text{H}^+] = 1 \times 10^{-6}$
   $[\text{OH}^-] = \dfrac{1 \times 10^{-14}}{1 \times 10^{-6}} = 1 \times 10^{-8}M$

b. pH $= 9.0$, $[\text{H}^+] = 1 \times 10^{-9}$
   $[\text{OH}^-] = \dfrac{1 \times 10^{-14}}{1 \times 10^{-9}} = 1 \times 10^{-5}M$
   c. pH $= 12.0$, $[\text{H}^+] = 1 \times 10^{-12}$
   $[\text{OH}^-] = \dfrac{1 \times 10^{-14}}{1 \times 10^{-12}} = 1 \times 10^{-2}M$

5. a. 4.0 b. 3.0 c. 9.0 d. 10.0

6. a. $[\text{H}^+] = 5.0 \times 10^{-6}$
   pH $= -\log(5.0) - \log(10^{-6})$
   $= -(0.70) - (-6)$
   $= 5.30$
   b. $[\text{H}^+] = 8.3 \times 10^{-10}$
   pH $= -\log(8.3) - \log(10^{-10})$
   $= -(0.92) - (-10)$
   $= 9.08$
   c. $[\text{H}^+] = \dfrac{1 \times 10^{-14}}{2.0 \times 10^{-5}} = 5 \times 10^{-10}$
   pH $= -\log(5.0) - \log(10^{-10})$
   $= -(0.70) - (-10)$
   $= 9.30$
   d. $[\text{H}^+] = \dfrac{1 \times 10^{-14}}{4.5 \times 10^{-11}} = 2.2 \times 10^{-4}$
   pH $= -\log(2.2) - \log(10^{-4})$
   $= -(0.34) - (-4)$
   $= 3.66$

7. a. $[\text{H}^+] = 1 \times 10^{-5}M$
   b. pH $= 5.80$
   $[\text{H}^+] = (10^{0.20})(10^{-6}) = 1.6 \times 10^{-6}M$
   c. pH $= 12.20$
   $[\text{H}^+] = (10^{0.80})(10^{-13}) = 6.3 \times 10^{-13}M$
   d. pH $= 2.64$
   $[\text{H}^+] = (10^{0.36})(10^{-3}) = 2.3 \times 10^{-3}M$

8. a. diprotic b. triprotic c. monoprotic

9. a. $2\text{K} + 2\text{H}_2\text{O} \rightarrow 2\text{KOH} + \text{H}_2$
   b. $\text{Ca} + 2\text{H}_2\text{O} \rightarrow \text{Ca(OH)}_2 + \text{H}_2$

10. a. $\text{KOH} \rightarrow \text{K}^+ + \text{OH}^-$
    b. $\text{Mg(OH)}_2 \rightarrow \text{Mg}^{2+} + 2\text{OH}^-$

11. $\underset{\substack{\text{acid} \\ \text{(hydrogen-} \\ \text{ion donor)}}}{\text{HNO}_3} + \underset{\substack{\text{base} \\ \text{(hydrogen-} \\ \text{ion acceptor)}}}{\text{H}_2\text{O}} \rightleftharpoons \underset{\substack{\text{conjugate} \\ \text{acid}}}{\text{H}_3\text{O}^+} + \underset{\substack{\text{conjugate} \\ \text{base}}}{\text{NO}_3^-}$

   $\underset{\substack{\text{base} \\ \text{(hydrogen-} \\ \text{ion acceptor)}}}{\text{CO}_3^{2-}} + \underset{\substack{\text{acid} \\ \text{(hydrogen-} \\ \text{ion donor)}}}{\text{H}_2\text{O}} \rightleftharpoons \underset{\substack{\text{conjugate} \\ \text{acid}}}{\text{HCO}_3^-} + \underset{\substack{\text{conjugate} \\ \text{base}}}{\text{OH}^-}$

12. a. H^+ is the Lewis acid; H_2O is the Lewis base.
    b. AlCl_3 is the Lewis acid; Cl^- is the Lewis base.

13. Boric acid; weakest acid has lowest K_a.

14. The strongest acid is oxalic acid. The weakest acid is carbonic acid.

15. Ammonia is a weak base because it has a small K_b. Sodium hydroxide is a strong base because it has a large K_b.

16. a. pH $= -\log[\text{H}^+] = -\log(9.86) - \log(10^{-4})$
    $= -(0.994) - (-4)$
    $= 3.006$

    b. $K_a = \dfrac{[\text{H}^+][\text{X}^+]}{[\text{HX}]}$
    $= \dfrac{(9.86 \times 10^{-4})(9.86 \times 10^{-4})}{0.199}$
    $= 4.88 \times 10^{-6}$

Chapter 19

1. **a.** $HNO_3 + KOH \rightarrow KNO_3 + H_2O$
   b. $2HCl + Mg(OH)_2 \rightarrow MgCl_2 + 2H_2O$
   c. $H_2SO_4 + 2NaOH \rightarrow Na_2SO_4 + 2H_2O$

2. Write the equation for the neutralization.
   $NaOH + HNO_3 \rightarrow NaNO_3 + H_2O$

   $0.20 \; \text{mol HNO}_3 \times \dfrac{3 \; \text{mol NaOH}}{1 \; \text{mol HNO}_3} = 0.20 \; \text{mol NaOH}$

3. Write the equation for the neutralization.
   $HCl + KOH \rightarrow KCl + H_2O$

   $0.025 \; \text{L KOH} \times \dfrac{1 \; \text{mol KOH}}{1 \; \text{L KOH}} \times \dfrac{1 \; \text{mol HCl}}{1 \; \text{mol KOH}}$
   $\times \dfrac{1 \; \text{L HCl}}{0.45 \; \text{mol HCl}} \times \dfrac{10^3 \; \text{mL HCl}}{1 \; \text{L HCl}} = 56 \; \text{mL HCl}$

4. Write the equation for the neutralization.
   $H_3PO_4 + 3NaOH \rightarrow Na_3PO_4 + 3H_2O$

   $0.0385 \; \text{L NaOH} \times \dfrac{0.15 \; \text{mol NaOH}}{1 \; \text{L NaOH}} \times \dfrac{1 \; \text{mol H}_3\text{PO}_4}{3 \; \text{mol NaOH}}$
   $= 0.00193 \; \text{mol H}_3\text{PO}_4$

   $\text{Molarity} = \dfrac{\text{mol}}{\text{L}} = \dfrac{0.00193 \; \text{mol}}{0.015 \; \text{L}} = 0.14M$

5. Divide the gram equivalent mass by the number of equivalents per mol.

   a. $1 \; \text{Gram equiv mass KOH} = \dfrac{56.1 \; \text{g}}{1} = 56.1 \; \text{g}$
   (1 equivalent per mol)

   b. $1 \; \text{Gram equiv mass HCl} = \dfrac{36.5 \; \text{g}}{1} = 36.5 \; \text{g}$
   (1 equivalent per mol)

   c. $1 \; \text{Gram equiv mass H}_2\text{SO}_4 = \dfrac{98.0 \; \text{g}}{2} = 49.0 \; \text{g}$
   (2 equivalents per mol)

6. $\text{Equiv} = \dfrac{\text{gram molecular mass}}{\text{gram equivalent mass}}$

   a. $\text{Gram equiv mass Ca(OH)}_2 = \dfrac{74.0 \; \text{g}}{2} = 37.0 \; \text{g}$

   $\text{Equiv Ca(OH)}_2 = \dfrac{3.7 \; \text{g}}{37.0 \; \text{g/equiv}} = 0.10 \; \text{equiv}$

   b. $\text{Gram equiv mass H}_2\text{SO}_4 = \dfrac{98.0 \; \text{g}}{2} = 49.0 \; \text{g}$

   $\text{Equiv H}_2\text{SO}_4 = \dfrac{189 \; \text{g}}{49.0 \; \text{g/equiv}} = 3.86 \; \text{equiv}$

   c. $\text{Gram equiv mass H}_3\text{PO}_4 = \dfrac{98.0 \; \text{g}}{3} = 32.7 \; \text{g}$

   $\text{Equiv H}_3\text{PO}_4 = \dfrac{9.8 \; \text{g}}{32.7 \; \text{g/equiv}} = 0.30 \; \text{equiv}$

7. Normality = molarity × number ionizable hydrogens
   a. $2M \; HCl = 2N \; HCl$
   b. $0.1M \; HC_2H_3O_2 = 0.1N \; HC_2H_3O_2$
   c. $0.3M \; H_3PO_4 = 0.9N \; H_3PO_4$
   (three ionizable hydrogens)
   d. $0.25M \; H_2SO_4 = 0.50N \; H_2SO_4$
   (two ionizable hydrogens)

8. Normality = equivalents/liter
   a. $20 \; \text{g NaOH} \times \dfrac{1 \; \text{equiv NaOH}}{40 \; \text{g NaOH}} = 0.50 \; \text{equiv NaOH}$

   $N = \dfrac{0.50 \; \text{equiv NaOH}}{1 \; \text{L}} = 0.50N$

   b. $4.9 \; \text{g H}_2\text{SO}_4 \times \dfrac{1 \; \text{equiv H}_2\text{SO}_4}{49.0 \; \text{g H}_2\text{SO}_4} = 0.10 \; \text{equiv H}_2\text{SO}_4$

   $N = \dfrac{0.10 \; \text{equiv}}{0.50 \; \text{L}} = 0.20N$

9. Find the number of equivalents by multiplying the volume of the solution by the normality of the solution.

   a. $0.55 \; \text{L} \times \dfrac{1.8 \; \text{equiv}}{1 \; \text{L}} = 0.99 \; \text{equiv}$

   b. $1.6 \; \text{L} \times \dfrac{0.50 \; \text{equiv}}{1 \; \text{L}} = 0.80 \; \text{equiv}$

   c. $0.25 \; \text{L} \times \dfrac{0.28 \; \text{equiv}}{1 \; \text{L}} = 0.070 \; \text{equiv}$

10. The number of equivalents must be equal before and after dilution.
    $N_1 \times V_1 = N_2 \times V_2$
    $V_1 = \dfrac{N_2 \times V_2}{N_1} = \dfrac{500 \; \text{mL} \times 0.20N}{4.0N} = 25 \; \text{mL}$
    Dilute 25 mL of $4.0N$ H_2SO_4 to 500 mL.

11. At the neutralization point, $N_A \times V_A = N_B \times V_B$.
    $V_B = \dfrac{N_A \times V_A}{N_B} = \dfrac{0.050N \times 75 \; \text{mL}}{0.20N} = 19 \; \text{mL}$

12. At the equivalence point, the number of equivalents of acid and base are equal.
    $N_A \times V_A = N_B \times V_B$
    $N_B = \dfrac{N_A \times V_A}{V_B} = \dfrac{0.40N \times 75 \; \text{mL}}{25 \; \text{mL}} = 1.2N \; \text{base}$

13. **a.** $HPO_4^{2-} + H^+ \rightarrow H_2PO_4^-$
    b. $H_2PO_4^- + OH^- \rightarrow HPO_4^{2-} + H_2O$

14. **a.** $C_2H_3O_2^- + H^+ \rightarrow HC_2H_3O_2$
    b. $NH_3 + H^+ \rightarrow NH_4^+$

15. Write the equation for the dissociation.
    $PbS \rightleftharpoons Pb^{2+} + S^{2-}$
    $K_{sp} = [Pb^{2+}] \times [S^{2-}] = 3 \times 10^{-28}$
    At equilibrium, $[Pb^{2+}] = [S^{2-}]$,
    Therefore: $[Pb^{2+}]^2 = 3 \times 10^{-28}$
    $[Pb^{2+}] = 2 \times 10^{-14}M$

16. **a.** Write the equation for the dissociation.
    $Ag_2S \rightleftharpoons 2Ag^+ + S^{2-}$
    $K_{sp} = [Ag^+]^2 \times [S^{2-}] = 8 \times 10^{-51}$
    When $[S^{2-}]$ is x, then $[Ag^+] = 2x$.
    Substitute into the equation for K_{sp}.
    $K_{sp} = (2x)^2(x) = 8 \times 10^{-51}$
    $4x^3 = 8 \times 10^{-51}$
    $x^3 = 2 \times 10^{-51}$
    $x = 1 \times 10^{-17}$
    $[Ag^+] = 2x = 2 \times 10^{-17}M$
    b. $[S^{2-}] = x = 1 \times 10^{-17}M$

17. $SrSO_4 \rightleftharpoons Sr^{2+} + SO_4^{2-}$
    $K_{sp} = [Sr^{2+}] \times [SO_4^{2-}] = 3.2 \times 10^{-7}$
    At equilibrium, $[SO_4^{2-}] = x$ and
    $[Sr^{2+}] = x + 0.10 \approx 0.10$.
    $K_{sp} = 0.10 \times x = 3.2 \times 10^{-7}$
    $x = 3.2 \times 10^{-6}M$

18. Calculate the solubility product of $PbCl_2$.
    $K_{sp} = [Pb^{2+}] \times [Cl^-]^2$
    $= 0.015 \times (0.050)^2 = 3.75 \times 10^{-5}$
    Since the calculated solubility product does not exceed the K_{sp} for the saturated solution (K_{sp} of $PbCl_2 = 1 \times 10^{-5}$), precipitation will occur.

Chapter 20

1. **a.** Write the equation to show the ions.
$2Na + S \rightarrow 2Na^+ + S^{2-}$
There has been a transfer of one electron from each of two sodium atoms.
The oxidation process is:
$2Na \rightarrow 2Na + 2e^-$ (loss of electrons)
The reduction process is:
$S + 2e^- \rightarrow S^{2-}$ (gain of electrons)
Na loses electrons and is oxidized; S gains electrons and is reduced. Therefore, Na is the reducing agent and S is the oxidizing agent.
b. Write the equation to show the ions.
$2K + Cl_2 \rightarrow 2K^+ + 2Cl^-$
There is a transfer of electrons from K to Cl_2.
Oxidation: $2K \rightarrow 2K^+ + 2e^-$ (loss of electrons)
Reduction: $Cl_2 + 2e \rightarrow 2Cl^-$ (gain of electrons)
K loses electrons and is oxidized; Cl_2 gains electrons and is reduced. K is the reducing agent; Cl_2 is the oxidizing agent.
c. Write the equation to show the ions.
$4Al + 3O_2 \rightarrow 4Al^{3+} + 6O^{2-}$
There is a transfer of electrons from Al to O_2.
Oxidation: $4Al \rightarrow 4Al^{3+} + 12e^-$ (loss of electrons)
Reduction: $3O_2 + 12e^- \rightarrow 6O^{2-}$ (gain of electrons)
Al loses electrons and is oxidized; O_2 gains electrons and is reduced. Al is the reducing agent; O_2 is the oxidizing agent.

2. **a.** Since chlorine is more electronegative than hydrogen, the electrons in the H—Cl bond will be shifted toward Cl and away from H. Therefore H_2 is oxidized and Cl_2 is reduced; H_2 is the reducing agent and Cl_2 is the oxidizing agent. **b.** Nitrogen is more electronegative than hydrogen. The electrons in the N—H bonds in NH_3 are shifted toward N and away from H. Therefore H_2 is oxidized and N_2 is reduced; H_2 is the reducing agent and N_2 is the oxidizing agent. **c.** Chlorine is more electronegative than sulfur. The electrons in the S—Cl bonds in SCl_2 are shifted toward Cl and away from S. Therefore S is oxidized and Cl_2 is reduced; S is the reducing agent and Cl_2 is the oxidizing agent. **d.** Oxygen is more electronegative than nitrogen. The electrons in the N—O bonds in NO_2 are shifted toward oxygen and away from nitrogen. Therefore N_2 is oxidized and O_2 is reduced; N_2 is the reducing agent and O_2 is the oxidizing agent. **e.** Fluorine is more electronegative than lithium. The electrons in LiF are completely transferred from Li to F_2 (LiF is an ionic compound). Therefore Li is oxidized and F_2 is reduced: Li is the reducing agent and F_2 is the oxidizing agent. **f.** Sulfur is more electronegative than hydrogen. The electrons in the H—S bonds in H_2S are shifted toward the sulfur and away from the hydrogen. Therefore hydrogen is oxidized and sulfur is reduced. Hydrogen is the reducing agent and sulfur is the oxidizing agent.

3. **a.** The oxidation number of oxygen is -2. Since there are five oxygen atoms and the sum of the oxidation numbers must be zero, the oxidation number of phosphorus must be $+5$.
$$\overset{+5\ -2}{P_4O_{10}}$$
check: $4(+5) + 10(-2) = 0$.

b. The oxidation number of hydrogen is $+1$. The sum of the oxidation numbers of the nitrogen and hydrogen must equal the ionic charge, $1+$. The oxidation number of nitrogen is -3.
$$\overset{-3\ +1}{NH_4^+}$$
check: $(-3) + (4)(+1) = +1$.

c. The oxidation number of sodium is $+1$ and the oxidation number of oxygen is -2. The oxidation number of chromium is $+6$.
$$\overset{+1\ +6\ -2}{Na_2Cr_2O_7}$$
check: $(2)(+1) + (2)(+6) + (7)(-2) = 0$.

d. The oxidation number of hydrogen is $+1$ and the oxidation number of oxygen is -2. The oxidation number of calcium is $+2$.
$$\overset{+2\ -2\ +1}{Ca(OH)_2}$$
check: $(+2) + (2)(-2) + (2)(+1) = 0$.

4. Assign oxidation numbers to each element. An oxidation number decrease indicates reduction; an increase indicates oxidation.
a.
$$\overset{0}{2Na} + \overset{0}{Cl_2} \rightarrow \overset{+1\ -1}{2NaCl}$$
Sodium is oxidized $(0 \rightarrow +1)$. Chlorine is reduced $(0 \rightarrow -1)$. Sodium (as Na) is the reducing agent; chlorine (as Cl_2) is the oxidizing agent.
b.
$$\overset{+1+5-2}{2HNO_3} + \overset{+1-1}{6HI} \rightarrow \overset{+2-2}{2NO} + \overset{0}{3I_2} + \overset{+1\ -2}{4H_2O}$$
Nitrogen is reduced $(+5 \rightarrow +2)$. Iodine is oxidized $(-1 \rightarrow 0)$. Iodine (as HI) is the reducing agent; nitrogen (as HNO_3) is the oxidizing agent.
c.
$$\overset{+1\ -2}{3H_2S} + \overset{+1+5-2}{2HNO_3} \rightarrow \overset{0}{3S} + \overset{+2-2}{2NO} + \overset{+1\ -2}{4H_2O}$$
Nitrogen is reduced $(+5 \rightarrow +2)$. Sulfur is oxidized $(-2 \rightarrow 0)$. Sulfur (as H_2S) is the reducing agent; nitrogen (as HNO_3) is the oxidizing agent.
d.
$$\overset{+2+6-2}{2PbSO_4} + \overset{+1\ -2}{2H_2O} \rightarrow \overset{0}{Pb} + \overset{+4\ -2}{PbO_2} + \overset{+1+6-2}{2H_2SO_4}$$
Lead is oxidized $(+2 \rightarrow +4)$. Lead is reduced $(+2 \rightarrow 0)$. Lead (as $PbSO_4$) is both the reducing agent and the oxidizing agent.

5. *Step 1:* Assign oxidiation numbers to all the atoms in the equation.
Step 2: Identify which atoms are oxidized and which are reduced.
Step 3: Use a line to connect the atoms that undergo oxidation. Use a separate line to connect the atoms that undergo reduction.
Step 4: Use appropriate coefficients to make the total increase in oxidation number equal to the total decrease in oxidation number.
Step 5: Do a final check to ensure the equation is balanced for both atoms and charge.
a. increase $(0 \rightarrow +3)$
$$\boxed{\text{oxidation}}$$
$$\overset{0}{Al} + \overset{0}{Cl_2} \rightarrow \overset{+3\ -1}{AlCl_3}$$
$$\boxed{\text{reduction}}$$
decrease $(0 \rightarrow -1)$

To balance the increase $(+3)$ and the decrease (-1) in oxidation numbers, three chlorine atoms must be reduced $(-3$ decrease) for each aluminum atom that is oxidized $(+3$ increase). To get 3Cl on the left-hand side of the equation, put a coefficient of $\frac{3}{2}$ in front of Cl_2.

$$(1)(+3) = +3$$
$$Al + \tfrac{3}{2}Cl_2 \rightarrow AlCl_3$$
$$(3)(-1) = -3$$

The fractional coefficient can be removed by multiplying each term in the equation by 2.

$2Al(s) + 3Cl_2(g) \rightarrow 2AlCl_3(s)$

This equation is now balanced.

b. decrease $(+5 \rightarrow -1)$

$$\boxed{\text{reduction}}$$
$$\overset{+1+5-2}{KClO_3} \rightarrow \overset{+1-1}{KCl} + \overset{0}{O_2}$$
$$\boxed{\text{oxidation}}$$
increase $(-2 \rightarrow 0)$

To balance the increase $(+2)$ and the decrease (-6) in oxidation number, three oxygen atoms must be oxidized $(+6$ increase) for each chlorine that is reduced $(-6$ decrease). To get 3 O on the right-hand side, put a coefficient of $\frac{3}{2}$ in front of O_2.

$$(1)(-6) = -6$$
$$KClO_3 \rightarrow KCl + \tfrac{3}{2}O_2$$
$$(3)(+2) = +6$$

The fractional coefficient can be removed by multiplying each term in the equation by 2.

$2KClO_3(s) \rightarrow 2KCl(s) + 3O_2(g)$

This equation is now balanced.

c. increase $(-3 \rightarrow +1)$

$$\boxed{\text{oxidation}}$$
$$\overset{-3+1}{PH_3} + \overset{0}{I_2} + \overset{+1-2}{H_2O} \rightarrow \overset{+1+1-2}{H_3PO_2} + \overset{+1-1}{HI}$$
$$\boxed{\text{reduction}}$$
decrease $(0 \rightarrow -1)$

To balance the increase $(+4)$ and the decrease (-1) in oxidation numbers, one phosphorus atom must be oxidized $(+4$ increase) for every four iodine atoms that are reduced $(-4$ decrease). To get 4 I on the left-hand side, put a coefficient of 2 in front of I_2; then put a coefficient of 4 in front of HI.

$$(1)(+4) = +4$$
$$PH_3 + 2I_2 + H_2O \rightarrow H_3PO_2 + 4HI$$
$$(4)(-1) = -4$$

To balance the oxygen and hydrogen (the spectator atoms), put a coefficient of 2 in front of H_2O.

$PH_3(g) + 2I_2(s) + 2H_2O(l) \rightarrow H_3PO_2(aq) + 4HI(aq)$

This equation is now balanced.

d. increase $(0 \rightarrow +5)$

$$\boxed{\text{oxidation}}$$
$$\overset{0}{Cl_2} + \overset{+1-2+1}{KOH} \rightarrow \overset{+1+5-2}{KClO_3} + \overset{+1-1}{KCl} + \overset{+1-2}{H_2O}$$
$$\boxed{\text{reduction}}$$
decrease $(0 \rightarrow -1)$

To balance the increase $(+5)$ and the decrease (-1) in oxidation numbers, one chlorine atom must be oxidized $(+5$ increase) for every five chlorine atoms that are reduced $(-5$ decrease). To get a total of 6 Cl on the left-hand side, put a coefficient of 3 in front of Cl_2; then put a coefficient of 5 in front of KCl.

$$(1)(+5) = +5$$
$$3Cl_2 + KOH \rightarrow KClO_3 + 5KCl + H_2O$$
$$(5)(-1) = -5$$

Finally, K, O, and H are balanced by inspection.

$3Cl_2(g) + 6KOH(aq) \rightarrow$
$$KClO_3(aq) + 5KCl(aq) + 3H_2O(l)$$

This equation is balanced.

e. decrease $(+5 \rightarrow +2)$

$$\boxed{\text{reduction}}$$
$$\overset{+5}{HNO_3} + \overset{-2}{H_2S} \rightarrow \overset{0}{S} + \overset{+2}{NO} + H_2O$$
$$\boxed{\text{oxidation}}$$
increase $(-2 \rightarrow 0)$

To balance the increase $(+2)$ and the decrease (-3) in oxidation numbers, two nitrogen atoms must be reduced $(-6$ decrease) for every three sulfur atoms that are oxidized $(+6$ increase). Put coefficients of 2 in front of HNO_3 and NO to balance the nitrogen atoms. Put coefficients of 3 in front of H_2S and S.

$$(2)(-3) = -6$$
$$2HNO_3 + 3H_2S \rightarrow 3S + 2NO + H_2O$$
$$(3)(+2) = +6$$

Finally, H and O are balanced by inspection.

$2HNO_3(aq) + 3H_2S(aq) \rightarrow 3S(s) + 2NO(g) + 4H_2O(l)$

This equation is balanced.

f. decrease $(+7 \rightarrow 0)$

$$\boxed{\text{reduction}}$$
$$\overset{+1+7-2}{KIO_4} + \overset{+1-1}{KI} + \overset{+1-1}{HCl} \rightarrow \overset{+1-1}{KCl} + \overset{0}{I_2} + \overset{+1-2}{H_2O}$$
$$\boxed{\text{oxidation}}$$
increase $(-1 \rightarrow 0)$

To balance the decrease (-7) and the increase $(+1)$ in oxidation numbers, one iodine atom (in KIO_4) must be reduced $(-7$ decrease) for every seven iodine atoms (in KI) that are oxidized $(+7$ increase). Put a coefficient of 7 in front of KI and a coefficient of 4 in front of I_2.

$$(1)(-7) = -7$$
$$KIO_4 + 7KI + HCl \rightarrow KCl + 4I_2 + H_2O$$
$$(7)(+1) = +7$$

Finally, balance K, H, O, and Cl by inspection.

$KIO_4(aq) + 7KI(aq) + 8HCl(aq) \rightarrow$
$$8KCl(aq) + 4I_2(s) + 4H_2O(l)$$

The equation is balanced.

6. a. This is a double-replacement reaction. There is no change in oxidation numbers, so this is not a redox reaction.

b. This is a decomposition reaction. There is no change in oxidation numbers.

c. $\overset{0}{Mg} + \overset{0}{Br_2} \rightarrow \overset{2+}{Mg}\overset{1-}{Br_2}$

This is a redox combination reaction. Magnesium is oxidized. It is the reducing agent. Bromine is reduced. Bromine is the oxidizing agent.

d. $\overset{-3+1+3-2}{NH_4NO_2} \rightarrow \overset{0}{N_2} + \overset{+1-2}{2H_2O}$

This is a redox decomposition reaction. Nitrogen (in NO_2^-) is reduced. It is the oxidizing agent. Nitrogen (in NH_4^+) is oxidized. It is the reducing agent.

e. $\overset{+1+5-2}{2KClO_3} \rightarrow \overset{+1-1}{2KCl} + \overset{0}{3O_2}$

This is a redox decomposition reaction. Chlorine is reduced. It is the oxidizing agent. Oxygen is oxidized. It is the reducing agent.

f. This is a double-replacement reaction followed by a decomposition reaction (of H_2CO_3). There is no change in oxidation numbers.

g. $\overset{+2-2}{CuO} + \overset{0}{H_2} \rightarrow \overset{0}{Cu} + \overset{+1-2}{H_2O}$

This is a redox reaction. Copper is reduced. It is the oxidizing agent. Hydrogen is oxidized. It is the reducing agent.

h. This is a decomposition reaction. There is no change in oxidation numbers.

i. $\overset{0}{Sb} + \overset{+1+5-2}{HNO_3} \rightarrow \overset{+5-2}{Sb_2O_5} + \overset{+2-2}{NO} + \overset{+1-2}{H_2O}$

This is a redox reaction. Antimony is oxidized. It is the reducing agent. Nitrogen is reduced. It is the oxidizing agent.

7. a. *Step 1:* Write the equation in ionic form.
Step 2: Write the oxidation and reduction half-reactions.
Step 3: Balance the half-reactions, using H_2O, H^+, or OH^- as needed.
Step 4: Add electrons to one side of each half-reaction to balance the charges.
Step 5: Multiply each half-reaction to make the electron changes equal.
Step 6: Add the half-reactions and subtract terms that are on both sides.
Step 7: Balance the spectator ions by inspection if they are included.

a. $ClO_3^- + I^- \rightarrow Cl^- + I_2$ (acid solution)

Oxidation: $I^- \rightarrow I_2$
Reduction: $ClO_3^- \rightarrow Cl^-$

Balance:

Oxidation: $2I^- \rightarrow I_2$
Reduction: $ClO_3^- + 6H^+ \rightarrow Cl^- + 3H_2O$

Add electrons.

Oxidation: $2I^- \rightarrow I_2 + 2e^-$
Reduction: $ClO_3^- + 6H^+ + 6e^- \rightarrow Cl^- + 3H_2O$

Multiply oxidation half-reaction by 3.

Oxidation: $6I^- \rightarrow 3I_2 + 6e^-$
Reduction: $ClO_3^- + 6H^+ + 6e^- \rightarrow Cl^- + 3H_2O$

Add the half-reactions and subtract terms.
$ClO_3^- + 6I^- + 6H^+ + \cancel{6e^-} \rightarrow$
$$Cl^- + 3I_2 + 3H_2O + \cancel{6e^-}$$

The balanced equation is:
$ClO_3^-(aq) + 6I^-(aq) + 6H^+(aq) \rightarrow$
$$Cl^-(aq) + 3I_2(s) + 3H_2O(l)$$

b. $C_2O_4^{2-} + MnO_4^- \rightarrow Mn^{2+} + CO_2$ (acid solution)

Oxidation: $C_2O_4^{2-} \rightarrow CO_2$
Reduction: $MnO_4^- \rightarrow Mn^{2+}$

Balance:

Oxidation: $C_2O_4^{2-} \rightarrow 2CO_2$
Reduction: $MnO_4^- + 8H^+ \rightarrow Mn^{2+} + 4H_2O$

Add electrons.

Oxidation: $C_2O_4^{2-} \rightarrow 2CO_2 + 2e^-$
Reduction: $MnO_4^- + 8H^+ + 5e^- \rightarrow Mn^{2+} + 4H_2O$

Multiply oxidation half-reaction by 5 and reduction half-reaction by 2.

Oxidation: $5C_2O_4^{2-} \rightarrow 10CO_2 + 10e^-$
Reduction: $2MnO_4^- + 16H^+ + 10e^- \rightarrow 2Mn^{2+} + 8H_2O$

Add the half-reactions and subtract terms:
$5C_2O_4^{2-} + 2MnO_4^- + 16H^+ + \cancel{10e^-} \rightarrow$
$$2Mn^{2+} + 10CO_2 + 8H_2O + \cancel{10e^-}$$

The balanced equation is:
$5C_2O_4^{2-}(aq) + 2MnO_4^-(aq) + 16H^+(aq) \rightarrow$
$$2Mn^{2+}(aq) + 10CO_2(g) + 8H_2O(l)$$

c. $Br_2 + SO_2 \rightarrow Br^- + SO_4^{2-}$ (acid solution)

Oxidation: $SO_2 \rightarrow SO_4^{2-}$
Reduction: $Br_2 \rightarrow Br^-$

Balance:

Oxidation: $SO_2 + 2H_2O \rightarrow SO_4^{2-} + 4H^+$
Reduction: $Br_2 \rightarrow 2Br^-$

Add electrons.

Oxidation: $SO_2 + 2H_2O \rightarrow SO_4^{2-} + 4H^+ + 2e^-$
Reduction: $Br_2 + 2e^- \rightarrow 2Br^-$

Since the electron changes are equal, add the half-reactions and cancel terms.
$Br_2 + SO_2 + 2H_2O + \cancel{2e^-} \rightarrow$
$$2Br^- + SO_4^{2-} + 4H^+ + \cancel{2e^-}$$

The balanced equation is:
$Br_2(l) + SO_2(g) + 2H_2O(l) \rightarrow$
$$2Br^-(aq) + SO_4^{2-}(aq) + 4H^+(aq)$$

d. $MnO_2 + H^+ + NO_2^- \rightarrow NO_3^- + Mn^{2+} + 2H_2O$

Oxidation: $NO_2^- \rightarrow NO_3^-$
Reduction: $MnO_2 \rightarrow Mn^{2+}$

Balance:

Oxidation: $NO_2^- + H_2O \rightarrow NO_3^- + 2H^+$
Reduction: $MnO_2 + 4H^+ \rightarrow Mn^{2+} + 2H_2O$

Add electrons.

Oxidation: $NO_2^- + H_2O \rightarrow NO_3^- + 2H^+ + 2e^-$
Reduction: $MnO_2 + 4H^+ + 2e^- \rightarrow Mn^{2+} + 2H_2O$

Since the electron changes are equal, add the half-reactions and cancel terms.
$MnO_2 + \overset{2}{\cancel{4}}H^+ + NO_2^- + \cancel{H_2O} + \cancel{2e^-} \rightarrow$
$$NO_3^- + Mn^{2+} + \cancel{2e^-} + \cancel{2H^+} + \overset{1}{\cancel{2}}H_2O$$

The balanced equation is:
$MnO_2(s) + 2H^+(aq) + NO_2^-(aq) \rightarrow$
$$NO_3^-(aq) + Mn^{2+}(aq) + H_2O(l)$$

e. $MnO_4^- + NO_2^- \rightarrow MnO_2 + NO_3^-$ (basic solution)

Oxidation: $NO_2^- \rightarrow NO_3^-$
Reduction: $MnO_4^- \rightarrow MnO_2$

Balance:

Oxidation: $NO_2^- + 2OH^- \rightarrow NO_3^- + H_2O$
Reduction: $MnO_4^- + 2H_2O \rightarrow MnO_2 + 4OH^-$

Add electrons.

Oxidation: $NO_2^- + 2OH^- \rightarrow NO_3^- + H_2O + 2e^-$
Reduction: $MnO_4^- + 2H_2O + 3e^- \rightarrow MnO_2 + 4OH^-$

Multiply the oxidation half-reaction by 3 and the reduction half-reaction by 2.

Oxidation: $3NO_2^- + 6OH^- \rightarrow 3NO_3^- + 3H_2O + 6e^-$
Reduction: $2MnO_4^- + 4H_2O + 6e^- \rightarrow 2MnO_2 + 8OH^-$

Add the half-reactions and cancel terms.

$$MnO_4^- + 3NO_2^- + \overset{1}{\cancel{4}}H_2O + \cancel{6OH^-} + \cancel{6e^-} \rightarrow$$
$$2MnO_2 + 3NO_3^- + \overset{2}{\cancel{3H_2O}} + \cancel{8}OH^- + \cancel{6e^-}$$

The balanced equation is:
$2MnO_4^-(aq) + 3NO_2^-(aq) + H_2O(l) \rightarrow$
$\qquad 2MnO_2(s) + 3NO_3^-(aq) + 2OH^-(aq)$

f. $Cl_2 \rightarrow ClO_3 + Cl$ (basic solution)

Oxidation: $Cl_2 \rightarrow ClO_3^-$
Reduction: $Cl_2 \rightarrow Cl^-$

Balance:

Oxidation: $Cl_2 + 12OH^- \rightarrow 2ClO_3^- + 6H_2O$
Reduction: $Cl_2 \rightarrow 2Cl^-$

Add electrons.

Oxidation: $Cl_2 + 12OH^- \rightarrow 2ClO_3^- + 6H_2O + 10e^-$
Reduction: $Cl_2 + 2e^- \rightarrow 2Cl^-$

Multiply the reduction half-reaction by 5.

Oxidation: $Cl_2 + 12OH^- \rightarrow 2ClO_3^- + 6H_2O + 10e^-$
Reduction: $5Cl_2 + 10e^- \rightarrow 10Cl^-$

Add the half-reactions and cancel terms.

$$\overset{6Cl_2}{\overline{5Cl_2 + Cl_2}} + 12OH^- + \cancel{10e^-} \rightarrow$$
$$2ClO_3^- + 10Cl^- + 6H_2O + \cancel{10e^-}$$

The balanced equation is:
$6Cl_2(g) + 12OH^-(aq) \rightarrow$
$\qquad 2ClO_3^-(aq) + 10Cl^- + 6H_2O(l)$

Chapter 21

1. The half-cell with the more positive reduction potential occurs as a reduction. In this cell Cu^{2+} is reduced and Al is oxidized. Because reduction takes place at the Cu^{2+} half-cell, this half-cell is the cathode.
The half-cell reactions are:

Oxidation: $Al(s) \rightarrow Al^{3+}(aq) + 3e^-$ (anode)
Reduction: $Cu^{2+}(aq) + 2e^- \rightarrow Cu(s)$ (cathode)

Before the reactions are added to give the overall cell reaction the electrons must cancel. Multiply each equation by a coefficient to give equal numbers of electrons.

$$2[Al(s) \rightarrow Al^{3+}(aq) + 3e^-]$$
$$\underline{3[Cu^{2+}(aq) + 2e^- \rightarrow Cu(s)]}$$
$$2Al(s) + 3Cu^{2+}(aq) \rightarrow 2Al^{3+}(aq) + 3Cu(s)$$

The standard cell potential can now be calculated.
$$E^\circ_{cell} = E^\circ_{red} - E^\circ_{oxid}$$

$$= E^\circ_{Cu^{2+}} - E^\circ_{Al^{3+}}$$
$$= +0.34V - (-1.66V)$$
$$= +2.00V$$

2. In this cell Ag^+ is reduced and Cu is oxidized. The Ag^+ half-cell is the cathode.
The half-cell reactions are:

Oxidation: $Cu(s) \rightarrow Cu^{2+}(aq) + 2e^-$ (anode)
Reduction: $Ag^+(aq) + e^- \rightarrow Ag(s)$ (cathode)

Before the reactions are added the electrons must cancel.

$$Cu(s) \rightarrow Cu^{2+}(aq) + 2e^-$$
$$\underline{2[Ag^+(aq) + e^- \rightarrow Ag(s)]}$$
$$Cu(s) + 2Ag^+(aq) \rightarrow Cu^{2+}(aq) + 2Ag(s)$$

The standard cell potential can now be calculated.
$$E^\circ_{cell} = E^\circ_{red} - E^\circ_{oxid}$$
$$E^\circ_{cell} = E^\circ_{Ag^+} - E^\circ_{Cu^{2+}}$$
$$= +0.80V - (+0.34V)$$
$$= +0.46V$$

3. Because the reaction is spontaneous as written Co^{2+} is reduced and Fe is oxidized.
The half-cell reactions are:

Oxidation: $Fe(s) \rightarrow Fe^{2+}(aq) + 2e^-$ (anode)
Reduction: $Co^{2+}(aq) + 2e^- \rightarrow Co(s)$ (cathode)

The standard cell potential can now be calculated.

$$E^\circ_{cell} = E^\circ_{red} - E^\circ_{oxid}$$
$$= E^\circ_{Co^{2+}} - E^\circ_{Fe^{2+}}$$

Substitute standard reduction potentials from Table 21·3.
$$= -0.28V - (-0.44V)$$
$$= +0.16V$$

4. a. The half-reactions are:

Oxidation: $Cu(s) \rightarrow Cu^{2+}(aq) + 2e^-$
Reduction: $2H^+(aq) + 2e^- \rightarrow H_2(g)$

The standard reduction potentials are found in Table 21·3.
$$E^\circ_{Cu^{2+}} = +0.34V$$
$$E^\circ_{H^+} = 0.00V$$
The standard cell potential is the difference of the potentials of the two cells.

$$E^\circ_{cell} = E^\circ_{red} - E^\circ_{oxid}$$
$$= E^\circ_{H^+} - E^\circ_{Cu^{2+}}$$
$$= 0.00V - (+0.34V)$$
$$= 0.34V$$

Since the standard cell potential is a negative number, the redox reaction is nonspontaneous as written.
b. The half-reactions are:

Oxidation: $Ag(s) \rightarrow Ag^+(aq) + e^-$
Reduction: $Fe^{2+}(aq) + 2e^- \rightarrow Fe(s)$

The standard reduction potentials are found in Table 21·3.
$$E^\circ_{Ag^+} = +0.80V$$
$$E^\circ_{Fe^{2+}} = -0.44V$$
The standard cell potential is the difference of the potentials of the two cells.

$$E^\circ_{cell} = E^\circ_{red} - E^\circ_{oxid}$$
$$= E^\circ_{Fe^{2+}} - E^\circ_{Ag^+}$$
$$= -0.44V - (+0.80V)$$
$$= -1.24V$$

Since the standard cell potential is a negative number, the redox reaction is nonspontaneous as written.

c. The half-reactions are:

Oxidation: $Cr(s) \rightarrow Cr^{3+}(aq) + 3e^-$
Reduction: $Zn^{2+}(aq) + 2e^- \rightarrow Zn(s)$

The standard reduction potentials are found in Table 21·3.

$$E^{\circ}_{Cr^{3+}} = -0.74V$$
$$E^{\circ}_{Zn^{2+}} = -0.76V$$

The standard cell potential is the difference of the potentials of the two cells.

$$E^{\circ}_{cell} = E^{\circ}_{red} - E^{\circ}_{oxid}$$
$$= E^{\circ}_{Zn^{2+}} - E^{\circ}_{Cr^{3+}}$$
$$= -0.76V - (-0.74V)$$
$$= -0.02V$$

Since the standard cell potential is a negative number, the redox reaction is nonspontaneous as written.

Chapter 22

1. $2 K + 2 H_2O \rightarrow 2 KOH + H_2$

2. $MgO + CO_2 \rightarrow MgCO_3$

Chapter 24

1. The reactant is radium-226 ($^{226}_{88}Ra$). The products are an alpha particle and radon, with an atomic number of 86 and a mass number 222 (226 − 4).

$$^{226}_{88}Ra \rightarrow {}^{222}_{86}Rn + {}^{4}_{2}He$$

2. In beta decay a neutron decomposes to give a proton and an electron (beta radiation). This means that the mass numbers of the reactant and product are equal, but the atomic number increases by 1.

$$^{210}_{82}Pb \xrightarrow{\text{same}} {}^{210}_{83}Bi + {}^{0}_{-1}e$$

(increased by 1)

3. The types of radiation in order of increasing penetrating power are: alpha radiation, beta radiation, and gamma radiation.

4.

Number of half-lives	Elapsed time (hr)	Mass of ^{56}Mn
0	0	1.00 mg
1	2.6	0.50 mg
2	5.2	0.25 mg
3	7.8	0.125 mg
4	10.4	0.0625 mg

After four half-lives there is 0.0625 mg of manganese-56 left.

5. A Geiger counter does not detect alpha particles; it is a beta radiation detector. Your hands may be contaminated.

Chapter 25

1.

```
    H  H  H  H  H
    |  |  |  |  |
H—C—C—C—C—C—H
    |  |  |  |  |
    H  H  H  H  H

    H  H  H  H  H  H
    |  |  |  |  |  |
H—C—C—C—C—C—C—H
    |  |  |  |  |  |
    H  H  H  H  H  H
```

2. $CH_3CH_2CH_2CH_2CH_3$, pentane
$CH_3CH_2CH_2CH_2CH_2CH_3$, hexane

3. a. propane
b. pentane
c. hexane

4. a. 2-methylbutane
b. 3-methylpentane
c. 3-ethylhexane

5. a.
$$CH_3CHCHCH_2CH_2CH_3$$ with CH_3 substituents
2,3-dimethylhexane

b.
$$CH_3CH_2CHCHCH_2CH_2CH_3$$ with CH_2CH_3
3-ethylheptane

c.
$$CH_3CHCHCHCH_2CH_2CH_2CH_3$$ with CH_3, CH_3, and CH_2CH_3
3-ethyl-2,4-dimethyloctane

6. Five structural isomers of molecular formula C_6H_{14} exist.

C—C—C—C—C—C
hexane

C—C—C—C—C with C
3-methylpentane

C—C—C—C—C with C
2-methylpentane

C—C—C—C with C and C
2,2-dimethylbutane

C—C—C—C with C and C
2,3-dimethylbutane

7. a. $CH_2=CHCH_2CH_2CH_3$,
1-pentene

b. (cis-2-hexene)
$$CH_3CH=CHCH_2CH_2CH_3$$
cis-2-hexene   trans-2-hexene

c.
$$(CH_3)_2C=CHCH_2CH_2CH_3$$
2-methyl-2-hexene

d.
$$(CH_3)_2C=C(CH_3)_2$$
2,3-dimethylbutene

8. Nonsuperimposable mirror images are **a, d, e,** and **f.**

9. The asymmetric carbon (circled) is found in **c.** and **d.**

c. $CH_3—\overset{H}{\underset{Cl}{C}}—OH$

d. $CH_3—\overset{H}{\underset{CH_2CH_3}{C}}—CHO$

10. a. 1-phenylpropane
b. 2-phenylpropane
c. 2,5-dimethyl-3-phenylhexane

Chapter 26

1. **a.** —OH, hydroxyl **b.** —COOH, carboxyl
   c. —NH₂, amino
2. **a.** bromobenzene **b.** chloroethane **c.** 3-chloro-1-butene
3. **a.**

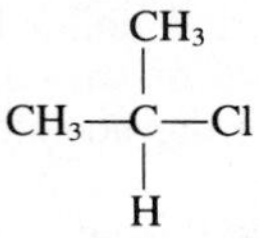

$$CH_3\!-\!\underset{\displaystyle \overset{|}{H}}{\overset{\displaystyle \overset{CH_3}{|}}{C}}\!-\!Cl$$

   c.

   b. ICH₂CCH₂CH₂CH₃ (with CH₃ above and CH₃ below)

4. **a.** butanol **b.** 2-propanol **c.** 2-methyl-1-butanol
5. **a.** primary **b.** secondary **c.** primary
6. **a.** 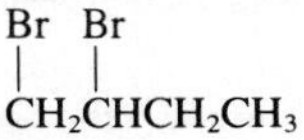 CH₂CHCH₂CH₃ (Br, Br)

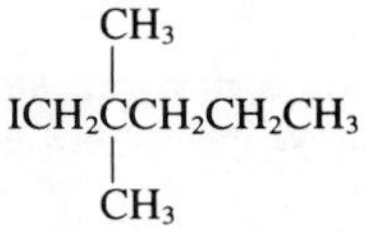

   b.  CH₃CHCHCH₃ (I, I)

   c. 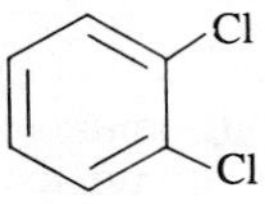 (benzene ring with Cl, Cl on adjacent positions)

   d.  CH₃CHCHCH₃ (H, H)

7. **a.**

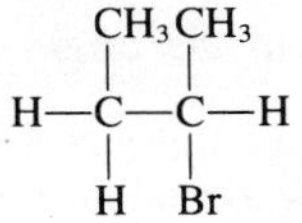

$$H\!-\!\underset{\displaystyle \overset{|}{H}}{\overset{\displaystyle \overset{CH_3}{|}}{C}}\!-\!\underset{\displaystyle \overset{|}{Br}}{\overset{\displaystyle \overset{CH_3}{|}}{C}}\!-\!H$$

   b. (benzene ring with Cl)

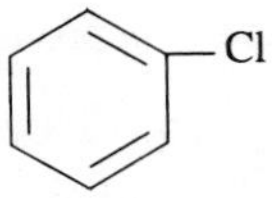

   c. (cyclohexane ring with Cl)

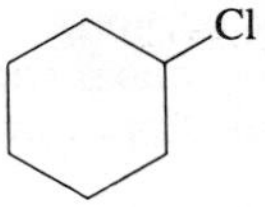

8. **a.** propanol **b.** 3-hexanone **c.** 3-methylpentanal
   d. butanone
9. **a.** CH₃CH₂CH₂COOH
   b. no reaction

   c.

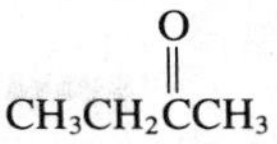

$$CH_3CH_2\overset{\displaystyle \overset{O}{\|}}{C}CH_3$$

   d. 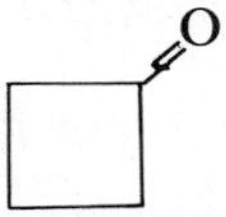 (cyclobutanone ring with =O)

10. **a.** propanol, CH₃CH₂CH₂OH

    b. 2-butanol, CH₃CH₂CHCH₃ (OH above)

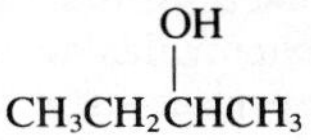

    c. 2-methyl-1-butanol, CH₃CH₂CHCH₂OH (CH₃ above)

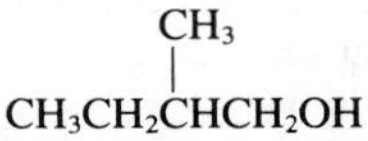

11. **a.** CH₃CH₂COONa, CH₃CH₂OH;

    b. CH₃COOK, (benzene ring)—OH

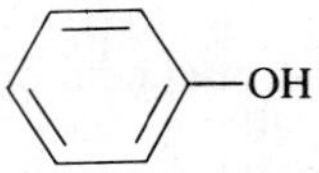

    c. CH₃CH₂COOH, CH₃CHCH₂OH (CH₃ below)

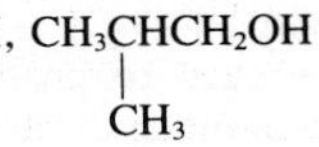

absolute zero the zero point on the Kelvin temperature scale, equivalent to $-273°C$; all molecular motion theoretically stops at this temperature. *2·11*

accuracy the closeness of a measurement to the true value of what is being measured. *2·2*

acid a compound containing hydrogen that ionizes to yield hydrogen ions (H^+) in water. *5·13*

acid dissociation constant (K_a) the ratio of the concentration of the dissociated form of an acid to the undissociated form; stronger acids have larger K_a values than weaker acids. *18·8*

acidic solution any solution in which the hydrogen-ion concentration is greater than the hydroxide-ion concentration. *18·2*

activated complex an unstable arrangement of atoms that exists momentarily at the peak of the activation energy barrier; it represents an intermediate or transitional structure formed during the course of a reaction. *17·1*

activation energy the minimum energy colliding particles must have in order to react. *17·1*

activity series of metals a table listing metals in order of decreasing activity. *7·5*

actual yield the amount of product that forms when a reaction is carried out in the laboratory. *8·6*

addition polymerization the process that occurs when unsaturated monomers add to each other, forming a polymer. *26·12*

addition reaction a reaction in which a substance is added at the double bond of an alkene or at the triple bond of an alkyne. *26·6*

alcohol an organic compound having an $-OH$ (hydroxyl) group; the general structure is R—OH. *26·4*

aldehyde an organic compound in which the carbon of the carbonyl group is joined to at least one hydrogen; the general formula is RCHO. *26·8*

aliphatic compound a hydrocarbon compound that does not contain a ring structure. *25·8*

alkali metal any metal in Group 1A of the periodic table. *12·10*

alkaline earth metal any metal in Group 2A of the periodic table. *12·10*

alkaline solution a basic solution. *18·2*

alkane a hydrocarbon containing only single covalent bonds; alkanes are saturated hydrocarbons. *25·1*

alkene a hydrocarbon containing one or more carbon–carbon double bonds; alkenes are unsaturated hydrocarbons. *25·5*

alkyl group a hydrocarbon substituent; methyl (CH_3) is an alkyl group. *25·3*

alkyl halide a halocarbon in which one or more halogen atoms are attached to the carbon atoms of an aliphatic chain. *26·2*

alkyne a hydrocarbon containing a carbon–carbon triple bond; alkynes are unsaturated hydrocarbons. *25·5*

allotrope a molecular form of an element that exists in two or more different forms in the same physical state; oxygen, O_2, and ozone, O_3, are allotropes of the element oxygen. *23·2*

alpha particle a positively charged particle emitted from certain radioactive nuclei; it consists of two protons and two neutrons and is identical to the nucleus of a helium atom. *24·2*

alpha radiation alpha particles emitted from a radioactive source. *24·2*

amide an organic compound having a nitrogen atom attached to the carbon of a carbonyl group; the general structures are

$$R-\overset{\overset{\displaystyle O}{\|}}{C}-NH_2 \qquad R-\overset{\overset{\displaystyle O}{\|}}{C}-\overset{\overset{\displaystyle H}{|}}{N}-R \text{ and } R-\overset{\overset{\displaystyle O}{\|}}{C}-\overset{\overset{\displaystyle R}{|}}{N}-R$$

26·13

amine an organic derivative of ammonia; the general structures are

$$R-\overset{\overset{\displaystyle H}{|}}{N}-H \qquad R-\overset{\overset{\displaystyle R}{|}}{N}-H \qquad R-\overset{\overset{\displaystyle R}{|}}{N}-R$$

26·13

amino acid an organic compound having amino ($-NH_2$) and carboxylic acid ($-COOH$) groups in the same molecule; proteins are made up of the 20 naturally occurring amino acids. *26·16*

amphoteric a substance that can act both as an acid and a base; water is amphoteric. *18·6*

amplitude the height of a wave from the origin to the crest. *11·6*

amorphous a term used to describe a solid that lacks an ordered internal structure; denotes a random arrangement of atoms. *9·8*

analytical chemistry the study of the composition of substances. *1·1*

anion any atom or group of atoms with a negative charge. *5·2*

anode the electrode at which oxidation occurs. *21·2*

antibonding orbital a molecular orbital whose energy is higher than that of the atomic orbitals from which it is formed. *14·7*

aqueous solution (*aq*) a solution in which the solvent is water. *15·6*

arene any member of a special group of unsaturated cyclic hydrocarbons. *25·8*

aromatic compound a name originally given to the arenes because many of them have pleasant odors. *25·8*

aryl halide a halocarbon in which one or more halogens are attached to the carbon atoms of an arene ring. *26·2*

asymmetric carbon a carbon atom that has four different groups attached. *25·7*

atom the smallest particle of an element that retains the properties of that element. *4·1*

atomic emission spectrum the pattern of frequencies obtained by passing light emitted by atoms of an element in the gaseous state through a prism; the emission spectrum of each element is unique to that element. *11·6*

atomic mass the weighted average of the masses of the isotopes of an element. *4·7*

atomic mass unit a unit of mass equal to one-twelfth the mass of a carbon-12 atom. *4·5*

atomic number the number of protons in the nucleus of an atom of an element. *4·4*

atomic orbital a region in space around the nucleus of an atom where there is a high probability of finding an electron. *11·3*

atmospheric pressure the pressure exerted by air molecules in the atmosphere surrounding the earth, resulting from collisions of air molecules with objects. *9·3*

Aufbau principle electrons enter orbitals of lowest energy first. *11·4*

Avogadro's hypothesis equal volumes of gases at the same temperature and pressure contain equal numbers of particles. *9·4*

Avogadro's number the number of representative particles contained in one mole of a substance; equal to 6.02×10^{23} particles. *6·2*

balanced equation a chemical equation in which mass is conserved; each side of the equation has the same number of atoms of each element. *7·2*

band of stability the location of stable nuclei on a neutron-vs-proton plot. *24·3*

barometer an instrument used to measure atmospheric pressure. *9·3*

base a compound that ionizes to yield hydroxide ions (OH^-) in water. *5·7*

base dissociation constant (K_b) the ratio of the concentration of the dissociated form of a base to the undissociated form. *18·8*

basic oxygen process the process by which most of today's high quality steel is made; pressurized oxygen and lime are used to remove impurities in pig iron. *22·10*

basic solution any solution in which the hydroxide-ion concentration is greater than the hydrogen-ion concentration. *18·2*

battery a group of voltaic cells that are connected together. *21·4*

Benedict's test a test commonly used to detect the presence of aldehydes. *26·10*

Bessemer process one of the first methods used for making steel; a blast of hot air is bubbled through molten pig iron in a huge egg-shaped converter. *22·10*

beta particles a fast-moving electron emitted from certain radioactive nuclei; it is formed when a neutron decomposes. *24·2*

beta radiation fast-moving electrons (beta particles) emitted from a radioactive source. *24·2*

binary compound a compound composed of two elements; $NaCl$ and Al_2O_3 are binary compounds. *5·9*

biochemistry the study of the composition and changes in composition of living organisms. *1·1*

blast furnace a towerlike furnace in which carbon is used to reduce iron ore to metallic iron. *22·9*

boiling point (bp) the temperature at which the vapor pressure of a liquid is just equal to the external pressure on the liquid. *9·7*

boiling point elevation the difference in temperature between the boiling points of a solution and of the pure solvent. *16·7*

bond dissociation energy the amount of energy required to break a bond between atoms; it is usually expressed in kJ per mol of substance. *14·12*

bonding orbital a molecular orbital whose energy is lower than that of the atomic orbitals from which it is formed. *14·7*

Boyle's law for a fixed mass of gas at constant temperature, the volume of the gas varies inversely with pressure. *10·6*

branched-chain alkane an alkane with one or more alkyl groups attached. *25·3*

bronze an alloy chiefly composed of copper and tin; zinc, lead or other metals may be added to give it special properties. *22·4*

Brownian motion the chaotic movement of colloidal particles, caused by collision with water molecules of the medium in which they are dispersed. *15·10*

buffer a solution in which the pH remains relatively constant when small amounts of acid or base are added; it consists of a solution of a weak acid and the salt of a weak acid (or a solution of a weak base with the salt of a weak base.) *19·6*

buffer capacity a measure of the amount of acid or base that may be added to a buffer solution before a significant change in pH occurs. *19·6*

calorie (cal) the quantity of heat that raises the temperature of 1 g of pure water 1°C. *2·12*

carbohydrate the name given to a group of monomers and polymers of aldehydes and ketones that have numerous hydroxyl groups attached; sugars and starches are carbohydrates. *26·14*

carbonyl group a functional group having a carbon atom and an oxygen atom joined by a double bond;

$$\underset{(—C—)}{\overset{O}{\|}}$$

it is found in aldehydes, ketones, esters, and amides. *26·8*

carboxyl group a functional group consisting of a carbonyl group attached to a hydroxyl group;

$$\underset{(—C—OH)}{\overset{O}{\|}}$$

it is found in carboxylic acids. *26·9*

carboxylic acid an organic acid containing a carboxyl group; the general formula is RCOOH. *26·9*

cast iron the product formed when pig iron is melted with scrap iron. *22·10*

catalyst a substance that speeds up a reaction without being used up. *7·1*

cathode the electrode at which reduction occurs. *21·2*

cathode ray a stream of electrons produced at the negative electrode (cathode) of a tube containing a gas at low pressure. *4·2*

cation any atom or group of atoms with a positive charge. *5·2*

cell potential the difference between the reduction potentials of two half-cells. *21·6*

Celsius temperature scale the temperature scale on which the freezing point of water is 0° and the boiling point is 100°. *2·11*

charge the load of raw materials added to a blast furnace. *22·9*

Charles' Law the volume of a fixed mass of gas is directly proportional to its Kelvin temperature if the pressure is kept constant. *10·7*

chemical equation an expression representing a chemical reaction; the formulas of the reactants (on the left) are connected by an arrow with the formulas for the products (on the right). *7·1*

chemical equilibrium a state of balance in which forward and reverse reactions are taking place at the same rate; no net change in the amounts of reactants and products occurs in the chemical system. *17·10*

chemical formula a shorthand method used to show the number and type of atoms present in the smallest representative unit of a substance; the chemical formula of ammonia, with one nitrogen and three hydrogens, is NH_3. *5·4*

chemical property the ability of a substance to undergo chemical reactions and to form new substances. *1·11*

chemical reaction the changing of substances to other substances by the breaking of old bonds and the formation of new bonds. *1·11*

chemical symbol a one or two letter representation of an element. *1·8*

chemistry the study of the structure, properties, and composition of substances, and the changes that substances undergo. *1·1*

cis **configuration** a term applied to geometric isomers; it denotes an arrangement in which the substituted groups are on the same side of the double bond. *25·6*

coefficient a small whole number that appears in front of a formula in a balanced chemical equation. *7·2*

colligative property a property of a solution that depends on the concentration of the solute particles but is independent of the nature of the particles; boiling point elevation, freezing point depression, and vapor pressure lowering are colligative properties. *16·7*

colloid a liquid mixture containing particles that are intermediate in size between those of a suspension and a true solution; these particles are evenly distributed throughout the liquid and do not settle out with time. *15·10*

combination reaction a chemical change in which two or more substances react to form a single new substance; also called a synthesis reaction. *7·3*

combined gas law a relationship describing the behavior of gases that combines Boyle's law, Charles' law, and Gay-Lussac's law. *10·9*

combustion reaction a chemical change in which oxygen reacts with another substance, often producing energy in the form of heat and light. *7·7*

common ion an ion that is common to both salts in a solution; in a solution of silver nitrate and silver chloride, Ag^+ would be a common ion. *19·8*

common ion effect a decrease in the solubility of a substance caused by the addition of a common ion. *19·8*

compound a substance that can be separated into simpler substances (elements or other compounds) only by chemical reactions. *1·7*

concentrated solution a solution containing a large amount of solute. *16·4*

concentration a measurement of the amount of solute that is dissolved in a given quantity of solvent; usually expressed as mol/L. *16·4*

condensed structural formula a structural formula that leaves out some bonds and/or atoms; the presence of these atoms or bonds is understood. *25·2*

conjugate acid the particle formed when a base gains a hydrogen ion; NH_4^+ is the conjugate acid of the base NH_3. *18·6*

conjugate acid-base pair two substances that are related by the loss or gain of a single hydrogen ion. Ammonia (NH_3) and the ammonium ion (NH_4^+) are a conjugate acid-base pair. *18·6*

conjugate base the particle that remains when an acid has donated a hydrogen ion; OH^- is the conjugate base of the acid water. *18·6*

continuous-chain alkane a hydrocarbon that contains any number of carbon atoms in a straight chain. *25·2*

conversion factor a ratio of equivalent units used to express the relationship between quantities expressed in different units. *3·3*

coordinate covalent bond a covalent bond formed when one atom contributes both bonding electrons. *14·4*

coordination number the number of ions of opposite charge that surround each ion in a crystal. *13·5*

covalent atomic radius half of the distance between the nuclei of two atoms in a hominuclear diatomic molecule. *12·4*

cracking the controlled process by which hydrocarbons are broken down or rearranged into smaller, more useful molecules. *25·9*

crystal a substance in which the atoms, ions, or molecules are arranged in an orderly, repeating, three-dimensional pattern called a crystal lattice. *9·8*

cyclic hydrocarbon an organic compound that contains a hydrocarbon ring. *25·8*

Dalton's atomic theory the first theory to relate chemical changes to events at the atomic level. *4·1*

Dalton's law of partial pressures at constant volume and temperature, the total pressure of a mixture of gases is the sum of the partial pressures of all the gases present. *10·5*

data the observations that are recorded during an experiment. *1·2*

de Broglie's equation an equation that describes the wavelength of a moving particle; it predicts that all matter exhibits wavelike motions. *11·10*

decomposition reaction a chemical change in which a single compound is broken down into two or more simpler products. *7·4*

dehydrogenation reaction a reaction in which hydrogen is lost. *26·10*

deliquescent a term describing a substance that removes sufficient water from the air to form a solution; the process occurs when the solution formed has a lower vapor pressure than that of the water in the air. *15·8*

denatured alcohol ethanol to which a poisonous substance has been added to make it unfit to drink. *26·5*

density the ratio of the mass of an object to its volume. *2·9*

dessicant a hygroscopic substance used as a drying agent. *15·8*

diffusion the tendency of molecules and ions to move toward areas of lower concentration until the concentration is uniform throughout the system. *10·12*

dilute solution a solution containing a small amount of solute. *16·4*

dimensional analysis a technique of problem-solving that uses the units that are part of a measurement to help solve the problem. *3·4*

dipole a molecule that has two electrically charged regions, or poles. *14·11*

dipole interaction a weak intermolecular force resulting from the attraction of oppositely charged regions of polar molecules. *14·13*

diprotic acid any acid that contains two ionizable protons (hydrogen ions); sulfuric acid (H_2SO_4) is a diprotic acid. *18·5*

disaccharide a carbohydrate formed from two monosaccharide units; common table sugar (sucrose) is a disaccharide. *26·14*

dispersion force the weakest kind of intermolecular attraction; thought to be caused by the motion of electrons. *14·12*

distillation a purification process in which a liquid is evaporated and then condensed again to a liquid; used to separate dissolved solids from liquids or liquids from liquids according to boiling point. *1·6*

double covalent bond a covalent bond in which two pairs of electrons are shared by two atoms. *14·2*

double-replacement reaction a chemical change that involves an exchange of positive ions between two compounds. *7·6*

dry cell a commercial voltaic cell in which the electrolyte is a moist paste. *21·3*

effloresce to lose water of hydration; the process occurs when the hydrate has a vapor pressure higher than that of water vapor in the air. *15·8*

effusion a process that occurs when a gas escapes through a tiny hole in its container. *10·12*

electrical potential the ability of a voltaic cell to produce an electric current. *21·6*

electrochemical cell any device that converts chemical energy into electrical energy or electrical energy into chemical energy. *21·1*

electrochemical process the conversion of chemical energy into electrical energy or electrical energy into chemical energy; in an electrochemical cell, all electrochemical processes involve redox reactions. *21·1*

electrode a conductor in a circuit that carries electrons to or from a substance other than a metal. *21·2*

electrolysis a process in which electrical energy is used to bring about a chemical change. *21·9*

electrolyte a compound that conducts an electric current in aqueous solution or in the molten state; all ionic compounds are electrolytes, but most covalent compounds are not. *15·9*

electrolytic cell an electrochemical cell used to cause a chemical change through the application of electrical energy. *21·9*

electromagnetic radiation a series of energy waves that travel in a vacuum at a speed of 3.0×10^{10} cm/s; includes radio waves, microwaves, visible light, infrared and ultraviolet light, x-rays, and gamma rays. *11·6*

electron a negatively charged subatomic particle. *4·2*

electron affinity the energy that accompanies the addition of an electron to a gaseous atom. *12·6*

electron configuration the arrangement of electrons around the nucleus of an atom in its ground state. *11·4*

electron dot structures a notation that depicts valence electrons as dots around the atomic symbol of the element; the symbol represents the inner electrons and atomic nucleus; also called Lewis dot structures. *13·1*

electronegativity the tendency for an atom to attract electrons to itself when it is chemically combined with another element. *12·8*

element a substance that cannot be changed into a simpler substance under normal laboratory conditions. *1·7*

elementary reaction a reaction in which reactants are converted to products in a single step. *17·4*

empirical formula a formula with the lowest whole-number ratio of elements in a compound; the empirical formula of hydrogen peroxide, H_2O_2, is HO. *6·10*

emulsion the colloidal dispersion of one liquid in another. *15·10*

endergonic a nonspontaneous chemical reaction that proceeds with the absorption of free energy; work must be done to make the reaction proceed. *17·8*

endothermic reaction a chemical change in which energy is absorbed; the energy content of the products is higher than the energy content of the reactants. *8·7*

end point the point in a titration at which neutralization is just achieved. *19·2*

energy the capacity for doing work; it exists in several forms, including chemical, nuclear, electrical, radiant, mechanical, and thermal energies. *1·9*

energy level a region around the nucleus of an atom where an electron is likely to be moving. *11·1*

enthalpy (H) the amount of heat that a substance has at a given temperature and pressure. *8·8*

entropy (S) a measure of the disorder of a system; systems tend to go from a state of order (low entropy) to a state of maximum disorder (high entropy). *17·5*

equilibrium constant (K_{eq}) the ratio of product concentrations to reactant concentrations with each raised to a power given by the number of moles of the substance in the balanced chemical equation. *17·11*

equilibrium position the relative concentrations of reactants and products of a reaction that has reached equilibrium; it indicates whether the reactants or products are favored in the reversible reaction. *17·10*

equivalence point the point in a titration at which the number of equivalents of acid and base are equal. *19·4*

equivalent (equiv.) one equivalent is the amount of an acid (or base) that can give one mole of hydrogen (or hydroxide) ions. *19·3*

ester a derivative of a carboxylic acid in which the —OH of the carbonyl group has been replaced by an —OR from an alcohol; the general formula is RCOOR. 26·11

ether an organic compound in which oxygen is bonded to two carbon groups; the general formula is R—O—R. 26·7

evaporation the vaporization of a liquid in an open container. 9·6

excess reagent a reagent present in a quantity that is more than sufficient to react with a limiting reagent; any reactant that remains after the limiting reagent is used up in a chemical reaction. 8.5

exergonic a spontaneous reaction that releases free energy. 17·8

exothermic reaction a chemical change in which energy is released in the form of heat; the energy content of the products is less than the energy content of the reactants. 8·7

experiment a carefully controlled, repeatable procedure for gathering data to test a hypothesis. 1·2

fatty acid the name originally given to continuous chain carboxylic acids first isolated from fats. 26·9

Fehling's test a test used to detect aldehydes. 26·10

fermentation the production of ethanol from sugars by the action of yeast or bacteria. 26·5

film badge a small radiation detector worn by persons who work near radiation sources; it consists of several layers of photographic film covered with black light-proof paper. 24·8

first order reaction a reaction in which the reaction rate is proportional to the concentration of only one reactant. 17·3

fission the splitting of a nucleus into smaller fragments, accompanied by the release of neutrons and a large amount of energy. 24·6

flux a material, usually lime, added during smelting to combine with undesirable material in an ore. 22·8

formula unit the lowest whole-number ratio of ions in an ionic compound; in magnesium chloride, the ratio of magnesium ions to chloride ions is $1:2$ and the formula unit is $MgCl_2$. 5·4

free energy the potential energy contained in a reaction system; the energy available to do work. 17·8

freezing point depression the difference in temperature between the freezing points of a solution and of the pure solvent. 16·7

frequency (ν) the number of wave cycles that pass a given point per unit of time. 11·6

froth flotation a method commonly used to enrich sulfide ores of copper and lead by separating ore from gangue. 22·8

fuel cell a voltaic cell in which a fuel substance undergoes oxidation to produce electrical energy. 21·5

functional group a specific arrangement of atoms in an organic compound that is capable of characteristic chemical reactions; the chemistry of an organic compound is determined by its functional groups. 26·1

fusion a reaction in which two light nuclei combine to produce a nucleus of heavier mass, accompanied by the release of a large amount of energy. 24·7

gamma radiation high energy electromagnetic radiation emitted by certain radioactive nuclei; gamma rays have no mass or electrical charge. 24·2

gangue the worthless sand and rock contained in freshly mined ore. 22·8

gas matter that has no definite shape or volume; it adopts the shape and volume of its container. 1·4

gas pressure a force resulting from the simultaneous collisions of billions of gas particles on an object. 9·3

Gay-Lussac's law the pressure of a fixed mass of gas is directly proportional to the Kelvin temperature if the volume is kept constant. 10·8

Geiger counter a gas-filled metal tube used to detect the presence of beta radiation. 24·8

geometric isomer an organic compound that differs from another compound only in the geometry of their substituted groups. 25·6

Gibbs free energy change (ΔG) the maximum amount of energy that can be coupled to another process to do useful work. 17·9

Graham's laws of effusion the rate of effusion of a gas is inversely proportional to the square root of its formula mass; this relationship is also true for the diffusion of gases. 10·12

gram a metric mass unit equal to the mass of 1 cm^3 of water at 4°C. 2·8

gram atomic mass (gam) the mass, in grams of one mole of atoms in a monatomic element; it is numerically equal to the atomic mass in amu. 6·3

gram equivalent mass of an acid the mass of one equivalent of an acid expressed in grams. 19·3

gram equivalent mass of a base the mass of one equivalent of a base expressed in grams. 19·3

gram formula mass (gfm) the mass of one mole of an ionic compound, equal to the formula mass expressed in grams; the expression may be used in broader sense to refer to a mole of any element, molecular compound, or ionic compound. 6·3

gram molecular mass (gmm) the mass of one mole of a molecular substance; it is equal to the formula mass expressed in grams. 6·3

ground state the lowest energy level occupied by an electron when an atom is in its most stable energy state. 11·9

group a vertical column of elements in the periodic table; the constituent elements of a group have similar chemical and physical properties. 5·1

half-cell the part of a voltaic cell in which either oxidation or reduction occurs; it consists of a single electrode immersed in a solution of its ions. 21·2

half-life ($t_{1/2}$) the time required for one-half of the atoms of a radioisotope to emit radiation and decay to products. 24·4

half-reaction an equation showing either the reduction or the oxidation of a species in an oxidation–reduction reaction. 20·7

half-reaction method a method of balancing a redox equation by balancing the oxidation and reduction half-reactions. 20·7

halide ions a negative ion formed when a halogen atom gains an electron. 13·3

halocarbon any member of a class of organic compounds containing covalently bonded fluorine, chlorine, bromine, or iodine. *26·2*

halogen any member of the nonmetallic elements in Group 7A of the periodic table. *12·15*

heat the energy that is transferred from one body to another because of a temperature difference. *1·9*

heat capacity the quantity of heat required to change an object's temperature by exactly 1°C. *2·13*

heat of combustion the heat absorbed or released during a chemical reaction in which one mole of a substance is completely burned. *8·8*

heat of condensation the heat, in calories, released when one gram of a gas condenses to a liquid at the liquid's boiling point. *9·10*

heat of fusion the heat, in calories, required to melt one gram of a solid at its melting point. *9·10*

heat of reaction the heat released or absorbed during a chemical reaction; equivalent to H, the change in enthalpy. *8·8*

heat of solidification the heat, in calories, given up as one gram of a liquid changes to a solid at the solid's melting point. *9·10*

heat of vaporization the heat, in calories, required to change one gram of a liquid to a gas at the liquid's boiling point (at atmospheric pressure). *9·10*

heat transfer the process in which heat moves from a warm object to a cooler object. *2·11*

Heisenberg uncertainty principle it is impossible to know both the velocity and the position of a particle at the same time. *11·10*

Henry's law at a given temperature the solubility of a gas in a liquid is directly proportional to the pressure of the gas above the liquid. *16·3*

hertz (Hz) the SI unit of frequency, equal to one cycle per second. *11·6*

heterogeneous mixture a mixture that is not uniform in composition; its components are readily distinguished. *1·6*

heterogeneous reactions a reaction carried out with a heterogeneous mixture of reactants. *17·2*

homogeneous mixture a mixture that is completely uniform in composition; its components are not distinguishable. *1·6*

homologous series a series formed by a group of compounds in which there is a constant increment of change in molecular structure from one compound in the series to the next. *25·2*

Hund's rule when electrons occupy orbitals of equal energy, one electron enters each orbital until all orbitals contain one electron with their spins parallel. *11·4*

hybridization a process in which several atomic orbitals (such as s and p orbitals) mix to form the same number of equivalent hybrid orbitals. *14·9*

hydration reaction a reaction in which water is added to an alkene. *26·6*

hydrocarbon an organic compound that contains only carbon and hydrogen. *25·0*

hydrogenation a process used to make solid shortenings and margarine; it involves adding hydrogen to an oil at high temperature and pressure in the presence of a catalyst. *23·1*

hydrogenation reaction a reaction in which hydrogen is added to a carbon–carbon double bond to give an alkane. *26·6*

hydrogen bond a relatively strong intermolecular force in which a hydrogen atom that is covalently bonded to a very electronegative atom is also weakly bonded to an unshared electron pair of an electronegative atom in the same molecule or one nearby. *14·13*

hydrogen-ion acceptor a base, according to the Brønsted-Lowry theory. *18·6*

hydrogen-ion donor an acid, according to the Brønsted-Lowry theory. *18·6*

hydrometer a device used to measure the specific gravity of a liquid. *2·10*

hydronium ion (H_3O^+) the positive ion formed when a water molecule gains a hydrogen ion; all hydrogen ions in aqueous solution are present as hydronium ions. *18·2*

hydroxide ion (OH^-) the negative ion formed when a water molecule loses a hydrogen ion. *18·2*

hydroxyl group the —OH functional group present in alcohols. *26·4*

hygroscopic a term describing salts and other compounds that remove moisture from the air. *15·8*

hypothesis a descriptive model used to explain observations. *1·2*

ideal gas constant (R) a term in the ideal gas law, which has the value 0.0821 (L × atm)/(K × mol). *10·10*

ideal gas law the relationship $P \times V = n \times R \times T$, which describes the behavior of an ideal gas. *10·10*

immiscible liquids that are insoluble in one another; oil and water are immiscible. *16·2*

inhibitor a substance that interferes with catalysis. *17·2*

inorganic chemistry the study of substances that do not contain carbon. *1·1*

intermediate a product of a reaction that immediately becomes a reactant of another reaction. *17·4*

International System of Units (SI) the revised version of the metric system, adopted by international agreement in 1960. *2·5*

ion an atom or group of atoms that has a positive or negative charge; cations are ions with a positive charge, and anions are ions with a negative charge. *5·2*

ionic bond the electrostatic attraction that binds oppositely charged ions together. *13·4*

ionic compound a compound composed of positive and negative ions. *5·3*

ionization energy the energy required to remove an electron from a gaseous atom. *12·5*

ionizing radiation radiation which has enough energy to produce ions by knocking electrons off some of the atoms it strikes. *24·8*

ion product constant for water (K_w) the product of the hydrogen ion and hydroxide ion concentrations of water; it is 1×10^{-14} at 25°C. *18·2*

isotopes atoms of the same element that have the same atomic number but different atomic masses due to a different number of neutrons. *4·6*

IUPAC system an internationally accepted system of naming compounds proposed by the International Union of Pure and Applied Chemistry (IUPAC). *25·2*

joule (J) the SI unit of energy; 4.184 joules equal 1.000 calorie. *2·12*

Kelvin temperature scale the temperature scale in which the freezing point of water is 273 K and the boiling point is 373 K; 0 K is absolute zero. *2·11*

ketone an organic compound in which the carbon of the carbonyl group is joined to two other carbons; the general formula is RCOR. *26·8*

kilogram (kg) the mass of 1 L of water at 4°C; it is the base unit of mass in the metric system. *2·8*

kinetic energy the energy an object has because of its motion. *1·9*

kinetic theory a theory explaining the states of matter, based on the concept that the particles in all forms of matter are in constant motion. *9·1*

law of conservation of energy energy is neither created nor destroyed in an ordinary chemical or physical process. *1·10*

law of conservation of mass mass can be neither created nor destroyed in an ordinary chemical or physical process. *1·12*

law of definite proportions in any chemical compound, the elements are always combined in the same proportion by mass. *5·3*

law of disorder it is a natural tendency of systems to move in the direction of maximum chaos or disorder. *17·5*

law of multiple proportions whenever two elements form more than one compound, the different masses of one element that combine with the same mass of the other element are in the ratio of small whole numbers. *5·5*

Le Châtelier's principle when stress is applied to a system at equilibrium, the system changes to relieve the stress. *17·12*

Lewis acid any substance that can accept a pair of electrons to form a covalent bond. *18·7*

Lewis base any substance that can donate a pair of electrons to form a covalent bond. *18·7*

limiting reagent any reactant that is used up first in a chemical reaction; it determines the amount of product that can be formed in the reaction. *8·5*

lime calcium oxide, CaO; it is sometimes called quicklime. *22·2*

lipid a member of a large class of relatively water-insoluble organic compounds; fats, oils, and waxes are lipids. *26·15*

liquid a form of matter that flows, has a fixed volume, and takes the shape of its container. *1·4*

liter (L) the volume of a cube measuring 10 centimeters on each edge (1000 cm^3); it is the common unprefixed unit of volume in the metric system. *2·7*

mass the amount of matter that an object contains; the SI base unit of mass is the kilogram. *1·3*

mass number the total number of protons and neutrons in the nucleus of an atom. *4·5*

matter anything that takes up space and has mass. *1·3*

melting point (mp) the temperature at which a substance changes from solid to a liquid; the melting point of water is 0°C. *9·8*

metal one of a class of elements that includes a large majority of the known elements; metals are characteristically lustrous, malleable, ductile, and good conductors of heat and electricity. *5·1*

metallic bond the force of attraction that holds metals together; it consists of the attraction of free-floating valence electrons for positively charged metal ions. *13·6*

metalloid one of a class of elements having properties of both metals and nonmetals. *5·1*

metallurgy the various procedures used to separate metals from their ores. *22·8*

meter the base unit of length in the metric and SI measurement systems. *2·6*

metric system the standards of measurement based on units of 10. *2·5*

millimeter of mercury (mm Hg) a unit of pressure; it is the pressure needed to support a column of mercury one millimeter high. *9·3*

miscible liquids that will dissolve in each other. *16·2*

mixture a combination of two or more substances that are not chemically combined. *1·6*

molal boiling point elevation constant (K_b) the change in boiling point for a 1 molal solution of a nonvolatile molecular solute. *16·10*

molal freezing point depression constant (K_f) the change in freezing point for a 1 molal solution of a nonvolatile molecular solute. *16·10*

molality (m) the concentration of solute in a solution expressed as the number of moles of solute dissolved in 1 kilogram (1,000 g) of solvent. *16·8*

molar mass an expression sometimes used in place of gram formula mass to refer to the mass of a mole of any element or compound. *6·4*

molar volume the volume occupied by one mole of a gas at standard temperature and pressure (STP); 22.4 L. *6·6*

molarity (M) the concentration of solute in a solution expressed as the number of moles of solute dissolved in 1 L of solution. *16·4*

mole (mol) the amount of a substance that contains 6.02×10^{23} representative particles of that substance; a gram formula mass of any substance. *6·2*

molecule a neutral chemically bonded group of atoms that act as a unit. *5·3*

molecular compound a compound that is composed of molecules. *5·3*

molecular formula a chemical formula that shows the actual number and kinds of atoms present in a molecule of a compound. *5·4*

molecular orbital an orbital resulting from the overlapping of atomic orbitals when two atoms combine. *14·7*

mole fraction the ratio of the moles of solute in solution to the total number of moles of both solvent and solute. *16·8*

Monel metal a strong, corrosion-resistant alloy of nickel and cobalt. *22·6*

monomer a simple molecule that combines to form the repeating unit of a polymer. *26·12*

monoprotic acid any acid that contains one ionizable proton (hydrogen ion); nitric acid (HNO$_3$) is a monoprotic acid. *18·5*

monosaccharide a carbohydrate consisting of one sugar unit; it is also called a simple sugar. *26·14*

network solid a substance in which all of the atoms are covalently bonded to each other. *14·14*

neutralization reaction a reaction in which an acid and a base react in an aqueous solution to produce a salt and water. *19·1*

neutral solution an aqueous solution in which the concentrations of hydrogen and hydroxide ions are 1.0×10^{-7} mol/L; it has a pH of 7.0. *18·2*

neutron a subatomic particle with no charge and a mass of one amu found in the nucleus of the atom. *4·2*

neutron absorption a process used in a nuclear reactor to slow down the chain reaction by decreasing the number of moving neutrons; this is done with control rods made of a material like cadmium that absorbs neutrons. *24·6*

neutron moderation a process used in a nuclear reactor to slow down the neutrons so they can be captured by the reactor fuel to continue the chain reaction; water and carbon are good moderators. *24·6*

noble gas any member of a group of gaseous elements in Group 0 of the periodic table; the s and p sublevels of their outermost energy level are filled. *12·3*

nonelectrolyte a compound that does not conduct an electric current in aqueous solution or in the molten state. *15·9*

nonmetal one of a class of elements that are not lustrous and are generally poor conductors of heat and electricity; nonmetals are grouped on the right side of the periodic table. *5·1*

nonpolar covalent bond a bond formed when the atoms in a molecule are alike and the bonding electrons are shared equally. *14·10*

nonspontaneous reaction a reaction that does not give products under the specified conditions. *17·7*

normal boiling point the boiling point of a liquid at a pressure of one atmosphere. *9·7*

normality (N) the concentration of a solution expressed as the number of equivalents of solute in 1 L of solution. *19·4*

nucleic acid a polymer of ribonucleotides (RNA) or deoxyribonucleotides (DNA) found primarily in cell nuclei; nucleic acids play an important role in the transmission of hereditary characteristics, protein synthesis, and the control of cell activities. *26·17*

nucleotide one of the monomers that make up DNA and RNA; it consists of a nitrogen-containing base (a purine or pyrimidine), a sugar (ribose or deoxyribose), and a phosphate. *26·17*

nucleus the dense central portion of an atom, composed of protons and neutrons. *4·3*

observation the noting and recording of facts and events. *1·2*

octet rule atoms react by gaining or losing electrons so as to acquire the stable electron structure of a noble gas, usually eight valence electrons. *13·2*

open-hearth method a process in which pig iron is converted to high quality steel in a shallow vessel, or hearth. *22·10*

ore a mineral used for commercial production of a metal. *22·8*

organic chemistry the study of compounds that contain the element carbon. *1·1*

oxidation a process that involves a complete or partial loss of electrons or a gain of oxygen; it results in an increase in the oxidation number of an atom. *20·1*

oxidation number a positive or negative number assigned to a combined atom according to a set of arbitrary rules; generally it is the charge an atom would have if the electrons in each bond were assigned to the atoms of the more electronegative element. *20·3*

oxidation number change method a method of balancing a redox equation by comparing the increases and decreases in oxidation numbers. *20·5*

oxidation–reduction reaction a reaction that involves the transfer of electrons between reactants during a chemical change. *20·0*

oxidizing agent the substance in a redox reaction that accepts electrons; in the reaction, the oxidizing agent is reduced. *20·2*

paramagnetic a term used to describe a substance that shows a relatively strong attraction to an external magnetic field; these substances have molecules containing one or more unpaired electrons. *14·6*

partial pressure the pressure exerted by each gas in a gaseous mixture. *10·5*

pascal (Pa) the SI unit of pressure. *9·3*

Pauli exclusion principle no more than two electrons can occupy an atomic orbital; these electrons must have opposite spins. *11·4*

peptide an organic compound formed by a combination of amino acids in which the amino group of one acid is united with the carboxylic group of another through an amide bond. *26·16*

peptide bond the bond between the carbonyl group of one amino acid and the nitrogen of the next amino acid in the peptide chain; the structure is

$$\underset{}{-\overset{\overset{\displaystyle O}{\|}}{C}-\overset{\overset{\displaystyle H}{|}}{N}-}.$$ *26·16*

percent composition the percent by mass of each element in a compound. *6·9*

percent yield the ratio of the actual yield to the theoretical yield for a chemical reaction expressed as a percentage; a measure of the efficiency of a reaction. *8·6*

period a horizontal row of elements in the periodic table. *12·2*

periodic law when the elements are arranged in order of increasing atomic number, there is a periodic repetition of their physical and chemical properties. *12·2*

periodic table an arrangement of elements into rows and columns according to similarities in their properties. *5·1*

pH a number used to denote the hydrogen-ion concentration, or acidity, of a solution; it is the negative logarithm of the hydrogen ion concentration of a solution. *18·3*

phase any part of a system with uniform composition and properties. *1·6*

phase change a change in the physical state of a substance. *9·9*

photoelectric effect a process in which electrons are ejected by certain metals when light of sufficient frequency shines on them. *11·8*

photon a quantum of light; a discrete bundle of electromagnetic energy that behaves as a particle. *11·8*

physical change an alteration of a substance that does not affect its chemical composition. *1·5*

physical chemistry the study of the theoretical basis of chemical behavior, relying on mathematics and physics. *1·1*

physical property a quality of a substance that can be observed or measured without changing the substance's chemical composition. *1·3*

pi bond (π bond) a bond in which the bonding electrons are most likely to be found in the sausage-shaped regions above and below the nuclei of the bonded atoms. *14·7*

pig iron the form of iron obtained from a blast furnace; it undergoes further treatment to produce steel. *22·9*

Planck's constant (h) a number used to calculate the radiant energy, E, absorbed or emitted by a body from the frequency of the radiation. *11·7*

polar bond a bond formed when two different atoms are joined by a covalent bond and the bonding electrons are shared unequally. *14·10*

polar covalent bond see polar bond. *14·10*

polar molecule a molecule in which one or more atoms is slightly negative, and one or more is slightly positive in such a way that the polarities do not cancel; water is a polar molecule. *14·10*

polyatomic ion a tightly bound group of atoms that behaves as a unit and carries a charge. *5·7*

polymer a very large molecule formed by the covalent bonding of repeating small molecules, known as monomers. *26·12*

polypeptide a peptide with more than 10 amino acid residues. *26·16*

polysaccharides a complex carbohydrate polymer formed by the linkage of many monosaccharide monomers; starch, glycogen, and cellulose are polysaccharides. *26·14*

positron a particle which has the same mass as an electron but has a positive charge. *24·3*

potential energy the energy stored in an object as a result of its position or composition. *1·9*

precipitate a solid that separates from a solution. *1·11*

precision the reproducibility, under the same conditions, of a measurement. *2·2*

product a substance formed in a chemical reaction. *1·11*

protein any peptide with more than 100 amino acid residues. *26·16*

proton a positively charged subatomic particle found in the nucleus of an atom. *4·2*

qualitative measurement a measurement which gives descriptive, nonnumeric results. *2·1*

quantitative measurement a measurement which gives definite, usually numeric results. *2·1*

quantum a small package, or unit, of electromagnetic energy; the amount of energy required to move an electron from its present energy level to the next higher one. *11·1*

quantum mechanical model the modern description, primarily mathematical, of the behavior of electrons in atoms. *11·2*

radiation the penetrating rays emitted by a radioactive source; also, the giving off of energy in various forms such as heat, light, or radiowaves. *24·1*

radioactive decay the process in which an unstable nucleus loses energy by emitting radiation. *24·1*

radioactivity the property by which an atomic nucleus gives off alpha, beta, or gamma radiation. *24·1*

radioisotopes isotopes that have unstable nuclei and undergo radioactive decay. *24·1*

rate law an expression relating the rate of a reaction to the concentration of the reactants. *17·3*

reactant a starting substance in a chemical reaction. *1·11*

reaction mechanism a series of elementary reactions that take place during the course of a complex reaction. *17·4*

reaction rate a measure of the number of atoms, ions, or molecules that react to form products in a specified time. *17·1*

redox reaction another name for an oxidation-reduction reaction. *20·0*

reducing agent the substance in a redox reaction that donates electrons; in the reaction, the reducing agent is oxidized. *20·2*

reduction a process that involves a complete or partial gain of electrons or the loss of oxygen; it results in a decrease in the oxidation number of an atom. *20·1*

reduction potential a measure of the tendency of a given half-reaction to occur as a reduction (gain of electrons) in an electrochemical cell. *21·6*

representative element an element that belongs to an A group in the periodic table; they are called representative elements because they illustrate the entire range of chemical properties. *5·1*

representative particle the smallest unit into which a substance can be broken down without a change in composition; the term refers to whether a substance commonly exists as atoms, ions, or molecules. *6·2*

resonance a phenomenon that occurs when two or more equally valid electron dot structures can be written for a molecule; the actual bonding is believed to be a hybrid or mixture of the extremes represented by the resonance forms. *14·5*

reversible reaction a reaction in which the conversion of reactants into products and the conversion of products into reactants occur simultaneously. *17·10*

roasting a process used to purify sulfide ores by heating the ore with air. *22·8*

salt bridge a tube containing a conducting solution used to connect half-cells in a voltaic cell; it allows the passage of ions from one compartment to another but prevents the solutions from mixing completely. *21·2*

salt hydrolysis a process in which the cations or anions of a dissociated salt accept hydrogen ions from water or donate hydrogen ions to water; solutions containing hydrolyzed salts may be either acidic or basic. *19·5*

saponification the process used to make soap; it involves the hydrolysis of fats or oils by hot aqueous sodium hydroxide. *26·15*

saturated compound an organic compound in which all carbon atoms are joined by single covalent bonds; it contains the maximum number of hydrogen atoms. *25·5*

saturated solution a solution containing the maximum amount of solute for a given amount of solvent at a constant temperature and pressure; in a saturated solution, the dissolved and undissolved solute are in dynamic equilibrium. *16·2*

scientific law a concise statement that summarizes the results of a broad variety of observations and experiments. *1·2*

scientific method a method of inquiry that involves observations, hypotheses, and experiments to formulate theories. *1·2*

scintillation counter a device that uses a specially coated surface called a phosphor to detect radiation; ionizing radiation striking this surface produces bright flashes of light (scintillations). *24·8*

self-ionization a term describing the reaction in which two water molecules react to give ions. *18·2*

semimetal a metalloid. *5·1*

sigma bond (σ bond) a bond formed when two atomic orbitals combine to form a molecular orbital that is symmetrical along the axis connecting the two atomic nuclei. *14·7*

significant figures all the digits that can be known precisely in a measurement, plus a last estimated digit. *2·3*

single covalent bond a bond formed when a pair of electrons is shared between two atoms. *14·1*

single-replacement reaction a chemical change in which atoms of an element replace atoms of a second element in a compound; also called a displacement reaction. *7·5*

skeleton equation a chemical equation that does not indicate the relative amounts of reactants and products. *7·1*

slag the fused waste material from a smelting process; it is usually composed of silicates. *22·8*

slaked lime calcium hydroxide, $Ca(OH)_2$. *22·2*

smelting a process in which impurities are removed from an ore by melting. *22·8*

solid matter which has a definite shape and volume. *1·4*

solubility the amount of a substance that dissolves in a given quantity of solvent at specified conditions of temperature and pressure to produce a saturated solution. *16·2*

solubility product constant (K_{sp}) an equilibrium constant that can be applied to the solubility of electrolytes; it is equal to the product of the concentration terms each raised to the power of the coefficient of the substance in the dissociation equation. *19·7*

solute dissolved particles in a solution. *15·6*

solution a homogenous mixture. *1·6*

solvation a process that occurs when an ionic solute dissolves; in solution, the ions are surrounded by solvent molecules. *15·7*

solvent the dissolving medium in a solution. *15·6*

specific gravity the ratio of the density of a substance to that of a standard substance (usually water). *2·10*

specific heat the quantity of heat required to raise the temperature of 1 g of a substance by 1°C. *2·13*

specific rate constant a proportionality constant relating the concentrations of reactants to the rate of the reaction. *17·3*

spectator ion an ion that does not change oxidation number or composition during a reaction. *20·7*

spectrum the range of wavelengths and frequencies making up light; the wavelengths are separated when a beam of white light passes through a prism. *11·6*

spontaneous reaction a reaction known to give the products as written in a balanced equation. *17·7*

standard atmosphere (atm) a unit of pressure; it is the pressure required to support 760 mm of mercury in a mercury barometer at 25°C; this is the average atmospheric pressure at sea level. *9·3*

standard cell potential the measured cell potential when the ion concentrations in the half-cells are 1.00 M, at 1 atm of pressure and 25°C. *21·6*

standard entropy (S^0) the entropy of a substance in its stable state at 25°C and 1 atm. *17·6*

standard heat of formation (H_f^0) the change in enthalpy (H) for a reaction in which one mole of a compound is formed from its constituent elements. *8·8*

standard hydrogen electrode an arbitrary reference electrode (half-cell) used with another electrode (half-cell) to measure the standard reduction potential of that cell; the standard reduction potential of the hydrogen electrode is assigned a value of 0.00 V. *21·6*

standard solution a solution of known concentration used in carrying out a titration. *19·2*

standard temperature and pressure (STP) the conditions under which the volume of a gas is usually measured; standard temperature is 0°C and standard pressure is 1 atmosphere (atm). *6·6*

stereoisomers an organic molecule having the same molecular structure as another molecule, but differing in the arrangement of the atoms in space. *25·7*

stoichiometry that portion of chemistry dealing with numerical relationships in chemical reactions; the calculation of quantities of substances involved in chemical equations. *8·0*

strong acid an acid that is completely (or almost completely) ionized in aqueous solution. *18·8*

strong base a base that completely dissociates into metal ions and hydroxide ions in aqueous solution. *18·8*

strong electrolyte a solution in which a large portion of the solute exists as ions. *15·9*

structural formula a chemical formula that shows the arrangement of atoms in a molecule or a polyatomic ion; each dash between two atoms indicates a pair of shared electrons. *14·1*

structural isomer a compound that has the same molecular formula as another compound but has a different molecular structure. *25·4*

sublimation the conversion of a solid to a gas without passing through the liquid state. *9·9*

substance a sample of matter having a uniform and definite composition; it can be either an element or a compound. *1·3*

substituent an atom or group of atoms that can take the place of a hydrogen atom on a parent hydrocarbon molecule. *25·3*

substitution the replacement of an atom or group of atoms by another atom or group of atoms. *26·3*

supercooled liquid a term sometimes used to describe glasses, transparent materials produced when an inorganic substance cools to a rigid state without crystallizing; their internal structure is intermediate between that of a crystalline solid and free-flowing liquid. *9·8*

supersaturated solution a solution that contains more solute that it can theoretically hold at a given temperature. *16·3*

surface tension an inward force that tends to minimize the surface area of a liquid; it causes the surface to behave as if it were covered by a thin skin. *15·2*

surfactant a surface active agent; any substance whose molecules interface with the hydrogen bonding between water molecules, reducing surface tension; soaps and detergents are surfactants. *15·2*

suspension a mixture from which some of the particles settle out slowly upon standing. *15·10*

temperature the degree of hotness or coldness of an object, which is a measure of the average kinetic energy of the molecules of the object. *2·11*

ternary compound a compound containing atoms of three different elements, with one or more polyatomic ions usually present; Na_2CO_3 and $Mg(OH)_2$ are ternary ionic compounds. *5·11*

tetrahedral angle a bond angle of 109.5° created when a central atom forms four bonds directed toward the corners of a regular tetrahedron. *14·8*

theoretical yield the amount of product that could form during a reaction calculated from a balanced chemical equation; it represents the maximum amount of product that could be formed from a given amount of reactant. *8·6*

theory a thoroughly tested model that explains why experiments give certain results. *1·2*

thermochemical equation a chemical equation that includes the amount of heat produced or absorbed during the reaction. *8·7*

titration a method of reacting a solution of unknown concentration with one of known concentration; this procedure is often used to determine the concentrations of acids and bases. *19·2*

trans **configuration** a term applied to geometric isomers; it denotes an arrangement in which the substituted groups are on opposite sides of the double bond. *25·6*

transition metal an element found in one of the B groups in the periodic table. *5·1*

transition state a term sometimes used to refer to the activated complex. *17·1*

transmutation the conversion of an atom of one element into an atom of another element by the emission of radiation. *24·5*

transuranium element an element in the periodic table whose atomic number is above 92. *24·5*

triglyceride an ester in which all three hydroxyl groups on a glycogen molecule have been replaced by long-chain fatty acids; fats are triglycerides. *26·15*

triple covalent bond a covalent bond in which three pairs of electrons are shared by two atoms. *14·2*

triprotic acid any acid that contains three ionizable protons (hydrogen ions); phosphoric acid (H_3PO_4) is a triprotic acid. *18·5*

Tyndall effect the visible path produced by a beam of light passing through a colloidal dispersion or suspension; the phenomenon is caused by the scattering of light by the colloidal and suspended particles. *15·10*

unit cell the smallest group of particles within a crystal that retains the geometric shape of the crystal. *9·8*

unsaturated a solution that contains less solute than a saturated solution at a given temperature and pressure. *16·2*

unsaturated compound an organic compound with one or more double or triple carbon–carbon bonds. *25·5*

unshared pair a pair of valence electrons that is not involved in bonding. *14·1*

vacuum a space where no particles of matter exist. *9·3*

valence electron an electron in the highest occupied energy level of an atom. *13·1*

van der Waals force a term used to describe the weakest intermolecular attractions; these include dispersion forces and dipole interactions. *14·13*

vapor a substance in the gaseous state that is ordinarily (at room temperature) a liquid or solid. *1·4*

vapor pressure the pressure produced when vaporized particles above the liquid in a sealed container collide with the container walls; when the container is saturated with vapor, a dynamic equilibrium exists between the gas and the liquid. *9·6*

vaporization the conversion of a liquid to a gas at a temperature below its boiling point. *9·6*

voltaic cell an electrochemical cell used to convert chemical energy into electrical energy; the energy is produced by a spontaneous redox reaction. *21·2*

volume the space occupied by matter. *2·7*

VSEPR theory valence-shell electron-pair repulsion theory; because electron pairs repel, molecules adjust their shapes so that valence-electron pairs are as far apart as possible. *14·8*

water of hydration water molecules that are an integral part of a crystal structure. *15·8*

wavelength (λ) the distance between two adjacent crests of a wave. *11·6*

wax an ester of a long-chain fatty acid and a long-chain alcohol. *26·15*

weak acid an acid that is only slightly ionized in aqueous solution. *18·8*

weak base a base that does not dissociate completely in aqueous solution. *18·8*

weak electrolyte a solution in which only a fraction of the solute exists as ions. *15·9*

weight a measure of the force of attraction between the earth and an object. *2·8*

zwitterion an internal salt of an amino acid. *26·16*

Index

The page on which a term is defined is indicated in **boldface** type.

Acknowledgments

Acknowledgments

Illustrations

Shirley Bortoli: Fig. 1.7, 1.12, 1.18, 2.5, 2.6, 2.9, 2.12, 2.15, 4.11, 4.14, 6.11, 7.20, 8.21, 9.6, 9.10, 9.12, 9.16, 11.13, 13.11, 13.13a, 14.14, 14.28, 14.29, 15.12, 15.15, 16.11, 16.16, 16.18, 17.7, 22.8, 22.15, 22.17, 24.3, 24.6, 25.14

Debby Morse: Fig. 4.6, 4.7, 4.8, 4.9, 5.9, 5.10, 5.13, 9.17, 11.3, 11.5, 11.6, 12.6, 12.9, 13.15, 14.2, 14.4, 14.5, 14.6, 14.7, 14.8, 14.9, 14.10, 14.13, 14.15, 14.16, 14.17, 14.18, 14.19a&b, 14.20, 14.21, 14.22, 14.23, 14.24, 14.25, 14.27, 14.32, 15.3 left, 15.7a&b, 18.4, 18.9, 18.12, 18.13, 19.2, 23.14, 23.15, 25.3, 25.9, 25.10, 25.13, 25.15, and pages 82, 140, 242, 314, and 567.

Masami Miyamoto: Fig. 10.4, 10.5, 10.6, 10.13, 10.17, 10.18, 10.23, 21.9, 21.10, 21.12, 21.15, 22.11, 22.21, 22.22, 22.23, 23.8, 23.12, 23.17, 23.22, 25.18, 25.19

Ignacio Gomez: 657T, 663

Precision Graphics: 652, 653, 654, 655, 657B, 659, 664

Periodic Table of the Elements

†This numbering system is used by the International Union of Pure and Applied Chemistry (IUPAC)

Metals

Nonmetals

State:
- Ⓢ Solid
- Ⓛ Liquid
- Ⓖ Gas
- Ⓝ Not found in nature

Key:
- Atomic number
- Electrons in each energy level
- Element symbol
- Average atomic mass

Example:
Ⓢ 11 Na Sodium 22.990 (2, 8, 1)

Group 1 (1A)	2 (2A)	3 (3B)	4 (4B)	5 (5B)	6 (6B)	7 (7B)	8 (8B)	9 (8B)	10 (8B)	11 (1B)	12 (2B)	13 (3A)	14 (4A)	15 (5A)	16 (6A)	17 (7A)	18 (0)
Ⓖ 1 H Hydrogen 1.0079 (1)																	Ⓖ 2 He Helium 4.0026 (2)
Ⓢ 3 Li Lithium 6.941 (2,1)	Ⓢ 4 Be Beryllium 9.0122 (2,2)											Ⓢ 5 B Boron 10.81 (2,3)	Ⓢ 6 C Carbon 12.011 (2,4)	Ⓖ 7 N Nitrogen 14.007 (2,5)	Ⓖ 8 O Oxygen 15.999 (2,6)	Ⓖ 9 F Fluorine 18.998 (2,7)	Ⓖ 10 Ne Neon 20.179 (2,8)
Ⓢ 11 Na Sodium 22.990 (2,8,1)	Ⓢ 12 Mg Magnesium 24.305 (2,8,2)											Ⓢ 13 Al Aluminum 26.982 (2,8,3)	Ⓢ 14 Si Silicon 28.086 (2,8,4)	Ⓢ 15 P Phosphorus 30.974 (2,8,5)	Ⓢ 16 S Sulfur 32.06 (2,8,6)	Ⓖ 17 Cl Chlorine 35.453 (2,8,7)	Ⓖ 18 Ar Argon 39.948 (2,8,8)
Ⓢ 19 K Potassium 39.098 (2,8,8,1)	Ⓢ 20 Ca Calcium 40.08 (2,8,8,2)	Ⓢ 21 Sc Scandium 44.956 (2,8,9,2)	Ⓢ 22 Ti Titanium 47.90 (2,8,10,2)	Ⓢ 23 V Vanadium 50.941 (2,8,11,2)	Ⓢ 24 Cr Chromium 51.996 (2,8,13,1)	Ⓢ 25 Mn Manganese 54.938 (2,8,13,2)	Ⓢ 26 Fe Iron 55.847 (2,8,14,2)	Ⓢ 27 Co Cobalt 58.933 (2,8,15,2)	Ⓢ 28 Ni Nickel† 58.71 (2,8,16,2)	Ⓢ 29 Cu Copper 63.546 (2,8,18,1)	Ⓢ 30 Zn Zinc 65.38 (2,8,18,2)	Ⓢ 31 Ga Gallium 69.72 (2,8,18,3)	Ⓢ 32 Ge Germanium 72.59 (2,8,18,4)	Ⓢ 33 As Arsenic 74.922 (2,8,18,5)	Ⓢ 34 Se Selenium 78.96 (2,8,18,6)	Ⓖ 35 Br Bromine 79.904 (2,8,18,7)	Ⓖ 36 Kr Krypton 83.80 (2,8,18,8)
Ⓢ 37 Rb Rubidium 85.468 (2,8,18,8,1)	Ⓢ 38 Sr Strontium 87.62 (2,8,18,8,2)	Ⓢ 39 Y Yttrium 88.906 (2,8,18,9,2)	Ⓢ 40 Zr Zirconium 91.22 (2,8,18,10,2)	Ⓢ 41 Nb Niobium 92.906 (2,8,18,12,1)	Ⓢ 42 Mo Molybdenum 95.94 (2,8,18,13,1)	Ⓝ 43 Tc Technetium (97) (2,8,18,14,1)	Ⓢ 44 Ru Ruthenium 101.07 (2,8,18,15,1)	Ⓢ 45 Rh Rhodium 102.91 (2,8,18,16,1)	Ⓢ 46 Pd Palladium 106.4 (2,8,18,18)	Ⓢ 47 Ag Silver 107.87 (2,8,18,18,1)	Ⓢ 48 Cd Cadmium 112.41 (2,8,18,18,2)	Ⓢ 49 In Indium 114.82 (2,8,18,18,3)	Ⓢ 50 Sn Tin 118.69 (2,8,18,18,4)	Ⓢ 51 Sb Antimony 121.75 (2,8,18,18,5)	Ⓢ 52 Te Tellurium 127.60 (2,8,18,18,6)	Ⓢ 53 I Iodine 126.90 (2,8,18,18,7)	Ⓖ 54 Xe Xenon 131.30 (2,8,18,18,8)
Ⓢ 55 Cs Cesium 132.91 (2,8,18,18,8,1)	Ⓢ 56 Ba Barium 132.33 (2,8,18,18,8,2)	Ⓢ 71 Lu Lutetium 174.97 (2,8,18,32,9,2)	Ⓢ 72 Hf Hafnium 178.49 (2,8,18,32,10,2)	Ⓢ 73 Ta Tantalum 180.95 (2,8,18,32,11,2)	Ⓢ 74 W Tungsten 183.85 (2,8,18,32,12,2)	Ⓢ 75 Re Rhenium 186.21 (2,8,18,32,13,2)	Ⓢ 76 Os Osmium 190.2 (2,8,18,32,14,2)	Ⓢ 77 Ir Iridium 192.22 (2,8,18,32,15,2)	Ⓢ 78 Pt Platinum 195.09 (2,8,18,32,17,1)	Ⓢ 79 Au Gold 196.97 (2,8,18,32,18,1)	Ⓛ 80 Hg Mercury 200.59 (2,8,18,32,18,2)	Ⓢ 81 Tl Thallium 204.37 (2,8,18,32,18,3)	Ⓢ 82 Pb Lead 207.2 (2,8,18,32,18,4)	Ⓢ 83 Bi Bismuth 208.98 (2,8,18,32,18,5)	Ⓢ 84 Po Polonium (209) (2,8,18,32,18,6)	Ⓢ 85 At Astatine (210) (2,8,18,32,18,7)	Ⓖ 86 Rn Radon (222) (2,8,18,32,18,8)
Ⓢ 87 Fr Francium (223) (2,8,18,32,18,8,1)	Ⓢ 88 Ra Radium 226.03 (2,8,18,32,18,8,2)	Ⓢ 103 Lr Lawrencium (260) (2,8,18,32,32,9,2)	Ⓝ 104* (261) (2,8,18,32,32,10,2)	Ⓝ 105* (262) (2,8,18,32,32,11,2)	Ⓝ 106* (263) (2,8,18,32,32,12,2)	Ⓝ 107* (264)	Ⓝ 109* (266)										

*Name not officially assigned

Lanthanide Series

Ⓢ 57 La Lanthanum 138.91 (2,8,18,18,9,2)	Ⓢ 58 Ce Cerium 140.12 (2,8,18,20,8,2)	Ⓢ 59 Pr Praseodymium 140.91 (2,8,18,21,8,2)	Ⓢ 60 Nd Neodymium 144.24 (2,8,18,22,8,2)	Ⓝ 61 Pm Promethium (145) (2,8,18,23,8,2)	Ⓢ 62 Sm Samarium 150.4 (2,8,18,24,8,2)	Ⓢ 63 Eu Europium 151.96 (2,8,18,25,8,2)	Ⓢ 64 Gd Gadolinium 157.25 (2,8,18,25,9,2)	Ⓢ 65 Tb Terbium 158.93 (2,8,18,27,8,2)	Ⓢ 66 Dy Dysprosium 162.50 (2,8,18,28,8,2)	Ⓢ 67 Ho Holmium 164.93 (2,8,18,29,8,2)	Ⓢ 68 Er Erbium 167.26 (2,8,18,30,8,2)	Ⓢ 69 Tm Thulium 168.93 (2,8,18,31,8,2)	Ⓢ 70 Yb Ytterbium 173.04 (2,8,18,32,8,2)

Actinide Series

Ⓢ 89 Ac Actinium (227) (2,8,18,32,18,9,2)	Ⓢ 90 Th Thorium 232.04 (2,8,18,32,18,10,2)	Ⓢ 91 Pa Protactinium 231.04 (2,8,18,32,20,9,2)	Ⓢ 92 U Uranium 238.03 (2,8,18,32,21,9,2)	Ⓢ 93 Np Neptunium 237.05 (2,8,18,32,22,9,2)	Ⓝ 94 Pu Plutonium (244) (2,8,18,32,24,8,2)	Ⓝ 95 Am Americium (243) (2,8,18,32,25,8,2)	Ⓝ 96 Cm Curium (247) (2,8,18,32,25,9,2)	Ⓝ 97 Bk Berkelium (247) (2,8,18,32,27,8,2)	Ⓝ 98 Cf Californium (251) (2,8,18,32,28,8,2)	Ⓝ 99 Es Einsteinium (254) (2,8,18,32,29,8,2)	Ⓝ 100 Fm Fermium (257) (2,8,18,32,30,8,2)	Ⓝ 101 Md Mendelevium (258) (2,8,18,32,31,8,2)	Ⓝ 102 No Nobelium (259) (2,8,18,32,32,8,2)